Circum-Pacific Council
for Energy and Mineral Resources
Earth Science Series

Volume 16

Energy and Mineral Potential of the Central American-Caribbean Region

Circum-Pacific Council
for Energy and Mineral Resources
Earth Science Series

H. Gary Greene, Editor-in-Chief
Florence L. Wong, Executive Editor

Volume 1
Tectonostratigraphic Terranes of the Circum-Pacific Region
Edited by D. G. Howell

Volume 2
Geology and Offshore Resources of Pacific Island Arcs–Tonga Region
Edited and Compiled by D. W. Scholl and T. L. Vallier

Volume 3
Investigations of the Northern Melanesian Borderland
Edited by T. M. Brocher

Volume 4
Geology and Offshore Resources of Pacific Island Arcs–Central and Western Solomon Islands
Edited by J. G. Vedder, K. S. Pound, and S. Q. Boundy

Volume 5A
The Antarctic Continental Margin: Geology and Geophysics of Offshore Wilkes Land
Edited by S. L. Eittreim and M. A. Hampton

Volume 5B
The Antarctic Continental Margin: Geology and Geophysics of the Western Ross Sea
Edited by A. K. Cooper and F. J. Davey

Volume 6
Geology and Resource Potential of the Continental Margin of Western North America and Adjacent Ocean Basins–Beaufort Sea to Baja California
Edited by D. W. Scholl, A. Grantz, and J. G. Vedder

Volume 7
Marine Geology, Geophysics, and Geochemistry of the Woodlark Basin–Solomon Islands
Edited by B. Taylor and N. F. Exon

Volume 8
Geology and Offshore Resources of Pacific Island Arcs–Vanuatu Region
Edited by H. G. Greene and F. L. Wong

Volume 9
Geology and Offshore Resources of Pacific Island Arcs–New Ireland and Manus Region, Papua New Guinea
Edited by M. S. Marlow, S. V. Dadisman, and N. F. Exon

Volume 10
Petroleum Resources of China and Related Subjects
Edited by H. C. Wagner, L. C. Wagner, F. F. H. Wang, and F. L. Wong

Volume 11
Geology of the Andes and Its Relation to Hydrocarbon and Mineral Resources
Edited by G. E. Ericksen, M. T. Cañas Pinochet, and J. A. Reinemund

Volume 12
Geology and Offshore Resources of Pacific Island Arcs–Solomon Islands and Bougainville, Papua New Guinea Regions
Edited by J. G. Vedder and T. R. Bruns

Volume 13
Terrane Analysis of China and the Pacific Rim
Edited by T. J. Wiley, D. G. Howell, and F. L. Wong

Volume 14
Geology and Offshore Mineral Resources of the Central Pacific Basin
Edited by B. H. Keating and B. R. Bolton

Volume 15
Basin Formation, Ridge Crest Processes, and Metallogenesis in the North Fiji Basin
Edited by L. W. Kroenke and J. V. Eade

Volume 16
Energy and Mineral Potential of the Central American-Caribbean Region
Edited by R. L. Miller, G. Escalante, J. A. Reinemund, and M. J. Bergin

Volumes 1, 3, and 5A through 13 available from AAPG, P. O. Box 979, Tulsa, OK 74101-0979.

Ralph L. Miller
Gregorio Escalante
John A. Reinemund
Marion J. Bergin
Editors

Energy and Mineral Potential of the Central American-Caribbean Region

Circum-Pacific Council
for Energy and Mineral Resources
Earth Science Series, Volume 16

With 181 Figures and 76 Tables

Springer

Ralph L. Miller (deceased)

Gregorio Escalante
Apartado 74, Pavas
San José, Costa Rica

John A. Reinemund
Marion J. Bergin
U.S. Geological Survey
12204 Sunrise Valley Drive
Reston, VA 22092
USA

Cover illustration: Momotombo Volcano, with the Momotombo geothermal power plant in the middle ground. Picture is taken from the northwest shore of Lake Managua, Nicaragua. Provided by Roger Arcia.

ISBN 3-540-58814-0 Springer-Verlag Berlin Heidelberg NewYork

CIP data applied for

Printed in Germany

Typesetting: Camera ready by editors
SPIN 10478865 32/3136 – 5 4 3 2 1 0 – Printed on acid-free paper

DEDICATION

This volume is dedicated to the memory of Ralph L. Miller in recognition of his role as technical editor for this volume, his help in organizing the CPCEMR-sponsored symposium in Costa Rica, and his long association with the geological community in Central America. Dr. Miller, who retired from the U.S. Geological Survey in 1979, participated in cooperative activities in Central America and adjacent areas over three decades as an adjunct to his work in the energy resources program of the Survey. His interest, knowledge, and contacts in Central America made him uniquely qualified to help develop the technical program and review the presentations of the symposium. During the two years prior to his death in October 1991, Dr. Miller devoted almost his entire time to the editorial work for CPCEMR. His efforts toward this volume and his participation in other CPCEMR activities in the past are remembered and most sincerely appreciated.

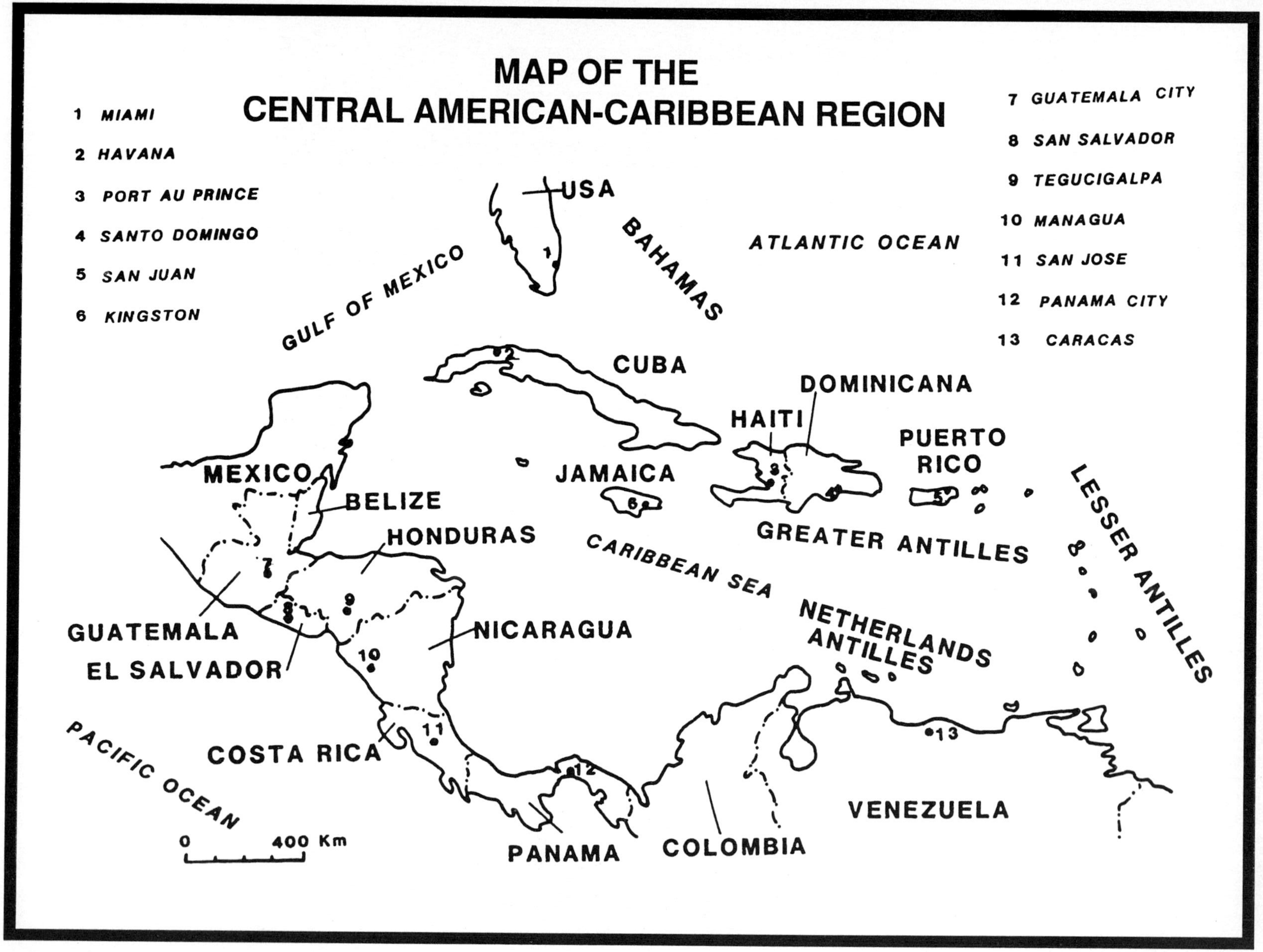
MAP OF THE
CENTRAL AMERICAN-CARIBBEAN REGION
1 MIAMI
2 HAVANA
3 PORT AU PRINCE
4 SANTO DOMINGO
5 SAN JUAN
6 KINGSTON
7 GUATEMALA CITY
8 SAN SALVADOR
9 TEGUCIGALPA
10 MANAGUA
11 SAN JOSE
12 PANAMA CITY
13 CARACAS
USA
BAHAMAS
ATLANTIC OCEAN
GULF OF MEXICO
CUBA
DOMINICANA
HAITI
PUERTO RICO
MEXICO
JAMAICA
BELIZE
HONDURAS
CARIBBEAN SEA
GREATER ANTILLES
LESSER ANTILLES
NETHERLANDS ANTILLES
GUATEMALA
NICARAGUA
EL SALVADOR
COSTA RICA
PACIFIC OCEAN
0 400 Km
PANAMA
COLOMBIA
VENEZUELA

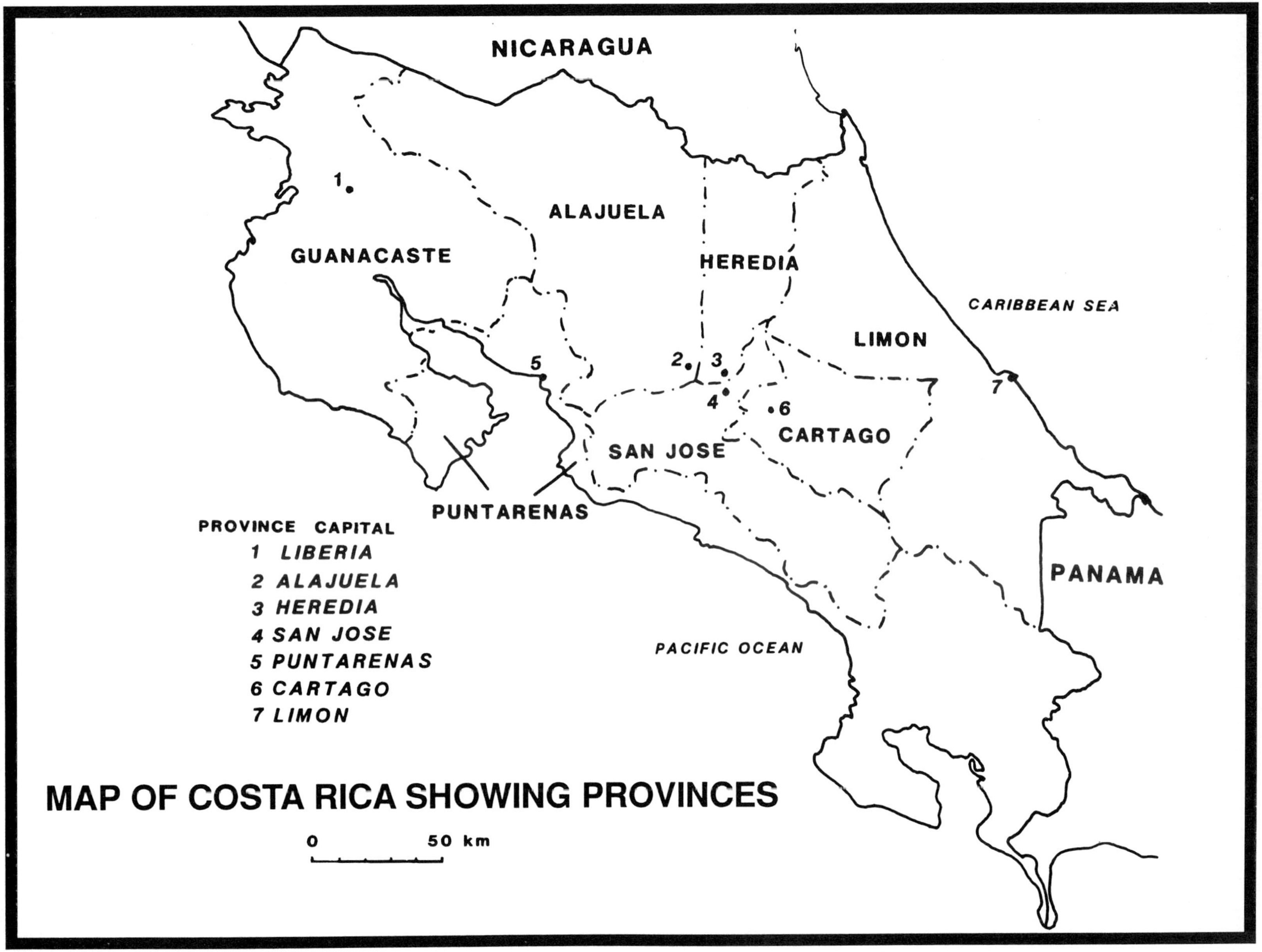

MAP OF COSTA RICA SHOWING PROVINCES

CIRCUM-PACIFIC COUNCIL FOR ENERGY AND MINERAL RESOURCES

FOREWORD

The Earth Science Series of the Circum-Pacific Council for Energy and Mineral Resources (CPCEMR) presents the results of scientific investigations that advance the knowledge of energy and mineral resources in the Pacific region and the environmental factors that affect resource utilization. The Series is part of the CPCEMR effort to encourage international cooperation and interest in the study of Pacific resources. It is intended to focus attention on new data and concepts based on offshore and onshore investigations. The Series includes reports from CPCEMR-sponsored symposia and workshops, as well as results of research projects that contribute to an understanding of the geological framework and resource potential of the Pacific Basin.

This volume in the Earth Science Series contains papers that were presented in a symposium on Energy and Mineral Potential of the Central American-Caribbean Region held in San Jose, Costa Rica, March 6-9, 1989. The symposium was hosted by the Costa Rica Ministerio de Recursos Naturales Energia y Minas and the Refinadora Cosstarricense de Petroleo, and had an attendance of over 350 participants from 24 countries. It was organized jointly by the CPCEMR and Costa Rican agencies to review the results of the extensive geological mapping, geophysical investigations, resources appraisal, and geohazard studies in this region in recent years. Formal papers and poster presentations dealt with regional geology; petroleum, coal, and geothermal resources; mineral resources and potential; hydrology and environment; geohazards; mapping technology; and remote sensing. Papers from the 120 presentations in the symposium are included in this volume.

On behalf of the CPCEMR, I thank the co-sponsoring and cooperating Agencies in Costa Rica for their help in organizing and conducting this symposium, and also the U.S. Geological Survey, U.S. Bureau of Mines, and U.S. Agency for International Development for their support and participation. I also thank Ralph L. Miller of the U.S Geological Survey (retired) and Gregorio Escalante, Consulting Geologist of Costa Rica, for their efforts in helping organize the technical program for the symposium and in the technical editing of the papers for this volume. Thanks also to John A. Reinemund and Merion J. Bergin of the U.S. Geological Survey for final editorial preparation of this volume following the death of Dr. Miller, to whom this volume is dedicated.

Michel T. Halbouty
Chairman

PREFACE

The symposium on the Energy and Mineral Potential of the Central American-Caribbean Region, which was held in San Jose, Costa Rica during March 6-9, 1989, attracted more than 350 participants from 46 countries, the largest group of earth scientists ever assembled in the area. Formal papers and poster presentations numbered 120, and five field trips gave opportunity for attendees also to view some of the geology and enjoy some of the spectacular scenery of Costa Rica.

For many years the scanty knowledge of the geology of the Central American and the Caribbean countries was mostly contained in publications of researchers from the United States and Germany. Results of work by the very small number of local geologists was largely confined to government files. Similarly, knowledge of the energy and mineral resources of the region was in foreign company files, with but an occasional published article by a foreigner, usually a university professor. Beginning about 25 or 30 years ago, interest by the local earth scientists in learning about the energy and mineral resources of their own countries arose more or less spontaneously in a number of places in the region. It was accelerated by the founding of a regional school of geology at the University of Costa Rica, which since then has graduated knowledgeable geologists from Costa Rica and from some of the other countries in Central America in ever increasing numbers. In recent years, much geologic mapping and other research have been accomplished in Central American, Greater Antillean, and northern South American countries. Recently, it seemed to leaders of the region and of the Circum-Pacific Council for Energy and Mineral Resources that the time was ripe for an assessment of progress to date, and of where future investigations might best be focused. As a result the Ministerio de Recursos Naturales, Energia y Minas de Costa Rica, the Refinadora Costarricense de Petroleo (RECOPE), and the Circum-Pacific Council for Energy and Mineral Resources decided to sponsor a symposium on resources of the region.

The Central American and Caribbean region comprises developing countries whose citizens contemplate the future with high expectations. The understanding of the potential resources of these countries and their orderly and fast development is, therefore, considered to be an indispensable requirement in the endeavors directed toward the attainment of a better life for many millions of people. A contribution toward the better knowledge of these resources, such as the present volume, becomes therefore a satisfying assignment for all involved. The task is even more enjoyable when we take note of the fact that fully trained and very capable local earth scientists are now in command of a number of important tasks in the development of the natural resources in these countries. This trend is clearly exemplified by noticing the large participation of local citizens of the Central American-Caribbean region as authors or co-authors of many of the papers which are included in this volume.

The program of the symposium was developed by affiliates of the Circum-Pacific Council and by a committee composed of representatives of the Costa Rican earth science organizations. Much of the non-technical work that is required for the success of a major meeting often goes unrecognized. We would also like to acknowledge here the major contributions to the success of this symposium made by Mary Stewart of the Circum-Pacific Council and Emilia Morales of RECOPE.

Subjects covered by the symposium and included in this volume as separate chapters are in order: background papers and papers of broad scope, petroleum resources, coal resources, geothermal resources, metallic minerals, industrial minerals, hydrology and environmental problems, and geologic hazards. Within the chapters our intent, not completely realized, was to have the lead-off paper as an overview of the theme of the chapter, followed by more circumscribed papers in a logical progression. Due to limitations of time for formal presentations, many excellent exhibits were displayed and discussed in poster sessions. Two of these resulted in papers that appear in this volume.

Papers at the symposium were presented in Spanish or in English, but the publication language is English. We have learned that people considered fluent in a language other than their own are not necessarily equally proficient in the written language not their own, particularly if it involves scientific writing. The translated papers have proved harder to edit. We hope that we have not, in the process of translating and editing, in places misinterpreted or distorted meanings of authors whose papers were submitted in Spanish.

We wish to thank our translators Oscar Salazar, Guatemala; Marco Baldioceda of Costa Rica; and Jean Weaver of the United States for their valued help with the Spanish. Weaver also corrected or redrafted many figures and prepared the two large maps on facing pages near the beginning of this volume. Miller's long association with the U.S. Geological Survey made it possible for him to call on the expertise of specialists, where the terminology and/or other technical aspects of parts of some manuscripts exceeded his ken. These authorities, with whom he has consulted, have been too numerous to list here, but thanks are owing to all of them. Escalante wants particularly to thank his secretary, Giselle Zamora, without whose support he probably could not have accomplished his commitment efficiently, and Miller wants to thank Harlene Padon, whose ability to understand his editing and often improve on it, has kept the flow of manuscripts through the technical approval stage reasonably continuous.

Finally, in this volume there are obviously many gaps in the coverage of the resources of a region as vast and with as many countries as Central America and the Caribbean. We do feel, however, that the papers here included do give a perspective of the current resources of energy and minerals and do provide a realistic appraisal of the potential for the future that will make this volume very useful to our science, to industry, and to the development of the region.

Ralph L. Miller
Deceased, U.S. Geological Survey

Gregorio Escalante
Consultant, Cost Rica

John A. Reinemund
Emeritus, U.S. Geological Survey

Merion J. Bergin
Emeritus, U.S. Geological Survey

OPENING REMARKS

I have the honor to address you in the name of the President of Costa Rica, Dr. Oscar Arias, who regrets not to be able to attend this opening session and who asked me to convey his greetings and welcome you to our country.

It is indeed a special occasion to have a select group of the international earth science community participating in this symposium. For us Costa Ricans, the technical papers that will be presented during the symposium are highly important, not only as scientific contributions, but also as guidelines for the development of our country.

Costa Rica is a small country, our geological resources are limited, and we need to improve our knowledge for the rational use and for the benefit of our people. The mineral resources are not completely known; aluminum laterite, known since the 1940s, is yet not explored, but new technologies will eventually be useful in providing a new source of income for our future generations. Gold mining in the past decades and at present represents the most important mining activity, which we hope will continue for several years. In non-metallic mineral activities we have made considerable progress, and we hope to expand this field and to balance the import of some minerals by local production.

In the field of energy, we have made a great advance in hydroelectric development and we are making an effort to develop our geothermal energy resources. Our potential in petroleum is very limited due to small sedimentary basins in fore-arc and back-arc tectonic conditions. However, we are also making an effort in basin evaluation for hydrocarbon accumulation.

Geographically, Costa Rica is located in a region of high geological risk, subject to frequent earthquakes, volcanic activity, and high rainfall, with the consequent results of flooding, mud flows, and landslides. Since the eruptions of Irazu volcano in 1963, which lasted two years, we have been able to organize a competent civil defense to cope with natural disasters. At present, this multi-institutional scheme works very effectively with the participation of several ministries, two national universities that are well equipped with seismic networks and volcanic activity monitoring experts, the national fire department, the International Red Cross, hospitals, civil guard, and other institutions. In this case, I am proud to say that we have one of the best systems in Latin America to cope with natural disasters.

In the basic earth sciences, the Universidad de Costa Rica and the Universidad Nacional, during the past 20 to 30 years, have advanced considerably in the training of young scientists as well as in research, despite the limited funds available for this purpose.

We have been managing our geological resources with the best of our possibilities and knowledge, and we have been trying to minimize the geological risk. However, we are also open to new ideas, new technologies, and advice on how to deal with these aspects of the earth sciences. We hope to learn from you, not only this time but also in future occasions, and we also expect that Costa Rica will be the host country again for scientific meetings similar to this one.

In the name of the Costa Rican people and our Government, I thank the Circum-Pacific Council for Energy and Mineral Resources, and especially its Chairman, Dr. Michel T. Halbouty, for holding this symposium in our country. We wish you a successful meeting and pleasant stay in Costa Rica. Thanks to all of you.

Jorge Manuel Dengo
Vice President, Costa Rica

Contributors

Fernando Alvarado
RECOPE
P.O. Box 4351-1000
San Jose, Costa Rica

Francisco J. Amezquita
Director General de Mineria
Secretaria de Estado de Industria y Comercio
Santo Domingo, Distrito Nacional
Dominican Republic

Allan Astorga
RECOPE
Gerencia de Produccion Primaria
P.O. Box 4351-1000
San Jose, Costa Rica

Lesmes E. Ballestero
RECOPE
P.O. Box 4351-1000
San Jose, Costa Rica

Fred Barnard
1835 Alkire Street
Golden, Colorado 80401

Rafael Barquero
Seccion de Sismologia e Ing. Sismica
Departamento de Geologia
Instituto Costarricense de Electricidad
Apdo. 10032
San Jose, Costa Rica

George Berman
Panama Canal Commission
Republic of Panama
APO Miami, Florida 34011-5000

Jevan Berrange
British Geological Survey
Overseas Directorate
Keyworth
Nottinghamshire NG12 5GG
United Kingdom

Kenneth Bolanos I.
RECOPE
P.O. Box 4351
San Jose, Costa Rica

A. A. Caicedo
Unidad de Desarrollo Geotermico
Instituto Nacional de Electrificacion
Torre Professional I
5° nivel, Centro Comercial Zona 4
Guatemala City, Guatemala

George Carpenter
Minerals Management Service
Offshore Resource Evaluation
Division MS 643
Reston, Virginia 22091

Arthur D. Cohen
Department of Geological Sciences
University of South Carolina
Columbia, South Carolina 29208

Thomas E. Daly
Resources Engineering and Management International, Inc.
5680 Greenwood Plaza Blvd., Suite 414
Englewood, Colorado 80111

Carlton E. Davis
Jamaica Bauxite Institute
Hope Gardens
P.O. Box 355
Kingston 6, Jamaica W.I.

Gabriel Dengo
Centro de Estudios Geologicos de America Central
Apartado 468
Guatemala City, Guatemala

Gregorio Escalante
Apartado 74, Pavas
San Jose, Costa Rica

Wilmer Flores
National Company of Electrical Energy
Apartado 99
Tegucigalpa, Honduras

Thomas Flynn
University of Nevada, Las Vegas
Environmental Research Center
Division of Earth Sciences
100 Washington Street, Suite 201
Reno, Nevada 89503

S. S. D. Foster
British Geological Survey
Casilla Postal 4337
Lima 100, Peru

Julio C. Franco A.
Ministerio de Energia y Minas
Direccion General de Hidrocarburos
Departamento Desarrollo Petrolero
Seccion de Geologia
Diagonal 17, 29-78 zona 11
01011 Guatemala, C. A.

Sue Goff
Los Alamos National Laboratory
EES-1, MS D462
Los Alamos, New Mexico 87545

Hernan Gomez-Mejia
Subdireccion de Docencia
Instituto Geografico "Augustin Codazzi"
A. A. 53754
Bogota, Colombia

Daniel B. Krinsley
2475 Virginia Avenue N.W.
Washington, D.C 20037

Michael J. Kozuch
Peace Corps
A. P. 3158
Tegucigalpa, Honduras

E. R. Landis
U.S. Geological Survey
P.O. Box 25046
Denver Federal Center
Denver, Colorado 80225

Matthew C. Larsen
U.S. Geological Survey
P.O. Box 4424
San Juan, Puerto Rico 00936

D. K. Larue
University of Puerto Rico
Department of Geology, R. U. M.
Mayaguez, Puerto Rico 00709-5000

Steve Ludington
U.S. Geological Survey
345 Middlefield Road MS 345
Menlo Park, California 94025

William D. MacDonald
Department of Geological Sciences
State University of New York
at Binghamton
Binghamton, New York 13901

Gregory E. McKelvey
U.S. Geological Survey
210 E. 7th Street
Tucson, Arizona 85701

Ernesto Martinez Tiffer
Instituto Nicaraguense de Energia
Direccion General de
Recursos Geotermicos
Aparado Postal 55
Managua, Nicaragua

Orlando Martino
Branch of Latin America
and Canada
Division of International Minerals
Bureau of Mines
U.S. Department of the Interior
2401 E Street, N.W.
Washington, D.C. 20241

S. J. Mathers
British Geological Survey
Keyworth, Nottingham NG12 5GG
United Kingdom

A. A. Meyerhoff
P. O. Box 4602
Tulsa, Oklahoma 74159

Walter Montero
Escuela Centroamericana
de Geologia
Universidad de Costa Rica
Apartado 35 2600 UCR
San Jose, Costa Rica

Sergio Mora C.
Department of Geology
Instituto Costarricense de
Electricidad
Apdo. 10032
San Jose, Costa Rica

Hadyn H. Murray
Department of Geology
Indiana University
Bloomington, Indiana 47405

Luis Gmo. Obando
RECOPE
P.O. Box 4351
San Jose, Costa Rica

James L. Pindell
Department of Earth Sciences
Dartmouth College
Hanover, New Hampshire 03755

Arturo Quesada Munoz
Escuela de Fisica
Universidad de El Salvador
San Salvador, El Salvador

Robert A. Renken
U.S. Geological Survey
Water Resources Division
6481-B Peachtree Industrial Blvd.
Doraville, George 30360

William H. Roberts III
HYDREXCO
11315 Bothwell
Houston, Texas 77024

Simon E. Rodriguez M.
ICGP National Committee
Ministry of Energy and Mines
Piso 5, Torre Oeste
Parque Central
Caracas, D.F.
Venezuela

Frederick R. Siegel
Department of Geology
George Washington University
Washington, D.C. 20052

David Sussman
Unocal Geothermal Division
3576 Unocal Place
P.O. Box 6854
Santa Rosa, California 95406

Denise Swink
Office of Planning and Environment
U.S. Department of Energy
Washington, D.C. 20585

Gary Thayer
Los Alamos National Laboratory
Los Alamos, New Mexico 87545

Linda K. Trocki
Los Alamos National Laboratory
Los Alamos, New Mexico 87545

Ed Van Eeckhout
Technical Coordinator
Central American Work
Earth and Space Sciences Division
Los Alamos National Laboratory
Los Alamos, New Mexico 87545

Agustin L. Viesca
Unocal Geothermal Division
P.O. Box 6854
Santa Rosa, California 95406

L. A. Wug
Ministerio de Energia y Minas
Direccion General de Hidrocarburos
Departamento Desarrollo Petrolero
Seccion de Geologia
Dagonal 17, 29-78 zona 11
01011 Giatemala, C. A.

Allen Zack
Caribbean District Chief
Water Resources Division
U.S. Geological Survey
GPO Box 4424
San Juan, Puerto Rico 00936

TABLE OF CONTENTS

Overview

Petroleum Resources

Coal and Peat Resources

Geothermal Resources

Industrial Minerals and Metallic Mineral Resources

Water Resources

Geologic Hazards

OVERVIEW

Circum-Pacific Council for Energy and Mineral Resources
Earth Science Series, Vol. 16
R. L. Miller, G. Escalante, J. A. Reinemund, and M. J. Bergin (Eds.)
Energy and Mineral Potential of the Central American-Caribbean Regions
Springer-Verlag, Berlin Heidelberg 1995

THE ROLE OF NATIONAL INSTITUTIONS AND INTERNATIONAL COOPERATION IN THE DEVELOPMENT OF THE GEOLOGICAL SCIENCES IN THE CENTRAL AMERICAN-CARIBBEAN REGION

Gabriel Dengo
Centro de Estudios Geológicos de América Central, Apartado 468, Guatemala City, Guatemala

ABSTRACT

The development of the geological sciences in the Central American-Caribbean region has been uneven as a result of different circumstances in each country. In spite of this, the knowledge of basic geology has advanced rapidly after World War II through the input from local, foreign, and international institutions, as well as from private companies.

Some historical aspects are: (1) the discovery of gold in the Río Yaque del Norte, Hispaniola in 1493, which initiated the Spanish search for minerals; (2) the inception of governmental mining offices with the Alcaldía Mayor de Minas, established in 1601 in Tegucigalpa; (3) the foundation of the Escuela de Minas in Medellín in 1887, which marked the beginning of professional training; and (4) the French Scientific Mission in El Salvador and Guatemala in 1865-1866, probably the first international cooperation in the region.

This paper presents the main aspects related to the local geological surveys, bilateral cooperation, and international programs in the land areas. The amount of work performed at sea is so overwhelming that this subject merits a separate presentation. Today, all countries in the region have national institutions in charge of geological mapping and supervision of mining and petroleum operations. Most of them have seismic networks to monitor this earthquake prone region.

Bilateral cooperation has played a very important role. Technical assistance from the United States, the Federal Republic of Germany, France, the Netherlands, Sweden, the U.S.S.R., the United Kingdom, and Japan has been of great importance. The assistance from the international organizations has also been considerable, particularly the programs of the UNDP, UNESCO, and OAS. These cooperative investigations covered a variety of subject matter including mineral exploration, groundwater, geothermal energy, seismology, training, and support of university centers.

The current Central American Geologic Map Project of the Circum-Pacific Council for Energy and Mineral Resources and of the Central American countries is the most recent case of international cooperation and coordinated regional effort.

INTRODUCTION

This paper is a summary of some salient aspects relating to the history of geological investigations in the Caribbean and Central America. It is a compilation of material presented more extensively in other papers such as Draper and Dengo (1990) on the Caribbean, Dengo (1988) on Central America, and Botero Arango (1978) on Colombia. It also includes information provided verbally by several colleagues. Therefore, this paper does not try to deal exhaustively with the topic or with all countries, but rather points out some important aspects of national and international cooperation. Several specific cases are mentioned instead of dealing separately with every country of the region.

Some of the circumstances that mark the inception of geological exploration in the region happened in the early years of the Spanish colonization, among them the casual discovery of placer gold in the Río Yaque del Norte, Hispaniola in 1493, which initiated the search for minerals by the Spaniards. This was followed by Columbus finding gold ornaments in use by the natives by when he landed in Cariari on the Caribbean coast of Costa Rica in 1502, during his fourth trip to the West Indies.

These events led to the intensive search for minerals, mostly gold, by the Spaniards and years later by the English, Dutch, French, and Portuguese, which resulted in the famous legend of El Dorado (Treasure City).

During the nineteenth century the geological knowledge of the region was augmented by the contributions from travelers and descriptions by scientific explorers like Alexander von Humboldt and several others who followed his steps. One case that is worth mentioning is that of Henty T. de la Beche who between 1823 and 1825 visited the plantations of his family in Jamaica, studied the geology of the island, and later published a geological description with a map. This is one of the oldest geological maps of any part of the Western Hemisphere and the first one of a part of the Caribbean region. De la Beche later was the founder of the British Geological Survey.

NATIONAL INSTITUTIONS

Mineral exploration and exploitation and, much later, petroleum exploration led to the establishment of governmental offices in charge of the regulation and fiscalization of these activities and, even later, of offices that may be grouped as Geological Surveys, although their names vary from country to country. Perhaps the first office of the kind was the Alcaldía Mayor de Minas, established in Tegucigalpa, Honduras in 1601.

It would be tedious to describe the history of the different national institutions that deal with the earth sciences in each of the countries of the region. For those persons interested in these matters, an account of how the Central American institutions developed and of those existing today was presented in some detail by Dengo (1988) and, in the case of Colombia, by Botero Arango (1978).

The present situation is that practically every country in the region has one or more offices that deal with geology, sometimes independently or as part of a larger institution that also deals with petroleum and mining concessions and fiscalization, as is the case of the Ministries of Energy and Mines in Venezuela, Guatemala, and Costa Rica, to mention some specific cases. However, in some countries like Guatemala the geological mapping is the responsibility of the Geographic Institute and there is a separate institution that deals with volcanology and seismology.

Two cases merit special mention, those of Colombia and Venezuela, which, besides being the largest countries in the region, also have a longer tradition of geological studies. This is particularly true for Colombia as the result of the foundation of the Escuela de Minas de Medellín in 1887. In Colombia the Instituto de Investigaciones Geológico-Mineras (INGEOMINAS) at present is probably the best equipped geological institution in the region both in human resources and physical equipment and one of the most active in research programs. This institute has contributed extensively to the knowledge of the geological resources of the country with continuous mapping programs and economic and other applied geological studies.

In Venezuela the Dirección de Geología (Ministry of Energy and Mines), which started as a small office under a different name in 1937, has also reached an excellent status in research, mapping, and thematic studies. Unfortunately, due to economic constraints in recent years its activities have declined, but it still is staffed by highly competent professionals.

Other cases that are worth mentioning are those of Cuba, the Dominican Republic, and Honduras. Cuba, which is the third country of the region in size and which has a very complex geology, has made a considerable advance in research and institutional development during recent decades. The geological research has been the product mainly of two institutions, namely the Centro de Investigaciones y Desarrollo del Petróleo (Ministry of Basic Industry) and the Instituto de Geología y Paleontología of the Academy of Sciences. The Dominican Republic and Honduras are almost parallel cases that have changed from almost inactive Direcciónes de Minas to well equipped and properly staffed aggressive organizations in geological studies and development of their mineral resources.

After these general remarks, with the exceptions of Colombia, Venezuela, and Cuba, it is necessary to state that one of the major problems that the other countries have faced has been the lack of national policies to study and map the basic geology as part of their development programs. This area is where international cooperation has been of utmost importance.

INTERNATIONAL COOPERATION

The international cooperation has been of different types, in some cases originating at the initiative of one or more individuals interested in studying the geology of certain areas, in other cases at the request of a particular country for assistance in specific fields such as mineral deposits, petroleum, ground water, and geothermal energy potential.

Perhaps the first international geological activity in the region, at least in Central America, was that of the French Scientific Mission in El Salvador and Guatemala, sent at the initiative of Napoleon III during 1865 and 1866 (Dollfus and de Montserrat, 1868). This Mission established the basis of the stratigraphic knowledge of the area and also produced one of the first geological maps of a part of the Western Hemisphere.

Much later, the U.S. Government sent a geological mission to the island of Hispaniola to undertake a geological reconnaissance, which also was the foundation of the basic geology of the Dominican Republic and Haiti (Vaughan et al., 1921). The individual studies during the last part of the nineteenth and early years of the twentieth centuries are not mentioned here because they are not considered international institutional cooperation. The geological contributions of those scientists are credited in the publication by Draper and Dengo (1990).

One of the more important types of cooperation has been the role played by several foreign universities, mostly by sending students to map areas and to write theses for advanced degrees. In these cases the national geological institutions also played a key role by providing local counterparts and logistic support. The first effort of this kind was started by L. M. R. Rutten of the University of Utrecht, The Netherlands, directed toward geological investigations in the Netherlands Antilles beginning in 1920 and followed by extensive studies in Cuba in the 1930s.

Shortly before the Second World War, Harry H. Hess started the Princeton Caribbean Research Project on the island of Tobago. The project was interrupted during the war years, but work in the region continued again in 1947 with mapping of the Caribbean Mountain System in northern Venezuela. It was also extended further to the island of Saint Martin and to the Dominican Republic and Puerto Rico. The studies in Venezuela continued for more than 20 years and were extended also to Colombia under the supervision of other Princeton professors.

One of Hess' students, Thomas W. Donnelly, started geologic quadrangle mapping in Guatemala, and then as Professor at Rice University and later at the State University of New York at Binghamton sent many of his students to map in Guatemala with the cooperation of the National Geographic Institute of that country. After this excellent start, the program broadened to include students from other universities, namely the University of Texas at Austin, Louisiana State University, and Dartmouth College. The University of Texas at Austin and Wesleyan University in Massachusetts have had mapping programs in Honduras. These have been continued by volunteers of the U.S. Peace Corps. In Nicaragua one thesis on regional volcanic geology was supported by the Universidad de Córdoba of Argentina. In recent years several French universities have sent students to Venezuela and Colombia, as well as to other countries of Latin America, to undertake geological studies for their advanced degrees. There must be still other cases of foreign universities that have had similar programs in Caribbean countries but of which the author is not aware.

International cooperation of another kind, specifically bilateral country technical assistance on projects sponsored by regional organizations, either through soft loans or non-reimbursable funds, has been considerable during the post-war years and continues at the present time.

One of the programs of regional coverage, which has produced excellent results in institutional building, technical training, and final products, was that of the Interamerican Geodetic Survey in topographic mapping, working in close collaboration with national cartographic institutions. This was not a geological program, but it has provided the topographic base maps on which most of the geological mapping of the countries has been done. It also provided a large amount of gravimetrical data obtained for geodetic purposes, which has been a useful geological tool, particularly in the interpretation of different models of the lithosphere.

Another American institution that has provided several kinds of technical assistance in the region is the U.S. Geological Survey. During the war years it made a regional study of the mineral resources of Central America. After the war it has made investigations in several geological disciplines throughout the region. Among the most important of these has been the mineral evaluation of Colombia in collaboration with INGEOMINAS on the project called Inventario Minero. In Venezuela and the Dominican Republic it has also participated in mineral deposit studies. In Central America the U.S. Geological Survey has participated with local institutions in several volcanological studies as well as in seismological and other studies related to major earthquakes such as the Managua earthquake in 1973 and the Guatemala earthquake in 1976. A different situation exists in Puerto Rico where, as a Commonwealth of the United States, the U.S. Geological Survey has worked extensively.

Another case of foreign technical assistance in the region is that of the geological mapping performed in El Salvador and Guatemala by the Bundesanstalt fur Geowissenschaften und Rohstoffe (BGR) of the Federal Republic of Germany in the 1960s. In El Salvador it produced a geological map of the entire country at a scale of 1:100,000, and in Guatemala a similar one of a fairly large area in the central part of the country. In more recent years the BGR initiated a similar mapping program in the Dominican Republic.

Other European countries have also provided valuable assistance to several of the countries of the region. Countries of the Socialist Block have performed geological studies in Cuba and Nicaragua. Sweden has provided assistance in mineral resource evaluation to Nicaragua over a period of several years, and has recently also established a regional center in Guatemala for geological hazard evaluation and prevention. A different category of assistance that needs to be mentioned is the geologic work of the United Kingdom, the Netherlands, and France in their former colonies. An important feature is their support in establishing local geological offices, particularly in Jamaica and Trinidad-Tobago. Besides European countries and the United States, Japan has provided technical cooperation to several countries in mineral and geothermal evaluation.

Several international organizations have had almost continuous projects in one country or another in the region. Among these organizations, the United Nations and some of its specialized agencies have played a very important role. In the 1960s and 1970s the UN supported mineral evaluation programs in all the Central American countries. It also supported ground water studies in Nicaragua, Costa Rica, and Guatemala and geothermal energy evaluation in El Salvador. Later it gave technical assistance in offshore petroleum exploration to Trinidad-Tobago, El Salvador, and, more recently, to Belize. Through UNESCO it has sponsored seismological studies in the Lesser Antilles, Central America, and northern South America. The Organization of American States (OAS) has provided assistance to many countries in training, seismology, and mineral inventories. Also, the Interamerican Development Bank has supported several studies related to energy resources and to mineral resource inventories, and the World Bank has supported similar studies related to petroleum exploration.

One important activity of a different sort because it has the participation of a working group of local geologists and a non-governmental organization, the Circum-Pacific Council for Energy and Mineral Resources, is the sponsorship of the ongoing Central American Geologic Map Project (scale 1:1,000,000). The Council was also the co-sponsor with the Costa Rican government of the 1989 symposium on Energy and Mineral Potential of the Central American-Caribbean Region. This publication results from papers delivered at that symposium.

This short and, by necessity, incomplete overview shows that the role of international cooperation is of utmost importance in the development of the geological sciences in the Caribbean-Central American region. However, it also shows that international cooperation, in order to be more effective, needs the support and active participation of national institutions. This situation should be recognized by the governments of the region in order to strengthen the technical and professional capabilities of their own geological surveys.

REFERENCES CITED

Botero Arango, G., 1978, Apuntes para una historia de las investigaciones geológicas en Colombia: Ciencia y tecnología en Colombia: Bogotá, Instituto Colombiano de Cultura, Editorial Escala, p. 147-169.

Dengo, G., 1988, Historia del desarrollo del conocimiento geológico de América Central: Anales de la Academia de Geografía e Historia de Guatemala, v. 62, p. 153-190.

Draper, G., and G. Dengo, 1990, History of geological investigation in the Caribbean region, *in* G. Dengo and J. E. Case, eds., The Caribbean region: Geological Society of America, Decade of North American Geology, v. H., p. 1-14.

Dollfus, A., and E. de Montserrat, 1868, Voyage geologique dans les Republiques de Guatemala et de El Salvador: Paris, Mission Scientifique au Mexique et dans l'Amerique Central, 539p.

Vaughan, T. W., W. Cooke, D. D. Condit, C. P. Ross, W. P. Woodring, and F. C. Calkins, 1921, A geological reconnaissance of the Dominican Republic: Washington, U.S. Geological Survey, Press of Gibson Brothers, Inc., 268p.

Circum-Pacific Council for Energy and Mineral Resources
Earth Science Series, Vol. 16
R. L. Miller, G. Escalante, J. A. Reinemund, and M. J. Bergin (Eds.)
Energy and Mineral Potential of the Central American-Caribbean Regions
Springer-Verlag, Berlin Heidelberg 1995

THE KNOWN AND POTENTIAL RESOURCE BASE FOR ECONOMIC DEVELOPMENT IN THE CENTRAL AMERICAN-CARIBBEAN REGION

Gregorio Escalante
Apartado 74, Pavas, San José, Costa Rica

ABSTRACT

The endeavors of local governments, of foreign and international institutions, as well as those efforts of private companies to evaluate the energy and mineral resources of the Central American and Caribbean countries, reveal a wide divergence in the results obtained for each individual country. These differences are due to factors or situations that have conditioned, stimulated, or limited the endeavors. Among these factors are the differences in the customs that characterize the Latin American countries, particularly those in Central America and the Caribbean, and also the negative attitudes of some governments toward the efforts of outsiders to study their resources. Of particular concern are those governments which have been unable to understand the importance of obtaining a potential resource base as a key factor in the economic development of their countries. Also pertinent to this problem has been the lack of native professional people and trained personnel able to carry out and give continuity to those programs that have been devised to evaluate the energy and mineral potential of the country. Of great significance to the success of such programs is the need for establishing and properly financing the governmental institutions that are to produce and eventually publish the data base required for planning the economic development of a country. Efforts to obtain an adequate data base for economic development in the countries of Central America, the area with which the author is familiar, are here considered.

Bilateral cooperation and international programs, especially those of a regional nature, recently completed or in progress in Central America as well as in the Caribbean, are now more likely to succeed than in the past. This optimistic viewpoint is based on the fact that most of these countries now have available native professionals capable of implementing the programs directed toward establishing the resource base for economic development and are able to lend continuity to the studies.

PROBLEMS OF THE PAST AND THE PRESENT

The economic development of the vast region covering Central America and the Caribbean has been very different for each country. This divergence derives from the varied customs that characterize every one of these countries and especially those that bear on the educational standards for each. Also important is the attitude of the governments that should be concerned with the programs that are designed to evaluate the potential resource base for each country. This attitude may be lacking in the necessary zeal.

For this paper, I shall rely mostly on my personal experience, especially that which I have gained from many years work as a geologist, particularly in Central America. I would like to narrate some anecdotes which I believe point out problems that, in many instances, have hampered some programs implemented by governmental or international institutions, and which were designed to properly evaluate the national resources. I also hope in this paper to be able to point out various factors or measures which should be considered by governments and international agencies so as to get the most value out of the efforts to establish the resource base for economic development of the countries of the Central American-Caribbean region.

From a personal point of view, I think that there are three conditions that have greatly limited the efforts of the countries to establish an adequate resource base for economic development. These conditions are:

1. The little importance and low priority that some governments have assigned to the study of a resource base for economic development, especially to those resources that involve the energy and minerals potential of the countries.

2. The scarcity of qualified personnel who are capable of understanding the true value of the information and who possess the necessary skills to properly catalog and use the information.

3. The lack of methods or means by which to divulge, within and outside the country, information that could be of great value to national and international agencies concerned with the economic planning for the countries.

The first concern is basically a problem of economics. The small countries that comprise the Central American-Caribbean region are, for the most part, very poor countries that cannot devote much of their limited economic resources to carrying out programs which do not produce immediate results. Sometimes, and quite unfortunately, this attitude may be influenced by the short-term need for producing a favorable political image. This situation is, fortunately, being solved with better planning and better organization of the geologic programs of the international agencies. A further contribution toward a solution of this problem is forthcoming with the emergence of governmental institutions dedicated to economic planning.

The second condition, mentioned previously, concerns the lack of qualified native personnel to properly direct or coordinate the programs leading to the investigation of a potential resource base of a country. This problem, which has been of major concern in the past, is slowly being solved with the availability of more capable personnel to direct and staff the programs and the recognition by the governments of the value of this kind of work.

Especially important to the success of resource evaluation is the need for obtaining the services of the personnel able to properly recognize and safeguard the valuable information with which they have been entrusted and who are endowed with the skills to prepare the reports and maps that summarize the work carried out by the government and by international and private institutions. A few examples from the past come to mind, and they vividly portray the problems deriving from the lack of capable personnel who were entrusted to safeguard this information but who were unable to recognize its value.

Many years ago in my own country, Costa Rica, a high government official permitted the destruction of an important collection of rock samples obtained from the drilling of more than ten deep wells by Union Oil Company of California and Gulf Corporation. In what used to be the parking lot of a government office entrusted with this valuable collection lay scattered pieces of rock, in hopeless disarray, which could have been reanalyzed using techniques that have been developed recently.

Soon after the Managua earthquake of 1972, I watched with horror how the flames consumed the rubble of a building where the Government of Nicaragua kept all the basic data on geology and mining for the entire country. Some years previously I also witnessed in Guatemala the confusion that followed the government's decision to open the geologic files to outsiders. For many days people of all sorts had the free run of the offices that contained valuable maps and reports. I cannot help but wonder how much of this collection of data was lost or has deteriorated as a result of the ill advised, uncontrolled access to the material.

My third point of concern is the need for making the best use of the basic information on the resource potential for each country by making it available to the people, the agencies, and the institutions which value it and need it in formulating their plans to develop the economies of the Central American and Caribbean countries. On this subject, unfortunately, I do not believe that we are making much headway. I do not believe that a nation will gain much by gathering information on its potential resource base if such is to be kept a state secret or a mysterious asset to be seen by just a few. Of course, any government has the right and obligation to declare the confidentiality of certain information on the resources of the country if it concerns national security or is in the interests of a future governmental commercial transaction. For example, I would be in agreement if the government treats as confidential information which could form part of a commercial promotion on the petroliferous potential of a particular area. I firmly believe, however, that some governments have been overzealous in keeping to themselves this valuable information. Examples of this unfortunate situation are very numerous. In fact, it is by giving out basic information on the potential resources of the country that the country will eventually gain the most in achieving a viable program for economic development.

ACCOMPLISHMENTS AND AN OPTIMISTIC VIEW OF THE FUTURE

I do not wish to present at this time a detailed description of the successful programs that have been carried out by numerous international and governmental institutions which have led to better understanding of the resource base now enjoyed by the countries of Central America and the Caribbean. This story of success was explained and reviewed during the symposium which generated this volume in the numerous papers presented or illustrated in the poster sessions by prominent scientists. The papers covered a number of disciplines and concerned the geology and resources of many countries.

I do wish, however, to refer briefly to some of the most important accomplishments in gaining a knowledge of the economic resources of the countries of Central America (Guatemala, Belize, El Salvador, Honduras, Nicaragua, Costa Rica, and Panama). This is the region with which I am personally best acquainted, particularly in the area of energy and mineral resources which are the concerns of my professional interests.

A review of the state of the knowledge on the geology, energy, and mineral resources of the Central American countries indicates a great advance, as exemplified by the quantity and quality of geological publications and maps which have been produced during the last decade. Of special interest is the status of geological mapping in the region. These maps incorporate much of the information required for economic planning. The Central American countries now possess many adequate geological maps, although extensive areas covered by jungle or of difficult access remain to be mapped. Noteworthy among those maps of a regional nature is the metalogenetic map of Central America, at a scale of 1:2,000,000, published in 1970 by the Central American Institute for Technology and Research

(Instituto Centroamericano de Investigación y Tecnología Industrial). The Geological Map of the Caribbean, prepared by the U.S. Geological Survey in 1980 at a scale of 1:2,500,000, places Central America in its regional geological framework, as had never before been done, by incorporating basic information on the marine geology. A new geologic map of Central America, at a scale of 1:1,000,000 (to be named geodynamic map), now in progress, will be the result of a joint effort by different institutions from the seven countries of Central America under the auspices of the Circum-Pacific Council. Progress in this regard has been outstanding, especially when consideration is given to the fact that just 25 years ago the most complete geologic map of Central America was that of Roberts and Irving, published in 1957 at a scale of 1:1,000,000, which was based mostly on the work of Sapper during the early part of this century.

Each of the countries of Central America has geological maps at different scales. Nicaragua has one of the whole country at a scale of 1:1,000,000 published in 1973 and has one in progress at a scale of 1:500,000. Those of Guatemala (1970) and Honduras (1974), also at a scale of 1:500,000, permit considerable detail. In 1976, Panama published a geologic map of the country at a scale of 1:250,000 and Costa Rica published in 1982 a geological map at a scale of 1:200,000. El Salvador has had since 1978 a geological map at a scale of 1:100,000, which remains today as the most complete and detailed national map coverage in the entire region. Numerous geological maps have been published covering parts of the Central American countries at the scale of 1:50,000. A number of these use as a base quadrangles or sheets which form part of the topographic mapping system of each country. Noteworthy among these maps are geological maps from Guatemala and Honduras which are the results of thesis work (Master's and Ph.D's) carried out by students from universities of the United States. This work, which has unfortunately been discontinued, produced very good results at a very low cost for the governments concerned. Maps on mineral resources, volcanism, seismicity, hydrogeology, and geophysics have been compiled and published for some of the Central American countries or for portions of these countries.

A definite trend toward a very rapid advance in the knowledge of the natural resources of Central America (and the Caribbean as well) is quite evident and dates from the last ten years or so. It is the result of ample cooperation of foreign and international institutions in development programs. This valuable international support has been the single most important factor in the accelerated progress. Applications of new technologies in these international programs, especially the use of satellite imagery, have also contributed greatly to the advance in the knowledge of national resources. An important factor in these advances concerns the positive attitude of the governmental personnel in many of the countries. They have come to recognize the value of the programs which are leading toward the understanding of the resources of the country. Another most significant factor in the search for a potential resource base for the economic development of the countries has been, in my view, the fast increase in the number of native professionals who have become available to direct, to work on, and to follow up the programs.

In 1965, during the celebration of a general assembly of the old Association of Geologists of Costa Rica, we became perturbed upon discovering to our amazement that the Board of Directors of the association was made up of half of the total members of the group! Today, the Colegio de Geólogos de Costa Rica, a professional society of geologists, as of this date counts 120 active members. In the case of Central America this has to a great extent been a felicitous result of the establishment of the Central American School of Geology in Costa Rica, which has already graduated more than 200 Central American geologists since it was inaugurated in 1970. I believe that the substantial increase in the number of native geologists in Central America, which includes an impressive list of specialists in different disciplines of the profession, has been the single, most important contributing factor to the recent advances in the knowledge of the energy and minerals potential. This becomes apparent when we take note of the large number of native professional geologists who have taken commanding jobs in governmental institutions and in private enterprises in these countries.

I feel quite optimistic about the future. We in Central America are advancing at a rapid pace in the knowledge of the potential of our natural resources. In 1965 during the First Meeting of Central American Geologists we had the participation of 25 geologists who presented eight papers. In the symposium on Energy and Mineral Potential of the Central American-Caribbean Region in San Jose (March, 1989) 60 papers on Central America were presented.

There is still much more to be done. Important information on the geology of Central America is still missing. We need to define the boundaries of the sedimentary basins of Central America and of the Caribbean as well. We need to obtain a better understanding of the facies changes in these basins and we require the establishment of more extensive programs to obtain radiometric datings, a better definition of the volcanic groups, and most importantly, we need to undertake the geologic mapping of extensive areas which today remain little known.

With regard to geologic mapping, I wish to indulge myself by bringing into consideration some thoughts on this matter and to present some ideas which I hope will contribute to an accelerated and a relatively inexpensive program directed toward expanding the basic knowledge of the resources of the countries of Central America.

I believe that the geologists of Central America are not making full use of a precious resource. They have at their disposal the basic topographic maps at a scale of 1:50,000, which are ideally suited for geological mapping and which give an accurate definition of those features necessary to the planning of hydroelectric projects and to the location of road systems. Such maps are ideally suited also to the search for mineral deposits, to determination of the potential for hydrocarbons, and to the recognition of areas that are at potential geological risk. The Central American countries have unfortunately made little use of this auspicious situation. I am fully convinced that a systematic geological mapping program at a scale of 1:50,000 is a very worthwhile undertaking. Therefore, I look forward with great expectation for the geologists of Central America to get out their geological picks and Bruntons and to start on this mapping endeavor. In all honesty I cannot understand why some governmental institutions make use of sophisticated electronic gadgetry, while the geology of large areas, some near main urban centers, remain today geologically unknown. What would the geologists of yesterday, those of us who suffered the rigors of making pace and compass surveys, not have given for the topographic maps that have now become available to the present generation of geologists!

Circum-Pacific Council for Energy and Mineral Resources
Earth Science Series, Vol. 16
R. L. Miller, G. Escalante, J. A. Reinemund, and M. J. Bergin (Eds.)
Energy and Mineral Potential of the Central American-Caribbean Regions
Springer-Verlag, Berlin Heidelberg 1995

GEODYNAMIC MAP OF COSTA RICA

Walter Montero P., Sergio Paniagua P., Siegfried Kussmaul, and Francois Rivier
Escuela Centroamericana de Geologia, Universidad de Costa Rica
Apartado 35, 2600 UCR, San José, Costa Rica

ABSTRACT

A geodynamic map of Costa Rica has been compiled from different sources. The map includes Plio-Pleistocene to Recent volcanic centers, magnitude Mb ≥ 4.5 earthquakes for the period 1964 to 1986, interplate and intraplate focal mechanism solutions for selected earthquakes, Quaternary and Recent faulting, Bouguer gravimetric anomalies with free air anomalies offshore, heat flow, and neotectonic stress directions.

INTRODUCTION

The Central American region is a very complex neotectonic area. The internal forces that exist in this area cause movements, develop structures, and result in a geologic environment on which we can get much direct or indirect information. These sources of information include seismicity, focal mechanisms, gravimetric anomalies, Quaternary volcanic centers, Quaternary faulting, geologic mapping, neotectonic stress directions, and heat flow data. Data on these geological and geophysical technologies are included on the simplified geodynamic map of Costa Rica (Figure 1) and are briefly described and interpreted in this paper. (The original geodynamic map at a scale of 1:750,000 will be printed by the Instituto Geográfico Nacional, San José, Costa Rica, and will be available on request to the senior author.)

SEISMICITY AND FOCAL MECHANISMS

The seismicity of Costa Rica shown on the geodynamic map (Figure 1) is based on the earthquake catalog of the National Geophysical Data Center/U.S. National Oceanic and Atmospheric Administration (NOAA) for the period 1964-1968. The catalog includes earthquakes of magnitude Mb (body wave) ≥ 4.5. The hypocenters are divided into two depth ranges: 0-69 km and 70-250 km. Focal mechanism solutions were compiled from published sources (Molnar and Sykes, 1969; Instituto Costarricense de Electricidad (ICE), Data from the Arenal seismic network, Interim Report; Dean and Drake, 1978; Montero and Ponce, 1979; Montero and Dewey, 1982; Burbach et al., 1984; Montero and Morales, 1984; Güendel, 1985, 1986; Alvarado et al., 1988; Red Sismológica National, ICE-Universidad de Costa Rica (UCR), Resumen de la sismicidad del año 1987, Internal Report). They are classified into three quality categories according to the restriction of the nodal planes: Class A earthquakes have both nodal planes well restrained (10° maximum variation on azimuth or inclination); Class B earthquakes have only one nodal plane well restrained; and Class C earthquakes have both nodal planes poorly restrained or the P polarization data used in the focal mechanism solution are not presented in the references. A nodal plane which depends on the P-wave arrival of only one station is considered as poorly restrained.

Most shallow seismicity has occurred along the Pacific coast of the country. It is caused by interaction of the Caribbean and Cocos Plates (Figure 2). The largest related earthquakes were the Golfito earthquake of April 3, 1983 (Ms (surface wave) = 7.3, Mb = 6.5), and the Samara earthquakes of August 23, 1978 (Ms = 7.0, Mb = 5.7 and Mb = 5.4). They have thrust type focal mechanisms with a slip vector trending N30°E. Deeper focus earthquakes (greater than 70 km) have been fewer. They are related to subduction of the Cocos Plate,

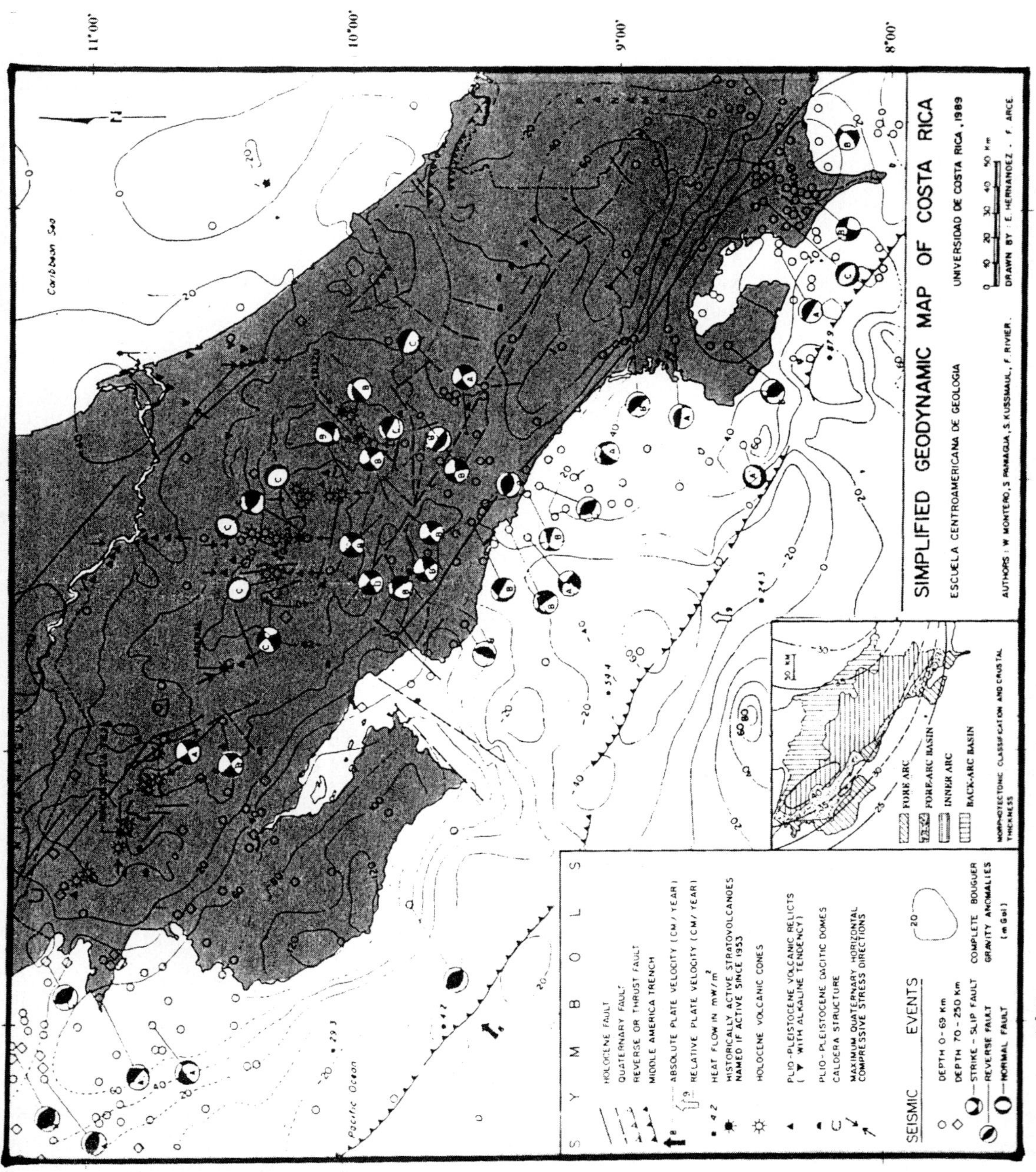

Figure 1. Map of Costa Rica showing plate motion, compressive stresses, faulting, gravity, volcanic centers, seismic events, and other features.

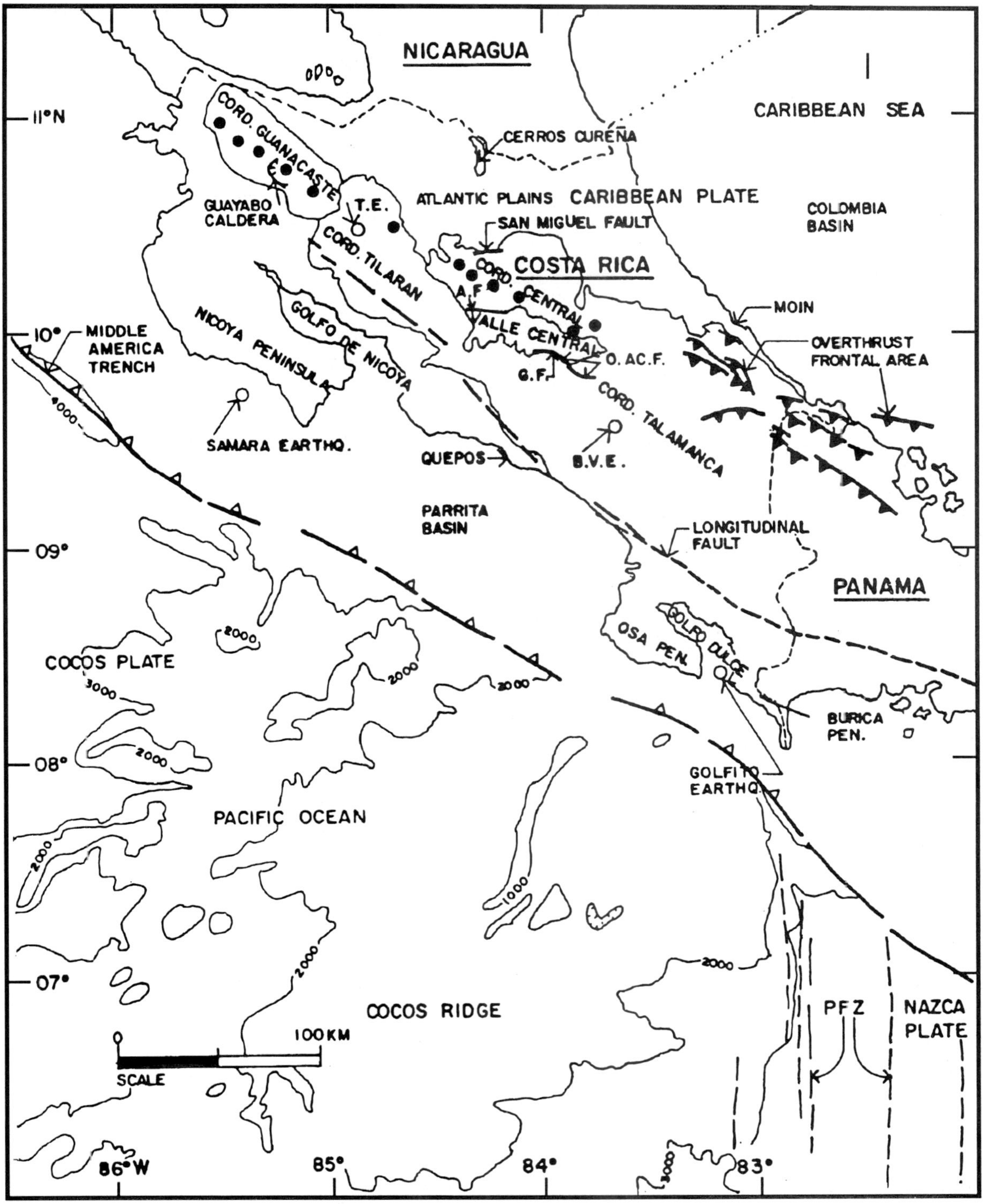

Figure 2. Generalized map showing principal geotectonic and volcanic features of Costa Rica. Morphotectonic and geographic terms used within the text are as follows: T.E. and B.V.E. refer to Tilaran and Buena Vista earthquakes, respectively. A.F., G.F., and O.AC.F. refer to the Alajuela, Guarco, and Orosi-Agua Caliente faults, respectively. PFZ refers to the Panama Fracture zone. Filled circles represent major Quaternary volcanos, unfilled circles are the epicenters of earthquakes mentioned in the text Unfilled teeth on the upper plate represent inferred zone of subduction, and filled teeth on the upper plate represent reverse or thrust fault. Based on Case and Holcombe (1980).

and most are located below the Quaternary volcanic cordilleras (the central and northwest regions of Costa Rica, Figure 1).

A normal faulting event with a hypocenter located on the inner side of the trench in the Golfo Dulce area (Figure 2) was interpreted by Molnar and Sykes (1969) as related to the bending of the Cocos lithosphere before it subducts below the continent. We propose that it is related to faulting induced by stresses associated with the tectonics of the triple junction area (Cocos-Caribbean-Nazca Plate interaction).

Strike-slip focal mechanisms were reported by Dean and Drake (1978) for the landward side of the Middle America Trench (Figure 1). They suggested that the faulting near 84°W, 10°N is related to alternating sinistral and dextral movements on a N30°E trending fracture in the oceanic lithosphere. A strike-slip fault event near 86°W and 10°45'N is spatially related to a transverse break in the Cocos Plate.

There are also shallow hypocenters along the inner arc. Damaging earthquakes which occurred in this region include the Tilarán earthquake of April 14, 1973 (Ms = 6.5, Mb = 5.7), and the Buena Vista earthquake of July 3, 1983 (Ms = 6.2, Mb = 5.9) (Figure 2). Both earthquakes have similar strike-slip focal mechanisms solutions with maximum horizontal compressive stress trending north.

GRAVIMETRIC ANOMALIES

The gravity anomalies shown on Figure 1 are from Ponce and Case (1987). The anomalies are free-air in the oceanic region and complete Bouguer in the continental area. The principal patterns shown by the gravity data moving from the Pacific to the Caribbean side begin with slightly positive free-air anomalies over oceanic basement on the marine side of the trench. There are strong local positive anomalies above the Cocos Ridge. Northwest-trending negative anomalies of over 70 mGal appear along the Middle America Trench caused by deepening of the basement by tectonic subduction. On the landward side of the trench and along the forearc, there are strong positive anomalies with values up to 140 mGal (Nicoya Peninsula). They decrease to the southeast over the Osa and Burica Peninsulas. These anomalies are related to the ophiolitic basement that crops out in these regions. The pattern is interrupted in the region south-southwest of Quepos where an 80 mGal amplitude negative anomaly is related to the Parrita basin. Between 84°10'W and 84°15'W there is a steep gravity gradient trending to the north that is related to a fracture zone in the Caribbean Plate. This interpretation is supported by seismicity (Morales and Montero, 1984) and by volcanic lineaments (Figure 1). There are northwest-trending negative anomalies over the inner arc. The anomalies over the Central and Talamanca Cordilleras are especially large. However, the latter one is not well constrained. The causes of the negative anomalies are the low density volcanic, plutonic, and sedimentary rocks that compose the cordilleras. In the back arc area, both slightly positive and negative anomalies reflect isostatic equilibrium.

QUATERNARY FAULTING

Quaternary faults (Figures 1 and 2) are taken from geological maps compiled by Tournon et al. (in preparation), Madrigal and Rojas (1980), Denyer and Montero (1988), Alvarado et al. (1988), and Kessel and Lowe (1987). Faults suspected or confirmed as Holocene are shown.

A prominent Holocene fault is the longitudinal fault of Costa Rica (Mora, 1979) which extends from Panama along the Pacific coast to the Gulf of Nicoya (Figures 1 and 2). It is a fault system that generally limits outcrops to the north from alluvial material to the south and separates two geomorphological zones. Berrangé and Thorpe (1988) interpreted this fault as a system of wrench faults with predominantly dextral strike-slip movement directly connected to the Panama fracture zone to the south (Figure 1).

On the Atlantic side of the Cordillera de Talamanca there is a frontal overthrust zone which extends offshore. A contact separates a very deep onshore sedimentary basin (> 8 km) with various lithologic repetitions due to overthrust tectonics from a thinner (3-4 km) marine basin, with simple Eocene paleostructures at Moin (Figure 2).

On the southern side of the Valle Central, the principal Quaternary faults are the Orosi-Agua Caliente (OACF) and the Guarco (G. E.) faults which are responsible for some historic earthquakes (Montero and Morales, 1988). On the northern side, the principal structure is the Alajuela thrust fault (A.F., Figure 2), a result of gravity sliding of the volcanic range impelled by the weight of the volcanic edifice and triggered by magma intrusions over the last 50,000 years (Borgia et al., 1987). A similar origin has been proposed for the San Miguel fault on the northern side of the Cordillera Central. Normal and strike-slip faulting transverse to the cordillera are also present (Alvarado et al., 1988).

In the Cordillera de Guanacaste, there are Quaternary faults with mainly north-south, and northwest-southeast directions; other minor structures (not shown on figures) trend east-west (Figure 1). They are located in the vicinity of Guayabo Caldera (ICE, 1983, Informe geo-vulcanológico, Proyecto Geotérmico Miravalles, Internal Report).

VOLCANIC STRUCTURES

Pliocene to Recent volcanism is concentrated in the northwestern part of Costa Rica. The location of the volcanos is based on the maps of Alvarado et al. (1980) and Barquero and Sáenz (1987) (Figures 1 and 2). The volcanos are classified into four groups using their record of activity (Alvarado, 1984; Paniagua and Soto, 1986), density of vegetation, morphology, and degree of dissection as criteria.

1. Stratovolcanos active within historic time. The volcanos of this group have built the Cordillera de Guanacaste

and the Cordillera Central. Both belts strike northwest-southeast and consist of structurally complex stratovolcanos. In the Cordillera de Guanacaste only Rincón de la Vieja has been active within the past one hundred years; tephra eruptions occurred in the years 1966-67, 1969-70, and 1983-84. At present only fumarolic activity persists. Reports of historic activity at other volcanos exist, but no signs of young tephra deposits have been found.

The Cordillera Central includes the volcanos Provenir-Platanar-Viejo, Poás, Barba, Irazú, and Turrialba. The volcanos Poás, Irazú, and Turrialba have erupted in the past two centuries and are now in the fumarolic stage. There are reports of historic eruptions of Viejo and Barba, but the reports are questionable. Poás had its last great eruption in 1953. Since 1953, small tephra eruptions occurred in 1961, 1964, 1978, and 1989. There is an alignment of volcanic centers in a northerly direction from the active crater, with a decrease in height and volume toward the north (Alvarado, 1984). Irazú had large tephra eruptions in the years 1963-65, with only minor fumarolic activity since that time.

Between the Cordillera de Guanacaste and the Cordillera Central is the Arenal volcano (Figure 1), the only conical stratovolcano in Costa Rica. Arenal volcano began its present phase of activity in 1968 with a pyroclastic flow. Since 1968 eruptive activity has been nearly continuous and has produced approximately 50 small aa-flows, the only ones of historic age in Costa Rica. The most typical eruptive behavior consists of phreatic and tephra explosions.

2. Holocene volcanos without historic reports of eruption, but with fairly well preserved craters. Most form part of the complex stratovolcano system of the Cordillera de Guanacaste and Cordillera Central. These inactive volcanos show various stages of erosion, but still preserve much of their bowl-shaped craters. Most of these large massifs have multiple craters and caldera collapses. Some show a migration of the volcanic activity in various directions. Along the southwestern flanks of the two volcanic chains there are extensive exposures of silicic pyroclastic flows. Some may be related to caldera subsidence of the volcanic massif, but for other pyroclastic flows, no likely source has been identified. This silicic volcanism generally predates the building of the composite stratovolcanos. Other Holocene (and Pleistocene) volcanos consist of small monogenetic cinder cones situated to the northeast of the Cordillera Central. They rise 20 to 130 m above their bases. Some are aligned in a northwest-southeast direction. Some have an alkaline character.

3. Pliocene to Pleistocene volcanic relicts are present in three different structural settings. One group forms the oldest cones of the two young volcanic chains. Between the volcanos Irazú and Barba are situated the Cerros Zurquí, which have been dated at 0.5 Ma (Bellon and Tournon, 1978). In addition to those relicts shown in Figure 1, there are probably relicts in the Cordillera de Talamanca in the southeastern part of Costa Rica, but little is known about the geology of this inaccessible mountain range. The second group is formed by isolated volcanic relicts to the southwest of the Guanacaste, Central, and Talamanca Cordilleras. They seem to be associated with formations of the Aguacate Group, which is of Miocene to Pliocene age. The third group is made up of widely scattered volcanos and cinder zones far to the northeast of the Cordillera Central. Two main concentrations are the Cerros Cureña (Figure 2), where volcanic relicts are aligned in north-south and northwest-southeast directions, and others are near the Caribbean Sea, where some alkaline lava and cinder cones of Pleistocene age (Bellon and Tournon, 1978) are aligned in north-south and northeast-southwest directions.

4. Dacitic domes. Six dacitic domes of early Pleistocene age (Bellon and Tournon, 1978) are located about 10 km southwest of Rincón de la Vieja, of which three are shown in Figure 2. They rise 120-370 m above their bases. There are also two isolated domes on the southwestern slope of the Cordillera de Tilarán and still others, probably of Pliocene age, in the Cordillera de Talamanca.

STRESS DIRECTIONS

The maximum horizontal compressive stress (MHCS) direction has been determined to be parallel to the trend of three or more aligned Quaternary volcanic centers. The MHCS data are taken from Montero and Morales (in press). Along the active volcanic arc, the MHCS is predominantly north-south, with variations to both the east and west. In the back arc area (inset map, Figure 1), the MHCS also has a similar north-south trend. The MHCS has also been determined from Class A focal mechanisms. The direction of MHCS is N30°E for Cocos-Caribbean interplate earthquakes. This is also the direction of relative movement between the plates as deduced from slip vectors (Molnar and Sykes, 1969; Dean and Drake, 1978). Focal mechanisms from intraplate shallow earthquakes along the inner arc show the MHCS trending north. This observed MHCS rotation of approximately 30° to the northeast, passing from the Pacific to the continental area, could reflect stresses related to the absolute movement of the Caribbean Plate or to the presence of local stress fields generated, for example, by subduction of the Cocos Ridge (Figure 2).

HEAT FLOW DATA

Heat flow values (mw/m^2) are from Ladd (1985). Measurements were made in various bathymetric areas in the Pacific Ocean, especially on the flanks of the Cocos Ridge, the Middle America Trench, and on the landward side of the Middle America Trench. Values are generally low, as expected. Locally high anomalous values over the Cocos Ridge suggest remnant magmatism.

MORPHOTECTONIC CLASSIFICATION AND THICKNESS OF THE CRUST

A morphotectonic division of Costa Rica modified from Weyl (1980) is presented on Figure 1, inset map. The morphotectonic elements (from southwest to northeast) include (1) fore arc: low hills with eroded slopes predominate, composed of Jurassic-Cretaceous ophiolitic basement (Dengo, 1962); (2) fore arc basins: discontinuous, elongated northwest-trending basins developed between the fore arc and the inner arc and composed of sedimentary rocks of Upper Cretaceous to Quaternary; (3) inner arc: northwest-southeast striking young mountainous terranes which are divided by the Valle Central into two principal arcs, the active volcanic cordillera to the northwest and the intrusive-sedimentary Miocene-Pliocene Cordillera de Talamanca to the southeast; (4) back arc area: the Atlantic plains and small basins in the Talamanca area; and (5) the Colombia basin (Figure 2) with turbidic and pelagic sediments.

Crustal thickness (in km) modified from Case, McDonald, and Fox (in press) is shown superimposed on the morphotectonic (inset) map (Figure 1). An increase in crustal thickness occurs from the oceanic area to the inner arc, where a 40-km depth to Moho has been deduced from seismic data in northwestern Costa Rica (Matumoto et al., 1977). Based on geological and geophysical data similar thicknesses are estimated for the Cordillera de Talamanca.

ACKNOWLEDGMENTS

The patient help of E. Hernandez during the compilation is acknowledged. Thanks are due to C. Nelson and J. Alt for English language proofreading, and to W. Addicott (Circum-Pacific Council) and H. Meyers (National Oceanic and Atmospheric Administration) for providing data used for the compilation.

REFERENCES CITED

Alvarado, G. E., 1984, Aspectos petrológicos-geológicos de los volcanes y unidades lávicas del Cenozoico Superior de Costa Rica: Escuela Centroamericana de Geología, Lic. Thesis, 183p.

Alvarado, G. E., S. Paniagua, R. Tejera, and S. Kussmaul, 1980, Mapa del vulcanismo Plio-Pleistoceno, escala 1:700,000: San José, Instituto Geográfico Nacional.

Alvarado, G. E., L. D. Morales, W. Montero, A. Climent, and W. Rojas, 1988, Aspectos sismológicos y morfotectónicos en el extremo occidental de la cordillera volcánica Central de Costa Rica: Revista Geológica de América Central, v. 9, p. 75-98.

Barquero, J., and R. Sáenz, 1987, Aparatos volcánicos de Costa Rica, escala 1:750,000: Heredia, Universidad Nacional, Observatorio Vulcanológico y Sismológico de Costa Rica.

Bellon, H., and J. Tournon, 1978, Contribution de la géochronométrie K-Ar a l'étude du magmatisme de Costa Rica, Amérique Central: Bulletin Société Géologique de France, v. 20, no. 6, p. 955-959.

Berrangé, J. P., and R. S. Thorpe, 1988, The geology, geochemistry, and emplacement of the Cretaceous-Tertiary ophiolitic Nicoya Complex of the Osa Peninsula, southern Costa Rica: Tectonophysics, v. 147, p. 193-220.

Borgia, A., J. Burr, W. Montero, L. D. Morales, and G. E. Alvarado, 1987, Is the central Costa Rica volcanic range extending?: EOS, Transactions of the American Geophysical Union, v. 16, p. 406.

Burbach, G. V., C. Frohlich, W. D. Pennington, and T. Matumoto, 1984, Seismicity and tectonics of the subducted Cocos Plate: Journal of Geophysical Research, v. 89, p. 7719-7735.

Case, J. E., and T. L. Holcombe, 1980: Geologic-tectonic map of the Caribbean Sea: U.S. Geological Survey Miscellaneous Investigations Series Map I-1100, scale 1:2,500,000.

Case, J. E., W. D. McDonald, and P. J. Fox, in press, Caribbean crustal provinces, seismic and gravity evidence, *in* G. Dengo and J. E. Case, eds., The Caribbean region: Boulder, Colorado, Geological Society of America, The geology of North America, v. H.

Dean, B. W., and C. L. Drake, 1978, Focal mechanism solutions and tectonics of the Middle America Arc: Journal of Geology, v. 86, p. 111-128.

Dengo, G., 1962, Estudio de la región de Guanacaste, Costa Rica: San José, Costa Rica, Instituto Geográfico Nacional, 112p.

Denyer, P., and W. Montero, 1988, Mapa geoestructural y sismos del Valle Central, escala 1:150,000: San José, Costa Rica, Instituto Geográfico Nacional.

Güendel, F., 1985, Enjambres sísmicos en el volcán Irazú, catálogo de temblores 1984: Heredia, Costa Rica, Universidad Nacional Autónoma, p. 100-104.

Güendel, F., 1986, Seismotectonics of Costa Rica: An analytical view of the southern terminus of the Middle America Trench: University of California at Santa Cruz, Ph.D. Thesis, 157p.

Kessel, R. H., and D. R. Lowe, 1987, Geomorphology and sedimentology of the Toro Amarillo alluvial fan in a humid tropical environment, Costa Rica: Geografiska Annales, v. 69A, p. 85-99.

Ladd, J. W., 1985, Heat flow, southern Mexico to Costa Rica margin, Middle America Trench off western Central America, *in* J. W. Ladd and R. T. Buffler, eds., Atlas 7: Woods Hole, Marine Science International.

Madrigal, R., and E. Rojas, 1980, Mapa geomorfológico de Costa Rica, escala 1:200,000: San José, SEPSA, Instituto Geográfico Nacional, 9 maps.

Matumoto, T., M. Ohtake, G. Latham, and J. Umaña, 1977, Crustal structure in southern Central America: Bulletin of the Seismological Society of America, v. 67, p. 121-134.

Molnar, P., and L. R. Sykes, 1969, Tectonics of the Caribbean and Middle America regions from focal mechanisms and seismicity: Geological Society of America Bulletin, v. 80, p. 1639-1684.

Montero, W., and J. W. Dewey, 1982, Shallow-focus seismicity, composite focal mechanisms, and tectonics of the Valle Central of Costa Rica: Bulletin of the Seismological Society of America, v. 72, p. 1611-1626.

Montero, W., and L. D. Morales, 1984, Seismotectónica y niveles de actividad de microtemblores en el suroeste del Valle Central, Costa Rica: Revista Geofísica, v. 21, p. 21-41.

Montero, W., and L. D. Morales, 1988, Zonificación sísmica del Valle Central de Costa Rica: San José, Memorias 4th Seminario de Ingeniería Estructural, p. 335-351.

Montero, W., and L. D. Morales, in press, Deformación y esfuerzos neotectónicos en Costa Rica: Universidad de Costa Rica, Editorial.

Montero, W., and L. Ponce, 1979, Localización hipocentral y mecanismo focal de algunos temblores fuertes de Costa Rica-Nicaragua: Geofísica Internacional, v. 18, p. 411-428.

Mora, S., 1979, Estudio geológico de una parte de la región sureste del Valle del General, Provincia de Puntarenas, Costa Rica: Escuela Centroamericana de Geología, Lic. Thesis, 3 v.

Morales, L. D., and W. Montero, 1984, Los temblores sentidos en Costa Rica durante 1973-1983 y su relación con la sismicidad del país: Revista Geológica de América Central, v. 1, p. 29-56.

Paniagua, S., and G. Soto, 1986, Reconociemiento de los riesgos volcánicos potenciales de la Cordillera Central de Costa Rica, América Central: Ciencia y Tecnología, v. 10, no. 2, p. 49-72.

Ponce, D. A., and J. E. Case, 1987, Geophysical interpretation of Costa Rica, *in* U.S. Geological Survey, Dirección de Geología, Minas e Hidrocarburos, Universidad de Costa Rica: U.S. Geological Survey Miscellaneous Investigations Series Map I-1865, p. 8-17.

Tournon, J., G. E. Alvarado, F. Rivier, and S. Mora, in preparation, Mapa geológico de Costa Rica, escala 1:500,000.

Weyl, R., 1980, Geology of Central America: Berlin, Gebrüder Borntraeger, 371p.

Circum-Pacific Council for Energy and Mineral Resources
Earth Science Series, Vol. 16
R. L. Miller, G. Escalante, J. A. Reinemund, and M. J. Bergin (Eds.)
Energy and Mineral Potential of the Central American-Caribbean Regions
Springer-Verlag, Berlin Heidelberg 1995

CENTRAL AMERICA AND THE NORTH CARIBBEAN: RESOURCE IMPLICATIONS INFERRED FROM PALEOMAGNETIC DATA AND PLATE TECTONIC HISTORY

William D. MacDonald
Department of Geological Sciences, State University of New York at Binghamton, Binghamton, New York 13901 USA

ABSTRACT

Jurassic paleomagnetic latitudes for northern South America are consistent with a post-Jurassic southward displacement of South America of about 15° relative to North America. This tectonic displacement allowed the development of the Caribbean Sea. It also had important consequences for the subsequent development of plate margins and all associated mineral and energy resources in the circum-Caribbean region. As an example of how plate tectonics can be incorporated into resource exploration strategies, the consequences of this post-Jurassic displacement are explored in terms of the disruption of Jurassic drainage systems of northern South America. It is hypothesized that this disruption may have resulted in the abandonment of an important Jurassic delta, a potential hydrocarbon reservoir. The possible sites for such a delta could be in widely scattered regions such as present-day Bahamas, south Florida-Yucatan, southern Gulf of Mexico-Chiapas, or the Nicaragua Rise-Honduras zone. Thus it is shown that geologic studies in South America can be important to exploration strategies in North America, the Greater Antilles, and Central America. The integration of a wider body of data, incorporating plate tectonic displacements, leads to a wider diversity of exploration possibilities.

This analysis also demonstrates the importance of accounting for accretion of terranes at subduction margins in paleogeographic reconstructions. Invoking this concept, a new method for fitting paleo-plate boundaries along the ultramafic lines of subsequent subduction zones is demonstrated. A satisfactory and novel Jurassic fit for South America against North America is obtained. This fit indicates a post-Jurassic rotation of South America relative to North America of 16.0° counterclockwise about a pole at 47°N, 03°E.

Finally, secondary effects of plate motions, such as fluid migration away from subduction zones, are evaluated in relation to examples of Caribbean mineral and hydrocarbon resources.

INTRODUCTION

The first conference on circum-Pacific energy and mineral resources in 1974 drew attention to the unique factors as well as some of the common denominators which characterize the resources of this great region (Halbouty, Maher, and Lian, 1976). Important among the characteristics of circum-Pacific resources especially are the plate tectonic settings. The majority of present-day global subduction zones, for example, occur here.

Numerous publications since 1974 have discussed the relationships between resources and plate tectonic settings (e.g., Fisher and Judson, 1975; Strong, 1976; Wright, 1977; Dickinson and Yarborough, 1978; Harding and Lowell, 1979; Bally et al., 1981; Mitchell and Garson, 1981; Woollard, 1981; Hutchinson, 1983; Kingston, Dishroon, and Williams, 1983; Sawkins, 1984; Price, 1989). With the focus now on the Caribbean-Central American region, the purpose of the present paper is to discuss briefly the effects of Cretaceous-Tertiary plate tectonics on resource distributions here. Except for geothermal resources and very young mineral deposits (epithermal veins; placers) for which present plate boundary activity can be important, most mineral and energy resources are associated with paleo-plate boundaries. Emphasis, therefore, is placed on the distribution of Caribbean paleo-plate margins and associated structures.

A correct understanding of paleogeography is important for the development of successful exploration strategies. Paleogeography and plate tectonic evolution are intimately associated because the paleogeographic reorganization of land and ocean requires motions on plate boundaries. Paleomagnetism is virtually the only geophysical science of value for paleo-plate reconstructions. An example of the use and limitations of paleomagnetism for Caribbean Jurassic paleogeography is provided. Finally, because the objectives of this publication are practical as well as conceptual, some suggestions are made for resource exploration, for both petroleum and mineral resources. First, a brief examination of resource associations with paleo-plate boundaries is in order.

RESOURCE DISTRIBUTION AND PALEO-PLATE BOUNDARIES

Plate boundaries are basically of three types: subducting, transforming, and rifting. Petroleum resources are associated with each type. Locally important hydrocarbon accumulations are associated with transform boundaries, as in the case of the pull-apart basins in California adjacent to the San Andreas fault (Crowell, 1974; Orwig, 1976). On a global scale, however, both energy and mineral resources are more typically associated with paleorift and paleosubduction zones than with paleotransform zones.

Paleorift zones are of two main types: (1) rifted continental margins, sometimes referred to as "Atlantic-type" margins, and (2) "failed" rift-zones or aulocogens. Paleosubduction zones are basically of three kinds: (1) intraoceanic or remnant-arc type, (2) intracontinental or "collision-type," and (3) continental margin or "Andean" type.

Petroleum resources are associated with both paleorift and paleosubduction zones. A variety of structural and stratigraphic traps characterize rifted continental margins, such as in northwest Australia (Johnstone, 1981) and eastern South America (de Figueiredo and Gabaglia, 1986), as well as failed rifts such as the Benue trough of Nigeria. Failed rifts are of interest even when relatively barren of hydrocarbons. Because they often control drainage systems, thereby channelling great amounts of clastic sediments to the sea, they may lead to great deltaic deposits with associated petroleum reservoirs, as in the Niger delta at the mouth of the Benue trough.

In subduction zones, the frontal accretionary prism is usually completely disrupted structurally and stratigraphically, and therefore is an unpromising exploration target for lack of adequate reservoirs (Montecchi, 1976). However, petroleum and associated products may migrate from this zone into adjacent forearc and retroarc basins where satisfactory reservoirs and traps occur. In the retroarc areas, migration may progress far into continental areas (Dickinson, 1974; Oliver, 1986).

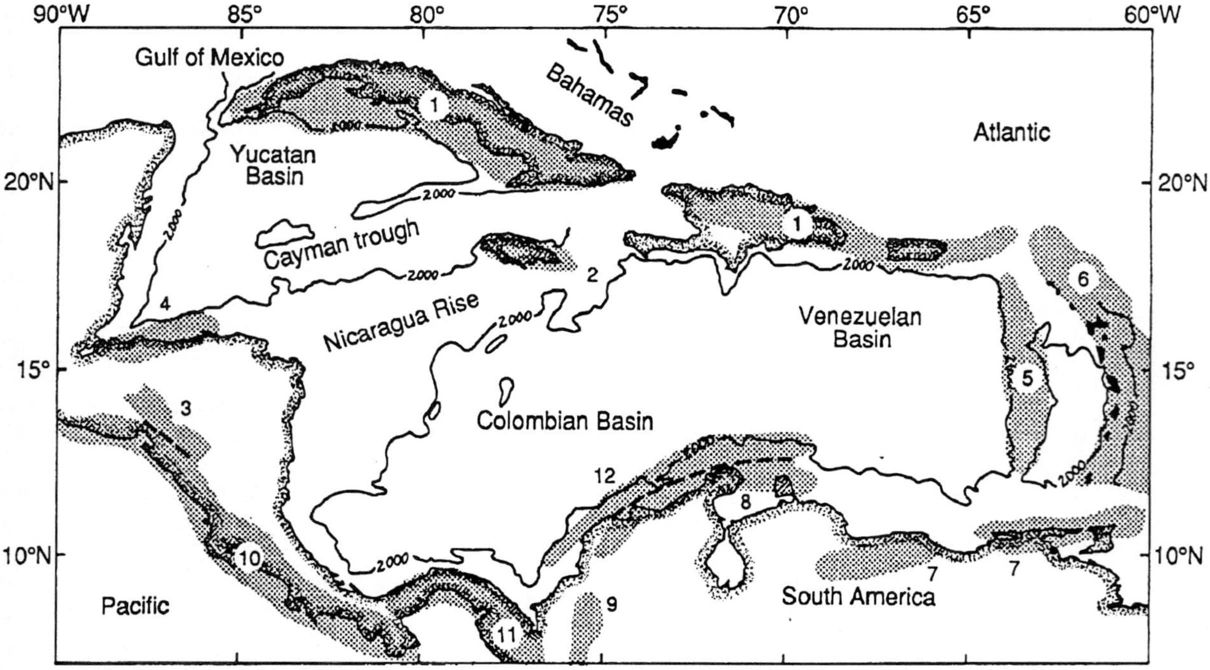

Figure 1. Major paleosubduction zones of the Caribbean region. See Table 2 for key.

Table 1. Principal metals and their paleo-plate boundary associations.

I. SUBDUCTION ZONE		
A. OPHIOLITE ASSOCIATION (oceanic crust accreted into subduction zone crust)		
CRUSTAL LEVEL	LITHOLOGY	MINERAL DEPOSITS
upper	sedimentary	Mn oxides, Pb-Zn sulfides
	mafic extrusive/	Fe-Cu sulfides
	intrusive	Au, Cu
		Pt (placer)
		Ni (sulfides/laterites)
lower	ultramafic	Cr podiform
B. FELSIC INTRUSIVE ASSOCIATION (diorite; porphyry) and related vein deposits		
EMPLACEMENT LEVEL	MINERAL DEPOSITS	TRENCH DISTANCE
shallow	Fe-Hg	near
	Cu-Au	
	Ag-Pb-Zn	
deep	Sn-Mo	far
II. RIFT ZONE		
TYPE		ORES
A. VEIN DEPOSIT		Pb, Zn, Cu
		Au, Ag tellurides
B. STRATIFORM MAFIC INTRUSIONS		Ni, Cr, Pt, Cu
III. TRANSFORM FAULT / FRACTURE ZONE		
A. MAFIC/ULTRAMAFIC ASSOCIATION		Cu, Ni, other

Mineral resources of diverse types are also associated with paleo-plate boundaries (Table 1). Created in both continental and oceanic rift zones (Woollard, 1981), these deposits may be subsequently uplifted and/or obducted into accretionary prisms at subduction zones. Those that are instead subducted may become incorporated into magmas and subsequently rise into the deformed rocks of the subduction zone.

CARIBBEAN PALEO-PLATE BOUNDARIES

From the preceding discussion, it is clear that paleosubduction zones are of primary importance especially in the search for mineral resources. Paleorift margins are, on the other hand, of special interest for petroleum exploration, and subduction zones secondarily so. With some notable exceptions, paleotransform settings are of less interest.

The major Cretaceous and Tertiary paleosubduction zones (Figure 1), paleorift zones (Figure 2), and paleotransform zones (Figure 3), compiled from many sources, are shown for the Caribbean region. This area has had an extremely active Cretaceous-Tertiary tectonic history. The names and approximate ages for these features are given in Tables 2, 3, and 4.

Many plate boundaries show mixed characteristics. For example, part of the east rifted margin of the Yucatan Peninsula or Maya block (1, Figure 2) seems likely to have suffered subsequent transform motion (7, Figure 3) and the subduction zone of northern South America (7, Figure 1) has seen subsequent transform motion (8, Figure 3), but undoubt-

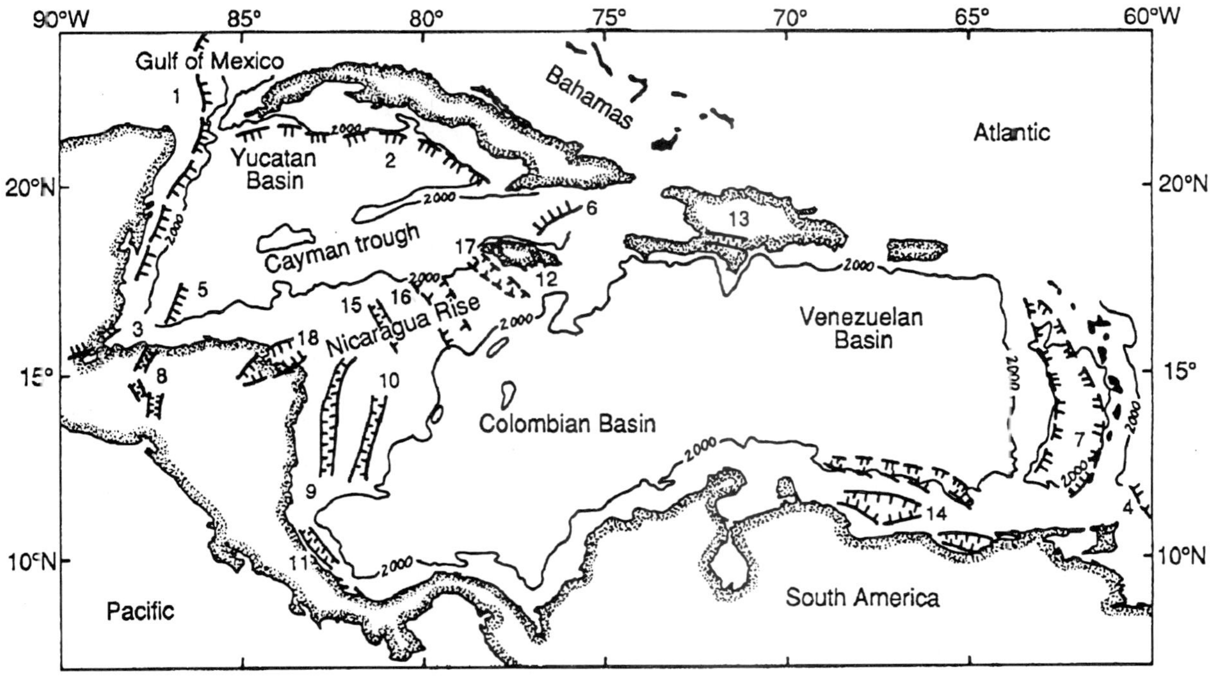

Figure 2. Major paleorift margins and rift zones of the Caribbean region. See Table 3 for key.

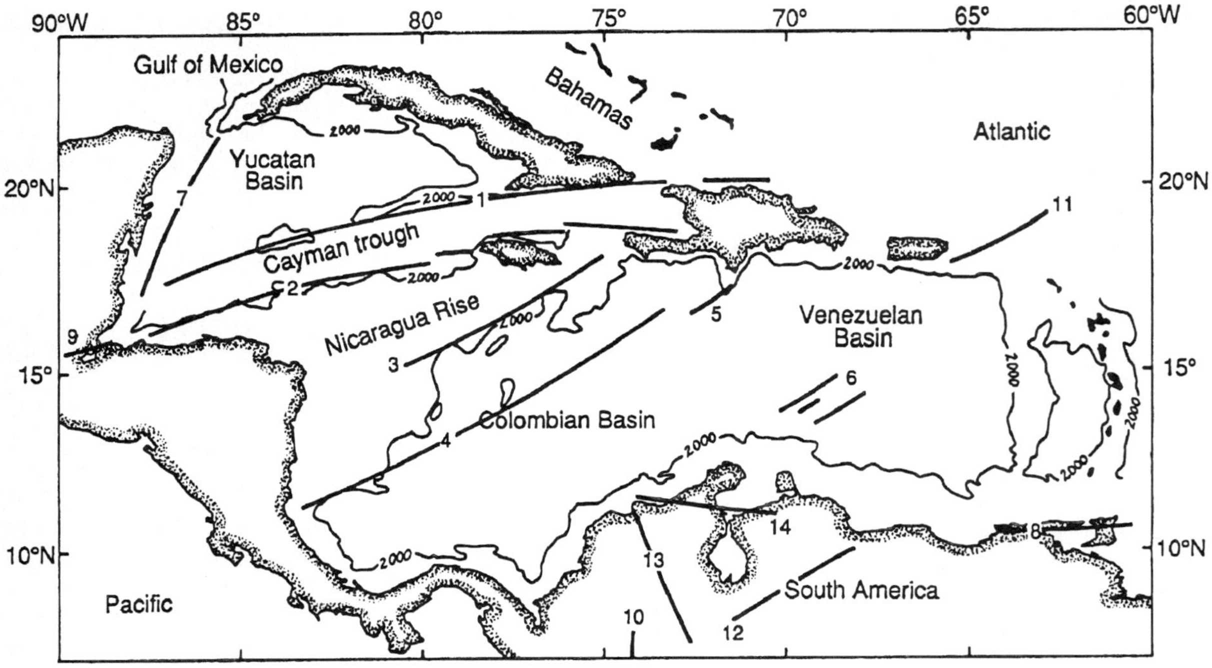

Figure 3. Major paleotransform boundaries and faults of the Caribbean region. See Table 4 for key.

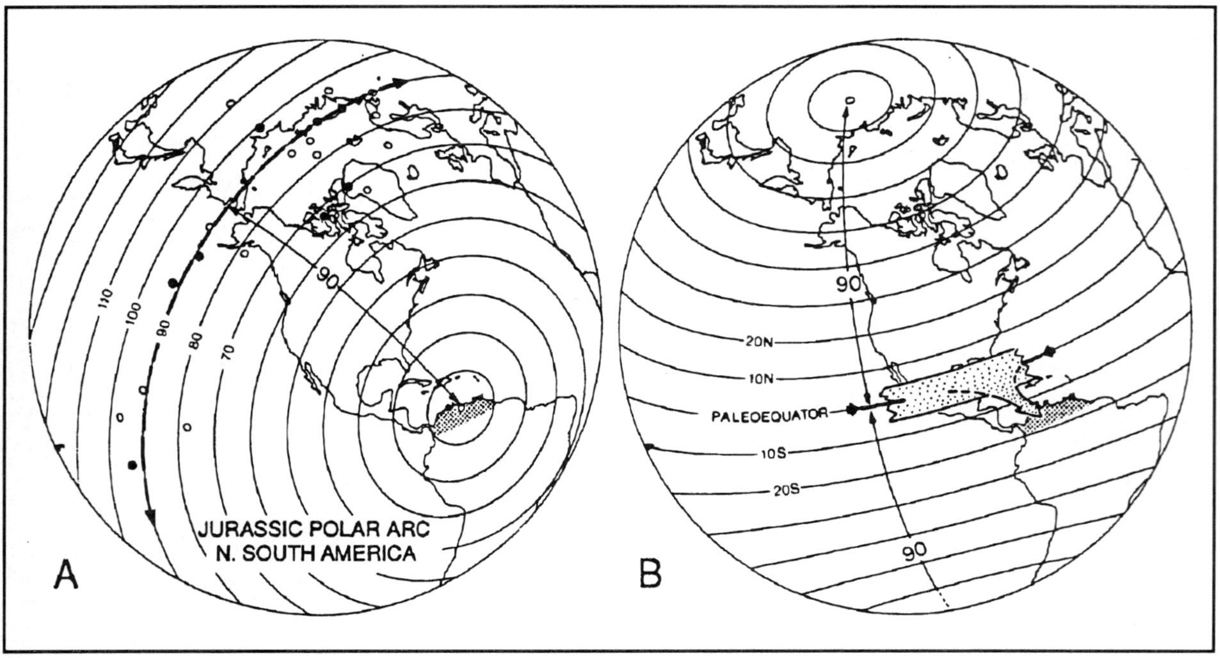

Figure 4A. Jurassic paleomagnetic poles of northern South America define a polar arc of 90° radius, indicating equatorial paleolatitudes for this region. **B.** Paleolatitudes at 10° intervals about a Jurassic pole for North America (Irving and Irving, 1982) suggest a 15° southward movement of South America since the Jurassic. After MacDonald (in press).

Table 2. Major Caribbean subduction zones (see Figure 1).

FEATURE	AGE
1. Greater Antilles	Cretaceous - E. Tertiary
2. Jamaica	Cretaceous - E. Tertiary
3. Honduran-Nicaraguan	E. Tertiary
4. Nombre de Dios	Cretaceous - Paleocene
5. Aves	Cretaceous
6. Lesser Antilles	Tertiary - Recent
7. Venezuela - Tobago	Cretaceous
8. Ruma	Cretaceous E. Tertiary
9. Cordillera Central	Cretaceous
10. Central American	Tertiary - Recent
11. Baudo	Tertiary - Recent
12. South Carihbean	Tertiary - Recent

edly also suffered pre-Cretaceous rifting, not considered in these figures. The Motagua zone of Guatemala (9, Figure 3), between the Maya and Chortis blocks, has changed through the sequence rifted margin, subduction zone, to transform zone. Therefore, the characteristics of these paleo-plate boundaries (Figures 1, 2, and 3) should be viewed as generally schematic.

PALEOGEOGRAPHY AND PALEOMAGNETISM

To interpret the plate tectonic movements which gave rise to the complex structures shown is a difficult task. This is especially so where the possibility of many smaller allochthonous terranes exists. There are few clues as to which landmass was where and when, and as to where and how extensive the intervening oceanic areas were. This is clearly of great importance to resource exploration. For example, consider the possibility that post-Jurassic rifting has isolated the Jurassic rifts of northern South America, perhaps initially an integrated drainage system following an aulocogen, from its terminal delta at the former continental margin. How can

Table 3. Major paleorift margins and zones (see Figure 2).

FEATURE		AGE
1.	Yucatan (N and E)	Early Mesozoic(?)
2.	South Cuba	Cretaceous?
3.	Guatemala (Tambor)	Cretaceous?
4.	Guyana	Cretaceous
5,6.	W and E Cayman	Early Tertiary
7.	Grenada	Early Tertiary
8.	Honduras depression	Tertiary - Recent
9.	E. Nicaragua	Tertiary
10.	San Andres	Tertiary - Recent
11.	Limon	Tertiary - Recent
12.	Jamaica	Tertiary
13.	Cul de Sac - Enriquillo	Tertiary - Recent
14.	Bonaire-Los Roques-Orchila-Cariaco	Tertiary - Recent
15.	Thunder	Tertiary - Recent
16.	Alice	Tertiary
17.	Walton	Tertiary
18.	Mosquitia	Jurassic

Table 4. Major paleotransform zones (see Figure 3).

FEATURE	AGE
1. North Cayman (Cayman Ridge)	Tertiary - Recent
2. South Cayman (Swan)	Tertiary - Recent
3. Pedro	Cretaceous
4. Hess	Cretaceous - Tertiary?
5. North Beata	Cretaceous
6. Venezuelan Basin 1,II,III	Cretaceous
7. E. Yucatan	Cretaceous?
8. El Pilar	Tertiary - Recent
9. Motagua	Tertiary - Recent
10. Dolores - Romeral	Cretaceous - Tertiary
11. Anegada	Tertiary
12. Bocono?	Tertiary
13. Santa Marta?	Tertiary
14. Oca?	Tertiary

the Jurassic paleogeography of South America be reconstructed in relation to Central America, North America, and the Greater Antilles, in order to evaluate the "abandoned" delta hypothesis? Numerous approaches in paleogeography are possible (e.g., Bullard, Everett, and Smith, 1965; Malfait and Dinkelman, 1972; MacDonald, 1976; Ladd, 1976; Pindell and Dewey, 1982).

Geologic intuition is an important basis, as are the use of marine magnetic anomalies, computer-fitted continental margins, paleomagnetically controlled paleolatitudes, and interpolations of plate rotations on the sphere. The many conceptual models published invoke numerous important assumptions, mainly implicit. Ideally, reconstruction should be evaluated by how well it accounts for known distributions of rock facies and faunas of the time (Scott, 1984); but these factors are very difficult to evaluate, even qualitatively.

Paleomagnetic data from both marine magnetic anomalies and oriented samples from land areas provide useful constraints on paleogeographic reconstructions in certain situations. However, when the uncertainties and assumptions are closely examined, including those of rock age, magnetization age, and paleomagnetic pole location, the confidence limits on the paleomagnetic data allow considerable latitude in paleogeographic interpretation. Paleomagnetic poles from land-based studies, for example, do not in themselves contain any information regarding paleolongitudes. This is especially important where many smaller terranes are involved, as in Central America and the north Caribbean. With these thoughts in mind, consider the problem of a Jurassic reconstruction for the North America-South America-Central America-Greater Antilles-Caribbean region.

Jurassic paleomagnetic poles for the northern Andes (Figure 4A) define a paleomagnetic polar arc, centered on the sampling area, and with radius of about 90°. For reference, small circles at 10° spacing are shown concentric about the sampling region. Paleomagnetic polar arcs of this type are relatively common from tectonically disturbed areas (MacDonald, 1980) and are primarily owing, directly or indirectly, to structural disturbances. Some amount of cross-arc and along-arc scatter in the polar arc is contributed by plate motion and polar wander. Figure 4A indicates that northern South America was approximately 90° from the Jurassic pole or, looked at another way, northern South America was approximately on the Jurassic equator.

Using a Jurassic pole for North America (Irving and Irving, 1982), the paleoequator for North America can be estimated (Figure 4B). This presumably indicates the approximate paleolatitude of northern South America in relation to North America. However, the situation is more complicated than it appears. There is still considerable uncertainty about the polar wander path for North America, some arguing for a large "cusp" in the Jurassic part of the curve, while others indicate that the North America Jurassic pole is simply not well determined (Steiner, Lawrence, and Beck, 1989). The uncertainties are more than 10°, i.e., greater than the north-south width of the Caribbean.

A JURASSIC RECONSTRUCTION

A Jurassic reconstruction for the Caribbean region is made in this paper by fitting northern South America to the Greater Antilles. The rationale and procedures are outlined below. The results are plotted in Figure 5.

Wilson (1966) has noted that rifted continental margins may subsequently become loci for subduction, possibly leading to a complete cycle of opening and closure of an oceanic region, as has apparently happened several times according to his interpretation of Appalachian structures. We note that a discontinuity in lithospheric thickness is created at rifted margins, resulting from the "withdrawal" of thicker continental lithosphere and its replacement by thinner oceanic lithosphere.

This discontinuity may focus lithospheric stresses and predetermine the location of subsequent subduction zones, responding, for example, to changes in plate dynamics. We assume that this is how the Greater Antilles developed, essentially in situ along the rift line away from which South America was withdrawn in post-Jurassic time.

Comparing subduction zones for the Greater Antilles (1, Figure 1) and northern South America (7, 8, and 9, Figure 1), we note the gross similarity in outline. The question arises as to how to match or "fit" these features in a least-squares sense, to develop the paleogeographic reconstruction.

Contours to be fitted were constructed along the "ultramafic line" which marks the approximate axis of the Greater Antilles. This line begins in southwest Puerto Rico, crosses Hispaniola obliquely, and runs along the axis of Cuba. The contour was continued along the rifted east margin of the Yucatan Peninsula, following the 2,000 m isobath, and ending in the Cayman trough vicinity. Although the use of the 1,000 m isobath (500 fm) was recommended by Bullard, Everett, and Smith (1965), the 2,000 m isobath was used here because it follows the Yucatan margin in a more regular way than does the 1,000 m isobath. The justification for using the "ultramafic line" is that ophiolites from Santa Elena Peninsula to Azuero Peninsula in Central America approximately mark the boundary between the Pacific and Caribbean Plates. Similarly, obducted ophiolites of the Bay of Islands complex and Hare Bay complex in Newfoundland are obducted onto the plate margin which marks the west limit of Iapetus. Thus, in subduction zones, ultramafic lines may represent significant lithospheric plate boundaries. The ultramafic line in the Greater Antilles is interpreted here as representing approximately the Jurassic margin away from which South America was rifted.

Similarly for northern South America, the fitting contour is also primarily an ultramafic line. Here, however, the presence of Paleozoic and Precambrian rocks can be used to interpolate the former continental margin where ultramafic rocks are lacking. That is, the edge or margin of the former continent, before modification by subduction complexes, is defined by the outer limit of pre-Jurassic, and mainly pre-Mesozoic, rocks (MacDonald, 1972). This contour begins near the Pacific coast south of Guayaquil, continues approximately up the west side of the Cordillera Central of Colombia, across to the northwest corner of the Santa Marta massif and across the central Guajira and Paraguana Peninsulas, then southeast down to follow approximately the north margin of South America along the Coast Range then north of Margarita and through Tobago; south from here it follows the 1,000 m isobath, ending near the Demerara rise.

The "ultramafic lines" were digitized at 1° intervals and are shown in Figure 5A. The best-fit pole is at 47°N, 03°E; the rotation angle is 16.0°. A triple scalar product fitting procedure, similar to that of Mckenzie and Sclater (1971), was used to fit the contours.

A detailed evaluation of this fit is not possible in this brief space. The fit is hypothetical and serves as a basis for discussing exploration strategies. The fit looks reasonably good. The largest overlap puts northeast South America (Margarita to Tobago) over the morphologically irregular southern tip of the Bahama platform. This overlap is balanced by several smaller gaps elsewhere along the contours.

RESOURCE DISTRIBUTIONS AND PALEOGEOGRAPHY

Now we return to the topic of resource distributions in the Caribbean-Central American region, and to the question of the usefulness of paleogeographic reconstructions in petroleum exploration. We are interested in evaluating the hypothesis that a Jurassic rift system, a possible aulocogen, funnelled sediments to a deltaic system which has subsequently been "abandoned" by rifting and the withdrawal of South America away to the south.

Both former rift margins were subsequently modified by subduction. To evaluate this hypothesis, the distribution of Jurassic rift deposits and volcanics of northern South America would have to be analyzed. Some of these are plotted schematically in Figure 5. More importantly, stratigraphic and sedimentological studies would have to be made to decipher current flow directions to interpret regional flow patterns. Ground-based studies would also have to be made in areas presumed to be peripheral to Jurassic South America. These include Guatemala, Cuba, and the Nicaragua-Honduras (Chortis) region which has to be relocated to allow the fit of Figure 5.

Some possibilities for the hypothetical Jurassic delta are (1) paleoflow crossed the trend of the present Greater

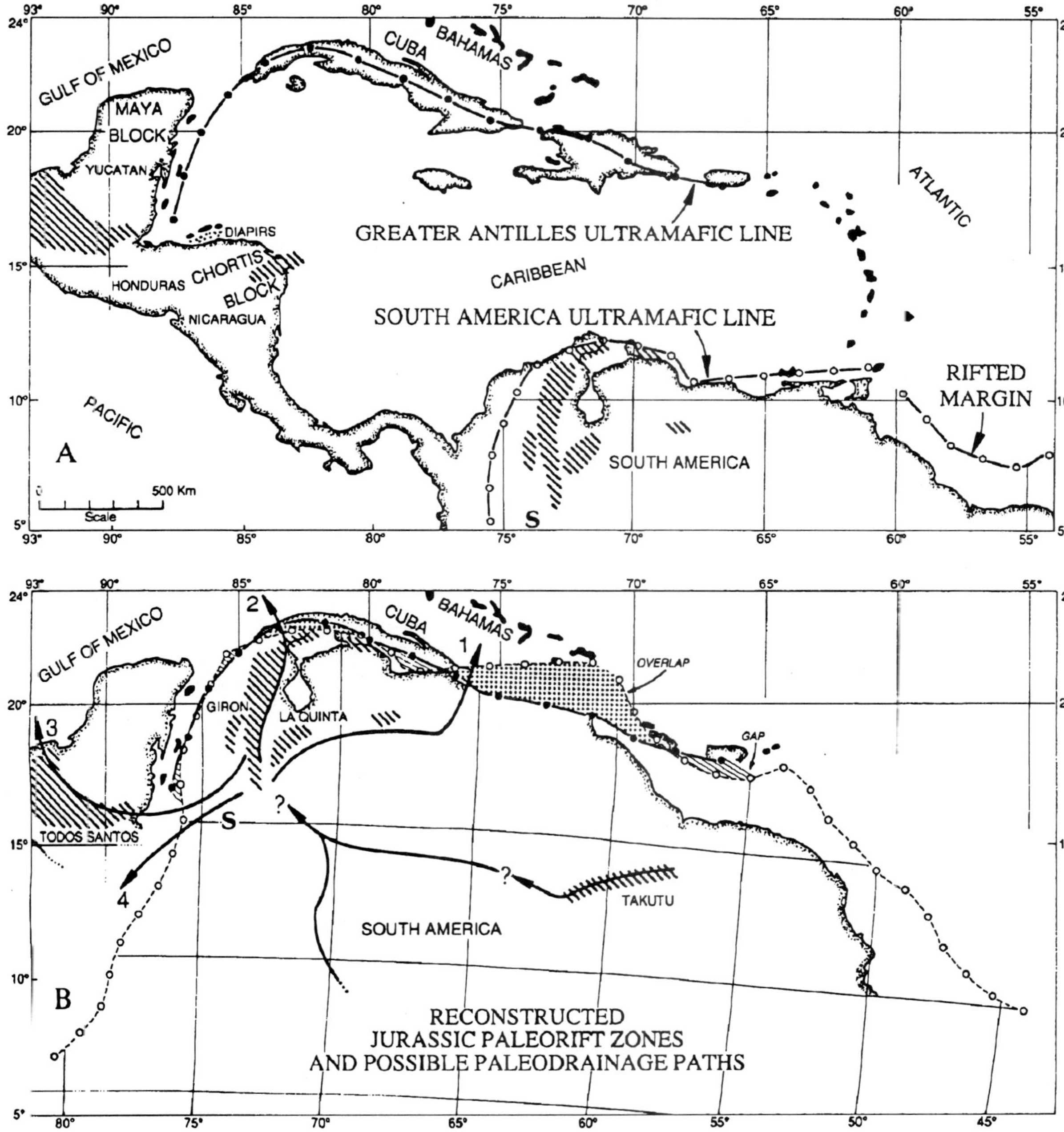

Figure 5A. Ultramafic "lines" in the Greater Antilles and northern South America mark approximate limits of crustal accretion along which a Jurassic reconstruction is made by fitting. **B.** A "best-fit" reconstruction as a basis for discussing possible paleoflow paths in Jurassic rifts in northern South America and the concept of an "abandoned" delta to the north.

Antilles and formed an "abandoned" delta now at the base of the Bahamas platform; (2) paleoflow was northward into south Florida or the southern Gulf of Mexico, with the "abandoned" deltaic sediments in those areas; (3) paleoflow was west and northwest across Guatemala and southern Mexico, with the "abandoned" delta in the southwest Gulf of Mexico; and (4) paleoflow was west across the relocated Chortis block, with "abandoned" deltaic deposits in Nicaragua or Honduras, perhaps under the Cenozoic volcanic apron, or in the Nicaragua Rise.

Other factors such as relief, drainage area, climate, duration of sediment accumulation, shelf width, the interaction of oceanic geostrophic currents at the continental margin, and the subsequent tectonic history affect delta development and preservation and need to be evaluated in this scenario for petroleum exploration.

TECTONIC SETTINGS IN RELATION TO OTHER EXPLORATION STRATEGIES

Numerous excellent summaries of Caribbean-Central American resource patterns are available (e.g., Bishop, 1980; Lee Moreno, 1980; Krushensky, Cargill, and Raines, 1987; Anonymous, 1988; Kesler, Levy, and Martin, in press).

Here we add some comments, emphasizing plate tectonic settings, especially for the north and west Caribbean regions.

Petroleum Aspects

Especially interesting in this region are the rift settings. The implications of several zones of diapirs and the potential of oil migration away from subduction zones also should be considered in evaluating petroleum possibilities.

Many smaller rifts occur in the Honduras-Nicaragua vicinity (Chortis block; see Mann and Burke, 1984), especially on the Nicaragua Rise. Two of these, the Honduras depression and the Limon rift (8 and 10, Figure 2) lie close to the Nicaraguan tectono-volcanic depression. This geothermal heat source trends through the Gulf of Fonseca, Lake Managua, Lake Nicaragua, and farther southeast. If we regard the Honduras depression as a smaller failed rift, which even today controls drainage in the area, the possibility arises that sediments deposited into the zone of subsidence at the Gulf of Fonseca might become thermally mature at a rate much faster than is conventional. Undoubtedly the same is true for the even larger sedimentary deposits (Anonymous, 1988) of the Limon rift, proposed by Christofferson (in press). This rift at present lies in an apparent retroarc position relative to the Central American subduction zone, near the southeast termination of the Nicaraguan volcanic depression.

For both areas, Fonseca and Limon, the possibility of reservoir rocks in ashflow tuffs (Katahira and Ukai, 1976) exists for strata younger than the onset of silicic volcanism in this region, approximately 32 m.y. ago.

Diapirs are widely distributed around the Caribbean and vicinity, with varying tectonic associations. Some are deep basin evaporites, possibly resulting from plate motions which closed off an arm of the sea; others are mud diapirs associated with accretionary prism deformation; and still others represent basin evaporites of local extent. The principal occurrences are in south Mexico (salt), north Cuba (salt), east of Barbados (mud), near Trinidad (mud), north of Colombia in the south Caribbean fold belt (mud), east of Panama in the deformed retroarc area (mud), and north of Honduras. The diapirs north of Honduras (Pinet, 1972) are controversial. As they are associated with sediments in a shallow-shelf setting, they represent possibly accessible petroleum-trap environments. The important questions are (1) are they evaporites, and, if so, salt; and (2) is there petroleum in the vicinity. We address these issues briefly.

On a regional basis, it seems justifiable to assume that these features are evaporite diapirs, and correlate with the Jurassic evaporites of the southern Gulf of Mexico. Or, possibly, they correlate with those of Cuba or with those of the nearby Mosquitia basin (Mills and Hugh, 1974). Parenthetically, the salt diapir of Cundinamarca, Colombia, lies in proximity to these other Jurassic evaporites in the Jurassic reconstruction of Figure 5.

However, a dilemma arises when one considers how Jurassic evaporites might survive off north Honduras adjacent to a Cretaceous subduction zone (4, Figure 1), which is presumed to have been associated with deformation and dioritic intrusion in the Cordillera Nombre de Dios. Because serpentinized ultramafics are known in the nearby Bay Islands, the possibility arises that the Honduran diapirs are serpentinites (Meyerhoff, 1973). Even so, perhaps serpentine diapirs, being penetrative and impermeable, can also generate petroleum traps. But is any oil available to be trapped?

The dilemma described above involves the problem of sources and of oil migration away from subduction zones. For example, the petroleum in the Orinoco tar sands has probably migrated a long way, possibly from beneath the overthrust Cretaceous rocks of the Caribbean ranges during subduction along the north margin of Venezuela (7, west, Figure 1). Thus in examining especially the distribution of the Cretaceous and early Tertiary subduction zones (Figure 1), the distribution of associated retroarc basins should be considered and possible oil migration pathways evaluated.

The immediate vicinity of the Cordillera Nombre de Dios in northern Honduras might be a poor place to search for Cretaceous oil because of the possible effects of Cretaceous heating there. However, that same circumstance could make the Cretaceous limestones of interior Honduras more inviting targets, if at sufficient depth and in the path of migrating oil. This is not a specific exploration recommendation, but rather an illustration of tectonic factors to consider in developing exploration strategies.

Economic Minerals Aspects

As Table 1 reveals, the principal plate boundary zones of interest in exploring for metallic mineral deposits is the subduction zone. The exception, as usual, proves the rule. In examining margin faults associated with the Honduras depression (8, Figure 2; Muehlberger, 1976), Kesler, Levy, and Martin (in press) note a high correlation with major limestone replacement and vein deposits. Evidently this is an aulocogen rather than a subduction-related setting. Relative to the present subduction setting, they note that precious metals epithermal vein deposits show a good correlation with the margins of the present tectonovolcanic Nicaraguan depression, which is parallel to and north of the present subduction axis (10, Figure 1).

Directly associated with paleosubduction zones are ophiolites and diorite-andesite intrusive and extrusive rocks. Both are of primary interest in mineral exploration.

Ophiolites are emplaced in linear belts, such as the Santa Elena-Nicoya-Osa-Burica-Azuero peninsular trend of Pacific Costa Rica. Although the primary mineralization of ophiolitic rocks is probably emplaced in oceanic crustal (rifting) environments, accretion and obduction in subduction zones make these deposits accessible. Subsequent enrichment of ophiolites may occur by laterization, forming Ni-laterites, or by lag-concentration to produce gold placers (Berrangé, 1987) in Costa Rica or platinum placers in Colombia. Ophiolite belts of the Caribbean should be thoroughly explored. Case (1980) provides a useful summary of Caribbean ophiolites. For finding and probing ophiolites in the shallow subsurface, Woollard (1981) offers some geophysical exploration clues. The possibility of finding Ni-laterites at shallow depth under unconformities should not be overlooked in assessing known ophiolite belts.

Finally, by taking tectonic factors into account, such as the subduction-release of oceanic metal deposits, perhaps combined with remobilization of older deposits in subsequent subduction cycles, the occurrence of metal deposits can be integrated into plate tectonic frameworks in a meaningful way (Petersen, 1980; Eriksen, 1976; Woollard, 1981; Sillitoe, 1988; Kesler, Levy, and Martin, in press). Kesler and coworkers have shown that it is possible to make tectonically important inferences from ore-deposit data, which in turn can be useful in exploration strategies. They point out, for example, that the southern limit of "continental" crust beneath the volcanic apron in Nicaragua can be deduced from lead-isotope ratios (Cumming and Kesler, 1976). Further, Kesler, Levy, and Martin (in press) note that the lead-isotope ratios in the Chortis block (Nicaragua-Honduras) are more consistent with those from southern Mexico than with those in the adjacent Maya block (Guatemala-Yucatan). This interpretation supports several paleogeographic reconstructions which would derive the Chortis block from south of Mexico.

CONCLUSIONS

Caribbean paleogeography is interesting from both academic and industrial points of view. A principal objective of this paper has been to show that scientific studies of structural and tectonic features can be used to develop practical energy and mineral exploration strategies. Thus, for example, a better understanding of the Jurassic polar wander path for North America can lead to better paleogeographic reconstructions for the Central America and Caribbean regions. By inference it is suggested that structural, stratigraphic, sedimentological, and paleomagnetic studies in northern South America can be important for locating petroleum and metalliferous provinces on the other side of the Caribbean. The integration of geologic information of many different kinds, from many different areas, will lead to a better understanding of the Earth's tectonic history. It will lead also to the practical exploitation of resources which have been created or redistributed by tectonic processes.

ACKNOWLEDGMENTS

S. Kesler and E. Christofferson kindly provided preprints of papers in press. The Circum-Pacific Council for Energy and Mineral Resources is thanked for their support.

REFERENCES CITED

Anonymous, 1988, The hydrocarbon potential of Costa Rica: San Jose, Refinadora Costarricense de Petróleo (RECOPE), Ministerio de Recursos Naturales, Energía y Minas, 7p.

Bally, A. W., A. B. Watts, J. A. Grow, W. Manspeizer, D. Bernouilli, C. Schreiber, and J. M. Hunt, 1981, Geology of passive continental margins: American Association of Petroleum Geologists, Education Course Note Series no. 19.

Berrangé, J. P., 1987, Gold in Costa Rica: Mining Magazine, v. 147, p. 402-407.

Bishop, W. F., 1980, Petroleum geology of northern Central America: Journal of Petroleum Geology, v. 3, p. 3-59.

Bullard, E. C., J. E. Everett, and A. G. Smith, 1965, Fit of continents around the Atlantic: London, Royal Society, Philosophical transactions, 258A, p. 41-75.

Case, J. E., 1980, Crustal setting of mafic and ultramafic rocks and associated ore deposits of the Caribbean region: U.S. Geological Survey Open File Report 80-304, 6 pl., 94p.

Christofferson, E., in press, Plate model of the collapsing Caribbean margin of Nicaragua: Transactions, Tenth Caribbean Geological Conference, Cartagena.

Crowell, J.C., 1974, Origin of late Cenozoic basins in southern California, *in* W. R. Dickinson, ed., Tectonics and sedimentation: Society of Economic Paleontologists and Mineralogists, Special Publication 22, p.190-204.

Cumming, G. L., and S. E. Kesler, 1976, Source of lead in Central America and Caribbean mineralization; Earth and Planetary Science Letters, v. 31, p. 262-268.

de Figueiredo, A. M. F., and G. P. R. Gabaglia, 1986, Sistema classificatorio applicado as bacias sedimentares Brasileiras: Revista Brasileira de Geociencias, v. 16, p. 350-369.

Dickinson, W. R., 1974, Plate tectonics and migration of petroleum, *in* Geological interpretations from global tectonics with applications for California geology and petroleum exploration: Bakersfield, San Joaquin Geological Society Short Course, p. 15-1 - 15-3.

Dickinson, W. R., and H. Yarborough, 1978, Plate tectonics and hydrocarbon accumulation: American Association of Petroleum Geologists, Continuing Education Course Note Series no. 1.

Eriksen, G. E., 1976, Metallogenic provinces of the southeastern Pacific region, *in* M. T. Halbouty, J. C. Maher, and H. M. Lian, eds., Circum-Pacific energy and mineral resources: American Association of Petroleum Geologists Memoir 25, p. 527-538.

Fisher, A. G., and S. Judson, eds., 1975, Petroleum and global tectonics: Princeton, Princeton University Press.

Halbouty, M. T., J. C. Maher, and H. M. Lian, eds., 1976, Circum-Pacific energy and mineral resources: American Association of Petroleum Geologists Memoir 25, 608p.

Harding, T. P., and J. D. Lowell, 1979, Structural styles, their plate-tectonic habitats, and hydrocarbon traps in petroleum provinces: American Association of Petroleum Geologists Bulletin, v. 63, p. 1016-1058.

Hutchinson, C. S., 1983, Economic deposits and their tectonic setting: New York, J. Wiley and Sons, 365p.

Irving, E., and G. A. Irving, 1982, Apparent polar wander paths Carboniferous through Cenozoic and the assembly of Gondwanaland: Geophysical Surveys, v. 5, p. 141-188.

Johnstone, M. H., 1981, The importance of continental fragmentation history to petroleum accumulation, *in* M. M. Creswell and P. Vella, eds., Gondwana five: Proceedings of the fifth international Gondwana symposium: Wellington, New Zealand, p. 329-334.

Katahira, T., and M. Ukai, 1976, Petroleum fields of Japan with volcanic-rock reservoirs, *in* M. T. Halbouty, J. C. Maher, and H. M. Lian, eds., Circum-Pacific energy and mineral resources: American Association of Petroleum Geologists Memoir 25, p. 276-279.

Kesler, S. E., E. Levy, and F. C. Martin, in press, Metallogenic evolution of the Caribbean region: Geological Society of America, Decade of North American Geology, Caribbean volume.

Kingston, D. R., C. P. Dishroon, and P. A. Williams, 1983, Hydrocarbon plays and global basin classification: American Association of Petroleum Geologists Bulletin, v. 67, p. 2194-2198.

Krushensky, R. D., S. M. Cargill, and G. L. Raines, eds., 1987, Proceedings of a workshop on development of mineral, energy and water resources and mitigation of geologic hazards in Central America: U.S. Geological Survey Circular 1006, 272p.

Ladd, J. W., 1976, Relative motion of South America with respect to North America and plate tectonics: Geological Society of America Bulletin, v. 87, p. 969-976.

Lee Moreno, J. L., 1980, Metalogenesis en Latinoamerica: International Union of Geological Sciences Publication 5, 361p.

MacDonald, W. D., 1972, Late Paleozoic tectonics of northern South America: Anais, Academia Brasileira de Ciencias, v. 44, supplement, p. 197-208.

MacDonald, W. D., 1976, Cretaceous-Tertiary evolution of the Caribbean: Transactions, Seventh Caribbean Geologic Conference Guadeloupe, p. 69-78.

MacDonald, W. D., 1980, Net tectonic rotation, apparent tectonic rotation, and the structural tilt correction in paleomagnetic studies: Journal of Geophysical Research, v. 85, p. 3659-3669.

MacDonald, W. D., in press, Survey of Caribbean paleomagnetism: Geological Society of America, Decade of North American Geology, Caribbean volume.

Malfait, B. T., and M. G. Dinkelman, 1972, Circum-Pacific tectonic and igneous activity and the evolution of the Caribbean plate: Geological Society of America Bulletin, v. 83, p. 251-272.

Mann, P., and K. Burke, 1984, Cenozoic rift formation in the northern Caribbean: Geology, v. 12, p. 732-736.

McKenzie, D. P., and J. G. Sclater, 1971, The evolution of the Indian Ocean since the Late Cretaceous: Geophysical Journal, Royal Astronomical Society, v. 24, p. 437-528.

Meyerhoff, A. A., 1973, Discussion: Geological Society of America Bulletin, v. 84, p. 2147-2152.

Mills, R. A., and K. E. Hugh, 1974, Reconnaissance geologic map of Mosquitia region, Honduras and Nicaragua, Caribbean coast: American Association of Petroleum Geologists Bulletin, v. 58, p. 189-207.

Mitchell, A. H. G., and M. S. Garson, 1981, Mineral deposits and global tectonic settings: New York, Academic Press, 405p.

Montecchi, P. A., 1976, Some shallow tectonic consequences of subduction and their meaning to the hydrocarbon explorationist, in M. T. Halbouty, J. C. Maher, and H. M. Lian, eds., Circum-Pacific energy and mineral resources: American Association of Petroleum Geologists Memoir 25, p. 189-202.

Muehlberger, W. R., 1976, The Honduras depression: Guatemala, Publicaciones Geologicas, Instituto Centroamericano de Investigacion y Tecnologia Industrial (ICAITI), v. 5, p. 43-54.

Oliver, J., 1986, Fluids expelled tectonically from orogenic belts: Their role in hydrocarbon migration and other geologic phenomena: Geology, v. 14, p. 99-102.

Orwig, E. R., 1976, Structural evolution of late Mesozoic and Cenozoic basins in western North America, *in* M. T. Halbouty, J. C. Maher, and H. M. Lian, eds., Circum-Pacific energy and mineral resources: American Association of Petroleum Geologists Memoir 25, p. 281-290.

Petersen, U., 1980, Metallogenesis in South America: Progress and problems, *in* J. L. Lee M., ed., Metalogenesis en Latinoamerica: International Union of Geological Sciences Publication 5, p. 249-274.

Pindell, J., and J. F. Dewey, 1982, Permo-Triassic reconstruction of western Pangea and the evolution of the Gulf of Mexico/ Caribbean region: Tectonics, v. 1, p. 179-211.

Pinet, P., 1972, Diapir-like features offshore Honduras; Implications regarding tectonic evolution of Cayman trough and Central America: Geological Society of America Bulletin, v. 83, p. 1911-1922.

Price, R., ed., 1989, Origin and evolution of sedimentary basins and their energy and mineral resources: American Geophysical Union Monograph 48, 216p.

Sawkins, F. J., 1984, Metal deposits in relation to plate tectonics: New York, Springer-Verlag, 325p.

Scott, R. W., 1984, Mesozoic biota and depositional systems of the Gulf of Mexico-Caribbean region: Geological Association of Canada Special Paper 27, p. 49-64.

Sillitoe, R. H., 1988, Epochs of intrusion-related copper mineralization in the Andes: Journal South American Earth Sciences, v. 1, p. 89-108.

Steiner, M., S. Lawrence, and M. Beck, 1989, A Jurassic pole position from the craton of North America: EOS, Transactions, American Geophysical Union, v. 70, no. 20, p. 600.

Strong, D. F., 1976, Metallogeny and plate tectonics: Geological Association of Canada Special Paper 14, 660p.

Wilson, J. T., 1966, Did the Atlantic close and then reopen?: Nature, v. 211, p. 676-681.

Woollard, G. P., 1981, Mineral deposits and plate tectonics: American Association of Petroleum Geologists, Studies in Geology 12, p. 507-516.

Wright, J. E., ed., 1977, Mineral deposits, continental drift, and plate tectonics: Stroudsburg, Pennsylvania, Dowden Hutchinson and Ross, Benchmark Papers in Geology, 417p.

Circum-Pacific Council for Energy and Mineral Resources
Earth Science Series, Vol. 16
R. L. Miller, G. Escalante, J. A. Reinemund, and M. J. Bergin (Eds.)
Energy and Mineral Potential of the Central American-Caribbean Regions
Springer-Verlag, Berlin Heidelberg 1995

THE STATUS OF MINERAL PRODUCTION IN THE CARIBBEAN BASIN COUNTRIES

Orlando Martino
Branch of Latin America and Canada, Division of International Minerals, Bureau of Mines,
U.S. Department of the Interior, 2401 E Street, N.W., Washington DC 20241 USA

ABSTRACT

This paper is designed to set the general stage for the technical papers on resources that follow. Any inquiry into the mineral potential of the Caribbean Basin needs to consider as a point of departure what mineral fuels (oil, gas, and coal), metallic minerals, and industrial minerals are already being produced in the Caribbean Basin countries.

The region of the Caribbean Basin is comprised of 25 countries and dependencies with a population of 187 million. This population represents about 47 percent, or almost one-half, of the total population of Latin America, which is approaching 400 million. The Caribbean Basin is characterized by a large diversity of ethnic, cultural, geological, and mineral endowment conditions.

Aside from Mexico and Colombia which are well endowed by nature, the other Caribbean Basin countries have a limited diversity of mineral output. This lack of diversity is especially apparent in the export sector. Some countries depend almost exclusively on exports of bauxite, gold, ferronickel, or petroleum.

By far the most important and most valuable mineral commodities produced today in the Caribbean region are the hydrocarbons—petroleum and natural gas. Of the 12 countries in all of Latin America that produce crude oil, seven are located in the Caribbean region. This Caribbean group produced 75 percent of Latin America's crude oil output in 1987. Of the seven producers, only Mexico and Venezuela are of world rank. As a nearby market, the United States has become highly dependent upon the Caribbean producers for its imports of crude oil and petroleum products. In recent years U.S. imports from the Caribbean have reached almost one-third of total U.S. oil imports, including shipments to the Strategic Petroleum Reserve.

Colombia has the largest resources of coal in Latin America, estimated at about 40 billion MT, and is now the largest producer and exporter of coal in all of Latin America. Colombia's rapid expansion of coal output dates from 1982. The Caribbean Basin is the only area in all of Latin America with installed capacity to exploit geothermal energy sources. This is not surprising when one considers the volcanism and seismic activity in the area.

The output of precious metals in the region is the second most valuable after crude oil, amounting to about $1.2 billion compared with $30 billion for oil. The production of gold is relatively widespread and not as concentrated in a few countries as is that of bauxite and iron ore. One-half of the countries in the region have been gold producers, while silver output has been more concentrated and only Colombia produces platinum. Colombia is the predominant gold producer in the region, followed in importance by Mexico, the Dominican Republic, and Venezuela.

In terms of value, bauxite ranks after gold and silver and iron ore among the metallic minerals produced in the region. Of the 25 countries or dependencies in the Caribbean region, only five have been significant bauxite producers in recent decades. In order of importance they are Jamaica, Guyana, Suriname, the Dominican Republic, and Haiti. Aluminum and ingot aluminum are also produced in the region. Venezuela is expected to become a significant producer and exporter of aluminum because of abundant bauxite reserves and low-cost energy.

Nickel, cobalt, and manganese are the chief strategic minerals produced in the Caribbean region. Cuba, the Dominican Republic, and Colombia are the major producers. Cuba has the largest reserves of nickel in the world and is the only significant producer of cobalt in the Caribbean.

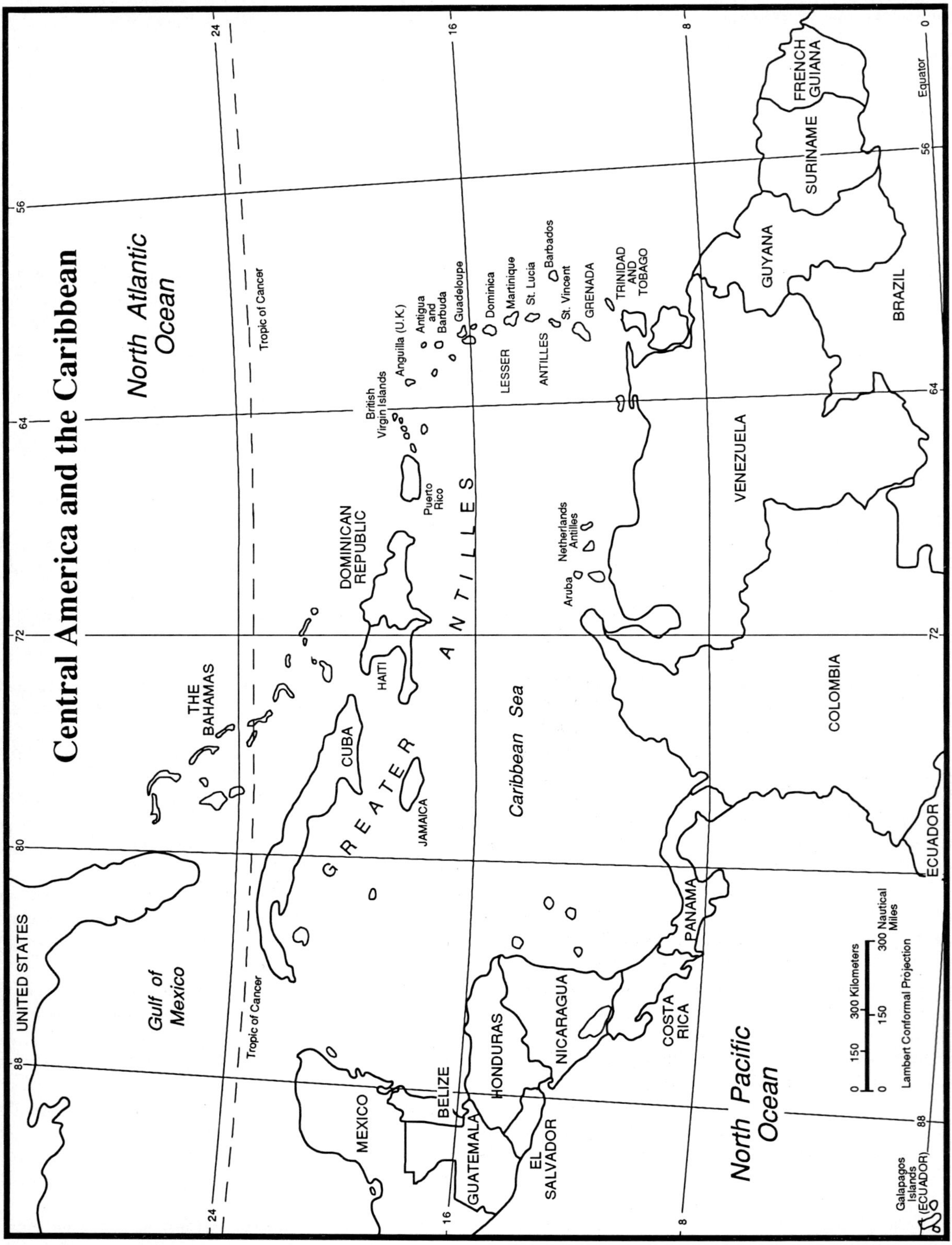

Figure 1. Central America and the Caribbean (U.S. Department of State).

Aside from Mexico and Colombia, the countries in the Caribbean Basin do not have as diversified a record as producers of industrial minerals. The most important industrial minerals produced in the region include cement, clays, gypsum, and salt. Of the 25 entities in the Caribbean, 19 produce cement. In practically all basin countries, materials are extracted to meet the domestic demands of the pervasive construction industry.

To summarize, mineral production in the Caribbean Basin is dominantly based on the energy minerals in the form of petroleum, natural gas, and coal. In addition, important geothermal sources of energy are being exploited. Of lesser value is the output of the precious metals—gold, silver, and platinum—and of bauxite, iron ore, nickel, cement, gypsum, and salt. The United States is highly dependent on the Caribbean Basin countries for its imports of energy minerals.

The text is augmented by numerous historical production tables and by a chronology of selected mineral developments.

INTRODUCTION

The region of the Caribbean Basin is comprised of about 25 countries (Figure 1) and dependencies with a population of 187 million. This population represents about 47 percent, or almost one-half, of the total population of Latin America, which is approaching 400 million. The Caribbean Basin is characterized by a large diversity of ethnic, cultural, geological, and mineral endowment conditions. Among the islands are tiny Aruba with 62,000 people and Cuba, the largest island, with 10 million people. On the mainland, which includes not only Central America but the northern part of South America—namely Colombia, Venezuela, Guyana, Suriname, and French Guiana—there is Belize with 168,000 people and Mexico, the largest mainland country, with 83 million people.

Except for Mexico and Colombia, most of the countries have a limited diversity of mineral output. Mexico, besides having a long and well-established mining tradition, produces some 50 minerals. Mexico is considered in this paper because its Yucatan Peninsula forms the western edge of the Caribbean Basin, but a separate paper would be required to deal with Mexico's highly diverse mineral output.

The resource papers that follow are oriented toward an examination of mineral prospects or mineral potential, including possibilities for development of new mineral resources in the basin countries. As a point of departure for such inquiry it is useful to present a report on the variety of important minerals currently being produced in the region. In this regard, tables are provided that give production data for the most important minerals that are being produced. Historical production data since 1970 are offered in each table to indicate production trends in each country. In the mineral fuels category, tables are included for crude oil, petroleum products, natural gas, coal, and geothermal energy. In the category of metallic minerals, the tables cover bauxite, alumina, aluminum, gold and silver, iron ore, crude steel, nickel, and lead and zinc. In the category of industrial minerals, the tables cover cement, clays, gypsum, and salt. These commodities represent the central core of mineral output from the Caribbean Basin countries. The data presented in this paper has been obtained from official reports of the country concerned or from files of the U.S. Bureau of Mines.

GENERAL PROFILE

Before proceeding with this detailed report by commodity, it would be interesting to offer a profile of the countries (Table 1) and a brief history of mineral developments in the region.

Long before the arrival of Columbus in the New World, the indigenous peoples had developed a metallurgy

Table 1. Caribbean Basin Countries — General Profile.

Country	Population	Area (Km^2)	Literacy (%)	Income per capita[1]
Cuba	10,260,000	110,860	96	1,800
Guatemala	8,454,000	108,890	57	1,376
Dominican Republic	6,716,000	48,440	70	1,400
Haiti	6,147,000	27,750	37	300
El Salvador	4,934,000	20,940	67	900
Honduras	4,679,000	112,088	60	780
Nicaragua	3,501,000	139,000	74	880
Costa Rica	2,791,000	50,900	90	2,010
Jamaica	2,391,000	10,960	73	1,704
Panama	2,274,000	77,080	86	2,550
Trinidad and Tobago	1,144,000	5,128	97	2,900
Guyana	988,000	214,970	91	670
Suriname	357,000	163,820	80	3,260
Martinique	344,920	1,100	70	4,300
Guadeloupe	336,350	1,780	70	3,200
Barbados	270,000	430	99	3,530
Bahamas	231,000	13,935	93	10,430
Netherlands Antilles	182,220	960	95	9,140
Belize	168,200	22,960	90	1,200
Dominica	94,190	750	80	1,100
French Guiana	92,040	91,000	73	3,400
Grenada	84,750	340	85	1,000
Aruba	62,130	193	95	6,900
SUBTOTAL	56,501,800			
Large Countries				
Mexico	83,039,000	1,967,183	88	2,420
Colombia	29,942,000	1,138,340	82	1,580
Venezuela	18,272,000	898,805	90	4,110
TOTAL	187,754,800			

[1] Source: Inter-American Development Bank, in 1986 U.S. dollars.

Table 2. Caribbean Basin Countries—Production of Crude Petroleum (thousand 42-gallon barrels).

COUNTRY	1970	1975	1980	1985	1986	1987
Mexico	177,599	261,540	708,454	960,114	886,092	927,333
Venezuela	1,353,420	856,364	793,397	613,581	653,563	603,594
Colombia	79,594	57,259	45,944	64,352	110,714	147,843
Trinidad and Tobago	51,047	78,613	77,616	64,259	61,435	56,621
Cuba[1]	1,330	1,000	1,819	5,771	e6,000	6,600
Guatemala	——	——	1,513	1,068	1,825	1,460
Barbados	——	123	305	679	559	510

[e]Estimated

[1]Cuba repts crude oil production in metric tons. A conversion to barrels was made by using a factor of 6.652.

based on gold found in placer deposits or in vein outcrops. The Indians learned to cast metal objects of two alloys that are unique to the Americas. One was an alloy of gold and platinum developed by the goldsmiths in Ecuador. The other uniquely American alloy was worked out by smiths in the region covering Colombia, Panama, and Costa Rica. It is a combination of copper and gold called "tumbaga" or "guanin" by the Indians. Shortly after the start of the Christian era, the goldsmiths of Colombia seem to have discovered the secret of casting objects of copper and gold by the lost-wax method.

During the four exploratory voyages of Columbus, the search for gold was an important activity along with the search for a passage to the Far East. Interest in gold and silver dominated mining for more than 300 years before independence from Spain and continued thereafter. Many mineral deposits in the Caribbean region show signs of having been worked by the Spaniards. Silver mining was especially fruitful in Mexico, where many bonanzas were discovered. The Spaniards were trained in lode as well as placer mining.

Where early mining in the Caribbean Basin has been dominated by gold and silver, mineral output began to diversify with the growing industrialization of the last half of the nineteenth century. The rapid growth of the iron and steel industry created a strong world demand for iron ore and manganese. Growth in the use of electricity at the turn of the century created a strong demand for copper. The phenomenal expansion in the use of the automobile created a strong market not only for steel, but also for lead and zinc, and for petroleum products for propulsion. Finally, the aviation and space age has stimulated the growth of output of bauxite and aluminum. Thus bauxite is the most recent major mineral to be exploited in the twentieth century, and the Caribbean has played a particularly significant role in this development.

Today when we look at the output of mineral commodities from the individual Caribbean Basin countries, we see a limited diversity, except for Mexico and Colombia which have been well-endowed by nature. This lack of diversity is especially apparent in the export sector. Some countries depend almost exclusively on exports of bauxite, gold, ferronickel, or petroleum. This survey begins with petroleum.

PETROLEUM

By far the most important and most valuable mineral commodities produced today in the Caribbean region are the hydrocarbons: petroleum and natural gas. Of the 12 countries in all of Latin America that produce crude oil, seven are located in the Caribbean region; namely, Mexico, Venezuela, Colombia, Trinidad and Tobago, Cuba, Guatemala, and Barbados in order of importance. This Caribbean group produced 75 percent of Latin America's crude oil output in 1987. Of the seven producers, only Mexico and Venezuela are of world rank. In 1981 Mexico surpassed Venezuela as the leading crude oil producer in all of Latin America and gained fourth place in the world after the U.S.S.R., the United States, and Saudi Arabia. However, by 1987, Mexico had slipped to sixth place, while Venezuela was in eighth place worldwide.

Statistics since 1970 show mixed production trends in this Caribbean group of seven (Table 2). Crude oil output by Mexico, Colombia, Cuba, and Barbados increased markedly between 1970 and 1987, while output from Trinidad and Tobago and Guatemala has been on a plateau. Crude oil output from Venezuela has been in a downward trend, such that by 1987 output was less than one-half of Venezuela's output in 1970.

In 1987, the total value of crude oil produced by these seven countries amounted to about $30 billion. Nonetheless, all the countries have suffered and are still suffering from the depressed world oil prices that in 1988 dropped to a low level of about $15 per barrel, compared with $35 per barrel in the early 1980s.

NATURAL GAS

All seven of the countries producing crude oil also produce natural gas, generally in association with the oil (Table 3). In Mexico, for example, 80 percent of the produced gas is associated with the oil output. However, Mexico and Colombia do have significant deposits that are primarily gas producers. As in the case of crude oil, Mexico and Venezuela are the dominant gas producers in the region; however,

Table 3. Caribbean Basin Countries—Production of Natural Gas[1] (in million cubic feet).

COUNTRY	1970	1975	1980	1985	1986	1987
Mexico	481,106	583,885	1,129,288	1,197,127	1,175,191	1,194,217
Venezuela	348,630	450,295	589,046	498,159	576,390	465,300
Colombia	46,731	65,905	118,534	[e]131,500	[e]131,700	[e]137,000
Trinidad and Tobago	66,687	62,990	128,800	124,197	[e]132,300	[e]123,000
Guatemala[2]	—	—	584	1,000	[e]1,100	600
Barbados	97	120	311	371	371	360
Cuba	—	—	630	244	[e]260	[e]240

[e]Estimated
[1]Marketable quantities
[2]Gross production. Guatemala began production of small quantities of crude oil and gas in 1976.

Mexico's output level is almost three times Venezuela's level. Since 1970, contrary to the case with oil, all the countries except Cuba show an increasing trend in gas output.

PETROLEUM PRODUCTS

Of the 25 countries and dependencies around the Caribbean Basin, 17 have constructed oil refineries to satisfy local energy demand and to obtain export earnings (Table 4). A number of these countries purchase foreign crude oil to refine a variety of products ranging from motor and aviation gasoline to residual fuel oil.

The Netherlands Antilles and the Bahamas are examples of small areas with considerable product capacity to enter export markets, especially the nearby United States. In this group of 17, the dominant producers of petroleum products are Mexico and Venezuela, followed by the Netherlands Antilles, Colombia, Cuba, Trinidad and Tobago, and the Dominican Republic. Refineries in El Salvador, Guatemala, Honduras, and Costa Rica are relatively smaller.

As a nearby market, the United States has become highly dependent upon the Caribbean producers for its imports of both crude oil and petroleum products. In recent years, U.S. imports from the Caribbean have reached almost one-third of the total U.S. oil imports (Table 5). Considering both crude oil and petroleum products, the most important suppliers to the United States in 1987 in order of magnitude were Venezuela, Mexico, Colombia, and Trinidad and Tobago. Mexico, however, was the chief supplier of crude oil to the United States, to which it allocates about 50 percent of its exports.

COAL

Only three countries in the region—Colombia, Mexico, and Venezuela—produce significant quantities of coal, pri-

Table 4. Caribbean Basin Countries — Petroleum Refinery Products (thousand 42-gallon barrels).

COUNTRY	1970	1975	1980	1985	1986	1987
Mexico	183,238	247,977	425,003	518,904	504,574	519,746
Venezuela	470,864	317,045	341,467	379,388	390,866	363,800
Netherlands Antilles[1]	336,017	221,228	214,350	20,000	58,400	83,600
Colombia	53,230	57,685	59,282	64,870	69,447	72,350
Cuba	32,648	42,824	44,337	46,020	46,500	46,500
Trinidad and Tobago	154,860	85,660	84,600	29,678	30,860	28,950
Dominican Republic	—	9,158	9,841	12,647	13,000	13,000
Jamaica	12,114	10,189	8,201	9,008	9,309	9,200
Panama	26,000	28,800	13,642	8,864	9,000	9,000
El Salvador	1,185	4,879	4,572	4,831	4,800	4,800
Martinique	—	3,673	3,990	4,300	4,938	4,800
Guatemala	4,687	5,555	5,381	4,926	3,821	3,800
Nicaragua	3,145	4,859	4,087	3,715	3,700	3,500
Costa Rica	1,999	1,993	3,781	3,045	3,000	4,594
Honduras	5,135	4,547	3,648	2,386	1,472	1,500
Barbados	932	1,179	1,364	1,500	1,500	1,400
Bahamas	25,500	79,446	67,880	6,000	—	—

[1]Includes Aruba

Table 5. United States Dependency on Imports of Petroleum from Latin America — 1986 (thousand 42-gallon barrels).

Country	Crude Oil	Petroleum Products[1]	Total Petroleum	% Share (Crude Oil)	% Share (Products)	% Share (Total Oil)
Venezuela	151,937.0	[2]137,532.0	289,469.0	32.8	54.80	40.6
Mexico	226,533.0	[3]28,457.0	254,990.0	49.0	11.30	35.7
Trinidad and Tobago	34,060.0	[4]11,477.0	45,537.0	7.5	4.50	6.4
Colombia	20,715.0	[5]11,106.0	31,821.0	4.4	4.40	4.4
Ecuador	23,408.0	4,836.0	28,244.0	5.1	1.90	3.9
Brazil	——	[6]18,254.0	18,254.0	N	7.40	2.5
Bahamas	——	[3]13,416.0	13,416.0	N	5.30	1.9
Peru	3,372.0	7,333.0	10,705.0	0.7	2.90	1.5
Netherlands Antilles	——	9,237.0	9,237.0	N	3.70	1.3
Argentina	——	7,734.0	7,734.0	N	3.00	1.1
Guatemala	1,783.0	——	1,783.0	0.4	N	0.3
Jamaica	——	706.0	706.0	N	0.30	0.1
Panama	——	698.0	698.0	N	0.30	0.1
El Salvador	——	467.0	467.0	N	0.20	——
Bolivia	268.0	——	268.0	0.1	N	???
Costa Rica	——	112.0	112.0	N	0.04	——
Sub-Total	462,076.0	251,401.0	713,477.0	100.0	100.00	100.0
Caribbean Basin Only	435,028.0	213,244.0	648,272.0	94.1	84.80	90.1
% Share from Latin America	30.3	33.7	31.4			
From Canada	208,163.0	86,254.0	294,413.0			
From Western Hemisphere	670,239.0	237,655.0	1,007,898.0			
% Share from Western Hemisphere	44.0	45.2	44.4			
Total U.S. Imports	1,524,970.0	746,425.0	2,271,395.0			

SOURCE: U.S. Department of Energy, Petroleum Supply Ann. May, 1987, DOE/EIA-340
N = Not Applicable
[1]Comprising LPG, unfinished oils, motor gasoline, jet fuel, kerosene, diesel oil, fuel oil, and other products.
[2]Mostly fuel oil, diesel oil, and motor gasoline.
[3]Mostly fuel oil diesel oil and LPG.
[4]Mostly fuel oil and diesel oil.
[5]Mostly fuel oil.
[6]Mostly motor gasoline and fuel oil.

Table 6. Caribbean Basin Countries — Production of Coal (thousand metric tons).

Country	1970	1975	1980	1985	1986	1987
Colombia						
Metallurgical	[e]700	[e]600	[e]700	618	630	759
Steam	[e]2,617	[e]3,200	[e]4,247	9,088	11,540	13,835
	3,317	3,800	4,947	9,706	12,170	14,594
Mexico						
Metallurgical	[e]1,300	2,906	3,051	2,890	2,594	2,330
Steam	[e]100	177	408	2,440	3,050	4,122
	1,400	3,083	3,459	5,330	5,644	6,452
Venezuela						
Steam	40	60	39	41	58	337

[e]Estimated

marily steam coal (Table 6). Colombia has the largest resources of coal in Latin America, estimated at about 40 billion MT, and is now the largest producer and exporter of coal in all of Latin America. Colombia's rapid expansion of coal output since 1982 comes primarily from the El Cerrejon deposit in the Guajira Peninsula just west of Lake Maracaibo. Exploitation of this deposit is by a joint venture of the Colombian Government and a foreign oil company. Colombia presently produces about 15 million MT of coal per year, but new rail and shipping port facilities have recently been especially constructed that will enable Colombia to export high-quality steam coal at the level of 25 million MT/yr. Most of Colombia's coal exports go to Europe.

Venezuela is engaged in a number of projects to exploit its sizeable reserves of steam coal and metallurgical coal, but combined output in 1987 was just over 300,000 MT/yr (Table 6). A large project in the Guasare region of Venezuela will be developed by Carbones del Zulia in a joint venture with energy companies from the United States and Italy. The planned production rate is 6.5 million tons per year. Costa Rica has identified significant coal deposits which are under evaluation and development near Puerto Limon.

Table 7. Caribbean Basin Countries—Geothermal Power./1

Country	Capacity (Megawatts)	Fields (in Latin America)
United States	2,830	
Philippines	948	
MEXICO	700	Cerro Prieto, Los Azufres
Italy	505	
Japan	215	
New Zealand	167	
Indonesia	142	
EL SALVADOR 2/	95	Ahuachapan
NICARAGUA 3/	70	Momotombo
Kenya	45	
Iceland	39	
Turkey	21	
China	15	
Soviet Union	11	
GUADELOUPE	4.2	La Bouillante
Others 4/	10.8	
B--Under Construction		
COSTA RICA	55	Miravalles 5/
C--Under Feasibility Study		
MEXICO 6/	50	Los azufres, La Primavera, Los Humeros
GUATEMALA	15	Zunil 7/
PANAMA	--	Valle de Anton, Chitre-Colobre 8/
ST. LUCIA	10	Soufriere

1/ As of early 1989 Based on data from David Sussman, UNOCAL-Geothermal Division.
2/ Expected increase to 115 MW by 1990 with new units at Berlin and Chipilapa.
3/ Second unit of 35 MW became operational in early 1989.
4/ Includes Taiwan (4 MW), Portugal - Azores (3 MW), Greece (2 MW), Argentina (1 MW), Australia (0 6 MW), and Zambia (0 2 MW).
5/ First unit of 55 MW under development for 1992 completion. Second unit of 55 MW has feasibility study completed. Its completion has been programmed for 1994.
6/ Ten units of 5 MW each to be installed at Los Azufres (25 MW), La Primavera (5 MW) and Los Humeros (20 MW) by 1990.
7/ Scheduled for startup in 1992. Second unit to be installed at Zunil or Amatitlan.
8/ Baru-Cerro Colorado field has been abandoned.

GEOTHERMAL ENERGY

The Caribbean Basin is the only area in all of Latin America with installed capacity to exploit geothermal energy sources. This is not surprising when one considers the volcanism and seismic activity in the area. Today four countries in the Caribbean region produce geothermal energy: Mexico, El Salvador, Nicaragua, and Guadeloupe (Table 7). Mexico was the first to develop geothermal energy, installing its first unit at its Cerro Prieto field in Baja California in 1973. By year-end 1987, Mexico ranked third worldwide in installed capacity after the United States and the Philippines, with a capacity of 655 MW. El Salvador came next in 1975 with its development of the Ahuachapán field and now has 95 MW of installed capacity. In early 1983, Nicaragua began production at a 35 MW unit in the Momotombo field. A second 35 MW unit was added in 1989.

Costa Rica has a very active geothermal program. At its Miravalles field an initial unit of 55 MW is currently under development and is scheduled for start-up in 1992. In addition, the feasibility study has been completed on a second unit for 55 MW, also in the Miravalles field and scheduled for completion in 1994.

Guatemala is currently developing its Zunil geothermal field for a 15 MW unit scheduled for completion in 1991.

METALLIC MINERALS

Bauxite

In terms of value, bauxite ranks after gold and silver and iron ore among the metallic minerals produced in the region. Of the 25 countries or dependencies in the Caribbean region, only five have been significant bauxite producers in recent decades (Table 8). In order of importance, they are Jamaica, Guyana, Suriname, the Dominican Republic, and Haiti. Venezuela became the sixth producer in 1987 when it started up its Los Pijiguaos mine in Bolivar State. After a lapse of two years, the Dominican Republic resumed production in 1987. Haiti ceased to produce bauxite in 1982.

Jamaica has continued as the dominant producer of bauxite in the Caribbean, but with respect to its rank in all of Latin America, Jamaica appears to have fallen behind Brazil in 1988. Commercial possibilities of bauxite in Jamaica were identified as early as 1943, and in 1957 Jamaica became the world's leading producer and exporter of bauxite. However, since the mid-1970s when output reached more than 15 million MT, Jamaica's bauxite output has been in a downtrend, amounting to 7.4 million MT in 1988. Today Jamaica ranks in fourth place worldwide after Australia, Guinea, and Brazil.

As the newest producer, Venezuela is actively developing its bauxite potential with plans to produce 3 million MT by 1990 and 6 million MT by 1992. Colombia and Costa Rica have identified significant bauxite deposits which are presently not exploited.

Alumina (Al_2O_3)

Several traditional bauxite producers such as Jamaica, Suriname, and Guyana have invested in facilities to produce alumina, the raw material of aluminum, to obtain value-added material for export. Venezuela, prior to initiating bauxite mining in 1987, was a major alumina producer in the region from bauxite imports, by taking advantage of its low-cost power. Guyana's production of alumina continued suspended in 1987.

Aluminum (Al)

Only three countries have installed plants to produce aluminum: Venezuela, Mexico, and Suriname, in order of

Table 8. Caribbean Basin Countries—Production of Bauxite, Alumina (Al_2O_3), and Primary Aluminum (Al).

Country	1970	1975	1980	1985	1986	1987
			PRODUCTION OF BAUXITE 1/ (Thousand metric tons)			
Jamaica	12,009	11,571	12,054	5,975	6,964	7,833
Guyana	4,417	3,250	3,052	1,675	1,466	2,200
Suriname 2/	6,022	4,928	4,864	3,000	3,847	1,200
Venezuela	—	—	—	—	217	
Dominican Republic	1 ,067	754	606	—	—	211
Haiti	632	522	312	—	—	—
Colombia 3/	—	—	—	—	—	—
			PRODUCTION OF ALUMINA (Thousand metric tons)			
Jamaica	1,689 4/	2,375	2,456	1,513	1,575	1,626
Venezuela	—	—	—	1,085	1,296	1,347
Suriname	1,014	1,148	1,329	1,000	1,471	370
Guyana	317	311	220	—	—	—
			PRODUCTION OF PRIMARY ALUMINUM (Metric tons)			
Venezuela	22,900	49,700	327,900	395,894	424,000	428,000
Mexico	33,955	40,090	42,601	42,744	37,016	60,200
Suriname 5/	55,000	42,000	55,000	29,000	29,000	9,000

1/Dry equivalent, gross weight.
2/Gross weight.
3/Production was registered in 1981-84 at the level of less than 1,000 tons per year.
4/Exports.
5/Represents exports for every year except 1970.

Table 9. Caribbean Basin Countries—Production of Gold.

Country	1970	1975	1980	1985	1986	1987
Colombia	201,518	311,321	510,439	1,142,385	1,285,878	853,468
Mexico	198,241	144,710	176,089	265,693	250,615	256,820
Dominican Republic	—	195,000	370,000	328,000	284,000	246,000
Venezuela	22,320	18,326	13,841	71,180	80,730	107,608
Nicaragua	115,173	83,457	59,984	25,316	28,664	30,486
Guyana	4,433	18,067	11,003	10,323	14,035	21,425
French Guiana	2,347	2,437	4,000	8,005	10,481	11,000
Costa Rica e/	500	18,000	18,000	1/15,997	1/11,600	1/9,645
Honduras	3,333	2,520	2,027	5,023	2,018	4,222
Suriname	1,137	141	350	500	600	700
El Salvador	2,301	8,713	2,492	—	—	—
Haiti e/	3,000	—	—	—	—	—

e/ Estimated
1/ Reported figure—gold from placer deposits and mines purchased by Banco Central. Actual production estimated to be at least twice that amount..

importance. Mexico produces aluminum entirely from imports of alumina since it has no economical bauxite deposits. Since 1970 Venezuela has aggressively expanded its aluminum output such that by 1988 its level of 500,000 MT was seven times the output by Mexico.

Venezuela's aluminum industry is expected to further expand output as one of the world's lowest cost producers because of abundant bauxite, inexpensive hydroelectric power, and basic infrastructure. Venezuela is also expected to become a significant exporter of aluminum. Venezuela is the region's only integrated aluminum producer.

Bauxite and aluminum producers in the region have enjoyed increased earnings from a sharp improvement in world prices. The average price of aluminum was $1.10/lb in 1988, compared with the depressed level of about $0.50/lb in 1985.

PRECIOUS METALS—GOLD, SILVER, AND PLATINUM

The output of precious metals in the region is the second most valuable after crude oil, amounting to about $1.2 billion compared with $30 billion for oil. The production of gold is relatively widespread and not as concentrated in a few countries as is that of bauxite and iron ore (Table 9). One-half

Table 10. Caribbean Basin Countries—Production of Silver (thousand troy ounces)

Country	1970	1975	1980	1985	1986	1987
Mexico	42,836	38,029	50,052.0	73,167	77,753	77,642
Dominican Republic	—	109	1,623.0	1,581	1,356	1,148
Honduras	3,816	3,802	1,766.0	2,765	1,745	747
Colombia	76	88	152.0	169	187	160
Nicaragua	216	435	164.0	30	25	29
Costa Rica e/	NA	3	1.6	2	2	2
El Salvador	154	176	146.0	—	—	—
Guatemala e/	NA	NA	10.0	—	—	—
Haiti e/	17	—	—	—	—	—

e Estimated.
NA Not available.

Table 11. Caribbean Basin Countries—Production of Iron Ore 1/ (Thousand metric tons).

Country	1970	1975	1980	1985	1986	1987
Venezuela	21,100	24,772	16,103	14,754	17,396	17,780
Mexico	4,354	5,054	7,631	7,820	7,298	7,523
Colombia	453	595	506	439	523	615
Guatemala	2	6	4	1	7	7

1/ Gross weight of marketable product.

Table 12. Caribbean Basin Countries—Production of Crude Steel (thousand metric tons).

Country	1970	1975	1980	1985	1986	1987
Mexico	3,881	5,272	7,156	7,367	7,175	7,571
Venezuela	928	2,074	3,182	3,055	3,467	3,722
Colombia	239	391	420	570	607	615
Cuba	140	240	304	401	416	420
Trinidad & Tobago	--	--	3	172	326	361
Dominican Rep.	NA	NA	NA	60	100	88
Guatemala	--	--	NA	10	17	21
El Salvador	NA	NA	NA	12	9	13
Honduras	--	--	--	19	7	7

NA Not Available..

of the countries in the region have been gold producers, but silver output has been more restricted, and only Colombia produces platinum.

Colombia is the predominant gold producer in the region, followed in importance by Mexico, the Dominican Republic, and Venezuela. Nicaragua, Costa Rica, and Honduras are much smaller producers. The major part of Colombia's gold output is from placer operations. Output exceeded 1 million troy ounces for the first time in 1985.

Colombia's output of platinum-group metals has been at the level of about 12,000 troy ounces per year.

As for silver production, Mexico is by far the preponderant producer (Table 10). With an output of almost 78 million troy ounces in 1987, Mexico extended its lead as the world's largest silver producer.

IRON ORE

The Caribbean region is modestly endowed with iron ore resources, in contrast to its abundance of energy resources. Venezuela, Mexico, and Colombia are the main iron ore producers, and very minor amounts are produced in Guatemala (Table 11).

Venezuela, the region's leading producer, is the only one that can be considered at a significant level on the world scale.

At present Venezuela is the only exporter of iron ore; the United States is an important export market. Venezuela's output level of about 20 million MT/yr of iron ore is far below the 125 million MT produced by Brazil, Latin America's first-place producer and exporter.

Table 13. Caribbean Basin Countries—Production of Nickel 1/ (metric tons).

Country	1970	1975	1980	1 985	1986	1987
Cuba 2/						
Oxide	——	NA	7,926	8,853	8,382	e/9,100
Oxide Sinter	18,500	NA	17,856	7,054	8,278	e/7,400
Sulfide	16,700	NA	16,812	16,180	16,863	e/17,700
Total	325,200	18,000	36,594	32,087	33,523	e/34,200
Dominican Republic 3/						
Ferronickel	——	30,000	16,552	25,394	21,878	32,521
Colombia 4/						
Ferronickel	——	——	——	11,800	18,600	20,700
Guatemala 5/						
Sinter	——	——	6,962	——	——	——

e/ Estimated.
NA Not available.
1/ Nickel content of products.
2/ Nickel content of the 3 nickel-cobalt compounds.
3/ The Bonao nickel mine started up in 1971.
4/ The Cerro Matoso Project started up in 1982.
5/ El Estor nickel mine was started up in 1977 and closed in 1980 because of adverse market conditions.

Table 14. Caribbean Basin Countries—Production of Lead and Zinc (metric tons).

Country	1970	1975	1980	1985	1986	1987
Mexico						
Lead	176,597	187,615	145,549	197,479	195,449	177,161
Zinc	266,400	228,851	238,231	291,938	278,065	271,480
Honduras						
Lead	15,965	23,263	13,315	21,250	12,558	4,573
Zinc	20,040	30,298	16,026	44,026	25,443	15,495
Colombia						
Lead	293	114	187	82	202	158
Zinc	156	7	152	2,000	6,000	1/
Guatemala						
Lead	500	100	100	——	——	——
Zinc 2/	——	300	——	——	——	——
Nicaragua 3/						
Lead	——	309	——	——	——	——
Zinc	——	6,327	——	——	——	——

1/ The new Equis Minc was shut down in Iate 1986 because of technical problems.
2/ Output wus first registered in 1971 at 506 tons. Next significant output was in 1981 at 2,996 tons of mined zinc.
3/ Nicaragua registered lead/zinc mine output between 1971-78 reaching a high in 1972 of 4,281 tons of lead and 17,495 tons of zinc.

CRUDE STEEL

Production of crude steel in the Caribbean Basin, from either pig iron or steel scrap, amounted to less than 13 million MT in 1987, representing more than 30 percent of total steel output in Latin America (Table 12). Mexico (the leader) and Venezuela are the dominant crude steel producers, followed at a much lower level by Colombia, Cuba, and Trinidad and Tobago.

Mexico and Venezuela have installed plants to produce sponge iron to take advantage of their ample resources of natural gas. A Mexican company invented its own direct-reduction process, known as

HyL, in the early 1950s. The HyL process is considered one of Latin America's foremost technological advances.

NICKEL

Nickel, cobalt, and manganese are the chief strategic minerals produced in the Caribbean region. Cuba, the Dominican Republic, and Colombia are the major producers of nickel (Table 13). Guatemala started up its El Estor nickel mine in 1977 and then closed it down in 1980 because of adverse market conditions. Cuba, the leading producer, produces nickel in the form of nickel-cobalt compounds, while the Dominican Republic and Colombia produce nickel as ferronickel.

Nickel producers have benefitted from increasing strong world demand for stainless steel, causing sharp increases in the price of nickel from a recent level of $2.00/lb to an average of $6.00/lb in 1988 and followed by another increase to $8.00/lb in early 1989. The most recent price increase was also due to supply disruptions during 1988.

LEAD AND ZINC

Mexico is by far the region's most important lead-zinc producer with world rank in both metals (Table 14). Output is from numerous mines scattered throughout the country. Mexico also has important large plants to refine lead and zinc. Honduras is the only other notable producer, but at a much lower level, primarily from the El Mochito mine.

The lead-zinc concentrate produced at El Mochito also contains cadmium, gold, and silver. Colombia, Guatemala, and Nicaragua have produced only minor amounts of lead and zinc in recent years.

The demand for zinc has been strong, causing prices to increase to $0.60/lb currently, compared with an average price of $0.38/lb in 1986. Lead prices have also been in an uptrend since 1985.

OTHER METALLIC MINERALS

Copper production is small in the Caribbean region. Only Mexico is a significant world producer from its large porphyry deposits at Cananea and La Caridad. Large copper deposits have been identified in Colombia and Panama (at Cerro Colorado) but remain undeveloped. The Rosita mine in Nicaragua has notable copper reserves.

Cobalt is produced by Cuba in its Moa-Nicaro nickel complex. Small amounts of chromite have been produced by Cuba and by Guatemala in its Jalapa district. Although antimony deposits are known in Guatemala, Honduras, and Nicaragua, only the first two countries have produced antimony in small quantities. Mexico is an antimony producer of world rank.

In Central America, manganese deposits are known in Costa Rica, Guatemala, Honduras, and Panama. A few of those in Costa Rica and Panama have been exploited in the past. Until about the 1940s, Cuba had an important level of manganese production in the Oriente Province. Manganese deposits are known in five areas in Panama and were studied in depth during World War II. Of the five areas, the Nombre de Dios area has been the most productive in high-grade ore. Today, only Mexico is a significant producer of manganese ore and ferromanganese alloys.

Guatemala has produced small quantities of tungsten. Mercury deposits have been reported in Guatemala and El Salvador, but only Honduras has ever reported any production.

Table 15. Caribbean Basin Countries—Production of Hydraulic Cement (thousand metric tons).

Country	1970	1975	1980	1985	1986	1987
Mexico	7,267	11,612	16,243	20,680	19,751	22,749
Venezuela	2,650	3,497	4,843	5,294	5,747	6,110
Colombia	2,757	3,091	4,351	5,394	6,011	5,965
Cuba	742	2,083	2,831	3,182	3,305	3,500
Guatemala	225	389	569	526	644	1,324
Dominican Rep.	492	578	1,015	1,007	1,066	1,100
El Salvador	167	332	520	450	443	606
Costa Rica	179	330	554	476	520	556
Honduras	151	298	445	348	350	400
Panama	180	278	565	305	336	350
Trinidad & Tobago	271	255	186	328	327	327
Jamaica	457	407	144	240	241	302
Barbados	——	——	——	215	199	205
Haiti	65	150	243	220	180	200
Martinique e/	NA	NA	180	200	200	200
Guadeloupe	NA	NA	183	173	181	190
Nicaragua	136	177	154	100	100	100
Suriname	——	31	69	50	50	50
Bahamas	833	381	472	——	91	——

e/ Estimated.

Table 16. Caribbean Basin Countries—Production of Gypsum (thousand metric tons).

Country 1/	1970	1975	1980	1985	1986	1987
Mexico	1,291	1,256	2,171	4,603	4,233	4,575
Colombia	189	200	262	250	295	302
Venezuela	e/100	164	117	189	257	247
Jamaica	283	240	105	179	117	176
Cuba e/	NA	NA	122	130	130	130
Dominican Republic	176	2/3/147	235	310	132	59
Guatemala	8	2/12	33	17	101	24
Honduras e/	9	1	23	22	22	22
Nicaragua	e/30	e/35	40	8	8	7
El Salvador e/	6	6	9	4	4	5
Trinidad & Tobago	4	NA	NA	NA	NA	NA

e Estimated.
NA Not Available.
1/Production in 1977 in Haiti was reported at 10,353 metric tons but general information has not been available to make reliable estimates of production for other years.
2/For mixing with cement only, other uses not available.
3/Exports.

Table 17. Caribbean Basin Countries—Production of Clays (thousand metric tons).

Country	1970	1975	1980	1985	1986	1987
Colombia						
Kaolin	93	105	786	1,041	1,155	1,221
Mexico						
Kaolin	79	120	271	282	276	151
Bentonite	65	33	176	268	136	130
Common	107	106	153	255	248	178
Fuller's Earth	24	38	57	58	52	49
Guatemala						
Kaolin	~	——	——	1	2	2
Bentonite	~	——	4	3	4	26
Unspecified	——	102	170	175	175	175
Panama						
Unspecified	NA	189	183	98	111	115
Suriname e/						
Common	3	NA	115	100	110	110
Venezuela						
Kaolin	——	15	22	19	15	30
Costa Rica						
Kaolin	——	1/	1/	1/	——	——

e/ Estimated.
NA Not Available.
1/ Less than 1,000 tons per year.

INDUSTRIAL MINERALS

Aside from Mexico and Colombia, the countries in the Caribbean Basin do not have as diversified a record as producers of industrial minerals. In practically all countries, materials are extracted to meet the domestic demands of the pervasive construction industry for sand and crushed stone and, in some cases, dimension stone and marble. The large bulk and low relative value of construction materials do not make it economically feasible to transport them long distances. The most important industrial minerals produced in the region include cement, clays, gypsum, and salt.

CEMENT

Of the 25 entities in the Caribbean, 19 produce cement (Table 15). As the most populous country, Mexico is by far the dominant cement producer. Practically all of the producers have increased their cement capacity in response to economic and population growth.

In 1986, Mexico gained sixth place as a world cement exporter and it replaced Canada as the leading cement supplier to the U.S. market. Mexico's cement exports to the United States expanded from 2 million MT in 1984 to 6 million MT in 1987. Other important cement exporters to the United States from the Caribbean Basin in 1987 included Venezuela (766,000 MT), Colombia (616,000 MT), Honduras (56,000 MT), and Panama (29,000 MT).

Caribbean cement producers have been successful in penetrating the U.S. market because of good quality and a large difference in price. The average import price is about $32/MT compared with $50/MT charged by U.S. cement producers. A number of U.S. producers have been forced to cease or reduce output because of this price competition from plants in the Caribbean region.

GYPSUM

Cement is mixed with three percent to five percent gypsum, which acts as a cement retarder. Calcined gypsum is also produced for prefabricated gypsum products. The dominant producer in the Caribbean Basin is, again, Mexico (Table 16). Mexico's sizeable exports of gypsum go mainly to the United States. U.S. imports of gypsum in 1987 amounted to almost 10 million MT and came mostly from Canada and Mexico. The price of crude gypsum has been level at about $7.00/MT, while calcined gypsum is valued at about $20.00/MT.

CLAYS

Colombia is the principal producer of kaolin in all of Latin America (Table 17). Kaolin is valued as a filler for paper and for refractories. Mexico is the only diversified clay producer with significant quantities of kaolin, bentonite, fuller's earth, and common clay. The United States imports about 35 percent of its requirements from Mexico.

SALT

Salt production in the region is about as widespread as that of cement and petroleum products. Mexico is the preponderant producer at a level of more than 6 million MT/yr from its large solar facilities in Baja California (Table 18). Although Mexico's salt output is only one-sixth of U.S. output, Mexico is the world's largest exporter of salt, primarily to Japan and the United States. Colombia has large underground rock salt operations at its Zipaquira mine.

OTHER INDUSTRIAL MINERALS

Mexico is also a producer of world rank of barite, celestite, fluorspar, amorphous graphite, natural sodium sulfate, and sulfur.

Much smaller quantities of barite for oil-drilling use are produced in Venezuela, Colombia, and Guatemala. Small amounts of sulfur are recovered from petroleum operations in Venezuela, the Netherlands Antilles, Colombia, Cuba, and Trinidad and Tobago. In all of Latin America, Mexico is the

Table 18. Caribbean Basin Countries—Production of Salt 1/ (thousand metric tons).

Country	1970	1975	1980	1985	1986	1987
Mexico	4,153	5,354	6,575	6,467	5,927	6,393
Bahamas	621	1,232	684	e/850	899	736
Colombia	762	925	838	730	728	655
Venezuela	265	290	243	339	511	500
Netherlands Antilles e/2/	——	480	400	350	350	350
Cuba	——	——	131	221	230	230
Jamaica e/	——	——	——	200	200	200
Dominican Rep. /e	37	40	56	47	54	55
St. Vincent e/	NA	50	50	50	50	50
Guatemala e/	——	9	10	17	39	37
Honduras e/	27	31	33	30	30	30
Nicaragua e/	14	12	20	15	15	15
Costa Rica e/	8	16	40	29	30	13
Panama 3/	7	29	19	16	10	10
El Salvador	31	23	2	3	3	3
Martinique	4/	163	NA	NA	NA	NA

e/ Estimated.
1/Primarily marine or solar salt with limited number of countries producing rock salt.
2/Includes Aruba.
3/Figure for 1980 refers to refined salt. Figures for 1985-87 represent sales.
4/Less than a thousand metric tons.

major sulfur producer; output is by the Frasch process from salt domes in the Isthmus of Tehuantepec. Guatemala was once a significant producer of mica. Deposits of quartz crystals in Central America have been exploited only in Guatemala. Costa Rica has promising deposits of diatomite of good quality and good deposits of pumice.

SUMMARY

On a value basis, mineral production in the Caribbean Basin is dominantly based on the energy minerals in the form of petroleum, natural gas, and coal. In addition, important geothermal sources of energy are being exploited. Of lesser value is the output of the precious metals—gold, silver, and platinum—and of bauxite, iron ore, nickel, cement, gypsum, and salt.

The United States is highly dependent on the Caribbean Basin countries for its imports of energy minerals.

ACKNOWLEDGMENTS

The statistical tables have been prepared with the collaboration of Ms. Ivette Torres, physical scientist, also in the Division of International Minerals.

CHRONOLOGY OF SELECTED MINERAL DEVELOPMENTS IN CARIBBEAN BASIN COUNTRIES

Pre-Columbian Period:

Ornamental use of gold and jade by indigenous peoples.

Salt was produced by the Mayans in the Yucatan Peninsula from solar facilities and was bartered for gold sheet and obsidian.

Native goldsmiths operating in a region covering Colombia, Panama, and Costa Rica produced a unique American alloy called "tumbaga," an alloy of gold and copper.

Diquis was the important goldworking center in Costa Rica.

1492 Discovery of the Americas by Christopher Columbus. First land sighted on October 12 was the Bahamian island of San Salvador, followed by discovery of Hispaniola (Haiti and Dominican Republic). Columbus engaged in placer gold mining at every opportunity.

1496 First permanent Spanish settlement in the new world at Santo Domingo founded by Bartolomeo, brother of Columbus.

1498 During his third voyage, Columbus discovered Trinidad and Venezuela, the first contact with the American mainland.

He noted greater abundance of gold ornaments. Natives called gold "guanin," which in many cases was really tumbaga.

1502 Columbus landed at Honduras on his fourth voyage. He was greatly interested in gold ornaments worn by the natives. First Spanish settlers of 1509 began gold exploration, but Olancho, the principal source of placer gold, was not reached until 1524.

1505 As early as this date African slaves were imported to dig in placer gold deposits and later worked in sugarcane fields.

1521 Conquest by Cortes of the Aztec empire opened up Mexico's large storehouse of precious metals, especially silver.

1535 The first mint of New Spain is established in Mexico City.

1848 The California gold rush stimulated interest in gold deposits of Central America.

1856 An English and French company began to mine the rich silver veins of the Concepción district of eastern Guatemala.

1879 Venezuela initiated small-scale oil development at Rubio, Tachira State.

1882 El Rosario mine in Honduras was considered one of the outstanding silver mines of the world, and has been in almost continuous operation since this date.

1883 Venezuela granted concession to Horace Hamilton, a U.S. citizen, to produce asphalt from Guanoco Lake, then the world's largest known deposit.

1889 Costa Rica constructed its first hydroelectric plant on Rio Anonos with 0.6 MW capacity.

1901 Mexico's first production of petroleum from the Tampico-Tuxpan embayment.

1903 The first modern steel works in Latin America was erected at Monterrey, Mexico.

1914 Discovery of Venezuela's first major commercial deposit of conventional oil at Mene Grande, east of Lake Maracaibo.

1915 Petroleum exploration began in the Limon Province of Costa Rica, but a German technician had searched for oil in Talamanca as early as 1874.

1918 Colombia discovered its Infantes oil field, the oldest currently producing. Cumulative production at year-end 1986 was 223 million bbl of oil.

1925 Discovery of the large El Pao iron ore deposit in Venezuela.

1932 Increase of the U.S. gold price to $32/oz stimulated mining and exploration. By 1938 the Neptune, La Luz, and La India mines in Nicaragua; Monte Cristo, Potosi, and San Sebastian mines in El Salvador; and Abangares and La Union mines in Costa Rica were in operation.

1938 Mexico nationalized its foreign-owned petroleum industry.

1943 Commercial possibilities of bauxite in Jamaica were identified.

El Mochito lead-zinc-gold-silver mine was purchased by the New York and Honduras Rosario Mining Company.

1947 Discovery of Cerro Bolivar, another large iron ore deposit, in Venezuela.

1949 The Instituto Costarricense de Electricidad (ICE) began operations in Costa Rica.

1950 Venezuela initiated production of iron ore from the El Pao deposit (Iron Mines Company of Venezuela), and output from the Cerro Bolivar deposit began in 1954 (Orinoco Mining Co.).

1951 Discovery of bauxite in Cerro El Chorro, Venezuela.

1957 Jamaica became the leading world producer of bauxite.

1958 Bauxite mining was initiated in the Dominican Republic based on a discovery by the Aluminum Company of America (ALCOA). Production was interrupted between 1982-86.

1960 Establishment of the Central American Common Market.

1967 Escuela Centroamericana de Geología in the University of Costa Rica was founded.

Venezuela began production of aluminum at Ciudad Guayana, Bolivar State. Construction was started in 1963.

Refinadora Costarricense de Petróleo (RECOPE) began production from a refinery at Port of Moin.

1969 Alumina Partners of Jamaica, Ltd. began production of alumina in May with the highest initial capacity plant of any such facility constructed worldwide up to that time.

1971 Falconbridge Dominicana initiated commercial exploitation of nickel laterites in Loma Peguera in Bonao, Dominican Republic. Exploration began in 1955 and construction in 1969.

1973 Mexico started up its 75 MW geothermal plant at Cerro Prieto, the first in Latin America.

1975 El Salvador became the first Central American country to construct and operate a geothermal electric power plant. Three units provide 95 MW of power. The first unit was started up in June 1975 and the third unit in November 1981 at the Ahuachapán field.

Mexico regained self-sufficiency in crude oil output and began its role as an important oil exporter. The largest increase in oil reserves occurred in 1978 to 45 billion bbl.

1976 Venezuela nationalized foreign oil properties after having nationalized iron ore operations in 1975.

1978 El Estor nickel project was brought into production near Izabal Lake, Guatemala, and then closed in 1980.

Coal was discovered in the Talamanca Valley of Costa Rica.

Venezuela installed two direct-reduction iron modules at Mantanzas, operated by C. V. G. Siderurgica del Orinoco C. A.

1979 Costa Rica initiated drilling in the geothermal field near Miravalles volcano, south of the border with Nicaragua.

1980 San Jose Agreement signed, under which Mexico and Venezuela supply petroleum at concessionary terms to certain Caribbean Basin countries (renewed yearly).

1981 Mexico surpassed Venezuela to become the major producer of crude oil in the Latin American region.

1982 Costa Rica enacted its new Mining Code in October with regulations issued in April 1984.

Colombia started up the Cerro Matoso ferronickel operation.

Colombia also initiated coal production from the Central Zone of the large El Cerrejon coal deposit.

1983 In August the U.S. Congress enacted the Caribbean Basin Economic Recovery Act (CBERA), providing for 12 years of duty-free imports except petroleum among other items excepted.

1984 Nicaragua became geothermal energy producer with a 35 MW plant near Momotombo volcano, northwest of Managua. Italy was the primary financial backer of the plant.

1985 Colombia became the leading producer and exporter of steam coal in Latin America.

1986 State-owned Mineral Nacional S. A. was created in Costa Rica to promote investment in the minerals sector by preparing exploration claims for public tender.

Colombia resumed exports of crude oil since its cut-off in 1975 as a result of new reserves discovered in the Caño Límon field.

1987 Costa Rica completed the design of its first 55 MW geothermal power unit at the Miravalles field. Completion is scheduled for 1992. In 1985 Costa Rica received a loan from the InterAmerican Development Bank for field development and power plant construction.

Bauxita Venezolana started up its Los Pijiguaos bauxite mine, expected to produce 3 million MT by 1990 and 6 million MT by 1992.

Guatemala signed a contract with a consortium of U.S.-Guatemalan firms for a feasibility study of a 15-MW geothermal plant at Zunil.

1988 Guatemala Accords signed in August to bring peace and stability to Central America.

Costa Rica began coal development operations at the El Indio mine west of Port Limon on the Caribbean Coast.

1989 Nicaragua completed installation of its second geothermal plant with 35 MW capacity at Momotombo field.

Circum-Pacific Council for Energy and Mineral Resources
Earth Science Series, Vol. 16
R. L. Miller, G. Escalante, J. A. Reinemund, and M. J. Bergin (Eds.)
Energy and Mineral Potential of the Central American-Caribbean Regions
Springer-Verlag, Berlin Heidelberg 1995

CIRCUM-CARIBBEAN SEDIMENTARY BASIN DEVELOPMENT AND TIMING OF HYDROCARBON MATURATION AS A FUNCTION OF CARIBBEAN PLATE TECTONIC EVOLUTION

James L. Pindell
Department of Earth Sciences, Dartmouth College, Hanover, NH 03755 USA

ABSTRACT

Sedimentary basin development in the Caribbean region is closely tied to the region's plate tectonic evolution, which involved (1) Jurassic rifting and passive margin development between North and South America, and (2) Late Cretaceous to Recent relative eastward migration of the Caribbean Plate from the eastern Pacific area to its present position between North and South America. Two primary stratigraphic suites of rock occur in the Caribbean region: (1) an autochthonous suite of Jurassic, Cretaceous, and Cenozoic passive margin sediments that was deposited along the rifted margins of North and South American basement after Jurassic continental breakup, and (2) an allochthonous suite consisting of oceanic crustal and magmatic arc rocks and overlying sedimentary units of the Caribbean Plate. The boundary of these suites coincides roughly with the limit of circum-Caribbean thrust belts of metamorphic/mafic rock above the formerly passive Proto-Caribbean shelf rocks. Most of the sediments of the allochthonous Caribbean Plate were deposited during the relative eastward translation between the Americas, and the older sections (Cretaceous) were deposited while the Plate was still in the eastern Pacific realm. Primary source rocks in both suites were deposited in the early Late Cretaceous, well after rifting and plate separation in the Proto-Caribbean Seaway between the Americas.

The timing of maturation of source material for hydrocarbons in the circum-Caribbean region is predicted herein to match the times of (1) highest heat flow or (2) most rapid subsidence and sediment accumulation in various basins around the Caribbean. Early maturation (Jurassic or Early Cretaceous) of Jurassic, Proto-Caribbean rift sediments, if local source material was present in them, may have occurred in areas of high heat flow related to rifting such as along some extensional basins of the northern Andes (e.g., Sierra Perija, Cordillera Merida, and Oriental). In contrast, well-known lower Upper Cretaceous source rocks of the Proto-Caribbean suite, deposited long after rifting, probably did not mature until a sufficient sedimentary overburden had accumulated above them. Throughout the primary circum-Caribbean basins (Guatemalan Sepur, southern Bahamas/Cuban, Venezuelan Maracaibo and Maturin), thick clastic units were deposited in foredeep basin settings above the previously passive (shelf) margins of the Proto-Caribbean Seaway. Development of the foredeep basins and rapid sedimentation (and predicted onset of hydrocarbon maturation) occurred in relation to the relative eastward advance of the Caribbean Plate from the Pacific, as the leading edge of the Caribbean Plate overthrust the former passive margins of the Proto-Caribbean Seaway. Peak maturation is predicted to have occurred in the latest Cretaceous to early Paleogene in southern Yucatan (Guatemalan Sepur), the Paleogene in northern Cuba/Bahamas and the Maracaibo area of Venezuela and Colombia, and in the Neogene and Quaternary in eastern Venezuela and Trinidad. In addition, however, Neogene intermontane basin sedimentation related to Andean uplift in northwest South America (e.g., Maracaibo), provided a second, Neogene to Present period of predicted maturation in the basins there. Hydrocarbon potential from the Cretaceous section of the allochthonous, Caribbean suite of rocks has yet to be realized.

INTRODUCTION

Neotectonic aspects of the Caribbean region have been well summarized by Mann and Burke (1984) and others. Concerning paleotectonics, the Caribbean is a region which historically has developed within a kinematic framework of the relative positions and motions of the North and South American continents (Ladd, 1976; Pindell et al., 1988). Most authors would agree that the oceanic portions as well as the partially continental Chortis "block" of the Caribbean Plate have arrived at their present positions after a long migration from the Pacific realm, relative to the Americas (Malfait and Dinkelman, 1972; Dickinson and Coney, 1980; Pindell and Dewey, 1982; Burke, 1988; Pindell and Barrett, in press). In that context, primary circum-Caribbean sedimentary basin development has been shown to be very closely tied to regional and local Caribbean plate tectonic evolution (Pindell et al., 1988). It is not the purpose of this paper to prove Pacific provenance for the Caribbean Plate; such arguments are summarized in Pindell (in press). Here, in the interests of brevity I will assume Pacific provenance and only briefly describe the regional plate tectonic evolution and outline the region's primary stratigraphic (basin) development as a function of relative motions and plate-boundary development between the North American, South American, and Caribbean Plates. Important periods of hydrocarbon source rock maturation are related to periods of basin development that pertain directly to plate tectonic evolution.

RELATIVE PLATE MOTIONS, JURASSIC RIFTING, SHELF DEVELOPMENT ALONG THE PROTO-CARIBBEAN SEAWAY, AND THE ORIGINS OF THE CARIBBEAN PLATE

The history of rifting and subsequent relative positions and motions between North and South America were outlined in Pindell et al. (1988), based on quantitative kinematic opening histories of the Central, Equatorial, and South Atlantic Oceans. In short, North and South America rifted apart

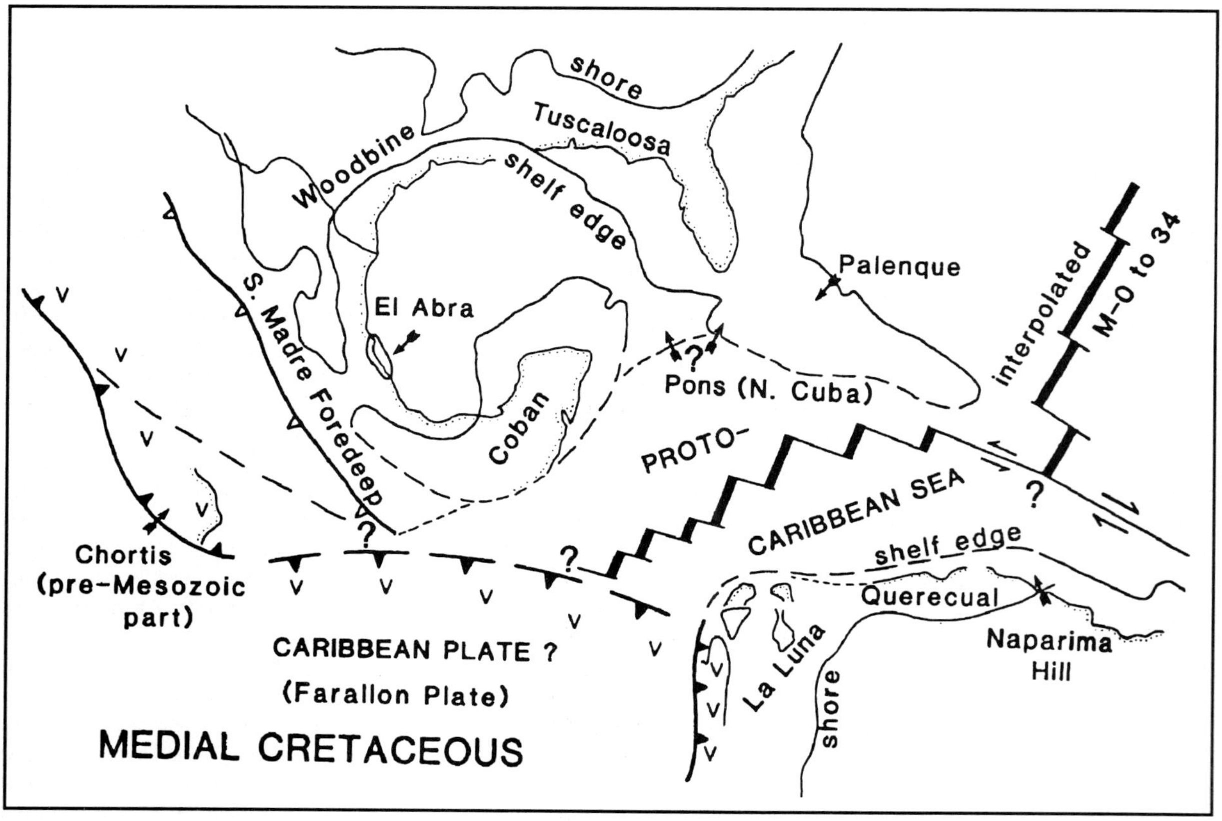

Figure 1. Median (early Late) Cretaceous paleogeography, showing limits of marine and shelf deposition, with key hydrocarbon source rock (La Luna, Querecual, Naparima Hill, Coban and other coeval stratigraphic units) identified along the shelves of the Gulf of Mexico and Proto-Caribbean Seaway. Mid-ocean ridge position is approximated from interpolation between anomalies M-O and 34 in the Central and South Atlantic oceans. Chortis Block is restored to its probable former position along the southwest Mexican margin, and the oceanic crust of the true Caribbean Plate (part of Farallon Plate in medial Cretaceous) is shown at the juncture of the Pacific realm prior to its relative eastward migration between North and South America. Late Cretaceous structural shortening in the Sierra Madre Oriental (100 to 150 km?) and approximate Neogene strike-slip offsets in northwest South America are palinspastically removed. Dotted shorelines (present day) are shown for reference.

during Late Triassic and Jurassic time and diverged to approximately their present relative positions by the Campanian (84 Ma). From the Campanian to the Eocene, very little or no continued divergent motion occurred, and since the Eocene very slow north-south convergence has occurred. Strain associated with the post-Eocene phase is probably seen mainly as the positive bathymetric expression of the Baracuda, Tiburon, and other fracture zones east of the Caribbean Sea in the western Central Atlantic. It is within this framework that the evolution of the Caribbean region has occurred, and I stress that little or none of the post-Campanian deformation in the Caribbean region has arisen from interactions between the North and South American Plates. Rather, such deformations are due mainly to the insertion of the allochthonous Caribbean Plate to its present position between the Americas, from the Pacific (subduction of Atlantic crust beneath the leading edge of the Caribbean Plate, and strike-slip offsets in the north and the south).

After Jurassic rifting and initial separation of the Americas, Atlantic-type continental shelves developed along the passive margins of the Gulf of Mexico and a widening seaway between the Yucatan Block and northern South America (Pindell, 1985). I will refer to this latter seaway as the Proto-Caribbean Seaway, as it formed an oceanic basin between the Americas that pre-dates the arrival of the true Caribbean Plate from the Pacific. It was along these passive margins of the Proto-Caribbean Seaway that the highest quality source rocks of the region were deposited in early Late Cretaceous time (Figure 1). High eustatic sea level at that time was responsible for the great advance of these sequences onto the continental margins.

The rifting and passive margin development along the margins of the Proto-Caribbean Seaway was the first primary phase of basin development in the evolution of the Caribbean region, and has *nothing* to do with the Caribbean Plate that we know today. Shelf sedimentary thicknesses of this phase were variable and hydrocarbon maturation may have occurred locally in areas of rapid subsidence, but depth of burial of the primary source rock interval along the shelves (i.e., the units shown on Figure 1) was largely insufficient to have caused maturity until the onset of tectonic interactions with the Caribbean Plate during its Late Cretaceous and Cenozoic relative eastward migration. One exception to this rule may have been along the western Magdallena Valley of Colombia; there, rapid uplift in the Central Andean Cordillera, as evidenced by many K-Ar "cooling" ages on plutonic and metamorphic rocks (Pindell and Barrett, in press), shed voluminous fine to sandy clastics eastward (Guasare and Oroque units), possibly providing a sufficient thermal blanket for the enhancement of maturation at the end of the Cretaceous and early Paleogene.

Turning to the Caribbean Plate proper, numerous examples of Jurassic and Early Cretaceous (pre-Aptian) rocks encircling the Caribbean Sea (basement of Costa Rica-Panama, Siquisique of Venezuela, La Rinconada metabasite of Margarita, Curacao Lava Formation, North Coast Schist of Tobago (Sharp and Snoke, 1988), basement rocks of La Desirade, Bermeja Complex of Puerto Rico, Duarte Complex of Hispaniola, ophiolitic basement of Cuba, and early volcanic rocks on Jamaica) and also thick pre-Coniacian layered reflectors in the Colombian basin (Stoffa et al., 1981) suggest that the crust of the Caribbean Plate is also Early Cretaceous to Jurassic in age. Because the Proto-Caribbean Seaway did not become wide enough by seafloor spreading to have accommodated the Caribbean Plate until the Late Cretaceous (Pindell et al., 1988), the oceanic crust of the Caribbean Plate must have formed somewhere in the eastern part of the Pacific Ocean, probably as part of the Farallon Plate (Figure 1). Unfortunately, not enough of the Farallon Plate is presently preserved to determine the exact origin of the Caribbean portion, nor how the Caribbean portion was tectonically isolated to form the Caribbean Plate. Despite these questions, the first signs of tectonic interaction between the North and South American (Proto-Caribbean) passive margins and rocks of the Caribbean Plate occurred in the Late Cretaceous, and thus from that time on, the former positions and motions of the Caribbean Plate relative to the Americas is fairly well constrained.

PRIMARY SOURCE ROCK HORIZONS IN THE CARIBBEAN REGION AND THE CONCEPT OF TWO DISTINCT SUITES OF ROCK

As mentioned above and shown in Figure 1, the early Late Cretaceous was a time of source rock deposition along the margins of the Proto-Caribbean Seaway. Likewise, the Caribbean Plate itself also possesses source rocks of that age, as shown by Deep Sea Drilling Program drill cores (Leg 15, Edgar and Saunders et al., 1973) from within the plate. General composite stratigraphic sections for both the Proto-Caribbean margins and the geographically separated Caribbean Plate (Figure 1) are shown in Figure 2. The lower Upper Cretaceous sections of both stratigraphies are of prime importance to the hydrocarbon potential of the Caribbean region.

As outlined below, the Late Cretaceous and Cenozoic history of tectonic development of the Caribbean region is primarily one of eastward-younging tectonic juxtaposition of the two distinct suites of rock shown in Figure 2. In all areas around the Caribbean, oceanic or arc-related rocks of the leading edge and flanks of the Caribbean Plate (Caribbean allochthons) have overthrust with "outward" vergence (direction of motion or thrusting) the "inward" facing passive margins of the Proto-Caribbean Seaway (Proto-Caribbean autochthons). At the boundary of these two suites of rock, orogenic foredeep basin sediments of various ages occur beneath and within the thrust zones. These sediments young to the east and serve to date the timing of tectonic interaction between the Americas and the relatively eastwardly migrating Caribbean Plate.

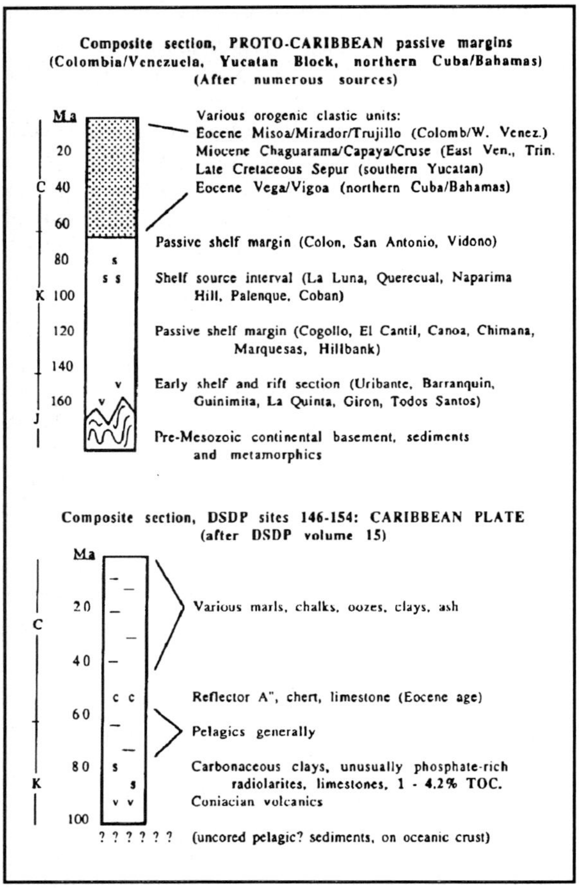

Figure 2. Generalized stratigraphic columns for the Proto-Caribbean passive margin shelves (from many sources) and the marine sediments of the true Caribbean Plate, from DSDP drilling. Ages of primary source rock sections (s) and volcanic sections (v) are indicated.

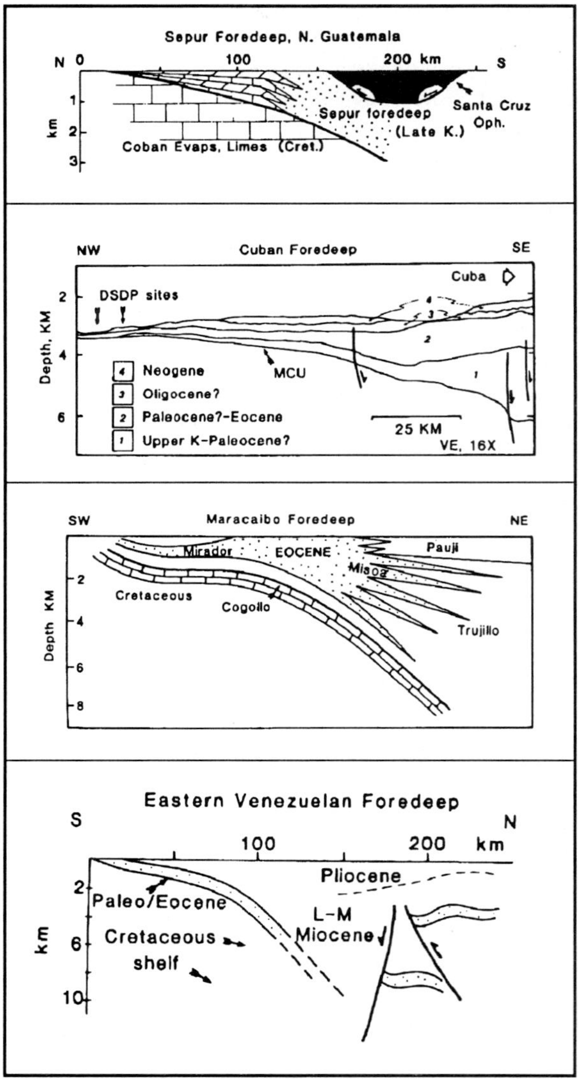

Figure 3. Rough cross sections of the four primary Proto-Caribbean margin foredeep basins, positions of basins and sections shown with heavy lines on Figure 4. These basins formed in response to overthrusting by the Caribbean Plate onto the passive margins of the Proto-Caribbean Seaway during the relative eastward migration of the Caribbean Plate from the Pacific to its present position between the Americas. Sepur Foredeep section (Campanian-Maestrichtian rapid subsidence) after Wilson (1974) and Rosenfeld (1980; 1981); Cuban Foredeep section (latest Cretaceous to early Paleogene rapid subsidence) after Angstadt, Austin, and Buffler (1985); Maracaibo Foredeep section (Eocene rapid subsidence; note that foredeep section is restored to Eocene datum, not the present) after Bockmeulen, Barker, and Dickey (1983); and Eastern Venezuelan Foredeep section (Neogene rapid subsidence) after Lamb and Sulec (1968); Case and Holcombe (1980).

FOREDEEP BASINS OF THE CARIBBEAN REGION AND THE RELATIVE EASTWARD ADVANCE OF THE CARIBBEAN PLATE

In Figure 3, the four primary foredeep basins of the Caribbean (former Proto-Caribbean continental shelves loaded by Caribbean overthrusts) are shown with the age of the thickest sedimentary sections identified (period of most rapid subsidence and, therefore, of overthrusting). The age of foredeep basin formation and sedimentation upon previously shallow water shelf areas progresses from west to east, beginning along southern Yucatan during the Campanian, in the Bahamas and northern Cuba in the early Paleogene, in the Maracaibo basin of Venezuela in the Eocene, and in the Eastern Venezuelan or Maturin basin (and Trinidad) in the Miocene (to younger in Trinidad). This is the second primary phase of basin formation in each of the areas mentioned.

In all cases, oceanic and/or arc-related rocks of the Caribbean Plate occur structurally above and "Caribbeanward" of these orogenic foredeep sediments, which in most cases are also incorporated into the thrusts, thereby constraining the timing of tectonic juxtaposition. I interpret this history of foredeep basin development as being caused by the onset of tectonic loading by the Caribbean Plate upon the passive margin and oceanic flanks of the Proto-Caribbean Seaway as the former migrated from the Pacific to its present position between the Americas, as no other driving mechanism for rapid subsidence of previously shallow-water shelf platforms appears plausible. In addition, the complete absence of older orogenic foredeep deposits along the Proto-Caribbean margins makes the possibility of allochthon emplacement at earlier times than shown in Figure 3 highly unlikely. Thus, the timing of progressive development of the primary Proto-Caribbean foredeep basins, as summarized in Figure 4, provides a first-order approximation of the former positions and rate of relative advance of the Caribbean Plate between North and South America. By associating the foredeep basins to the position of the Caribbean Plate across a typical wavelength of foredeep basin shape, the rate of Caribbean/American relative motion appears to be on the order of 1.5 to 2.5 cm/yr back to Campanian time (total migration of about 1,500 km over roughly 75 Ma). It is interesting to note that these rates are far slower than convergence rates suggested for the Farallon Plate relative to North America (Engebretson, 1982; Engebretson, Cox, and Gordon, 1985; Pindell et al., 1988) during Late Cretaceous and Cenozoic time. The difference is taken up by subduction of Farallon crust along the Chortis and Costa Rica-Panama arcs (western boundary of the Caribbean Plate), the latter of which began its magmatic history in the Late Cretaceous at about the time convergence rates dramatically increased (Lundberg, 1983; Pindell and Barrett, in press).

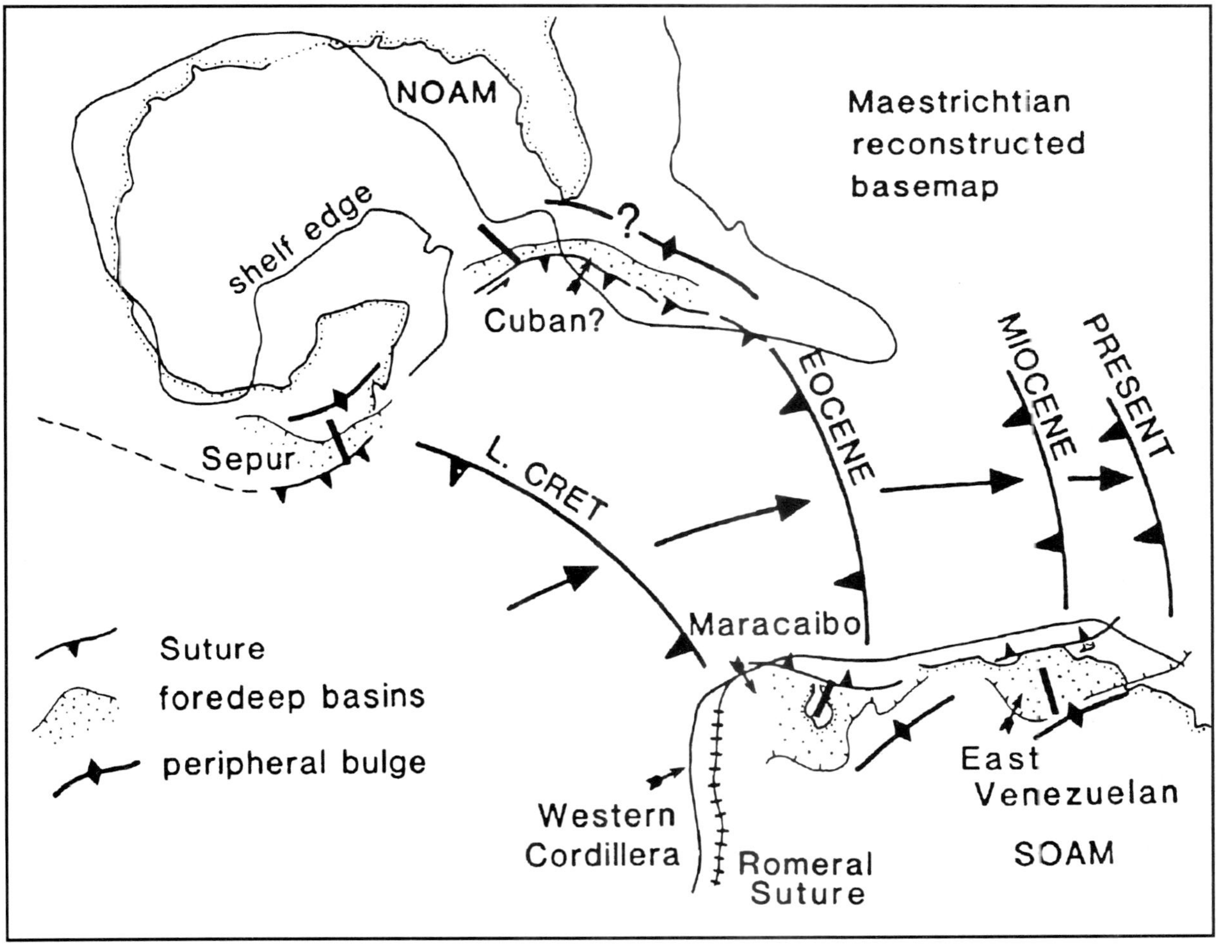

Figure 4. Map of four primary Proto-Caribbean foredeep basins and positions of cross-sections of Figure 3 (heavy lines), summarizing the history of relative eastward migration of the Caribbean Plate between the Americas as discussed in text. North and South America are plotted in their Maestrichtian relative positions, and the blocks of northwest South America are restored to their pre-Neogene relative positions prior to Andean deformation (see Figure 6). A net rate of about 2 cm/yr relative motion between the Caribbean and the American plates is suggested by the analysis of the timing of foredeep basin development as a function of Caribbean Plate displacement.

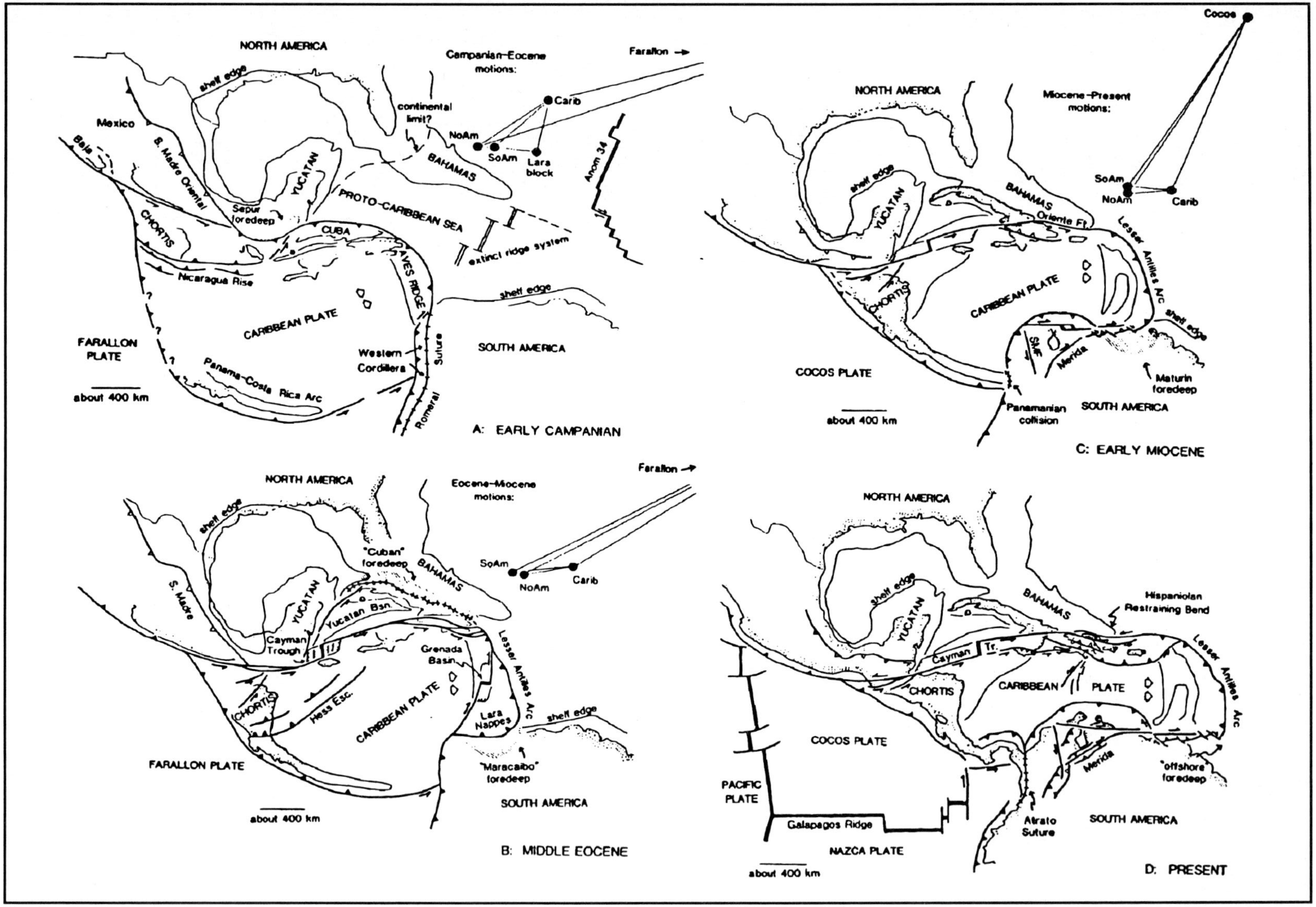

Figure 5. Four-stage depiction of the evolution of the Caribbean region, after Pindell et al. (1988). The four primary foredeep basins are shown (Sepur, Cuban, Maracaibo, and East Venezuelan or Maturin) that developed at the onset of tectonic interaction between the American or Proto-Caribbean passive margins and the relatively eastward advancing Caribbean Plate. Strike-slip tectonics have developed at each of these areas subsequent to the initial emplacement of the allochthonous masses of Caribbean rock .

Figures 5A-D summarize in map form the motion history of the Caribbean Plate relative to the greater "American" plate after seafloor spreading had ceased in the Proto-Caribbean Seaway, as well as other aspects of Caribbean evolution. Of particular note in this history is the opening of the Yucatan, Grenada, and Cayman Trough basins during probably Paleogene time (Figure 5B). I suggest that the Yucatan and Grenada basins formed by the expansion (intra-arc rifting) of the Greater Antillean Arc as the Caribbean Plate entered the Proto-Caribbean Seaway after having passed through the Yucatan-Colombian Central Cordillera "bottleneck" (Pindell and Barrett, in press). Basal sedimentary sections of these deep, presumably oceanic basins within the older Caribbean crust may have been deposited in isolated, anoxic conditions. Having been deposited on highly rifted or oceanic crust, heat flow into these sediments should have been high during and shortly after deposition, presumably during the Paleogene.

OUTLINE OF THE NEOGENE OROGENIC DEVELOPMENT OF NORTHWEST SOUTH AMERICA

In addition to the general tectonic development of the Caribbean region outlined above, a second phase of orogenesis has been superposed upon the northwest corner (Andean Chains) of South America mainly in Neogene time. There, Miocene marine sediments of the Perija, Merida, and Central Cordilleras of Venezuela and Colombia indicate young, rapid deformation, uplift and erosion (Irving, 1975). Several attempts have been made to unravel the net palinspastic deformation of northwest South America (Dewey and Pindell, 1985, 1986; Schubert, 1984; Mann and Burke, 1984; others), but Neogene strike-slip motions of about 65 km, 100 km, and 150 km on the Oca, Bocono, and Santa Marta Faults, respectively, seem to be tentatively accepted by many workers.

In Figure 6, a three-stage depiction of the neotectonic deformation of northwest South America is shown, following the methodology of Dewey and Pindell (1985), which restores the magnitudes of motion along the Oca, Bocono, and Santa Marta Faults mentioned above. Primary implications of this reconstruction are (1) that the Guajira and Paraguana Peninsulas become retracted to the southwest from their presently protruding positions northward into the Caribbean Sea, thereby allowing the southeastern corner of the Caribbean Plate to have migrated past northwest South America during the Paleogene as shown in Figure 5B, and (2) that total shortening associated with uplift, primarily within the Eastern Cordillera and at the flanks of the Central and Western Cordilleras, approaches 150 km, such that wide areas of former shelf deposition have been overthrust beneath the thrust fronts of the Andean Chains. The tectonic loading associated with uplift of the Andean Chains has induced a regional subsidence in the intermontane basinal areas, such that the basins have received enormous quantities of sediment since the Miocene. Thus, in northwest South America a third primary phase of basin development (subsidence and sedimentation) has occurred. Examples are the 4 to 6 km of Neogene in the Llanos basin of Colombia (Case and Holcombe, 1980) and the several kilometers of Neogene section in the southern Maracaibo basin (Bockmeulen, Barker, and Dickey, 1983). Concerning the Maracaibo, the original dip of the Eocene foredeep basin phase was to the north toward the Caribbean overthrust zone (Figure 3), whereas the Neogene sedimentary burial phase (due to uplift of the Merida and northern Eastern Cordilleras to the south) has affected mainly the southern part of the basin, such that downdip in the Neogene section is toward the south.

PREDICTIONS FOR LOCAL TIMING OF HYDROCARBON MATURATION THROUGHOUT THE CARIBBEAN REGION BASED ON TECTONIC EVOLUTION AND BASIN DEVELOPMENT

From the foregoing discussion of evolution and basin development in the Caribbean region, some predictions for the timing of hydrocarbon source rock maturation may be made. First, rifting and formation of the passive margins of the Proto-Caribbean Seaway may have produced some local areas where heat flow was sufficient and sedimentary deposition and overburden occurred sufficiently fast to have driven maturation of hydrocarbon source material in possible Jurassic, rift-related sediments of the Proto-Caribbean suite of rocks. However, the best known source rocks of the Caribbean province (Proto-Caribbean) are of early Late Cretaceous age (Figure 2), and the heat flow from Jurassic rifting by that time had probably diminished such that maturation of Upper Cretaceous or younger source-bearing horizons could not have occurred without either secondary heat input or rapid sedimentary burial by a rift-independent mechanism.

Second, all four foredeep basins shown in Figure 3 occupy a position above previously shallow-water shelf sequences of the Proto-Caribbean suite which possess source material of early Late Cretaceous age, especially in northern South America. Source rock quality in the northern South American basins may have been enhanced by upwelling (Ekman Transport produces a net northerly surface flow along a north-facing continental margin in the northern hemisphere within the easterly trade winds). Assuming typically low heat flow values for a 100 million-year-old rifted margin (Middle Jurassic to Late Cretaceous), depth of burial during the Late Cretaceous in each case was insufficient to have triggered maturation until the inception of foredeep basin subsidence and deposition, as shown in Figure 3. Therefore, peak maturation and hydrocarbon generation in the foredeep basins can be predicted to roughly coincide with the emplacement of the Caribbean allochthons in those areas. Specifically, maturation is expected to have occurred during the latest Cretaceous to early Paleogene in the Guatemalan Sepur basin of southern Yucatan, the Paleogene in northern

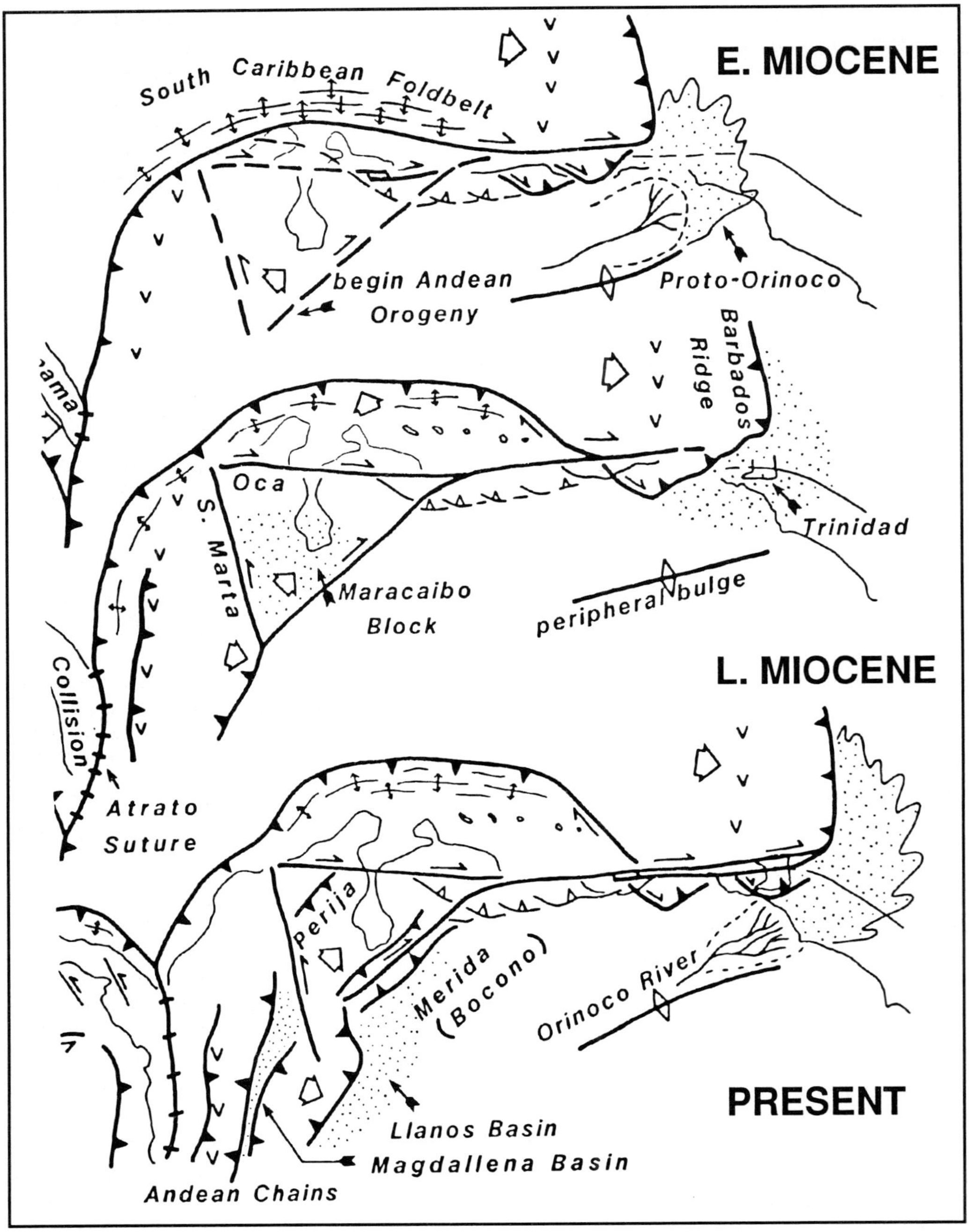

Figure 6. Three-stage outline of the Neogene structural development of northwestern South America, caused largely by the collision of the Panama arc along the Atrato Suture (Pindell and Dewey, 1982) during eastward motion of the Caribbean Plate (and Panama) relative to South America, modified after Dewey and Pindell (1985; 1986). East-west shortening and northeastward tectonic escape of the Maracaibo Block (Mann and Burke, 1984) is primarily responsible for this development. Neogene uplift of Andean Ranges has caused associated subsidence and rapid sedimentation in intermontane basins, leading to the prediction of a Neogene phase of hydrocarbon maturation. Wide arrows indicate approximate motion directions at associated fault zones. Note the transition along northern South America from a mainly strike-slip boundary to a compressional boundary as a result of the northward component of the relative motion of the Maracaibo Block. Note also the foredeep development of the Eastern Venezuelan and Trinidadian basins (controlling Orinoco drainage) as a function of the partial loading (compressional component) of the Caribbean Plate upon the northeastern corner of the South American crust.

Cuba/Bahamas and the Maracaibo area of Venezuela and Colombia, and the Neogene and Quaternary in eastern Venezuela and Trinidad. In addition, however, the Neogene phase of intermontane basin deposition and subsidence in northwest South America associated with Andean uplift provided a second period of potential maturation in the basins there. Likewise, similar effects on maturation may have occurred during the mainly Neogene structural uplift and associated subsidence of overthrust sections in Chiapas and northern Guatemala.

Concerning the allochthonous Caribbean suite of rocks, little or no production has yet to be realized from the Cretaceous source rock section. Deep water prevents drilling in most areas, but the accretionary prisms at the Muertos, South Caribbean, Sinu (western Colombia), and north Panamanian Foldbelts (see Figures 5D and 6) should be comprised at least partially of the Cretaceous sections of the Caribbean stratigraphy, and should not be written off.

CONCLUSIONS AND SUMMARY

Sedimentary basin development in the Caribbean region is closely tied to the region's plate tectonic evolution. Two primary suites of rock occur in the Caribbean region: (1) an autochthonous suite of Jurassic, Cretaceous, and Cenozoic passive margin sediments that was deposited along the rifted margins of North and South America (Proto-Caribbean Seaway) after continental breakup in Jurassic time, and (2) an allochthonous suite of mainly oceanic crust (except Chortis Block) and magmatic arc rocks and overlying sediments of the Caribbean Plate that has migrated from the Pacific to its present position between the Americas. Primary source rock horizons in both suites are of early Late Cretaceous age. The Proto-Caribbean source rocks are well known and highly productive, but hydrocarbon potential from the Caribbean source rocks has yet to be realized.

A third category of rocks unrelated to the others occurs in the small oceanic basins that have developed by extensional plate boundary processes within the Caribbean Plate (Grenada, Yucatan, and Cayman Trough) during regional (mainly Paleogene) evolution. Unknown source rocks may occur at depth in these basins, particularly as the basins were probably sites of restricted, possibly anoxic circulation during their development. Only the Cayman Trough has continued to develop into the Neogene and Quaternary.

The timing of hydrocarbon source rock maturation for the Proto-Caribbean suite is predicted to match times of highest heat flow or most rapid subsidence and burial during basin development. Late Jurassic or Early Cretaceous maturation of source material in Jurassic rift-related sediments may have occurred locally where rifting and sediment accumulation was most extreme, but heat flow pertaining to Jurassic rifting probably had diminished to a level insufficient for maturation without significant overburden by the time the best-known source rock sections (early Late Cretaceous) were deposited. Therefore, maturation of the Late Cretaceous source horizons was delayed until the onset of rapid deposition in foredeep basins that were tectonically-driven by plate interactions between the Caribbean Plate and North and South America. Like foredeep basin development, the onset of hydrocarbon source rock maturation progressed from west (Late Cretaceous) to east (Neogene-Recent) across the Caribbean. The Neogene uplift and deposition associated with the Andean Chains, however, has produced a second period of maturation in northwest South America.

ACKNOWLEDGMENTS

I thank John Dewey, Edward Robinson, and Gary Karner for assistance with data synthesis and discussions on the conclusions and predictions presented herein. I also thank Michel Halbouty and the members of the Circum-Pacific Council for inviting the presentation of this paper. Ralph Miller and Art Snoke provided helpful comments on the manuscript.

REFERENCES CITED

Angstadt, D. M., J. A. Austin, Jr., and R. T. Buffler, 1985, Early Late Cretaceous to Holocene seismic stratigraphy and geologic history of southeastern Gulf of Mexico: American Association of Petroleum Geologists Bulletin, v. 69, p. 977-995.

Bockmeulen, H., C. Barker, and P. A. Dickey, 1983, Geology and geochemistry of crude oils, Bolivar Coastal Fields, Venezuela: American Association of Petroleum Geologists Bulletin, v. 67, p. 242-270.

Burke, K., 1988, Tectonic evolution of the Caribbean: Annual Reviews, Earth and Planetary Science, v. 16, p. 201-230.

Case, J. E., and T. Holcombe, 1980, Geologic-tectonic map of the Caribbean region: U.S. Geological Survey Miscellaneous Investigations Series Map I-1100, scale 1:2,500,000.

Dewey, J. F., and J. L. Pindell, 1985, Neogene block tectonics of Turkey and northern South America: Continental applications of the finite difference method: Tectonics, v. 4, p. 71-83.

Dewey, J. F., and J. L. Pindell, 1986, Neogene block tectonics of Turkey and northern South America: Continental applications of the finite difference method: Reply: Tectonics, v. 5, p. 703-705.

Dickinson, W. R., and P. J. Coney, 1980, Plate-tectonic constraints on the origin of the Gulf of Mexico, *in* R. H. Pilger, ed., The origin of the Gulf of Mexico and the early opening of the central North Atlantic: Baton Rouge, Louisiana State University, p. 27-36.

Edgar, N. T., and J. B. Saunders, et al., 1973, Initial reports of the Deep Sea Drilling Project, Leg 15: Washington, U.S. Government Printing Office.

Engebretson, D. C., 1982, Relative motion between continental and oceanic plates in the Pacific basin: Stanford University, Ph.D. dissertation, 211p.

Engebretson, D. C., A. Cox, and R. G. Gordon, 1985, Relative motions between oceanic and continental plates in the Pacific basin: Geological Society of America Special Paper 206, 59p.

Irving, E. M., 1975, Structural evolution of the northernmost Andes, Colombia: U.S. Geological Survey Professional Paper 846, 47p.

Ladd, J. W., 1976, Relative motion of South America with respect to North America and Caribbean tectonics: Geological Society of America Bulletin, v. 87, p. 969-976.

Lamb, J. L., and J. A. Sulec, 1968, Miocene turbidites in the Carapita Formation of eastern Venezuela: Arima, Trinidad and Tobago, Caribbean Printers, Transactions, 4th Caribbean Geological Conference, p. 111-120.

Lundberg, N., 1983, Development of forearcs of intraoceanic subduction zones: Tectonics, v. 2, p. 51-61.

Malfait, B. T., and M. G. Dinkelman, 1972, Circum-Caribbean tectonic and igneous activity and the evolution of the Caribbean Plate: Geological Society of America Bulletin, v. 83, p. 251-272.

Mann, P., and K. Burke, 1984, Neotectonics of the Caribbean: Reviews of Geophysics and Space Physics, v. 22, p. 309-362.

Pindell, J. L., 1985, Alleghenian reconstruction and the subsequent evolution of the Gulf of Mexico, Bahamas and Proto-Caribbean Sea: Tectonics, v. 4, p. 1-39.

Pindell, J. L., in press, Arguments for a Pacific origin for the Caribbean Plate: Christianstadt, St. Croix, Transactions, 12th Caribbean Geological Conference.

Pindell, J. L., and S. F. Barrett, in press, Geologic evolution of the Caribbean: A plate-tectonic perspective, *in* J. E. Case and G. Dengo, eds., Decade of North American Geology, v. H: Geological Society of America.

Pindell, J. L., and J. F. Dewey, 1982, Permo-Triassic reconstruction of western Pangea and the evolution of the Gulf of Mexico/Caribbean region: Tectonics, v. 1, p. 179-212.

Pindell, J. L., S. C. Cande, W. C. Pittman, D. B. Rowley, J. F. Dewey, J. LaBrecque, and W. Haxby, 1988, Plate-kinematic framework for models of Caribbean evolution: Tectonophysics, v. 155, p.121-138.

Rosenfeld, J. H., 1980, The Santa Cruz ophiolite, Guatemala, Central America: Santo Domingo, Dominican Republic, Transactions, 9th Caribbean Geological Conference, p. 451-452.

Rosenfeld, J. H., 1981, Geology of the western Sierra de Santa Cruz, Guatemala, Central America, an ophiolite sequence: State University of New York at Binghamton, Ph.D. dissertation, 313p.

Schubert, C., 1984, Basin formation along the Bocono-Moron-El Pilar Fault System, Venezuela: Journal of Geophysical Research, v. 89, p. 5711-5718.

Sharp, W. D., and A. Snoke, 1988, Tobago, West Indies: Geochronological study of a fragment of a composite Mesozoic oceanic island arc: Abstracts with Programs, Geological Society of America, v. 20, p. A60.

Stoffa, P. L., A. Mauffret, M. Truchan, and P. Buhl, 1981, "Sub-B" layering in the southern Caribbean: The Aruba gap and Venezuela basin: Earth and Planetary Science Letters, v. 53, p. 131-146.

Wilson, H. H., 1974, Cretaceous sedimentation and orogeny in nuclear Central America: American Association of Petroleum Geologists Bulletin, v. 58, p. 1348-1396.

Circum-Pacific Council for Energy and Mineral Resources
Earth Science Series, Vol. 16
R. L. Miller, G. Escalante, J. A. Reinemund, and M. J. Bergin (Eds.)
Energy and Mineral Potential of the Central American-Caribbean Regions
Springer-Verlag, Berlin Heidelberg 1995

THE HUMAN AND ECONOMIC LOSSES OF SELECTED NATURAL DISASTERS IN LATIN AMERICA, 1970-1987

Daniel B. Krinsley
2475 Virginia Avenue, N.W., Washington, D.C. 20037 USA

ABSTRACT

Natural disasters are particularly devastating to the social and economic fabric of developing countries which lack sufficient resources and infrastructure to cope with these events. The coincidence of poverty along hurricane tracts, floodplains, volcanic arcs and seismic zones further exacerbates the problem and creates regional pockets of chronic stress.

Deaths resulting from natural disasters are a shock to the communities in which they occur, but these losses to the work force plus the costs of caring for the injured and homeless create a continuing economic drain in the most vulnerable countries. A significant measure of the economic loss resulting from a natural disaster is the total replacement costs to the economy of a country, as a percentage of its Gross National Product, for the year of the disaster. This method factors-in the relative economic abilities of countries to respond to disasters and presents a more balanced view of the actual economic impact of a natural disaster in a particular country.

Graphic representation of the human and economic losses from natural disasters aids in the development of a classification system for comparing the impacts of these events. Such a system should facilitate the work of planners and engineers before replacement funds are allocated to construction projects in hazardous areas.

INTRODUCTION

The human and economic losses associated with many natural disasters are a serious blow to any community but are frequently devastating when they strike already impoverished countries. Natural disasters cannot be predicted with any accuracy as yet, but their repetitive nature in many places and their geologic and climatic history provide sufficient data for prudent anticipation of future events.

Planning for future disasters in vulnerable places by relocating people and structures is severely limited by the size of the population, the funds and land available, and political considerations. The resistance of people to leave their land, in spite of repeated disasters, is well-documented. The anticipated effects of future disasters can be mitigated through engineering procedures in places where high real estate values can justify the expense. Planning and funding for either avoiding or ameliorating the effects of disasters are generally absent in those countries that are most vulnerable.

Human losses include the dead, injured, affected and homeless, and become part of the staggering economic costs of a disaster. Destroyed buildings and roads are easier to assess, but damage to fields, crops, irrigation systems, communication and transportation nets, and their impacts on peripheral economies are much more difficult to measure. Quantifying the economic losses of a disaster is difficult in a well-developed country, and becomes elusive in poorer countries with little or no relief infrastructure. For these countries, the United Nations may provide a team of specialists who assess the damage and recommend the appropriate scale of relief.

An important perspective of the economic loss resulting from a natural disaster is to calculate the total replacement costs to the economy of a country, as a percentage of its Gross National Product (GNP) for the year of the disaster. GNP is the total value of goods and services produced in a country during a particular period of time, usually one year. GNP is made up of consumer and government purchases, private domestic and foreign investments, and the total value of exports. This information can be graphed on logarithmic paper, because of its broad range in data, with other important items such as the number of persons killed during a disaster. The aggregation of these data extends the parameters of the disaster.

The graphic representation and analysis of the human and economic dimensions of natural disasters provide a tool for developing a classification scheme for the relative risks and costs of natural disasters for the affected countries and for those of similar vulnerability. Such classification should facilitate more focused planning with future replacement funds tied to relocation and engineering safety specifications, and to scientifically approved site selection.

AN OVERVIEW OF NATURAL DISASTERS

With the exception of floods and landslides, there is no evidence that natural disasters have occurred with increasing frequency during the last two decades, but we seem to be increasingly aware of these disasters because of the speed of modern communication satellites with video transmissions. The startling explosion of human populations into many high risk urban areas and the increasing degradation of the environment accentuate the impacts of natural disasters.

During the last 20 years, 2.8 million persons have died and 820 million have been affected by all kinds of natural disasters (National Research Council et al., 1987). Losses to property damage are estimated at US $25-100 billion as a result of these disasters, while actual, total losses to the economies and social organizations of the devastated areas are much higher.

A recent review of the economic impacts of windstorms, floods, and earthquakes during the period 1980-1985 was conducted by Zupka (1988) utilizing data from the United Nations Disaster Relief Organization (UNDRO), the Office of Foreign Disaster Assistance (OFDA) of the U.S. State Department, the Swiss Reinsurance Company's economic studies on natural disasters and major losses, and the Munich Reinsurance Company's list of greatest natural disasters. In this review, a disaster involved 10 or more deaths and/or US $1 million or more of damage. A summary of his conclusions follows.

Windstorms include hurricanes in the Atlantic, Caribbean, and eastern Pacific; typhoons in the western Pacific; cyclones in the Indian Ocean; and tornados and other storms that affect all countries in varying degree. These types of disasters accounted for 49 percent of the recorded events, 33 percent of the dead, and 25 percent of the damage.

Floods are intensified by increasing the runoff through farming, deforestation, and urbanization. Because of these man-made alterations to the earth's surface, flooding has increased in frequency and in severity. Floods were 38 percent of the recorded events and accounted for 36 percent of the deaths and 28 percent of the damage.

Earthquakes provide a serious risk to at least 35 countries (National Research Council et al., 1987), and the risk is greatest for the moderate (6-7) and large (7-8) magnitude earthquakes because of their greater frequency when compared to the great earthquakes of magnitude 8 and above. Earthquakes were responsible for 13 percent of the disasters and caused 31 percent of the deaths and 47 percent of the damage.

There is tremendous suffering caused by these catastrophes, particularly among the injured and the homeless. Floods in densely populated floodplains and deltas regularly devastate entire communities. Earthquakes cause more dam-

Table 1. The human and economic costs of categories of thirteen natural disasters in Latin America (1970-1987).

CATEGORIES OF NATURAL DISASTERS	APPROXIMATE DEATH TOLL	PERCENT	COST—BILL. OF U.S. DOLLARS	PERCENT	TOTAL GNP[1]	COST OF DISASTERS AS % TOTAL GNP
EARTHQUAKES AND ASSOCIATED LANDSLIDES (7)[2]	115,300	78.2	11.31	69.4	216.13	5.2
FLOOD/DROUGHT CYCLES (2)	400	0.3	2.84	17.5	24.23	11.7
HURRICANES (2)	9,400	6.3	1.29	7.9	5.94	21.7
FLOOD (1)	300	0.2	.64	3.9	11.57	5.5
VOLCANIC ERUPTION (1)	22,000	15.0	.21	1.3	38.38	0.5
TOTALS	147,400	100.0	16.29	100.0	296.25	

[1]The total Gross National Product (GNP) is the sum of the GNPs for the countries affected in each category in the year in which the event occurred.
[2]Number in parenthesis indicates the number of events in each category of disaster. Thirteen natural disasters were studied in this report.
SOURCES: Compiled by D.B. Krinsley from Office of U.S. Foreign Disaster Assistance (1988); Natural Research Council et al. (1987); Economic Commission for Latin America and the Caribbean (1987); and World Bank (1987).

Table 2. Approximate Death Toll
Selected Natural Disasters In Latin America, 1970-1987.

Country	Event	Year-Month - Day[1]	Approximate Death Toll
1. Peru	Earthquake/ Landslide	70-5-31	70,000
2. Guatemala	Earthquake	76-2-4	24,000
3. Colombia	Volcanic Eruption	85-11-13	22,000
4. Nicaragua	Earthquake	72-12-23	10,000
5. Mexico	Earthquake	85-9-19	9,000
6. Honduras	Hurricane	74-9-18	8,000
7. Dominican Republic	Hurricane	79-8-00	1,400
8. El Salvador	Earthquake	86-10-10	1,100
9. Ecuador	Earthquake	87-3-5	1,000
10. Peru	Flood/Drought	83-1-00	400
11. Ecuador	Flood	82-11-00	300
12. Chile	Earthquake	85-3-5	200
13. Bolivia	Flood/Drought	83-1-00	?
TOTAL			147,400

[1] Date indicates the reported status of the event. 00 indicates that the exact day is unknown.
Sources: Compiled by D.B. Krinsley from Office of U.S. Foreign Disaster Assistance (1988) and National Research Council et al. (1987).

Table 3. Total Cost in Billions of U.S. Dollars
Selected Natural Disasters in Latin America, 1970-1987

Country	Event	Year-Month - Day[1]	Cost in Billions of U.S. Dollars
1. Mexico	Earthquake	85-9-19	4.10
2. Peru	Flood/Drought	83-1-00	2.00
3. Nicaragua	Earthquake	72-12-28	1.80
4. Chile	Earthquake	85-3-5	1.50
5. Guatemala	Earthquake	76-2-4	1.31
6. El Salvador	Earthquake	86-10-10	1.07
7. Ecuador	Earthquake	87-3-5	1.00
8. Bolivia	Flood/Drought	83-1-00	.84
9. Dominican Republic	Hurricane	79-3-00	.83
10. Ecuador	Flood	82-11-00	.64
11. Peru	Earthquake/ Landslide	70-5-31	.53
12. Honduras	Hurricane	74-9-18	.46
13. Colombia	Volcanic Eruption	85-11-13	.21
TOTAL			16.29

[1] Date is start of event. 00 indicates that the exact day is unknown.
Sources: Compiled by D. B. Krinsley from Economic Commission for Latin America and the Caribbean (1987) and Office of U.S. Foreign Disaster Assistance (1988) .

age to the built environment than other types of disasters but generally affect fewer people except when they strike urban centers such as Mexico City or Managua.

ANALYSIS OF SELECTED NATURAL DISASTERS IN LATIN AMERICA, 1970-1987

In view of the geographic area of interest of this symposium, an analysis was made of 13 major natural disasters in Latin America during the period 1970-1987. Data concerning the human and economic losses from natural disasters is difficult to obtain and when available it must be evaluated. Additionally, preliminary estimates are continually revised. Some of the data presented in this report have been revised twice since the report was initiated in June 1988. The natural disasters chosen for study are those that are not only important because of their dimensions, but because of the fact that the data on them are considered reliable.

The number of deaths from natural disasters was obtained from reports by the Office of U.S. Foreign Disaster Assistance (1988) and from the National Research Council et al. (1987). Table 1 summarizes the human losses associated with these 13 natural disasters. Earthquakes and associated landslides caused 78 percent of the 147,400 deaths and of that number the Peruvian event (Table 2) caused 70,000 or 48 percent of all deaths. The Guatemalan earthquake and Colombian volcanic eruption with accompanying lahars were equally devastating (Table 2). The hurricanes in Honduras and the Dominican Republic and the El Niño phenomena in Ecuador, Peru, and Bolivia took fewer lives but resulted in widespread economic dislocation and losses (see Table 3).

Estimating the economic losses (Table 1) associated with natural disasters is extremely difficult because (1) very few countries have the organizational infrastructure to measure and record the actual extent of the economic damage; (2) preliminary figures are frequently unreliable; (3) later revised figures are often distorted by inflation or changes in exchange rates; and (4) countries and national organizations have different criteria for measuring losses and these criteria are constantly being revised. After these problems are controlled, though never solved, it is important to consider the total economic damage as a percentage of that country's GNP during the year in which the disaster occurred. For example, the Mexico City earthquake of 1985 cost at least US $4.1 billion (Table 3) and was the most damaging of all the 13 disasters examined. However, Mexico's GNP was much greater than any of the others and the cost of its disaster as a percentage of its GNP was 2.4 percent, which made it twelfth on the list (Table 4). Thus Mexico's capacity to deal with disaster, in economic terms, is greater than for most of the countries examined.

The Economic Commission for Latin America and the Caribbean (ECLAC) is the foremost United Nations organization to provide a specialized team effort to measure and record the extent of damage as a result of a natural disaster in Latin America. An example of its expertise is provided by its report on the earthquake of March 5, 1987, in Ecuador (Economic Commission for Latin America and the Caribbean, 1987). That report, prepared at the request of the government of Ecuador, identifies the economic and social sectors that were most severely damaged by the disaster and that needed priority attention.

A summary of the damage (in millions of U.S. dollars) caused by that disaster is as follows:

	Total	Direct	Indirect
Total	1,001.06	186.21	814.85
Social infrastructure	30.17	30.17	---
Health	1.95	1.95	---
Education	7.55	7.55	---
Housing	20.67	20.67	---
Economic infrastructure	41.97	21.93	20.04
Transport	38.16	18.41	19.75
Electricity	3.81	3.52	0.29
Productive sectors	900.38	123.90	776.48
Agriculture	11.96	2.23	9.73
Hydrocarbons	888.42	121.67	766.75
Other sectors	28.54	10.21	18.33
Public, private buildings	3.00	3.00	---
Historical heritage	7.54	7.21	0.33
Emergency expenses	18.00	---	18.00

Direct costs involve replacing the property destroyed and repairing the property damaged. Indirect costs include crops that could not be harvested and oil deliveries that could not be made because of the ruptured trans-Ecuadoran oil pipeline. It should be emphasized that other subsequent costs such as increased expenditures for loans, decreased revenues, reduction in exports and increase in imports are significant factors in later computations of the cost of each disaster.

For the purposes of this paper, the total cost provided by ECLAC was used in Table 3. Other sources included Office of U.S. Foreign Disaster Assistance (1988) and the Embassy of Chile (personal communication, January 30, 1989). Earthquakes, the most prevalent of the 13 disasters, cost US $11.31 billion or 69 percent of the total cost of all (Table 1). On the other hand, the two devastating hurricanes cost approximately 8 percent of the total cost but represented almost 22 percent of the costs of the disasters as a percent of the total (combined) GNP of the two poor countries (Tables 1, 3, and 4).

Table 4. National Economic Impact
Selected Natural Disasters in Latin America, 1970-1987.

Country	Event	Year-Month - Day[1]	Cost of Disaster as a Percent of GNP
1. Nicaragua	Earthquake	72-12-23	209.3
2. Honduras	Hurricane	74-9-18	46.5
3. Guatemala	Earthquake	76-2-4	31.8
4. Bolivia	Flood/Drought	83-1-00	28.2
5. El Salvador	Earthquake	86-10-10	26.7
6. Dominican Republic	Hurricane	79-8-00	16.8
7. Ecuador	Earthquake	87-3-5	9.7
8. Peru	Flood/Drought	83-1-00	9.4
9. Chile	Earthquake	85-3-5	8.6
10. Peru	Earthquake/ Landslide	70-5-31	7.7
11. Ecuador	Flood	82-11-00	5.5
12. Mexico	Earthquake	85-9-19	2.4
13. Colombia	Volcanic Eruption	85-11-13	0.5

[1] Date is start of event. 00 indicates that the exact day is unknown.
Sources: Compiled by D. B. Krinsley from Economic Commission for Latin America and the Caribbean (1987); World Bank (1987); and Office of U.S. Foreign Disaster Assistance (1988).

Once the total cost of a disaster is reliably established, it is necessary to ascertain the GNP of the affected country for the year in which the disaster occurred in order to derive a percentage of the cost of the disaster to the country's GNP. The value of GNP for the particular country and specific disaster year was examined in (1) International Financial Statistics Yearbook 1987, of the International Monetary Fund; (2) Inter-American Development Bank, 1988 report; and (3) World Tables 1987, of the World Bank. There were significant differences in measuring GNP among the countries and among the above sources. There was the problem of converting the local currency to U.S. dollars for compatibility. This entailed a consideration of several exchange rates. Then there was the problem of devalued currencies, inflation, and deflation. These problems were exacerbated by the passage of time. After many conversations with members of the Brookings Institution, the Inter-American Development Bank, and the World Bank, the author received an extensive computer printout from Mr. John C. O'Connor, the Chief of the Socio-Economic Data Division of the International Economics Department of the World Bank. Among the items specifically requested and obtained was a world list of countries and their GNPs from 1965 through 1987, as computed by the World Bank utilizing the "Atlas" conversion factor. "The 'Atlas' conversion factor for any year is the average of the exchange rate for that year, and the exchange rates for the two preceding years, which have been adjusted for differences in relative inflation between the country and the United States. This three-year average smooths fluctuations in prices and exchange rates for each country" (World Bank, 1988b). These data, which are the standard for the World Bank, satisfied the requirements of this study and have been used for all GNP computations.

Table 4 reflects the national economic impact of the studied disasters by listing their costs as a percentage of their countries' GNP during the year in which they occurred. These data significantly supplement those in Table 3 and provide a more realistic and meaningful understanding of the true economic impact of these disasters. Although the Nicaraguan earthquake cost US $1.8 billion, which is in itself a terrible blow to such a poor country, the fact that this sum represented 209 percent of its GNP that year is quite devastating financially. Such a disaster places unbearable financial burdens on future generations and further complicates that country's serious political difficulties.

Though not as startling as the Nicaraguan example, the disasters in Honduras, Guatemala, Bolivia, El Salvador, and the Dominican Republic (Table 4) were very serious blows to weak economies. All of the 13 countries except Colombia had GNP per capita growth rates of less than 0 percent annually during the period 1980-1987 (World Bank, 1988a), and Colombia's GNP per capita growth rate was less than 0.9 percent annually during that period.

CLASSIFICATION OF NATURAL DISASTERS

There are many organizations that utilize risk warning systems for the impending dangers of avalanches, landslides, volcanic eruptions, floods, storms, tsunamis, and forest fires. Some major faults and volcanos are carefully instrumented for the slightest movements. All of these systems employ categorizations of relative risk such as slight, moderate, high, and likely within a stated period. Abundant records exist on the frequency and intensity of earthquakes in populous regions, hence their risk of recurrence. The Richter Scale is the standard by which all earthquakes are measured and their magnitude understood.

Once a disaster has occurred, its inclusion in an organization's records depends upon the criteria for inclusion established by that organization. The Office of Foreign Disaster Assistance (OFDA) of the U.S. Department of State established the following criteria (Office of Foreign Disaster Assistance, 1988):

1. Earthquake and volcanic disasters are included if the number of people killed is at least six, or the total number killed and injured is 25 or more, or at least 1,000 people are affected, or damage is US $1 million or more.

2. Weather disasters (except droughts) are included if the number of people killed and injured totals at least 50, or 1,000 or more are homeless or affected, or damage is at least US $1 million.

3. Drought disasters are included if the number affected is substantial.

None of the major relief organizations have published any systems of categorization of natural disasters which would attempt to classify the human and economic losses in terms of national impact. Such systems might provide more focus to organizations providing emergency relief and subsequent economic rehabilitation.

Figure 1 was prepared by recording the number of deaths and the costs of the natural disaster as a percent of the

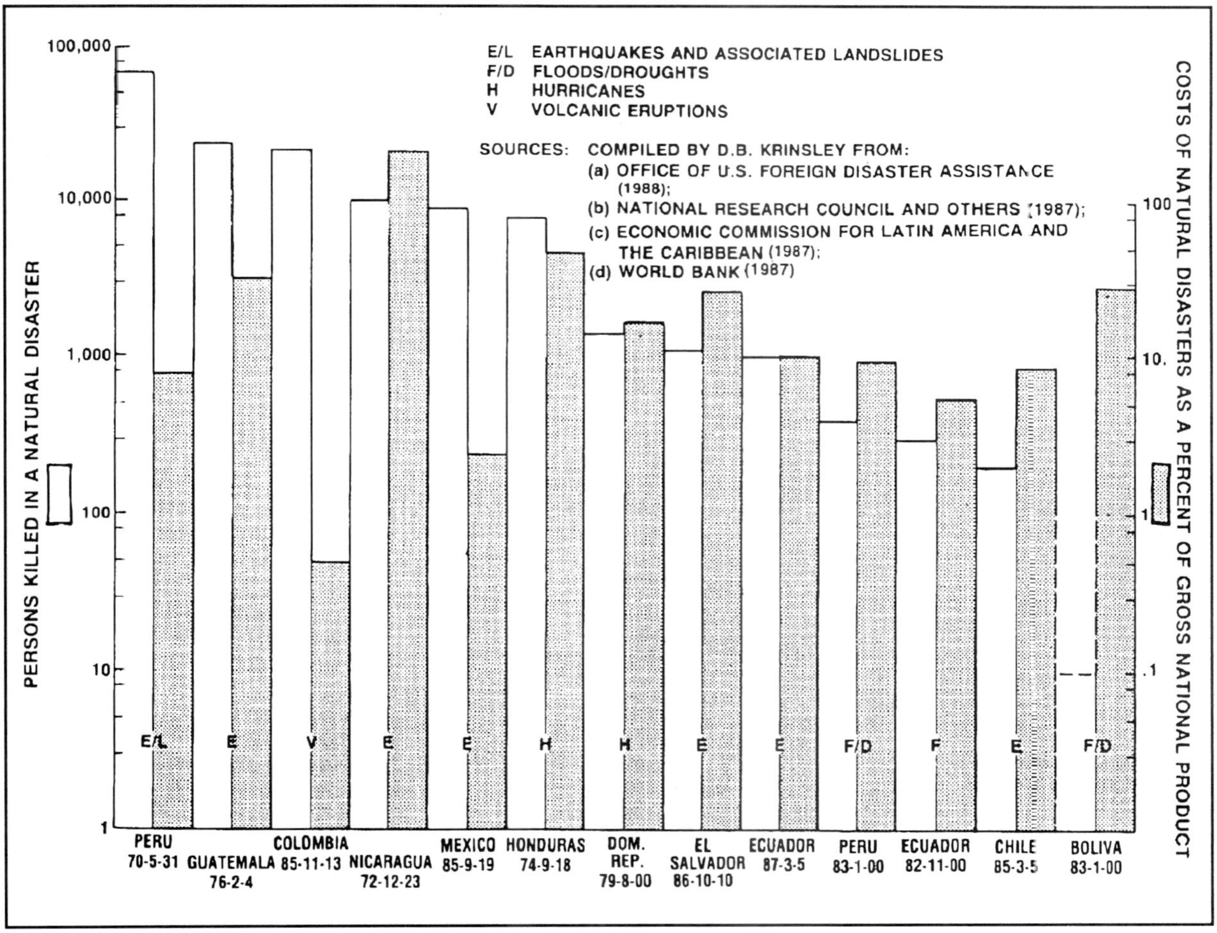

Figure 1. The human and economic costs of natural disasters.

GNP during the year of the disaster. Because of the extreme ranges in values these data were plotted on semilogarithmic paper. In spite of the admittedly small population, a perusal of the figure suggests the following categories of disasters:

Category	Deaths		Cost/GNP(%)
A	more than 10,000	and/or	more than 100.
B	1,000 to 10,000	and/or	10. to 100.
C	100 to 999	and/or	1. to 9.
D	less than 100	and/or	less than 1. *

* Almost any U.S. natural disaster involving less than 100 fatalities and less than US $45 billion would fall into this category because the U.S. GNP is so large ($4.5 trillion in 1987).

Notations based on the data provided in Tables 2 and 4 make it possible to assign the Guatemala earthquake to Category A with 24,000 deaths and a cost as a percent of GNP of 31.8, or A24.-31.8. Further notations could include data on the injured, affected, and homeless, and tables can be constructed to meet the specific needs of relief organizations. A coherent standardized and widely accepted system, however, is necessary so that everyone knows and understands the dimensions of each disaster and can respond quickly and effectively to match need with relief. Such classification facilitates data handling and processing and clarifies the national impact of these disasters. It will also encourage more emphasis on planning, avoidance, mitigation, and location of facilities at sites that are physically safe as well as economically feasible before large sums are allocated for rebuilding.

ACKNOWLEDGMENTS

This study was greatly facilitated through the kindness and generosity of several economists, statisticians, and engineers who provided materials, suggestions, and guidance. I am particularly indebted to Mr. Frederick M. Cole and Mr. John M. Harrington, Jr., of the U.S. Department of State; Mr. Isaac Cohen, Mr. Walter M. Cabrera, and Mr. J. Roberto Jovel of ECLAC; Mr. John C. Elac, Mr. Uziel Nogueira, and Mr. Michael Anderson of the Inter-American Development Bank; Mr. John C. O'Connor and Ms. Rosalinda Quintanilla of the World Bank; and Mr. Jorge Valenzuela and Mr. Cristian Hohlberg R. of the Embassy of Chile.

REFERENCES CITED

Economic Commission for Latin America and the Caribbean, 1987, The natural disaster of March 1987 in Ecuador and its impact on social and economic development: United Nations, LC/G. 1456, 56p.

Inter-American Development Bank, 1988, Economic and social progress in Latin America, 1988 report: Washington: 604p.

International Monetary Fund, 1987, International financial statistics yearbook: Washington, v. XL, 744p.

National Research Council et al., 1987, Confronting natural disasters—An international decade for natural hazard reduction: Washington, National Academy Press, 60p.

Office of U.S. Foreign Disaster Assistance, 1988, Disaster history—Significant data on major disasters worldwide, 1900-present: Washington, Agency for International Development, 217p.

World Bank, 1987, World tables 1987—The fourth edition: Washington, 521p.

World Bank, 1988a, The World Bank atlas 1988: Washington, 29p.

World Bank, 1988b, World development report 1988: New York, Oxford University Press, 307p.

Zupka, D., 1988, Economic impact of disasters: UNDRO News, January/February, p. 19-22.

Circum-Pacific Council for Energy and Mineral Resources
Earth Science Series, Vol. 16
R. L. Miller, G. Escalante, J. A. Reinemund, and M. J. Bergin (Eds.)
Energy and Mineral Potential of the Central American-Caribbean Regions
Springer-Verlag, Berlin Heidelberg 1995

CENTRAL AMERICAN RESOURCE STUDIES

Ed Van Eeckhout and Bill Laughlin
Los Alamos National Laboratory, Los Alamos, New Mexico 87545, USA

ABSTRACT

The Los Alamos National Laboratory has been involved in geothermal and mineral resource studies in Central America over the last four years. These studies have concentrated on finding mineral resources in Costa Rica (using geochemical prospecting), peat in Costa Rica and Panama, geothermal resources in Honduras and Guatemala, and developing existing geothermal fields in Costa Rica (Miravalles), El Salvador (Ahuachapán), and Guatemala (Zunil). We have been assisted in these efforts by the counterpart electrical utilities in each country, the U.S. Geological Survey, Lawrence Berkeley Laboratory, and private consultants. The work has been funded by the U.S. Agency for International Development (USAID) at a level of \$8.0 million over calendar years 1985-1987.

Significant achievements to date include the release of geochemical atlases for quadrangles in Costa Rica, the identification of Platanares as the site with greatest geothermal energy potential in Honduras, a geothermal assessment of the Tecuamburro volcano in Guatemala, and use of state-of-the-art geothermal well logging and direct current resistivity equipment in the region. This equipment was used in well logging and geophysical surveys at the Miravalles geothermal field in Costa Rica, the Ahuachapán geothermal field in El Salvador, and the Zunil geothermal field in Guatemala.

INTRODUCTION

Los Alamos National Laboratory has been working with five Central American countries to assist in the development of their energy and mineral resources. Since 1985, mineral resources in Costa Rica, peat resources in Costa Rica and Panama, geothermal energy resources in Honduras and Guatemala, and geothermal field development in El Salvador and Costa Rica have been topics of study. This paper presents an overview of these studies; within these proceedings are papers that deal with specific aspects of each, and they will be duly noted.

The studies described herein were funded by the U.S. Agency for International Development (USAID), and were administered through its Regional Office of Central American Programs (ROCAP) in Guatemala City. Mr. Carl Duisberg was the project manager. Due credit for assistance for much of this work must go to our counterparts in Central America; in Costa Rica, the Instituto Costarricense de Electricidad (ICE) (for the geothermal work), Minera Nacional, S. A. (MINASA) and the University of Costa Rica (UCR) (for the minerals work), and the Refinadora Costarricense de Petróleo, S. A. (RECOPE) (for the peat studies); in El Salvador, the Comisión Hidroeléctrica del Río Lempa (CEL); in Guatemala, the Instituto Nacional de Electrificación (INDE); in Honduras, the Empresa Nacional de Energía Eléctrica (ENEE); and in Panama, the Instituto de Recursos Hidraulicos y Electrificación (IHRE). Many personnel from these organizations helped greatly in our efforts. Thanks must also be given various contractors who worked on these projects: the U.S. Geological Survey (USGS), Lawrence Berkeley Laboratory (LBL), and various private consultants.

Los Alamos involvement in the region dealt first with mineral assessment in Costa Rica and peat assessment in Costa Rica and Panama. We later became more heavily involved in geothermal resource assessment in Honduras and Guatemala, and geothermal field development at Miravalles in Costa Rica and Ahuachapán in El Salvador. For the purposes of this paper, it is convenient to discuss these four "components" (mineral resource assessment, peat resource assessment, geothermal resource assessment, and geothermal field development) in the order mentioned above.

MINERAL RESOURCE ASSESSMENT

A major metallic mineral resource assessment was initiated in Costa Rica by Los Alamos, the USGS, MINASA and personnel from the UCR. The USGS and Los Alamos utilized two different approaches in assessing the mineral potential of a large part of the country. Los Alamos concentrated on the generation of geochemical atlases based upon the analysis of stream sediment samples while the USGS produced a mineral resource assessment folio for Costa Rica.

The USGS folio (Mineral Resource Assessment of the Republic of Costa Rica, USGS Miscellaneous Investigations Series Map I-1865) is comprised of three parts. The first part is a series of newly compiled country-wide maps for mineral resource assessment. Considerable new data was collected to generate the physiographic, geologic, mineral occurrence and a variety of geophysical maps which are included in this section. Accompanying text describes the geology of Costa Rica, the more prominent geophysical anomalies, and identifies areas that are of interest in exploring for various types of mineral deposits. The second part of the folio has the supporting data used to generate the mineral resource assessment maps. These data include additional maps, figures, and explanatory text. The third part of the folio has the results of detailed investigations of the Costa Rican "gold belt" by the USGS and the counterpart organizations are given.

Los Alamos published two geochemical atlases for Costa Rica (LA-10965-MS and LA-10965MS Addendum 1) which cover the San José, Golfito and Quepos quadrangles. Work is in progress on the atlas for the southwestern portion of the Talamanca quadrangle (at a scale of 1:200,000). These atlases are based on the analytical results of stream sediment sampled at a density of 1/20 km^2 over the four quadrangles.

Before beginning the stream sediment sampling program, Los Alamos conducted a pilot study in the San José and Talamanca quadrangles. This study, which became a M.Sc. thesis project for a Costa Rican geology student at the Colorado School of Mines (Arauz, 1986), determined the best sampling procedures and density for a tropical environment. Wet sieving techniques were developed to maximize sampling efficiency and the sample density of 1/20 km^2 was selected as optimal. Upon completion of the pilot study Los Alamos trained the counterpart geologists and students in the newly adopted sampling procedures and the field work began.

After collection, the stream sediment samples were dried and shipped to Los Alamos for analysis. The Los Alamos nuclear reactor was used to determine the abundances of 46 elements by neutron activation analysis, and a commercial geochemical laboratory was used for four additional elements.

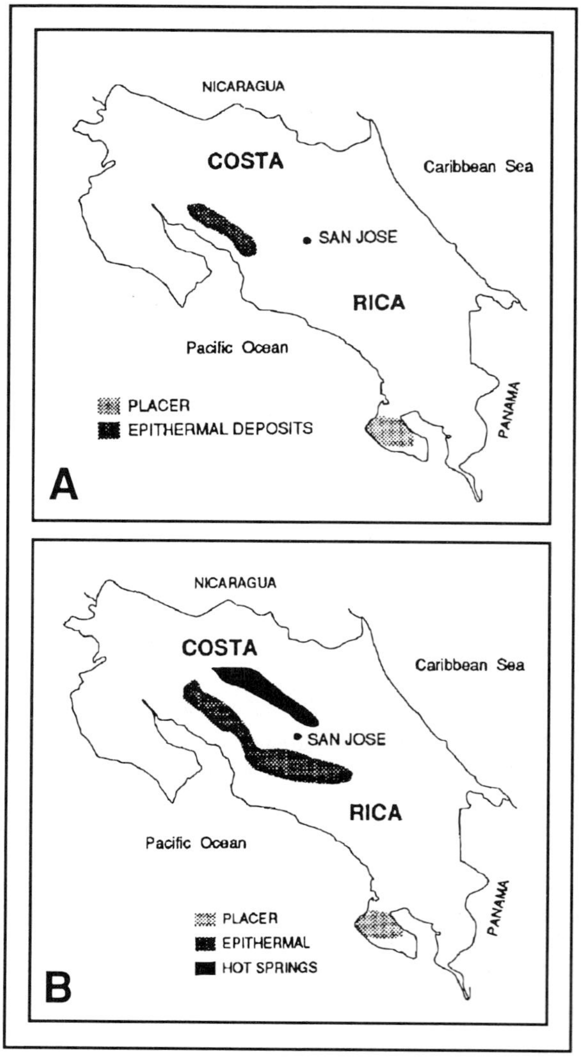

Figure 1. (a) Known gold districts before Los Alamos work. (b) Potential areas for gold after Los Alamos work.

The combined results of the USGS and Los Alamos assessments indicate that the Costa Rican "gold belt" is much larger than previously thought, extending a considerable distance to the southeast (Figures 1a and 1b). The results also indicated a high potential for hot spring-type gold deposits. At the request of the Costa Rican government all results of the assessments were made public at an international gold conference held in San José in the fall of 1987. Specially prepared promotional brochures giving geochemical and geological data on the most promising prospects were distributed at this conference as were the USGS folio and the Los Alamos atlases. Release of this vast amount of new data has greatly stimulated interest in gold development in Costa Rica and MINASA has received several proposals for more detailed evaluations of the prospects.

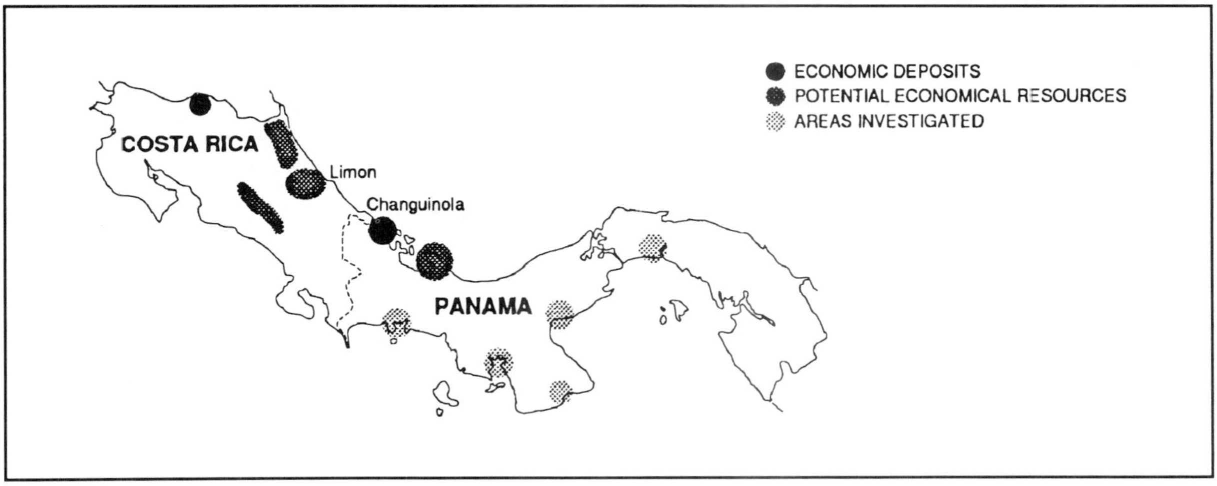

Figure 2. Areas of Costa Rica and Panama evaluated for potential peat deposits.

Training of counterpart geologists and students has been an important part of the entire assessment program. The USGS taught two formal courses in economic geology and both Los Alamos and the USGS performed on-the-job training in all aspects of modern mineral assessment and exploration. More than 150 geologists or students received some type of training.

PEAT RESOURCE ASSESSMENT

Los Alamos initiated an effort to find and evaluate peat deposits in Costa Rica and Panama, and were successful in identifying world class deposits in both countries. A detailed discussion of this work is contained in Thayer, Ramirez, and Ramirez (this volume). Suffice it to say that certain prospects were identified that would certainly be of interest to the counterparts involved. Those prospects are noted in Figure 2.

GEOTHERMAL RESOURCE ASSESSMENT

Geothermal resource assessments have also been an important part of the Los Alamos Central American Program. Assessment work began in Honduras in 1985 and has continued there until the present time; in Guatemala it began in 1988.

Assessment work in Honduras began with a reconnaissance investigation of six previously identified areas that appeared to have geothermal potential. Results of these geological and geochemical investigations are reported in Eppler, Fakundiny, and Ritchie (1986), Goff et al. (1987), and Goff et al. (1988). The major conclusions of these reconnaissance-scale investigations were that three sites, Platanares, Azacualpa, and San Ignacio, appeared to have the highest potential for electricity generation and that a fourth site, Pavana, had high potential for direct-use of the geothermal resource. Detailed geologic investigations were consequently focussed on these four sites. Results of these detailed studies are given in Heiken et al. (1987), Eppler et al. (1987a), Aldrich et al. (1987) and Eppler et al. (1987b) for Platanares, Azacualpa, San Ignacio and Pavana, respectively. Detailed geochemical work was also done at the four sites and these results are presented in Goff et al. (1987) and Goff et al. (1988). Based upon the results of these studies, Platanares clearly had the highest potential for electricity generation with a calculated reservoir temperature of 225°C. Surface flow rates were also high at Platanares.

The encouraging results at Platanares led to the decision to proceed with geophysical surveys and gradient drilling at this site. Telluric profiling and gravity and self-potential surveys were performed and three gradient wells were drilled. Results of the drilling are given in Goff et al. (1988). Two of the three gradient wells encountered 160°C water at relatively shallow depths (<650 m) and the third, non-flowing well yielded a geothermal gradient of 139°C/km (Figure 3 shows the artesian flow encountered at one well). All data from Platanares were integrated and a recommendations report (Laughlin, 1988) and an economics report (Trocki, 1989) submitted to ENEE. These reports concluded that it is economically very attractive to begin the immediate development of the shallow reservoir while funding is sought for development and testing of the deeper 225°C reservoir. Los Alamos is now working closely with ENEE in seeking these funds. A more detailed discussion of the Honduras work is given by Goff et al. (this volume).

In Guatemala, Los Alamos and the USGS have been working closely with INDE to complete a geothermal prefeasibility study of the Tecuamburro area in the eastern part of the country. Two field trips have been made and numerous rock and fluid samples have been collected for geology, geochronological studies, and reservoir geothermometry. To date, results have been very exciting:

Figure 3. Artesian flow encountered at Platanares, Honduras.

Figure 4. The caliper tool used to log wells at Ahuachapán.

reservoir temperatures as high as 298°C have been calculated and the geochronological results indicate that volcanism has continued as recently as 3,000 years before present. Gradient drilling at Tecuamburro began in late 1989.

GEOTHERMAL FIELD DEVELOPMENT

Los Alamos efforts under this category have concentrated on development of existing geothermal fields, well logging, geochemical brine sampling and analyses, reservoir engineering, and geophysics. The geothermal fields receiving attention to date include Miravalles, in Costa Rica, and Ahuachapán, in El Salvador.

A large share of the funding was used to purchase a state-of-the-art well logging truck and downhole tools for the region. This equipment is patterned after the equipment developed by Los Alamos for the Hot Dry Rock Project in the United States, which allows the monitoring of downhole temperatures, pressures, flows, and well diameters, plus the collection of brine samples at depth. These tools are calibrated for use at temperatures up to 300°C in very caustic environments. One of these tools, a caliper tool, is shown in Figure 4. Detection of calcite buildup in certain wells at Miravalles (Dennis et al., 1989b) and recharge zones at Ahuachapán (Dennis et al., 1989a) are typical results observed. This equipment is being provided to the region and will be operated and maintained by a consortium involving CEL, ICE, INDE, ENEE, Los Alamos, and USAID. The details of this consortium are currently being worked out.

A second aspect of this component, geochemistry brine sampling and analysis, is linked to the well logging (downhole samples are collected as part of the well logging program). Analyses to date have yielded interesting conclusions on recharge scenarios at Miravalles (Grigsby et al., 1989) and surface water encroachment at Ahuachapán (Dennis et al., 1989a).

A third aspect of this component, reservoir engineering, has been performed principally by a sister national laboratory, LBL, and has been concentrated at Ahuachapán. The focus of this effort has been to build a data base of past temperatures, pressures, drawdown, and geology in the field and then model various scenarios for development. This study will show how best to handle the current trend of dropping temperatures in the field (by perhaps using reinjection) and the direction of future drilling efforts. This work is on-going and initial results were presented at the Fourteenth Geothermal Reservoir Engineering Conference at Stanford in January 1989 (proceedings to be published).

The final aspect, geophysics, has dealt primarily with direct current resistivity studies at Ahuachapán. Greater powered equipment was built to extend the commercially available depth ranges so that the geologic structure could be seen better. The results of this work are still being analyzed and should be forthcoming soon.

Training has also been an important part of this component. Los Alamos and its contractors have made great efforts to transfer the ability to perform and analyze each of the aspects mentioned above to Central American counterparts. Several training sessions have been held, and certain personnel within the organizations of CEL, ICE, INDE, and ENEE have been trained in the techniques used.

REFERENCES CITED

Aldrich, M. J., D. Eppler, G. Heiken, W. Flores, N. Ramos, and A. Ritchie, 1987, San Ignacio (La Trembladora) geothermal site, Departamento de Francisco Morazan, Honduras, Central America: Los Alamos National Laboratory Report LA-10696-MS, 27p.

Arauz, A., 1986, Evaluation of stream sediments in areas of known mineralization, San Jose and Talamanca quadrangles, Costa Rica: Los Alamos National Laboratory Report LA-10840-T.

Dennis, B., F. Goff, E. Van Eeckhout, and R. Hanold, 1989a, Results of investigation at the Ahuachapán geothermal field, El Salvador, Part 1: Well logging and brine geochemistry: Los Alamos National Laboratory Report, to be published.

Dennis, B., R. Lawton, J. Kolar, and A. Alvarado, 1989b, Results of investigation at the Miravalles geothermal field, Costa Rica, Part 1: Well logging: Los Alamos National Laboratory Report LA-11510-MS, 18p.

Eppler, D., R. Fakundiny, and A. Ritchie, 1986, Reconnaissance evaluation of Honduran geothermal sites: Los Alamos National Laboratory Report LA-10685-MS, 15p.

Eppler, D., S. Baldridge, F. Perry, W. Flores, J. R. Paredes, and R. Finch, 1987a, Geology of the Azacualpa geothermal site, Departamento de Comayagua, Honduras, Central America: Los Alamos National Laboratory Report LA-10687-MS, 32p.

Eppler, D., G. Heiken, K. Wohletz, W. Flores, J. R. Paredes, and W. A. Duffield, 1987b, Geology of the Pavana geothermal area, Departamento de Choluteca, Honduras, Central America: Los Alamos National Laboratory Report LA-10864-MS, 17p.

Goff, F. E., A. H. Truesdell, C. O. Grigsby, C. J. Janik, L. A. Shevenell, J. R. Paredes, J. W. Gutierrez, P. E. Trujillo, Jr., and D. A. Counce, 1987, Hydrogeochemical investigations of six geothermal sites in Honduras, Central America: Los Alamos National Laboratory Report LA-10634-MS, 168p.

Goff, F. E., A. W. Laughlin, H. D. Rufenacht, F. Goff, G. Heiken, A. I. Adams, J. Musgrave, H. Planner, and N. Ramos, 1988, Exploration geothermal gradient drilling, Platanares, Honduras, Central America: Los Alamos National Laboratory Report LA-11349-MS, 29p.

Grigsby, C., F. Goff, P. E. Trujillo, Jr., D. Counce, B. Dennis, and J. Kolar, 1989, Results of investigation at the Miravalles geothermal field, Costa Rica, Part 2: Downhole fluid sampling: Los Alamos National Laboratory Report, to be published.

Heiken, G., D. Eppler, K. Wohletz, W. Iores, N. Ramos, and A. Ritchie, 1987, Geology of the Platanares geothermal site, Departamento de Copán, Honduras, Central America: Los Alamos National Laboratory Report LA-10634-MS, 24p.

Laughlin, A. W., 1988, Recommendations report for the Platanares geothermal site, Department of Copán, Honduras: Los Alamos National Laboratory Report LA-11410-MS, 73p.

Thayer, G., O. Ramirez, and A. Ramirez, 1990, Economics of selected energy applications of peat in Panama and Costa Rica, this volume.

Trocki, L., 1989, An economic prefeasibility study of geothermal energy development at Platanares, Honduras: Los Alamos National Laboratory Report, 49p., in pres

PETROLEUM RESOURCES

Circum-Pacific Council for Energy and Mineral Resources
Earth Science Series, Vol. 16
R. L. Miller, G. Escalante, J. A. Reinemund, and M. J. Bergin (Eds.)
Energy and Mineral Potential of the Central American-Caribbean Regions
Springer-Verlag, Berlin Heidelberg 1995

CARIBBEAN REGION: PERSPECTIVES ON PETROLEUM

Arthur A. Meyerhoff
P. O. Box 4602, Tulsa, OK 74159 USA

Anthony E. L. Morris
P. O. Box 64610, Los Angeles, CA 90064 USA

Irfan Taner
3336 East 32nd Street, Suite 210, Tulsa, OK 74135 USA

ABSTRACT

The Caribbean region unambiguously illustrates several principles of petroleum geology, the most important of which is that tectonic stability in the continental platform environment is essential to the generation, accumulation, and preservation of petroleum. Platform blocks account for 99.99 percent of the region's cumulative production and proved oil and gas reserves, 95 percent of its oil resources (oil still to be discovered), and 87 percent of its gas resources (gas still to be discovered). However, a platform can be too stable, as illustrated by large parts of the Yucatan and Florida-Bahamas platforms where production and reserves are nil. In contrast, the mobile-belt environment ultimately leads to the destruction of generated petroleum, unless the mobile belt becomes stabilized by accretion to an already-existing platform. Use of these principles in the Caribbean suggests that future exploration should be concentrated almost entirely in stable blocks. However, the search for future reserves should not be limited to rocks of Late Jurassic through Pleistocene ages, as it has been in the past, but expanded to include a concerted search in late Paleozoic rocks whose platform facies are well developed in at least two large areas of the Caribbean region.

INTRODUCTION

The volume of generated and trapped petroleum that is preserved in a region is related closely to that region's tectonic stability. However, a long-stable region is not destined automatically to become a petroleum province, because the potential to generate, accumulate, and trap commercial volumes of petroleum also must be present. Nevertheless, it is a fact that the preferred habitat of the commercial petroleum field is the stable foreland platform.

In sharp contrast to the stable platform, its antithesis, the mobile belt, provides an environment intrinsically hostile to petroleum preservation. In fact, many mobile belts are totally unsuited even for the generation of petroleum. However, if a mobile belt stabilizes by accretion to a platform, it gradually will acquire the characteristics of the stable platform, and the petroleum potential of the post-accretionary sedimentary section will increase accordingly.

The Caribbean region (Figure 1) provides clear examples of these broad principles. This region—with an area of 8.5 million km^2—is essentially an assemblage of two mobile belts with their branches sandwiched between two large cratons, North America in the northwest and South America in the southeast. The important petroleum accumulations are on the platform blocks. They account for 99.99 percent of the region's cumulative oil and gas production, 99.99 percent of the region's proved oil and gas reserves, 95 percent of its oil resources (oil still to be discovered), and 87 percent of its gas resources (gas still to be discovered). Table 1, based on tables 2 and 3 in Morris et al., 1989b, shows that not only is the platform environment the Caribbean region's preferred petroleum habitat, but also the platform areas of

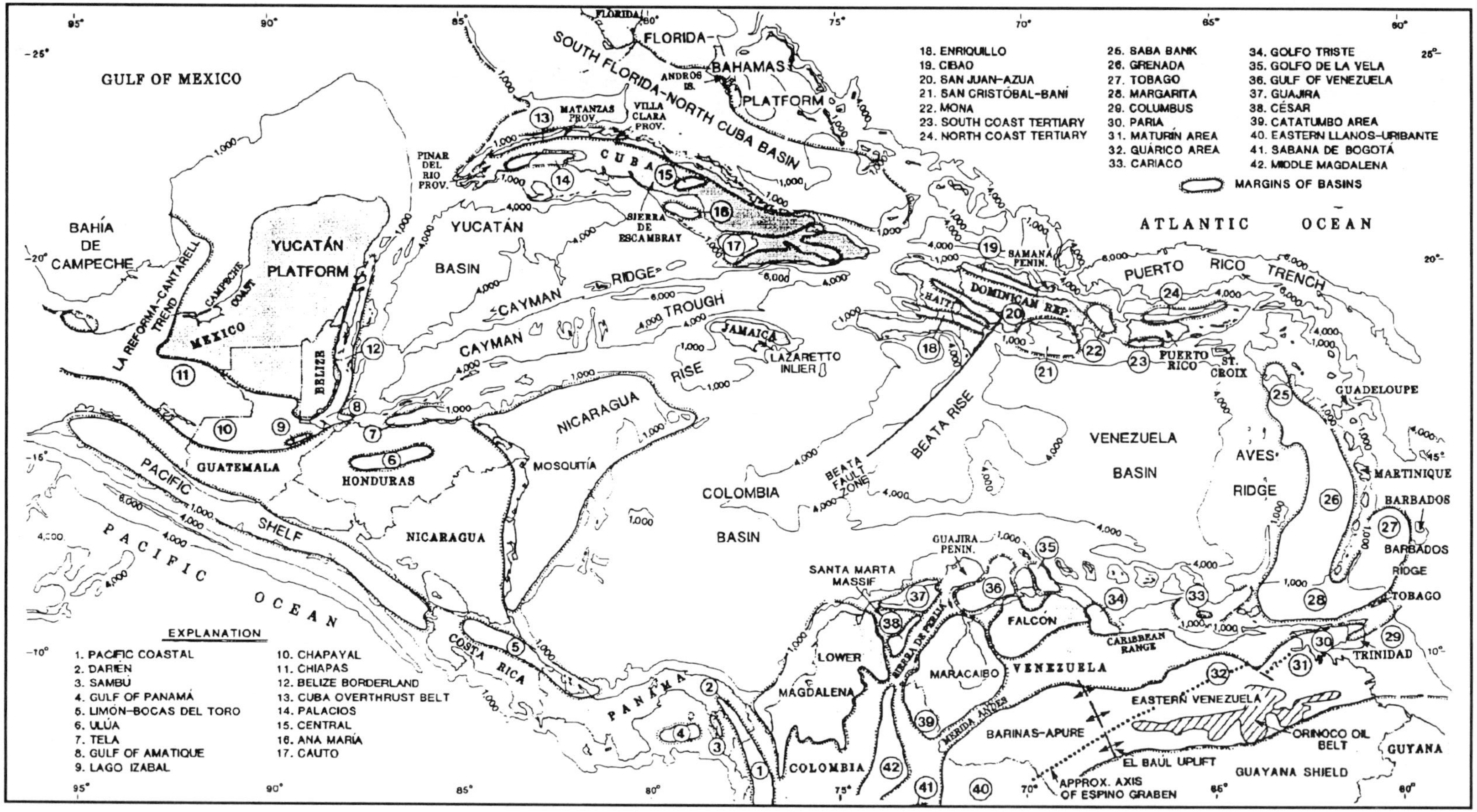

Figure 1. Index map, Caribbean region, showing major physiographic features mentioned in text, together with all basins used in the compilation of Tables 1, 2, 3, and 4.

Table 1. Distribution of Caribbean petroleum by tectonic environment (as of end-1986). Based on Tables 2 and 3 in Morris et al. (1989b).

	OIL			HEAVY OIL			GAS		
	Cumulative Production	Proved Reserves (Millions of Barrels)	Resources	Cumulative Production	Proved Reserves (Millions of Barrels)	Resources	Cumulative Production	Proved Reserves (Trillions of Cubic Feet)	Resources
Platforms									
Guatemala	14.3	23	514	--	10	10	--	30	490
Colombia	2,347.9	3,631	1,852	29.2	43	70	3,840	10,057	5,699
Venezuela	40,447.0	21,409	47,490-103,390	1,716.1	276,200	61,370-101,370	49,570	48,623	101,750-169,750
Trinidad and Tobago	2,333.0	750	3,500	218.4	111	800	5,155	6,075	12,060-15,060
Subtotals	45,142.2	25,813	53,356-109,256	1,963.7	276,364	62,250-102,250	58,565	64,785	119,999-190,999
Mobile Belts									
C. America	0.008	--	815	--	--	--	--	--	4,600
Colombia	--	--	75	--	--	--	--	--	1,315
Venezuela	--	--	90	--	--	--	--	--	1,000-2,000
Trinidad and Tobago	--	--	225	--	--	--	--	--	15,000-20,000
Barb. Ridge	4.3	1	250	--	--	--	7	1	250
Puerto Rico	--	--	1,150	--	--	--	--	--	1,500
Hispaniola	--	--	45	--	--	--	--	--	230
Cuba	50.0	34	200	0.08	--	0.02	28	340	250
Subtotals	54.3	35	2,850	0.08	--	0.02	35	341	24,145-30,145
TOTALS	45,196.5	25,848	56,206-112,106	1,963.8	276,364	62,250-102,250	58,600	65,126	144,144-221,144

Venezuela have most of that petroleum. Venezuela's share of cumulative oil and gas production in the Caribbean is 89 percent and 85 percent, respectively; of proved reserves, 83 percent and 75 percent; and of resources, 85 percent and 71 percent. Venezuela has, in addition, 97 percent of the region's heavy oil resources. Thus Venezuela alone has an assured ultimate recovery of 340 billion bbl of oil and 98.3 Tcf of gas, with a possible ultimate recovery of 449 to 545 billion bbl of oil and 200 to 268 Tcf of gas. These figures place Venezuela very close to the forefront of the world's petroleum-producing nations.

We did not include the La Reforma-Cantarell petroleum fields of southern Mexico within the Caribbean region because they are more closely related to the evolution of Yucatan and the Gulf of Mexico (Viniegra, 1981). Even if we had included them, Venezuela's preeminence would be unchanged, as the estimated ultimate recovery from them of 20 billion bbl would leave Venezuela's total share of the Caribbean's conventional oil still more than 60 percent, and its 97 percent share of the heavy oil unchanged.

Nor did we include the southern Florida fields within the region, even though they are an intrinsic part of the geological history of the Florida-Bahamas platform. Like much of Yucatan, the history of this region is more closely related to the evolution of the Gulf of Mexico. Moreover, the inclusion of these minor fields would only change our percentage values less than 0.002.

Table 1 shows that the mobile belts are poor places in which to prospect for petroleum, although it does show that some areas within the mobile belts (e.g., the northern South American margin) have the potential for rather large volumes of gas. However, this volume of gas—if present—will be in sediments accumulated after accretion of the mobile belt to the platform margin.

FUTURE TRENDS IN EXPLORATION AND DISCOVERY

The preceding premises suggest that future exploration in the Caribbean on the same scale as during the first six or seven decades of this century is unlikely, as the region has been rather thoroughly culled for the obvious traps, and companies searching for the unexpected, or "subtle," traps

Table 2. Petroleum production, reserves, and resources, Caribbean region, 1986 (excluding Orinoco Heavy Oil Belt).*

PROVINCE	PRODUCTION*			RESERVES AND RESOURCES*			
Basin	Barrels per Day (bopd)	Cumulative Production		Proved (1986)		Future Potential Reserves+	
		Oil (Mil. bbl.)	Gas (bcf)	Oil (Mil. bbl.)	Gas (bcf)	Oil (Mil. bbl.)	Gas (bcf)
CENTRAL AMERICA							
Chapayal	6,200	14.3	--	23	30	463.5	190
Gulf of Amatique-Belize borderland	--	--	--	--	--	50	200
Lago Izabal, Tela, Ulúa	--	--	--	--	--	--	100
Mosquitía	--	--	--	--	--	500	2,000
Pacific Shelf	--	--	--	--	--	50	1,000
Límon-Bocas del Toro	--	0.008	--	--	--	100	1,500
Gulf of Panamá	--	--	--	--	--	10	200
Sambú-Darién	--	--	--	--	--	5	100
NORTHERN S. AMERICA							
Upper Magdalena	46,030	113.8	36.0	149.3	55	110	30
Middle Magdalena	95,000	1,666.0	1,953.0	2,000	2,600	342	444
Lower Magdalena-Caribbean Platform	1,340	69.8	772.0	68.7	778	70	1,000
Guajira	--	--	564.0	--	6,2000	50	3,000
Eastern Llanos	135,100	91.3	42.0	1,000	--	1,100	825
Pacific-Atrato	--	--	--	--	--	50	1,200
César	--	--	--	--	--	25	115
Sabana de Bogotá	--	--	--	--	--	60	370
Maracaibo							
Venezuela	1,372,300	30,716	25,227	16,495	24,493	28-51,000	30-50,000
Colombia	7,451	407	473	413	424	120	130
Falcón	240	108	64	1	1	300	500
Golfo de Venezuela	--	--	--	--	--	3-26,000	35-60,000
Golfo de Vela	--	--	--	--	--	500	1,000
Golfo Triste	--	--	--	--	--	90	100
Barinas-Apure	36,500	555	5	160	8	200	150
Guárico	5,860	311	1,394	59	3,477	300	5-8,000
Maturín	367,000	8,757	22,880	4,690	24,121	15-25,000	30-50,000
Trinidad-Paria	77,740	1,773	655	250	75	300	60
Columbus	91,260	560	4,500	500	6,000	3,200	12-15,000
Cariaco-Margarita	--	--	--	--	--	90	1-2,000
Tobago	--	--	--	--	--	225	15-20,000
EASTERN CARIBBEAN							
Grenada (S. of 16°)	--	--	--	--	--	200	1,000
Saba Bank-Northern Grenada	--	--	--	--	--	200	2,000
Tobago	--	--	--	--	--	500	10,000
Barbados Ridge	1,499	4,27	6.8	1	1	250	250
GREATER ANTILLES							
South Coast Tertiary Basin	--	--	--	--	--	--	1
North Coast Tertiary Basin	--	--	--	--	--	400	500
Mona	--	--	--	--	--	750	1,000
San Cristóbal-Baní	--	--	--	--	--	10	10
Enriquillo	--	--	--	--	--	5	10
San Juan-Azua	--	.013	--	--	--	20	200
Cíbao	--	--	--	--	--	1	10
Cauto	--	--	--	--	--	--	10
Ana María	--	--	--	--	--	--	10
Central	700	3.1	--	1	1	50	50
Palacitos	--	--	--	--	--	--	5
Overthrust Belt	17,000	50.0	27.9	34	340	200	250
South Florida-North Cuba (excluding U.S.)	--	--	--	5	5	20	20

*All figures are for end-1986. Reliability of numbers varies. Sources: AAPG (1947-1987); Oil and Gas Journal; International Petroleum Encyclopedia (1975-1987); DeGolyer and MacNaughton's Twentieth Century Petroleum Statistics (1976-1987); U.S. Department of Energy, Energy Data Report; World Natural Gas (1977-1983); Venezuela, Ministerio de Energía y Minas, Memoria y Cuenta (1976-1985); González de Juana and others (1980); Semenovich and Namestrikov (1981); Sotsialista Industriya (1984), in Petroconsultants, Review of Soviet Oil, v. 19, n. 5, p. 88-93).

+Future potential reserves = Proved + Unproved + Speculative (11th World Petroleum Congress).

are not reaping significant rewards. A second, perhaps more important reason is that the exploration of the first 60 or 70 years of the century was conducted under very liberal laws that permitted huge financial rewards for an aggressive internationally oriented group; today, laws are complex and inhibiting and the potential profit to a private company has been greatly reduced. Moreover, private exploration is forbidden in the two most promising countries, Mexico and Venezuela.

On the basis of the data summarized in Tables 1, 2, and 3, we believe that future exploration in the Caribbean region should be concentrated, and will be most successful, in the unmetamorphosed sections of the stable platforms, and in the young, relatively undisturbed sequences of the recently created (by accretion) neocratonic areas. With these ends in mind, we summarize the basic geological facts of the Caribbean that we believe must be taken into account for successful exploration in this region.

Paleozoic

Exploration for Paleozoic objectives has been nil, even though Pennsylvanian and Permian marine sections are worthy objectives in several areas. These include southern Mexico (Thompson and Miller, 1944), Guatemala (Walper, 1960; Meyerhoff, 1967), Belize (Dixon, 1956), Colombia (Bürgl, 1973), and Venezuela (Compañía Shell-Creole, 1964; Bowen, 1972).

Meyerhoff et al. (1989) recognized two principal late Paleozoic mobile belts in the region. The first, the Northern Caribbean mobile belt, extends at least from western and south-central Cuba (Hatten et al., 1988; Tikhomirov et al., 1988) to the southern slope of the Puerto Rico Trench (Morris et al., 1989a) via the Samaná Peninsula of Hispaniola (Joyce and Aronson, 1983). Permian and older radiometric dates have been obtained from mafic to ultramafic eugeosynclinal assemblages in Cuba and the Dominican Republic (Joyce and Aronson, 1983; Hatten et al., 1988; Tikhomirov et al., 1988). (Joyce and Aronson consider their own 250-Ma date from the Samaná Peninsula to be erroneous, but it comes from rocks which they themselves correlated with the Escambray mafic-ultramafic rocks of south-central Cuba where Hatten et al. (1988) recorded 255-Ma dates.) The Paleozoic rocks are adjacent to or directly overlie 952-903 Ma rocks (Somin and Millan, 1977; Renne et al., 1989). The westward extent of the Northern Caribbean mobile belt is unknown, but Meyerhoff et al. (1988) suggested that it splits into two branches, one along the eastern margin of the Yucatan platform and the other along the western edge of the Florida-Bahamas platform. Only the former can be corroborated at present.

The second, or Southern Caribbean, mobile belt strikes eastward from Honduras (Horne et al., 1976) via the Nicaragua Rise to Jamaica (Green Bay Metamorphics, Lazaretto Inlier; Morris et al., 1989a). Morris et al. (1989a) suggested that the Nicaragua Rise originally was attached to northwestern South America and that, beginning in Middle Jurassic time, it was rotated counterclockwise to its present position. If one accepts this interpretation, then it follows that the eastern part of the Southern Caribbean mobile belt underlies the Caribbean Range of coastal Venezuela, from which Paleozoic mafic, ultramafic, and granitic rocks have been dated radiometrically together with marine strata that have been dated paleontologically (MacDonald, 1972; Urbani, 1983; Feo-Codecido et al., 1984; Benjamini et al., 1987). Irving (1975) described a coeval Paleozoic mobile belt paralleling Colombia's Pacific margin. Meyerhoff et al. (1989) suggested that this is a branch of the Southern Caribbean mobile belt.

In both northern Central America (Chiapas, Guatemala, Belize) and northwestern South America (Sierra de Perijá, Mérida Andes), thick marine Pennsylvanian and Permian platform sequences border the coeval mobile belts, presenting a full range of source rocks, reservoirs, and seals. In southern Venezuela, the largely undeformed Espino graben runs west-east from the Colombian Llanos to Trinidad, paralleling the Southern Caribbean mobile belt. In southeastern Mexico and Guatemala, similar platform-type late Paleozoic sections underlie the Chapayal basin (Figure 1). One of us (Meyerhoff) has studied in detail probable Permian shelf carbonates in the bottom 244 m of the Manabique-1C well (total depth, 4,230 m) in the Gulf of Amatique, Guatemala, an eastern extension of the Chapayal basin. Thin sections revealed extensive oil staining and some vuggy porosity associated with tests of fusulinids. This may be the only drilled Paleozoic in this part of the Caribbean region.

Recently, similar Pennsylvanian and Permian facies were found in unmetamorphosed pebbles in the Middle to Late Jurassic San Cayetano Group of western Cuba (A. Pszczolkowski, Polish Academy of Sciences, written communication, July 17, 1988). Pszczolkowski identified *Parafusulina* sp., *Schwagerina* sp., *Tetrataxis* sp., *Rhabdomeson* sp., and other taxa; one pebble had Permian (probably Leonardian) faunas. The fauna, like those of northwestern South America and northern Central America, belongs to the same warm-water faunal realm as coeval assemblages in western Texas, a fact that has important implications for plate-tectonic reconstructions of the Florida-Bahamas platform. The important point for petroleum exploration is that unmetamorphosed late Paleozoic strata, nearly identical with equivalent and highly prospective strata in northwestern South America and Central America, are now known to have been deposited also in, or near, western Cuba, offering the possibility that these strata may have been preserved somewhere in the subsurface. Thus late Paleozoic marine strata are promising objectives in several large areas of the Caribbean region.

Tectogenesis (orogeny) at the end of the Paleozoic deformed all parts of the two mobile belts, extending into the adjacent platforms. However, not enough information is available to enable us to judge the severity and full extent of this tectogenesis.

Table 3. Very heavy and extra-heavy crude oils and bitumens, Caribbean region: Production, reserves and resources, end-1986 estimates*.

Country Basin or Area	Name or Nearest Community	API Gravity	Production (1,000 bbl) Cumul.	Production (1,000 bbl) Yearly	Reserves and Resources (1,000 bbl) Proved	Reserves and Resources (1,000 bbl) Future Potential	Remarks
Guatemala							1. Xan field (disc. 1951) will be developed when the price is right.
Chapayal	Xan, NW Petén	13.4°	None	None	10,000	10,000	
Honduras							1. Used locally for caulking
Ulua	Comayagua	n.a.	None	None	--	--	2. Noncommercial.
Panamá							1. Several local "tar pits" exploited
Sambú-Darién	--	n.a.	None	None	--	--	2. Small deposit
Colombia							1. Disc. 1963. Being expanded
Middle Magdalena	Cocorna	12.0°-13.5°	29,000	680	Included in next column	100,000	2. To use steam-scale in future 3. More than 10% mineral matter 4. 300 million bbl in-place; could be 4.4 billion
	Nare	13.0°	239	102	Included in next column	13,000	1. Being developed; Will use steam soak 2. 148 million bbl in-place
Venezuela							
Officina Zone	Adás	9-16°	8,400	--	20,000	75,000	--
	Lejos	8-14°	14,500	--	33,500	65,000	--
	Merey	10-18°	244,800	--	14,000	25,000	--
	Miga	8-18°	10,600	--	30,000	50,000	1. Disc. 1957
	Melones	8-15°	81,300	--	121,500	200,000	1. Disc. 1955
	Ostra	11-21°	64,000	--	37,000	50,000	--
	Oveja	10-21°	216,000	--	62,000	75,000	--
	Oleos	10-18°	30,000	--	10,000	20,000	1. Disc. 1955
	Yopales Sur	10-21°	32,000	--	35,000	125,000	--
	SUBTOTAL	--	701,600	--	363,000	685,000	--
Templador Zone	Pilón	13°	127,100	--	226,000	--	1. Disc. 1937
	Jobo	9-15°	251,000	--	736,000	--	1. Disc. 1953
	Uracoa	16°	3,000	--	68,000	--	--
	Morichal	7-12°	168,500	--	490,000	--	1. Disc. 1958
	SUBTOTAL	--	549,600	--	1,520,000	--	--
Maracaibo	Boscán	10°	665,800	--	1,000,000	--	1. Disc. 1946
	Los Claros	10.4°	41,800	--	7,000	--	1. Disc. 1957
	Urdaneta	12-13°	11,800	--	1,710,000	--	--
	SUBTOTAL	--	719,400	--	2,717,000	--	--
Orinoco Oil Belt (post 1978 exploration)		5-17°	5,000+	1,000+	267,000,000	60,000,000-100,000,000	1. In-place oil estimated to be 700-1,200 billion bbl
Trinidad and Tobago							1. 60 million bbl in place
Eastern Venezuelan Basin	Asphalt Lake	1-2°	minor	minor	9,000	--	2. Exploitation began in 1888
	Guapo	13.5°	38,964	816	16,327	--	1. Disc. 1913
	Palo Seco-Los Rajos-Erin	12.0°	179,483	4,256	85,132	800,000	1. Disc. 1926
Eastern Caribbean							
	Barbados, Charnocks	12-18°	None	None	--	--	1. Described in early literature as "Barbados tar"
Greater Antilles							
	Enriquillo basin	1°	None	None	--	--	1. Noncommercial; see Maurrasse and others (1983)
	Overthrust Belt (Cuba)	9.5°	80	0.1	20	--	1. Guanabo field, disc. 1957 2. See text

*Sources: Same as Table 2, plus: Brodermann and others (1945), Marrero (1964), Ayaleto and Louder (1974), Maurasse and others (1983), Khayan (1984), Meyer and Schenk (1988), Zamora and Zambrano (1984), Fiorillo (1987), and Rambarran (1987).
+As of end-1986.

Table 4. Summary of Sedimentary Basins.*

	REGION Basin	Area (km 2) Onshore	Offshore	Total	Ave. Sed. Thickness (m)	Sediment Volume (km^3) Onshore	Offshore	Total	Reservoirs Sediment Column Present	Position Column	Lithologies	Source Rocks	Principal Trap Type	No. of Fields **
	CENTRAL AMERICA													
10,11	Chapayal (Chiapas)	20,000	--	20,000	4,000	80,000	--	80,000	U Penn-Eo	LK	dolo	LK ls	Faulted anticline	5
8, 12	Gulf of Amatique - Belize Borderland	1,000	34,000	35,000	2,000(n) 4,000(s)	4,000	76,000	80,000	U Penn-Mio		--	--	--	--
7,9	Lago Izabal, Tela, Ulúa	16,000	5,000	21,000	2,000	60,000	10,000	70,000	U Jur-Tert	--	--	--	--	--
+	Mosquitía	6,000	124,000	130,000	3,500	15,000	415,000	430,000	K-rec	--	--	--	--	--
+	Pacific Shelf	4,000	76,000	80,000	4,000	15,000	305,000	320,000	UK-Plio	--	--	--	--	--
5	Limón-Bocas del Toro	6,000	6,000	12,000	3,800	22,000	23,000	45,000	UK-rec	Oligo,	Frac lava, Frac ls, tuffac ss	Tert	Faulted anticline	2-non-c
4	Gulf of Panamá	--	8,000	8,000	2,500	--	20,000	20,000	Eo-rec	--	--	--	--	--
3	Sambú	1,000	--	1,000	3,000	20,000	--	20,000	IK-Plio	--	--	--	--	--
2	Darién	6,000	--	6,000	3,300	20,000	--	20,000	UK-Plio	--	--	--	--	--
	NORTHERN SOUTH AMERICA													
N.S.	Upper Magdalena	16,000	--	16,000	6,000	96,000	--	96,000	Tr-Pleis	M+UK, Mio	K ss, ls; Mio ss	M + UK	Faulted anticline, pinchouts	7
42	Middle Magdalena	30,000	--	30,000	8,000	240,000	--	240,000	Tr-Pleis	UK, Paleo, Eo,	Ss, UK	M + UK	Faulted anticline, pinchouts, truncation	26
+	Lower Magdalena - Caribbean margin	47,000	40,000	87,000	6,000(on) 9,000(off)	280,000	360,000	640,000	UK-Pleis	L	ss,	Paleo, Oligo	Strat: reefs, facies changes, pinchouts	11
37	Guajira	10,000	21,000	31,000	3,500	35,000	75,000	110,000	UK-Plio	Eo?, Oligo, L	ss,	UK, Eo	Strat: reefs, ss, pinchouts	3
40	Eastern Llanos - Uribante	195,000	--	195,000	7,500	1,450,000	--	1,450,000	Pz-Pleis	UK, Paleo, Eo	ss	UK?	Faulted anticline	22
1	Pacific Coastal	63,000	7,000	70,000	6,000	280,000	40,000	320,000	MK-Pleis	--	--	--	--	--
38	Cesar	9,000	--	9,000	4,000	36,000	--	36,000	Tr-Plio	--	--	--	--	--
41	Sabana de Bogotá	9,000	--	9,000	7,500	65,000	--	65,000	Cambr.-Pleis	--	--	--	--	--

Table 4. Continued.

	REGION Basin	Area (km 2) Onshore	Offshore	Total	Ave. Sed. Thickness (m)	Sediment Volume (km³) Onshore	Offshore	Total	Reservoirs Sediment Column Present	Position Column	Lithologies	Source Rocks	Principal Trap Type	No. of Fields **
+, 39	Maracaibo - Catatumbo	33,000	14,000	47,000	5,500	180,000	77,000	257,000	U Jur-rec	Bc, MK, UK, Paleo, Eo,	ign, ls, ss	L-MK(?)	Faults, anticline	11-Col 40-Ven
+	Falcón	35,000	--	35,000	8,250	275,000	--	275,000	UK-rec	Mio	ss	L+M Mio	Pinchouts, onlap, fault closure, anticline	10
36	Golfo de Venezuela	--	32,650	32,650	5,000	--	160,000	160,000	U Jur-rec	--	--	--	--	--
35	Golfo de Vela	--	3,900	3,900	4,200	--	16,400	16,400	Mio-rec	Bc,	ign, ls, ss	Mio	Fault closure	4
34	Golfo Tristw	--	3,000	3,000	4,100	--	12,000	12,000	Eo-rec	--	--	--	--	--
+	Barinas-Apure	90,000	--	90,000	2,750	245,000	--	245,000	Cambrian - recent	M+UK, Eo	ss	M-UK	Fault closure	10
+	Eastern Venezuela													
32	Guárico	23,000	--	23,000	2,750	63,000	--	63,000	Cambrian - recent	M+UK(?), Oligo, Mio	ss	UK?, Oligo, Mio	Fault closuire	16
31	Maturín	79,700	--	79,700	5,100	400,000	--	400,000	LK-rec	M+UK, Eo, Oligo, Mio, Plio	ss	UK?, Oligo, Mio	Fault closure	132
+	Orinoco Heavy Oil Belt	54,000	--	54,000	850	43,350	--	43,350	Cambrian - rec	M+UK, Eo, Oligo, Mio	ss	UK?, Oligo, Mio	Fault closure, onlap	33
30	Trinidad - Paria	4,000	2,000	6,000	2,100	8,400	4,200	12,600	U Jur-rec	UK, Eo, Mio,	ss	UK?, Oligo, Mio	Fault closure, anticline	45
29	Columbus	--	35,700	35,700	6,500	--	230,000	230,000	UK-rec	Mio, Pleis	ss	Mio, Plio, Pleis	Fault closure, shale swells, shale diapirs, anticline	12
33	Cariaco	--	23,000	23,000	6,250	--	145,000	145,000	K-rec	Mio, Pleis	ss	Mio?	Fault closures, onlap	4
27,28	Margarita - Tobago	--	66,700	66,700	3,600	--	240,000	240,000	K-rec	Mio	ss	Mio?	Fault closures, onlap	4-Ven 2-Trin
	EASTERN CARIBBEAN													
26	Grenada (S. OF 16°)	--	36,000	36,000	3,000	--	108,000	108,000	K-rec	--	--	--	--	--

Table 4. Continued.

	REGION Basin	Area (km 2) Onshore	Offshore	Total	Ave. Sed. Thickness (m)	Sediment Volume (km^3) Onshore	Offshore	Total	Reservoirs Sediment Column Present	Position Column	Lithologies	Source Rocks	Principal Trap Type	No. of Fields **
25	Saba Bank-Northern Grenada	--	18,000	18,000	3,500	--	63,000	63,000	UK-rec					
27	Tobago	--	63,000	63,000	4,000	--	250,000	250,000	K-rec	--	--	--	--	--
+	Barbados Ridge	400	18,800	19,200	3,000	1,200	56,400	57,600	Eo-Pleis	L Eo	cgl, ss	unk	sh seal of flysch, chaos breccia	1
	GREATER ANTILLES													
22	South Tertiary	750	1,750	2,500	1,750	1,000	3,400	4,400	Oligo-Plio					--
24	North Coast Tertiary	1,500	15,000	16,500	3,000	2,500	47,500	50,000	Eo-Plio	--	--	--	--	--
23	Mona	--	3,500	3,200	3,200	--	11,200	11,200	Eo?=Pleis?	--	--	--	--	--
21	San Cristóbal-Baní	14,000	--	14,000	4,000	56,000	--	56,000	MK-Plio	--	--	--	--	--
18	Enriquillo	6,400	1,000	7,400	6,000	37,400	7,000	44,400	MK-Pleis	--	--	--	--	--
20	San Juan-Azua	5,000	--	5,000	6,000	30,000	--	30,000	MK-Plio	Mio	ss	unk	Faulted anticline	2-non-c
19	Cibao	8,000	--	8,000	7,000	56,000	--	56,000	MK-Plio	--	--	--	--	--
17	Cauto	10,700	5,000	15,700	4,000	42,800	20,000	62,800	UK-Plio	--	--	--	--	--
16	Ana María	--	4,000	4,000	3,200	--	12,800	12,800	UK-rec	--	--	--	--	--
15	Central	1,500	--	1,500	3,000	--	4,500	4,500	LK-Mio	MK, UK	frac tuff, reef, ls	UK	drape anticline, reef strat traps	4
14	Palacios	1,500	1,500	3,000	1,800	2,700	2,700	5,400	L Jur-Mio	--	--	--	--	--
13	Overthrust Bel (Cuba)	35,000	15,000	50,000	3,300	115,000	50,000	165,000	Tr-Plio	U Jur, LK, MK, UK	frac ls, tuff, ss, serp	U Jur	faulted anticline	14
+	South Florida-North Cuba	11,400	490,000	501,400	10,000	100,000	4,900,000	5,000,000	Tr-rec	MK, LK	ls,	MK, U Jur	strat (reef)	10-Fla 1-Cub

*Data are through 1986.
**The definition of "field" differs greatly (e.g., three horizontally separated accumulations on a single structure would be considered to be one field by some, and three by others). Sources include AAPG Bulletin (foreign development issues), Oil and Gas Journal, and International Petroleum Encyclopedia for various years.

16 Numbers for basins on Figure 1
\+ Basins named on Figure 1
N.S. Not shown; outside of map area

Triassic-Jurassic

A mixed red-bed and volcanic sequence deposited in nonmarine intermontane grabens characterizes the Triassic through Early or Middle Jurassic section of large parts of the Caribbean region. Marine strata for this time interval are sparse except in parts of Colombia, northern Venezuela, and in Pinar del Río Province, western Cuba.

The red-bed unit is known by a variety of names; Todos Santos, Eagle Mills, Newark, Girón, and La Quinta are familiar to most persons active in the Caribbean. Equivalent red beds and volcanic rocks have been identified in recent years in Jamaica (Morris et al., 1989a), the western Bahamas (Tator and Hatfield, 1975), and several parts of Cuba (M. Iturralde-Vinent, written communication, May 16, 1988; Iturralde-Vinent, 1989). Although this section is not considered to be a prime petroleum-exploration target, it cannot be dismissed altogether. Similar continental sequences produce petroleum in many parts of the world (e.g., China, Argentina, Bolivia, Sudan) and, in fact, a gas show was recorded in the red beds at the base of the Great Isaac-1 well (5,355-5,440 m), western Bahamas (B. A. Tator, oral communication, May 23, 1976).

Late Jurassic-Middle Eocene

Strata of this time interval are widespread in the Caribbean region. Except for a few oil fields that produce from fractured basement, all production in the Caribbean region is from Late Jurassic or younger beds. Important offshore production in the Cantarell complex of southern Mexico's Campeche Shelf comes from Middle Jurassic (Callovian) and younger carbonates (Viniegra, 1981), but in the Caribbean region as defined here, the oldest producing beds are fractured deep-water Kimmeridgian to Tithonian limestones in northwestern Cuba.

Facies favorable for petroleum generation and accumulation, Callovian to Kimmeridgian and similar lithologically to coeval strata that are productive in southern Mexico, are present in western Cuba but untested.

As during the late Paleozoic, two mobile-belt systems dominated the Caribbean. The Northern Caribbean mobile belt extended from the Pacific coast of Mexico (Carafantan, 1981) through Guatemala (Wilson, 1974) via the eastern Yucatan margin (Pyle et al., 1973), the Greater Antilles (including Cuba; Khudoley and Meyerhoff, 1971; Knipper and Cabrera, 1974; Pardo, 1975), and the Aves Ridge (Meyerhoff et al., 1989). The Northern and Southern Caribbean mobile belts joined south of the Aves Ridge. A branch of the northern mobile belt underlay the present northern margin of the newly emplaced Nicaragua Rise, the site of the modern Cayman Trough, and the Cayman Ridge, rejoining the main part of the mobile belt in southeastern Cuba or Hispaniola.

The Southern Caribbean mobile belt paralleled the present Pacific coast of Colombia, swung to an east-west direction north of the Santa Marta and Lake Maracaibo massifs, and continued eastward through Trinidad-Tobago and the Barbados Ridge. Late in Cretaceous time, the modern arc-trench system of Middle America-Colombia began to extend southeastward from Guatemala to modern Colombia, where it joined the Southern Caribbean mobile belt.

The platform areas adjacent to the Northern Caribbean mobile belt were the sites mainly of carbonate-platform deposition (e.g., Florida-Bahamas platform, Yucatan platform, Nicaragua Rise). In contrast, adjacent to the Southern Caribbean mobile belt, terrigenous-clastic deposition predominated except on the Maracaibo massif where—and adjacent to which—important carbonate deposition took place, becoming predominant northward toward the present Guajira Peninsula (González de Juana et al., 1980). Most of the Caribbean's Cretaceous oil was generated adjacent to the Southern Caribbean mobile belt, both in Colombia and Venezuela (Figure 1, Tables 1-4).

The only other Cretaceous production within the region is (1) from the Chapayal basin of Guatemala (Morris et al., 1989b), (2) the Cuba overthrust belt (Semenovich and Namestnikov, 1981), and (3) the Sunniland trend in Florida. In Guatemala and Florida, production is from Lower Cretaceous shallow-water carbonates interbedded with evaporites. In Cuba, production in the north coast fields between Pinar del Río and Matanzas Provinces is from the same deep-water facies that produces petroleum in the underlying Late Jurassic strata. Elsewhere, as in the Central basin of Cuba (Figure 1, Table 4), production is from rudist reefs, deep-water carbonate lentils, and deep-water to neritic sandstones of Campanian through Maastrichtian ages. Some production is from Mesozoic tuff and from fractured mafic volcanics and serpentinites of the mobile belt where these have been thrust over structures containing petroleum in fractured deep-water carbonates. None of the fields in these three areas will produce more than 35 million bbl. Most have recoverable reserves of 10 million bbl or less.

Tectogenesis of the Caribbean Mesozoic mobile belts commenced perhaps as early as Albian or Cenomanian time, and extensive evidence of intra-Cretaceous unconformities has been described (e.g., Khudoley and Meyerhoff, 1971; Wilson, 1974; Weyl, 1980). However, the main tectogenesis to affect the region in Mesozoic time commenced during the Campanian, and continued through Maastrichtian and Paleocene times, culminating generally by middle or late Eocene time (Khudoley and Meyerhoff, 1971; Wilson, 1974; Irving, 1975); González de Juana, 1980; Weyl, 1980).

Late Eocene-Holocene

Tectogenesis has continued intermittently until the present day. The Northern Caribbean orogenic belt underlay the Cayman Trough, as it still does. The Cuban segment became tectonically quiescent. In Hispaniola, the orogenic belt east of the longitude of the Beata fault zone (ca. 71°W) was considerably less active after middle Eocene time. The Aves Ridge portion was abandoned in favor of a new position farther east, and thus the Lesser Antilles magmatic arc was born. Like its Aves Ridge predecessor, the Lesser Antilles mobile belt rejoins the Southern Caribbean mobile belt just off South America.

The Southern Caribbean mobile belt has remained approximately in its present position since middle Eocene, shifting laterally from time to time to occupy adjacent but closely parallel positions. It was severely deformed by middle Miocene tectogenesis, unlike the Northern Caribbean belt which was deformed during middle Miocene only in central and western Hispaniola. Possible reasons for these changes in tectonic patterns have been considered by numerous Caribbean specialists (e.g., Butterlin, 1977; Pindell and Dewey, 1982; Meyerhoff et al., 1989).

Basin Summary

Case and Holcombe (1980) noted the extreme tectonic segmentation that took place during the Campanian-middle Eocene and younger tectogeneses; the many basins and trends that developed are reflected on Figure 1, and summarized on Table 4. Of the 47 basins and trends listed, 16 formed during Late Jurassic to Early Cretaceous, or were already in existence; seven, perhaps more, began to form during Campanian to Paleocene; 25 are Eocene or younger. (Table 4 gives the age range of possible objectives in the stratigraphic section, but does not give the time when the basin or trend first formed.) Of the oil and gas produced and proved (Tables 1-3), 99.99 percent of each is in the 16 "older" basins—those that formed in (or already were in existence by) Late Jurassic to Early Cretaceous, and which have remained tectonically stable. Of the total reserves proved in the "older" basins, we note that more than 50 percent are in rocks of Tertiary age, and almost all of these are in Colombia, Venezuela, and Trinidad. These statistics demonstrate clearly the importance of tectonic stability.

Yucatan and Florida-Bahamas Platforms

However, one should note also those sedimentary basins, provinces, and trends that are *not* listed on Table 4. Foremost among these are the huge Yucatan and Florida-Bahamas platforms. These two areas, unified by a common history, both sedimentary and tectonic, are the most stable continental blocks within the Caribbean region. In fact, the two platforms may be considered as having been *too* stable. One needs only to contrast the La Reforma-Cantarell productive trend along the southwestern margin of the Yucatan platform with the rest of the platform. Here, in the southwest off the coast of Campeche, a series of north-south trending fault blocks was active during deposition of the Late Jurassic-Early Cretaceous reservoirs and source beds, a fact which also implies that an elevated geothermal gradient characterized the La Reforma-Cantarell trend (Viniegra, 1981) during Late Jurassic-Cretaceous time. In contrast, the lack of similar trends elsewhere on the Yucatan platform, as well as on the Florida-Bahamas platform, implies that geothermal gradients were low outside of the La Reforma-Cantarell trend. Subsurface data from the wells drilled confirm the existence, through Cretaceous and Cenozoic, of low geothermal gradients in large areas of both platforms. The same data confirm the absence of structure (e.g., fault-block movements). (A notable exception may be the eastern margin of the Yucatan platform, where Dillon and Vedder (1973) have documented the presence of a north-south trend of parallel faults that may have been active as early as Late Jurassic.) However, drilling of the platforms outside of the Florida mainland has not been sufficient to ascertain whether other La Reforma-Cantarell trends might, or might not, exist. Presumably, seismic work has been done to provide indications of underlying structure in several areas, but in the final analysis only direct evidence from the subsurface can tell the story.

Despite this negative evaluation, Sunniland-type (Albian of southwestern Florida) fields will continue to be discovered on the Florida-Bahamas platform and, possibly, on the Yucatan platform.

Maracaibo Platform and Eastern Venezuela Basin

Mild tectonism has long affected the stable platform areas of Venezuela. The Maracaibo platform, on a salient of the Guayana shield (González de Juana, 1980), has been subjected to strong vertical motions on the order of thousands of meters, as well as to an elevated geothermal gradient. Similarly, the Eastern Venezuela basin has undergone major vertical movements. In fact, those movements have taken place intermittently since Pennsylvanian time, and perhaps earlier (Feo-Codecido et al., 1984). Again by implication, times of vertical movements were also times of higher geothermal gradient. The fact that the Cretaceous-Cenozoic Eastern Venezuela basin is underlain by an even older depression might also have provided the conditions for the entrapment and preservation of the large volumes of organic matter needed to generate the immense in-place volume of 700 to 1,200 billion bbl of heavy oil (Meyer and Schenk, 1988), of which 267 billion bbl are recoverable (Table 3).

CONCLUSIONS

1. Petroleum in very large volumes is most likely to be preserved in tectonically stable blocks.

2. However, such blocks must be the sites of some tectonic activity, accompanied by an elevated geothermal

gradient, if accumulated organic matter is to be converted to petroleum, migrated, and trapped. A stable block that has not undergone a history of mild tectonic movements and elevated heat flow is no better, or worse, than a highly active mobile belt in terms of the volume of ultimately recoverable petroleum.

3. Mobile belts that are stabilized by accretion to stable blocks also can accumulate large volumes of reserves in the post-accretionary stratigraphic sequence.

4. Future exploration in the Caribbean region, to be successful, must be concentrated in stable and newly stabilized geological provinces. Some of these provinces, even though considered to be "old" or "mature" in terms of exploration potential, still are capable of revealing "pleasant surprises," as recent activity in eastern Venezuela is proving.

5. Deliberate exploration of late Paleozoic shallow marine facies should result in numerous discoveries, perhaps even some giant fields.

REFERENCES CITED

American Association of Petroleum Geologists Bulletin, various issues, 1947-1987.

Ayaleto-E., M., and L. W. Louder, 1974, The geology and exploration potential of the heavy oil sands of Venezuela (the Orinoco Heavy Oil Belt), *in* L. V. Hills, ed., Oil sands of the future: Canadian Society of Petroleum Geologists Memoir 3, p. 1-18.

Benjamini, C., R. Shagam, and A. Menéndez-V., 1987, (Late?) Paleozoic age for the "Cretaceous" Tucutunemo Formation, northern Venezuela: Stratigraphic and tectonic implications: Geology, v. 15, no. 10, p. 922-926.

Bowen, J. M., 1972, Estratigrafía del precretáceo en la parte norte de la Sierra de Perijá, *in* Cuarto Congreso Geológico Venezolano, Memoria, t. II: Caracas, Ministerio de Minas e Hidrocarburos, Dirección de Geología, Boletín Geológico, Publicación Especial no. 5, p. 729-761.

Brodermann, J., F. Villoch, and A. Andreu, 1945, Yacimientos asfaltíferos de Cuba: La Habana, Dirección de Montes, Minas y Aguas, Boletín de Minas no. 19, 191p.

Bürgl, H., 1973, Precambrian to middle Cretaceous stratigraphy of Colombia: privately published by C. G. Allen and N. R. Rowlinson, Bogotá, 214p.

Butterlin, J., 1977, Géologie structurale de la région des Caraïbes (Mexique-Amérique Centrale-Antilles-Cordillère Caraïbe): Paris, Masson, 259p.

Carfantan, J.-C., 1981, Evolución estructural del sureste de México: Paleogeografía e historia tectónica de las zonas internas Mesozóicas: Universidad Nacional Autónoma de México, Instituto de Geología, Revista, v. 5, no. 2, p. 207-216.

Case, J. E., and T. L. Holcombe, 1980, Geologic-tectonic map of the Caribbean region: U.S. Geological Survey, Miscellaneous Investigations Series, Map I-1100, scale, 1:2,500,000, 3 sheets.

Compañía Shell de Venezuela, and Creole Petroleum Corporation, 1964, Paleozoic rocks of Mérida Andes, Venezuela: American Association of Petroleum Geologists Bulletin, v. 48, no. 1, p. 70-84.

DeGolyer and McNaughton, 1976-1987, Twentieth century petroleum statistics.

Dillon, W. P., and J. G. Vedder, 1973, Structure and development of the continental margin of British Honduras: Geological Society of America Bulletin, v. 84, no.8, p. 2713-2732.

Dixon, C. G., 1956, Geology of southern British Honduras, with notes on adjacent areas: Belize, Government Printer, 85p.

Feo-Codecido, G., F. D. Smith, Jr., N. Aboud, and E. de Di Giacomo, 1984, Basement and Paleozoic rocks of the Venezuelan Llanos basins, *in* W. E. Bonini, R. B. Hargraves, and R. Shagam, eds., The Caribbean-South American plate boundary and regional tectonics: Geological Society of America Memoir 162, p. 175-187.

Fiorillo, G., 1987, Exploration and evaluation of the Orinoco Oil Belt, *in* R. F. Meyer, ed., Exploration for heavy crude oil and natural bitumen: American Associations of Petroleum Geologists Studies in Geology no. 25, p. 103-115.

González de Juana, C., J. M. Iturralde de Arozena, and X. Picard-Cadillat, 1980, Geología de Venezuela y de sus cuencas petrolíferas: Caracas, Ediciones Foninves, 2 v., 1031p.

Hatten, C. W., M. Somin, G. Millán, P. Renne, R. W. Kistler, and J. M. Mattinson, 1988, Tectonostratigraphic units of central Cuba, *in* L. Barker and J. Gordon, eds., Review, 11th Caribbean geological congress transactions, 20th to 26th July, 1986, Bridgetown, Barbados: Bridgetown, Ministry of Finance, Energy Division, Paper 36, p. 1-12.

Horne, G. S., G. S. Clark, and P. Pushkar, 1976, Pre-Cretaceous rocks of northwestern Honduras: Basement terrane in Sierra de Omoa: American Association of Petroleum Geologists Bulletin, v. 60, no.4, p. 566-583.

International Petroleum Encyclopedia, 1975-1987.

Irving, E. M., 1975, Structural evolution of the northernmost Andes, Colombia: U.S. Geological Survey Professional Paper 846, 47p.

Iturralde-Vinent, M., 1989, Características generales del magmatismo de margen continental de Cuba (abs.), *in* Geología '89, Primer Congreso Cubano de Geología, Programa y Resúmenes: La Habana, Sociedad Cubana de Geología, p. 207.

Joyce, J., and J. Aronson, 1983, K-Ar age dates for blueschist metamorphism on the Samaná Peninsula, Dominican Republic: Tenth Caribbean Geological Conference, Cartagena, 1983, preprint, 8p.

Khayan, M., 1984, Proposed classification and definitions of heavy crude oils and tar sands, *in* R. F. Meyer, J. C. Wynn, and J. C. Olson, eds., The future of heavy crude and tar sands: New York, McGraw-Hill, Inc., Coal Age Mining Information Sources, p. 7-11.

Khudoley, K. M., and A. A. Meyerhoff, 1971, Paleogeography and geological history of Greater Antilles: Geological Society of America Memoir 129, 199p.

Knipper, A. L., and R. Cabrera, 1974, Tectónica y geología histórica de la zona de articulación entre el mio- y eugeosinclinal y del cinturón hiperbásico de Cuba, *in* Contribución a la geología de Cuba: La Habana, Academia de Ciencias de Cuba, Instituto de Geología y Paleontología, Publicación Especial no. 2, p. 15-77.

MacDonald, W. D., 1972, Late Paleozoic tectonics of northern South America: Rio de Janeiro, Academia Brasileria das Ciências, Anais, v. 44 (Suplemento), p. 197-208.

Marrero, L., 1964, Venezuela y sus recursos: Caracas, Cultural Venezolana, S. A., 700p.

Maurrasse, F., J. Robert, J. Claude, and A. Loctamar, 1983, Geologic characteristics of the Massanga rock asphalt deposit, Southern Peninsula of Haiti (abs.), *in* X Caribbean Geological

Conference, Cartagena, 1983, Program and abstracts of papers: Bogotá, INGEOMINAS, p. 52.

Meyer, R. F., and C. J. Schenk, 1988, An estimate of world resources of heavy crude oil and natural bitumen, *in* R. F. Meyer, ed., UNITAR/UNDP International conference on heavy crude and tar sands, 4th, 1988, Edmonton: Edmonton, Alberta Oil Sands Technology and Research Authority, p. 73-83.

Meyerhoff, A. A., 1967, Future hydrocarbon provinces of Gulf of Mexico-Caribbean region: Gulf Coast Association of Geological Societies Transactions, v. 17, p. 217-260.

Meyerhoff, A. A., A. E. L. Morris, and I. Taner, 1989, Tectonic evolution of the Caribbean region (abs.), *in* New concepts in global tectonics, a symposium, Smithsonian Institution and Texas Tech University, July 20-21, 1989, Washington, D. C., Program with abstracts: Lubbock, Texas Tech University Press, in press.

Morris, A. E. L., I. Taner, H. A. Meyerhoff, and A. A. Meyerhoff, 1989a, Tectonic evolution of the Caribbean region: Alternative hypothesis, *in* The geology of North America, v. H., The Caribbean region: Geological Society of America, Decade of North American Geology, in press.

Morris, A. E. L., A. A. Meyerhoff, I. Taner, R. Bueno-Salazar, and G. A. Young, 1989b, Energy resources of the Caribbean region, *in* The geology of North America, v. H., The Caribbean region: Geological Society of America, Decade of North American Geology, in press.

Oil and Gas Journal, various issues.

Pardo, G., 1975, Geology of Cuba, *in* A. E. M. Nairn and F. G. Stehli, eds., The ocean basins and margins, v. 3., The Gulf of Mexico and the Caribbean: New York, Plenum Press, p. 553-615.

Pindell, J., and J. F. Dewey, 1982, Permo-Triassic reconstruction of Pangea and the evolution of the Gulf of Mexico/Caribbean region: Tectonics, v. 1, no. 2, p. 179-211.

Pyle, T. E., A. A. Meyerhoff, D. A. Fahlquist, J. W. Antoine, J. A. McCrevey, and P. C. Jones, 1973, Metamorphic rocks from northwestern Caribbean Sea: Earth and Planetary Science Letters, v. 18, p. 339-344.

Rambarran, H., 1987, Significance of gravity segregation in heavy-oil distribution, Forest 'A' reservoir, Parrylands field, Trinidad, West Indies, *in* R. F. Meyer, ed., Exploration for heavy crude oil and natural bitumen: American Association of Petroleum Geologists Bulletin Studies in Geology no. 25, p. 281-291.

Renne, P. R., J. M. Mattinson, C. W. Hatten, M. Somin, G. Millán, and E. Linares, 1989, Confirmation of late Proterozoic age for the Socorro Complex of north-central Cuba from 40Ar/39Ar and U-Pb dating (abs.), *in* Geología '89, Primer Congreso Cubano de Geología, Programa y Resúmenes: La Habana, Sociedad Cubana de Geología, p. 118.

Semenovich, V. V., and Yu. G. Namenistnikov, 1981, Neftegazonosnyye basseyny sotsialisticheskikh stran Evropy (NRB, VNR, GDR, PNR, SRR, SSSR, ChSSR i SGRYu) i respubliki Kuba: Moscow, Sovet Ekonomicheskoy Vzaimopomoshchi, 400p.

Somin, M. L., and G. Millán, 1977, Sobre la edad de rocas metamórficas cubanas: Academia de Ciencias de Cuba, Instituto de Geología y Paleontología, Informe Científico-Técnico, no. 2, 11p.

Sotsialista Industriva, 1984, *in* Petroconsultants, Review of Soviet oil, v. 19, no. 5, p. 88-93.

Tator, B. A., and L. E. Hatfield, 1975, Bahamas present complex geology: Oil and Gas Journal, v. 73, no. 43, p. 172-176; no. 44, p. 120-122.

Thompson, M. L., and A. K. Miller, 1944, The Permian of southernmost Mexico and its fusulinid fauna: Journal of Paleontology, v. 18, no. 6, p. 481-504.

Tikhomirov, I. N., E. de los Santos, A. L. Vtulochkin, A. Bruto, A. V. Dovbnya, E. Linares, B. A. Markovskiy, V. A. Trofomov, and G. Furrazola, 1988, Novyye dannyye po geologii Kuby: Sovetskaya Geologiya, no. 1, p. 58-66.

U.S. Department of Energy, 1977-1983, Energy data report: world natural gas.

Urbani, F., 1983, Las rocas graníticas del área de Las Trincheras-Mariara, Estado Carabobo: Geología y edad: Acta Científica Venezolana, v. 34 (Suplemento 1), p. 93.

Venezuela, Ministerio de Energía y Minas, 1976-1985, Memoria y cuenta.

Viniegra-O., F., 1981, Great carbonate bank of Yucatan: Journal of Petroleum Geology, v. 3, no. 3, p. 247-278.

Walper, J. L., 1960, Geology of Cobán-Purulhá area, Alta Verapaz, Guatemala: American Association of Petroleum Geologists Bulletin, v. 44, no. 8, p. 1273-1315.

Weyl, R., 1980, Geology of Central America: Berlin, Gebrüder Borntraeger, 371p.

Wilson, H. H., 1974, Cretaceous sedimentation and orogeny in nuclear Central America: American Association of Petroleum Geologists Bulletin, v. 58, no. 7, p. 1348-1396.

Zamora, L., and G. Zambrano, 1984, Occurrences of heavy and extra-heavy crude oil deposits in Venezuela, other than the Orinoco Oil Belt, *in* R. F. Meyer, J. C. Wynn, and J. C. Olson, eds., The future of heavy crude and tar sands: New York, McGraw-Hill, Inc., Coal Age Mining Information Services, p. 159-168.

Circum-Pacific Council for Energy and Mineral Resources
Earth Science Series, Vol. 16
R. L. Miller, G. Escalante, J. A. Reinemund, and M. J. Bergin (Eds.)
Energy and Mineral Potential of the Central American-Caribbean Regions
Springer-Verlag, Berlin Heidelberg 1995

SEDIMENTARY BASINS OF COSTA RICA: LATE MESOZOIC-CENOZOIC EVOLUTION AND HYDROCARBON POTENTIAL

Allan Astorga, Jose A. Fernandez, Guillermo Barboza, Lolita Campos, Jorge Obando, Alvaro Aguilar, and Luis Obando
RECOPE, Gerencia de Produccion Primaria, P. O. Box 4351-1000, San Jose, Costa Rica

ABSTRACT

The Upper Cretaceous/Cenozoic basins of Costa Rica are underlain by oceanic crust and located along the western margin of the Caribbean Plate. A primitive intraoceanic arc formed along the western margin of the Caribbean Plate since Albian time, with associated anoxic pelagic sedimentation. This arc marked the beginning of subduction of the Pacific Plate in this region and of the first elements of the land bridge joining North and South America. Continued subduction of the Pacific Plate from Campanian until the present created a convergent marginal tectonic setting with, from southwest to northeast, a trench, an outer arc, a forearc region, a volcanic arc, and a backarc region. Localized platforms of shallow water carbonates, overlain by extensive cyclical turbidite systems, characterized the sediments of most of the basins up to the Eocene, when a major tectonic compressive event affected the area. Also about this time, a major sinistral east-west fault system (Costa Rica Transcurrent Fault Zone) divided the region into north and south segments, and transtensional or pull apart basins developed along the fault system. During this pre-Miocene time, the north segment was emergent and subject to erosion, whereas in the south segment sedimentation continued. From Miocene on, shallow marine deposition was widespread in both segments. In the north segment, extension in the backarc region created a structural depression filled with Neogene sediments. The south segment experienced compressional tectonics due to north-directed movement of the Panama block. Later, in the Pleistocene, northeasterly-directed shallow subduction of the thick Cocos Ridge resulted in additional thrusting and the development of a foreland basin.

Organic-rich, deep-water, thermally mature source rocks occur in several basins with high TOC (total organic content) values and adequate maturation levels. Potential carbonate reservoirs in some of these basins include shallow water carbonates, rudistic reef buildups, and grainstone accumulations. Potential clastic reservoirs include turbidite sandstones, regressive deltaic deposits, and near-shore porous sandstones. Potential petroleum traps are diverse and numerous in all of the basins. Seismic and stratigraphic well data recently acquired and interpreted in a joint RECOPE-PCIAC (Petro-Canada International Assistance Corporation) program added significantly to the understanding of the Costa Rican basins.

INTRODUCTION

Published geological studies on Costa Rica referring to its evolution have focused principally on tectonic or petrologic aspects (Dengo, 1962a, 1962b, 1968, 1983; Lloyd, 1963; Henningsen, 1966; Pichler and Weyl, 1975; Stibane, Schmidt-Effing, and Madrigal, 1977; Galli-Olivier, 1979; Kuypers, 1980; Weyl, 1980; Baumgartner et al., 1984). The evolution of the sedimentary basins has been outlined by Seyfried, Astorga, and Calvo (1987). The present study on the tectonic-stratigraphic evolution of the sedimentary basins of Costa Rica endeavors to (1) outline the different basins, (2) establish the tectonic framework, (3) analyze the nature of the sedimentary fill, (4) define the structural styles, (5) analyze the paleogeographic evolution, and (6) evaluate the oil potential.

This study evolves from geological information published on Costa Rica and the Caribbean region and from the geological-geophysical data base obtained by the national oil company of Costa Rica (RECOPE S.A.) during the last nine years. The information includes surface and subsurface geology, seismic surveys in onshore and offshore areas, gravimetry and aeromagnetometry, and other methods such as geochemistry, petrology, sedimentology, and remote sensing.

TECTONIC FRAMEWORK

Costa Rica forms part of the intraoceanic insular mature arc of southern Central America, which developed since the Cretaceous on the southwestern margin of the Caribbean Plate (Figure 1a). The principal tectonic features recognized about this margin (Mann and Burke, 1984) include (a) the Middle America Trench, (b) the Panama Fracture Zone, (c) the Hess Escarpment, (d) the North Panamanian Deformed Belt, and (e) the shallow subduction of the Cocos Ridge beneath the south part of Costa Rica. These tectonic features are associated with two generations of sedimentary basins in Costa Rica (Figure 1b): the first generation corresponds to the classical intraoceanic island arc-basin system, i.e., the trench-slope, the forearc, and backarc basins, and second generation basins formed since Oligocene time from local tectonic phenomena of the transtensional type.

In a general sense, it has been established that related movements of the North American and South American Plates have displaced the Caribbean Plate eastward from the Pacific in Cretaceous time (Malfait and Dinkelman, 1972; Pindell and Dewey, 1982; Anderson and Schmidt, 1983; Pindell et al., 1988; Eva et al., 1989; among others). In addition to this displacement, the Caribbean Plate has been internally deformed by systems of regional "strike slip" faults (Burke, Fox, and Sengor, 1978; Wadge and Burke, 1983; Mann and Burke, 1984).

Four important geodynamic events, developed in the Caribbean Plate, have been relevant to the geological evolution of Costa Rica: (1) the Early Cretaceous subduction of oceanic crust beneath the west margin of the Caribbean Plate, (2) the thickening and buoyancy of the Caribbean Plate in Santonian-Campanian time (Burke, Fox, and Sengor, 1978), (3) the major convergence of the North and South American Plates starting at the end of the late Eocene-early Oligocene, which intensified the narrowing of the Caribbean Plate, and (4) the collision of the Panama Block with South America in the middle Miocene-early Pliocene (Mann and Burke, 1984).

STRATIGRAPHIC SEQUENCES

Based on regional angular unconformities and lithologic discontinuities, the sedimentary deposits of the Costa Rican basins are divided into four stratigraphic megasequences named A, B, C, and D (Figure 2).

Megasequence A is considered to be the economic basement of the region. Its age extends from Early Jurassic

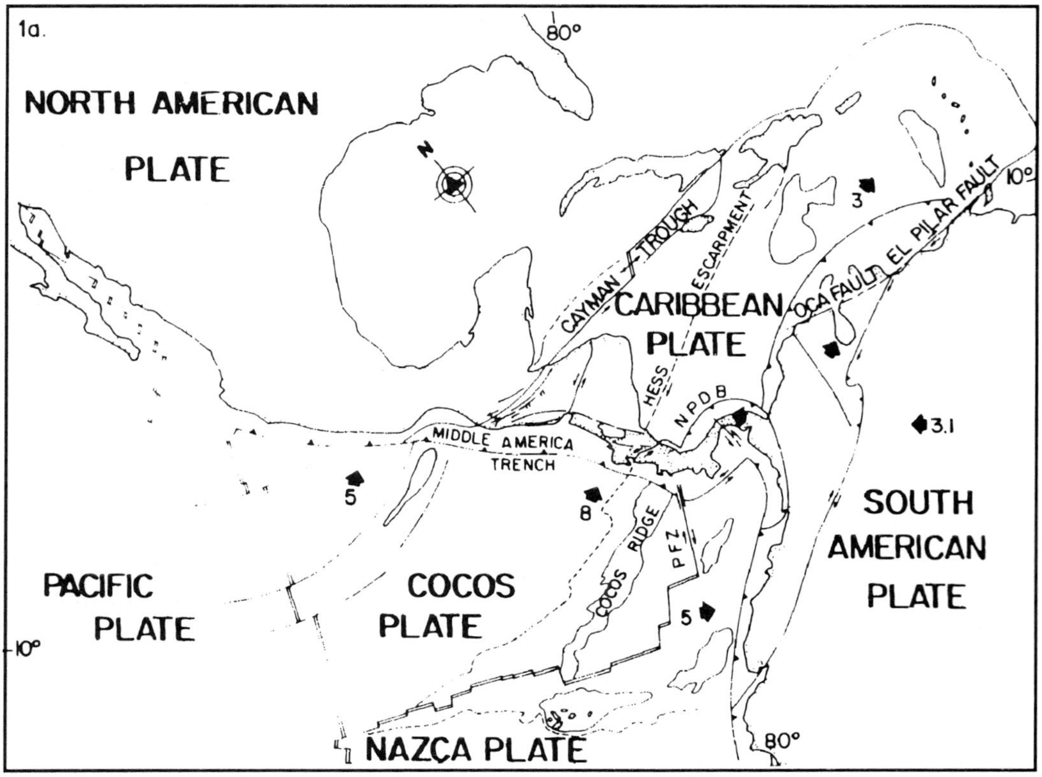

Figure 1. (a) Tectonic map of Central American region: PFZ (Panamanian Fracture Zone), NPDB (North Panamanian Deformed Belt).

(Dewewer et al., 1985) to Late Cretaceous and locally to the Paleogene (Baumgartner, 1987). Its lower part is ophiolitic in character, whereas the upper part consists of primitive island arc basalts, associated with pelagic siliceous Cretaceous sediments (Loma Chumico Formation, cf. Astorga, 1987, 1988), which include organic-rich oil shales.

Megasequence B, Late Cretaceous to Oligocene in age, has more than 3,000 m of sediments. They include deep-water turbidites, local occurrences of neritic limestone, volcaniclastic deposits, and rudistic biostromes that grade laterally to upper Campanian neritic limestones. Different levels of neritic sandstones, interbedded with volcaniclastics, and carbonate turbidites with deep marine fans culminate the sequence. Megasequence B is separated from megasequence A by an upper Santonian angular unconformity.

Megasequence C, late Oligocene-Miocene in age (late Tertiary), includes principally shallow marine siliciclastic sediments (lithic sandstones and shales) with deltaic affinities, neritic limestones, and volcaniclastic deposits in the northern part of the country, whereas in the southern region the deposits are lower Miocene clastic deep-water turbidites, covered by clastic sediments and in places by shallow marine carbonate deposits. Thickness of the sequence reaches a maximum of 4,000 m, and averages 2,000 m. Separating megasequence B and C is an angular unconformity.

The Plio-Quaternary megasequence D, the youngest deposits of the sedimentary basins, is formed mainly of continental sediments (tephritic, alluvial, and lacustrine) and locally by clastic marine sediments. It is discordant over the C sequence and its thickness ranges from a few hundred meters to more than 3,000 m.

TECTONIC DEVELOPMENT

Tectonic evolution of the island arc of Costa Rica and the deformation style of its basins have been controlled by (a) the geodynamics of the Caribbean Plate (compressive phases) and (b) the internal adjustment processes between the blocks which form the arc (local tectonic features).

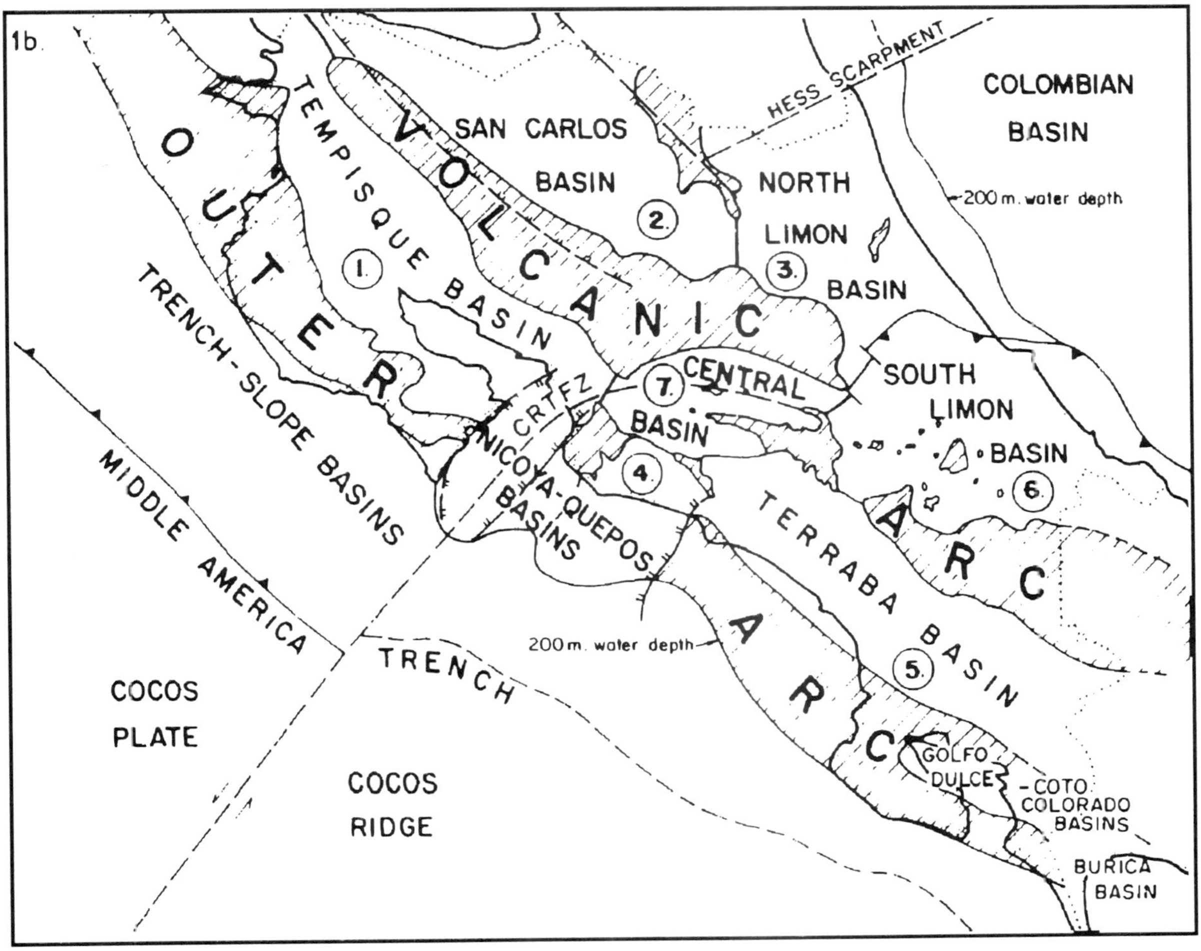

Figure 1. (b) Morphotectonic model of Costa Rica and delineation of the sedimentary basins: CRFZ (Costa Rica Transcurrent Fault Zone).

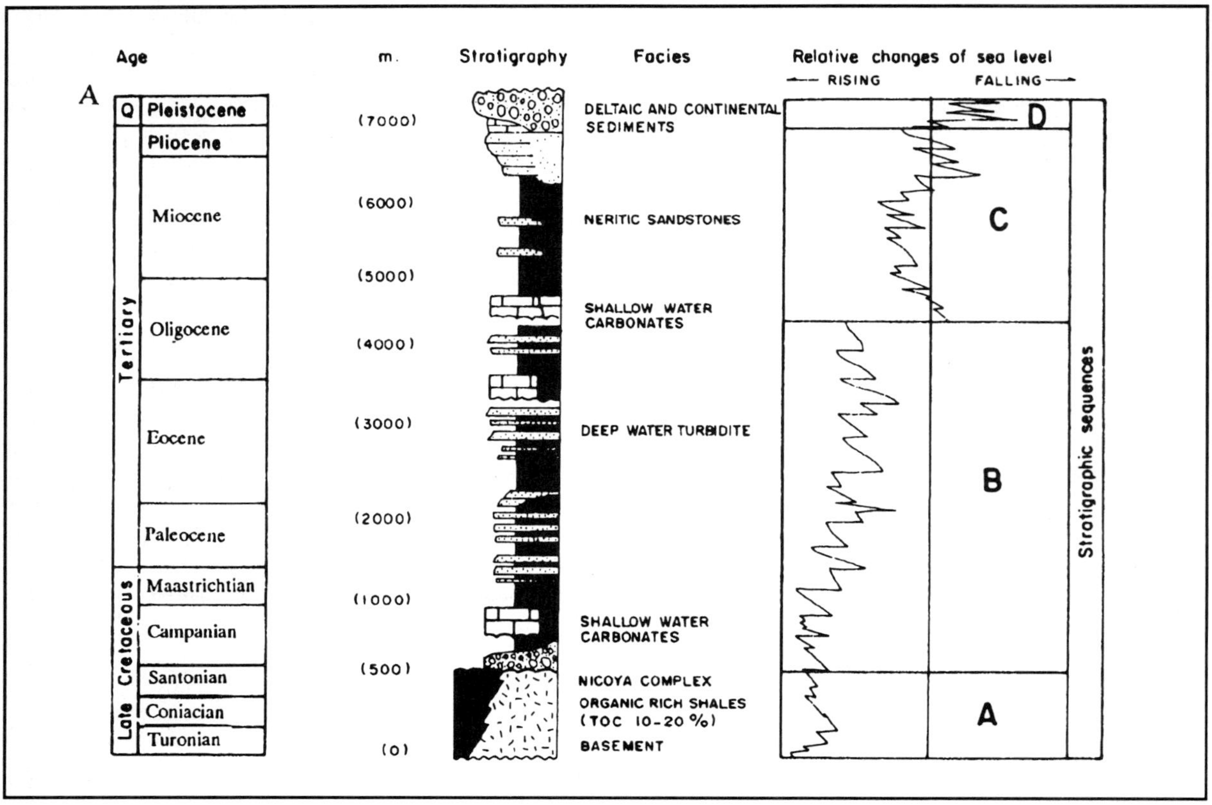

Figure 2. (a) Generalized Stratigraphic column of Costa Rica and megasequences based on lithologic discontinuities and angular unconformities, related to relative sea level changes.

Compressive Phases

The four deformative phases (D1, D2, D3, and D4) defined by Gursky (1988) for northwestern Costa Rica are extrapolated to the rest of the country, using geological and geophysical evidence.

Phase D1, with north northwest-south southeast directed stress, developed during the Santonian (Kuypers, 1980; Astorga, 1988). During this phase, folds of magnitudes up to several kilometers in length were developed, with northeast-southwest and east-west oriented axes. Their genesis is associated with the convergence of the North and South American Plates (Gursky, 1988).

Phase D2 (in the Campanian), with compressional stress direction north northeast-south southwest, produced cortical telescoping and shortening, with thrust sheets (Kuypers, 1980; Bourgois et al., 1984; Seyfried, Astorga, and Calvo, 1987; Meschede, Frisch, and Sick, 1988), which lifted sections of the basement (megasequence A), over which were deposited platform carbonates (Calvo, 1987). The cause of this compression is attributed to the strong coupling of the oceanic Farallon to the Caribbean Plate (Burke, Fox, and Sengor, 1978).

Phase D3, with north-south to north northwest-south southeast directed stress, generated compressive east-west trending structures (reverse faults and folds) during the middle-late Eocene. This compression affected a great part of the Caribbean Plate and occurred due to the convergence of the North and South American Plates.

During the late Miocene and early Pliocene, phase D4 developed, with compressional stress direction northeast-southwest. It resulted in northwest-southeast trending folds and the general emergence of the island arc. The intensity of this deformation is variable: thrust displacements attain several kilometers in southern Costa Rica, whereas in the north, displacements are only hundreds of meters. This tectonic phase was caused by the collision of the Panama Block with Colombia (Mann and Burke, 1984; Eva et al., 1989).

Local Tectonic Features

Apart from the development of the four compressive phases, three local tectonic events have affected the evolution of the island arc and are responsible for the second generation basins. The first is the sinistral "strike slip" fault system

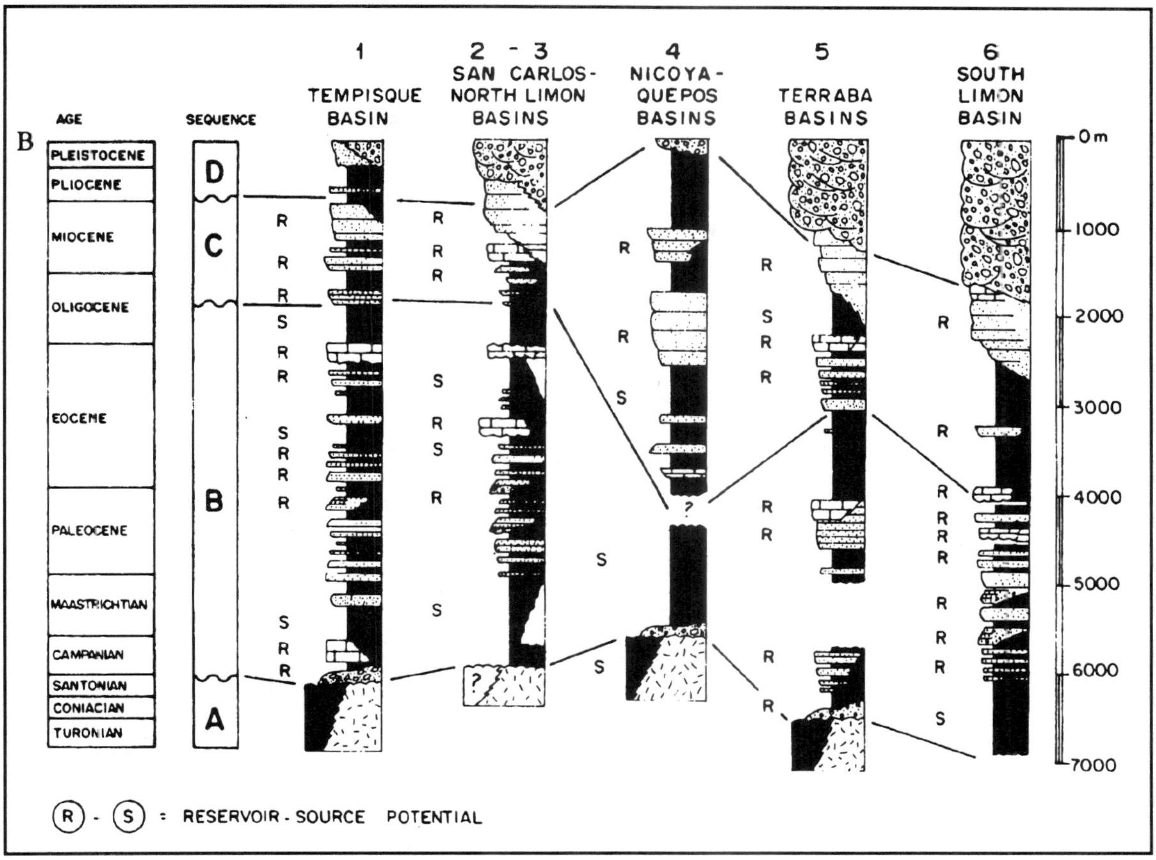

Figure 2.(b) stratigraphic columns of some basins indicated in Figure lb and their reservoir and source rock potential.

which, since the Oligocene, has divided Costa Rica into a north and south segment. The two segments are separated by a major east-west trending strike-slip fault system designated here has the Costa Rica Transcurrent Fault Zone (Figure 3). This major strike-slip fault zone is terminated on the north-east by a series of northeasterly-directed thrust faults (North Panamanian Deformed Belt, Figure 1). During this tectonic regime, which is dominated by the fault system, the transtensional Central Valley and Nicoya-Quepos basins were formed and also the Limón Sur compressional basin on the Caribbean side of the country. The second feature of the post-Miocene tectonism is the graben or rift development in the backarc region of the San Carlos basin (Figure 1). It developed as an extension of the Nicaragua and Honduras depressions (Mann and Burke, 1984). The third feature is the shallow subduction of the Cocos Ridge (Figure 1) in the south segment, which began one million years ago (Lonsdale and Klitgord, 1978). This process intensified the deformation of the basins and the uplift of the magmatic arc, and was accompanied by a cessation of volcanism in this region. In addition, the Panama Fracture Zone (Lonsdale and Klitgord, 1978) opened, which resulted in the Golfo Dulce transtensional basin (cf. Obando and Baumgartner, 1986; Barritt and Berrangé, 1987).

Structural Styles of the Basins

Regional and local deformation and the resultant structural styles are illustrated on Figures 3 and 4 and are discussed here. Four compressive phases can be recognized in the Tempisque "forearc" basin (Figure 1b). However, phase D3 is the one whose structural style is most easily recognized because of its open and concentric folds up to several kilometers in length. Phase D4 is identified by thrust faults and associated fold trends, northwest-southeast oriented, overturned toward the southwest.

In the Térraba basin (Figure 1b) phase D4 originates an overthrust belt and associated folds of northwest-southeast orientation and vergence (overturning). The folds tend to be closed and overturned to the southwest. Their lengths are measured in kilometers, and the faults show displacements of the same order of magnitude (Mora, 1979). The shallow subduction of the Cocos Ridge accentuated this structural style and influenced its development.

The structural style of the Limón Sur basin is compressive with two phases of stress (D3 and D4) being identified. Middle Eocene thrusts are succeeded by a northwest-southeast oriented thrust system. Due to these compressive stresses a typical foreland basin is developed with associated thrust belt, foredeep, and foreland components.

The Limón Norte basin has a distinctive structural style typical of a passive margin (Atlantic) type basin (Figures 4 and 5). Based on a study of the reflection seismology, two systems of faults can be recognized, namely a series of steeply dipping down-to-the-basin normal faults in the Oligocene-Miocene succession and a Miocene-Pliocene listric (curved downward-flattening) fault system associated with prograding beds with deltaic characteristics.

The Nicoya, Quepos, and Valle Central transtensional basins have structural styles typical of transtensional basins, such as horst and graben systems, tilted blocks, flower structures, and peripheral normal faulting (Figures 4 and 5; Seyfried, Astorga, and Calvo, 1987; Rivier and Calvo, 1988; Barboza and Zucchi, in press), as well as some other structures associated with fault lines, such as en echelon folds and synthetic and antithetic tear faults (Figure 4; Bergoing, Malavassi, and Jiminez, 1980; Denyer and Montero, 1989).

In the San Carlos extensional basin, both extensional and compressional structural styles are present: a system of northwest-southeast trending normal faults aligned with the Nicaragua depression, and a later system of east-west trending compressional reverse faults and folds. This later system is believed to be due to strike-slip faults that are steeply dipping and have strike-slip seismic characteristics such as faults splaying upward (flower structures).

Finally, in the Burica, Golfo Dulce, and Coto-Colorado basins in southern Costa Rica (Figure 1), two superimposed structural styles are also present. The first one, of a compressive type, is associated with phase D4. It presents northwest-southeast oriented overthrusts and folds overturned toward the southwest. The second style is extensional normal faults

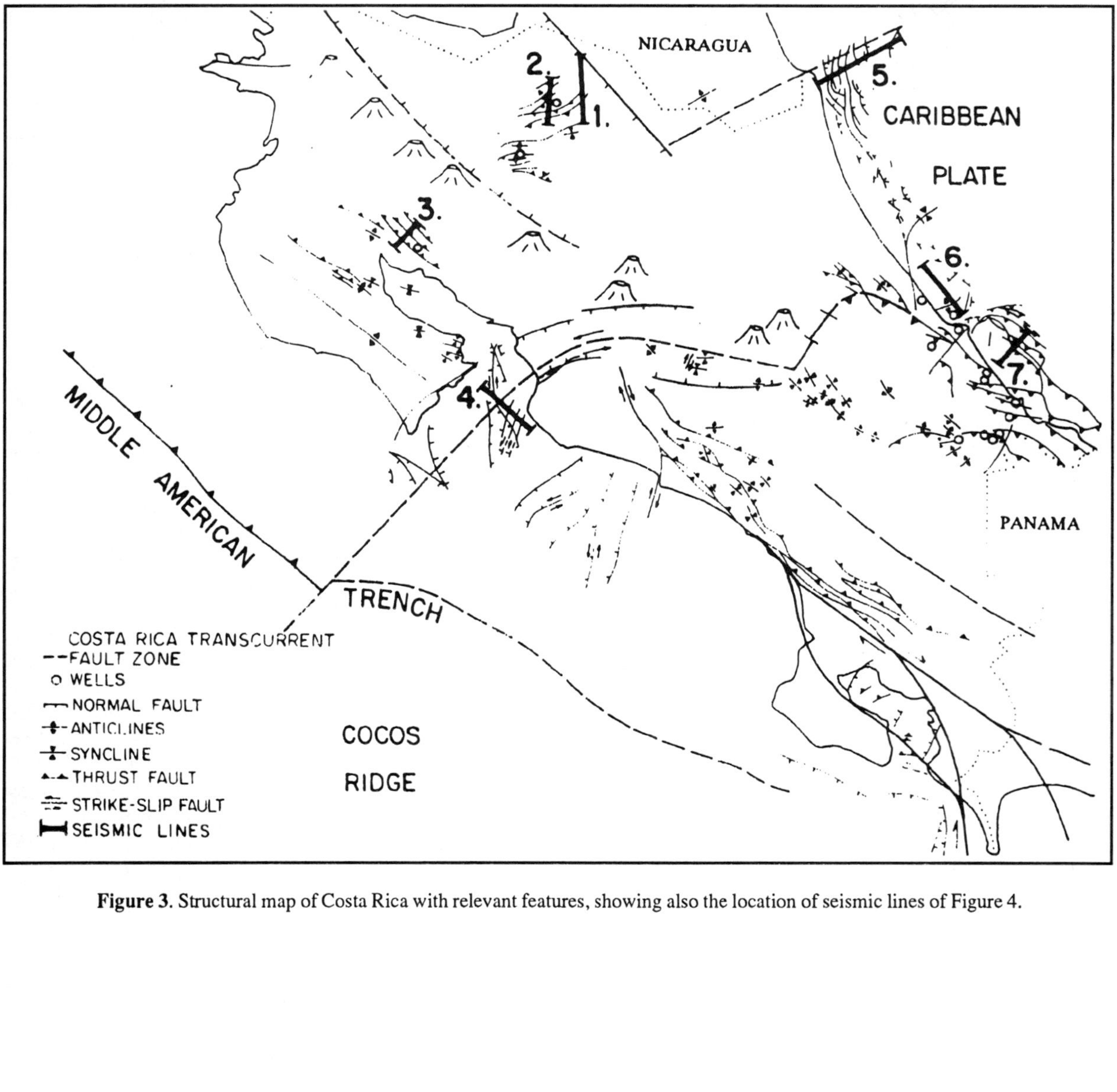

Figure 3. Structural map of Costa Rica with relevant features, showing also the location of seismic lines of Figure 4.

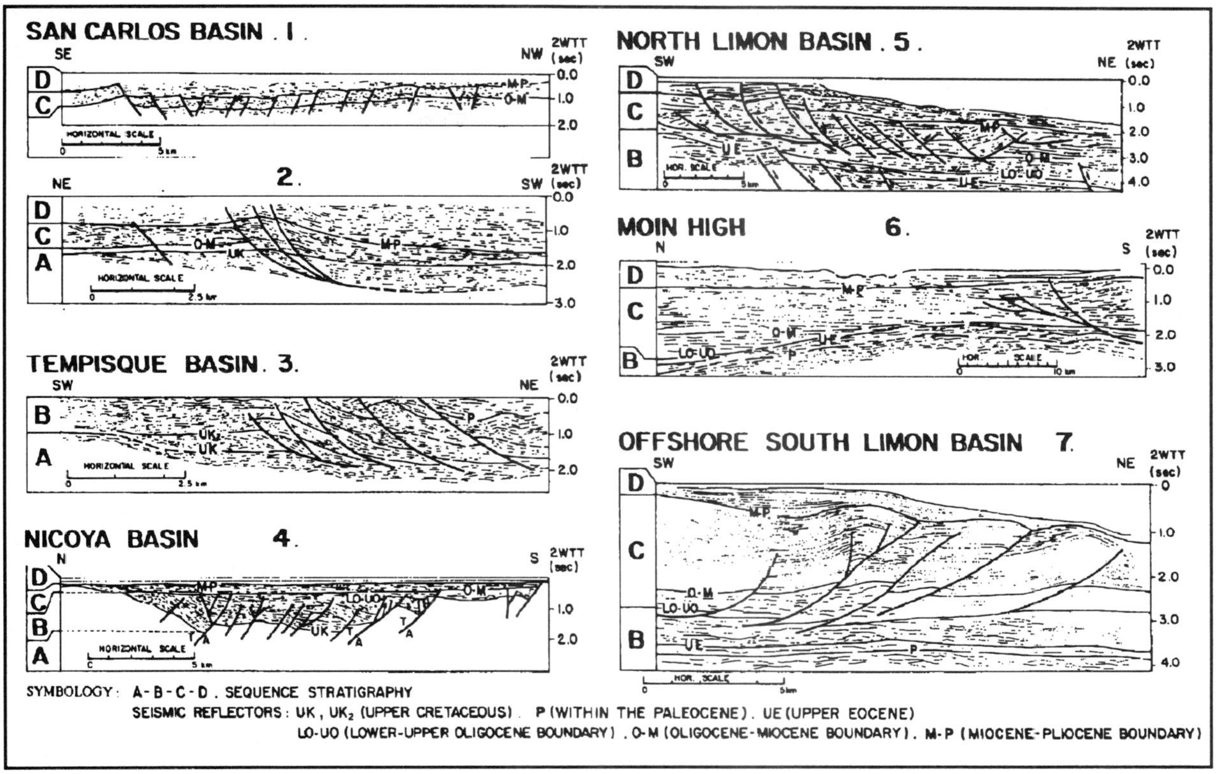

Figure 4. Seismic interpretation of key sections for some of the basins.

of Quaternary age with a west southwest-east northeast direction. They are part of a "strike-slip" system and are possibly still active.

TECTONIC STRATIGRAPHIC EVOLUTION

The tectonic-stratigraphic evolution of Costa Rica's sedimentary basins is separated into four great evolutionary episodes (Figure 5): (a) Late Cretaceous, (b) early Tertiary, (c) late Tertiary, and (d) Plio-Quaternary.

Late Cretaceous

During the Cretaceous and up to the Santonian (Late Cretaceous), a primitive island arc volcanism developed, concomitantly with extensive and thick deposits of bituminous shales, resulting from sedimentation in anoxic environments (Astorga, 1987; 1988). During the basal late Campanian, there was a partial lifting of the basement in the outer arc, and the deposition of a carbonate platform (Calvo, 1987), which interfingers seaward with turbidites and pelagites. Toward the end of the Campanian, an incipient inner volcanic arc developed, with its activity gradually increasing into the Maastrichtian. In the early Maastrichtian (latest Cretaceous), a quick subsidence opened the forearc basin and the carbonate platform was partially buried.

Early Tertiary

From the end of the Cretaceous to the middle of the late Eocene, sedimentation in the slope, forearc, and backarc basins consisted of thick sequences of clastic turbidites (megasequence B). In the middle of the late Eocene the region was affected by compressive stresses (D3), which uplifted the marginal parts of the basins and resulted in structural highs such as the Moin and Parasmina highs in the backarc area. Extensive carbonate ramp-type sediments were deposited during a period of volcanic quiescence toward the end of the Eocene.

Late Tertiary

Between the latest Eocene and into early Oligocene, a new tectonic-stratigraphic evolutionary phase took over in the region. Costa Rica's island arc became segmented by a sinistral "strike-slip" fault system which generated transtensional basins in the central part of the country (Costa Rica Transcurrent Fault Zone). During the Oligocene, the rotation of the southern segment originated the uplifting of

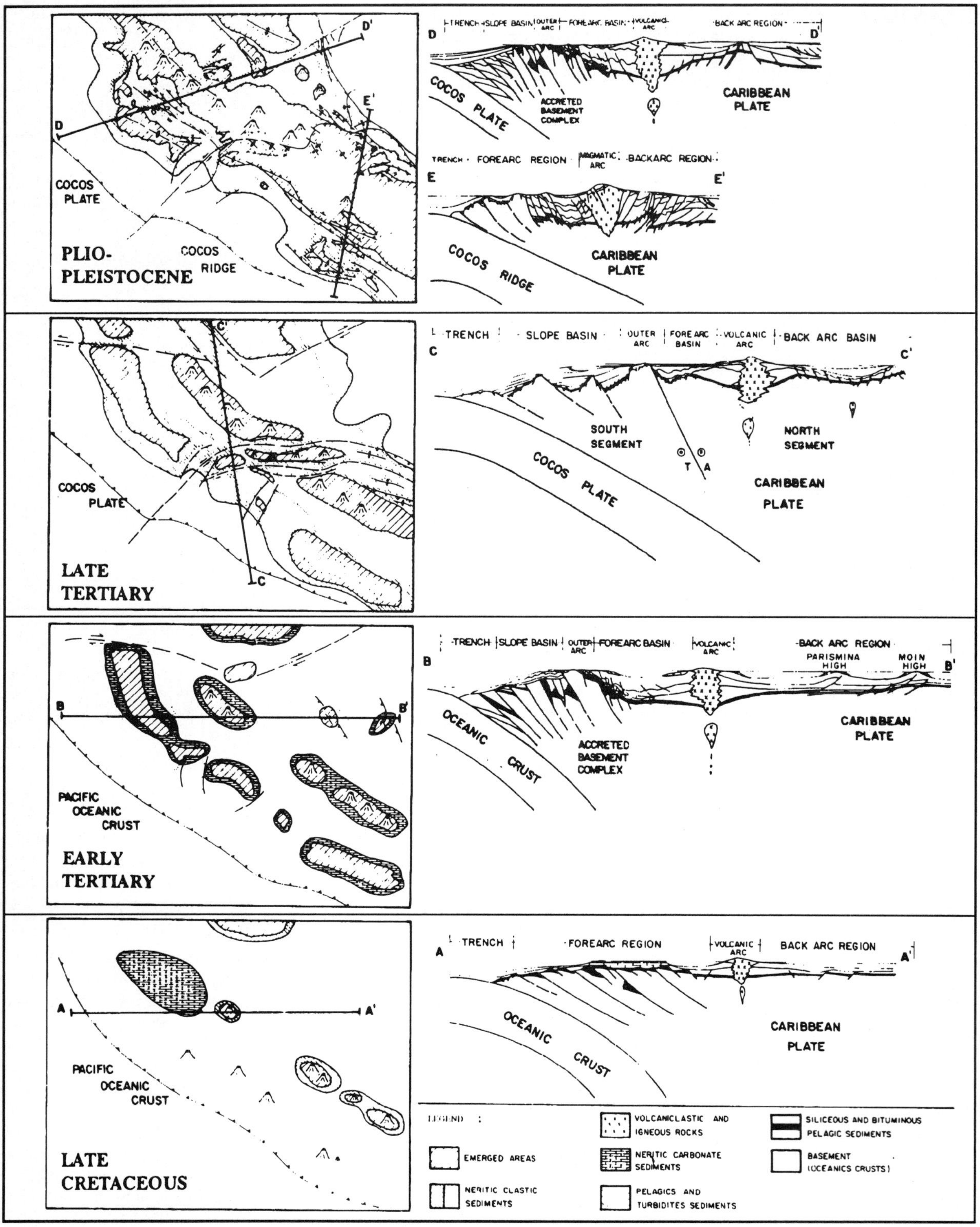

Figure 5. Schematic tectonostratigraphic and paleogeographic evolution of Costa Rica.

the northern segment, subjecting it to erosion. Meanwhile, in the southern segment, sedimentation of clastic turbidites continued and culminated in the early Miocene. The transtensional basins of the central region of Costa Rica were filled in mainly by clastic marine sediments (Rivier and Calvo, 1988). In the Miocene, shallow marine clastic sedimentation prograded in the northern segment. In the "back arc" region of San Carlos an embryonic "rifting" system (graben) was generated, which has continuity with the Lake Nicaragua depression (Mann and Burke, 1984). This graben was partially filled by marine clastic and carbonate sediments from the middle of the late Miocene (cf. Obando, 1986; Calvo and Bolz, 1987) to the Pliocene. In the southern segment, sedimentation was clastic marine during the period. Toward the end of the Miocene to middle Pliocene, a new compressive phase occurred. This induced a strong uplift of the internal arc and of the Talamanca Range caused by acid and intermediate intrusions in the southern segment. During this same time period andesitic volcanism began in the northern segment, followed by intrusions of intermediate composition in the volcanic arc and backarc regions (Obando, 1986; Kussmaul, 1987).

PLIO-QUATERNARY

During the Pliocene, in the northern segment shallow marine clastic sedimentation continued in the San Carlos graben, followed by Quaternary volcaniclastic and marginal marine deposits. In the Limón Norte basin, beginning in the late Miocene, a deltaic system developed in the region. The basin was filled with sediments that have deltaic affinities and which prograde northeastward toward the Colombia basin. During the Pliocene, in the southern forearc and backarc regions thick and coarse shallow marine to continental clastic sediments ("fan and braided delta" to "alluvial fans") were deposited. In the present Golfo Dulce-Burica region a foredeep basin was formed and a thick prism of shallow and continental marine sediments was deposited during the Pliocene (Corrigan, 1986). In this same region, during the Pleistocene, the opening of the Panama Fracture Zone produced a right lateral strike-slip fault system, which borders the Térraba basin thrust, and also a "pull apart" basin which is still active (cf. Barritt and Berrangé, 1987).

HYDROCARBON POTENTIAL

This study has identified several basins that are deemed to have petroleum potential. A stratigraphic column for each of these basins is illustrated in Figure 2b and potential source and reservoir beds are noted. Structural styles and potential trapping components are illustrated on the seismic profiles shown in Figure 4.

Of special interest and discussed here are four areas, namely the Limón Sur fold and thrust belt, the San Carlos area, the prograding delta-like succession of the Limón Norte basin and the interpreted build-up on the Moin High, and the Nicoya-Quepos basin complex.

In the Limón Sur basin more than 7 km of section, which includes deltaic and near-shore sandstone depositional systems as well as ramp-type (continental slope type) carbonates, is present. Clean porous quartzitic sandstones of Pliocene age eroded from the granodioritic uplifted pluton, the Talamanca Range, provide reservoir potential. Oil seeps in the area and mature source rocks with TOC (total organic content) values up to 1.5 percent enhance the prospects of this basin. The seismic section (Figure 4-7) illustrates a faulted anticline prospect.

The reservoir and source potential of the San Carlos and Limón Norte basins are illustrated in the stratigraphic section (Figure 2-3). Deep-water fans, carbonate ramps, and calcarenite shoreface deposits are typical of the depositional systems in this region. In the Limón Norte basin, as illustrated on Figure 4-5, prograding delta-like features associated with down-to-the-basin listric faults characterize the offshore region. Ramp-type carbonates on the flanks of the Moin High offer an additional prospect.

In the San Carlos basin east-west folds, thrusts, and en echelon structures as well as an interesting "flat spot" shown on Figure 4-2 are attractive features.

ACKNOWLEDGMENTS

We wish to thank the Primary Production Management of RECOPE S.A. for permission to publish this paper, also Petro-Canada International Assistance Corporation for its cooperation, Dr. Andrew Baillie (Calgary) for his valuable assistance, and geologist Giovanni Bottazzi for his collaboration. Lastly, we would like to thank all those who have in one way or another assisted in the preparation of the manuscript.

REFERENCES

Anderson, T. H., and V. A. Schmidt, 1983, The evolution of Middle America and the Gulf of Mexico-Caribbean Sea region during Mesozoic time: Geological Society of America Bulletin, v. 94, p. 941-966.

Astorga, A., 1987, El Cretácico Superior y el Paleógeno de la vertiente Pacífico de Nicaragua meridional y Costa Rica septentrional: Origen, evolución y dinámica de cuencas profundas relacionadas al margen convergente de Centroamérica: San José, Universidad de Costa Rica, Tesis de licenciatura, 250p.

Astorga, A., 1988, Geodinámica de las cuencas del Cretácico Superior-Paleógeno de la región "forearc" del sur de Nicaragua y norte de Costa Rica: Revista Geológica de América Central, v. 9, p. 1-40.

Barboza, G., and H. Zucchi, in press, Sismoestratigrafía de las cuencas "forearc" del Pacífico central y sur en Costa Rica.

Barritt, S., and J. P. Berrangé, 1987, Interpretation of a gravity survey of the Osa Peninsula and environs, southern Costa Rica: Overseas Geology and Mineral Resources, p. 64.

Baumgartner, P. O., 1987, Tectónica y sedimentación del Cretácico Superior en la zona pacífica de Costa Rica (América Central), *in* C. Barbarin, H.-J. Gursky, and P. Meiburg, eds., El Cretácico de México y América Central: Actas Facultad de Ciencias de la Tierra, Universidad Autónoma de Nuevo Leon, v. 2, p. 251-260.

Baumgartner, P. O., C. R. Mora, J. Butterlin, J. Sigal, G. Glacon, J. Azema, and J. Bourgois, 1984, Sedimentología y paleogeografía de Cretácico y Cenozoico del litoral pacífico de Costa Rica: Revista Geológica de América Central, v. 1, p. 57-136.

Bergoing, J. P., E. Malavassi, and R. Jimenez, 1980, Síntesis geológica del Valle Central de Costa Rica: San José, Instituto Geografico Nacional, 2 parts.

Bourgois, J., J. Azema, P. O. Baumgartner, J. Tournon, A. Desmet, and J. Aubouin, 1984, The geologic history of the Caribbean-Cocos Plate boundary with special reference to the Nicoya ophiolite complex (Costa Rica) and D.S.D.P. results (Legs 67 and 84 off Guatemala): A synthesis: Tectonophysics, v. 108, p. 1-32.

Burke, K., P. Fox, and A. M. C. Sengor, 1978, Buoyant ocean floor and the evolution of the Caribbean: Journal Geophysics Resources, v. 83, p. 3949-3954.

Calvo, C., 1987, Las calizas neríticas de la vertiente pacífica del norte de Costa Rica y sur de Nicaragua: Epocas y sistemas de sedimentación asociades con la apertura y evolución del margen convergente de la América Central meridional: San José, Universidad de Costa Rica, Tesis de licenciatura.

Calvo, C., and A. Bolz, 1987, La secuencia de Venado, un estuario lagunar tropical del Mioceno Medio, San Carlos, Costa Rica: Revista Geológica de América Central, v. 6, p. 1-24.

Corrigan, J. D., 1986, Geology of the Burica Peninsula, Panamá-Costa Rica: Neotectonic implication for the southern Middle America convergent margin: Austin, University of Texas, Masters Thesis, 152p.

Dengo, G., 1962a, Estudio geológico de la región de Guanacaste, Costa Rica: San José, Instituto Geográfico Nacional, ix and 112p.

Dengo, G., 1962b, Tectonic-igneous sequence in Costa Rica, *in* Petrologic studies, A. F. Buddington volume: Geological Society of America, p. 133-161.

Dengo, G., 1968, Estructura geológica, historia tectónica y morfología de América Central: México, Centro Regional de Ayuda Técnica, México, Buenos Aires, 55p., 2nd ed., 1973.

Dengo, G., 1983, Mid America: Tectonic setting for the Pacific margin from southern Mexico to northwestern Colombia: Guatemala, Centro de Estudios Geológicos de América Central, iv, 90p.

Denyer, P., and W. Montero, 1989, Mapa geo-estructural y sismos del Valle Central, Costa Rica: San José, Instituto Geográfico Nacional, 1 map, 23p.

Dewewer, D., J. Azema, J. Tournon, and A. Desmet, 1985, Dècouverte de matèriel du Lías-Dogger infèrieur, dans la pèninsule de Santa Elena (Costa Rica, Amèrique Centrale): Paris, Academia de Ciencias de Paris, v. 300, p. 759-764.

Eva, A. W., K. Burke, P. Mann, and G. Wadge, 1989, Four-phase tectonostratigraphic development of the southern Caribbean: Marine and Petroleum Geology, v. 6, p. 9-21.

Galli-Olivier, C., 1979, Ophiolite and island-arc volcanism in Costa Rica: Geological Society of America Bulletin, v. 90, p. 444-452.

Gursky, M., 1988, Análisis tectónico de la Península de Nicoya (Costa Rica) y su significado para el desarrollo estructural-geodinámico de América Central meridional: Revista Geológica de América Central, v. 8, p. 19-75.

Henningsen, D., 1966, Die pazifische küstenkordillere (Cordillera Costeña) Costa Ricas und ihrer stellung innerhalb des südzentralamerikaanischen Gebirges: Stuttgart, Geotekto Forsch., v. 23, p. 3-66.

Kussmaul, S., 1987, Petrología de las rocas intrusivas Neógenas de Costa Rica: Revista Geológica de América Central, v. 7, p. 83-112.

Kuypers, E., 1980, The geologic history of the Nicoya ophiolite complex, Costa Rica, and its geotectonic significance: Tectonophysics, v. 68, no. 3/4, p. 233-255.

Lonsdale, P., and K. D. Klitgord, 1978, Structure and tectonic history of the eastern Panama basin: Geological Society of America Bulletin, v. 89, p. 981-999.

Lloyd, J., 1963, Historia tectónica del orógeno sur centroamericano, *in* The backbone of the Americas, tectonic history from pole to pole; A symposium: American Association of Petroleum Geologists Memoir 2, 28p.

Malfait, B. T., and M. G. Dinkelman, 1972, Circum-Caribbean tectonics and igneous activity and the evolution of the Caribbean Plate: Geological Society of America Bulletin, v. 83, p. 251-272.

Mann, P., and K. Burke, 1984, Neotectonics of the Caribbean: Reviews of Geophysics and Space Physics, American Geophysical Union, v. 22, no. 4, p. 309-362.

Mora, S., 1979, Estudio geológico de una parte de la región sureste del Valle del General, Provincia de Puntarenas, Costa Rica: San José, Universidad de Costa Rica, Tesis de licenciatura, 200p.

Meschede, M., W. Frisch, and M. Sick, 1988, Interpretación geoginámica de los complejos ofiolíticos de Costa Rica: Revista Geológica de América Central, v. 8, p. 1-17.

Obando, J., and P. O. Baumgartner, 1986, Estratigrafía y tectónica de la cuenca de Golfo Dulce y Charco Azul: Jornadas Geológicas de Costa Rica, v. 2, p. 15-16.

Obando, L., 1986, Estratigrafía de la Formación Venado y rocas sobreyacientes (Mioceno-Reciente) Provincia de Alajuela, Costa Rica: Revista Geológica de América Central, v. 6, p. 73-104.

Pichler, H., and R. Weyl, 1975, Magmatism and crustal evolution in Costa Rica (Central America): Stuttgart, Geologie Rundschau, v. 62, p. 357-396.

Pindell, J. L., and J. F. Dewey, 1982, Permo-Triassic reconstruction of western Pangea and the evolution of the Gulf of Mexico-Caribbean region: Tectonics, v. 1, p. 179-212.

Pindell, J. L., S. C. Cande, W. C. Pittman, III, D. B. Rowley, J. F. Dewey, J. L. La Brenque, and W. Haxby, 1988, A plate-kinematic framework for models of Caribbean evolution, *in* C. R. Scotese and W. W. Sager, eds., Mesozoic and Cenozoic plate reconstructions: Tectonophysics, v. 155, p. 121-138.

Rivier, F., and C. Calvo, 1988, Terciario de sur del Valle Central: Sección estratigráfica del Cerro Caraigres, Provincia de San José, Costa Rica: Revista Geológica de América Central, v. 9, p. 61-74.

Seyfried, H., A. Astorga, and C. Calvo, 1987, Sequence stratigraphy of deep and shallow water deposits from an evolving island arc: The Upper Cretaceous and Tertiary of Central America: Facies, v. 17, p. 203-214.

Stibane, F. R., R. Schmidt-Effing, and R. Madrigal, 1977, Zur stratigraphisch-tektonischen Entwicklung der Halbinsel Nicoya (Costa Rica) in der zeit von oberkreide bis Unter-Tertiät: Giessen, Giessener Geologische Schriften, v. 12, p. 315-358.

Wadge, G., and K. Burke, 1983, Neogene Caribbean Plate rotation and associated Central American tectonic evolution: Tectonics, v. 2, p. 633-643.

Weyl, R., 1980, Geology of Central America: Berlin, Gebruder Borntraeger, viii, 372p.

Circum-Pacific Council for Energy and Mineral Resources
Earth Science Series, Vol. 16
R. L. Miller, G. Escalante, J. A. Reinemund, and M. J. Bergin (Eds.)
Energy and Mineral Potential of the Central American-Caribbean Regions
Springer-Verlag, Berlin Heidelberg 1995

AN INTEGRATED GEOLOGICAL AND GEOPHYSICAL INTERPRETATION OF THE SAN CARLOS BASIN, COSTA RICA

Lesmes E. Ballestero C., Mihaela M. Dobrinescu, and Gerardo Jager C.
Refinadora Costarricense de Petróleo, S. A. (RECOPE), Apartado 4351, San José, Costa Rica

Ian R. Mayers
Petro-Canada International Assistance Corporation (PCIAC), P. O. Box 2844, Calgary, Alberta T2P 3E3 Canada

ABSTRACT

Hydrocarbon exploration began in Costa Rica with Sinclair Corporation's five well program of 1916-1923. Since then geological, geochemical, potential field, land, and marine seismic surveys have been made with 49 stratigraphic and exploration wells drilled, the majority of these in the Limon Sur basin. Some intriguing non-commercial oil and gas shows were encountered.

The San Carlos basin in northern Costa Rica was considered prospective and made an objective of the 1986-1989 RECOPE-PCIAC exploration/technical assistance program. Building on sparse geological information and new potential field interpretations, 600 km of dynamite, multifold reflection and 50+ km of refraction seismic surveys were conducted there. These data were interpreted, along with older information, and drill hole locations were selected. These stratigraphic wells at Pataste No. 1 and Tonjibe No. 1 (Figure 1) were finished in November, 1987, and January, 1988, respectively. They significantly increased the understanding of the basin.

INTRODUCTION

The San Carlos basin, with an area of 8,000+ km^2, lies northeast of the Guanacaste Cordillera (inset map, Figure 1, northwest part of volcanic arc), which is continuing to grow by active volcanism. The underlying Cocos Plate's subduction zone here dips to the northeast at > 25°. The basin forms part of the Central American back arc region, and is separated from the Limon Norte basin to the east by a basement ridge. From Early Cretaceous it has been intermittently filled with 6 km of primarily fluvial and marine rocks of flysch and molasse facies. Potential reservoirs include dolomitized limestones and also turbidites. Basin geochemical analyses of drill samples identified a predominantly terrestrial kerogen within the Tertiary sediments. This type of kerogen is prone to gas and condensate generation, but oil and gas seeps do occur in the region. These indicate the presence of a richer, unidentified source rock, perhaps within the underlying Cretaceous strata. The high geothermal gradient (4.3°C/100 m) suggests hydrocarbon generation beginning at a shallow depth of 2.0-2.5 km.

Structurally, the San Carlos basin is bounded by northwest-aligned normal faults which began forming in the Cretaceous as a result of northeast-southwest-directed back arc extension. During the Neogene, east-trending structures were formed by north-south compression of the Tertiary strata. Future exploration should concentrate along the resulting east-west structural trends or on the basin's flanks, where migration pathways and traps are focused. The presence of "bright spot" anomalies also indicate stratigraphic prospects may exist elsewhere in the basin.

TECTONIC FRAMEWORK AND REGIONAL GEOLOGY

The Middle American Trench subduction zone parallels the southwest coast of Costa Rica. At this convergent

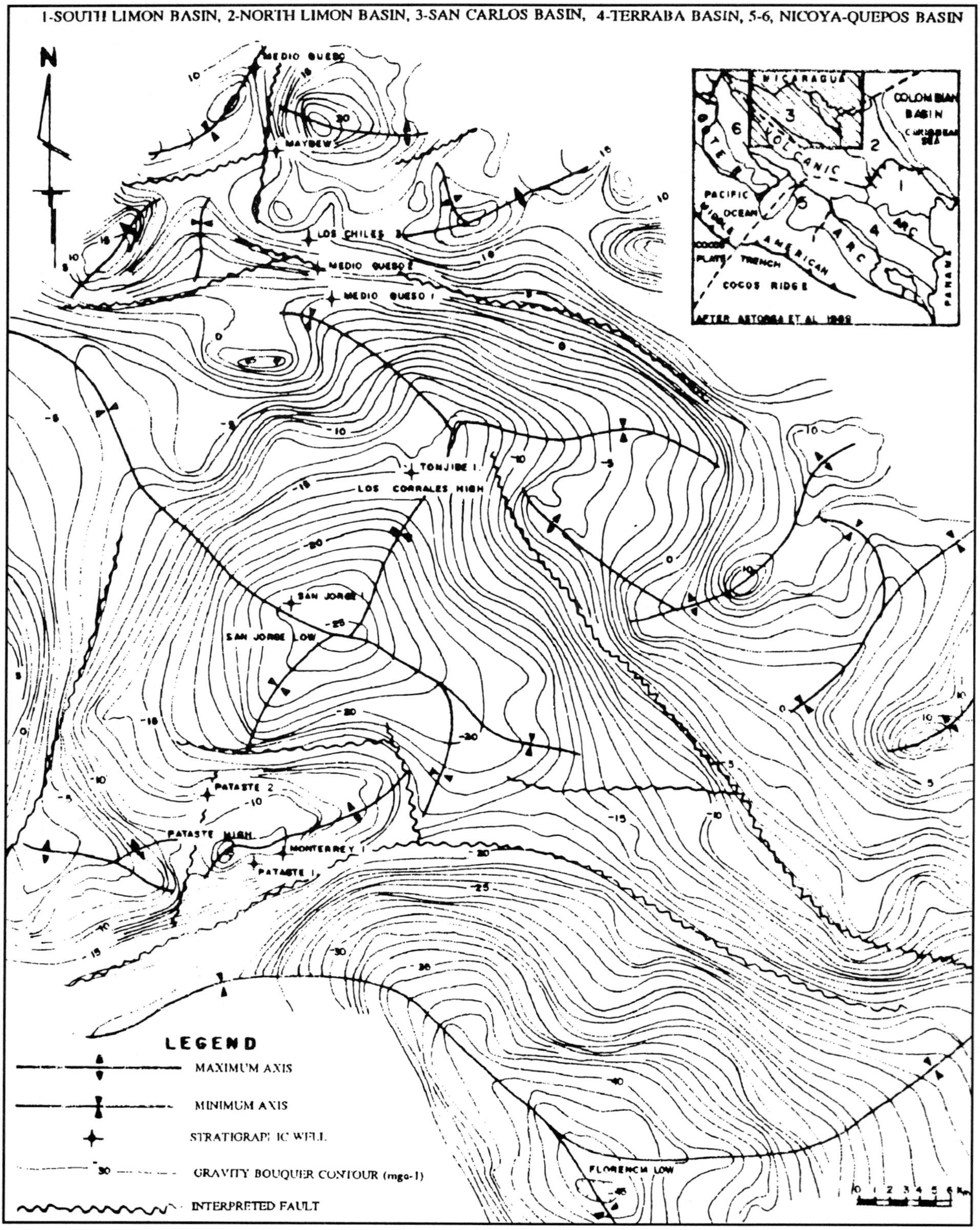

Figure 1. Maps showing location of San Carlos Basin (inset map) and Bouguer gravity anomaly map of the basin and contiguous areas. Also shows locations of wells drilled for stratigraphic information and interpreted faults.

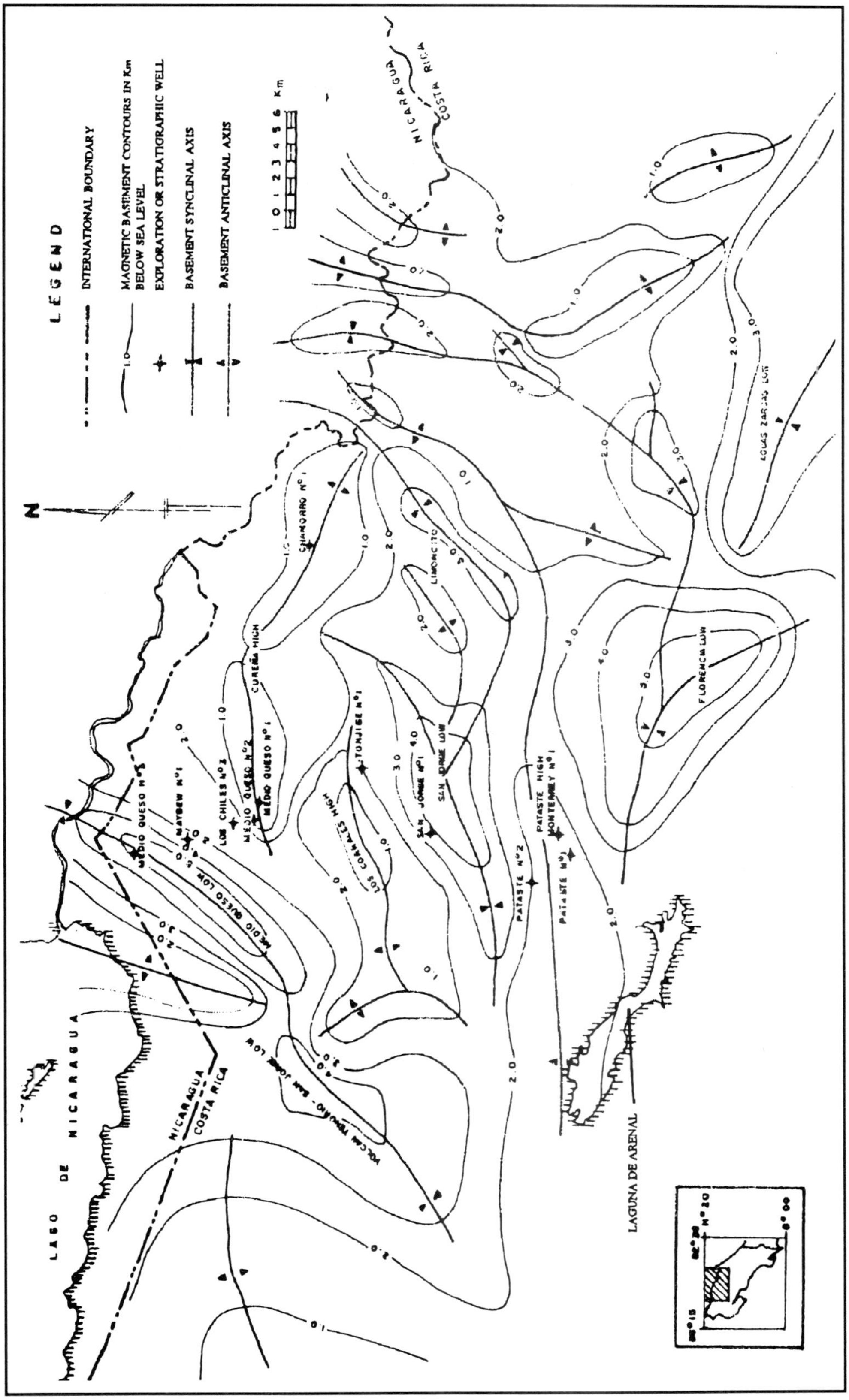

Figure 2. Magnetic map of the San Carlos Basin area. Also shows location of wells drilled for stratigraphic information.

margin the Cocos Plate is consumed beneath the Caribbean Plate at 9 cm/yr down a low angle (average dip < 25°) Benioff Zone (Mora, 1981). Forces in the subduction zone have created two northwest-southeast-trending 20 to 50-km wide ranges separated by an east-west-striking rift, the Valle Central. The northern range, the Tilarán-Talamanca Range, consists of active and dormant volcanos. During the early Tertiary it was the locus of mafic eruptions. Since the Pliocene the volcanism has become more acidic. This volcanism supplied the magmas that are thought to rise to the surface along a northwest-trending fault zone. The fault zone may also be the southwestern boundary of the San Carlos-Limon Norte (back arc) basin (inset map, Figure 1). The basin is probably cut by a suite of deep vertical faults aligned northeast-southwest (Montero and Dewey, 1982; WAI, 1988).

The Costa Rican back arc basin has been the site of relatively continuous volcanism and sedimentation since the Paleocene which resulted in a Tertiary sequence 6-10 km thick immediately to the northeast of the volcanic arc. It is subdivided into the intensely thrust and wrench faulted and folded Limon Sur sub-basin and the less deformed Limon Norte and San Carlos sub-basins located to the northwest of the northeast-trending "Puerto Limon-Moin Ridge." The Limon Sur sub-basin is relatively well understood, having been studied through surface geology mapping, exploration drilling, and considerable seismic coverage. It will not be discussed further except to assist in the general understanding of the back arc region.

The Limon Norte and San Carlos sub-basins cover an (onshore) area in Costa Rica of 14,500 km^2. Much of the surface area is covered by young volcanics and alluvium deposits except along the banks of the Rio San Juan and in the San Carlos-Venado area (Calvo and Bolz, 1987).

Geophysical studies have shown that the sub-basin area is divided into the San Carlos basin to the northwest and the Limon Norte basin to the southeast by a basement ridge that runs northeast-southwest beneath the San Carlos Plain. In this paper we will consider only the 8,000 km^2 of the San Carlos basin.

THE SAN CARLOS BASIN—EXPLORATION ACTIVITY

Potential Field Data Mapping

Gravity and Magnetics

The basin was surveyed by RECOPE in 1982 using a La Coste-Romberg meter with an in-line station spacing of 0.3 km and an average distance between polygons of 5.0 km. The Bouguer gravity anomaly map (Figure 1) shows that the basin consists of two major troughs. The larger, Florencia Low (to the south of the east-west-trending Pataste High), is estimated to be filled by >5 km of post-Cretaceous sediment. The northern San Jorge Low is estimated to be 3 km deep. The Pataste High is thought to be caused by a hypabyssal andesite intrusion. The major structural trends of these anomalies run either west northwest-east southeast or east northeast-west southwest. The interpreted fault trends are predominantly west and west-southwest.

In 1983 Aeroservice surveyed the basin using an optically-pumped magnetometer flown at 762 m ASL (above sea level) in a 5 km grid with control lines 20 km apart. Again the data indicate that the basin is composed of two troughs, the Florencia Low (in the south) and the San Jorge Low, separated by the large, upper Miocene-lower Pliocene andesitic Pataste High (Obando, 1986) (Figure 2). The northeast-trending Los Corrales High extending into the San Jorge Trough is believed to be produced by the same serpentine body that was encountered at the bottom of the PCIAC-RECOPE Tonjibe No. 1 well. The Cureña High is a high-frequency sequence of anomalies, paralleling the Nicaraguan border, produced by lava flows of the upper Miocene-lower Pliocene Cureña Formation.

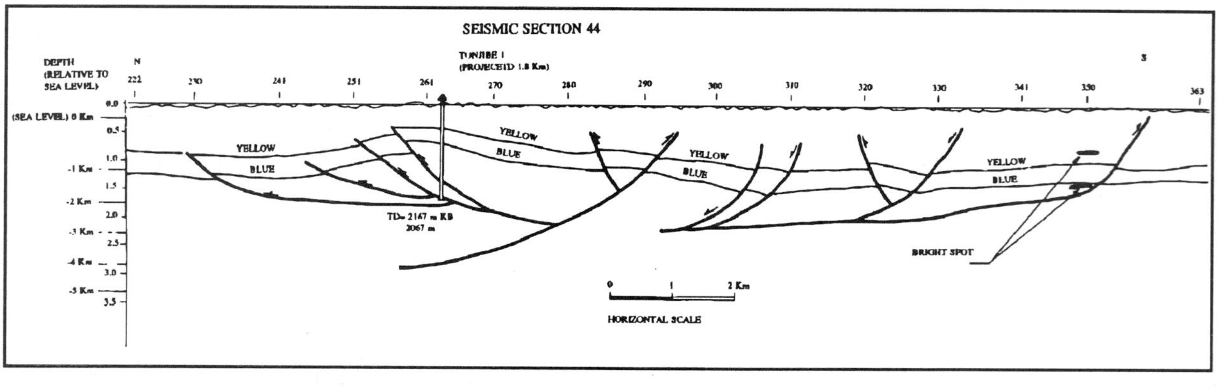

Figure 3. Interpretation of north-sout seismic section from the San Jorge Trough in the south across the Los Corrales High toward the northern edge of the basin. The lines designated yellow line and blue line are seismic reflectors indentified in the text.

The Well Data Base

The San Carlos basin has been drilled by two deep wells, PCIAC-RECOPE Pataste No. 2 (TD 1,979 m) and PCIAC-RECOPE Tonjibe No. 1 (TD 2,168 m) (Figure 2). Both of these had a comprehensive set of logs, including velocity surveys. In addition, nine shallower RECOPE core holes have been drilled and logged in the area: Chamorro No. 1 (TD 321 m), Mayhew No. 1 (TD 132 m), Medio Queso No. 1 (TD 401 m), Medio Queso No. 2 (TD 355 m), Medio Queso No. 3 (TD 632 m), Medio Queso No. 4 (TD 132 m), Monterrey No. 1 (TD 620 m), San Jorge No. 1 (TD 550 m), and Pataste No. 1 (TD 320 m).

The Seismic (Refraction and Reflection) Data Base

The seismic data base used in this study consists primarily of 520+ km of 1,200-1,500 percent dynamite reflection data of variable (poor to very good) quality. In addition, 50+ km of refraction data were interpreted and used.

Interpretation

Potential Field Depth to Basement Map (Figure 2)

This map is derived from both sets of data using the slope method and the simple and weighed average (in the space domain) method. Bodies estimated to be at depths equal to or less than 1 km are inferred to be intrusives and lava flows; those at >2 km are believed to be basement. Two major structural trends show up in the data: (1) northwest-southeast trends related to the late Mesozoic-early Tertiary opening of the basin, under east-west extension, and (2) northeast-southwest trends produced later, in the upper Tertiary, by northwest-southeast compression. The three major troughs are the Florencia-Aguas Zarcas Low, the San Jorge Low, and the Volcan Tenorio-Medio Queso Low. The major basement highs are the Pataste and Los Corrales Highs. It is estimated, from this map, that the San Carlos basin covers some 7,850 km^2, of which about 5,000 km^2 has the Mesozoic basement covered by more than 2 km of sediment.

Seismic Section No. 44 (Figure 3)

This north-south interpreted seismic section (s. p. 221-366 only) passes 1.8 km to the east of PCIAC-RECOPE Tonjibe No. 1 well, from the San Jorge Trough in the south, across the Los Corrales High (crest at s. p. 259) toward the northern edge of the basin. It shows the complex reverse and normal faulting that occurred as a result of wrench movements under north-south compressional forces. The Yellow (Y) Reflector is an early Pliocene alluvial shale and the Blue (B) Reflector, a late Miocene sandstone/volcanic contact in the Tonjibe No. 1 well. At s. p. 350, at reflection times of 0.8 sec and 1.5 sec there are two "bright spots" in small anticlines on the upthrust side of a reverse fault which may represent gas accumulations. The tectonism that deformed the strata was probably recurrent during the late Tertiary, with the most recent episode occurring in the late Pliocene.

The Tonjibe No. 1 well drilled into the Los Corrales High (Figure 2). It encountered a shallowing upward sequence of Holocene, Pliocene, and Miocene marine to non-marine sediments down to 1,700 m. Below that, primarily non-marine, Lower Cretaceous sediments were drilled. Below 1,819 m the well entered serpentinized and chloritized tuffs and basalts, representing economic basement. It appears that the Los Corrales basement high acted as local focusing ramp, over which the Tertiary strata were upwarped and faulted during the late Tertiary north-south compression.

Contour Map of Dix Interval Velocities (Figure 4)
(At a 2.0 sec time slice, below the +200 m ASL datum, or -2,500 m)

Throughout the back arc basins a generally good correlation has been noted between (carefully edited) Dix Interval Velocities (DIV) (calculated velocities between two parallel seismic reflectors) and the check shot/VSP (vertical shot point) interval velocities. Smoothed DIV contour plots in the plane of the seismic sections and in map form frequently improved the structural interpretation in areas where reflector patterns were not easily distinguished or correlated. The smoothed DIV plots also facilitated reasonably accurate depth mapping and regional interpretation. For example, a DIV contour map at a 2.0 sec time slice shows the higher velocities and fault densities associated with the Pataste High igneous rocks and the lower velocity zones delineate the clastic depocenters.

Blue Reflector Isochron Structure Map

The Blue Reflector surface of Figure 5 is one of the few reflectors that can be fairly continuously mapped across most of the basin. It represents a time transgressive event, which at Pataste No. 2 was correlated with the contact between a middle Miocene (Venado Formation) limestone overlying a lava. At Tonjibe No. 1 it corresponds to a sandstone over volcanics contact dated as being within the late Miocene. The Venado Formation limestone was absent at Tonjibe No. 1. The maximum mapped time on this reflector was 2.3 sec in the Florencia Low and the minimum time was 0.6 sec over the Pataste High.

Structurally, the map shows the dominant east-west-trending, reverse faulting and folding (with probably a wrench component) which was produced by north-south-directed, late Tertiary compressional forces. The major structural elements are the Los Corrales High (drilled by Tonjibe No. 1), Pataste High (drilled by Pataste No. 2), Florencia Low, and the San Jorge Low.

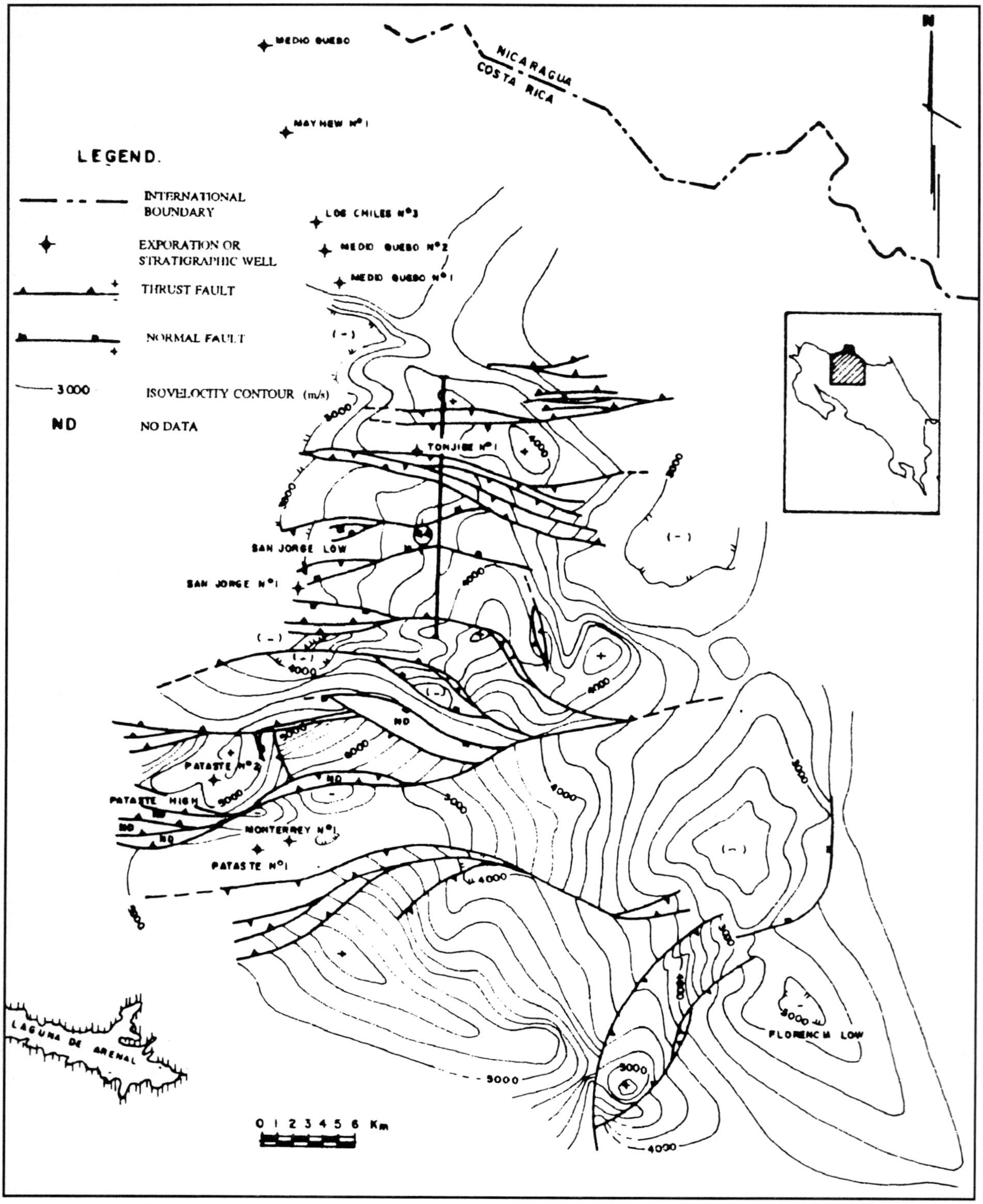

Figure 4. Contour map of Dix interval velocities (calculated velocities between two parallel seismic reflectors), showing structural interpretations.

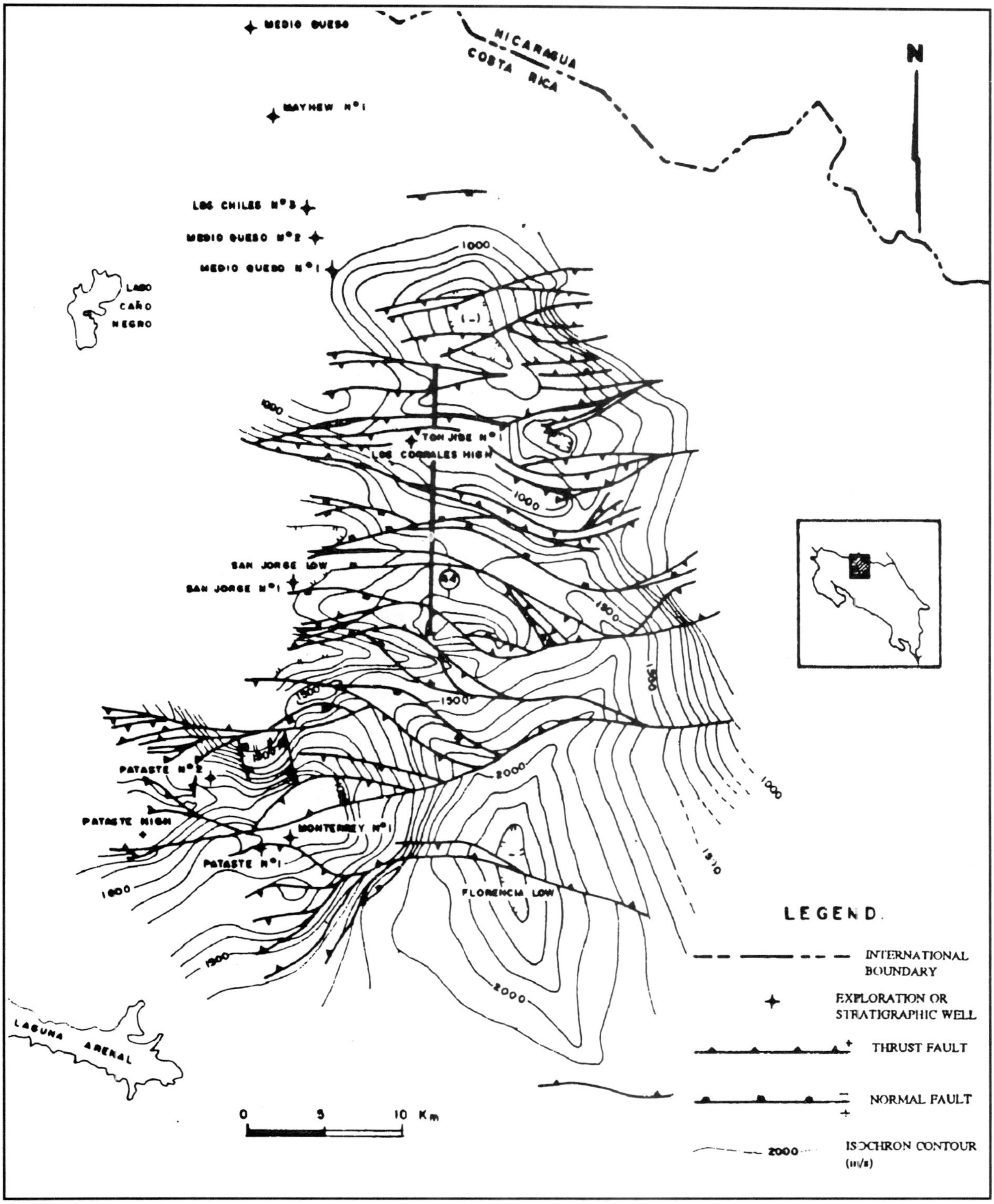

Figure 5. Structural interpretation of the San Carlos Basin based on the nearly continuously mapped Blue reflector (Figure 3).

Geothermal Data Analysis and Source Rock Potential

The geothermal data used in this project are derived from the PCIAC-RECOPE stratigraphic wells and from Pataste No. 2 and Tonjibe No. 1 deep wells. The calculation of the geothermal gradient of these wells was based on the maximum temperature taken at the beginning of the first run of the well logging at each interval. It has been considered that the resting period before logging of the wells was sufficient for the recovery and stabilization of the temperature. Consequently, there was no need for temperature corrections to calculate the geothermal gradient. The Pataste No. 2 well had an estimated geothermal gradient of 2.89°C/100 m down to 1,580 m KB and 7.04°C/100 m down to TD at 1,979 m KB. The weighted average gradient for the entire well was 3.74°C/100 m. The higher gradient coincided with a plagioclase rich, volcanic lava.

The Tonjibe No. 1 well has calculated gradients of 4.96°C/100 m (50-1,400 m KB) and 4.38°C/100 m (1,400-2,168 m KB). The weighed average for the entire well is 4.75°C/100 m. Below 1,400 m the strata drilled were predominantly non-marine volcanics. The overall weighted gradient for both wells is estimated to be 4.27°C/100 m, a high value but typical of back arc basins.

Geochemical studies of possible source rocks encountered in the Pataste No. 2 and Tonjibe No. 1 wells indicate that the Tertiary sediments contain predominantly gas prone kerogen (structured terrestrial, biodegraded terrestrial, miospores, and marine amorphous). These will begin generating hydrocarbons at Ro 0.7 percent (TAI = 2 to 2+) and under the prevailing high geothermal gradients these conditions should begin to occur between depths of 2.0-2.4 km. As the (potential field) basement map (Figure 2) indicates that there is an area of about 5,000 km^2 with Tertiary sediment deeper than 2 km, there is evidently considerable gas potential in the basin. The Cretaceous basin here also contains the oil prone marine shales of the Loma Chumico Formation. This indicates that the basin may have a significant oil potential.

SUMMARY AND CONCLUSIONS

Using a substantial data base of potential field, seismic, well, surface geology, and other proprietary information obtained by RECOPE (before and during the PCIAC 1986-1989 technical assistance program), a series of multidiscipline maps, etc., have been produced. This work has improved the geological understanding of the basin and allowed RECOPE to quantitatively upgrade its knowledge and evaluation of the basin's potential hydrocarbon resources.

The San Carlos basin formed in the Cretaceous by east-west-aligned, back arc extension. Its sedimentary sequence consists of shallowing upward bathyal, continental shelf, deltaic, and alluvial facies which were deformed during the Tertiary by north-south compression.

Potential reservoirs are dolomitized limestones, turbidites, shallow marine sandstones, and deltaic/alluvial clastics. Potential traps are expected in the Tertiary strata in compressional and wrench-derived structures, onlapping porous sequences along the flanks of the basin, and in stratigraphic traps of various types.

The Tertiary source rocks encountered contain primarily gas prone, terrestrial kerogen which, given the prevailing high geothermal gradient, should begin to generate hydrocarbons below 2.0-2.4 km depth. However, the basement Nicoya Complex is known to contain shales with 10-20 percent TOC elsewhere, and may have generated the oil observed in the basin's surface seeps.

Dix Interval Velocities have proven to be a good approximation of velocity survey values of the exploration wells. Here, and in the other Costa Rica back arc basins, this has permitted reasonably accurate depth mapping and has also assisted the structural interpretations.

RECOMMENDATIONS FOR FUTURE SAN CARLOS BASIN EXPLORATION

1. In future exploration all existing and new data should be completely analyzed, interpreted, and used to further improve the San Carlos basin geological model.

2. The known structural and stratigraphic prospects should be rigorously evaluated for their oil and gas potential and then ranked.

3. A new, more focused (on the more prospective areas as indicated above) RECOPE-operated exploration program consisting primarily of microgravity surveys and multifold seismic surveying tied to deep and shallow stratigraphic wells should be carried out. In the lava flow covered, seismic "blind zones," magnetotelluric surveys should be conducted and interpreted in order to obtain essential, supplementary data.

4. Once these data have been integrated, RECOPE should re-evaluate and re-rank the San Carlos basin's hydrocarbon plays and prospects. With this improved assessment and a new law covering exploration for and development of hydrocarbons in Costa Rica, RECOPE should be in a good position in 1991 to implement its strategic plan for exploration in the San Carlos basin in cooperation with the international oil industry.

ACKNOWLEDGMENTS

The authors would like to thank the management of RECOPE's Primary Production Division and PCIAC for their support and for their permission to submit this paper for publication. We gratefully acknowledge the practical assistance and advice given by Dr. A. D. Baillie, international geological consultant, and L. Piedra B., D. Badilla R., and A. Jimenez A. of RECOPE during the preparation of the paper.

REFERENCES CITED

Calvo, C., and A. Bolz, 1987, La secuencia de Venado, un estuario lagunar tropical del Mioceno medio, San Carlos, Costa Rica: Revista Geológica America Central, v. 6, p. 1-24.

Mora, C. S., 1981, Clasificación morfotectónica de Costa Rica: San José, Instituto Geográfico Nacional de Costa Rica, Informe Semestral, v. 2, p. 35-55.

Montero, W., and J. W. Dewey, 1982, Shallow-focus seismicity, composite focal mechanisms and tectonics of the Valle Central, Costa Rica: Bulletin of the Seismological Society of America, v. 71, p. 1611-1626.

Obando, L. G., 1986, Estratigrafía de la Formación Venado y rocas sobreyacentes (Mioceno-Reciente), Provincia de Alajuela, Costa Rica: Revista Geológica America Central, v. 5, p. 73-104.

WAI Staff, 1988, Synthesis report on the hydrocarbon potential of Costa Rica: Houston, Proprietary report prepared for RECOPE but now available for purchase, 200+p.

Circum-Pacific Council for Energy and Mineral Resources
Earth Science Series, Vol. 16
R. L. Miller, G. Escalante, J. A. Reinemund, and M. J. Bergin (Eds.)
Energy and Mineral Potential of the Central American-Caribbean Regions
Springer-Verlag, Berlin Heidelberg 1995

PRELIMINARY TECTONIC OUTLINE OF NORTHERN GUATEMALA

Julio C. Franco A., Luis F. López, and Miguel A. Carballo H.
Ministerio de Energía y Minas, Dirección General de Hydrocarburos , Departamento de Desarrollo Petrolero, Sección de Geología, Diagonal 17, 29-78 zona 11, 01011 Guatemala Ciudad, Guatemala

ABSTRACT

Due to the importance of understanding the structural styles of northern Guatemala and their relations to the occurrence of hydrocarbons, the preliminary tectonic-structural framework is herein presented, with emphasis on the Petén basin. This basin, the largest in the country, is important because the exploration for hydrocarbons has resulted in commercial discoveries.

Geologic, seismic, gravimetric and magnetometric data have resulted in the recognition of the principal structural elements that make up the tectonic environment of the zone. They have been used to prepare a theoretical model that explains the evolution of the Petén basin from the Late Jurassic to the Tertiary.

A series of extensional movements mark the beginning of the basin's evolution, followed by salt related phenomena, compressive events, and wrench faulting. This sequence of tectonic events has induced groups of structural features that are regionally associated with the movement of the North American and Caribbean Plates, and with the opening of the Gulf of Mexico.

The structural style of the basin is principally due to the tectonic movements from the Cretaceous to the present.

INTRODUCTION

The Mesozoic sedimentary basin of the Petén, located in the department of the same name in northern Guatemala, covers an area of approximately 50,000 km^2. Geologically, it extends north and west into Mexico (Figure 1). To the east, with the exception of the Maya Mountains, it covers the major part of Belize. Its southern limit is the zone of concurrent left lateral faulting referred to as Polochic. Traditionally, it has been divided in two sub-basins; southern Petén (or Chapayal) and northern Petén (or Paso Caballos), which are separated by the structural high called La Libertad Arc. The accumulations of hydrocarbons that are known in this area are not evenly distributed. Rather they are concentrated in the southern part which has better developed structures.

Since 1930, a large quantity of information has been obtained from the Petén's exploration history. From this information, which includes surface geologic maps, gravimetric, aerial magnetometric, and seismic surveys and well data, a series of structural and stratigraphic interpretations have been generated that give a better understanding of the area's geologic history. However, doubts still exist as to the origin and evolutionary sequence of some of the features.

The purpose of this paper is to describe the tectonic processes that are responsible for the development of the Petén basin and their relation to the structural features that constitute its structural framework both in the concept of time and space. This should define a preliminary model that will serve as the criterion for the selection of possible exploration and should form the basis for a complete study to guide future development. The study is restricted to the northern zone of Guatemala because it is there where the major part of the information is available and because of its importance in the exploration and exploitation of hydrocarbons. Comparison is made between the surface and subsurface structural features and a rough draft is made of their genetic relationship with the regional tectonics so as to explain their tectonic evolution from the Late Jurassic to the Tertiary.

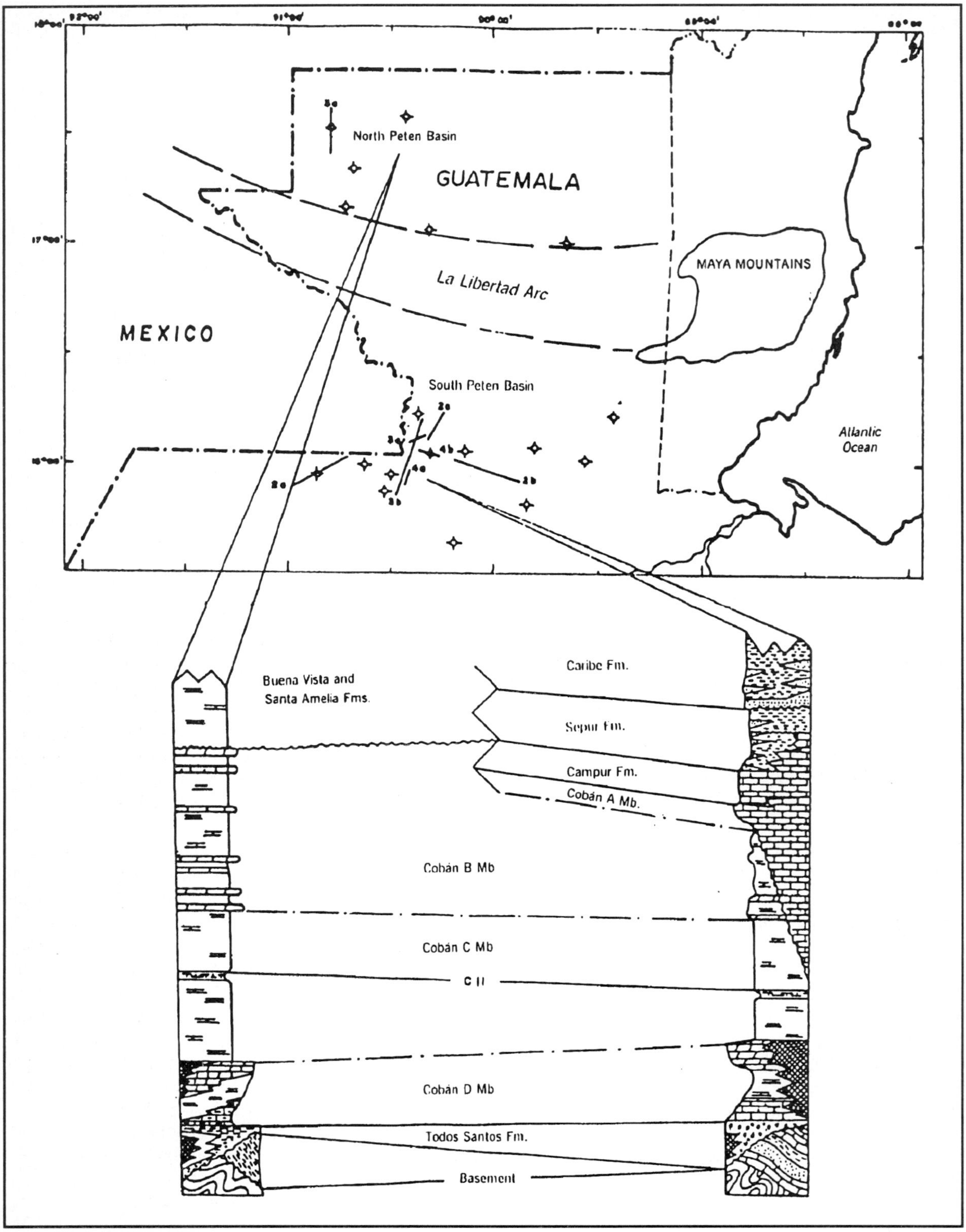

Figure 1. Locality map and stratigraphy of the Petén basin, Guatemala.

GEOLOGY OF THE PETÉN BASIN

Stratigraphy

The sedimentary sequence of the Mayan Block (Dengo, 1983) that occupies the present portion of the Petén basin lies on metasediments equivalent to rocks of the Santa Rosa Group (Dollfus and De Mont-Serrat, 1868) of pre-Permian age (Figure 1).

The initial rifting, existent during the opening of the Gulf of Mexico, controlled the sedimentation of the Late Jurassic-Early Cretaceous that resulted in the sedimentation of "redbeds" of the Todos Santos Formation (Sapper, 1894) (Table 1). During this extension and subsidence, which is common to passive margins, the sedimentation was continued into the Late Cretaceous deriving in the carbonate platform deposits of Member D of the Cobán Formation (Wug et al., this volume). The combined phenomena associated with the extensional phase has been designated Phase I (Norconsult, 1987).

The collision of the volcanic arc of the ancestral Greater Antilles with the southern part of the Yucatan block marks the beginning of Phase II. This was characterized by the upheavals and folds of an evaporitite-carbonatite sequence of the members B and C of the Cobán Formation (Wug et al., this volume).

During the Jurassic, uplift of the La Libertad Arc greatly controlled the sedimentation, dividing the Petén basin into two depocenters, north and south.

Table 1. Summary of tectonic-stratigraphic events of the Petén basin.

TIME (m.y.)	PERIOD	PHASE	STYLE	EVENT	STRATIGRAPHY
	Pliocene	III	TRANSCURRENCE	Folding and transcurrence in the southern and western margins of the basin.	Caribbean
	Miocenc				
	Oligocene				— Major Discordance —
50	Eocene	II	COMPRESSION	Laramidic folding in the west of the basin (culmination in the Eocene)	Santa Amelia - Cambio
	Paleocene				— Discordance —
	Ku			Collision with the Antillean Arc in the south. Uprising and extensive folding in the Peten Basin	Sepur Campur Coban A
					— Major Discordance —
100	Kl	I	EXTENSION	Passive platform	Coban B Coban C Coban D Todos Santos
150	J			Rifting	

During the Late Cretaceous, under conditions of compressional stresses, the formation of a major depocenter in the southern Petén basin resulted in the sedimentation of carbonates in the Cobán Formation (Member A) (Wug et al., this volume) and the Campur Formation (Vinson, 1962). This tecto-sedimentary regime (Phase II) continued to the end of the Late Cretaceous, at which time the southern block (the ancestral Chortis Block) emerged to become a source of clastic turbidites named the Sepur Formation by Sapper (1899). The Laramide orogeny that ended in the early part of the Oligocene took place during this phase. The strong unconformity at the base of the Tertiary marks the upper limit of the compressional phase (Table 1).

In the northern part of the basin, mild subsidence without significant tectonism resulted in deposition of the Santa Amelia and Buena Vista formations (Figure 1) (Vinson, 1962).

The tectonism (Phase III) that followed the compressional phase was characterized by left lateral faulting, sedimentation of the clastic-Caribe Formation (Vinson, 1962), and erosion. This phase began in the Miocene and continues to date. It is very closely tied to the transcurrent left-lateral faulting that forms the boundary between the North American and Caribbean Plates.

Regional Structural Framework

Features of the tectonic patterns that characterize the Petén basin are observed at the surface. One of the outstanding ones is the northwest-southeast orientation of the folds and faults in the western part of the basin with concave structures toward the northeast. Also perceivable is the influence of the transcurrent faulting in the orientation of the folds in the proximity of the zone of the Polochic Fault and, thirdly, the halotectonic (salt related) phenomena in the southwestern part of the basin.

The aeromagnetic data identify the La Libertad Arc as the barrier between the great depocenters of shallow basins to the northeast, and the deeper basins to the north and south (Figure 1). When this data is analyzed, the relationship between the arc and the Maya Mountains becomes apparent, as well as the westward extension of the arc as far as the Sierra del Lacandon. In the central part of the basin, the arc loses its aeromagnetic definition due to the presence of magnetic matter in the south which belongs to the horst of the Maya Mountains and its associate graben of the Sarstun Portal.

STRUCTURAL STYLES

This subject describes the tectonic evolution of the basin. It has been divided in four sub-sections, each one corresponding to a structural style. The mechanics that are responsible for each of the styles will not be considered, but rather a description of same and their relation with the present configuration of the basin.

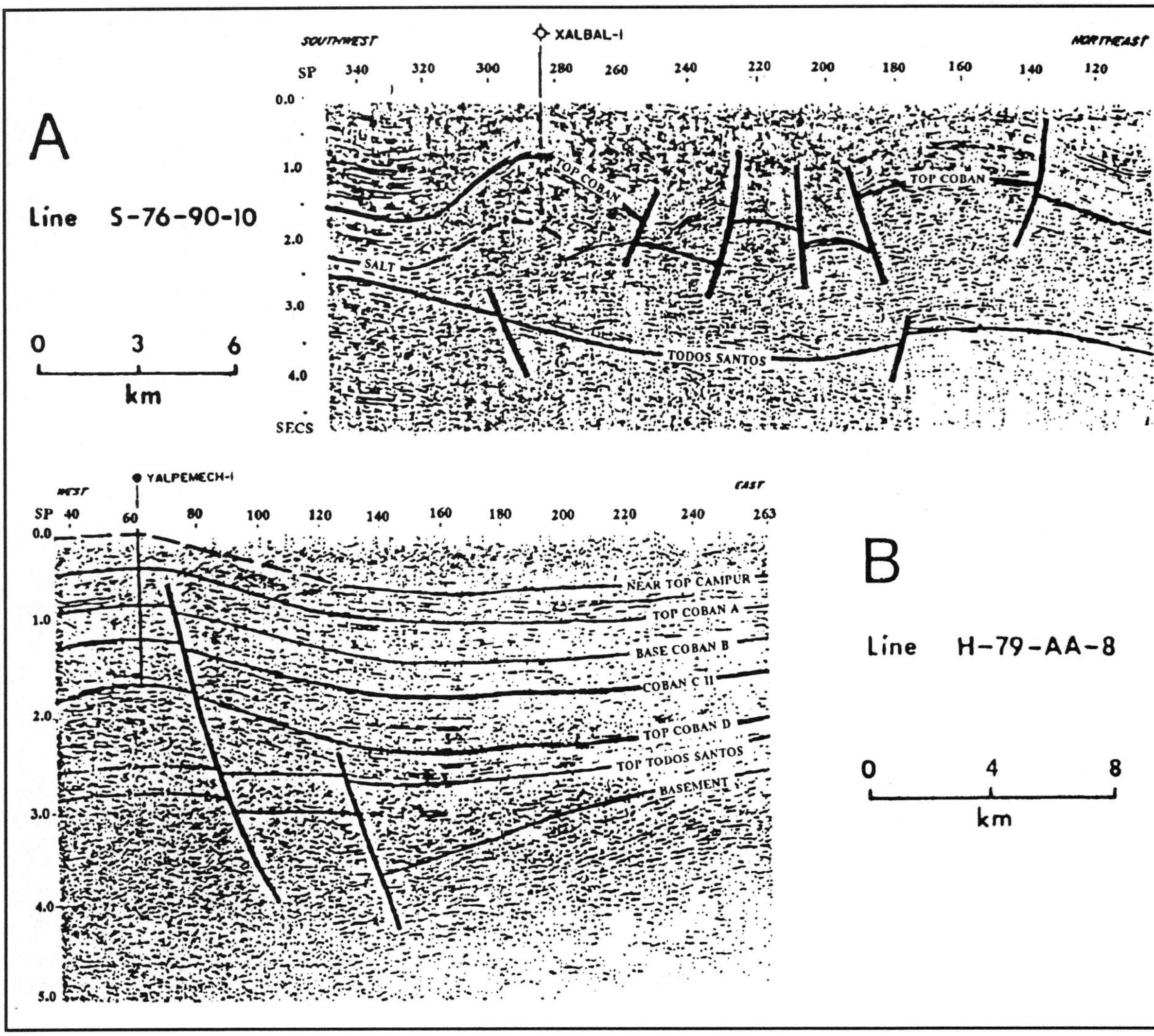

Figure 2. Seismic sections showing features of the following structural styles: (a) extensional and compressive; (b) extensional; and (c) extensional.

Extensional Style

The initial extensional movements of the Petén basin are bound to the expansion of the Atlantic's oceanic floor during the final phase of the opening of the Gulf of Mexico during the Tithonian. Such opening was increased after the accumulation of salt deposits, with acceleration of the expansion of the continental crust under the Gulf's central part and with the beginning of the separation of the Yucatan platform from the North American Plate, moving toward the south up to its present position (Salvador, 1987). During this period, the greater part of the basin remained as an emerged terrestrial mass, excepting the southwestern end (southern basin), where the sedimentation of red beds, limestones, shales, and evaporites of environments of superficial platform and engulfings (its) was in progress. In Figure 2, evidences of the initial rifting of the southern basin are shown. In Figure 2a, the high angle normal faults (65-70°) displace the top of the Todos Santos Formation (JTS). Present here is their apparent convergence in depth and inclination in direction of the regional dip of said formational top, which clearly shows a concave form between said faults. In Figure 2b, the normal faults of variable displacement and extension are also at high angle (60-70° East) with a certain tendency to converge at depth. Its dips are contrary to those shown regionally by sediments underlying the top of the JTS. It should be noted that prior to the faulting, the basement configuration corresponded to an arc, which allowed for a greater accumulation of sediments in the central part. In Figure 2c, the extensional style is again observed, where the two faults of lesser extension toward the

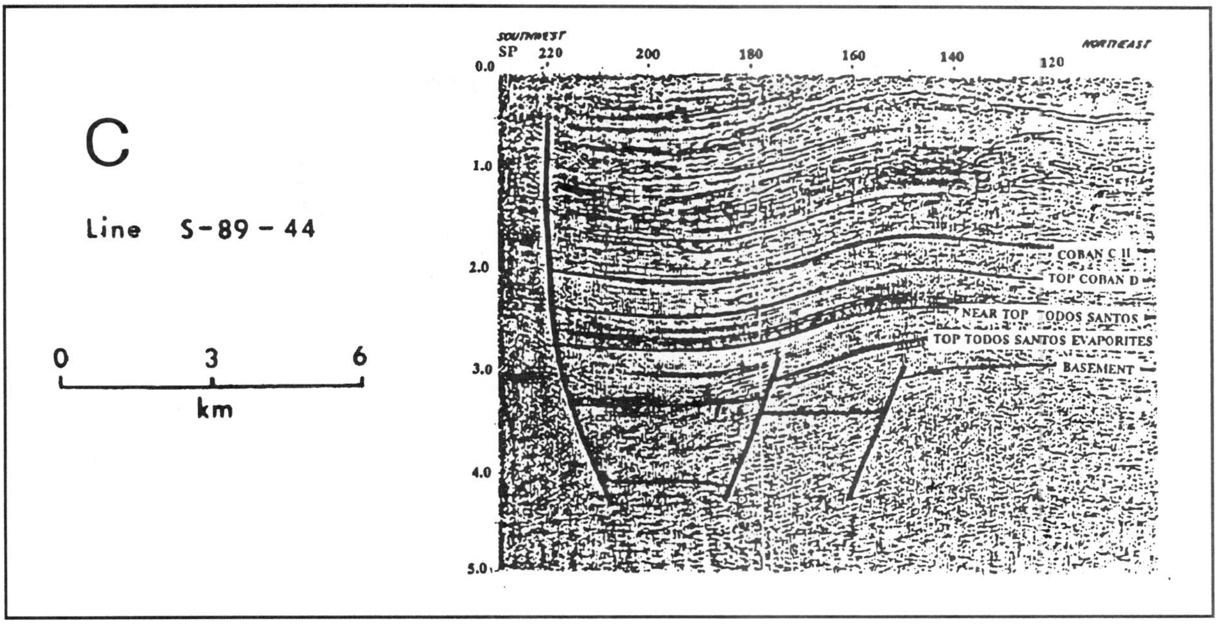

Figure 2. Continued.

northeast affect the evaporites at the top of JTS. When the three sections are observed jointly, it can be seen that the basement has an undulatory form. In the northern basin (Figure 3a), this type of extensional movement has not been observed, which indicates that this zone is under a relative tectonic calm.

Compressive Style

This style is one of the most dominant in the basin and presents a complex structural history. This structure begins in the Neocomian and continues to the end of the Tertiary. Figure 3b shows a seismic section located in the southern basin through the Rubelsanto and Tierra Blanca petroleum fields. The compressive style has originated a combination of disharmonic folds with normal and inverted faults. Noteworthy are the antithetic semi-parallel faults in the Rubelsanto area. All the geological formations from the Upper Jurassic (JTS) until the Paleocene (Sepur Formation) are affected by these phenomena. The combination of these structures and faults are present between Las Mercedes and Tierra Blanca, resulting in a structural configuration in the form of a positive flower (Harding, 1983). In Figure 2a, the compressive style can also be observed above the initial rifting (Xalbal area), which corresponds to a symmetrical anticline which contrasts with those of Figures 3b and 3c, where the Tierra Blanca structure appears faulted and slightly overturned towards the north.

Halotectonic Style

The effects of the halokinetic and probably halotectonic phenomena are best represented in the southern basin, where there are non-perforating domes with outstanding topographic manifestations such as the Tortugas Saline Dome. In the northern basin, the only incipient salt movements that have been detected are in the Sierra del Lacandon (southwest end). Figure 4a is a seismic section perpendicular to the Rubelsanto structure, where the salt movement with its separation surface in contact with Todos Santos Formation may be seen. The salt anticline slightly overturned toward the northeast is non-perforating with outstanding topographical manifestations such as the Tortugas Salt Dome. In the northern basin, only incipient salt movements have been detected in Sierra del Lacandon (southwest end). Figure 4a is a seismic section perpendicular to the Rubelsanto structure, where the movement of the salt may be seen with its surface separation in contact with the Todos Santos Formation. The saline anticline, slightly overturned toward the northeast, is non-perforating. The considerable tapering of the salt is evident in the southwest end, but its swelling expansion toward the northeast indicates that the direction of its movement has been toward this end. The inverted faults converging in the surface do not affect the salt or are not detected by seismic reflection. The non-perforating structure shown in Figure 4b is different from the former one, and from its thickness it can be inferred that the movement has been toward the east. This structure is affected by the listric fault that limits the cuneiform mass in

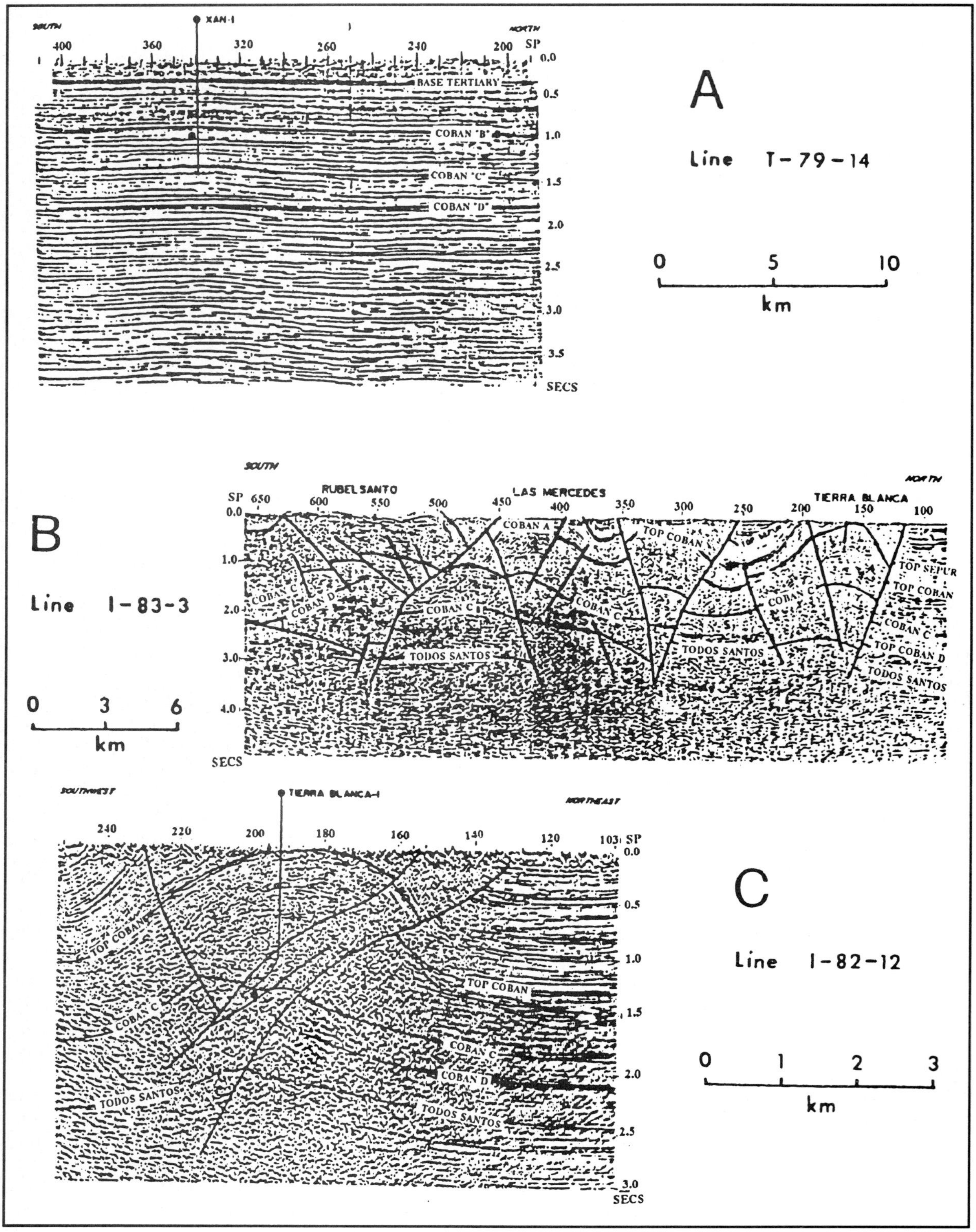

Figure 3. Seismic sections showing features of the following structural styles: (a) without evidence of extension; (b) compressive; and (c) compressive.

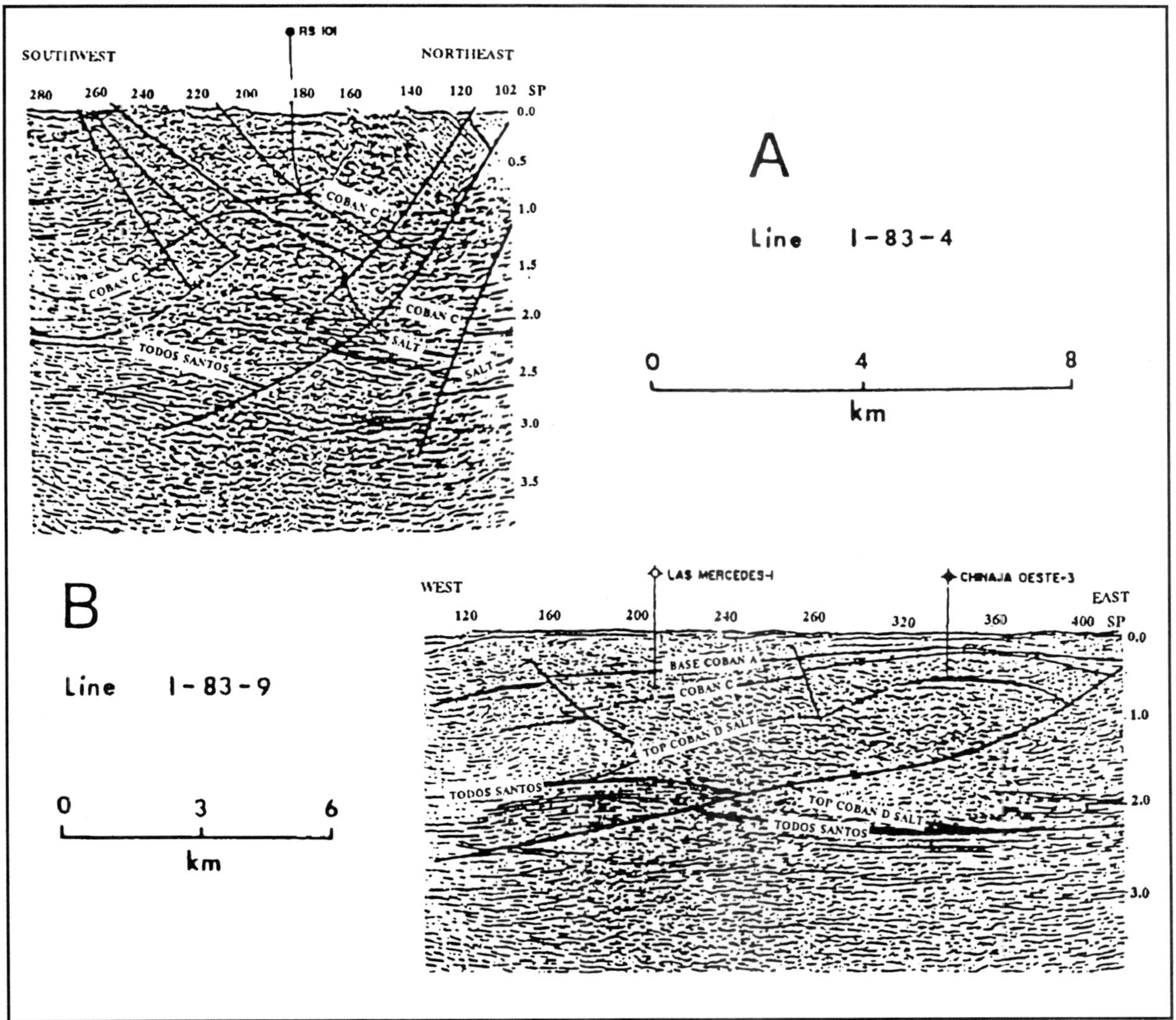

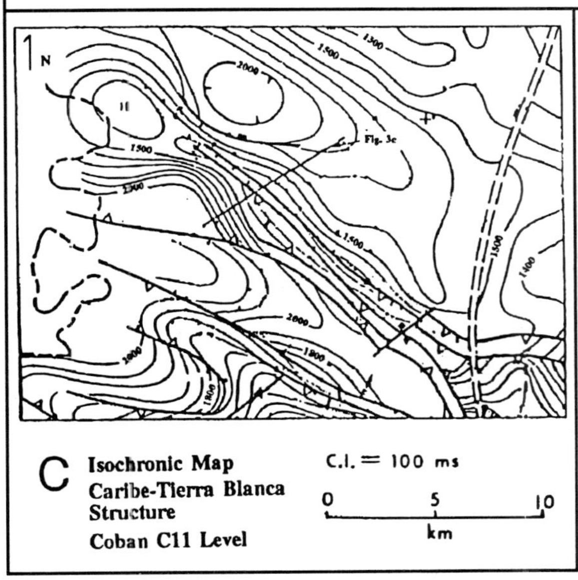

Figure 4. Seismic sections and isochronic map of the following structural styles: (a) halotectonic phenomena; (b) halotectonic phemonena; and (c) transcurrency.

the central part of the section. Noteworthy is the tapering between the Cobán horizons C-11 and the Cobán base A, east of the Chinaja West-3 well which, in a certain form, evidences the vertical pressure of the salt on the overlying stratum.

Transversal Faulting

Although this type of sub-vertical fault with left lateral displacement is difficult to grasp in the seismic sections, it should be mentioned because it is a tectonic event, which associated with other types of deformation mechanisms, has stirred complexity and orientation changes in the Petén basin

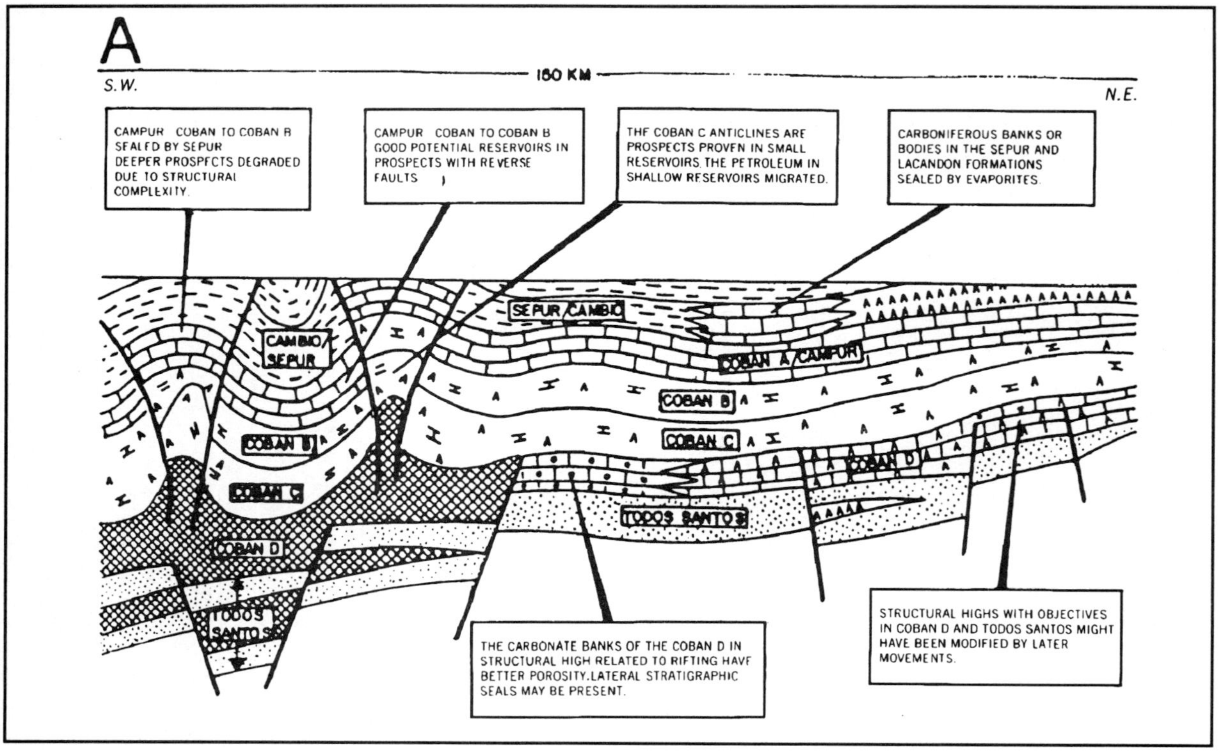

Figure 5. (a) Structural interpretation and entrapment types of the south Petén basin (Norconsult, 1957).

(south) structures. The transversal faults that are spectacular are those oriented east northeast-west southwest in the central portion of the basin. Their age possibly ranges between middle Tertiary and the Pliocene. During this period, the axis of maximum stress that occurred during the Late Neocomian Jurassic was vertical, rotated 90° in a vertical plane with orientation northeast-southwest along which the faults originated. These are responsible for the structural modification of the Laramide folded belts which, in the surface, changes from a north northwest-south southeast bearing in the central-west portion of the basin, to the east northeast-west southwest bearing in the south. This change in the alignment of the structures is quite evident in the SLAR images and in surface structural maps.

Probably these are ancient faults deeply settled in the basement and formed during the Late Paleozoic, which were reactivated during the Tertiary (D. R. Ray, 1983, Preliminary report concerning the structural and tectonic developments in the Petén area, northern Guatemala, internal report to Texaco Exploration Guatemala Inc.). Their bearing coincides with the alignment of La Libertad Arc. In a lesser scale, there are also lateral displacement faults, well located, that are present principally in the south. In Figure 4c, two families of this type of small faults can be seen: one bearing northwest-southeast, which dominates the area, and the other northeast-southwest which is more recent and smaller. The evidence of the horizontal displacement is found in the southeast end, where the small faults displace those of greater extension in left direction. This type of faulting is clearly present in the southern basin, while in the northern one the faulting produced before the Laramide orogeny is of a normal type with dips to the south which afterward were inverted during the said orogeny.

Petroleum Potential of Structural Styles

Associated with the structural styles derived from the various aforementioned structural tecto-sedimentary phases, there are stratigraphic and structural entrapments that are important from the point of view of petroleum (Figure 5). Pillars, anticlines, faults, traps associated with saline tectonics, and overthrusts exist in the southern Petén basin (Figures 2, 3, 4, and 5), and changes of facies, bioconstructions, stratigraphic thinnings, and diagenetic entrapments exist in the northern Petén basin (Figures 3a and 5).

CONCLUSIONS

Three tecto-sedimentary phases are recognized which have originated four different structural styles: extensional, compressive, halotectonic, and transcurrent. The conjunction

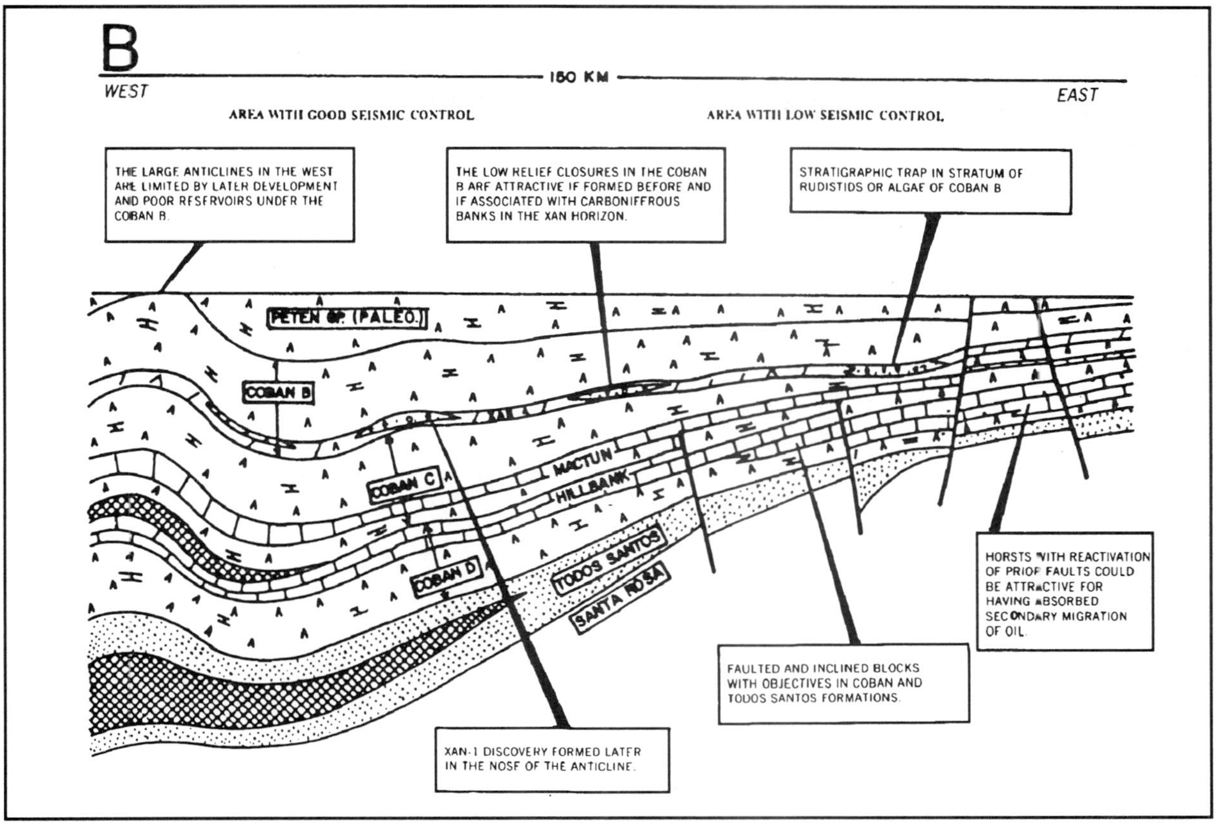

Figure 5. (b) Structural interpretation and entrapment types of the north Petén basin (Norconsult, 1957).

of these styles, with petroliferous characteristics of the lithological succession, has generated conditions for the generation, accumulation, and entrapment of hydrocarbons.

The description of the various structural styles and their chronological and spatial location can support future work for the formulation of hypotheses on the mechanics of deformation and genesis of the various existing styles.

There is a probability of petroleum entrapments at greater depths of the basin, derived from the extensional phase.

REFERENCES CITED

Dengo, G., 1983, Mid-America tectonic setting for the Pacific margin from southern Mexico to northwestern Colombia: Centro de Estudios Geológicos de América Central, Guatemala, 96 p.

Dollfus, A., and E. De Mont-Serrat, 1868, Voyage geoligique dans les Républiques de Guatemale et de Salvador: Paris, France, Imp. Impériale, p. 1-539.

Harding, T. P., 1983, Divergent wrench fault and negative flower structure, Andaman Sea, in A. W. Bally, ed., Seismic expression of structural styles: American Association of Petroleum Geologists Studies in Geology No. 15, v. 3.

Norconsult, A. S., 1987, Petroleum geology of Guatemala: Final report of the Technical Assistance in Petroleum Geology and Geophysics Project, Guatemala (GUA/83/007): United Nations Development Program.

Salvador, A., 1987, Late Triassic-Jurassic paleogeography and origin of Gulf of Mexico basin: American Association of Petroleum Geologists Bulletin, v. 71, p. 419-451.

Sapper, K., 1894, Grundzüge der physikalischen geographie von Guatemala: Gotha, Germany, Peterm. Mitt. Erg. -H. 113.

Sapper, K., 1899, Gber gebrigsbau und boden des Nördlichen Mittelamerikia: Gotha, Germany, Peterm. Mitt. Erg. -H 127.

Vinson, G. L., 1962, Upper Cretaceous and Tertiary stratigraphy of Guatemala: American Association of Petroleum Geologists Bulletin, v. 46, p. 425-456.

Wug, L. A., R. E. Matias, C. A. De León, and J. S. De La Cruz, this volume, La Formación Cobán en la cuenca Petén, Guatemala.

Circum-Pacific Council for Energy and Mineral Resources
Earth Science Series, Vol. 16
R. L. Miller, G. Escalante, J. A. Reinemund, and M. J. Bergin (Eds.)
Energy and Mineral Potential of the Central American-Caribbean Regions
Springer-Verlag, Berlin Heidelberg 1995

THE COBÁN FORMATION IN THE PETÉN BASIN, GUATEMALA

L. A. Wug, R. M. Matias, C. A. De Leon, and J. S. De la Cruz
Ministerio de Energía y Minas, Dirección General de Hidrocarburos, Departamento Desarrollo Petrolero,
Sección de Geología, Diagonal 17, 29-78 zona 11, 01011 Guatemala, C.A.

ABSTRACT

The Cobán Formation of Cretaceous age is known from well data in both North and South Petén subbasins. These subbasins are separated by the east-west trending Arco de la Libertad. Four lithologic subdivisions have been recognized in the wells, which from the top down have been called Cobán A, B, C, and D. This paper formalizes these subdivisions, here designated as members, and provides lithologic descriptions of each member together with its stratigraphic relations to adjacent members. The Cobán has an average thickness of about 16,000 ft. It was deposited in a shallow water platform environment. The Cobán is overlain by the Campur Formation of Late Cretaceous age and underlain by the Todos Santos Formation of Jurassic age.

INTRODUCTION

The present work consists of a clarification of the stratigraphy of the Cobán Formation in the subsurface of the Petén basin, and includes technical data related to its formation. The purpose is to give formal definitions that will help in its identification, mapping, and interpretation in the subsurface. The information consulted consists mainly of unpublished technical material available at the Dirección General de Hidrocarburos.

THE PETÉN BASIN

In Guatemala, three sedimentary basins are present, which differ from each other both structurally and stratigraphically: Pacific to the south, Amatique to the northeast, and Petén to the north.

The Petén basin is divided into North Petén or Paso Caballos and South Petén or Chapayal, which are separated by the anticline Arco de la Libertad, which has an east-west trend. It is believed, however, that both subbasins were intimately linked during the greater part of their history. This basin has an area of approximately 60,000 km^2 and covers most of northern Guatemala. It is a lithologic element that developed during the Mesozoic-Cenozoic when approximately 30,500 ft of sediments were deposited on Permian limestones and shales. The basal unit of this sequence is composed of Jurassic clastic sediments, salt, and other evaporites; on this lies a thick transgressive sequence of Cretaceous carbonates and evaporites, capped by clastic regressive sediments (Figure 1).

THE COBÁN FORMATION

The Cobán Formation (Cretaceous: Barremian-Cenomanian) is a part of the stratigraphic sequence of the Petén basin. In the North Petén subbasin it is overlain unconformably by the Buena Vista and Santa Amelia (Tertiary) Formations. In the South Petén subbasin it is overlain conformably and sometimes gradationally by the Campur Formation (Santonian-Maestrichtian), and overlies the Todos Santos Formation in an apparent gradational contact (Jurassic).

Historical Background

The first one to use the term Cobán Formation was Sapper (1899), who gave this name to a thick sequence of

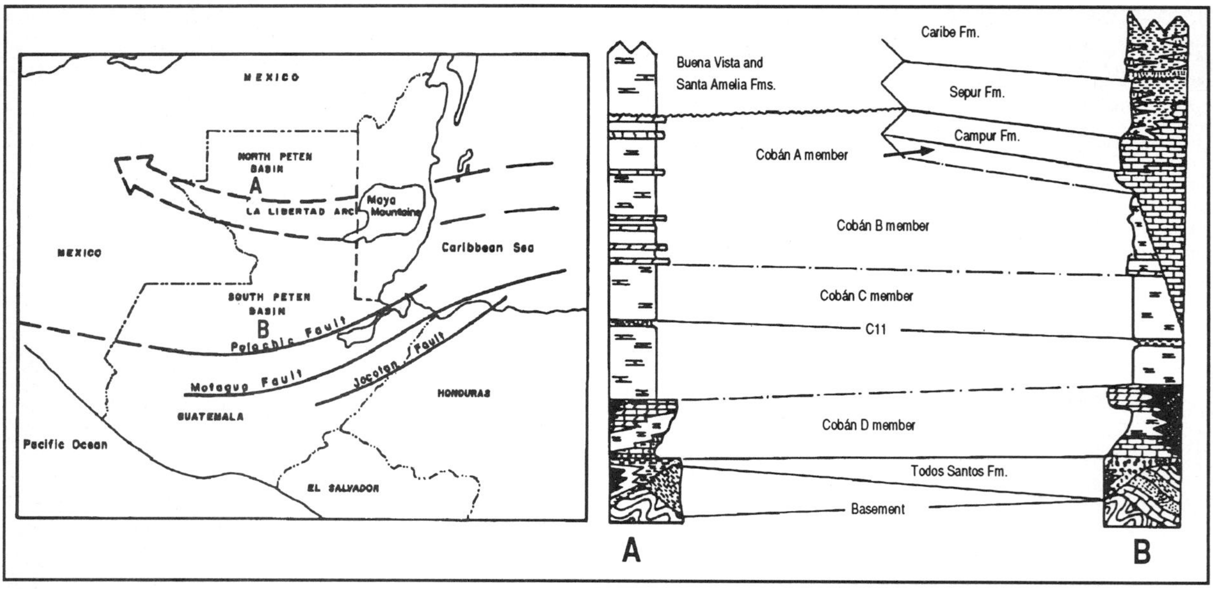

Figure 1. Stratigraphic columns showing generalized lithology and relations of the Cobán Formation and overlying and underlying formations for both the North Petén subbasin and the South Petén subbasin.

Cretaceous carbonates and evaporites located near the town of Cobán, Alta Verapaz, central Guatemala. At the beginning of oil exploration, the first descriptions of the Cobán Formation were obtained from deep wells. Several names and ages were designated, including local divisions, and mentioning some correlations with nearby formations. In 1976 Leigh and Buis (1976) divided the formation in the Rubelsanto area (South Petén subbasin) into three members, naming them from the top down A, B, and C.

Later, Shenandoah Chisec No. 1 well (Shenandoah Guatemala, Inc., 1978, unpublished final well report) drilled a saline interval underlying the Cobán C Member which could be correlated with a halite sequence drilled initially in the Tortugas area. It has been named Chisec Salt.

The Getty Oil Yaxa No. 1X well (Getty Oil Guatemala, 1980, unpublished final well report) used the divisions of Leigh and Buis for the sequence of Cobán A-B-C and, in addition, drilled an interval which showed the top of a carbonaceous unit that lacks the anhydritic intercalations of Cobán C, and which underlies it. It was named Cobán D.

Elf Aquitaine Guatemala (1983, unpublished geologic synthesis of Guatemala) indicated that the Chisec Salt of the Shenandoah well is equivalent in age to Cobán D of the Getty well and to the interval named Hillbank Dolomite of the Esso Standard 1963 well, Petén Itza No. 1 (Esso Standard Guatemala, 1963, unpublished well report).

The divisions of the Cobán proposed by Leigh and Buis with the addition of Cobán D of Getty Oil are now generally accepted and are in common usage. Type sections are here proposed to facilitate identification and to unify the nomenclature as used in the subsurface.

DIVISIONS OF THE FORMATION

It is here proposed to formally divide the Cobán Formation in the subsurface into four members, Cobán A, B, C, and D, based on the relative positions, the thicknesses, and the number of layers of carbonates related to evaporites. The basal unit of the sequence is Cobán D.

Cobán A Member

The well Rubelsanto No. 3, drilled by Shenandoah Guatemala, Inc. (1976, unpublished final well report) (coordinates 15°58′34′′N, 90°23′06′′W), has been designated as the type section of the Cobán A Member. It is 2,037 ft thick in the well, but in a nearby surface section it is 2,052 ft thick. Lithologically it consists of limestone and peletoidal dolomitized limestone, dolomite and calcareous dolomite, calcareous lutites and calcarenties, with sparse breccia and conglomerate, and with milliolids and fragments of other fossils. Diagnostic fossils in the type section are absent. These rocks were deposited in a shallow marine environment of lagoonal type behind a barrier bar and they transgress toward the platform.

Regional Aspects

The Cobán A Member is present in the South Petén subbasin and absent in the North Petén subbasin except on the north flank of the Arco de la Libertad. Its average thickness is 2,000 ft with a minimum in the San Roman No. 2 well of

594 ft (Elf Aquitaine Guatemala, 1982, unpublished final well report), whereas in the Las Casas No. 1X well (Getty Oil Guatemala, 1981, unpublished final well report) the maximum thickness exceeds 5,000 ft (Figure 2). Its upper contact is with the Campur Formation (Santonian-Maestrichtian), which in places is gradational through a transition zone. Its lower contact is unconformable on different parts of the underlying members B and C, varying from an apparent gradation with the topmost Cobán B down to the lowest part of the C Member (C12).

Investigation of outcrop samples and well samples shows the presence of *Heterohelix, Hedbergella, Spiroplectamina, Pithonella spherica, P. ovalis, Rh. laurinensis, Nummoluculina, Murciella cuvillieri, Omphalocyclus* cf. *macrophorus, Pararotalia* cf. *schambi, Raadhoovenia* cf. *saletina, Whiteinella paradubia, Strepalveolina mexicana*, and *Pseudorapydionina*. On the basis of studies of the fauna, the assigned age is Turonian-Senonian.

Cobán B Member

For a type section of this member the well Rubelsanto No. 3 (see Member A) is selected, where the thickness is 3,195 ft (interval 2,052 ft to 5,247 ft). It consists of massive anhydrites with intercalations of carbonates and in some places the presence of clays. The upper contact is with the member A and the lower contact is with the member C. It was deposited in an environment of the coastal sabkha type, with variations between the low tide and high tide sediments in a transgressive regime of continuous regional subsidence.

Regional Aspects

The member Cobán B is divided into 12 units (B0 to B11). It is partially absent in the South Petén subbasin and in the Arco de la Libertad due to the pre-Cobán A erosion. In the North Petén subbasin, the complete section underlies a cover of Tertiary sediments. In the Rubelsanto area, early wells recognized 11 lithologic units (B1-B11). Later, other wells of the area identified another lithologic unit (B0) at the top. The clearest separation of the lithologic units (B0-B11) marks the beginning of tectonic instability (Elf Aquitaine Guatemala, 1983, unpublished geologic synthesis of Guatemala).

The thickness of the Cobán B Member is variable, reaching up to 6,000 ft in a complete section (Figure 3). It consists of alternate anhydrites and carbonates, with some intercalations of clay and halite. The carbonate content varies

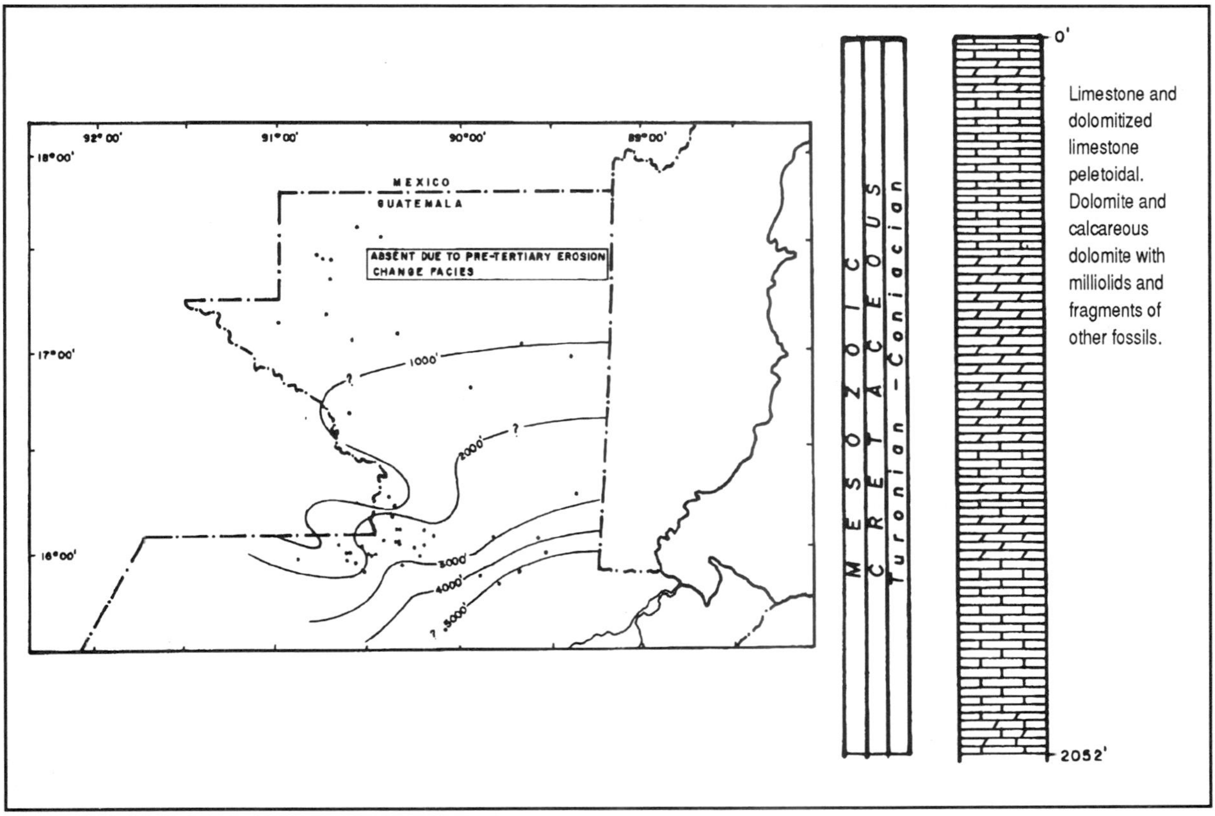

Figure 2. Stratigraphic column showing generalized lithology (right) and thickness (left) of Cobán A Member.

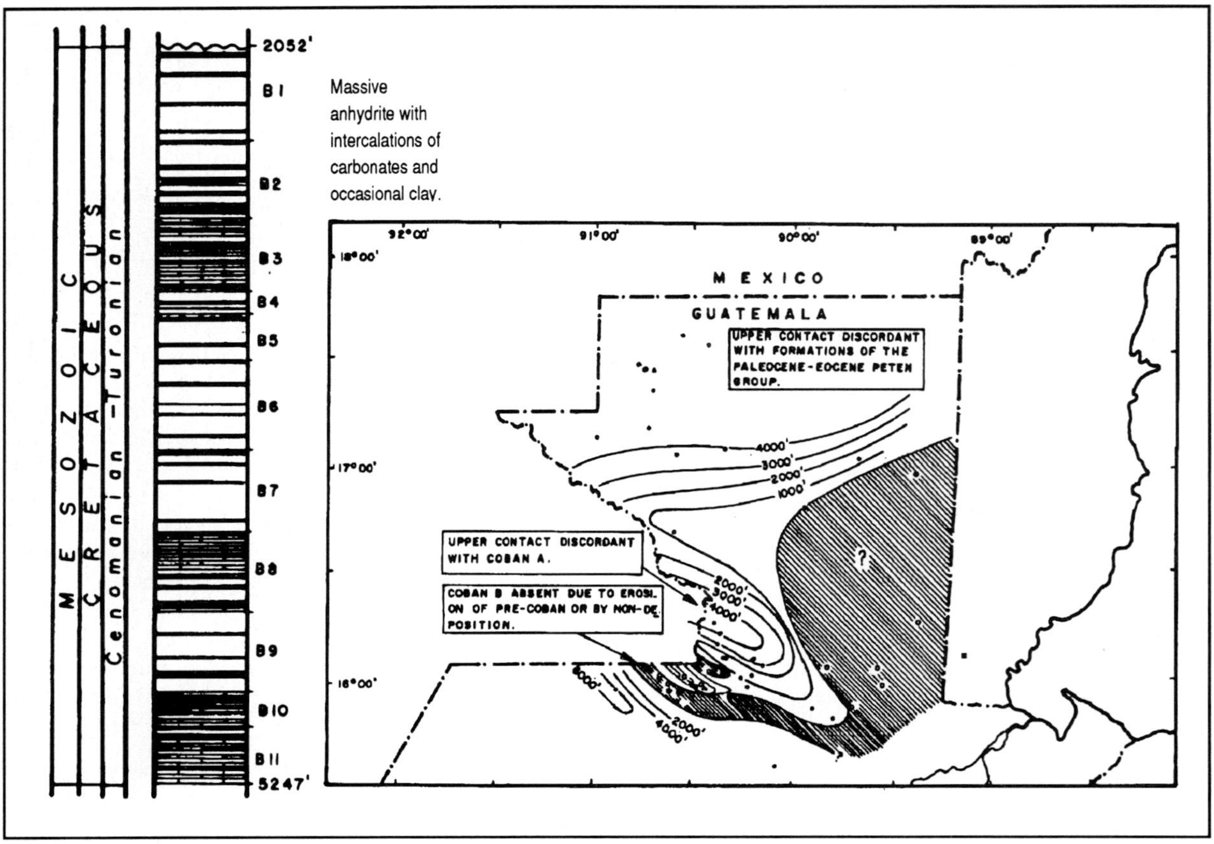

Figure 3. Stratigraphic column showing generalized lithology (left) and thickness (right) of Cobán B Member.

between 30 and 50 percent of the member. Where not overlain by Cobán A, its upper contact is unconformable with the Tertiary formations of the Petén Group (Buena Vista and Santa Amelia). Its lower contact is conformable with Cobán C.

In the North Petén subbasin it has been named "Zona de Xan" (Texaco, 1981, Xan No. 1 well, unpublished final well report), the interval that correlates with the Cobán B8-B11 of the South Petén subbasin. The Cobán B Member was deposited in a shallow platform environment. Its age is probably between Cenomanian and Senonian. The fauna present in the type section includes *Glomospira* sp., *Nummoloculina heimi*, and milliolids.

Cobán C Member

The Cobán C Member consists of 19 units, with the C19 unit at the base of the sequence. The proposed type section is in the San Diego No. 1 well (Hispanoil Guatemala, 1982, unpublished final well report) (coordinates 16°03´35.506´´N, 90°06´11.134´´W) in the depth interval 6,355 ft-12,887 ft, a thickness of 6,532 ft). It consists of massive beds of anhydrite with intercalations of carbonates and clay. Its upper and lower contacts are with the members B and D, respectively (Figure 4). Its depositional environment is similar to that of the member Cobán B.

Regional Aspects

Individual units of Member C of the Cobán Formation have lateral continuity. Total thickness is variable and exceeds 7,000 ft when the section is complete. It consists of anhydrite (80 percent) with intercalations of dolomite and limestone (20 percent). In some intervals the presence of clays is significant (C11) and beds of halite have been reported in the North Petén subbasin.

Its upper contact is conformable with the Cobán B11 in a normal sequence. Where the complete member is present, the contact is with the Cobán A Member. In the South Petén subbasin, because of the effect of the unconformity at the base of Cobán A, its topmost beds range from C3 through C12. Its lower contact is conformable with Cobán D. Lithologically the units C1 to C10 consist of massive anhydrite with intercalations of dolomite. From C11 to C12 anhydrite predominates with intercalations of clay and dolomite. Because of the presence of layers of clay, this horizon is an excellent seismic

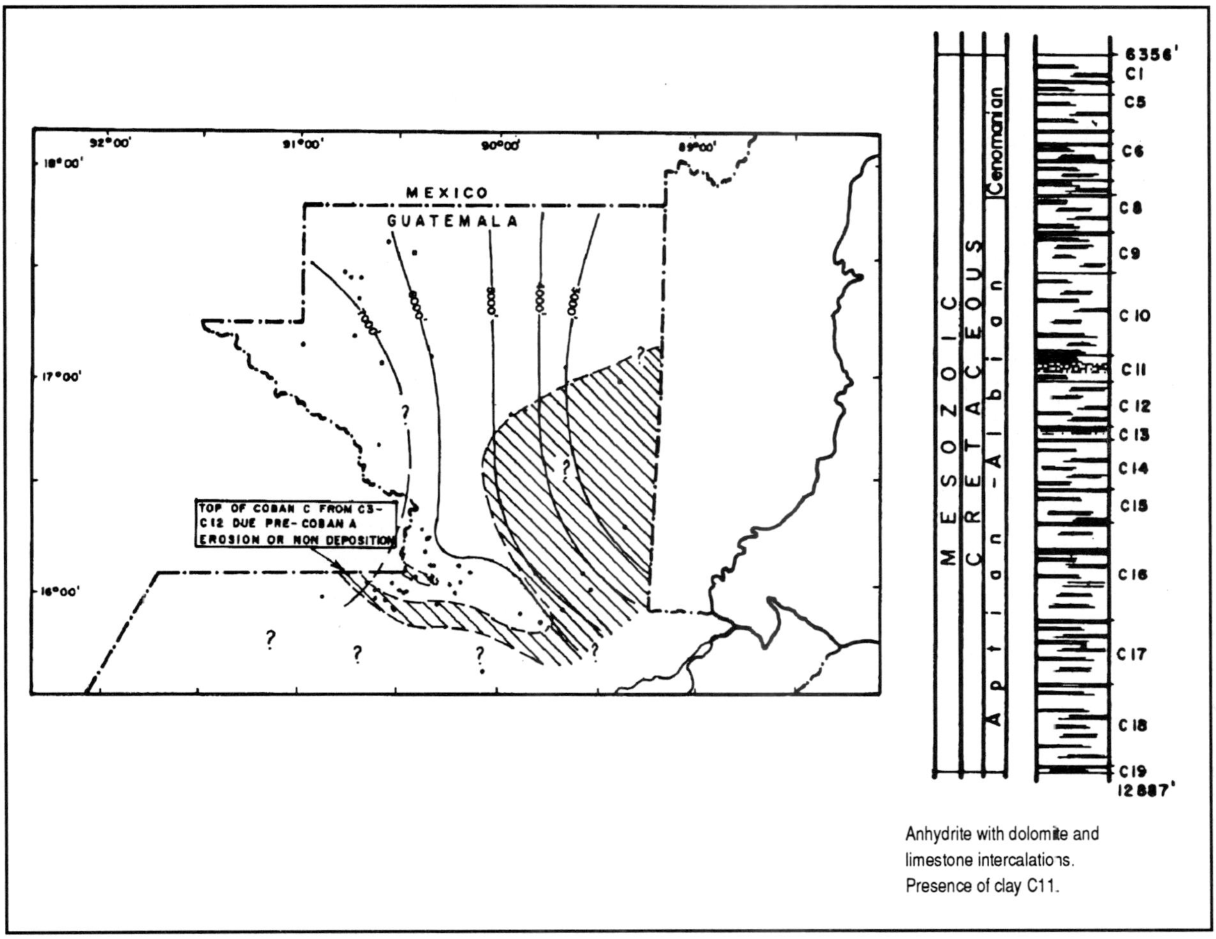

Figure 4. Stratigraphic column showing generalized lithology (right) and thickness (left) of Cobán C Member.

marker, mappable throughout the basin because of its acoustic impedance.

Units C13 to C19 consist of anhydrite with intercalations of limestone and dolomite. The C13 carbonate, originally known in the South Petén subbasin as "first limestone marker" and in the North Petén subbasin as "Caliza Mactun," has an approximate thickness of 70 ft. It was deposited on a broad shallow water platform. It has been assigned an age of late Aptian to Cenomanian, based on the fauna present in the type section: *Ovalveolina reichel, Nautiloculina, Thaumatoporella parvovesiculifera, Tritaxis, Cuneolina, Cacinella irregularis, Marinella lugeoni*, stromatolites, and lamelibranchs.

Cobán D Member

The Cobán D Member is divided into three units: upper, middle, and lower. Like the C Member, the type section is located in the San Diego No. 1 well (Hispanoil Guatemala, 1982, unpublished final well report) in the 12,887 ft to 15,866 ft interval, a thickness of 2,979 ft. It consists of massive carbonates and anhydrites with intercalations of thin carbonates and clays. The base of Cobán D is probably in unconformable contact with the Todos Santos Formation. The upper contact is with the Cobán C19 (Figure 5). Its sedimentary environment is similar to that of members B and C.

Regional Aspects

The Cobán D Member has been penetrated in full by few wells. Its thickness varies between 750 ft (Getty Oil Guatemala, 1981, Tzuncal No. 1X well, unpublished final well report) and 2,979 ft (Hispanoil Guatemala, 1982, San Diego No. 1 well, unpublished final well report). The lithology of the upper part of the member consists of massive limestone, the uppermost part of which has been dolomitized (the dolomitized zone varies in thickness). Norconsult (1987)

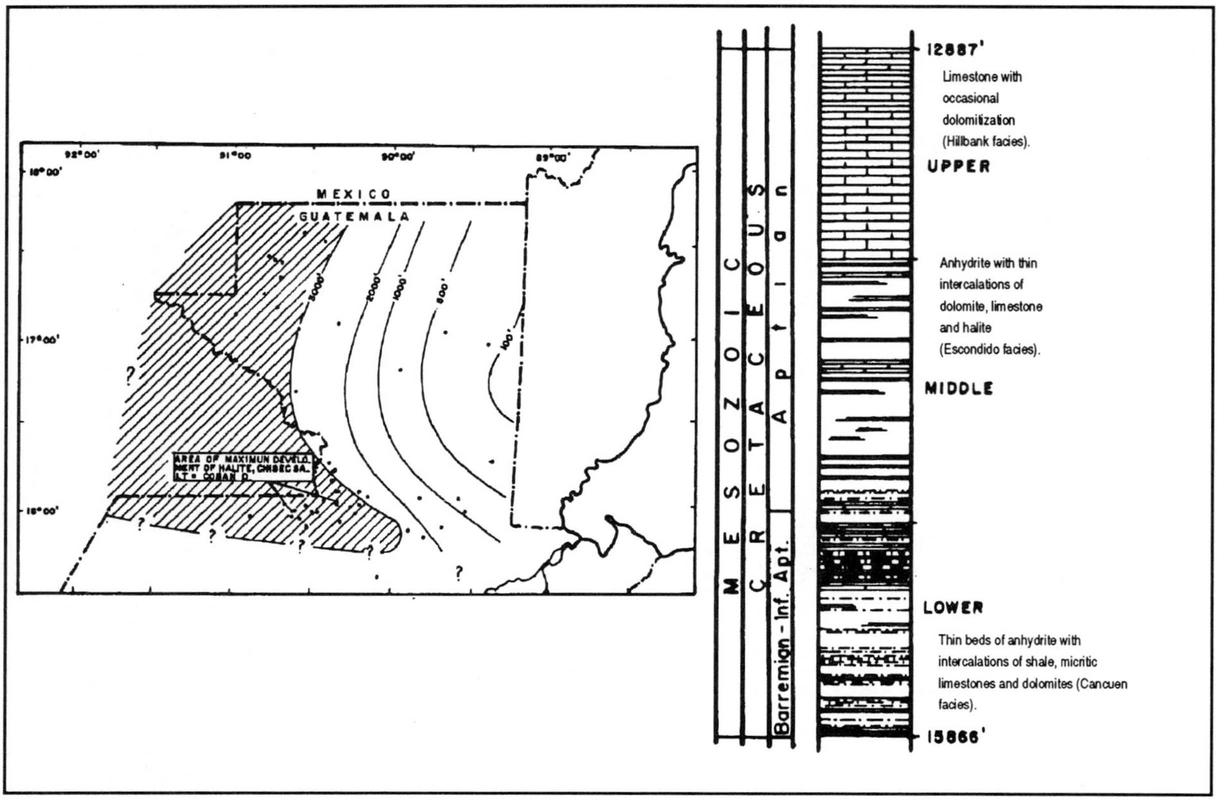

Figure 5. Stratigraphic column showing generalized lithology (right) and thickness (left) of Cobán D Member.

names this interval the Hillbank facies. The middle Cobán D Member consists lithologically of anhydrite with intercalated thin layers of dolomite, limestone, and halite similar to those found in the Cobán C Member. Norconsult (1987) named this sequence Escondido facies. The lower Cobán D Member consists of lutite, limestone, and dolomite intercalated with thin layers of anhydrite. The lower part is characterized by its major content of lutite, which diminishes going upward. Norconsult (1987) named it Cancuen facies because the Cancuen No. 1X well (Getty, 1980, unpublished final well report) shows an abundance of lutite. The rocks of this facies overlie the clastics of the Todos Santos Formation of Jurassic age.

A lateral equivalent of the Cobán D Member is named Sal (salt) Chisec, which consists predominantly of halite with minor intercalations of anhydrite, limolitic clay, traces of dolomite, fossiliferous limestone, sandstone, pyrite grains, and carbonaceous fragments. In the South Petén subbasin its thickness is hard to estimate, but may be between 1,000 ft and 1,500 ft. A thickness of 4,920 ft was drilled in the Nueve Cerros No. 1 well (Shenandoah Guatemala, Inc., 1976, unpublished final well report) without reaching its base, but this is probably not its true drilled thickness because of flowage. The Sal Chisec underlies Cobán C19. The lateral transition of Cobán D to Sal Chisec in the South Petén subbasin is very abrupt. In the North Petén subbasin, the salt intertongues with the upper part of Cobán D, but pinches out to the east. According to Norconsult (1987), lateral changes in thickness of the salt seem to be gradual because seismic evidence detected no abrupt changes. On the basis of microflora studies, the Cobán D Member has been assigned a Barremian-Aptian age. Its upper contact is with the Cobán C Member and its lower with the Todos Santos Formation (red beds). The fauna present in the type section of the Cobán D Member includes *Pseudocyclammina hedbergi*, *Marssonella* cf. *Oxycona*, *Tritaxia*, *Ophthalmidium*, *Lenticulina*, *Arenobulimina*, *Glomospira*, *Orbitolinopsis*, *Haplophragmium*, bryozoans, gastropods, and remnants of crinoids.

CONCLUSIONS

The Cobán Formation is present in the entire Petén basin in the subsurface, and has an approximate thickness of 5,000 m (16,450 ft). Its lithology consists predominantly of carbonates (limestone and dolomite) and anhydrite with sporadic thin strata of clay, lutite, and halite except in its upper part where it consists totally of carbonates.

The depositional environment of the Cobán Formation is predominantly that of a shallow platform. On the basis of the relative positions, the thicknesses, and the number of layers of carbonates with relation to the evaporites, the formation has been divided into four members named Cobán A (Turonian-Senonian), B (Cenomanian-Senonian), C (upper Aptian-Cenomanian), and D (Barremian-Aptian). For each one the type locality is a well where the lithologies are characteristic and are recognizable over a wide area. The lower contact of the Cobán Formation is with the Todos Santos Formation (Jurassic). Its upper contact is with the Campur Formation (Upper Cretaceous) in the South Petén subbasin and with the Tertiary formations of the Petén Group in the North Petén subbasin. When the sections are complete, the contacts of the members are conformable, but in most of the basin the base of Cobán A is a regional unconformity.

REFERENCES CITED

Leigh, R. A., and O. I. Buis, 1976, Exploración petrolera en Guatemala: Acapulco, Third Congreso Latino-Americano de Geología.

Norconsult, A. S., 1987, Reporte final geología petrolera de Guatemala: v. 2, 113p., unpublished report.

Sapper, C., 1899, Ueber gebingsbau und boden der nördlichen Mittelamerika: Petermanns Mitteilungen Erganzungsneft Band 27, Ergenzungsheft-H 127, vi and 119 p. mit kartan, 25 profilen.

Circum-Pacific Council for Energy and Mineral Resources
Earth Science Series, Vol. 16
R. L. Miller, G. Escalante, J. A. Reinemund, and M. J. Bergin (Eds.)
Energy and Mineral Potential of the Central American-Caribbean Regions
Springer-Verlag, Berlin Heidelberg 1995

THE PETROLEUM POTENTIAL OF THE NETHERLANDS ANTILLES

Thomas E. Daly
Resources Engineering & Management International Inc.,
5680 Greenwood Plaza Boulevard, Suite 414, Englewood CO 80111 USA

ABSTRACT

Seismic surveys conducted in the offshore areas of the Netherlands Antilles indicate that there is a good possibility for the presence of hydrocarbons in commercial quantities in at least two specific areas. Integrated interpretations of these surveys show the presence of thick Cretaceous and Tertiary sediments and large potential traps. Some of these traps exceed 100 km^2 in size. Geochemical studies and other evidence indicate the presence of source rocks and there is good evidence for the presence of adequate to good reservoir rocks.

The two exploratory wells that have been drilled in the Netherlands Antilles have been dry holes, but the information provided by those wells helps to define a geological model that shows several interesting areas for new exploration. The wells have proven the presence of very thick Eocene reefs in addition to thick sandstones deposited in a submarine fan environment. The wells also indicate the presence of source rocks.

When the well data is integrated with the seismic data, there is excellent evidence for the presence of extensive Cretaceous reefing as well as several very large, faulted Cretaceous anticlines. These large potential traps within the thick Cretaceous section have not yet been tested. Additional exploration activity is in progress which is expected to test some of these large traps, and activity in nearby areas is expected to stimulate additional exploration activity. Presently available data suggest that those factors necessary for the accumulation of hydrocarbons are all favorable in the Netherlands Antilles.

INTRODUCTION

The Netherlands Antilles is made up of a group of five islands in the Caribbean Sea. As the name would suggest, the islands are former colonies of The Netherlands, and together with the island of Aruba, they form part of the Kingdom of The Netherlands. The five islands are located in two widely separated areas (Figure 1) with St. Maarten, Saba, and St. Eustatius located in the northeastern part of the Caribbean while Bonaire and Curacao are located in the southern Caribbean just north of Venezuela. A considerable amount of petroleum exploration has been conducted in both areas, and presently available geological and geophysical data suggest that those elements which are necessary for the accumulation of hydrocarbons are probably present in both areas.

CURACAO-BONAIRE AREA

The islands of Curacao and Bonaire are located approximately 70 km north of Venezuela, and the international boundary between Venezuela and the islands has been agreed by treaty. Although no petroleum exploration wells have been drilled in the Curacao-Bonaire offshore area, more than 5,000 km of seismic profiles have been recorded offshore from these islands. In general, the quality of the seismic data is fair to good, and more than three-fourths of the data is available for purchase through Western Geophysical Company of Houston, Texas. The seismic profiles show a considerable amount of block faulting at depth and there is excellent evidence for reverse faulting in the upper portion of the section.

Figure 1. Index map showing the location of the five islands which make up the Netherlands Antilles.

Reflection Identification

Three excellent reflections are present over much of the Curacao-Bonaire offshore area. These seismic surveys have not been tied to La Vela field, which is approximately 100 km southwest of Curacao, so horizon identification is somewhat subject to question. Nevertheless, it is believed that the three reflections correspond to specific stratigraphic units identified in the wells at La Vela field. The deepest horizon is considered to be the top of the pre-Eocene. In all probability, it represents economic basement in this area. This reflector shows very extensive block faulting.

There is an excellent reflection which is believed to be the equivalent of the Cauderalito Member of the Agua Clara Formation at La Vela. This formation is early Miocene in age. The unit varies from 100 to 200 m in thickness and is probably a carbonate as it is at La Vela.

The shallowest horizon mapped is considered to be the equivalent of the Socorro Formation at La Vela. The Socorro at La Vela is composed of sandstones and interbedded mudstones and is middle Miocene in age. This formation varies from 800 to 1,700 m in thickness in La Vela Bay (Vasquez, 1975). The Socorro Formation is reported to contain 17 sandstone bodies in the old Cumarebo field in northern Venezuela and they vary from less than 6 m to more than 50 m in thickness.

Curacao Area

An interpretation of the seismic data at Cauderalito depth shows the presence of a number of very large structures offshore the island of Curacao (Figure 2). Structure "A", for example, covers almost 100 km^2 and has over 300 m of vertical closure. Structure "E" has in excess of 1,300 m of closure against a major fault and covers more than 150 km^2. This interpretation shows the presence of numerous potential traps for hydrocarbons in offshore Curacao.

The fact that La Vela field, which is located within the same sedimentary basin as offshore Curacao, has produced both oil and gas would indicate that the basin does contain adequate source rocks. Furthermore, the hydrocarbon accumulations at La Vela also prove the presence of adequate cap rocks within the basin. Both cap rocks and source rocks are expected to be present in offshore Curacao.

The major unknown in the Curacao area at this point in time is the presence or absence of suitable reservoirs. Since La

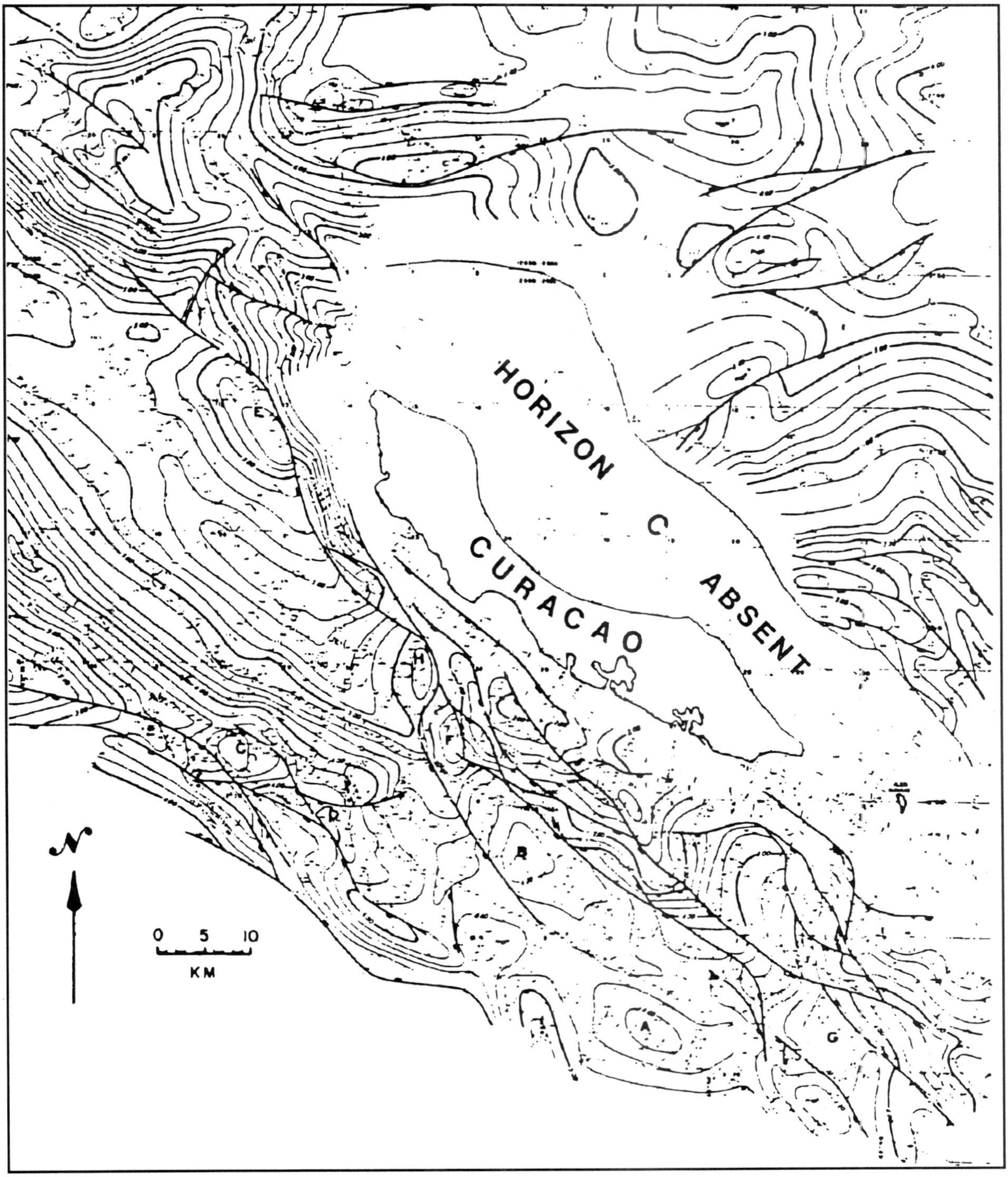

Figure 2. Structural interpretation of Horizon C (Cauderalito) offshore Curacao.

Vela field is 75 to 100 km from the most prospective areas in offshore Curacao, any projection of similar reservoir quality in the Curacao area would be highly speculative.

One of the major factors that has delayed petroleum exploration in offshore Curacao is the depth of water. It varies between 900 m and 1,300 m over the most prospective areas.

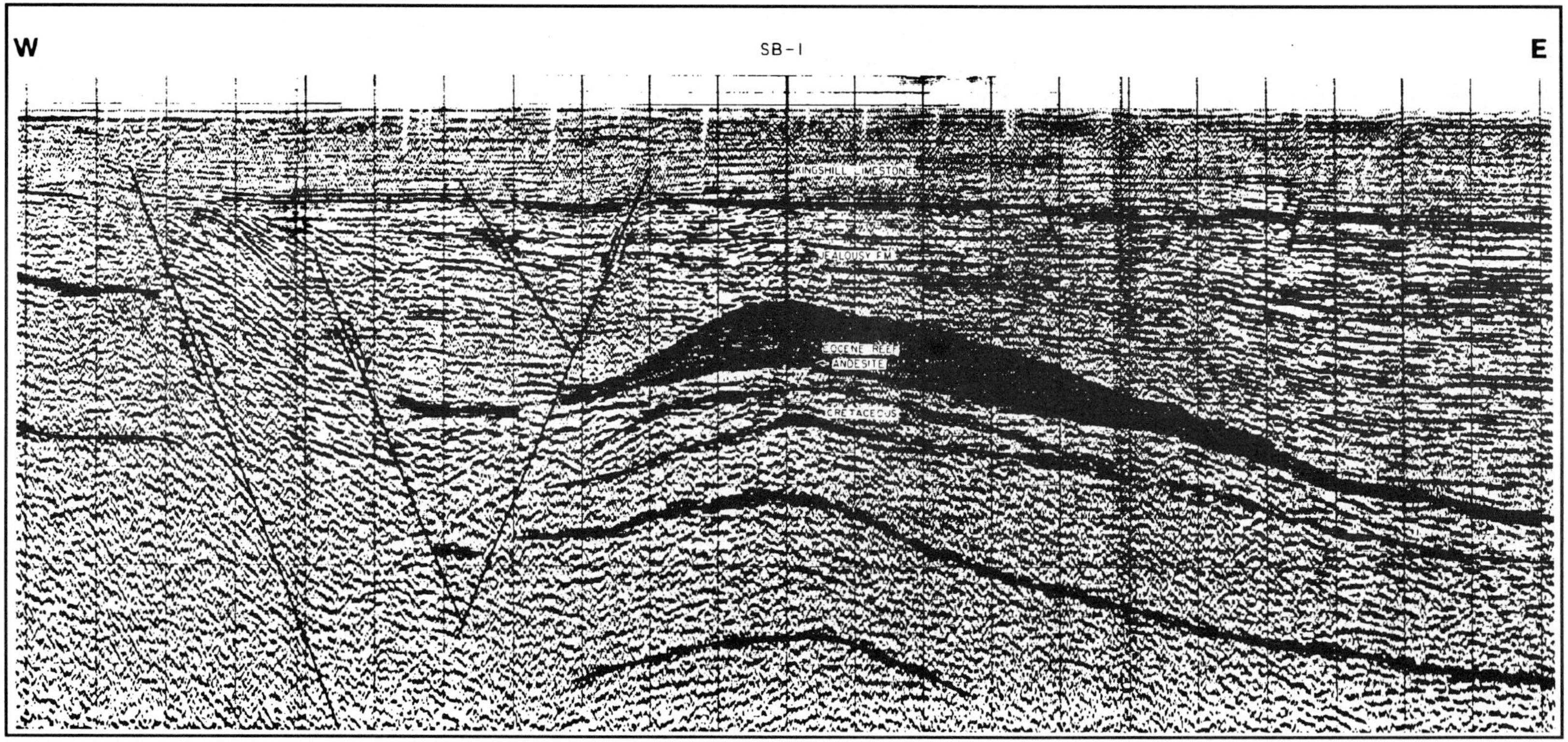

Figure 3. Seismic line S-15 in Saba Bank area showing the well location and the thick Eocene carbonate in the Saba Bank No. 1 well.

However, recent deep water activity in the Campos basin of Brazil and in the Gulf of Mexico would suggest that these water depths are no longer the barrier that they were considered to be just a few years ago.

Bonaire Area

The seismic data in the Bonaire area (Figure 1) show a thinner sedimentary section than in the Curacao area. The areas of thicker sediments are also areas of deep water, so Bonaire's problems are similar to Curacao's problems in this respect. The structural interpretation shows at least two large structural anomalies in offshore Bonaire. If future exploration in the Curacao area is successful, it is expected that some company will almost certainly be interested in testing these structures in offshore Bonaire.

Saba Bank Area

The portion of the Netherlands Antilles located in the northeastern corner of the Caribbean also has considerable petroleum potential. The area of specific interest is a large shallow water area called Saba Bank located to the southwest of the small island of Saba (Figure 1). The total area covered by the Saba Bank is in excess of 4,100 km^2 where the water depth is less than 200 m.

A considerable amount of petroleum exploration has been conducted in this area with more than 3,900 km of seismic profile being recorded by various parties. In addition, two dry holes have been drilled on Saba Bank. The first well drilled on Saba Bank (SB-1) was drilled by a group of companies with Marathon Oil Company as operator. The objective of this well was a very large, well defined carbonate buildup which is shown on seismic profile S-15 (Figure 3). Subsequent studies have shown this to be an Eocene reef. This reefal buildup is associated with a very large structure which covers approximately 150 km^2 and is located in the south-central part of Saba Bank (Figure 4). The sedimentary section penetrated by this well consisted of 969 m of carbonate followed by 918 m of what Marathon called a volcaniclastic series followed by 921 m of Eocene reef and that is followed by 131 m of andesite. The strong reflection at the base of the Eocene reef is generated by the top of the andesite.

Age dating has shown the upper carbonate unit to be from late Miocene to Recent in age. Consequently, this unit is partially equivalent to the Kingshill Limestone which outcrops on the island of St. Croix, located approximately 140 km west of Saba Bank. The volcaniclastic section is predominantly a siltstone-mudstone sequence with interbedded sands and a few thin limestones. This interval is partially equivalent to the Jealousy Formation which also outcrops on St. Croix and is lower Oligocene to middle Miocene in age.

The lower carbonate unit in SB-1 was for the most part micritic limestone, but there were a number of 10 to 20 m zones with excellent porosity development. Age dating showed this unit to be middle to upper Eocene in age.

Saba Bank No. 1 had only minor gas shows, but source rock studies showed the presence of fair to good but immature source rocks within the Jealousy Formation. In spite of its proximity to the volcanic islands arc, this well had a very low temperature gradient. Both the temperature gradient and reflectance studies indicated that the top of the "oil window" would be expected at a depth of approximately 3,300 m or well below the total depth of SB-1 which is bottomed at a depth of 2,975 m in Eocene andesite.

The deep oil window indicated by the studies of SB-1 led to the drilling of Saba Bank No.2 by Petrofina and others in 1982. The basic idea behind this second well was to move down dip from SB-1 in search of a second Eocene reef at a depth below 3,300 m where the surrounding source rocks would be mature. The drill site was located approximately 15 km east-northeast of SB-1 on a smaller, deeper structure than the structure at SB-1. The results of SB-2 were surprising because it encountered sandstones and siltstones where carbonates had been expected. Only 15 m of carbonate was encountered near the bottom of this well. Saba Bank No. 2 is also bottomed in Eocene andesite at 4,231 m. Well logs indicated the presence of hydrocarbons in the sands which were approximately 80 m thick, but a test recovered only mud, brackish water, and a puff of gas during reverse circulation. Later studies showed the presence of large amounts of smectite clays in the sands. These clays swell tremendously in fresh water and Fina now believes that the 250 barrels of mud lost in this formation while drilling badly damaged these sands and that the water recovered on the test was filtrate water. Fina engineers consider the test to be nonconclusive.

Robertson Research studied SB-2 and reported that the sands and shales recovered in cores from 3,421 to 3,428 m and 3,786 to 3,795 m indicate an outer submarine fan environment with pelagic shales and fine-grained turbidites. The limestone near the base of the well was deposited in an outer shelf to upper slope environment. This shows that during Eocene deposition while the thick carbonate was being deposited in SB-1 in a middle to inner shelf environment, the SB-2 carbonate was deposited in a deep water environment. This relationship suggests that the area to the west of the arcuate fault system may have been a positive area during Eocene deposition.

This idea is substantiated by source rock studies in both wells which report the presence of reworked Cretaceous nannofossils in the Jealousy interval. In addition, both wells show the presence of reworked light brown cuticle, fusinite, and wood fragments which independent source rock studies indicate probably originated in a Cretaceous coal seam. To the north and west of the wells the seismic lines show a major arcuate zone of faulting which was active during Tertiary time. They also show a platform area in the western half of the Saba Bank. Every seismic line that passes from the platform area into the Tertiary basin where the wells were drilled shows well defined progradation from the platform area toward the southeast

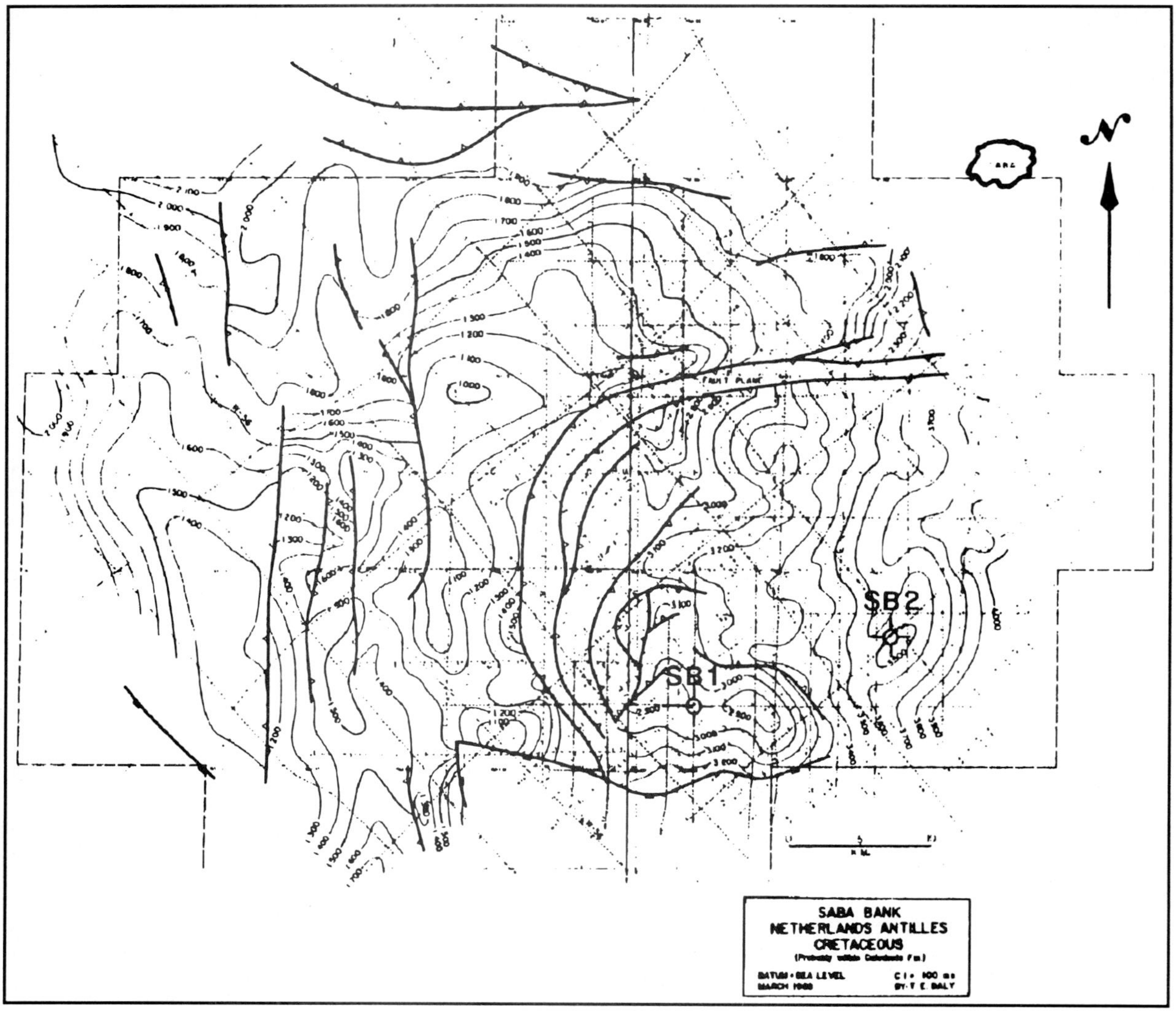

Figure 4. Structural interpretation of Horizon D, a seismic reflection from within the Cretaceous section.

indicating that the platform to the west of the fault zone was a positive area. As this land mass was being eroded, it served as the source of the sediments during Jealousy deposition. Consequently, it was also the source of the reworked Cretaceous nannofossils found in both wells as well as the source of the reworked Cretaceous coal material. This reworked material shows that the Cretaceous at Saba Bank was deposited in a shallow water environment, which is in sharp contrast to the thick Cretaceous sediments that crop out on St. Croix. These outcrops are reported to have been deposited in a deep water environment (Whetten, 1966).

Because the platform area to the west of the arcuate fault system was serving as the source of the Jealousy sediments, it follows that the Kingshill Limestone in the platform area is presently resting unconformably on eroded Cretaceous. The seismic data in the platform area also provide some interesting clues as to the type of Cretaceous that was deposited there. Seismic line W-56 (Figure 5) is almost entirely within the platform area.

This line shows the presence of two fairly strong reflectors below the base of the Kingshill Limestone, which would also be the top of the eroded Cretaceous. The deepest reflection would indicate at least 3,000 m of Cretaceous sediments below approximately 900 m of Kingshill Limestone. Note that this line shows an interesting reversal centered near shot point 900 at a time of 0.900 sec and strongly suggests a probable carbonate buildup in this area. The carbonate appears to be at least 400 m thick, and the crest of the buildup coincides with the crest of the huge structure located immediately to the west of the major arcuate fault system (Figure 4). This particular structure, which is cut off at both ends by a major fault, downthrown to the west, and which splits at its north end, covers approximately 300 km^2

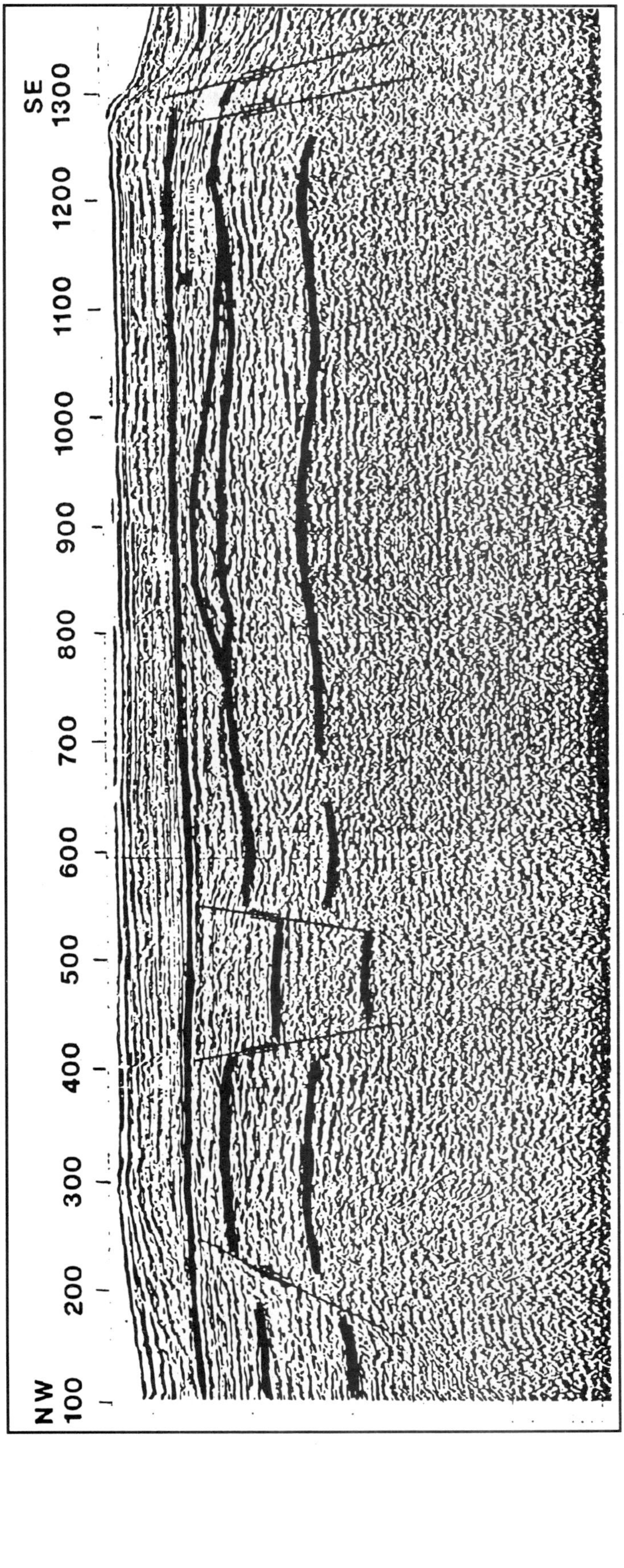

Figure 5. Seismic line W-56 in the Saba Bank platform area showing excellent evidence of a thick reefal buildup within the Cretaceous section.

and has about 850 m of vertical closure. A 50 m oil column over a structure of this size would contain several billion barrels of oil.

Seismic interval velocity analyses suggest that the lithology above and below the Cretaceous reefal development is probably shale indicating the probable presence of both source rocks and cap rocks. It is postulated that the reefal development may be the age equivalent of limestones reported in the Caledonia Formation on St. Croix (Lidz, 1988). If that interpretation is correct, there may have been as much as 3,500 m of Upper Cretaceous sediments overlying this reefal buildup prior to uplift and erosion during Jealousy deposition. That, in turn, would suggest that any associated source rocks would have reached maturity.

A second very large structure is mapped in the western portion of the Saba Bank area on the downthrown side of a north-south fault, the westernmost of the three parallel faults (Figure 4). This structure covers about 200 km^2 and has in excess of 360 m of vertical closure. Additional seismic data was recently recorded in the platform area by Aladdin Petroleum Corporation in an effort to better define the indicated Cretaceous reefing. However, that data is proprietary information and has not been released for publication. It is possible to report that the results of the survey were good and Aladdin is very excited about the area. To summarize, the Saba Bank is a shallow water area containing very large structures, a thick sedimentary section, apparent Cretaceous reefing, and there is reasonable evidence to suggest the presence of mature source rocks and cap rocks. It is expected that a third well will be drilled in this area in the near future to test these Cretaceous reefs.

CONCLUSIONS

Good quality seismic data recorded in the offshore areas of the Netherlands Antilles show the presence of a number of large structural traps for potential hydrocarbon accumulation. In addition, there is excellent evidence indicating the presence of thick Cretaceous reefs in the Saba Bank area. There is geological evidence suggesting the presence of adequate source rocks and cap rocks and the probable presence of reservoirs in both the Saba Bank and Curacao areas. In short, it appears that all of the prerequisites for the accumulation of hydrocarbons are present in the Netherlands Antilles, and it is expected that future exploration will confirm the presence of such accumulations.

ACKNOWLEDGMENTS

The author wishes to thank the Government of the Netherlands Antilles for permission to publish this paper and Mr. Clark Gomes Casseres and Mr. Henry Parisius for their review and very helpful comments.

REFERENCES CITED

Lidz, B. H., 1988, Upper Cretaceous (Campanian) and Cenozoic stratigraphic sequence northeast Caribbean (St. Croix, U.S. Virgin Islands): Geological Society of America Bulletin, v. 100, p. 282-298.

Vasquez, E., 1975, Results of exploration in La Vela Bay: Proceedings, Ninth World Petroleum Congress, v. 3.

Whetten, J. T., 1966, Geology of St. Croix, U.S. Virgin Islands, Caribbean geological investigations: Geological Society of America Memoir 98, p. 177-239.

Circum-Pacific Council for Energy and Mineral Resources
Earth Science Series, Vol. 16
R. L. Miller, G. Escalante, J. A. Reinemund, and M. J. Bergin (Eds.)
Energy and Mineral Potential of the Central American-Caribbean Regions
Springer-Verlag, Berlin Heidelberg 1995

SEDIMENTARY BASINS AND PETROLEUM POTENTIAL OF PUERTO RICO

D. K. Larue
University of Puerto Rico, Department of Geology, R.U.M. Mayaguez, Puerto Rico 00709-5000

R. Torrini, Jr.
Woodward-Clyde, 5055 Antioch, Overland Park, Kansas 66203 USA

ABSTRACT

The petroleum potential of Puerto Rico is poorly known, owing to a lack of exploration effort. Studies to date have concluded that pre-Eocene basement rocks on the island possess little petroleum potential, although, ironically, the only free oil and gas that has been discovered to date was found in these rocks. Of four middle Tertiary basins on Puerto Rico, the North Coast, San Juan, North Mona, and South Coast, only the San Juan and North Coast basins hold particular promise. The North Mona and South Coast basins contain apparently too little stratal thickness to allow petroleum maturation, given what is known about temperature/depth gradients on Puerto Rico. However, more seismic coverage is needed in the San Juan, South Coast, and North Mona basins. The North Coast and San Juan basins contain up to perhaps 6 km of middle Tertiary and younger strata, probably enough to allow for petroleum maturation. However, sedimentary-organic facies of these deep basinal rocks are essentially unknown because they are not observed in outcrop nor penetrated by drilling. Until these deep basinal facies are explored, the petroleum potential of Puerto Rico cannot be properly assessed.

INTRODUCTION

This work addresses the petroleum potential of Puerto Rico through a description of some of the basins onshore and offshore of the island. Meyerhoff et al. (1983) presented an earlier view of the petroleum potential of Puerto Rico which is updated here with new references, data, and interpretations.

CRETACEOUS-EARLY TERTIARY HISTORY OF PUERTO RICO

Puerto Rico (Figure 1) consists of several sequences separated by major unconformities (for example, Mattson, 1960, 1967). Basement rocks on the island of Puerto Rico include the Bermeja Complex, an Upper Jurassic to Lower Cretaceous dismembered ophiolite-like sequence in southwestern Puerto Rico, and a Lower Cretaceous island arc sequence in central Puerto Rico (Mattson and Pessagno, 1971, 1979). Joyce (1985) suggested that these different basement types are the cores of two discrete tectonostratigraphic terranes. Sedimentary rocks in the Bermeja Complex of southwestern Puerto Rico consist mostly of ribbon cherts and contain essentially no organic material. Lower Cretaceous volcanic sandstones and shales of central Puerto Rico (Mattson, 1967) contain little organic carbon and are overmature (Hayes et al., 1985).

In the Late Cretaceous, island arc volcanism on Puerto Rico resulted in the deposition of thick piles of volcanogenic strata. In southwest Puerto Rico, a complicated arc platform accumulated thick sequences of shelfal and basinal carbonates and volcanic flows (Mattson, 1960). A deep summit basin between southwest Puerto Rico and central Puerto Rico (Figure 1) filled with several kilometers of submarine fan and related facies, consisting of recycled volcanic detritus mixed with carbonate materials. In central and eastern Puerto Rico,

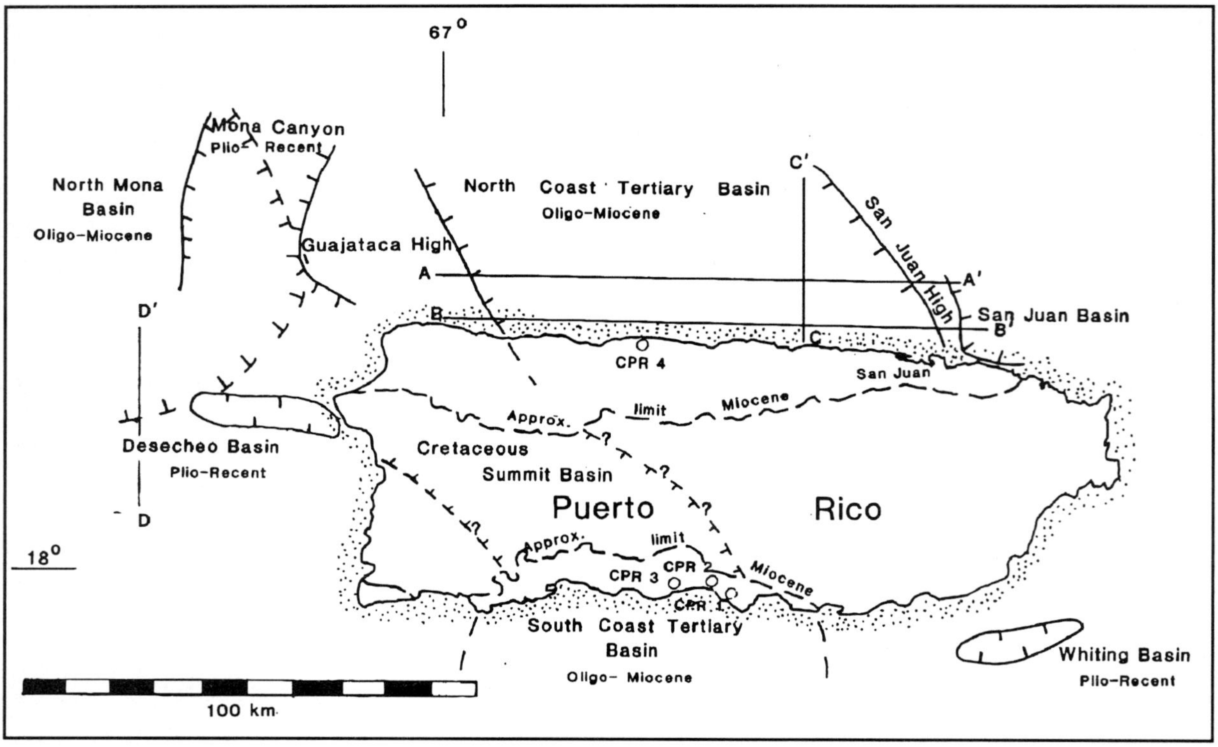

Figure 1. Location map of basins on Puerto Rico. In part after Meyerhoff et al. (1983), Western Geophysical Company of America and Fugro, Inc. (1973), and work in preparation by the authors. The Whiting basin and Desecheo basin are not described in the text. They are actively subsiding half-grabens filled with presumably immature Plio-Recent sediments.

shallow water and transition facies to the basinal facies are present, and thus define the other side of the basin (Mattson, 1967). The source rock potential of these Upper Cretaceous rocks was studied by Hayes et al. (1985) who concluded that the rocks were dominantly overmature, although some samples yield oil window maturation indices. Organic carbon in the samples is significant (however, more recent studies by the present authors have shown that the Hayes et al. (1985) total organic carbon values (TOC) were off by as much as an order of magnitude because of poor analytical techniques by the company they employed). Hayes et al. (1985) were only partly successful in determining the type of kerogens present in the rocks. Most kerogens appear to be derived from terrestrial plants, although some marine organic matter may be indicated. The only free petroleum on Puerto Rico has been found associated with rocks of Late Cretaceous age (Glover, 1971), and in Eocene rocks from the Cerrillos damsite, near Ponce in south-central Puerto Rico (A. Soto, personal communication, 1989).

Lower Tertiary sedimentary rocks on Puerto Rico (most dated as Eocene) consist mostly of volcanogenic turbidites and associated shales with minor limestones. The environment of deposition is thus quite similar to that of the Upper Cretaceous rocks. Hayes et al. (1985) performed a few analyses of petroleum potential of these Eocene rocks, and concluded that organic carbon content was low, and maturation state was variable, in part associated with proximity to Eocene plutons.

In the middle Eocene, volcanism stopped on Puerto Rico (and much of the Greater Antilles), and the rocks were deformed, uplifted, and eroded. Probably the most prominent stratigraphic break on Puerto Rico is between the highly deformed and metamorphosed Cretaceous through Eocene rocks and flat-lying middle Tertiary and younger rocks of the overlap sequence. The origin of this stratigraphic break is debatable. Pindell and Dewey (1982) and Joyce (1985) have argued that in the middle Eocene Puerto Rico collided with the Bahama bank. However, the presence of a relatively youthful slab under western Puerto Rico and the eastern Dominican Republic suggests that if there was a collision event, it was followed by further subduction of oceanic crust (Schell and Tarr, 1978). Such a "collisional" event followed by renewed subduction might occur during the accretion of an aseismic oceanic ridge to the leading edge of an arc. Such aseismic ridges are known to be present in the subducted sea floor beneath the Puerto Rico trench (McCann and Sykes, 1984).

The petroleum potential of the Jurassic through Eocene rocks on Puerto Rico is severely limited by a lack of potential source rocks, general high maturity, poor reservoirs, and complicated deformation such that potential structural traps are difficult to define.

MIDDLE TERTIARY BASINS

There are essentially four middle Tertiary basins onshore and offshore of Puerto Rico: the North Coast, South Coast, North Mona, and San Juan basins (Meyerhoff et al., 1983; Western Geophysical Company of America and Fugro, Inc., 1973). These basins are defined principally using seismic reflection studies. Only the North Coast and South Coast basins are well-exposed on land, and tilted so that their stratigraphy can be studied in part at the outcrop. The San Juan basin is only marginally exposed onshore, and the stratigraphy of the North Mona basin is known only through dredge haul and submersible studies. As discussed below, a limited amount of borehole and drilling information is available.

North Coast Basin

The North Coast basin is an approximately rhomboid-shaped depression, with a maximum sedimentary fill of more than 6 km of middle Tertiary rock (Figure 1). The basin can be subdivided into a number of sub-basins. Based on seismic reflection studies, two basin-filling associations are recognized. The lowermost unit was deposited within the confines of a fault-bound basin, whereas the overlying unit is apparently not cut by basin-defining faults. However, the overlying unit is also a basinal facies in that stratigraphic units thicken into the basin center, thus indicating continued basinal subsidence (Figure 2). We refer to the lower unit as the basinal association and the upper unit as the sag association (that is, the basin was sagging during deposition, though not actively subsiding through fault motion).

The stratigraphy of the basin fill is known through surface studies on land and by correlation between a seismic line and a well (CPR 4; see Briggs, 1961) drilled in 1960 along that seismic line (Meyerhoff et al., 1983; Moussa et al., 1987). Exposed in outcrop on Puerto Rico and from drillhole information, the following stratigraphy is recognized. The lowermost unit is the upper Oligocene San Sebastian Formation (about 320 m in well CPR 4), representing deposition in a dominantly clastic coastal plain/deltaic setting. Overlying the San Sebastian Formation is a mostly Miocene reefal limestone unit, the Lares Limestone (504 m); a chalky limestone, the Cibao Formation (305 m); another reefal unit, the Aguada limestone (150 m); and the upper Miocene/Pliocene Aymamon limestone (400 m). Moussa et al. (1987) disagree with the stratigraphic nomenclature used by Briggs (1961), and instead substitute the Los Puertos limestone for the Aguada limestone, and divide the Aymamon into the "Aymamon" limestone and the overlying upper Miocene/Pliocene Quebradillas limestone. Detailed arguments concerning such problems in stratigraphic interpretation are presented in Moussa et al. (1987; and references therein).

Based on our studies, the break between the sag association and the basinal association seems to correspond with the onset of Lares Limestone deposition, corresponding approximately to be base of the Miocene. Birch (1986) has argued that the subsidence history of CPR 4 can be modelled best as a function of the square root of time, indicating a thermal subsidence. Birch argues that such thermal subsidence was provided by the cooling Eocene island arc.

Petroleum Potential of the North Coast Basin

It is extremely difficult to establish the petroleum potential of the North Coast basin because so little is known about source rocks in the area, and the geothermal gradient is not known with any precision (see below). Moreover, most of the basin is in water depths greater than one kilometer, such that the definition of an economic basin becomes more restricted. Not only is there little information about source rocks, there is essentially no information concerning the deep basinal stratigraphy. It is possible that these deep sub-basins contain almost any lithology, from continental to deep marine, from carbonate to clastic. The San Sebastian Formation as exposed on Puerto Rico and in CPR 4 is thought to represent a condensed sequence of the basinal facies, but it is unclear what facies changes occur between the platforms and the basins. In fact, based on analysis of the seismic reflection lines, it is likely that the San Sebastian Formation is dominantly carbonate toward the north, and is only siliciclastic near the mountainous source region. Potential traps in the North Coast basin are dominantly reefal and stratigraphic inasmuch as there is little structural deformation (Figure 2).

San Juan Basin

The San Juan basin (Figure 2) occurs to the east of the North Coast basin and the San Juan High (de Jong, 1974). It contains on the order of three seconds of strata (approximately 6 km), but it is poorly studied, and few seismic reflection studies have been made in the area. Because the deep part of the basin occurs relatively close to shore, exploration will not be as restricted by water depths as in the North Coast basin. Additional seismic coverage is desperately needed in this basin.

North Mona Basin

The North Mona basin occurs in the offshore of Puerto Rico in the Mona Passage. Only its southern and eastern margins have been defined by seismic reflection studies

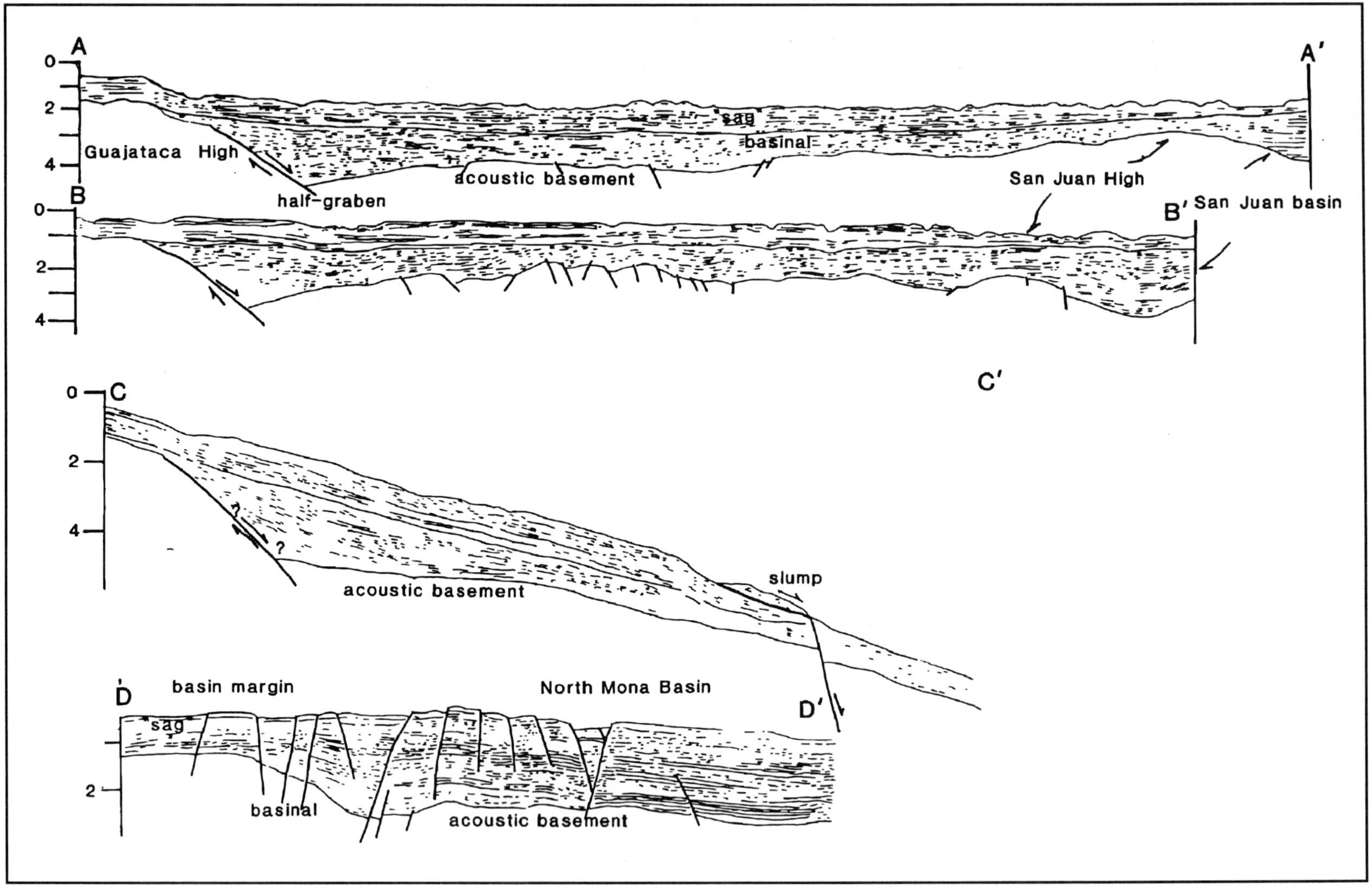

Figure 2. Line drawings of multichannel seismic lines presented in Western Geophysical Gompany of America and Fugro, Inc. (1973). Lines located in Figure 1. Vertical scale is two way travel time (sec). Mean interval velocity for rocks is about 2,573 m/sec (Western Geophysical Gompany of America and Fugro, Inc. (1973)), indicating two seconds of basin fill represents about 2.5 km of strata.

(Figure 2). Maximum thickness of basinal strata is on the order of 2.5 km thickness where seismic data are available. The stratigraphy of the basin is known only through comparison with the North Coast basin, and meager information from dredge hauls and submersible observations in the Mona Canyon (Heezen et al., 1985; Perfit et al., 1980; Fox and Heezen, 1975; Weaver, Smith, and Seiglie, 1975), which support a middle Tertiary age. The seismic stratigraphy indicates history of basin filling similar to the North Coast basin, a basinal association overlain by a sag association. The Mona Passage is cut by a number of faults (Figure 2) trending mostly east-west or north-south, such that potential structural traps (fault cut-outs, rollovers) are abundant. The active Mona Canyon graben is centered on the eastern margin of the North Mona basin, and seems to represent reactivation of a middle Tertiary basin-defining fault system. That is, the Mona Canyon is an actively subsiding submarine graben that cuts the margin of the older, inactive North Mona basin.

The petroleum potential of the basin rests on finding out whether the unstudied portion of the basin is thick enough to be capable of maturing any organic-rich source rocks.

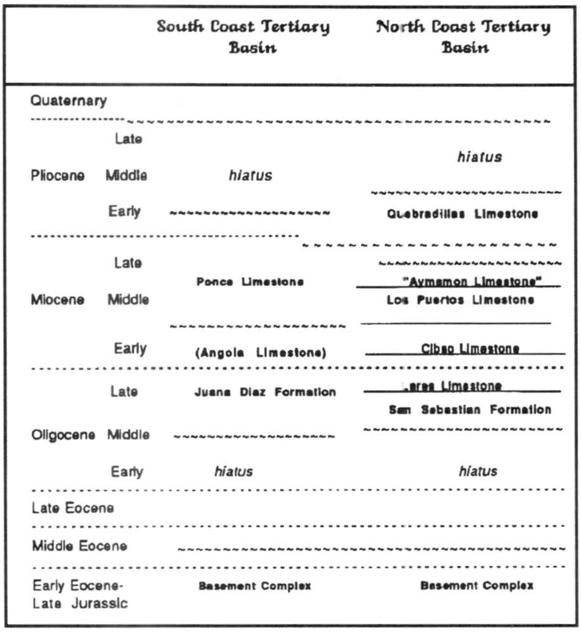

Figure 3. Comparative stratigraphy on the North and South Coast Tertiary basins of Puerto Rico. Modified from Meyerhoff et al. (1983) using data presented by Seiglie and Moussa (1984) and Moussa et al. (1987).

South Coast Tertiary Basin

The South Coast Tertiary basin represents an ideal basin to study because normal faulting has resulted in significant tilting of basinal strata permitting study in outcrop, and three boreholes (CPR 1, 2, 3; Figure 1) and offshore seismic data are available. However, based on these studies, the basin does not seem to be hot enough or deep enough to have generated any significant quantities of petroleum.

The stratigraphy of the South Coast basin is grossly comparable to that of the North Coast basin (Figure 3). The lowermost middle Tertiary unit is the Juana Diaz Formation, an upper Oligocene/lower Miocene clastic unit that fines upward into a reefal facies and chalk facies (grossly comparable to the Lares Limestone), and is overlain unconformably by the Miocene Ponce Limestone, a reefal/shelfal facies (Frost et al., 1983). The three wells were drilled into a gravity low, probably representing a sub-basin depocenter, and penetrated approximately 1.3 km of middle Tertiary section (Todd and Low, 1976). Favorable facies for petroleum sources are observed in outcrop, such as organic-rich Oligocene turbiditic sandstones and mudstones (Moussa, 1987). Preliminary work by our group indicates TMAX values (the temperature at which most kerogens thermally crack during step-wise heating in the laboratory) from Rock-Eval pyrolysis studies to be significantly low for petroleum generation (for an explanation of Rock-Eval pyrolysis, see Espitalie, Madec, and Tissot, 1980). Although what is known about the South Coast basin does not indicate good petroleum potential, only the on land portion and northern offshore portion of the basin have been studied (Figure 1); the southern margin of the basin is not defined owing to lack of seismic coverage. It is possible that a deeper part of the basin exists to the south, and depths capable of petroleum production are reached in the basin. In this scenario, updip migration of petroleum to shallower parts of the basin is possible. Although there are no data to refute such an argument, there is little evidence to indicate it is correct. Seismic reflection work is needed to define the southern and eastern margins of the South Coast basin.

CONCLUSIONS

The onshore and offshore petroleum potential of Puerto Rico is poorly known. We conclude, as did Meyerhoff et al. (1983), that the pre-middle Tertiary rocks have low petroleum potential (although ironically they contain the only proven petroleum seeps). The middle Tertiary basins are largely unexplored, especially their basinal facies. The greatest petroleum potential lies in the thick deposits of the North Coast Tertiary basin and the little-known San Juan basin. Because the North Mona and South Coast Tertiary basins contain only a few kilometers of middle Tertiary strata, it is unlikely that they have enjoyed sufficient thermal maturation to allow for petroleum formation. More seismic coverage is needed to define the extent of the San Juan, North Mona, and South Coast basins.

ACKNOWLEDGMENTS

This study was supported by grants to the University of Puerto Rico by the Puerto Rico Electric Power Authority. We are grateful for their continued support. We also acknowledge the donation of the multichannel seismic lines from the Puerto Rico Electric Power Authority and the Department of Natural Resources to the Department of Geology. Finally, the Department of Geology has benefited greatly from an N.S.F. grant entitled "Experimental Program to Stimulate Competitive Research."

REFERENCES CITED

Birch, F. S., 1986, Isostatic, thermal and flexural models of the subsidence of the north coast of Puerto Rico: Geology, v. 14, p. 427-429.

Briggs, R. P., 1961, Geology of Kewanee Interamerican Oil Company Test Well Number 4CPR, Northern Puerto Rico, *in* Oil and gas possibilities of northern Puerto Rico: San Juan, Puerto Rico Mining Commission, p. 1-23.

de Jong, A., 1974, Evaluation of hydrocarbon prospects of the island of Puerto Rico: Western Geophysical Company of America and Fugro, Inc., unpublished report: San Juan, Department of Natural Resources, 20 p.

Espitalie, J., M. Madec, and B. Tissot, 1980, Role of mineral matrix in kerogen pyrolysis: Influence on petroleum generation and migration: American Association of Petroleum Geologists Bulletin, v. 64, p. 59-66.

Fox, P. J., and B. C. Heezen, 1975, Geology of the Caribbean crust, *in* A. E. M. Nairn and F. G. Stehli, eds., The ocean basins and margins, v.3, The Caribbean and the Gulf of Mexico: New York, Plenum Press, p. 421-466.

Frost, S. H., et al., 1983, Oligocene reef tract development Southwestern Puerto Rico: Sedimenta IX: The Comparative Laboratory Division of Marine Geology and Geophysics Rosenthiel School of Marine and Atmospheric Science, University of Miami, 143p.

Glover, L., III, 1971, Geology of the Coamo area, Puerto Rico, and its relation to the volcanic arc-trench association: U.S. Geological Survey Professional Paper 636, 102p.

Hayes, J., D. K. Larue, J. Joyce, and J. H. Schellekens, 1985, Puerto Rico: Reconnaissance study of the maturation and source rock potential of an oceanic arc involved in a collision: Journal of Marine and Petroleum Geology, v. 3, p.151-161.

Heezen, B. C., W. D. Nesteroff, M. Rawson, and R. P. Freeman-Lynde, 1985, Visual evidence for subduction in the western Puerto Rico trench, *in* A. Mascle, ed., Caribbean Geodynamics Symposium, Paris, March 5-8, 1985: Paris, France, Editions, Technip, p. 287-304.

Joyce, J., 1985, High pressure-low temperature metamorphism and the tectonic evolution of the Samana Peninsula, Dominican Republic, Greater Antilles: Ann Arbor, Michigan, University Microfilms, 250p.

Mattson, P. H., 1960, Geology of the Mayaguez area, Puerto Rico: Geological Society of America Bulletin, v. 71, p. 319-362.

Mattson, P. H., 1967, Cretaceous and lower Tertiary stratigraphy in west-central Puerto Rico: U.S. Geological Survey Bulletin 1254-B, 35p.

Mattson, P. H., and E. A. Pessagno, Jr., 1971, Caribbean Eocene volcanism and Atlantic Ocean Layer A: Science, v. 87, p. 293-304.

Mattson, P. H., and E. A. Pessagno, Jr., 1979, Jurassic and Early Cretaceous radiolarians in Puerto Rican ophiolite—Tectonic implications: Geology, v. 7, p. 440-444.

McCann, W. R., and L. R. Sykes, 1984, Subduction of aseismic ridges beneath the Caribbean Plate: Implications for the tectonics and seismic potential of the northeastern Caribbean: Journal of Geophysical Research, v. 89, p. 4493-4519.

Meyerhoff, A. A., E. W. Krieg, J. D. Cloos, and I. Taner, 1983, Petroleum possibilities of Puerto Rico: Oil and Gas Journal, v. 81, no. 51, p. 113-120.

Moussa, M. T., G. A. Seiglie, A. A. Meyerhoff, and I. Taner, 1987, The Quebradillas Limestone (Miocene-Pliocene), northern Puerto Rico, and tectonics of the northeastern Caribbean margin: Geological Society of America Bulletin, v. 99, p. 427-439.

Perfit, M. R., B. C. Heezen, M. Rawson, and T. W. Donnelly, 1980, Chemistry, origin, and tectonic significance of metamorphic rocks from the Puerto Rico Trench: Marine Geology, v.34, p. 125-156.

Pindell, J., and J. F. Dewey, 1982, Permo-Triassic reconstruction of western Pangea and the evolution of the Gulf of Mexico/Caribbean region: Tectonics, v. 1, no. 2, p. 179-211.

Schell, B. A., and A. C. Tarr, 1978, Plate tectonics of the northeastern Caribbean Sea region: Geologie en Mijnbouw, v. 57, p. 319-324.

Seiglie, G. A., and M. T. Moussa, 1984, Late Oligocene-Pliocene transgressive-regressive cycles of sedimentation in northwestern Puerto Rico: American Association of Petroleum Geologists Memoir 36, p. 89-96.

Todd, R., and D. Low, 1976, Smaller foraminifera from deep wells on Puerto Rico and St. Croix: U.S. Geological Survey Professional Paper 863, 32p.

Weaver, J., A. Smith, and G. Seiglie, 1975, Geology and tectonics of the Mona Passage: Transactions of the American Geophysical Union, v. 56, p. 451.

Western Geophysical Company of America and Fugro, Inc., 1973, Geological-geophysical reconnaissance of Puerto Rico for siting of nuclear power plants: San Juan, The Puerto Rico Water Resources Authority, 127p.

Circum-Pacific Council for Energy and Mineral Resources
Earth Science Series, Vol. 16
R. L. Miller, G. Escalante, J. A. Reinemund, and M. J. Bergin (Eds.)
Energy and Mineral Potential of the Central American-Caribbean Regions
Springer-Verlag, Berlin Heidelberg 1995

AN OIL AND GAS ASSESSMENT OF THE U.S. CONTINENTAL SLOPE IN THE GULF OF MEXICO

George Carpenter and Pulak Ray
Minerals Management Service, Offshore Resource Evaluation Division (MS 643), Reston, Virginia 22091, USA

ABSTRACT

Exploration and development decisions in high cost deep-water areas can be greatly facilitated by a rigorous assessment of the resource potential of the area of interest. The Minerals Management Service, an agency of the U.S. Department of the Interior, has recently completed an assessment of undiscovered, economically recoverable oil and gas on the Continental Slope of the Gulf of Mexico. The area assessed includes the Flexure Trend, a high-potential play with exceptionally favorable petroleum geology. The Plio-Pleistocene Flexure Trend extends seaward from roughly the shelf break and includes the deep-water Green Canyon field. Traps tend to be large, reservoir sands have high porosity and permeability, and organic-rich source rocks generate both oil and gas. These qualities allow profitable production at current low prices even in very deep water.

The resource assessment was conducted using a drilling simulation model, which uses risk analysis to "drill" user-defined prospects. The model can function under uncertain conditions regarding the presence of oil and gas because geologic and engineering input data (acreage, pay thickness, and oil, gas, and condensate recovery factory) are entered as ranges of values which are randomly sampled by the model. The model integrates these data with a multi-level risking hierarchy and a user-defined economic scenario to generate estimates of undiscovered, economically recoverable hydrocarbons. Model outputs are estimates of remaining resources, which are given as probability distributions for four commodities: oil, associated and nonassociated gas, solution gas, and condensate.

INTRODUCTION

Development in deep water is expensive. Industry, therefore, tries to be as well informed as possible about potential amounts of economically recoverable oil and gas in these areas prior to exploratory drilling and development. A rigorous resource assessment of an area of interest can provide a framework for decisions on future exploration and development activities. As a precursor to drilling in a frontier area, a carefully crafted assessment can highlight areas of good potential or, conversely, can indicate whether or not exploration is likely to be economically worthwhile.

Resource estimation methodologies cover the range from art to science, depending on the quality and quantity of data available to the evaluator. The method used by the Minerals Management Service (MMS) relies on the Probabilistic Resource Estimates Offshore (PRESTO) computer model, which integrates geological, geophysical, geochemical, engineering, and economic data with a multi-level risking hierarchy to generate estimates of undiscovered economically recoverable oil and gas (Carpenter, 1985). The model, now in its third generation (PRESTO III), develops estimates for numerous economic scenarios and can provide projections of conventionally recoverable resource estimates from small, uneconomic fields not normally included in the database.

The following discussions of the PRESTO III resource assessment model, the geology of the deep-water Gulf of Mexico, and the resulting resource estimates demonstrate an analytical approach that is applicable to any offshore area. This approach is usable under a wide range of uncertainty as to the nature and distribution of potential hydrocarbon accumulations and to whether or not hydrocarbons are present at all.

THE PRESTO III MODEL

Before drilling occurs, the only quantifiable information about the existence of hydrocarbons in a trap is based on inference derived from geophysical data and extrapolations of regional tectonic and depositional histories. The degree of uncertainty regarding the presence of oil and gas is also ill-defined. Thus, any resource estimates developed under such conditions are likely to be inexact because uncertainties in the underlying geologic assumptions translate directly into uncertainties in the estimates derived from them.

The PRESTO III grouped-prospect play analysis model can generate resource estimates under a broad range of geologic and economic uncertainty. The model uses the same types of data used by industry to locate and define oil and gas prospects. The data are interpreted and converted to maps and cross sections that provide the basis for a set of program inputs that numerically model all identified prospects in the area of study. (Procedures for adding resources attributed to plays and prospects that have not been conventionally identified will be discussed later.)

PRESTO III is a computer-based drilling simulation model that "drills" prospects by comparing a random number (0<RN<1) to the prospect risk (Cooke, 1985). If the random number is greater than the risk, the drilling is considered to be successful (a hit) for that play and/or prospect on that trial. This process is repeated for up to 1,000 successful trials, and a hit history is compiled in program memory. The model then calculates resource volumes for successful trials using the following listed variables:

1. areal extent (acres),
2. zone net pay thickness (feet),
3. oil recovery factor (stock tank barrels/acre-feet),
4. gas recovery factor (thousand cubic feet/acre-feet),
5. gas share, the proportion of the zone pay thickness consisting of gas, if a mixture,
6. solution gas-to-oil ratio (standard cubic feet/stock tank barrel), and
7. condensate yield (stock tank barrels/million cubic feet of gas).

These variables are entered as ranges of values (distributions) because they are only rarely precisely known. In addition to the distributed variables, a number of single-point inputs are required by PRESTO III. Four estimates of risk at the zone, prospect, play, and area levels are used in the model to ensure that the probability of occurrence of hydrocarbons is adequately assessed. Economic screens in the form of minimum economic field sizes (MEFS) and minimum play and area reserves are also required. These inputs provide economic thresholds at the prospect, play, and area (usually a geologic province or administrative planning area) levels below which profitable development is not feasible. The economic screens consider water depth, distance from shore, presence or absence of infrastructure, depth to pay, and other economic and engineering constraints. The final single-point values required by the model are estimates of the probability that each zone in a given prospect contains all gas or all oil.

PRESTO includes a provision for assessing dependencies among the input variables through the use of a correlation matrix. For instance, if pay thickness is observed to decrease as an area increases, the strength or degree of the dependency can be numerically quantified by a correlation coefficient. The correlation coefficient controls the Monte Carlo sampling process by forcing the pairing of dependent variables. In the example given, thin pays would more likely be paired with large areas (i.e., the randomness of the sampling process is reduced owing to additional knowledge concerning the variable relationships).

The model calculates conditional resource estimates by prospect from the variables on the condition that recoverable oil and gas exist in the area being analyzed. To begin the calculation process, the model refers to the previously stored hit history and selects those prospects that have passed the drilling simulation test. Monte Carlo or Latin Hypercube sampling techniques are then used to sample a single point from each variable distribution for each zone in every prospect that has "hit." The selected points are entered into the equations listed below to compute resource volumes for each of the four commodities assessed by PRESTO III.

1. Volume of oil, thousands of barrels = (acres) (thickness) (1-gas share) (oil recovery factor) (.001).
2. Volume of nonassociated and associated gas, million cubic feet = (acres) (thickness) (gas share) (gas recovery factor) (.001).
3. Volume of condensate, thousands of barrels = (condensate yield) (nonassociated and associated gas) (.001).
4. Volume of solution gas, million cubic feet = (gas-to-oil ratio) (oil, thousands of barrels) (1,000).

The sequence of randomly sampled points and resource calculation for each trial is repeated in the program up to 1,000 times. The prospect-specific result for each trial is compared to the MEFS for that prospect and, if the calculated resource exceeds it, the value is stored; if less, the value is set to zero. Similarly, the aggregated resource estimates for the play and area are compared to play and area economic screens.

When the program has completed the specified number of trials, the viable prospect resource volumes (those which have survived both the drilling simulation and the economics screen) are sorted, ranked, and recorded as a distribution of conditional resources for the play. An example of a conditional resource distribution with sample values is shown in Figure 1. The median (50th percentile) and the mean (average of all non-zero trials) values are shown on the figure. For simplicity, the MMS ordinarily reports only the 5th (high) and 95th (low) values and the mean.

PRESTO calculates economically recoverable resource estimates. Therefore, the model must factor in the possibility that a prospect, play, or area may contain hydrocarbons but in subeconomic volumes. Input risks are geologic risks (i.e., are

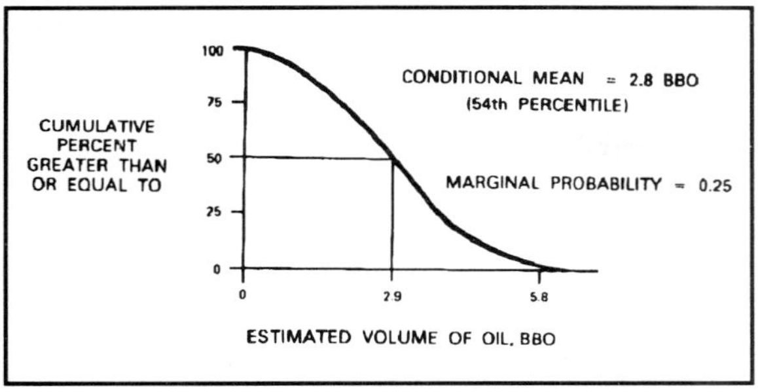

Figure 1. An example of a conditional resource distribution. (BBO is a billion barrels of oil.

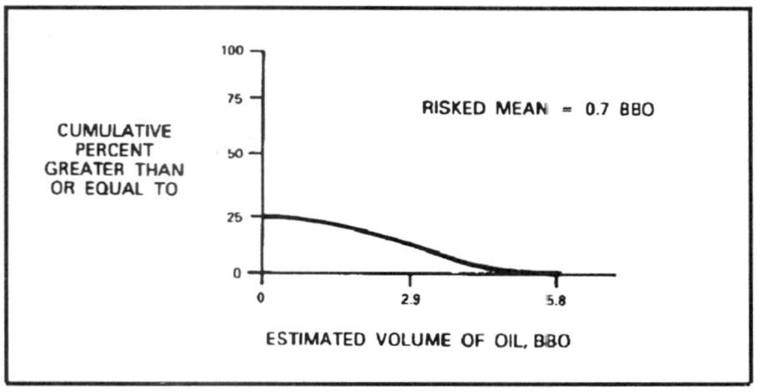

Figure 2. An example of a risked resource distribution.

hydrocarbons present in any amount?). In order to reflect the condition that economic quantities of oil and/or gas may not be present, the input risk is adjusted by the model to account for subeconomic resource amounts between zero and the economic volume screens. The final (output) risk is thus the geologic risk plus the economic risk, the complement of which is derived from the ratio of successful trials to total trials in the run.

Conditional resource estimates calculated by PRESTO are predicated on the condition that economic resources exist in the area as modeled, when in reality this condition is not often met in most offshore planning areas.

Thus, the probability that oil and gas may not exist is often incorporated into the resource assessment of an area. Such estimates are said to be risked. An example of a risked distribution curve is shown in Figure 2. The following example illustrates the essential difference between conditional and risked estimates. Assume that two areas have been assessed, each with very different conditional resource volumes and marginal probabilities (MP = 1 - risk, i.e., the complement of the risk).

Conditional Mean Estimate (106) bbl MP		Risking	Risked Mean Estimate (106 bbl)
Area 1	1,000	0.1	1,000 x 0.1 = 100
Area 2	125	0.8	125 x 0.8 = 100

Note that Area 1 has a far larger resource potential than Area 2 if oil is present in the area. However, the probability of oil occurrence is considerably lower in Area 1, so that when the chance that each area is dry is considered, the risked resources for each area are identical. Risking removes the condition that resources are present, which can vary widely by area. Risked estimates are thus directly comparable (Cooke, 1985).

The above discussion generally shows how the PRESTO III model calculates resource estimates for identified prospects. Since it is most unlikely that all potential traps in any given area have been identified (or are even identifiable) with seismic data, PRESTO III includes a number of provisions for restoring resources of possible missing prospects and plays to the database.

The model includes a subroutine termed the unidentified play option. The option allows the evaluator to assess the resource potential of unidentified plays in one of three ways, depending on the degree of certainty regarding the nature of the postulated plays. The first method, which assumes a general knowledge of the geological attributes of the play, requires the evaluator to directly specify zone and prospect variables. The second method extrapolates resources from those calculated by PRESTO III for the identified prospects in the area. The third, and least certain of the methods, requires the user to enter a probability distribution for prospect size in barrels of oil equivalent. All three methods require an average prospect risk, a play risk, and a distribution for the number of presumed prospects that compose the play. The result is a probability distribution of estimated resources for the play, which is then aggregated with the resources calculated for all identified prospects in the area.

Another subroutine of PRESTO III (DIST) provides an assessment of resources contributed by unmapped small prospects. The DIST methodology is an extension of the observations of Ivanhoe (1976), Folinsbee (1977), Coustau (1980), and others who showed that field size and field rank have a quantifiable correlation. A semilog graph of field size versus field rank for any play will be a straight line if all fields in the play have been discovered. Since total discovery is a rare event, most such graphs at some point will show a downward deflection beyond which smaller size classes are progressively less well represented in the field size distribution (Figure 3). Pecora and Cooke (1989) show that an extrapolation of the straight line portion of the curve to some pre-selected minimum value "discovers" those small fields that, by reason of economics or data coverage, are absent from most field size distributions. Resource estimates for prospects smaller than those normally evaluated (sizes below the deflection point) are calculated by integrating the area under the extrapolated part of the curve (Figure 3). The procedure is reasonable as long as its use is limited to areas of significant geologic uniformity (e.g., plays and provinces having an uncensored distribution of field sizes with similar geology).

A final, rather basic, method of accounting for potential resources in unidentified prospects is to simply add a certain

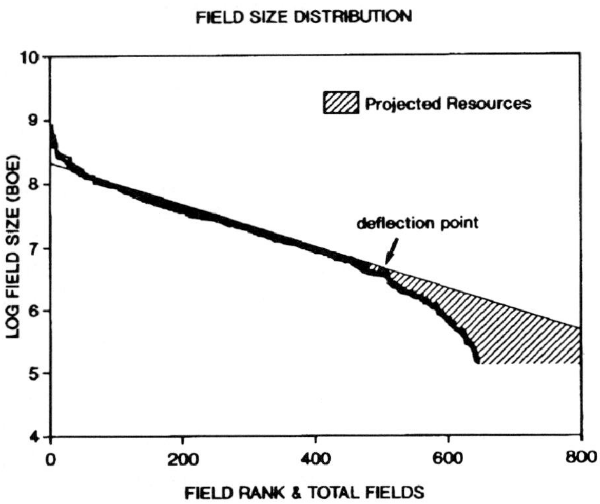

Figure 3. A semilog field size distribution. The hachured area is a projection of resources in fields smaller than the field size at the deflection point. (BOE is barrels of oil equivalent.)

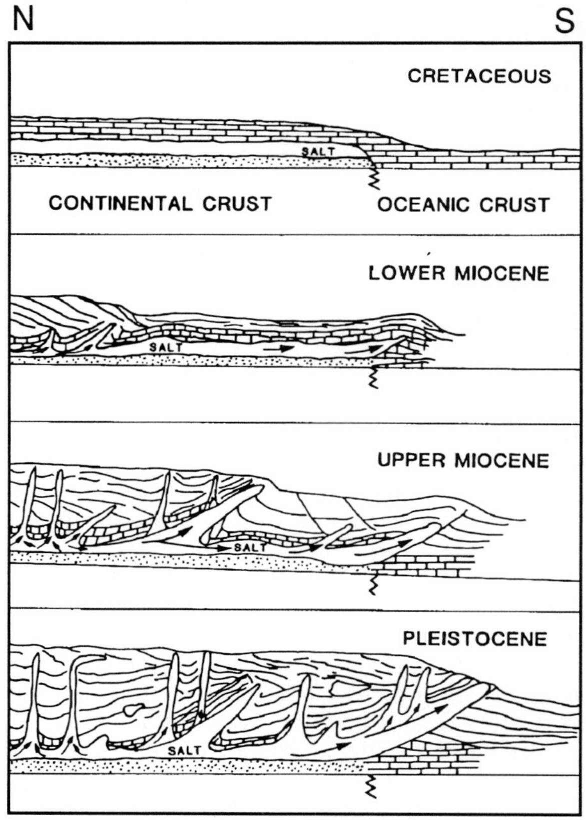

Figure 4. Salt flow and evolution of various structures related to salt tectonics (Ray, 1988).

number of phantom prospects to the identified prospect database. This method is most useful in areas of similar geology but with unevenly distributed data coverage. In such cases, the prospect density in a mapped area is assumed to be equivalent to that in an unmapped area, and an appropriate proportion of similar prospects are added to the preexisting database.

Most of the data processed by the PRESTO III model have a geological basis. A thorough understanding of the geology of a study area will optimize the results of any resource assessment effort. The following discussion describes the geology of the continental slope, including the highly promising Flexure Trend, and illustrates how geologic information is integrated into the PRESTO model.

GEOLOGY AND MODELING

The Continental Slope of the northern Gulf of Mexico is an area of high hydrocarbon potential. Exploration on the upper slope has been quite successful, particularly in the Plio-Pleistocene Flexure Trend. Although more than 70 fields have thus far been discovered, the remaining potential of the Flexure Trend is believed to be excellent.

The Flexure Trend is characterized by intense episodic salt movement, which includes both vertical and lateral components. Figure 4 illustrates possible mechanisms responsible for the unique structural attributes of the Trend. Prograding Miocene sediments forced the underlying mobile Jurassic salt ahead of the depocenter because of sediment loading. This process accelerated in the Pleistocene when large volumes of glacially derived sediments were supplied by the ancestral Mississippi River. The vector resulting from the horizontal forces on the salt generated by overburden pressure and the upward positive buoyancy of the salt moves the salt along the landward-dipping thrust planes (Ray, 1988). In some cases, when the updip location of the depocenter changes, salt flow ceases and the tongue of salt is separated as an isolated salt lens. Salt flow uplifts overlying sediments forming topographic highs that become efficient sediment traps because the high forms a barrier normal to the provenance. Most production in the Flexure Trend originates from this trap type. In deeper water farther down the slope, the structural and stratigraphic framework is significantly different from the Flexure Trend. Tertiary sediments tend to thin considerably in deeper water areas. The most attractive exploration target may thus be limestone and/or dolomite reefs and reef talus of Mesozoic age. Stratigraphic traps may also be important.

Most of the petroleum potential in the Gulf of Mexico is in the central and western sections coincident with the greatest accumulations of Cenozoic sediments. Traps in deep water in this area (and indeed on parts of the shelf as well) tend to be discontinuous, of highly variable thickness, and scattered vertically throughout the Cenozoic section. Modeling trap dimensions and production characteristics under such conditions requires some generalizing. The geologic attributes of specific traps are based on seismic data, nearby wells, appropriate analogs, and the analyst's intuition.

Inputs to the model are developed by an interdisciplinary team of subject matter specialists. Net pay, productive acreage, and the probability that the prospect is dry (risk) are estimated by geologists. Petroleum engineers base estimates of oil and gas recovery factors on projected reservoir conditions and also determine costs and economic screens. Economists develop price paths based on projected rates of financial return, cost projections, inflation, and price changes.

The procedures and parameters used to numerically model a prospect are nearly identical to those used to calculate volumetric oil and gas reserves in known fields, except that the greater uncertainty inherent in an undiscovered trap is compensated through the use of ranges of values. In cases where the uncertainty is high, the ranges may be very wide and the risk high.

ESTIMATES OF UNDISCOVERED OIL AND GAS

The estimated volumes of undiscovered, economically recoverable hydrocarbons listed below are based on an oil price of $18 per barrel and a gas price of $1.80 per thousand cubic feet, as of January 1, 1987. They represent our judgment on the deep-water potential of the central and western Gulf of Mexico in terms of economically recoverable oil and gas.

	Conditional			Risked			Marginal
	Mean 1	95% 2	5%	Mean	95%	5%	Probability
OIL	4.12	1.10	7.69	4.11	1.06	7.66	.997
GAS	43.95	12.48	82.97	43.95	12.02	82.69	—

1. Oil given in billions of barrels; gas, in trillions of cubic feet.
2. The upper and lower bounds of the range of estimated resources are reported by the 95th and 5th percentiles of the output distribution.

The remaining resources in the deep-water Gulf of Mexico are considerable. Calculated on the basis of risked mean estimates, it is estimated that fully 70 percent of remaining economically recoverable oil and gas in the Gulf of Mexico will be found seaward of the shelf break. Note also that the associated risk is very low (0.003), indicating that the presence of commercial oil and gas in the area is a virtual certainty.

CONCLUSIONS

The resource estimation methodology used in the Gulf of Mexico is applicable to any area, whether or not large amounts of geological and geophysical data are available. This assessment provides an example of the type of approach upon which important decisions regarding exploration and development can be based.

REFERENCES CITED

Carpenter, G. B., 1985, PRESTO, a program for estimating oil and gas resources: Geological Society of America Bulletin, v. 69, no. 9, p. 1434.

Cooke, L. W., 1985, Estimates of undiscovered, economically recoverable oil and gas resources for the Outer Continental Shelf as of July 1984: Minerals Management Service OCS Report MMS 85-0012, 45 p.

Coustau, H., 1980, Habitat of hydrocarbons and field size distribution: A first step towards ultimate reserve assessment, in Assessment of undiscovered oil and gas: Bangkok, United Nations ESCAP, CCOP Publication 10, p. 180-194.

Folinsbee, R. E., 1977, World's view from Alph to Zipf: Geological Society of America Bulletin, v. 88, p. 897-907.

Ivanhoe, L. F., 1976, Evaluating prospective basins: Oil and Gas Journal, December 6, p. 154-156; December 13, p. 108-110; December 20, p. 82-84.

Pecora, W. C., and L. W. Cooke, 1989, A new methodology to assess the conventionally recoverable undiscovered petroleum resource base of OCS lands: 21st Annual OTC Conference, May, 1989, Houston, Texas, in press.

Ray, P. K., 1988, Lateral salt movement and associated traps on the Continental Slope of the Gulf of Mexico: Transactions, Gulf Coast Association of Geological Societies, v. 38, p. 217-223.

Circum-Pacific Council for Energy and Mineral Resources
Earth Science Series, Vol. 16
R. L. Miller, G. Escalante, J. A. Reinemund, and M. J. Bergin (Eds.)
Energy and Mineral Potential of the Central American-Caribbean Regions
Springer-Verlag, Berlin Heidelberg 1995

THE PERPETUAL MYSTERY OF PETROLEUM MIGRATION

William H. Roberts III
HYDREXCO, 11315 Bothwell, Houston, TX 77024 USA

ABSTRACT

As petroleum explorationists, we live with the ultimate embarrassment. We have no tangible proof of how gas and oil come to be in the places where we find them. Many pages of our journals are devoted to "upstream" organic sources and "downstream" reservoir traps. We have some reasonable ideas of where those places are, but in between is that no-man's land of migration. I think the mystery can be solved by viewing the entire petroliferous system, upstream and downstream, as a three-dimensional, water-driven, fluid continuum. If we can accept the idea that the colloidal admixtures of hydrocarbons we know as gas and oil are assembled in the downstream traps from water-borne organic derivatives (e.g., acids, salts, hydrocarbons, etc.), the bulk migration of pre-formed, separate-phase gas and/or oil becomes unnecessary. Let us admit that the movement of significant quantities of gas and/or oil over significant distances has never been factually observed or proven to occur, even for the largest accumulations. An analogy to mineral deposits, which likewise do not move in bulk form, is suggested.

The concept of a water-driven petroliferous continuum raises questions about some popular assumptions (e.g., temperature windows, episodic fluid movement, buoyancy, differential entrapment) which characterize our literature. For some of these, alternatives based on field observations can be offered. Although we are dealing with a noisy, complicated system, I think we can simplify and improve our understanding of it. It could make finding the remaining gas and oil reserves easier and less expensive.

The reason we have not understood petroleum migration may be very simple. I am convinced that petroleum, as such, does *not* actually migrate under natural conditions. The movement of pre-formed, non-wetting, separate-phase hydrocarbon fluids through generally water-wet (hydrophyllic) sedimentary porous media is unnatural, illogical, and unproven. Most importantly, it is not even needed for explaining the occurrence of known gas and oil deposits, large or small. There is a more logical explanation.

For over a hundred years now, oil seekers have had the idea (especially in the Western nations) that oil is primarily formed in one place and secondarily collected in another place. Therefore, they insist that the pre-formed hydrocarbon mixtures must have moved from one place to the other, sometimes over great distances. Because it is usually found in the uppermost available reservoir space, and most oils are lighter than water, buoyancy is assumed to have been the principal cause of movement. I find that nearly all gas and oil deposits around the world are located in, or near, basinal hydrologic discharge areas. In most cases, especially for the larger deposits, there is clear evidence of a significant upward vertical component of water movement. It is this upward movement which carries hydrocarbons into the uppermost spaces where they are unloaded from the waters. The unloading results from depressuring and cooling due to the upward progress and possibly salting-out due to brine filtration. The anticlinal convergence of rising aquifers, of course, is an ideal focal mechanism for deep-water-discharge (DWD) leading to hydrocarbon accumulation; hence the validation of the "anticlinal theory," which has been and still is very useful in oil-finding despite being misinterpreted.

For a trap to function as a collector of hydrocarbons, it must be able to receive organic-rich waters and simultaneously discharge those waters, minus some organic freight. To prevent the loss of accumulating hydrocarbons, a reservoir must be covered by a fine-grained, water-wet membrane essentially impervious to non-wetting fluids.

In a continuously water-filled sedimentary basin, the placement, the composition, and the duration of gas and oil deposits as well as mineral deposits are almost totally con-

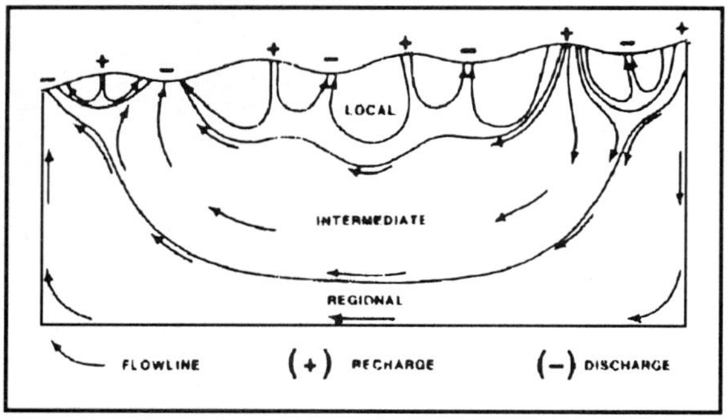

Figure 1. Schematic profile: three orders of basinal water movement (from Toth, 1962).

trolled by water. The amount and the activity of the water is more than adequate to provide that control. There is a relatively simple concept whereby hydrocarbons and other organic derivatives in one place can be accommodated in transient waters, transported, and separated from the water in another place. And I'd like to help you understand the major controls of basinal water traffic through geologic time.

Accumulation of hydrocarbons, as of minerals, logically depends *not* on buoyancy but on *forced draft* (gravity-driven) aqueous transport. Figure 1 is a schematic profile suggesting different depth levels of basinal water exchange related to topographic elevations. It is assumed, of course, that three-dimensional focusing effects caused by the subsurface rock framework and the placement of more and less permeable lithofacies will also influence patterns of basinal water movement. For either petroleum or mineral exploration we should be mainly concerned with the focused, convergent, generally upward water movement leading to (and through) the places where gas and oil deposits are found. It should be noted that the focusing of this deep-water-discharge (DWD) can occur above as well as below the level of the gas or oil reservoirs. In fact, as will be shown later, any local weakness in the "cover" may be of major importance in allowing the spent, previously organic-rich, waters to escape.

The vehicular capacity of basinal waters for hydrocarbons and other organic derivatives is not large in terms of unit volumes, but is exceedingly large in terms of dimension, continuity, and time. Basinal water exchange begins with the earliest compaction effluent. In most cases, the water exchange driven by compaction will at some point in basinal history be mixed and blended with waters driven by meteoric recharge and discharge, which will continue through the entire history of any given basin.

Hydrocarbon accumulation can begin very early in basin history. For example, Figure 2 shows a 5,000-year old carbon-dated sand lens containing a significant concentration of carbon-dated hydrocarbons 12,000 to 16,000 years old, probably coming from a slightly deeper level in this very young and watery deltaic section. Could this be an incipient "stratigraphic trap"? This work in the Pedernales area of the Orinoco Delta, Venezuela, included measurement of water pressures in the core holes. In these young sands and clays, pressures up to 16 pounds per square inch in excess of hydrostatic pressure were recorded. Clearly, complications of early water traffic are caused by differential compaction.

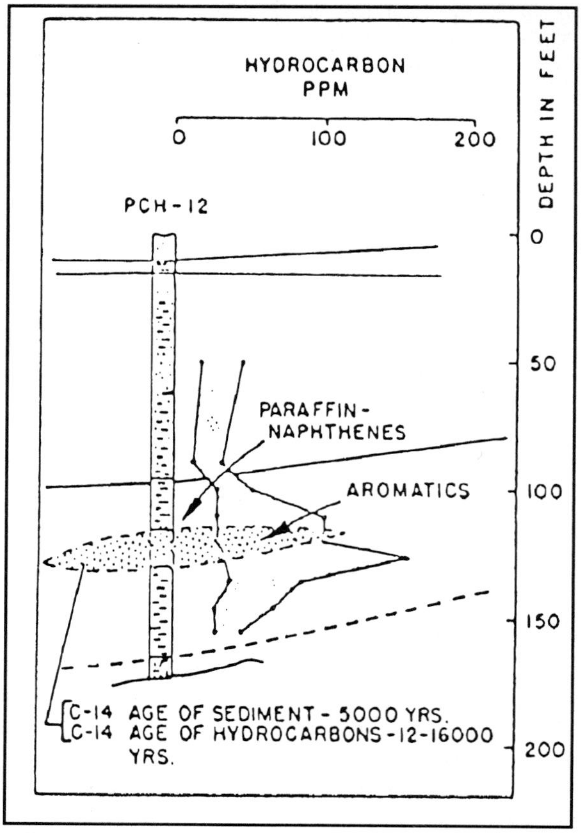

Figure 2. Early hydrocarbon accumulation at Pedernales, Venezuela (from Kidwell and Hunt, 1958).

Three reasonable modes by which hydrocarbons and/or precursors may move, or be moved, from source to deposit are diagrammed in Figure 3. The primary organic source is non-specific. Probably the movement is not restricted to one mode or one geologic episode. The effectiveness of each mode will depend on such variables as pressure, temperature, source quality, lithology, and water chemistry. Aqueous reworking from site to site is believed pertinent to the question of source or "source rock." Obviously, we should know more about the organic content of basinal waters as well as organic-rich rocks. Within the chain of repositories depicted by Figure 3, the organic enrichment of any fine-grained, non-reservoir rock is believed more likely to be secondary than primary. At the same time, in a geologic context every organic or mineral concentration in porous media is susceptible to leaching or otherwise reworking and re-locating. In the role of source rocks, then, all concentrations tend to

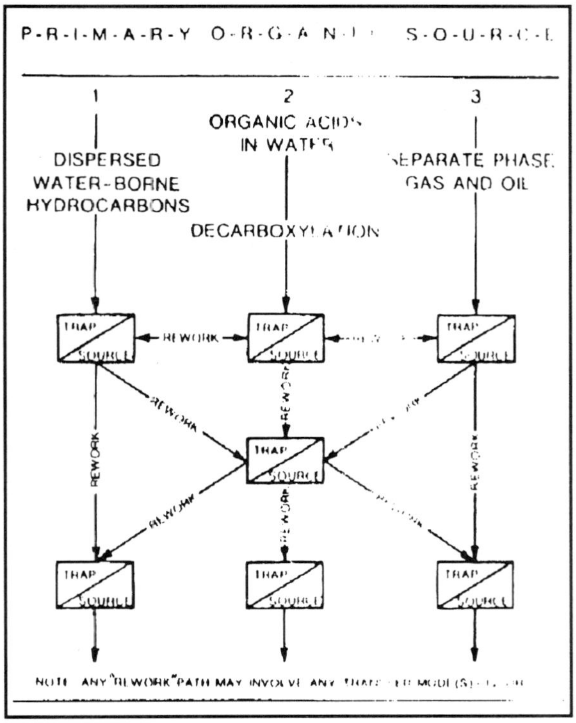

Figure 3. Schematic diagram of source-to-trap and trap-to-trap transfer of organic matter (from Roberts, 1979).

rework, migrate, and rebuild in the general direction of water movement.

In my view, there is considerable room for argument about the relation between progressive maturation of "upstream" organic sources and the selective functions of "downstream" hydrocarbon collectors. The terms "upstream" and "downstream" refer simply and respectively to where the organic components are coming from and where they are going, all of which happens in a three-dimensional context of water movement. As a counterpart to the organic maturation in the source area, which may or may not impose a source imprint on the composition of whatever it is that moves downstream, I wish to emphasize the discriminative functions which probably imprint the composition of resident hydrocarbon admixtures (natural gases and crude oils) in downstream reservoir traps. It seems equally logical to apply the term "maturation" to progressive changes in composition imposed by ongoing processes of entrapment. It should not be confused with effects of thermal maturation in upstream source areas.

It is significant that the variety of hydrocarbon admixtures in reservoirs is virtually infinite, and that among the heavier mixtures no two are exactly alike. None are in thermodynamic equilibrium. Clearly, in the context of geologic time, no trap for hydrocarbons is completely isolated from water exchange and chemical change.

Because lighter, less complex hydrocarbons (or precursors) are more easily accommodated in waters, they are apt to preempt early trap space. Gradually and more slowly, however, heavier and less water-sensitive hydrocarbons will be arriving in the trap as the lighter ends tend to be winnowed or washed through the boundaries of the trap with the partially spent waters. If the functions of entrapment continue effectively in this fashion, processing essentially the same organic-rich aqueous feedstock, a trapped hydrocarbon admixture will progressively evolve, or mature, from light toward heavy. Inasmuch as trap function is very rate-sensitive (i.e., sensitive to rates of water exchange), the deeper trap with lower rates of water exchange will mature more slowly, retaining the light ends longer, whereas the shallower trap is more apt to mature toward a heavier mixture in a relatively short time (other conditions being equal). Such a progression is almost the reverse of the sequence expected from thermal maturation of upstream organic matter (i.e., kerogen). The proximity of residence-time controls in the downstream area, however, is apt to have a more direct effect on the composition of entrapped gases and oils than would the maturation of a distant source.

The bottom line of the above argument is that the composition of a resident (trapped) hydrocarbon admixture at any given geologic moment is expected to reflect the collective residence times of its components. How that relates to progressive maturation (thermal or otherwise) in the upstream source area is a matter for serious further study.

Hydraulic connections between "upstream" areas of organic enrichment (i.e., sources) and "downstream" areas for hydrocarbon accumulation (i.e., products) are easily defined by basic subsurface geology, plus potentiometric and hydrothermal data. The notion of specific aquifers, however, as discrete or exclusive conduits for the movement of organic components does not make geologic sense. Although basinal water movements between areas of hydrologic recharge and discharge are focused in considerable degree by structural, stratigraphic, and lithofacies configurations, within the context of geologic time and space there are no absolute barriers (aquicludes) to water movement. Extensive cross-formational flow is common. The real world of sedimentary transmissibility to water is, in fact, truly three-dimensional.

Given the continuous supply of potentiometric energy in a natural hydrologic system, the dominant regional control of basinal water traffic is exerted by the placement and dimensions of more or less transmissible facies units. Because those units are fixed in position, the facies-guided traffic pattern will be continued through geologic time. The major long-term control is lithofacies. Within that context, it must be realized that facies units occur in a variety of sizes, shapes, and transmissibilities. It is also important to recognize the shunting principle, whereby a local, highly transmissive unit will attract greater water exchange than the ambient facies. Such a condition (e.g., sand lens, bioherm) can create or destroy a gas or oil deposit, depending on the organic content, the water exchange rate, and the nature of the reservoir boundary.

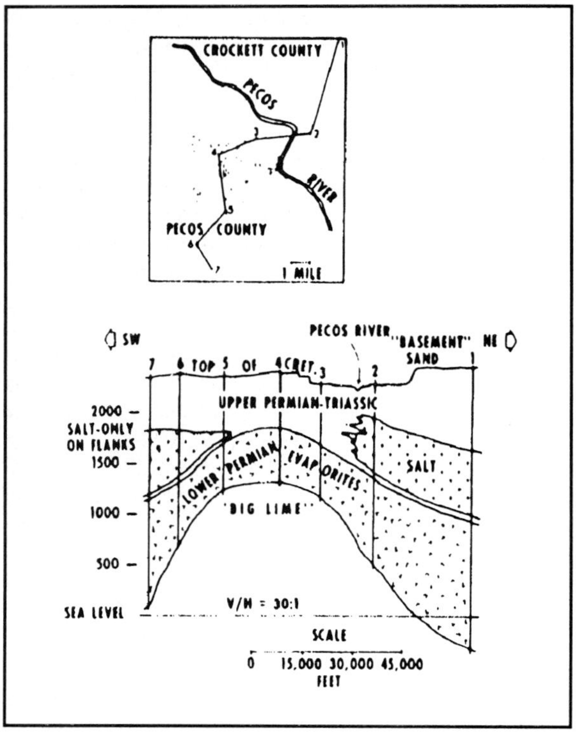

Figure 4. Simplified structure profile, Yates field, Pecos County, Texas (after McCoy, 1934).

The work of progressively gathering hydrocarbons to form and maintain the gas or oil deposits where they are presently found requires a lot of water over geologic time. Based on data from several oil-rich, fully compacted basins with gross average void space of 10 percent, the static volumetric ratio of hydrocarbons to water for a total basin is only a few hundred parts per million. That meager oil-water ratio actually represents countless dispersed oil accumulations, of all sizes, from stains to "shows" to giant fields. Of great importance is the extensive contact area between the non-wetting fluids (gases and oils) and the ubiquitous water. Not only are the gas and oil concentrations spread out three-dimensionally in the basinal aqueous continuum, but within the gas and oil deposits per se (or stains, or shows), there is "irreducible" interstitial water, some free and some bound to the mineral surfaces. Obviously this adds tremendously to the interfacial contact between the hydrocarbons and the water in a basin. The inherent mobility of most of this water prevents the accumulation of any water-sensitive material from being truly permanent in any geologic context.

The aforesaid static basinal water volume, exceeding basinal hydrocarbon volume by several orders of magnitude, is yet hardly a "drop in the bucket" compared to the full aqueous history of a typical petroliferous basin. It doesn't begin to account for the total initial compaction effluent, much less the perpetual water exchange by basinal recharge and discharge, which could exceed present water volume by many additional orders of magnitude. The total exposure of organic matter to water during basinal history thus appears more than adequate to explain the distribution of gas and oil deposits at strategic points within the water traffic system. That basinal water exchange continues to and through the

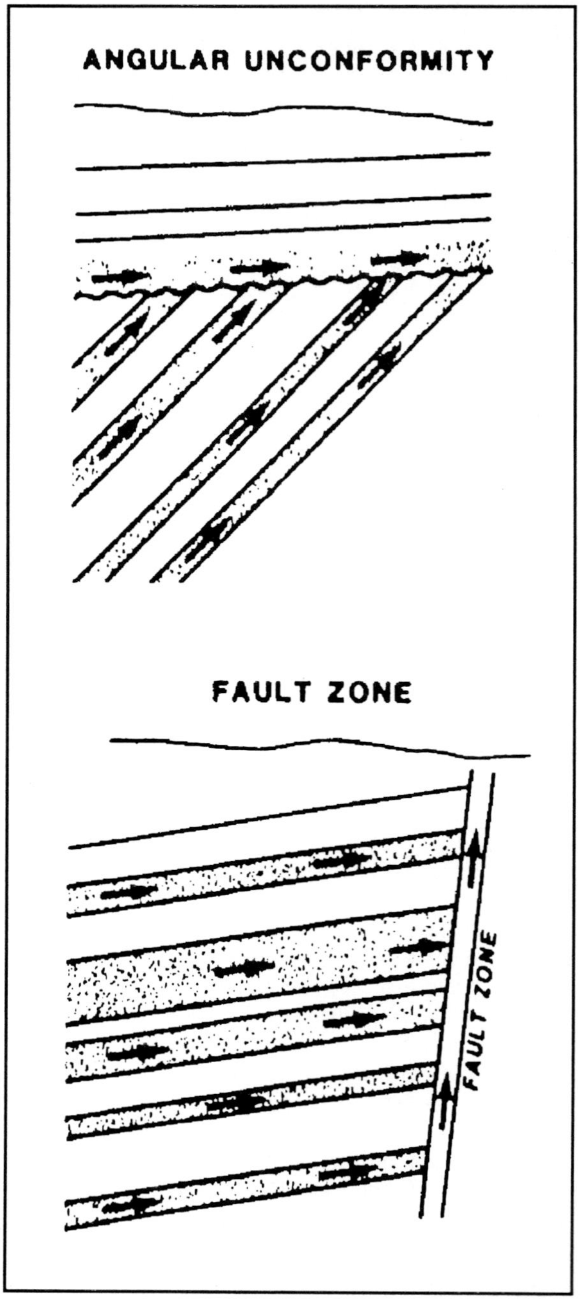

Figure 5. Schematic profile: manifold flow through unconformity or fault (from Roberts, 1986).

present, progressively reworking all water-sensitive materials, organic as well as mineral, at geologically significant rates. It cannot be stopped. Therefore, current observations of the role played by water in the ongoing functions of hydrocarbon entrapment, retention, or loss can simulate the history of existing gas and oil deposits.

Proof of the foregoing argument is found in field evidence of upwardly focused water movements passing through gas and oil-bearing reservoir configurations. In my experience, the clearest evidence is provided by the world's largest existing oil accumulations. Although the heavy oils of the Athabasca and Orinoco supertraps are well qualified examples, they are unwieldy in dimension. So the Yates field of West Texas, Figure 4, is better suited for illustration. This simplified cross-section shows the Lower Permian "Big Lime" oil reservoir to be partially covered by obviously thinned evaporites. The removal and thinning of salts in the cover by solution and the proximity of important surface drainage (Pecos River), fed in part by the discharging waters, are clearly indicative of deep-water-discharge (DWD). The rate of solution and collapse in the cover of this multi-billion barrel field is unknown. It appears, however, that in the near geologic future the oil reservoir will be exposed to the atmosphere, leading to extensive seepage of what oil is then left.

A major factor in the three-dimensional gathering of organic-rich source waters is the manifolding of principal aquifers, caused by angular unconformities and fault zones as depicted in Figure 5. Manifolding not only increases flow rate in a common, upward-leading aquifer, but it tends to blend waters and their organic freight from several stratigraphic intervals. Such stratigraphic and structural manifolding is found in nearly all major oil-productive regions of the world. Moreover, it commonly contributes to the lineal DWD known to be associated with the principal rivers in sedimentary areas. In fact, very few major oil fields have been found far from important surface drainage.

In conclusion, the problems of petroleum migration seem to be mostly in the minds of men. More attention should be paid to the materials, the conditions, and the forces which operate visibly and perpetually today in petroliferous basins. The movement of petroleum components or precursors from site to site in a basinal aqueous continuum is assumed to be perpetual, with no clear beginning or ending. This is migration, but not mystery. It's there for all to see today and tomorrow.

REFERENCES CITED

Kidwell, A. L., and J. M. Hunt, 1958, Migration of oil in recent sediments of Pedernales, Venezuela, *in* Habitat of oil: Tulsa, American Association of Petroleum Geologists, p. 790-817.

McCoy, A. W., 1934, An interpretation of local structural development in Mid-Continent areas associated with deposits of petroleum, *in* Problems of petroleum geology: Tulsa, American Association of Petroleum Geologists, p. 581-627.

Roberts III, W. H., 1979, A basic undocumented re-think on petroleum origin: Houston Geological Society Bulletin, November, p. 6-8.

Roberts III, W. H., 1986, Deep water discharge: Key to hydrocarbon and mineral deposits, *in* B. Hitchon, S. Bachu, and C. Sauveplane, eds., Proceedings, 3rd Canadian-American conference on hydrogeology: Dublin, Ohio, National Water Well Association.

Toth, J., 1962, A theory of groundwater motion in small drainage basins in central Alberta, Canada: Journal of Geophysical Research, v. 67, no. 11, p. 4375-4387.

Circum-Pacific Council for Energy and Mineral Resources
Earth Science Series, Vol. 16
R. L. Miller, G. Escalante, J. A. Reinemund, and M. J. Bergin (Eds.)
Energy and Mineral Potential of the Central American-Caribbean Regions
Springer-Verlag, Berlin Heidelberg 1995

SUMMARY OF WORKING GROUP SESSION ON PETROLEUM RESOURCES

Gerardo Jager
RECOPE

Many of the countries of the Caribbean and Central American region have had a long history of exploration with varying success. But now, with new technologies and an expanded data base, it is appropriate to explore for new prospects and to reinvestigate old basins with modern technology.

We are located in a region surrounded by major oil provinces of the Gulf of Mexico, northern Venezuela, and Colombia. It is essential to extrapolate geological knowledge from these regions and to develop from them exploration models to encourage the expansion of exploration in our region.

We recognize the large risks involved in the exploration of new basins, and value both the scientific expertise and the investment from energy companies in providing the setting for future exploration.

At this symposium we have learned of new interpretations of the plate tectonic evolution of the Caribbean-Central American region. Incorporation of these new ideas which have originated from diverse sources and from skilled workers is essential to new exploration efforts. Additional funding for research and workshops in new plate tectonic theory is strongly encouraged.

In order to ensure success in energy prospecting it is necessary to greatly expand our drilling program, utilizing all new techniques available to us. The Nicoya complex has known middle Cretaceous source rocks of high quality and thick Late Cretaceous shale-turbidite sequences that have not yet been tested for organic content. This is the biggest potential source rock area in Costa Rica, and it includes rocks of Aptian and Albian age that are recognized world-wide as promising source rocks. Although known middle Cretaceous carbonate reservoirs are lacking, siliciclastic reservoirs may occur in structural and stratigraphic traps. These would be the prime Mesozoic targets in Costa Rica.

In the late Miocene, there was a northern Pacific climatic-paleooceanographic event that resulted in deposition of organic-rich diatomaceous rocks (Monterrey Formation), the source rock for several giant oil fields. It is suggested that Pacific basin-Caribbean countries investigate the California literature to see if Monterrey source rock, reservoir rock, and seals may apply as models for the Pacific coastal region of Central America.

We have learned about areas having promise of petroleum accumulations in great water depths. Solutions to production problems are already being carried out in the Gulf of Mexico today, but at high risk and great cost. Whereas this may be a future target elsewhere, the countries of Central America can not now be concerned with these prospects. Rather, we will focus our immediate concern on finding small fields in known basins that can be economic, especially for small companies with low overhead. A major discovery in northern Guatemala provides an example of a major company leasing the discovery well out and a small company taking it over and producing the oil and developing a small field for the country and the company. This is an example of why a country should encourage companies of all sizes.

A major inhibition to our knowledge of the Central American-Caribbean province and the application of its geology to exploration for resources is the lack of availability of unpublished data from both independent researchers and energy companies to the scientific community at large. Efforts need to be made to acquire and make available this knowledge, perhaps through an annual Caribbean-Central American newsletter that will interface with all geologists and economic developers interested in this area. It is also of importance that all earth scientists interested in the Caribbean province meet frequently, for example at the Caribbean Conference held every three years, and equivalent Central American meetings.

The members of all working groups emphasize the importance of expanding our role in education throughout the Caribbean Province and recognize the tremendous strides in geological education accomplished at the University of Costa

Rica in the past 20 years. The excellence of the Costa Rican geologic education system is witnessed by the large number of regional geologists participating in this symposium. We believe that other Central American and Caribbean nations need more and better trained geologists and pledge ourselves to assisting in that endeavor. We hope that students from these countries will be supported by government and industry scholarships to attend the regional school of geology at San Jose.

We are very cognizant of the tremendous amount of planning and work by participants at this conference and we are enthusiastic about the results of this positive force. Let's direct it together into the search for energy resources here in Costa Rica and throughout the Caribbean and Central American region. Good luck to you all in your adventures.

COAL AND PEAT RESOURCES

Circum-Pacific Council for Energy and Mineral Resources
Earth Science Series, Vol. 16
R. L. Miller, G. Escalante, J. A. Reinemund, and M. J. Bergin (Eds.)
Energy and Mineral Potential of the Central American-Caribbean Regions
Springer-Verlag, Berlin Heidelberg 1995

COAL IN THE CENTRAL AMERICAN-CARIBBEAN REGION

E. R. Landis and J. N. Weaver
U.S. Geological Survey, Denver, Colorado, USA

M. D. Carter
U.S. Geological Survey, Reston, Virginia, USA

G. H. Wood, Jr. (Deceased)

ABSTRACT

Coal is reported to be present in almost all the countries that border the Caribbean Sea. All the countries of Central America are known to contain coal except Belize. All the countries of northern South America contain coal except the Netherlands Antilles and Aruba. All the countries of the Greater Antilles contain coal, but among the island nations of the Lesser Antilles, only Anguilla has been reported to contain coal or coaly material.

The known coal deposits range in age from late Paleozoic to mid-Tertiary and occur in geologically complex areas. Most of the coals are probably of subbituminous rank and of early Tertiary age. The ash and sulfur contents of the coals have a broad range but most of the coals can probably be utilized in environmentally acceptable manners for a variety of purposes through use of presently available technology.

With presently available information, estimates of total reserves and recoverable quantities of coal in the region are not possible. However, application of geologic inference and speculation indicates that the total quantity of original coal-in-place in the region may be more than 100 billion MT. Considerable exploration and development will be needed to allow detailed evaluation of the coal resources of the region.

INTRODUCTION

This report is an expansion of the oral presentation at the symposium on the Energy and Mineral Potential of the Central American-Caribbean Region held in San Jose, Costa Rica, March 6-9, 1989. It is intended to provide an overview of the coal situation in the Central American-Caribbean region without in-depth treatment of particular aspects. It is intended that the references cited will provide, or lead to, specific information not presented in this general treatment.

Much of the information discussed in this report is derived from or guided by the exhaustive literature research of Wood and Bour (1988), and Wood et al. (written communication, 1984). Regional and sub-regional syntheses and summaries such as those by Bohnenberger and Dengo (1978); Olive (1978); Orndorff (1985); Landis and Weaver (1987); and Trocki (1988) were extremely useful. Many available country-specific studies are cited as appropriate.

GEOLOGIC SETTING

Regional Tectonics and Stratigraphy

The known coal-bearing rocks of the region are largely within the boundaries of the Caribbean Plate. Some areas that are known to contain coal, such as Cuba, and other areas where coal-bearing rocks may be present, such as the south-eastern part of Venezuela, fall within the North and South American Plates. The Caribbean Plate is seismically active

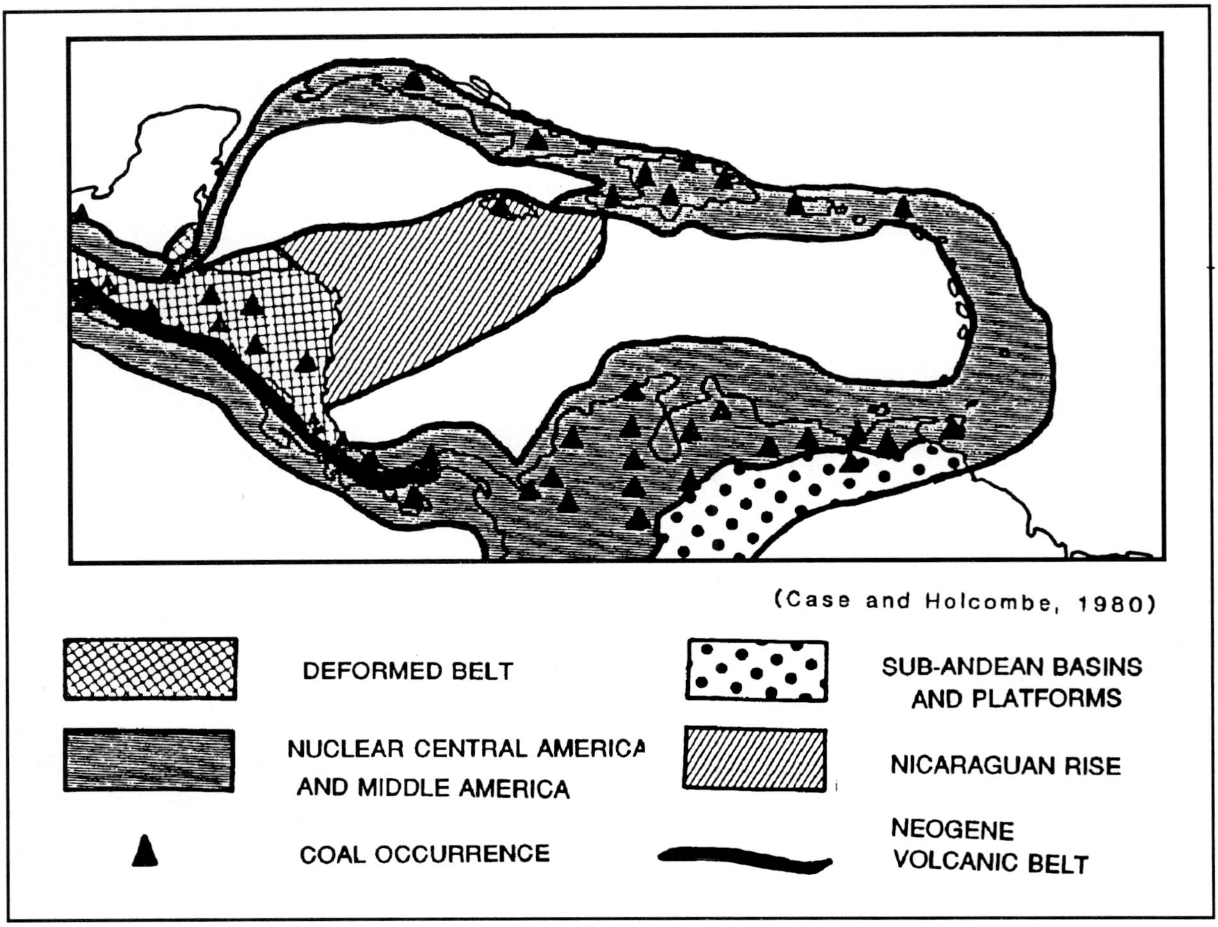

Figure 1. Grouped tectonic-geomorphic provinces of the Central American-Caribbean region.

(Tarr, 1974), particularly along its southwestern and eastern edges.

Case and Holcombe (1980) divided the Caribbean Plate and adjoining part of abutting plates into tectonic-geomorphic provinces. For the purposes of this report the provinces can be assembled into four groups. Figure 1 shows the four groups of provinces along with generalized locations of known and reported coal deposits of the region. One of the groups comprises the nine deformed belts that Case and Holcombe (1980) recognized. Most of the known and reported coal deposits of the region are within this group of deformed belts, along with the bulk of the land masses of the region.

The three other groups shown on Figure 1 are made up of provinces, as defined by Case and Holcombe (1980), that have been relatively stable compared to the deformed belts. All three groups are known or believed to contain coal. Nuclear Central America and the less-deformed part of Middle America are combined in one group, the Nicaraguan Rise is included because of coal-bearing rocks in Jamaica, and the Sub-Andean basins and platforms between the Andean uplift and the Guyana Shield are shown on Figure 1 as a group.

The amount of tectonic deformation that the coal-bearing rocks in the deformed belts have undergone ranges from areas of locally complex folding and faulting, such as the Greater Antilles, to other areas, such as the Andean belt of Colombia and Venezuela, which "has involved extreme vertical uplift and local depression, accompanied by block faulting, local folding and thrusting, and considerable strike-slip faulting during the late Cenozoic. These later Cenozoic belts were superimposed on belts of major Late Cretaceous to Eocene deformation. These, in turn, were superimposed on various styles and degrees of Mesozoic and Paleozoic deformation..." (Case and Holcombe, 1980). Structural deformation of the coal-bearing rocks in the relatively stable areas commonly takes the form of block faulting and open, minor folding, without the repetitive extremes shown in many parts of the deformed belts.

Both the total thickness and the geologic-age span of the sedimentary rocks of the Central American-Caribbean region range widely. As pointed out by Weyl (1980), it has long been recognized that Central America can be divided into two parts that differ in geologic history. The division can be made on the basis of the type of crust, continental or oceanic, that underlies the rock sequence present in different parts of Central America. This same general division can be made throughout the Central American-Caribbean region.

In general, areas underlain by continental crust have a recorded geologic history through Phanerozoic time and contain rock units of Paleozoic, Mesozoic, and Cenozoic age. In contrast, most of the areas underlain by oceanic crust contain only rocks of Cretaceous age and younger. Puerto Rico and Hispaniola in the northeastern part of the Caribbean plate are underlain by oceanic crust (Case and Holcombe, 1980) and the oldest sedimentary rocks on the island are shown as late Mesozoic in age (Blesch, 1967; Garrison et al., 1972; Case and Holcombe, 1980). In contrast to areas underlain by oceanic crust, the Andean deformed belt of Colombia and Venezuela, which is underlain by continental crust (Case and Holcombe, 1980), has a sedimentary-rock sequence that includes rocks of all three eras of Phanerozoic time (Cediel et al., 1976; Martin, 1978; Case and Holcombe, 1980).

Most of the rocks exposed on, or at shallow depths within, the present land areas of the region are of Cenozoic age because (1) older rocks were never deposited in some areas; (2) many of the rocks older than Cenozoic age that were deposited have been removed during subsequent repetitive erosional episodes; and (3) large parts of the present land areas have been and still are the sites of deposition of rocks of Cenozoic age.

Local Structural and Stratigraphic Considerations

The bulk of the coal occurrences in the Central American-Caribbean region fall within deformed belts (Figure 1) and the local structural setting ranges from relatively simple folding and faulting related to middle and late Cenozoic tectonism to very complex structures resulting from several tectonic episodes, each of which further complicates an already structurally complex situation.

The coal-bearing rocks in relatively stable areas such as nuclear Central America and part of Middle America are generally not as deformed as the coal-bearing rocks of the deformed belts. The amount of deformation is only a matter of degree, however. The older sedimentary rocks in the relatively stable areas can be strongly folded and faulted while only the very youngest rocks, those of Pliocene and Pleistocene age, are not deformed. As a minimum, all but the very youngest rocks have been displaced by faults, isolated in outliers by erosion, or concealed by younger rocks.

As pointed out in the discussion on regional stratigraphy, the bulk of the sedimentary rocks are Cenozoic in age. In the areas of continental crust, the oldest dated rocks are of Precambrian age and are composed of metasedimentary, metavolcanic, and metaigneous rocks of varying metamorphic grade. Sedimentary and metasedimentary rocks of Paleozoic age are present in several areas underlain by continental crust, such as Guatemala, Colombia, and Venezuela. Sedimentary rocks of Mesozoic age commonly are present both in areas underlain by continental crust and in areas underlain by oceanic crust.

Coal or coaly material is reportedly present in rocks of late Paleozoic age in Guatemala (Wood and Bour, 1988). The oldest coal known to be present in unmetamorphosed sedimentary rocks in northern Central America is of Early Jurassic to Late Jurassic age (Bohnenberger and Dengo, 1978). In most parts of the Central American-Caribbean region, the next youngest coal-bearing rocks are of mid-Tertiary age; but in parts of Colombia the coal-bearing rocks are of latest Cretaceous age (Suescun-Gomez, 1978), and coal is reportedly present in rocks of Cretaceous age in Guatemala (Bohnenberger and Dengo, 1978). Coal is common in rocks of Oligocene, Miocene, and earliest Pliocene age throughout most of the region, but the Miocene-age rocks contain far more coal than the rocks of any other epoch or period, and almost certainly contain more coal than all the other age units combined.

Depositional Environments

Coal is potentially present in only about 4 percent of the total volume of sedimentary rocks in the crust of the Earth (Fettweis, 1979), and recognition and understanding of the restricted range of occurrence of coaly rocks have become increasingly important in modern coal exploration, development, and utilization planning (Ethridge and Flores, 1981; Galloway and Hobday, 1983; Rahmani and Flores, 1984).

Within the Central American-Caribbean region, coals have been deposited in most of the possible environments because of the diverse geologic history of the region. Coals deposited in alluvial, fluvial, and coastal environments generally differ in such factors as the thickness, lateral continuity, areal size, and quality factors (Cavaroc and Flores, 1981; Flores and Ethridge, 1981). However, the possible range of these characteristics is wide, and detailed study of the sedimentary facies associated with the coals is usually necessary before it is possible to derive reliable predictions of the probable characteristics of the coal deposits. With the exception of some coal fields in Colombia and Venezuela and the investigations underway in Costa Rica, very few in-depth studies have been made in the Central American-Caribbean region.

RESOURCE POTENTIAL

Table 1 is a comparison of reserve and resource estimates for the Central American-Caribbean region. Although the bases, parameters, and objectives of the estimates are vastly dissimilar, Table 1 does allow two conclusions: (1) most reserve and, particularly, resource estimates are conser-

Table 1. Comparison of reserve and resource estimates, Central American-Caribbean Region.

Source	Areas Included	Reserves (million MT)	Resources [1] (million MT)
IGC, 12th (1913)	Honduras	5.0	5
	Colombia	——	27,500
	Venezuela	——	[2]5+
	TOTAL	5.0	27,010+
Herrera (1965)	Honduras	1.0	——
	Colombia	12.5	——
	Venezuela	342.0	——
	TOTAL	355.5	——
Suescun-Gomez et al. (1978) and Olive (1978)	Columbia	——	10,000
	Venezuela	——	1,000+
	TOTAL	——	11,000
Greene, Gallagher, and Wilson (1980)	Colombia	996.0	8,318
	Venezuela	978.0	1,630
	Other Latin Amer.	——	39
	TOTAL	1,974.0	9,987
World Energy Conference (1983)	Colombia	2,073.0	7,990
	Venezuela	309.0	15,058
	TOTAL	2,382.0	23,048
Wood and Bour (1988) and J. Weaver	Central America	——	1,069
	Colombia, Venezuela and Trinidad	——	133,990
	N. Caribbean Countries	——	157
	TOTAL	——	[3]135,216

[1]Includes reserves
[2]"Inexhaustible amount"
[3]Includes identified/hypothetical and speculative resources

Table 2. Past and projected production and consumption (Greene, Hallagher, and Wilson, 1980). (million MT)

	1975		2000	
	Production	Consumption	Production	Consumption
Colombia	3.0	3.0	31.9	[1] 19.5
Venezuela	.1	.1	16.0	[2] 19.2

[1]Projects exports of 12.4 x 10^6, largely steam coal.
[2]Projects imports of 1.2 x 10^6 metallurgical coal and 2.0 x 10^6 steam coal.

vative; and (2) in most parts of the Central American-Caribbean region, much exploration and development are required before definitive resource evaluation will be possible.

PRODUCTION AND UTILIZATION

The earliest commercial utilization of coal in the region probably was for fueling steamships (IGC, 1913). Suescun-Gomez (1978) reported that in Colombia the annual production was "consumed totally in Colombia for production of electrical and mechanical energy, for the manufacture of coke, and for domestic use."

Table 2 shows production and consumption for Colombia and Venezuela in 1975, and projections for the year 2000, according to Greene et al. (1980). Plans for Colombia to supply 10 percent of the world's steam coal market by the end of the century have been reported (International Mining, 1986). In 1987, Colombia exported 8.3 million MT of coal from the Cerrejon coal field, and plans are being made to both increase the amount of coal exported from Cerrejon and to develop for export the resources of other coal areas in the northern part of the country (International Mining, 1988a).

In Venezuela, plans are underway to develop at least two coal fields for production of steam coal for export (U.S. Bureau of Mines, 1988), and other steam coal and coking areas are undergoing technical and economic studies (International Mining, 1988b; Coal Age, 1987).

In addition to the steadily increasing exploration, development, and mining activities in Colombia and Venezuela, Costa Rica has expressed increasing interest in utilizing indigenous coal to replace imported petroleum. Costa Rica recently became the first developing country to reach agreement with the U.S. Department of Energy (DOE) for participation in the DOE Clean Coal Technology Program (Coal News, 1989).

SUMMARY

Fifteen of the nations or territories of the Central American-Caribbean region are reported to contain deposits of coal. Some of the coal deposits have been known and exploited for more than a century, but the vast bulk of the deposits remain unstudied, unexplored, and unused.

Most of the known coal deposits are in geologically deformed parts of the region. Deformation ranges from intense in areas with major folding and thrust faulting in which recovery is difficult, to relatively minor in areas of block faulting and minor folding in which recovery is much easier.

The few definitive studies that are available indicate that the following broad generalizations may be confirmed by future research: (1) the coal-bearing rocks deposited in synclines marginal to the shield areas of northern South America are thick (1,000 m or more), extensive (200 km or more in length), and contain many coal beds (as many as 20, each more than 1 m thick); (2) the coal-bearing rocks deposited elsewhere in the region are thin (a few hundred meters or less), of limited extent (commonly a few tens of kilometers or less), and contain only a few coal beds more than 1 m thick; and (3) because of the active geologic history of much of the region both the physical situation of the coal-bearing rocks and the quality characteristics of the coals have been extensively modified in most areas where coal-bearing rocks were deposited, yet most of the changes should be predictable once we understand the geologic history of the deposits.

Estimates of the quantity of coal in the region range from engineering probabilities of a few million metric tons of available reserves to geologic speculations of more than a hundred billion metric tons of original coal-in-place.

The known coals of the region range in rank from high-volatile bituminous to lignite and range in age from late Paleozoic(?) to Pliocene. However, most of the coals of the region are probably of subbituminous rank and are of Miocene age. Both ash and sulfur contents of the coals from different deposits range from low to high. Most, and perhaps almost all, of the coals are usable with presently available technology. Utilization of clean coal technology should allow a wide range of beneficial usage with environmentally acceptable results.

REFERENCES CITED

Blesch, R. R., 1967, Preliminary geologic map of the Dominican Republic: Oficina de Planification de la Republica Dominica, 1 sheet, scale 1:250,000.

Bohnenberger, O. H., and G. Dengo, 1978, Coal resources in Central America, *in* F. E. Kottlowski, A. T. Cross, and A. A. Meyerhoff, eds., Coal resources of the Americas: Geological Society of America Special Paper 179, p. 65-72.

Case, J. E., and T. L. Holcombe, 1980, Geologic-tectonic map of the Caribbean region: U.S. Geological Survey Miscellaneous Investigations Series Map I-1100, 3 sheets, scale 1:2,500,000.

Castillo M. R., 1984, Geología de Costa Rica, una sinopsis: San Jose, Editorial de la Universidad de Costa Rica, 182p.

Cavaroc, V. V., Jr., and R. M. Flores, 1981, Facies control of Upper Cretaceous Cleary and Gibson Coal Members near Gallup, New Mexico, abs.: American Association of Petroleum Geologists Bulletin, v. 65, p. 556-557.

Cediel M., F., G. Ujueta L., and C. Cacares G., 1976, Mapa geológico de Colombia: Servicio Geológico Nacional, 2 sheets, scale 1:1,000,000.

Coal Age, 1987, U.S. coal producers look to South America, as well: Coal Age, June, p. 12-14.

Coal News, 1989, DOE signs clean coal pact with Costa Rica: National Coal Association, Coal News, June 5.

Ethridge, F. G., and R. M. Flores, eds., 1981, Recent and ancient nonmarine depositional environments: Models for exploration: Society of Economic Paleontologists and Mineralogists Special Publication 31, 349p.

Fettweis, G. B., 1979, World coal resources: Methods of assessment and results: Amsterdam, Elsevier, Developments in economic geology, v. 10.

Flores, R. M., and F. G. Ethridge, 1981, Nonmarine deposits and the search for energy resources and minerals, *in* F. G. Ethridge and R. M. Flores, eds., Recent and ancient nonmarine depositional environments: Models for exploration: Society of Economic Paleontologists and Mineralogists Special Publication 31, p. 1-17.

Galloway, W. E., and D. K. Hobday, 1983, Terrigenous clastic depositional systems: New York, Springer-Verlag, 423p.

Garrison, L. E., R. G. Martin, Jr., H. L. Berryhill, Jr., M. W. Buell, Jr., H. R. Ensminger, and R. K. Perry, 1972, Preliminary tectonic map of the eastern Greater Antilles region: U.S. Geological Survey Miscellaneous Geologic Investigations Map I-732, scale 1:500,000.

Greene, R. P., and J. M. Gallagher, eds., and C. L. Wilson, Project Director, 1980, Future coal prospects: Country and regional assessments: World coal study (WOCOL): Ballinger, Cambridge, Massachusetts Institute of Technology, 577p.

Herrara, A. O., 1965, Los recursos minerales de America Latina: Biblioteca de America: Editorial Universitaria de Buenos Aires, 127p.

International Geological Congress, 12th, 1913, The coal resources of the world: Toronto, Canada, Morang and Co., 3v., 1266p.

International Mining, 1986, Coal news—Colombia: London, International Mining, April, p. 67.

International Mining, 1988a, Coal news—Colombia: London, International Mining, April, p. 94.

International Mining, 1988b, Coal news—Venezuela: London, International Mining, February, p. 41-42.

Landis, E. R., and J. N. Weaver, 1987, Coal resources of Central America: U.S. Geological Survey Open-File Report 87-365, 5p.

Martin F., C., 1978, Mapa tectonico Norte de America del Sur: Ministerio de Energía y Minas, Republica de Venezuela, 2 sheets, scale 1:2,500,000.

Olive, W. W., 1978, Coal deposits of Latin America, *in* F. E. Kottlowski, A. T. Cross, and A. A. Meyerhoff, eds., Coal resources of the Americas: Geological Society of America Special Paper 179, p. 57-64.

Orndorff, R. C., 1985, Annotated bibliography of coal in the Caribbean region: U.S. Geological Survey Open-File Report 85-110, 29p.

Rahmani, R. A., and R. M. Flores, eds., (1984), Sedimentology of coal and coal-bearing sequences: International Association of Sedimentologists Special Publication 7, 412p.

Suescun-Gomez, D., 1978, Coal deposits in Colombia, *in* F. E. Kottlowski, A. T. Cross, and A. A. Meyerhoff, eds., Coal resources of the Americas: Geological Society of America Special Paper 179, p. 49-56.

Tarr, A. C., 1974, World seismicity map: U.S. Geological Survey, scale 1:39,000,000.

Trocki, L., 1988, An energy atlas of five Central American countries: Los Alamos National Laboratory, LA-11205-MS, 126p.

U.S. Bureau of Mines, 1988, Mineral industries of Latin America: U.S. Bureau of Mines Mineral Perspective, 134p.

Weyl, R., 1980, Geology of Central America: Berlin-Stuttgart, Gebrüder Borntraeger, 2nd ed., 371p.

Wood, G. H., Jr., and W. V. Bour, III, 1988, Coal map of North America: U.S. Geological Survey Special Geologic Map, 2 sheets, scale 1:5,000,000.

World Energy Conference, 1983, Survey of energy resources: London, World Energy Conference, 148p.

Circum-Pacific Council for Energy and Mineral Resources
Earth Science Series, Vol. 16
R. L. Miller, G. Escalante, J. A. Reinemund, and M. J. Bergin (Eds.)
Energy and Mineral Potential of the Central American-Caribbean Regions
Springer-Verlag, Berlin Heidelberg 1995

EXPLORATION, DEVELOPMENT, AND UTILIZATION OF COAL IN COSTA RICA

Oldemar Ramirez, Kenneth Bolaños, and Luis Malavassi
Division de Recursos Carboniferos, RECOPE, P. O. Box 4351, San José 1000, Costa Rica

ABSTRACT

Costa Rica has undertaken a program of investigation of its coal resources because of the cost of importing foreign oil to meet much of its energy needs. In 1982 it was decided to investigate the potential of each of the eight known coal occurrences. Reconnaissance studies determined that three had coal beds thick enough to warrant further investigation. Prefeasibility studies at the two better known deposits (Uatsi and Zent) have been completed, and a total of 50 million MT of proved and probable reserves of coal is calculated. This figure includes a small number of tons estimated for the third area, Venado. At Zent, feasibility studies have been completed and a mine design for a part of the deposit has been prepared.

The coal beds are in the thick Gatun Formation, which is of Miocene age. In rank the coal ranges from lignite to subbituminous A.

The first usage of the coal will probably be in the cement industry, but the country is considering the feasibility of other uses in the industrial and residential sector as a replacement for imported Bunker C fuel.

INTRODUCTION

Imported hydrocarbons constitute one of the principal sources of energy on which Costa Rica to a large degree is dependent. Variations in the international price of petroleum impinge directly on the Costa Rican economy. An important part of the national energy program in recent years has been the evaluation of national resources such as evaluation of the national petroleum potential, coal investigation and development, search for and appraisal of peat deposits, and consideration of the potential of biomass, among others. This new strategy looks forward to the use of alternate sources of energy in order to lessen the imbalance of foreign exchange caused by consumption of traditionally imported hydrocarbons. Use of alternate fuels would also create employment and promote building of infrastructure adapted to development of a new national industry.

THE COALS

A coal deposit with a potential on the order of 6 million MT had in 1982 already been identified in an area in Lower Talamanca. This was recognized as constituting a possible and attractive alternative source of energy. With this in mind, in that year the Government of the Republic began a comprehensive program for investigation and development of the coal resources in the entire country, using a systematic approach that would produce results in a short time. Because of its earlier work on coal, RECOPE was entrusted with the implementation of this program directed toward the investigation of the areas throughout the country where coal is known to exist.

Outcrops of coal have been reported in Costa Rica ever since the end of the last century. As the result of a library search, eight different localities with coal potential have been

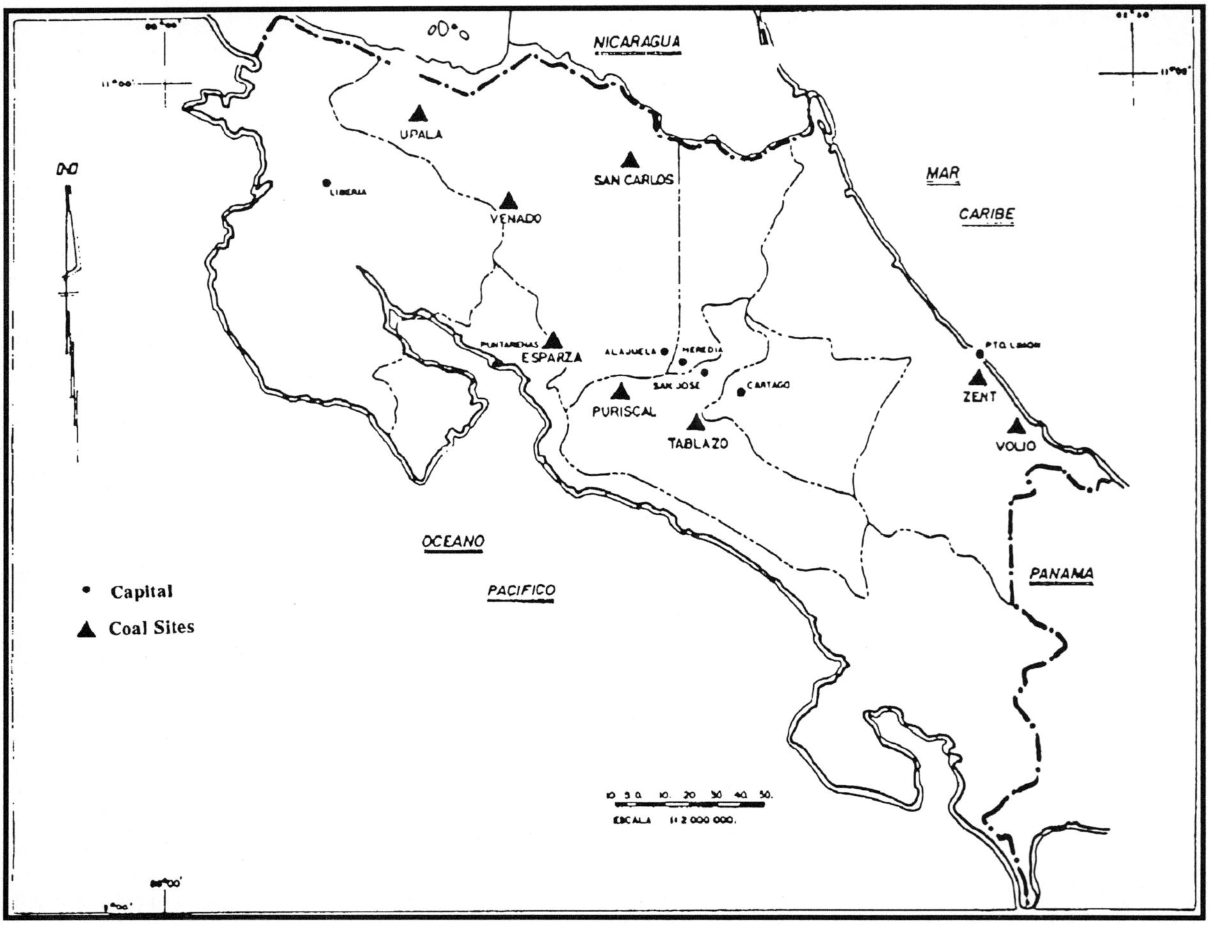

Figure 1. Location of the eight known coal occurrences in Costa Rica.

Table 1. Coal reserves of Costa Rica (millions of metric tons)

AREA	R. Probada	R. Probable	R.	Total
UATSI (Talamanca)	19.3	9.1	4.1	32.5
ZENT (Limon)	5.6	5.4	3.8	14.8
VENADO (San Carlos)	2.0	——	——	2.0
TOTALES	26.9	14.5	7.9	49.3

identified: Venado, Rio San Carlos, and Upala in the Province of Alajuela; Esparza in the Province of Puntarenas; El Tablazo and Purizal in the Province of San Jose; and Zent and Uatsi in the Province of Limon (Figure 1). Once reconnaissance of these eight areas had been accomplished, priority for more thorough studies was given to the areas of Venado, Zent, and Uatsi, where the results of the reconnaissance investigations were more significant and where thicknesses of some of the beds, as much as 7 m, have been measured. The coal is of good quality. Based on work to date, a total potential of 50 million MT of coal has been calculated (Table 1). Once the respective stages of prefeasibility studies had been completed, the greatest coal resources appeared to be concentrated in the areas of Zent and Uatsi, so that the following stages of feasibility and mine design studies have proceeded.

The coal in these areas is in the Gatun Formation of Miocene age. Peat accumulated in a reducing environment in sea marginal lakes. In spite of the fact that the coal occurrence at Zent is tens of kilometers from that of Uatsi, from the point of view of coal formation and age, the two correlate. Not so from the point of view of coal rank. The coals of the Limon sedimentary basin increase in rank in a southern direction, ranging from lignite to subbituminous A with some areas of bituminous coal in the Uatsi coal field. The chemical composition of seven coal samples from the Zent deposit are shown in Table 2. At the present time (1989), feasibility studies of a part of the Zent area have been completed, and a mine design for production of the coal has been prepared, so as to supply

Table 2. Analyses of samples.

Sample Number Laboratory-Field-		Proximate Analysis: Moisture	Ash	Volatile Matter	Fixed Carbon	Ultimate Analysis: Hydrogen	Carbon	Nitrogen	Sulfur	Oxygen	Heating Value (Btu/lb)	Forms of Sulfur: Sulfate	Pyritic	Organic	Equilibrium Moisture	Free Swelling Index	Hardgrove Grindability	Ash Fusion Temperatures (°F): Initial Deformation	Softening Temperature	Fluid Temperature	Apparent Specific Gravity	Apparent Rank	
MR26187	(1)	40.33	22.47	21.14	16.36	6.72	26.09	0.52	2.47	41.73	4,622	0.00	1.45	1.02	23.58	0.0	---	2,210	2,340	2,480		lig B	U-6 bed
PCU-1-M-3	(2)		37.46	35.26	27.28	3.74	43.51	0.88	4.12	10.29	7,708	0.00	2.41	1.71									C.92 m thick, 130 m depth
	(3)			56.38	43.62	5.98	69.52	1.41	6.59	16.44	12,326	0.00	3.85	2.74									
	(4)	23.58	28.63	26.94	20.85	5.49	33.25	0.66	3.15	28.82	5,890	0.00	1.85	1.3								sub C	
L10220485	(1)	30.1	11.0	10.1	48.8	---	---	---	2.24	---	7,321											lig A	U-6 bed
PCU-3-M-1	(2)		15.8	14.4	69.8	---	---	---	3.20	---	10,472												0.60 m thick, 40 m depth
	(3)			17.1	82.9				3.80	---	12,429												
MR27096	(1)	25.56	35.63	22.99	15.83	5.21	23.77	0.38	5.22	29.79	4,160	0.12	3.30	1.80	24.07	0.0	58	2,300	2,420	2,480		lig B	U-3 bed
PCU-1-M-7	(2)		47.85	30.88	21.27	3.16	31.93	0.50	7.02	9.54	5,587	0.17	4.43	2.42									0.55 m thick
	(3)			59.22	40.78	6.06	61.23	0.96	13.46	18.29	10,717	0.33	8.50	4.63									
	(4)	24.37	36.34	23.45	16.15	5.09	24.24	0.39	5.32	28.62	4,240	0.12	3.37	1.84								lig A	
MR27097	(1)	26.00	21.56	28.90	23.60	6.17	37.95	0.70	1.48	32.20	6,610	0.01	0.89	0.58	26.92	0.0	46	2,370	2,450	2,470		sub C	Y-7 bed
PCV-1-M-1	(2)		29.05	39.05	31.90	4.41	51.29	0.95	2.00	12.30	8,933	0.01	1.20	0.79									
	(3)			55.04	44.96	6.22	72.29	1.34	2.82	17.33	12,591	0.01	1.69	1.12									
	(4)	26.92	21.23	28.54	23.31	6.23	37.48	0.69	1.46	32.91	6,530	0.01	0.88	0.57								sub C	
L-5 (1981	(1)	26.67	8.14	36.03	29.16	3.85	47.66	0.85	0.89	11.94	8,275	---	---	---	25.0	---	41	---	---	---	1.3	sub C	Y-9 bed
	(2)		11.18	49.13	39.77	5.25	65.00	1.16	1.22	16.27	11,285												Outcrop sample, 1.1 m thick
	(3)	---	---	---	---	---	---	---	---	---	---												Chlorine 0.80 %
	(4)	26.00	8.21	36.36	29.43	3.78	48.1	0.86	0.9	38.15	8,350											sub C	
MR26190	(1)	34.39	13.76	32.19	19.66	7.23	38.06	0.64	2.28	38.03	6,914	0.00	0.75	1.53	---	0.0	---	2,150	2,230	2,260		lig A	Uncorrelated bed
PCU-1-M-1	(2)		20.97	49.07	29.96	5.16	58.01	0.97	3.48	11.41	10,538	0.00	1.15	2.33									C.45 m thick, 64 m depth
	(3)			62.09	37.91	6.53	73.40	1.23	4.40	14.44	13,336												
MR26191	(1)	32.63	14.71	29.68	22.98	6.80	37.84	0.65	2.47	37.53	6,768	0.00	0.58	1.89	---	0.0	---	2,190	2,310	2,410		lig A	Uncorrelated bed
PCU-1-M-2	(2)		21.84	44.05	34.11	4.67	56.17	0.97	3.67	12.68	10,046	0.00	0.87	2.80									C.69 m thick, 89 m depth
	(3)			56.36	43.64	5.97	72.86	1.24	4.70	16.23	12,853	0.00	1.11	3.59									

the demand for coal in the industrial sector, especially the cement industry, starting in 1989. For its part, the Uatsi coal project is in the feasibility study stage. The coal may be produced after additional exploration over a period of one or two more years.

UTILIZATION OF THE COAL

One of the cement factories of the country has already made changes that will enable it to substitute coal for approximately 25 percent of its fuel requirement without many additional changes. Tests on coal utilization were made at the cement plant in the final months of 1988 with positive results. In the first trial, a coal substitution of 11 percent was attained. One result was perfect control of problems that have occasionally arisen with use of other solid fuels. The objective is to attain total substitution of coal for petroleum in the next few years.

Other avenues for consumption of the coal are using it as a substitute for Bunker C fuel in the industrial and residential sector. These will be considered in a short time. These usages will require treatment of the coal at the mine site to prepare it for the most efficient transportation and utilization. Much work will be needed to develop the technology for storing, crushing, washing, and drying of the coal.

CONCLUSION

The project on the investigation and development of coal resources of Costa Rica, a pioneer endeavor in the Central American region, has been successful. It should bring enormous benefit to the country in a short time.

Circum-Pacific Council for Energy and Mineral Resources
Earth Science Series, Vol. 16
R. L. Miller, G. Escalante, J. A. Reinemund, and M. J. Bergin (Eds.)
Energy and Mineral Potential of the Central American-Caribbean Regions
Springer-Verlag, Berlin Heidelberg 1995

COAL RESOURCES OF THE BAJA TALAMANCA AREA OF COSTA RICA

Kenneth Bolaños I. and Manuel Salas P.
RECOPE, P. O. Box 4351, San José 1000, Costa Rica

ABSTRACT

A total of 17 million MT of subbituminous C and B coal had been evaluated to be present in the three sub-areas of the southern part of Baja Talamanca coal field (Uatsi Project); 8.7 million MT are classified as proven (measured) reserves. Within the northern sub-area of the Carbon Volio River basin, 2.6 million MT, with a thickness of 1.5 m, are present in the V-9 coal seam. Around 2 million MT are classified as proven reserves in an area of 1.2 km^2. The coal-bearing sequence is folded in an asymmetric, much faulted anticline. Some 3.1 million MT, with an average thickness of 0.95 m, in the V-7 of the eastern sub-area of the Carbon Volio River basin, cover an area of 1.7 km^2. Approximately 1.9 million MT of the total coal estimates are proven reserves. The coal-bearing member dips 10-20°. Underground mining will be needed to recover most of these resources. In the western sub-area of the Carbon Volio River basin, 11.3 million MT are estimated to be present in the V-1 and V-9 Nueva coal seams. The lower bed (V-1), with a thickness of 0.95 m, has 5.1 million MT, of which 1.8 are proven reserves. The upper bed, 140 m higher than the V-1 bed, has 6.2 million MT with an average thickness of 1.5 m, of which 2.9 million MT are proven. The V-9 Nueva coal bed has an overburden ratio of 20:1 or less in an area of 0.4 km^2, where coal can be recovered by open pit methods. Proven reserves are calculated within a radius of 0.25 km from outcrops or boreholes. Total reserves are calculated within a radius of 0.5 km from known points of observation. A total of 37 measured outcrops and 20 boreholes were used to derive these resource figures.

INTRODUCTION

The work conducted during the project was an Exploration Stage I study of the Uatsi project area in the southern part of the Baja Talamanca (Rio Carbon Volio) coal field. This phase of exploration included compilation of information acquired through drilling and borehole geophysical logging to allow correlation of coal beds of potential interest, to estimate resources in the probable category, and to obtain a preliminary understanding of the quality of the coal and its potential utilization. During this stage, 20 exploratory drill holes were completed, samples from drill cuttings and cores were collected and described, coal samples were selected for analysis, and geophysical logs were made and interpreted. A report on the results of the study, comprising 120 pages of text and 30 illustrations, was prepared with the assistance of other RECOPE personnel. The report will provide guidance during subsequent exploration phases in both the Uatsi project area and adjoining parts of the Baja Talamanca coal field.

COAL RESOURCES

The potentially economic coal beds in the Uatsi project area are located in the middle member of the Gatun Formation of Miocene age (Figure 1). The formation in this area consists of (1) a dominantly sandy lower member at least 270 m thick, (2) a middle member divided into a coaly lower part as much as 320 m thick, and a dominantly sandy upper part as much as 600 m thick, and (3) a sandy and conglomeratic upper member as much as 450 m thick. The Gatun Formation is underlain, possibly conformably, by the Uscari Formation, a dominantly fine to very-fine grained clastic sequence at least 750 m thick. The Suretka Formation of Pliocene and/or Pleistocene age, which unconformably overlies the Gatun Formation in the Baja Talamanca coal field, ranges from 200 to 1,000 m thick, but has been eroded from the Uatsi project area.

Structurally, the area is complex. The coal-bearing rocks are folded and faulted; displacements range from small

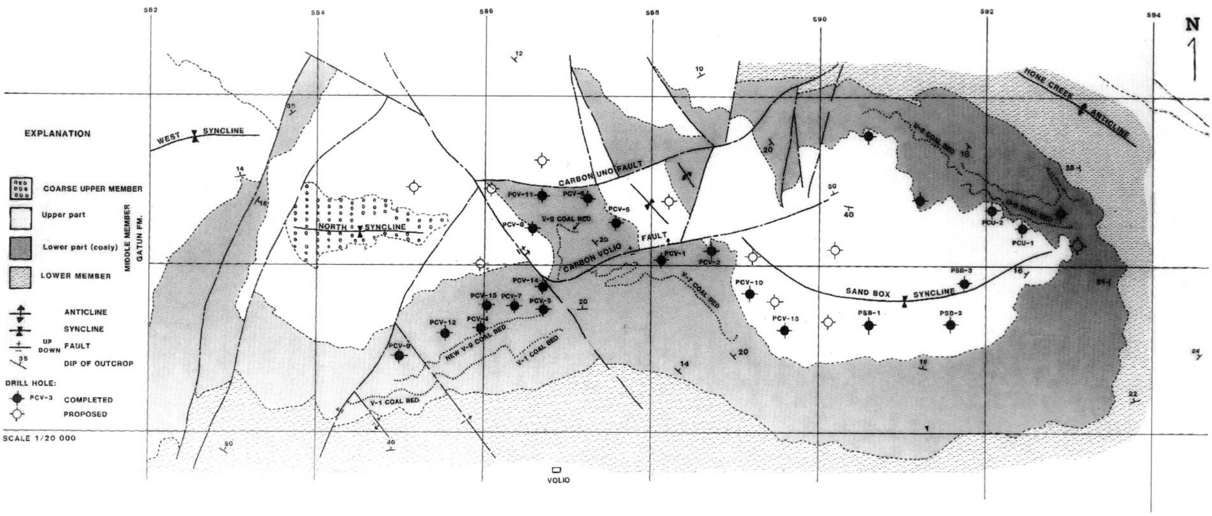

Figure 1. Geologic map, Uatsi Project, with completed and proposed drill sites.

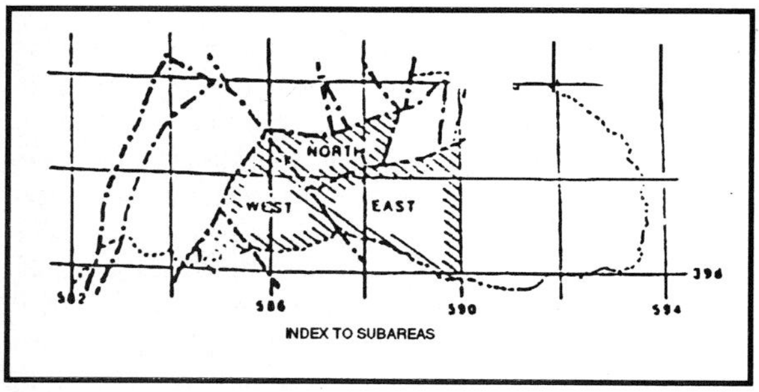

Figure 2. Boundaries of areas for wich coal resources have been evaluated. Also shows major faults.

to large, and the net effect is division of the area into fault-bounded blocks (Figure 2). Correlation of the rock sequences and coal beds between blocks is difficult. For coal resource purposes, the structurally-defined blocks are treated as individual entities.

Within the fault-bounded blocks, resources were estimated for four coal beds for which correlation within the individual blocks is reasonably assured. Other coal beds are present, but their correlation is uncertain based upon the current geologic information, and no resources were estimated for these beds (Figure 1).

A total of about 17 million MT of subbituminous C coal, of which about 8.7 million MT is in the measured category, is estimated to be present in three sub-areas (Figure 1) of the southern part of the Baja Talamanca coal field. The boundaries of each sub-area are either faults or outcrops. In the sub-area North, an area of about 1.2 km^2, about 2.6 million MT are estimated to be present in one coal bed (V-9), which has an average thickness of about 1.6 m (Figure 2). About 2 million MT of the estimated total are in the measured category. The coal-bearing rocks are folded into an asymmetric anticline, and underground mining would almost certainly be required for recovery.

In an area of about 1.7 km^2 in the sub-area East, about 3.1 million MT of coal are estimated to be present in one bed (V-7), with an average thickness of about 0.9 m. About 1.9 million MT of the estimated total are in the measured category. The coal-bearing rocks dip about 10° in a gently-folded sequence. Underground mining would be required for recovery of the coal, though a limited amount might be recovered through surface mining along the outcrop.

In the sub-area West, a total of about 11.3 million MT of coal is estimated to be present in two coal beds (V-1 and New V-9). The lowest bed (V-1), which has an average thickness of about 0.9 m in the area of estimation (3.2 km$^{2)}$, contains an estimated total of about 5.1 million MT of which about 1.8 million MT are in the measured category. The upper bed (New V-9), which occurs about 140 m above the lower bed, is estimated to contain a total of almost 6.2 million MT of coal with an average thickness of about 1.5 m in an area of about 2.6 km^2. Of this total, about 2.9 million MT are in the measured category. The lower coal bed (V-1) would have to be recovered mainly by underground mining, but some surface mining might be done along the outcrop. The upper coal bed (New V-9) is recoverable by surface mining in part of the area of about 0.4 km^2 with an overburden factor of less than 20 to 1.

The combined estimates in millions of metric tons are:

ASSURANCE CATEGORY	ESTIMATE I
Proven (measured)	8,704
Probable (indicated)	4,146
Possible (inferred)	4,210
TOTAL	17,060

In summary, with the available data, a minimum of almost 13 million MT of measured and indicated coal are estimated to be present in an area of about 6 km^2. Additional exploration is required to increase the estimated resources, both by adding to our knowledge about the four coal beds represented in the estimates to date, and by allowing estimation of resources in beds not included in the foregoing estimates. Most of the estimated resources must be recovered by underground mining, but some part, undefined as yet, could be recovered by surface mining.

COAL QUALITY

Coal sample analyses, 10 by RECOPE and six by USGS contractors, show a range in coal rank from lignite B to high volatile C bituminous. Most analyses are in the subbituminous C category. The coal does not appear to have any coking properties; the four coal samples for which equilibrium moisture (inherent moisture by American Society for Testing and Materials [ASTM] definition) was determined have an average of 25 percent inherent moisture; the five coal samples for which ash fusion temperatures were determined had an average initial ash deformation temperature of 2250°F, an average ash softening temperature of 2350°F, and an average fluid temperature of 2400°F. Three Hardgrove Grindability Indexes that were determined ranged from 41 to 58. Average sulfur content on as-received basis of 13 samples was 1.7 percent, but the range was from 0.28 to 5.22 percent. Forms of sulfur were determined for five samples; pyritic sulfur was dominant in three samples, and organic sulfur in the other two. The ash content of 15 samples on an as-received basis averaged 13.8 percent, and ranged from 6.3 to 35.6 percent (Table 1).

Obviously, many more analyses of representative samples were needed for adequate characterization of the coals of the Uatsi project area, but at this time it appears that the coal is most likely to be of subbituminous C rank with medium sulfur and ash contents. This characterization should be used with caution because of the small number of samples analyzed to date.

CONCLUSION

Further exploration efforts should be designed to yield needed additional information in parts of the coal field already partially explored in the Uatsi project area and to begin the new exploration needed for assessment of the coal resources of the remainder of the Baja Talamanca coal field. Exploration Stage II activities are underway in the Uatsi project area to provide data needed for some planning and evaluation of utilization potential. Some of the Exploration Stage II activities will also provide information needed for areas outside the Uatsi project area.

Table 1. Analyses of samples. L-prefix samples by RECOPE, remainder by USGS contractor. Condition (1) as received; (2) dry; (3) dry, ash-free; (4) changed to equilibrium moisture.

SAMPLE NUMBER Laboratory - Field -		Proximate Analysis				Ultimate Analysis					Heating Value — Btu / lb.	Forms of Sulfur			Equilibrium Moisture	Free Swelling Index	Hardgrove Grindability	Ash Fusion Temperatures (°F)			Apparent Specific Gravity	Apparent Rank	COMMENTS
		Moisture	Ash	Volatile Matter	Fixed Carbon	Hydrogen	Carbon	Nitrogen	Sulfur	Oxygen		Sulfate	Pyritic	Organic				Initial Deformation	Softening Temp.	Fluid Temperture			
L03310185	(1)	24.0	8.1	--	--	--	--	--	--	--	9,041												New V-9 bed
PCV-14-M-2	(2)	--	10.1	--	--	--	--	--	--	--	11,210												1.6 m thick,
	(3)	--	--	--	--	--	--	--	--	--	12,388												50 m depth
L04190285	(1)	27.9	7.6	11.9	52.6	--	--	--	0.28	--	10,732											hv Cb	New V-9 bed
PCV-12-M-1	(2)		10.5	16.5	79.9	--	--	--	0.39	--	14,884												0.45 m thick,
	(3)			14.9	66.0	--	--	--	0.35	--	16,637												33 m depth
L19070575	(1)	14.8	6.3	38.1	40.8	--	--	--	0.81	--	8,892											sub B	New V-9 bed
PCV-7-M-1(A)	(2)		7.6	45.7	46.7	--	--	--	0.97	--	10,661												0.30 m thick,
	(3)			57.9	42.1	--	--	--	1.02	--	11,270												20.7 m depth
L16290485	(1)	24.4	19.3	34.5	21.8	--	--	--	0.40	--	9,290											hv Cb	New V-9 bed
PCV-7-M-1(B)	(2)		25.6	45.6	28.8	--	--	--	0.50	--	12,289												0.92 m thick,
	(3)			61.3	38.7	--	--	--	0.70	--	16,501												21 m depth
L12290485	(1)	25.6	9.7	42.1	22.6	--	--	--	0.60	--	8,577											sub B	New V-9 bed
PCV-7-M-1(C)	(2)		13.0	56.6	30.4	--	--	--	0.84	--	11,529												-.40 m thick,
	(3)			65.1	34.9	--	--	--	0.97	--	13,255												22 m depth
L18070585	(1)	12.2	57.2																				New V-9 bed
PCV-7-M-1(D)	(2)																						Not coal, 0.45 m thick,
	(3)																						22.4 m depthj
L01310185	(1)	23.6	11.5	24.7	40.2	--	--	--	--	--	7,772												V-1 bed
PCV-7-M-2	(2)		--	--	--	--	--	--	--	--	9,637												0.61 m thick,
	(3)			--	--	--	--	--	--	--	11,192												167 m depth
L21070585	(1)	18.1	7.7	38.8	35.4	--	--	--	0.77	--	8,698											sub C	V-9 bed
PCV-7-M-2(A)	(2)		9.4	47.4	43.2	--	--	--	0.94	--	10,620												0.285 m thick,
	(3)			63.9	36.1	--	--	--	1.04	--	11,722												79.3 m depth
L22070585	(1)	15.2	9.9	41.5	33.4	--	--	--	2.06	--	7,466											sub C	V-9 bed
PCV-7-M-2(B)	(2)		11.7	48.9	39.4	--	--	--	2.43	--	8,806												0.49 m thick,
	(3)			65.3	34.7	--	--	--	2.75	--	9,968												79.6 m depth

Table 1, Con't. Analyses of samples. L-prefix samples by RECOPE, remainder by USGS contractor. Condition (1) as received; (2) dry; (3) dry, ash-free; (4) changed to equilibrium moisture.

SAMPLE NUMBER Laboratory - Field -		Proximate Analysis				Ultimate Analysis					Heating Value — Btu / lb.	Forms of Sulfur			Equilibrium Moisture	Free Swelling Index	Hardgrove Grindability	Ash Fusion Temperatures (°F)			Apparent Specific Gravity	Apparent Rank	COMMENTS
		Moisture	Ash	Volatile Matter	Fixed Carbon	Hydrogen	Carbon	Nitrogen	Sulfur	Oxygen		Sulfate	Pyritic	Organic				Initial Deformation	Softening Temp.	Fluid Temperture			
M226187	(1)	40.03	22.47	21.14	16.36	6.72	26.09	0.52	2.47	41.73	4,622	0.00	1.45	1.02	23.58	0.0	- -	2,210	2,340	2,480		lig B	U-6 bed
PCU-1-M-3	(2)		37.46	35.26	27.28	3.74	43.51	0.88	4.12	10.29	7,708	0.00	2.41	1.71									0.92 m thick,
	(3)			56.38	43.62	5.98	69.58	1.41	6.59	16.44	12,326	0.00	3.85	2.74									130 m depth
	(4)	23.58	28.63	26.94	20.85	5.49	33.25	0.66	3.15	28.82	5,890	0.00	1.85	1.3								sub C	
L10220485	(1)	30.1	11.0	10.1	48.8	- -	- -	- -	2.24	- -	7,321											lig A	U-6 bed
PCU-3-M-1	(2)		15.8	14.4	69.8	- -	- -	- -	3.20	- -	10.472												0.60 m thick,
	(3)			17.1	82.9				3.80	- -	12,429												40 m depth
M227096	(1)	25.5	35.63	22.99	15.83	5.21	23.77	0.38	5.22	29.79	4,160	0.12	3.30	1.80	24.07	0.0	58	2,300	2,420	2,480		lig B	U-3 bed
PCU-1-M-7	(2)		47.85	30.88	21.27	3.16	31.93	0.50	7.02	9.54	5,587	0.17	4.43	2.42									0.55 m thick
	(3)			59.22	40.78	6.06	61.23	0.96	13.46	18.29	10,717	0.33	8.50	4.63									
	(4)	24.07	36.34	23.45	16.15	5.09	24.24	0.39	5.32	28.62	4,240	0.12	3.37	1.84								lig A	
M227097	(1)	26.00	21.50	28.90	23.60	6.17	37.95	0.70	1.48	32.20	6,610	0.01	0.89	0.58	26.92	0.0	46	2,370	2,450	2,470		sub C	V-7 bed
PCV-1-M-1	(2)		29.05	39.05	31.90	4.41	51.29	0.95	2.00	12.30	8,933	0.01	1.20	0.79									
	(3)			55.04	44.96	6.22	72.29	1.34	2.82	17.33	12,591	0.01	1.69	1.12									
	(4)	26.92	21.23	28.54	23.31	6.23	37.48	0.69	1.46	32.91	6,530	0.01	0.88	0.57								sub C	
L-5 (1981)	(1)	26.67	8.14	36.03	29.16	3.85	47.66	0.85	0.89	44.94	8,275	- -	- -	- -	26.0	- -	41	- -	- -	- -	1.3	sub C	V-9 bed
	(2)		11.10	49.13	39.77	5.25	65.00	1.13	4.22	16.27	11,285												Outcrop sample,
	(3)	- -	- -	- -	- -	- -	- -	- - -	- -	- -	- -												1.1 m thick,
	(4)	26.00	8.21	36.36	29.43	3.78	48.1	0.86	0.9	38.15	8,350											sub C	Chlorine 0.00%
M226190	(1)	34.39	13.94	32.18	19.66	7.23	38.006	0.64	2.28	38.03	6,914	0.00	0.75	1.53	- -	0.0	- -	2,160	2,230	2,260		lig A	Uncorrelated bed
PCU-1-M-1	(2)		20.97	49.07	29.96	5.16	58.01	0.97	3.48	11.41	10,538	0.00	1.15	2.33									0.45 m thick,
	(3)			62.09	37.91	6.53	73.40	1.23	4.40	14.44	13,336												64 m depth
M226191	(1)	32.63	14.71	29.68	22.98	6.80	37.84	0.65	2.47	37.53	6,768	0.00	0.58	1.89	- -	0.0	- -	2,190	2,310	2,410		lig A	Uncorrelated bed
PCU-1-M-2	(2)		21.84	44.05	34.11	4.67	56.17	0.97	3.67	12.68	10,046	0.00	0.87	2.80									0.69 m thick,
	(3)			56.36	43.64	5.97	71.86	1.24	4.70	16.23	12,853	0.00	1.11	3.59									89 m depth

Circum-Pacific Council for Energy and Mineral Resources
Earth Science Series, Vol. 16
R. L. Miller, G. Escalante, J. A. Reinemund, and M. J. Bergin (Eds.)
Energy and Mineral Potential of the Central American-Caribbean Regions
Springer-Verlag, Berlin Heidelberg 1995

A MINING DESIGN PROJECT FOR THE ZENT COAL, LIMON, COSTA RICA

Rogelio Samuels and Gladys Cubilla
Refinadora Costarricense de Petróleo S.A., P. O. Box 4351, San José 1000, Costa Rica

ABSTRACT

The Zent coal deposit is one of two in Costa Rica that are being explored to determine their potential for production of coal. The deposit is in the northern foothills of the Talamanca Range, 32 km southeast of Puerto Limon.

Two coal beds, each more than 1 m thick, are in the thick Rio Banano Formation, which consists of fine, medium, and coarse-grained sandstone, conglomerate, siltstone, and shale, and is of middle Miocene age. The coal beds are separated by 9 m of fine-grained sandstone and siltstone. Chemical analyses of many samples indicate that the coal is of lignite A rank, has a low sulfur content, and an average ash content of 12 percent.

A semi-detailed mine design for strip mining the two coals has been prepared, covering an area of 0.4 km^2. Exploration of this area includes numerous boreholes. Aspects of the mine that are discussed include structure of enclosing rocks, overburden, highwall construction, surface and ground water problems, and mineable reserves.

The first sign of exploitation will cover a seven year interval during which 1.02 million MT will be mined. An identified market for the coal is the cement industry.

The material presented in this paper is considered preliminary information. More work will be needed before production can begin.

INTRODUCTION

Nine coal occurrences have been identified in Costa Rica. Based on reconnaissance studies, two of these, Zent and Uatsi (also called Volio), have been selected by the Costa Rican Government for preliminary exploration to determine if they may have potential to augment the energy resources of the country. This paper deals with the Zent coal deposit, and includes information on the geologic setting, the coals, and the mining design.

The Zent deposit is located in southeastern Costa Rica in the northern foothills of the Talamanca Range and 32 km southeast of Puerto Limon. Local relief is about 100 m. Access to the exploration site is via an improved gravel road from the paved coastal highway to the camp site. The coals have been found over an area of 268 km^2, but exploration studies to date have been concentrated within an area of 0.4 km^2, which is described here.

The hydrology of the area is simple. Regional stations were used to determine the precipitation, which averaged 3,500 mm (138 in) per year over a 10 year period. The distribution of the precipitation and intensity of the rains is not known. The area is poorly drained. Two intermittent streams, Quebrada Veinticino and Quebrada Jean, are the only drainages in the area. Average flows are 0.15 m^3/day and 0.019 m^3/day, respectively.

GEOLOGY

The coal beds are in the Rio Banano Formation, which consists of interbedded fine, medium, and coarse-grained sandstone with varying amounts of calcareous cement, siltstone, shale, and conglomerate. It is of middle Miocene age. In a part of the formation, 100 m thick, four coal beds are present, of which two are over one meter thick (Figure 1).

Structure of the region is quite complex and not yet well understood. In the area of current exploration, however, the coal-bearing beds are part of a homocline, in which the beds strike N75°W and dip 10° northeast. Two normal faults

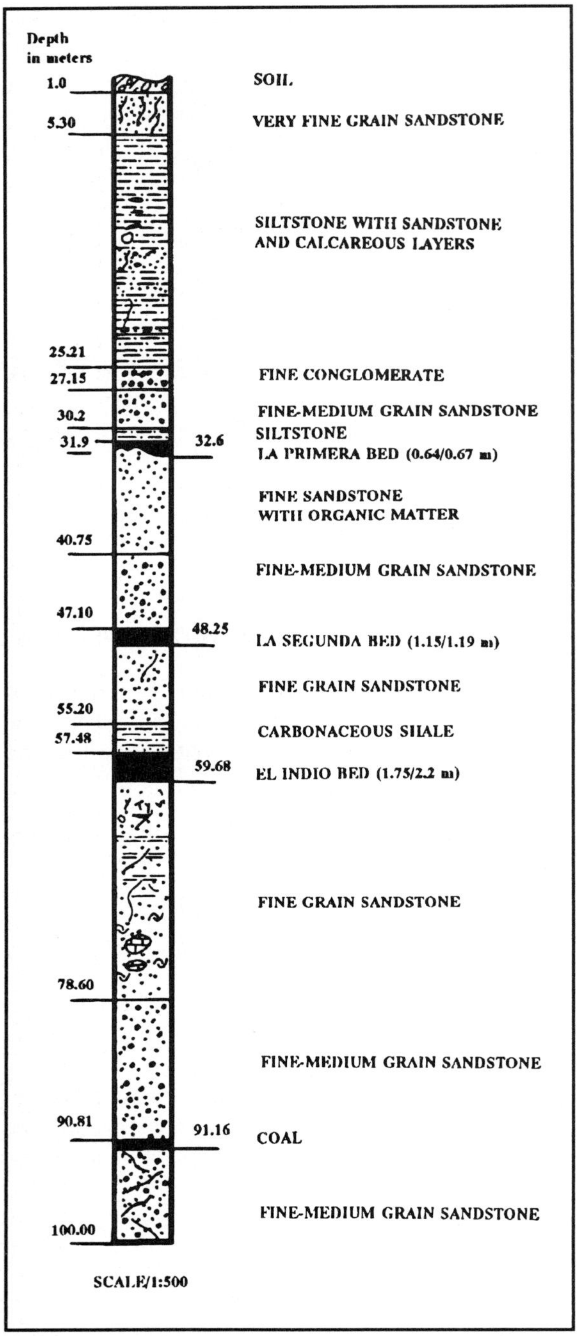

Figure 1. Typical lithologic column for the coal-bearing part of the Rio Bonano Formation, which is of middle Miocene age.

rupture the homocline (Figure 2). Fault No. 1 strikes N40°E, and dips southeast at 60°. Vertical displacement is 15 to 20 m, with a fault breccia zone about 1 m wide. Fault No. 2 strikes N30°E, dips southeast at 55°, and has a vertical displacement of 40 m. Its breccia zone is 7 m wide. As discussed later, these faults divide the exploration area into three sectors (Figure 2), based on average thickness of overburden.

A system of partings strikes N70°W, dips 10° northeast, and is concordant with the bedding. The partings are best developed in a sandstone unit 10 m thick, where they are filled with iron oxide. There are minor fractures less than 30 cm in length. These, it is believed, will have little significance in the development of the future mine.

THE COAL BEDS

In the drilled area four coals are present in a 60-m thick sequence (Figure 1). They have been named, from the top down, the Primera bed, which averages 0.65 m thick, the Segunda bed, which is 15 m deeper and averages 1.17 m thick, and the El Indio bed, which is 9 m below the Segunda and averages 1.8 m thick. A thin coal 31 m below the El Indio bed and also the Primera bed has no part in the mine plans.

Numerous samples of the coal have been analyzed. Table 1 lists the average and maximum and minimum values for a large number of samples. They show that in rank the average of sampled coals, following the standards of the American Society for Testing and Materials (1982), is in the lignite A range. In quality the coals are low sulfur. The ash content has a wide range with an average of about 12%. This is a little above the low ash grade, less than 10%, which is considered most desirable by industry.

Reserves of coal amounting to 14.9 million MT in the La Segunda and El Indio beds combined have been estimated for an area of 5.6 km^2. Of these reserves 5.58 million MT are considered to proven reserves and 9.36 million MT are probable reserves. Exploration studies to date have been concentrated in the small area of 0.4 km^2 shown in Figure 2. In this area 1.02 million MT of coal will be mined in stage 1 (seven years).

MINE DESIGN

Studies directed toward exploitation of the coal are in a preliminary stage. In the first stage, the area of concentrated exploration is shown in Figure 2. It lies on the nose of a ridge, the crest of which slopes southeast, as shown by 150 m and 100 m contours. The part being explored in the first stage is 0.4 km^2 in size and is shown by a dashed line in Figure 2. Within that area, three sectors have been delineated. Sector 1, the largest, includes the area southeast of Fault No. 1. The wedge-shaped area between Faults No. 1 and No. 2 constitutes Sector 2, and the remainder of the exploration area which is northwest of Fault No. 2 makes up Sector 3. The sectors are distinguished from each other in that the physical conditions that will be encountered in removal of overburden will be different. The area of proposed mining is further divided into seven blocks, based on estimated production per year for seven years (Figure 2). Production will be undertaken

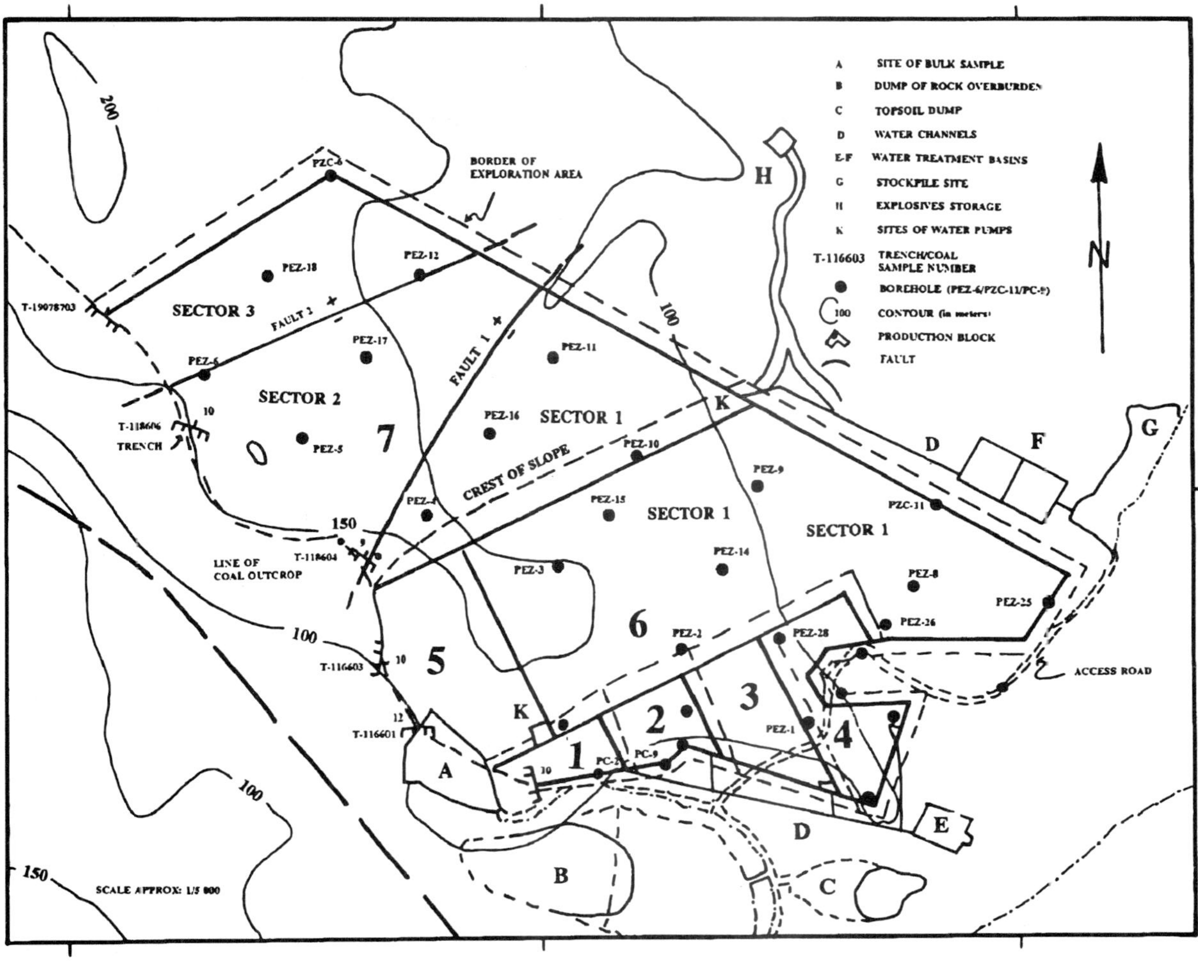

Figure 2. Mine design for the first stage (seven years) of operation. Sectors 1, 2, and 3 are separated by Faults 1 and 2. The large numbers 1-7 are the planned areas of strip mining of coal for each of the years of the first stage.

in the order of numbers, with numbers 1 to 4 being the smallest area. Also shown in Figure 2 are the locations of the exploratory boreholes and the outcrop of the coal beds (La Segunda and El Indio) 9 m below. Along the outcrop the beds have been exposed in trenches as shown by the rake symbols, with dip of beds measured (10°) in two places.

The slopes of the ridge and the joint (fracture) systems control the plan of the pit. The front face of the highwall will trend N68°E, approximately at right angles to the direction of the ridge crest. The highwall will slope back at about 58°. The adjoining lateral mine face approximately at right angles to the front face will cut diagonally across the ridge and will trend at about N23°W. It also will slope back at about 58°. It is difficult to visualize these relations in a two-dimensional diagram, but Figure 3 shows some of these relations. The height of each step in overburden removal is controlled partly by the topography and partly by the physical characteristics of the overburden. Figure 4 diagrams these steps in profile.

Of the overburden, about 90% is unconsolidated and easily stripped, but 10% will require explosives. Economic studies have indicated that the maximum allowable thickness of overburden to be stripped is 50 m, which will average about 20 m^3 of overburden per metric ton. Mining will begin at the southeast end of the mine where the beds crop out and will proceed downdip till an overburden thickness of 50 m is reached. Stripping on the lateral faces will proceed simultaneously. The total estimated "in situ" volume of overburden to be stripped in the first seven years (first stage) of operation is 13.41 x 10^6 m^3. Two dump sites for the overburden for the first two years of mining have been planned. The first dump (Figure 2C) will be for topsoil, and the second (Figure 2B) for rock, both coming from stripping of Blocks 1 and 2. Topsoil from the dump will be used to restore the mined area to its original contour after completion of the mining. Having the dump very near the mine site will also lessen transportation costs. Site of a coal stockpile (Figure 2G) is also shown.

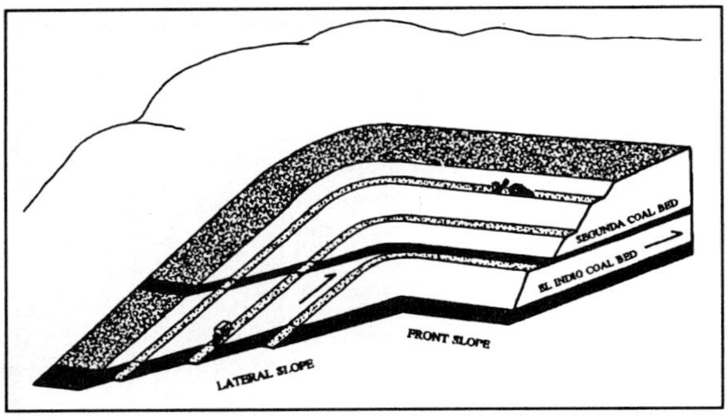

Figure 3. Schematic representation of the proposed surface mine at Zent. The ramps (dotted pattern) will be used to access the coal as well as the overburden. The arrows indicate the main direction of mining effort. (Diagram modified by Hettinger and Weaver, USGS).

Table 1. Coal Quality (as received).

COMPONENT		AVG.	MAX.	MIN.
Air Dry Loss	(%wt)	28.078	40.52	9.85
Sulfur	"	0.98	2.15	0.44
Fixed Carbon	"	19.24	27.74	11.48
Ash	"	12.15	20.50	6.03
Moisture	"	41.36	49.91	28.61
Volatile Matter	"	27.24	37.19	18.71
Cal.				
Value (Kcal/Kg) Btu/lb)		5684	7511	4123

ASH COMPOSITION

Silicon Dioxide (SiO_2)	46.46
Aluminium Oxide (Al_2O)	30.50
Iron Oxide (Fe_2O_3)	4.87
Phosphorus Pentoxide (P_2O5)	0.31
Calcium Oxide (CaO)	7.14
Magnesium Oxide (MgO)	1.69
Sodium Oxide (Na_2O)	0.02
Potassium Oxide (K_2O)	0.23
Sulfur Trioxide (SO_3)	6.17
Manganese Dioxide (MnO_2)	0.01

The water table is at a depth of 15 m in the southern part of the area. To the north near Quebrada Jean (not shown on Figure 2), it reaches the surface at an elevation of 75 m. A gradient for the water table of about 12° to the northeast is indicated. The recharge area is located in the higher country to the northwest, at elevations ranging from 250 to 300 m, and the zones of discharge are to the east at elevations of 75 m. In the final days of mining at this site, with maximum stripped area exposed, the combination of rainwater and ground water entering the mine pit is calculated to be on the order of 1369 m^3/d. This will require an adequate system for pumping, draining, and treatment of the water before it is released into the natural drainage. Sites of water treatment basins are shown in Figure 2 (E and F), as is pumpsite (Figure 2K), just north of Block 1. In addition, a drainage system around the mine pit will be needed to collect the surface water that otherwise would run into the pit. This water will be led into the natural streams without undergoing any treatment.

RESERVES AND PRODUCTION

The mineable reserves considered for exploitation during the first stage (seven years) are 1.02 million MT of coal. Based on potential market studies, this should be adequate to meet the national demand for those years. First usage will probably be by the cement industry.

The mining design here described is preliminary. More detailed work is required including, in particular, the areas adjacent to Faults 1 and 2. The extended studies will include the establishment of monitoring systems and production control. Some of these studies are already underway.

REFERENCE

American Society for Testing and Materials, 1982, Annual book of ASTM standards, gaseous fuels, coal and coke, atmospheric analysis.

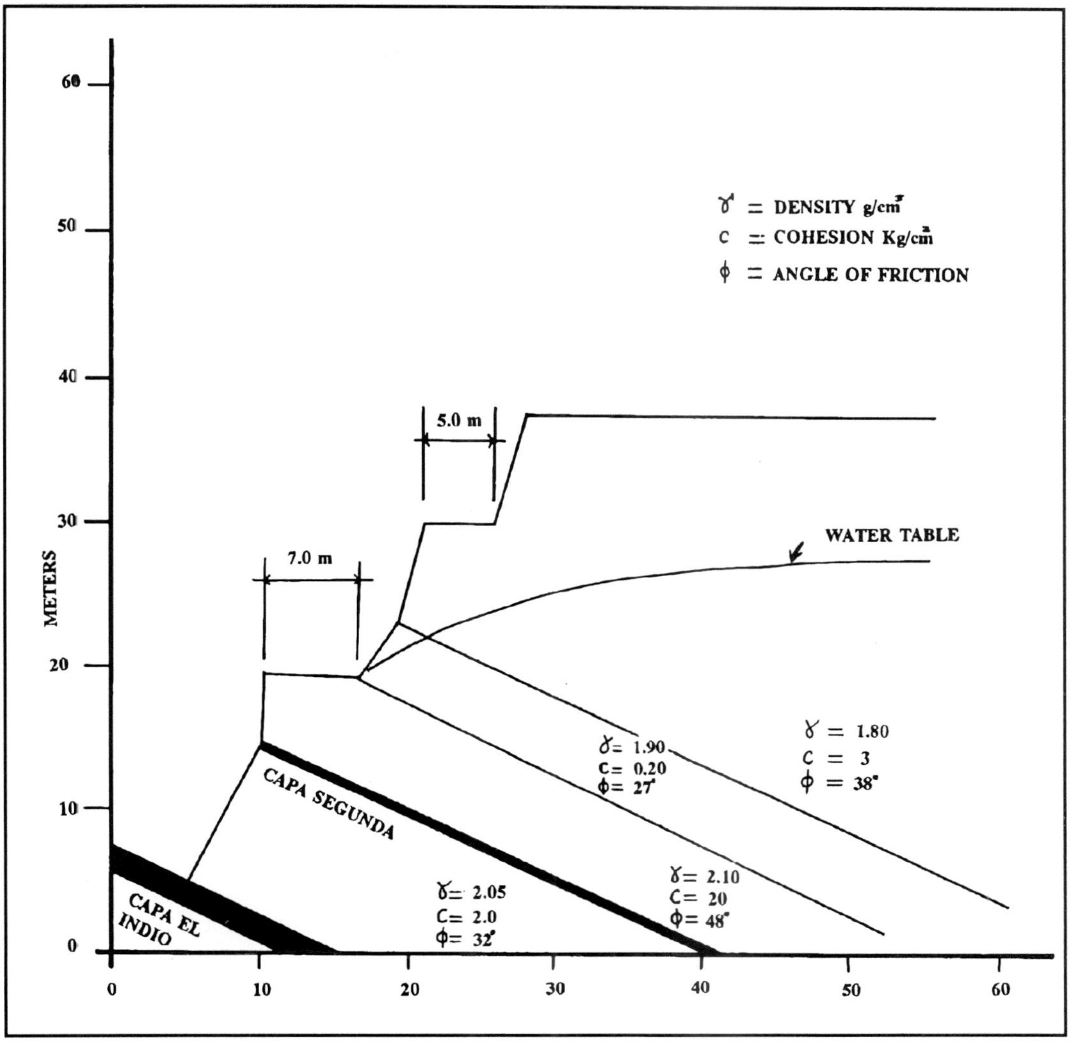

Figure 4. Typical profile of highwall of Zent mine, showing physical characteristics of different layers of the overburden of the coal beds.

Circum-Pacific Council for Energy and Mineral Resources
Earth Science Series, Vol. 16
R. L. Miller, G. Escalante, J. A. Reinemund, and M. J. Bergin (Eds.)
Energy and Mineral Potential of the Central American-Caribbean Regions
Springer-Verlag, Berlin Heidelberg 1995

UTILIZING COAL IN THE CEMENT INDUSTRY IN COSTA RICA

Fernando Alvarado and Rafael Yglesias
Refinadora Costarricense de Petróleo (RECOPE), Apartado 4351, San José, Costa Rica

ABSTRACT

Costa Rica has produced cement since 1978. There are three cement plants in the country, one of which operated only from 1978 to 1983. Total production in 1988 was 618,000 MT, almost all of which was used domestically.

Basic ingredients in making Portland Cement are limestone (or other source of lime), silica (sand), and alumina (shale or clay). All of these are available within Costa Rica. Traditionally bunker fuel oil has been the source of energy for the cement kilns, but alternatives are being sought by RECOPE because of the cost of the imported oil. Coal has seemed a potential substitute for oil because recent exploration has located nine coal deposits. One of these has been extensively drilled and has large enough proved reserves to consider production. Unfortunately, the moisture content of the coal is 41 percent. Tests have shown that the practical limit of direct substitution of the wet coal for bunker oil at the cement kilns is 13 percent. Drying the coal with hot gas is being considered, but the problems involved have not yet been fully resolved. Additional work by RECOPE is needed, in the hope that some of the coal that is now available may become a partial substitute for bunker fuel oil in the cement plants.

INTRODUCTION

At the start of the Portland Cement industry about the middle of the last century, coal represented the major energy source that was in use in industrialized countries. This continued until the end of the 1950s when the technical advantage and the lower cost of hydrocarbons took over. Coal was first replaced by bunker oil and later in some places by natural gas. In Costa Rica, imported bunker oil has traditionally been used as the combustible for the cement industry. In 1988, this industry accounted for 47 percent of the total bunker oil consumed annually (321,000 bbl). The cost is about 50 percent of the total cost of producing the cement. Bunker oil provides 65 percent of the total energy required to produce the cement, with electrical energy accounting for 33 percent of the remainder. The total production of cement in Costa Rica in 1988 was 618,000 MT, almost all of which was used domestically. The small quantity exported went to Caribbean markets.

THE CEMENT INDUSTRY IN COSTA RICA

In the manufacture of Portland Cement, the feed stock consists of calcium carbonate, usually in the form of limestone, silica, and alumina, usually in the form of clay or shale. Fine grinding and mixing follow either on a wet or dry basis. The feed is introduced into the upper end of a long, gently inclined rotary kiln, commonly 200 to 400 ft (about 60 to 120 m) long, and a source of heat is blown into the lower end. Temperatures up to 1,500°C are attained. As the feed moves slowly down the kiln, complex chemical reactions take place, with fusion and agglomeration into nodules called clinker, which exits the kiln. The clinker is mixed with about 5 percent

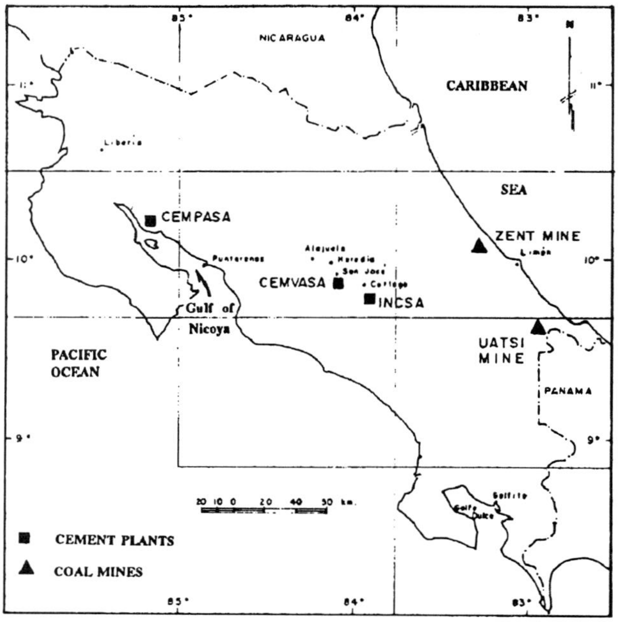

	INDUSTRIA NACIONAL DE CEMENTO S.A. INCSA	CEMENTOS DEL PACIFICO S.A. CEMPASA	CEMENTOS DEL VALLE S.A. CEMVASA
OWNERSHIP	SWISS (49%) or COSTARICAN (51%) CAPITAL CAPITAL	COSTARICAN GOV.	COSTARICAN GOV.
	1964	1980	1978
LOCATION	AGUA CALIENTE, CARTAGO PROV.	COLORADO, ABANGARES, GUANACASTE PROV.	PATARRA, SAN JOSE PROV.
PRODUCTION CAPACITY CEMENT/TONS PER YEAR	650,00	450,000	160,000
PRODUCTION PER YEAR CEMENT/TONS	360,000	258,000	NO PRODUCTION SINCE 1983
DISTANCE FROM SAN JOSE (Km)	24	170	15
DISTANCE FROM ZENT MINE (Km)	143	289	134
DISTANCE FROM UATSI MINE (Km)	233	379	224
FUEL USED	BUNKER-FUEL C, OIL PALM FRUIT SHELL, COKE, COAL	BUNKER FUEL C	BUNKER FUEL C
COAL SUBSTITUTION TESTED (%)	13	—	—
COAL SUBSTITUTION PLANNED (%)	AT LEAST 40	—	—

Figure 1. Cement industry in Costa Rica.

of gypsum to control setting time of the cement, and ground very fine to become Portland Cement. The process is monitored and very closely controlled at every stage to assure the uniformity of the product. Small deviations can adversely affect the setting and binding qualities of the cement.

Costa Rica has three cement plants (Figure 1): (1) Industria Nacional de Cemento, S.A. (INCSA), privately owned, started in 1964 and has an installed capacity of 650,000 MT/yr of cement; (2) Cementos del Pacifico, S.A. (CEMPASA), a governmental enterprise, started in 1980, and currently has an installed capacity of 450,000 MT/yr; and (3) Cementos del Valle, S.A. (CENVASA), a governmental enterprise, in operation between 1978 and 1983.

The production system of the two cement plants that are in operation in Costa Rica is similar (Figure 2). Both operate on the dry basis. In 1984 INCSA finished the construction of a precalcination tower that features two injection burners, one for bunker oil and the other for solid combus-

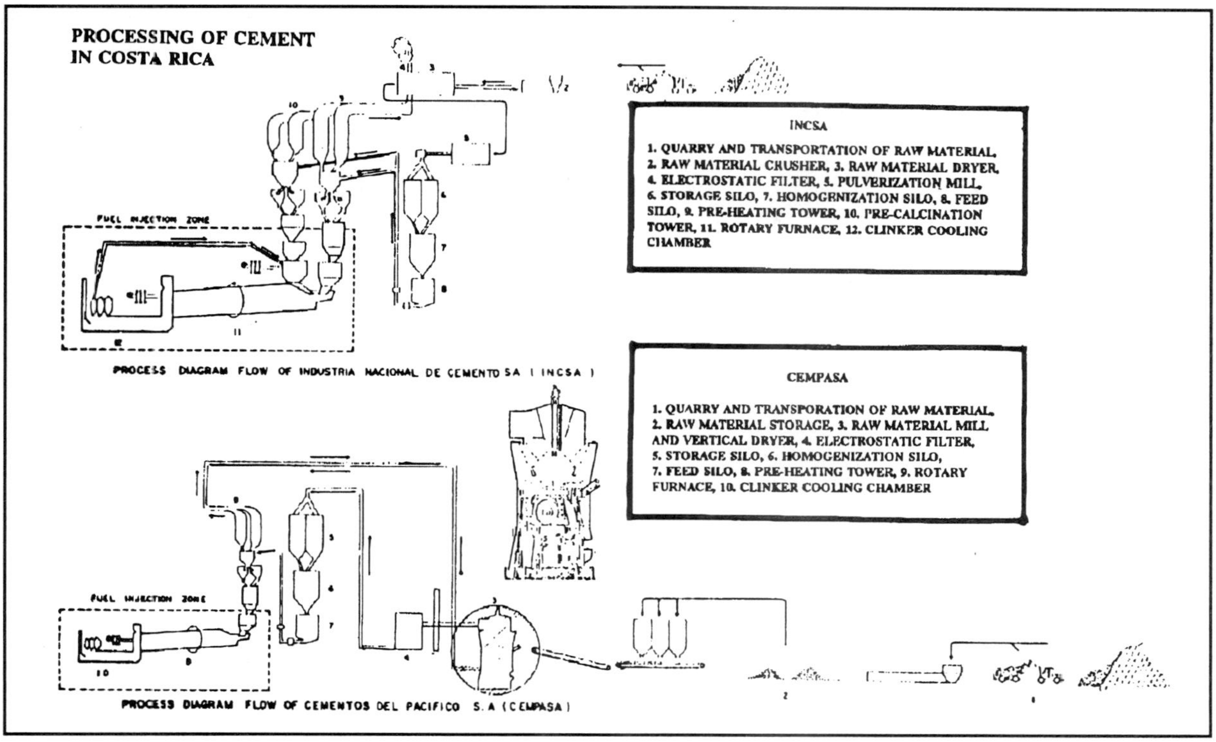

Figure 2. Processing of cement in Costa Rica.

tibles including coal. This addition will permit an increase in the plant's production of clinker to 1,800 MT/d. CEMPASA does not have a precalcination tower, but because of the increase in the demand for cement, construction of one is planned for 1992, which will have a burner with the capability of accepting solid combustibles including coal. When this precalcination tower is installed in CEMPASA, its clinker production will increase from 1,250 to 2,000 MT/d.

INCSA has become the leader in Costa Rica in the investigation and experimentation of combustibles that can be substitutes for bunker oil. The burner installed at the base of INCSA's precalcination tower allows utilization of various solid combustibles such as African palm husks, imported coke, corn cobs, and coal. All of these potential substitutes for the bunker oil have favorable and unfavorable aspects for their utilization. African palm husks are available only at a maximum of 8,000 MT/yr, which amounts to a heating level of only 20 percent of the bunker oil now in use. The coke that has been imported has a high sulfur content (greater than 3 percent), which has introduced some problems into the process because of the formation of clinker in the precalcination tower. INCSA has stockpiled close to 3,000 MT of the coke hoping to dilute it with other combustibles of low sulfur content. Corn cobs have a low calorific value and one must burn great quantities to obtain appreciable results.

RECOPE, a state enterprise, is charged with the exploration for, refining, and distribution of coal, oil, and gas in the country. It has recently located and quantified a total of 50 million MT of coal, primarily on the country's Caribbean coast. Of this total, 15 million MT represent the proven reserves of the Zent deposit. The age of the coal is middle Miocene. In rank it is subbituminous C, and it has an average calorific value of 5,385 kcal/kg (dry base). Sulfur content is less than 1 percent, ash is 12.15 percent, and moisture 41.36 percent.

Coal represents one of the most viable alternatives as a substitute for bunker oil in the cement industry. The most important economic aspects that can result from the substitution of coal for bunker oil in the cement industry are:

1. Coal can lower the consumption of bunker oil produced by refining imported petroleum in the hope of increasing the exportation of the bunker oil to the international market. For each barrel of bunker oil that Costa Rica can export, it can purchase 0.65 bbl of high-grade oil on the Venezuelan or Mexican market.

2. It can contribute to the development of a new energy industry in the country, utilizing its own coal resources. This could also lower the dependence of the country on imported oil, and it would also result in the development of a new work force.

3. It could lower the cost of cement, because the cost of domestic coal would be lower than that of imported bunker oil by about 54 percent.

4. It could help stabilize the cost of energy combus-

tibles because the cost of coal will not be as tightly linked to the fluctuation of the international prices of bunker oil and other grades of imported oil.

All of the above mentioned benefits are contingent on the substitution of the bunker oil by coal in the manufacture of cement. At present the practical upper limit at which coal can be substituted is 13 percent, due to the fact that the moisture content of the coal is so high.

INCSA now has the capacity to use up to 4 MT/hr of solid combustibles. During the tests to substitute 100 percent coal, however, it was only possible to process 2 MT/hr because of the moisture content of the coal. Thus, we now know that the drying of the coal is fundamental to successfully compete with other solid combustibles.

It will require a large quantity of hot gases to dry the coal. Unfortunately, the hot gases, even though vented to some degree, can pick up the clinker that is in dust form and by mixing the dust with the burning coal might contaminate and dilute the feedstock by as much as 50 percent. INCSA is now making small changes in the process, such as the installation of a cyclone, in order to overcome the dilution of the coal in the drying process.

RECOPE is also conducting tests on dryer beds to reduce the moisture content of the coal to as little as 15 percent. This will increase the potential for substitution up to 40 percent of the total bunker oil now in use.

THE FUTURE OF COAL IN THE CEMENT INDUSTRY

The following developments involving use of coal in the cement industry are anticipated.

1. Increase by INCSA of its capacity so as to use 8 MT/hr of solid combustibles.
2. Changes in the operations at INCSA so as to divert the drier gas from the clinker.
3. Increased production capacity at CEMPASA with the construction of a precalcination tower enabling it to handle solid combustibles.
4. Continuing tests that RECOPE is currently performing involving a bed for drying the coal.
5. Construction of an experimental oven to dry coal.

Circum-Pacific Council for Energy and Mineral Resources
Earth Science Series, Vol. 16
R. L. Miller, G. Escalante, J. A. Reinemund, and M. J. Bergin (Eds.)
Energy and Mineral Potential of the Central American-Caribbean Regions
Springer-Verlag, Berlin Heidelberg 1995

COAL DISTRICTS OF VENEZUELA

Simón E. Rodríguez M.
IGCP National Committee, Ministry of Energy and Mines, Piso 5, Torre Oeste, Parque Central, Caracas, D. F., Venezuela

ABSTRACT

Ten separate coal districts have so far been recognized in Venezuela. Those in western Venezuela, near the Colombian border, are of Paleocene or Eocene age. Others in the northern part of the country near the Caribbean are of Miocene age. In the Guasare district, 20 coal beds have been identified with thicknesses from 0.5 m to 13 m. Coal is being mined from this district at the rate of about 700,000 MT annually, but plans call for a build-up to 6 million MT, all of which will be for export. Each of the other nine mining districts is described briefly.

INTRODUCTION

To date, 10 main coal districts have been located, outlined, and partially evaluated in Venezuela. These districts are Guasare, Northern Zulia State; Santa Rosa, Southern Zulia State; Lobatera, Táchira State; Rubio, Táchira State; Santo Domingo, Táchira State; Pedregal, Falcón State; Taguay, southern Aragua State; Sabana Grande, northern Guárico State; Guanape, northwestern Anzoátegui State; and Naricual, northeastern Anzoátegui State. The total of different types of coal reserves in Venezuela, according to publications of the Ministry of Energy and Mines, amount to approximately 9,169,690,000 MT.

DISTRICTS

The Guasare district, the most important of the country, is located in the northern Perijá Mountains, Zulia State, close to the Guasare and Cachirí Rivers (Figure 1). Geologically, the coal seams are within the Marcelina Formation of Paleocene age. More than 20 seams, with thicknesses from 0.5 m to 13 m, have been studied, correlated, and evaluated. Total reserves of the district are on the order of 8,489,000,000 MT. Actually CARBOUZULIA, a company of the National Oil Company PDVSA, mines about 700,000 MT/yr. from the Paso Diablo open pit mine. The final planned production of the mine will be 6 million MT/yr., totally for export.

The Santa Rosa district is located in the foothills of the southern Perijá Mountains, Zulia State, between the Oro and Tucuco Rivers. The deposits, also within the Marcelina Formation, are under intensive exploration and evaluation by CORPOZULIA, the local development corporation, and a Venezuelan-American company, which will be responsible for the mining. Production may have started in this district before publication of this paper, at a planned rate of 300,000 MT/yr. Total reserves amount to 300 million MT.

Táchira State, in the southwestern Venezuelan Andes, is rich in good thermic and coking coal. Several local districts, separated by regional faults, constitute large tectonic blocks, each one characterized by its own structural features. The main identified districts are the Lobatera, Rubio, and Santo Domingo.

The Lobatera district is located about 20 km north of San Crístobal, the capital city of the State. Geologically, the coal seams are within the Carbonera Formation of Eocene age. Three main coal seams have been studied and evaluated. Regional reserves amount to 36 million MT, mainly of thermic type. At the present time, coal is being mined by the open pit method at a rate of 200,000 MT/yr.

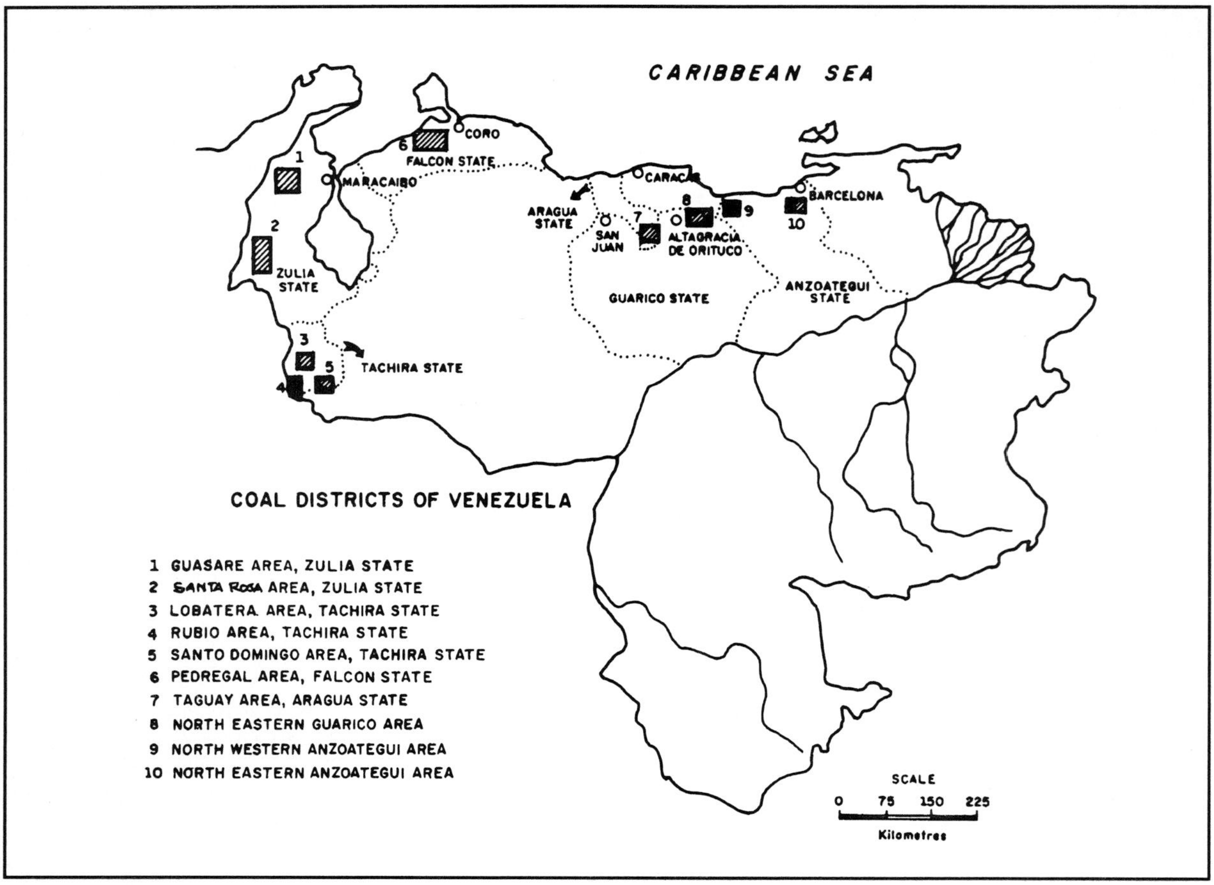

Figure 1. Coal districts of Venezuela.

The Rubio district contains probably the best coking coal of the country. The seams are located about 15 km west of San Crístobal. They are in the Los Cuervos Formation of Paleocene age. The seams have been highly affected by regional tectonism, so the quality of the coal has been improved. The whole area is under evaluation by a local development corporation, CORPOSUROESTE. Reserves probably amount to as much as 30 million MT, based on the geological studies. All mining will have to be underground.

The Santo Domingo coal district is the most important of the three in Táchira State (Figure 1). The district is located about 35 km southeast of San Crístobal. The coal seams dip 30-60°. They are within the Carbonera Formation. The deposits were evaluated in the 1970s by the Regional Development Corporation in cooperation with the Polish Government. More than 280 million MT of thermic coal were quantified.

In northern Venezuela all the coal districts are associated with Miocene rocks cropping out in Falcón, Aragua, Guárico, and Anzoátegui States. These rocks are present in the Falcón-Zulia sedimentary basin and also in the large Eastern Venezuelan basin. They are rich in oil and gas.

The Pedregal district is located about 50 km west of Coro, the capital city of Falcón State, in a region that is characterized by a semi-desert climate. The coal seams are within the Cerre Pelado Formation of Miocene age. They were studied between 1983 and 1985 by the Ministry of Energy and Mines, and about 40 million MT of good quality thermic coal were estimated. The entire district is now being drilled by a Venezuelan-American company for additional coal data. Production estimated at a rate of 300,000 MT/yr. is predicted. The coal will be exported using the Puerto Cumarebo facilities.

The Taguay district in northern Aragua State has also been studied by the Ministry of Energy and Mines. The coal seams are within the Quiamare Formation of Miocene age. They are flat-lying levels and are covered by about 15 m of clay overburden. The district probably contains more than 25 million MT of high sulfur thermic coal. The seams will be evaluated and eventually mined by a private company under contract with CORPOCENTRO, the local development corporation.

The Sabana Grande district, northern Guárico State, contains an important reserve of Miocene coal in the Naricual and Quebradon Formations. More than 10 seams have been identified during regional investigations by the Ministry of Energy and Mines. No activities are presently being carried out in this district.

The Guanape district, northwestern Anzoátegui State, is rich in coal seams. About 17 economic coal seams in the Naricual Formation have been outlined regionally and a reserve of more than 27 million MT has been calculated. About 400,000 MT/yr. of thermic coal are now being mined from the Fila Maestra district and are exported through the Guanta Port facilities in northeastern Anzoátegui State.

The Naricual district in northeastern Anzoátegui State contains more than 35 million MT of thermic coal, also in the Naricual Formation. More than six economic seams have been investigated and the area was underground mined until 1975. All mining activities ceased after that year because of technical and economic problems. The underground mine has a production capacity of about 350,000 MT/yr.

Circum-Pacific Council for Energy and Mineral Resources
Earth Science Series, Vol. 16
R. L. Miller, G. Escalante, J. A. Reinemund, and M. J. Bergin (Eds.)
Energy and Mineral Potential of the Central American-Caribbean Regions
Springer-Verlag, Berlin Heidelberg 1995

STATUS OF COAL EXPLORATION IN THE ATLANTIC COAST OF COLOMBIA

Hernán Gómez-Mejía
Subdirección de Docencia, Instituto Geografico "Agustin Codazzi"
(IGAC), A.A. 53754, Bogotá, Colombia

ABSTRACT

There are 36 coal basins in Colombia, grouped into seven principal coal zones. The most important of these are the César, Cerrejón, and San Jorge zones, all located in the northern part of the country. A feasibility study to locate a 600-MW thermal plant required that the authors investigate and explore the coal resources of the area. This work led to the discovery of the new field, El Descanso, with proven reserves at this date of 1,000 million MT of coal.

Presented herein are the results of the exploration and evaluation of fields in the César and Cerrejón zones, including the geological structures, the concession blocks that have been granted, and data on the reserves and quality of the coal deposits. This report includes information on the exploration of and the characteristics of the coal sequences in the four most important coal deposits in the César area: Boquerón, El Descanso, La Jagua, and La Loma, and the northern and central blocks of the Cerrejón basin. Also included are data on the lesser known San Jorge zone which contains the Alto San Jorge-San Pedro basin.

The heating power of the César coal is 26,100 kj/kg (11,200 BTU/lb); that of the Cerrejón coal 28,000 kj/kg (12,000 BTU/lb). The coal at San Jorge averages 8,500 BTU/lb. Total demonstrated reserves (proved plus probable reserves) in the César area are 2,068.3 million MT of coal; in Cerrejón, 3,667.4 million MT; and in Alto San Jorge-San Pedro, 1,679.6 million MT, for a grand total of 7,425.3 million MT. The production capacity of coal from the Colombian Atlantic coast is 50,700 kt/yr.

INTRODUCTION

The information contained in this paper is based on unpublished reports of investigations by the author and on unpublished reports by various companies and governmental agencies that have studied the coal deposits of Colombia, among them Siderurgica de Medellin S.A. (SIMESA), Instituto Nacional de Investigaciones Geológico-Mineras (INGEOMINAS), Ministerio de Minas y Energía (MINMINAS), Compañia Minera Marathon Ltda., Lobo Guerro Gomez Geología Ltd., Consultores Unidos Ltd., Interconexión Eléctrica S.A. (ISA), Colombia Coal Company (CARBOCOL), Carbones del Caribe Ltda., and Intercor, a subsidiary of Exxon.

THE ATLANTIC COASTAL COAL BASINS

There are 36 identified coal basins in Colombia (Figure 1), grouped into seven principal coal zones that are considered by INGEOMINAS and the Estudio Nacional de Energía to be the areas most favorable for development of a modern mining industry in the course of the next 20 years. Three of the principal coal zones are near the Atlantic coast, namely César, Cerrejón, and San Jorge, of which the most important are the first two.

The Alto San Jorge-San Pedro basin in Cordoba Department has two deposits: in the west, Alto San Jorge, explored by CARBOCOL; and in the east, San Pedro-Uré, explored by Carbones del Caribe Ltda. and CARBOCOL

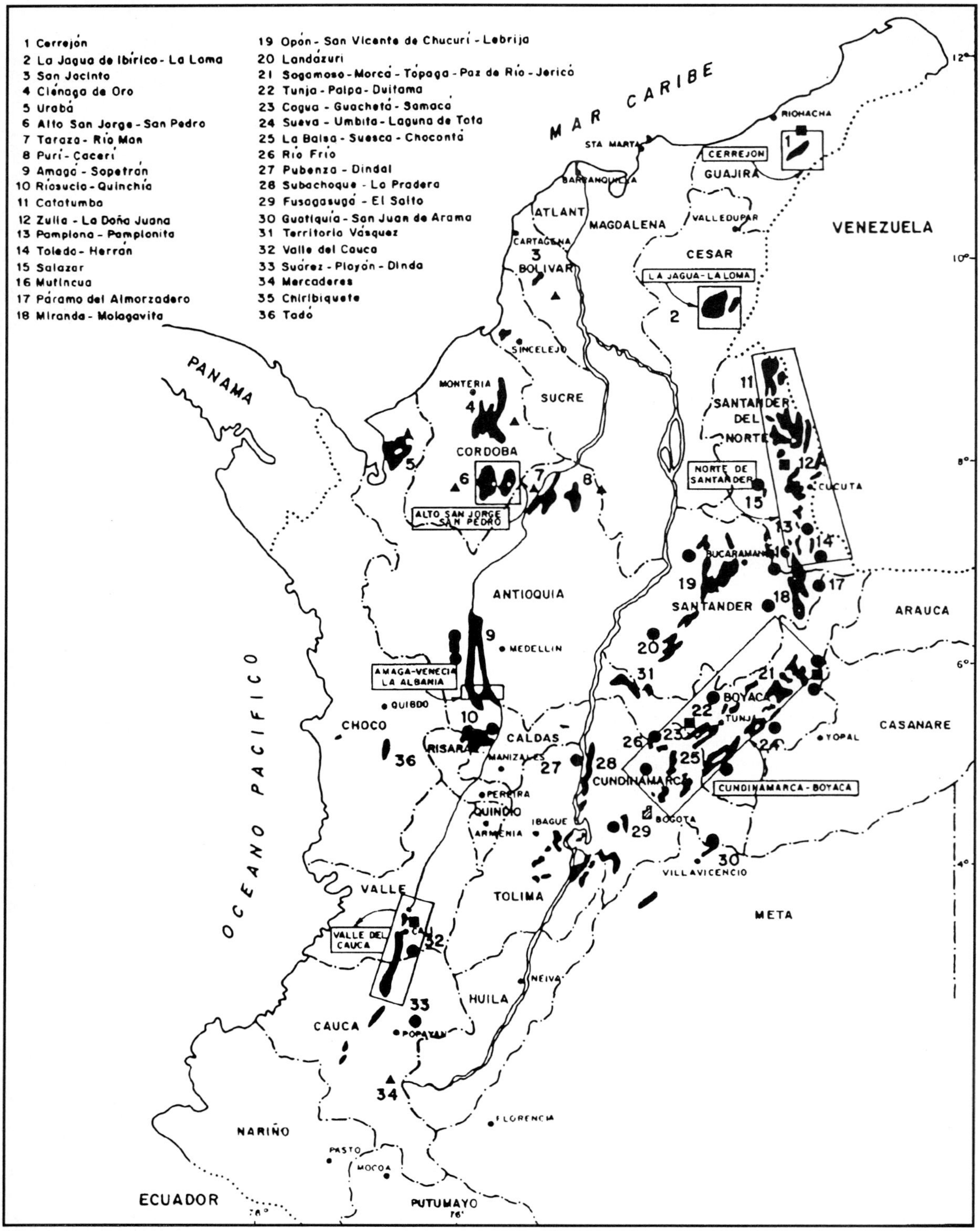

Figure 1. Colombian coal basins. Large circles = prospects, small circles = cities and towns, triangles = showings, and squares = deposits.

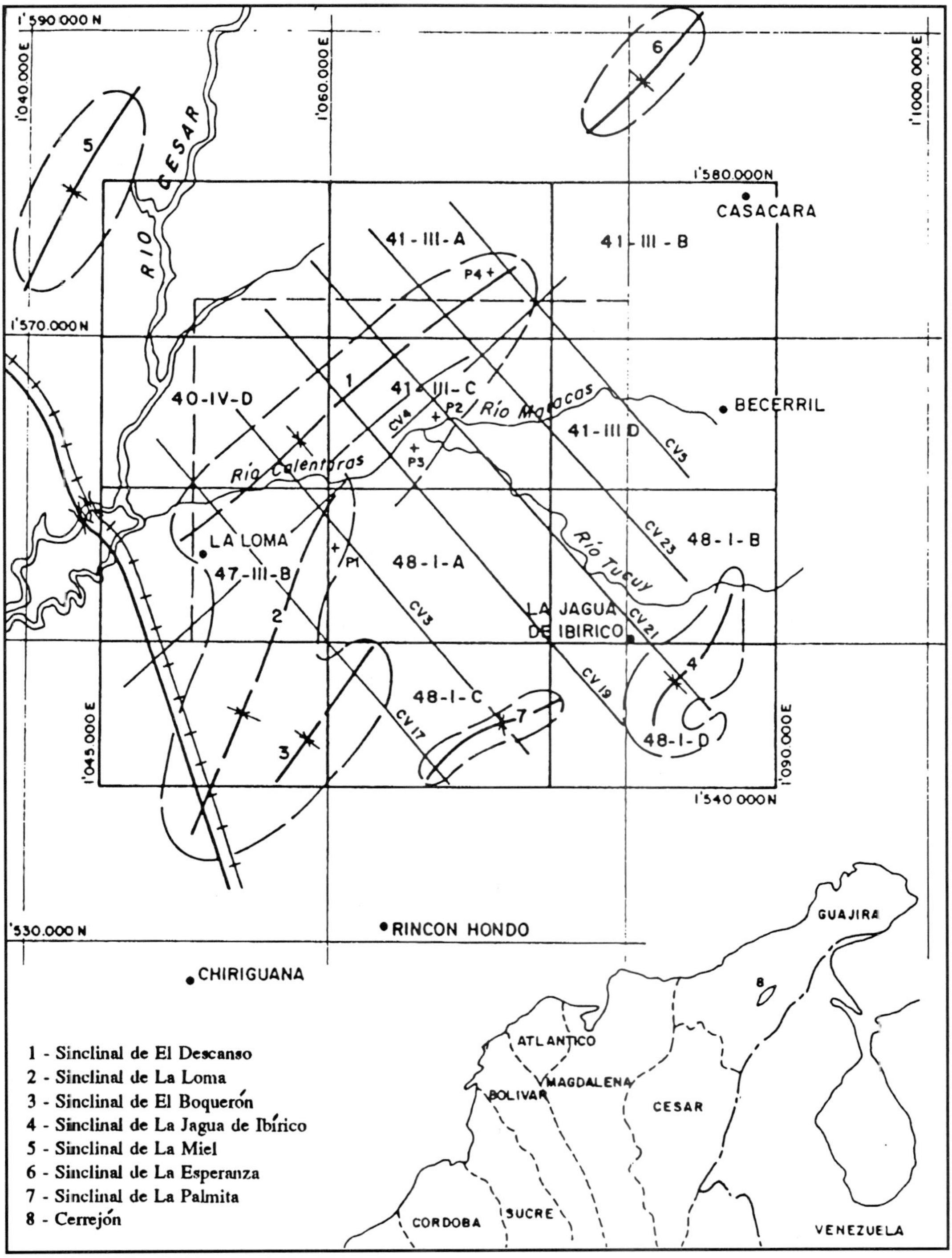

Figure 2. Location of principal coal-bearing struetures in the Cesar region. Based on unpublished CARBOCOL report.

Table 1. Coal beds of the Boquerón deposit.

Coal Bed	Thickness in Meters Variation	Average	Average Interval between seams (meters)
Aquila	1.36 - 2.26	1.76	13
Peric	0.91 - 2.28	2.25	16
Monito	0.30 - 0.91	0.55	10
Caballero	0.40 - 1.58	1.42	12
Gran Madre	7.07 - 8.38	7.90	6
Donecilla	391 - 1.45	1.28	4
Cazador	1.40 - 2.21	1.60	19
Babilla	0.50 - 1.07	0.86	16
Rouzaud	2.22 - 5.11	4.27	22
Borrego	2.81 - 6.21	4.82	13
Alanito	0.24 - 1.19	0.91	14
Diablito	0.70 - 2.95	2.42	
		30.04	145

(Figure 1, no. 6). Data on the Alto San Jorge-San Pedro basin is included in some of the tables, but the basin is not described in this report.

CÉSAR ZONE

The César coal zone in César Department (Figure 2) has four important coal fields: Boquerón, explored by SIMESA-Greenley; El Descanso, explored by ISA-CARBOCOL; La Jagua, explored by Compañia Minera Marathon Ltda.; and La Loma, explored by SIMESA-Greenley and ISA-CARBOCOL. The La Loma and La Jagua deposits are near the towns of the same names (Figure 2, upper map).

Boquerón-La Loma Deposit

SIMESA-Greenley Company explored the Boquerón deposit, which is in the Boquerón Syncline (Figure 2, no. 3), up to the feasibility stage. The syncline is a large structure that plunges 3° to the southwest, with gently dipping limbs between 8° and 15°. The coal-bearing zone has a length of 10 km and an average width of 1.5 km. The stratigraphic section in the Boquerón Syncline includes twelve coal beds 1 m to 8 m thick within 175 m of the lithologic sequence, as shown in Table 1. The carbonaceous sequence is covered by Quaternary alluvium which varies between 14 m and 75 m in thickness. The coaly unit is in the upper part of the Los Cuervos Formation (Table 2). This part of the formation consists of 450 m to 700 m of claystone and sandstone with intercalations of carbonaceous beds.

Another coal-bearing structure in the César basin is the La Loma Syncline (Figure 2, no. 2), located to the northwest of the Boquerón Syncline. The Cuesta Formation, which overlies the Los Cuervos Formation (Table 2), surfaces in the northern part of the syncline. The outcrop belt of the Cuesta is 25 km long and averages 3 km wide. On the western side of this structure, the SIMESA-Greenly Company found seven coal beds, presumably in the underlying Los Cuervos Formation. The beds dip steeply between 45° and 60°. ISA-CARBOCOL found the same coaly sequence on the eastern side of the syncline.

La Jagua Deposit

In 1978 Compañia Minera Marathon Ltda., through SIMESA, entered into a joint venture with St. Clair to carry

Table 2. Stratigraphy of La Loma-Lagua.

Age		Surficial Deposits/ Bedrock Formations	Thickness (in meters)	Lithologic Descriptions
QUATERNARY		Alluvium	0 - 70	Sandstone, quartzose, clear to gray, fine to coarse grain, and claystone, yellow or gray. Contains rounded and angular quartzite clasts.
		Terraces	0 - 120	Conglomerate, and sandstone, conglomeratic, brown quartzose
TERTIARY	EOCENE-MIOCENE	Cuesta	0 - 2,000	Basal conglomerate, and sandstone, brown or yellow. Overlain by gray claystone. ANGULAR UNCONFORMITY
	UPPER PALEOCENE	Los Cuervos T-3	450 - 700	Claystone, gray-blue sandy, claystone and sandstone, gray, fine grained. Numerous coal seams.
		Los Cuervos T-2	250 - 350	Sandstone, gray, fine to coatse grained, quartzose, somewhat ferruginous. Laminated, with cross bedding, resistant.
		Los Cuervos T-1 Lower part	900	Argillite, gray and black, interbedded with somewhat micaceous sandstone. Coal beds in upper part.
	L. PALEO.	Barco	not known	Argillite and sandstone, with thin intercalations of coal.
CRETACEOUS	UPPER	Molino	500 - 590	Argillite, black, and limestone, a little sandstone.

Table 3. Coal beds of the La Jagua deposit.

COAL BED	Average Thickness (meters)	
	Block A	Block B
1	2.27	2.21
2	5.51	4.67
3	1.19	1.62
4	1.90	1.97
5	2.40	1.89
6	5.89	4.84
7	2.21	1.51
8	1.17	0.69
9	2.17	2.10
10	1.05	0.89
11	0.99	0.57
12	1.27	1.55
13	2.45	3.14
14	0.57	0.97
TOTAL	31.04	28.62

Based on unpublished reports by Compañia Minera Marathon Ltd.

out exploration in the La Jagua Syncline (Figure 2, no. 4) up to the feasibility stage. Geologically, the deposit has a similar stratigraphic and coaly sequence to the Boquerón deposit. The upper unit of the sequence known as Los Cuervos T-3 is the most important: in a 210-m interval there are 14 coal seams (Table 3).

El Descanso Deposit

In January 1982, Lobo Guerro Gomez Geología Ltd. on behalf of Consultores Unidos Ltd. and ISA carried out the exploration of 1,300 km^2 in the La Loma-La Jagua area for the purpose of locating a new coal deposit for the operation of a large-capacity thermal power plant. After a photogeological study with field checking and an interpretation of seismic profiles were completed, the area was reduced to 300 km2 for exploration drilling. In May 1983, an exploratory hole was drilled in the Descanso Syncline (Figure 2, no. 1) by ISA-CARBOCOL. A coal-bearing sequence was penetrated beneath a Quaternary alluvial cover. These beds have been traced for 35 km. On the southeast limb of the syncline, they dip to the northwest at an average angle of 3°. Drill hole P4, located near the northeast end of the syncline as now known, reached a depth of 277.5 m after penetrating 13 coal seams, which have the thicknesses shown in Table 4.

CERREJÓN BASIN

The Cerrejón basin is located in Guajira Department (Figure 1, no. 1) and is divided into three exploration blocks (Figure 3). The northern or "B" block is under the jurisdiction of CARBOCOL and Intercor; the central block has been explored by CARBOCOL and is in the exploitation stage; and the southwestern or "A" block remains unexplored.

Table 4. Coal beds of the El Descanso deposit.

Number of Coal Bed	Thickness (meters)	Name of Coal Bed
2	1.32	Margarita
4	1.60	Turpial
5	0.93	Delirio
6	3.30	Platanal
9	3.44	Mpchuelo
13	4.26	Lucía
15	0.78	Angel II
16	5.90	Olgana
18	2.69	Zorro I
19	1.37	Zorro II
20	0.76	Zorro III
21	1.39	Zorro IV
22	1.17	Zorro V
TOTAL	28.91	

Table 5. Coal beds of the Cerrejón Central Block. (average thickness of coal beds in meters)

BED	Sector I	Sector II	Sector III	Sector Patilla
1	1.3	1.1	——	——
2	1.3	1.0	——	——
3	1.8	1.9	1.7	——
4	1.7	1.3	1.3	1.4
5	2.5	2.9	4.0	2.4
6	0.8	0.8	0.8	0.6
7L	2.8	2.9	2.6	1.8
7U	0.9	1.1	2.1	1.6
8L	1.9	3.1	1.8	1.3
8U	4.5	4.8	5.2	1.6
9L	5.9	2.6	3.1	3.6
9X	0.7	1.2	1.1	--
9U	1.3	4.3	4.2	5.0
9C	0.8	——	——	——
10	1.8	2.1	1.7	3.2
11	4.8	4.9	3.4	4.7
12	2.7	3.2	2.9	2.7
Omitida	1.4	1.4	1.1	1.7
13L	2.0	1.9	1.6	2.6
13U	5.2	8.4	6.3	10.7
14	7.0	4.2	2.6	1.8
15	4.0	5.0	5.1	4.5
16	4.9	5.0	6.1	2.8
17L	4.8	4.9	3.7	7.0
17U	2.6	3.8	2.8	3.4
18	0.5	0.8	1.2	0.8
F	6.5	3.7	4.2	3.9
E	4.8	5.4	4.3	5.4
D	3.3	3.3	——	4.2
C	3.6	7.3	3.9	5.6
B	5.5	5.4	3.2	4.3
A	2.1	3.4	3.5	5.4
TOTALS	95.7	103.1	85.5	94.0

Cerrejón Central Block

The Cerrejón basin is in Guajira Department (Figure 2, inset map, no. 8). The Cerrejón Central Block (Figure 3) has

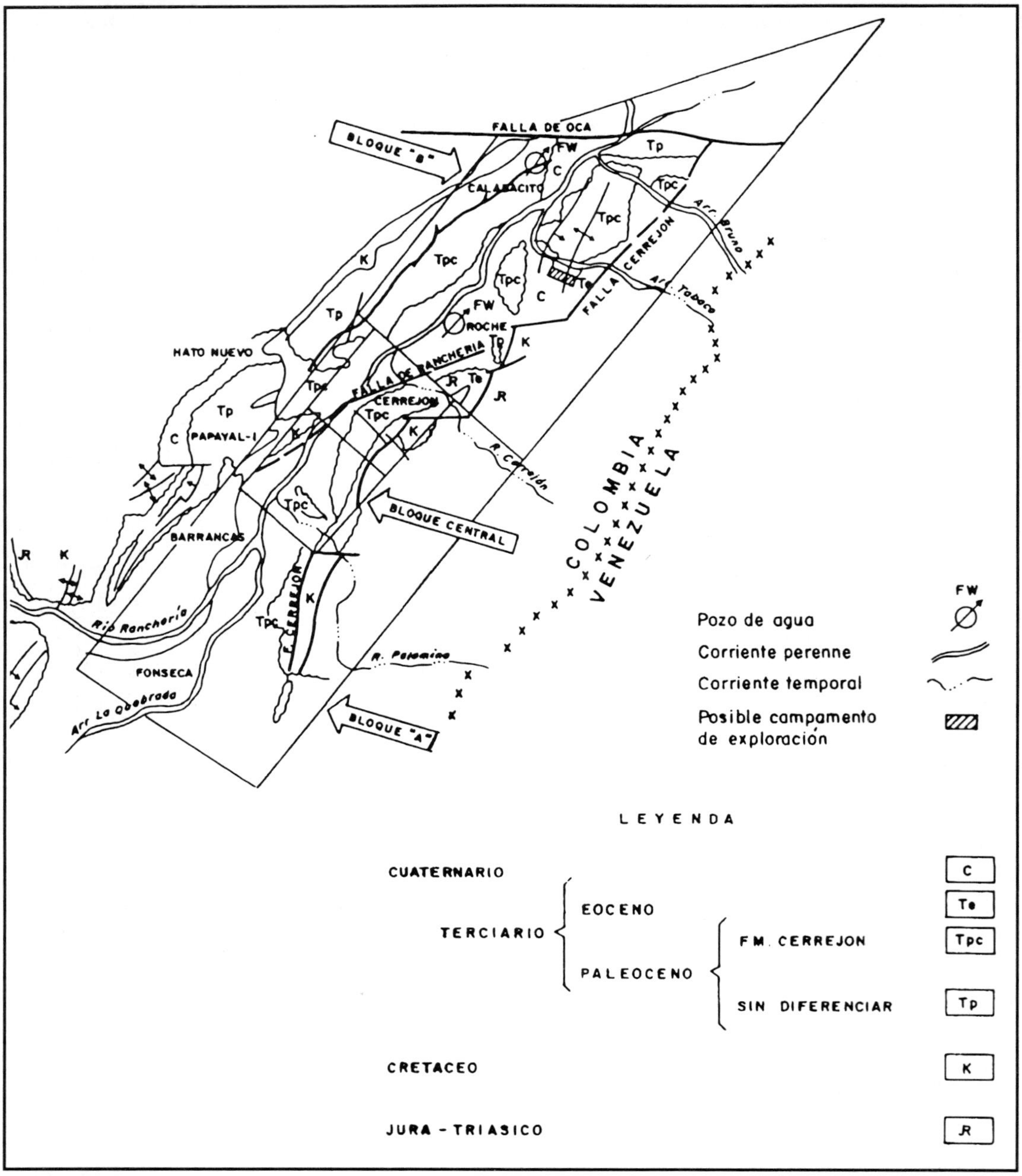

Figure 3. Ceneralized geologie map of the Cerrejón area. Pozo de agua = water well, Leyenda = explanation, falla = fault, Corriente perenne = perrenial river (stream), Corriente temporal = intermittent river (stream).

been divided into four sectors known as I, II, III, and the northern or Patilla sector. On the north it adjoins block B of the CARBOCOL-Intercor concession. At that border, the coal-bearing Cerrejón Formation is nearly 1,000 m thick. It is overlain by about 20 m of alluvium. More than 30 coal beds have been identified in the Cerrejón Central Block, as shown in Table 5.

Cerrejón North "Block B"

The evaluation of this block by Peabody Co. began in 1971. In 1973, Intercor entered into an association contract with Empresa Colombiana de Petroleos (ECOPETROL). Around 1975, MINMINAS granted concession No. 389 to Intercor for exploration of an area of 70,000 hectares in the

Table 6. Quality of Atlantic coast coals.

	Cerrejón Basin Northern	Cerrejón Basin Central	César Zone	Alto San Jorge-San Pedro Basin
Heat Content (kj/kg)	27,725	28,425	26,1000	18,400 - 22,600
Moisture Content %	9.2	8.2	11.8	17.0 - 13.0
Ash %	8.0	8.3	7.5	17.0 - 12.0
Volatile Matter %	34.9	34.0	38.0	33.7
Fixed Carbon %	47.9	49.5	46.5	32.3 - 36.7
Sulfur %	0.60	0.66	0.57	1.50

Based on feasibility studies available at Ministerio de Minas y Energía (MINMINAS).

block. It was not until 1980, however, that drilling to determine reserves was undertaken. Between January 1, 1980, and July 1, Intercor drilled 652 wells for a total of 177,100 m, of which 142,600 m were rotary drilled and 34,500 m were cored. The coal is being exploited but at the same time the drilling program has been continued to establish the dimensions and reserves of the deposit. Up to 1988 an estimated 150,000 linear meters had been drilled. The coal belt is now known to extend approximately 50 km along the Río Ranchería. About 40 m of coal are present to a depth of 300 m. Average thickness of the coal beds is about 3.3 m.

QUALITY OF THE COAL

The coals from the Atlantic coast area of Colombia are of bituminous rank, ranging from bituminous A through C. None have coking quality. Those of the César zone and the Cerrejón basin are of better quality than those of the Alto San Jorge-San Pedro basin, as shown in Table 6. Table 7 is a more complete set of anaylses for coal from the Boquerón, La Jagua, La Loma, and El Descanso fields in the César area and from the Cerrejón field.

COAL RESERVES AND PRODUCTION

The three coal zones discussed here (César, Cerrejón, and San Jorge) have proved reserves of 3,958.9 million MT to a depth of 150 m in some areas, 200 m in others. There are additional probable reserves totalling 3,466.4 million MT, for a grand total of 7,425.3 million MT. Coal mining in the Atlantic coastal area began in 1982, when the Cerrejón (Figure 1, no. 1) and San Pedro-Bijao (Figure 1, no. 6) mines were put in production. Since then, production has increased significantly from 90 kt/yr in 1982 to 10,000 kt/yr in 1988. The distribution of these reserves and current and projected production figures are summarized in Table 8.

Table 7. Physical and chemical characteristics of coals in the César1 and Cerrejón areas.

	Boquerón	Descanso	La Jagua	La Loma	Cerrejón
Volatile Matter %	34.4 - 36.8	31.3 - 36.9	33.6 - 40.0	41.8 - 43.5	31.5 - 36.5
Fixed Carbon %	43.0 - 47.1	46.3 - 54.8	42.3 - 46.0	50.9 - 55.2	45.0 - 52.5
Ash %	3.7 - 10.96	1.7 - 4.1	0.6 - 5.1	1.4 - 6.2	5.5 - 11.5
Moisture %	10.5 - 12.5	5.89 - 9.1		10.3 - 14.3	8.3 - 10.3
Sulfur %	0.2 - 1.7*	0.3 - 1.04	0.4 - 3.9*	0.3 - 1.9*	0.4 - 0.8
Phosphorous	0.15 - 0.85	0.05 - 0.58		0.01 - 0.08	
Index of Swelling					0 - 1.5
Calorific Value*	13,5000 - 14,100 BTU 7,500 - 7,833 kc/kg	11,340 - 12,402 BTU 6,300 - 6,890 kc/kg	9,927 - 13,433 BTU 5,515 - 7,462 kc/kg	13,810 - 13,854 BTU 7,672 - 7,696 kc/kg	11,548 - 12,249 BTU 6,416 - 6,805 kc/kg
Grindability Index*	40 - 55	45.4 - 49.1		47 - 49	48.3
Rank (apparent)	Bituminous	Bituminous	Bituminous	Bituminous	Bituminous
Int'l Classification			711-800		
Fusion Temperature °F	2,400 - 2,700	2,360 - 2,680		2,460 - 2,840	2,300 - 2,700
Slagging Factor	0.2 - 0.5				
Fouling Factor	0.04 - 0.48				
Ratio base/acid					0.7

1 The César area contains the Boquerón, Descanso, La Jagua, and La Loma basins.
* Anomalous data from defined beds located in the sequence

Table 8. Coal reserves and production capacity in the Atlantic coast.

Zone/Mine	Proved Reserves [1] (million	Probable Reserves 1986 (million	Total Reserves (million	Developed Capacity (kt/yr)	Planned Future Capacity [2] (kt/yr)
César [3]					
Boquerón	179.5	—	179.5	—	6,000
El Descanso	600.0	390.0	990.0	—	9,000[4]
La Jagua	209.0	274.8	483.8	40	7,000
La Loma	90.0	325.0	415.0	—	—
Cerrejón					
Central (I, II, III)	340.5	—	340.5	500	5,000
Central (Patilla)	336.9	—	336.9	—	3,500
Norte	1,600.0	1,400.0	3,000.0	6,100	15,000
Alto San Jorge-San Pedro					
Alto San Jorge	—	726.0	726.0	—	—
San Pedro Bijao	20.0	—	20.0	250	900
San Pedro Palmeras	515.0	—	515.0	—	1,500[4]
San Pedro Pica-Pica	68.0	350.6	418.6	—	2,800
TOTAL	3,958.9	3,466.4	7,425.3	6,890	50,700

1 Reserves "in situ" to a maximum depth of 200 m.
2 Annual production determined by feasibility studies and planned or projected for 25 to 30 years of production.
3 Other promising areas in La Loma and El Descanso have not been included.
4 Estimate of the consultants.

Circum-Pacific Council for Energy and Mineral Resources
Earth Science Series, Vol. 16
R. L. Miller, G. Escalante, J. A. Reinemund, and M. J. Bergin (Eds.)
Energy and Mineral Potential of the Central American-Caribbean Regions
Springer-Verlag, Berlin Heidelberg 1995

OPPORTUNITIES FOR COLLABORATING IN UTILIZATION OF CLEAN COAL TECHNOLOGIES

Denise F. Swink
Office of Planning and Environment, U.S. Department of Energy, Washington, D.C. 20585 USA

ABSTRACT

The paper calls attention to the importance of coal in energy security and in balance of trade opportunities. It contains information relative to the current state of research and development of clean coal technologies by the U.S. Department of Energy (DOE), and it outlines the objectives of these programs. It also discusses the ongoing efforts of the DOE directed toward promoting international bilateral collaboration in the research, demonstration, development, and utilization of clean coal technology.

COAL—SYMBOL OF ENERGY STRENGTH

Coal's image as the world's "fuel of last resort" is being replaced by a vision of coal as a symbol of energy strength. In countries having supplies of coal, it offers energy security, and it can also result in an improved balance of trade through reduced consumption of more expensive imported oil. Coal utilization is now increasingly linked with clean coal technologies. The link between increased use of coal and environmental damage is being broken, not only in perception but also in reality. Coal plants will soon compete efficiently and economically with liquid and gaseous fuels in the modular power generating market. Coal can also find new markets in a future in which a coal plant is no longer just a combustion facility, but a "refinery" tailoring its outputs of solid, liquid, and gaseous fuels to a wide range of uses.

It is an objective of the United States to extend the use of new coal technologies beyond its borders. This paper focuses on the research and development of clean coal technology by the U.S. Department of Energy (DOE). The agency is actively promoting bilateral collaboration in the utilization of these new technologies in other countries.

Many countries are facing the same energy problems, such as increasing electrical demand, declining reserve margins, and aging plants. Challenges are also occurring relative to the capability for suppliers to provide, or for purchasers to obtain, adequate financing on acceptable terms for the needed investment in goods and services for energy projects. Clean coal technologies for power generation can respond to these situations by being more efficient, and more reliable. They can provide modular increments of power to better meet peak load demand, and they can also provide prefabricated equipment to decrease architectural and engineering costs and to shorten construction periods.

U.S. COAL TECHNOLOGY RESEARCH AND DEVELOPMENT PROGRAM

The DOE for the last eight years has been managing an intensive research and development effort focused on clean coal technologies. The intent is to develop a suite of technologies that can be used singularly or in combination to effectively respond to the unique characteristics of coal at any particular plant site. An array of technologies is being pursued: precombustion technologies such as coal cleaning and preparation; combustion technologies such as fluidized-bed and slagging combustion; conversion technologies such as gasification, fuel cells, and magnetohydrodynamics; and post-combustion technologies to clean stack gases. The research and development objectives for all these technologies include increases in technical and economic performance in the use of coal. Technical performance for power generation, which is currently about 30 to 35 percent efficient, is targeted at 40 to 50 percent conversion efficiency and environmental performance at 50 to 99 percent of stack emissions.

The non-electric utility applications of coal use that have been pursued include:

1. Direct and indirect liquefaction of coal for transportation fuels and chemical feedstocks.

2. Synthesis gas production for direct substitute of natural gas and feedstock purposes.

3. Direct coal firing of gas turbines and diesels for cogeneration and locomotive applications.

4. Mild gasification for production of char and liquid products such as high-density jet fuels.

5. Advanced combustors and fluidized-bed combustors for light industrial, commercial, and residential applications (for example, in hospitals, hotels, apartment buildings, schools and other institutions). There are hundreds of fluidized-bed combustors operational in the U.S. industrial sector now.

6. Super-clean coal-water slurries integrated with all of these technologies.

U.S. COAL TECHNOLOGY DEMONSTRATION PROGRAM

As the research and development program continues and feeds the technology commercialization and demonstration path, the U.S. has now invested in a cost-shared technology demonstration program with industry for which the government will cost-share up to $2.5 billion.

In 1988, the DOE ran a second clean coal competition for this demonstration program. Sixteen first-class projects valued at more than $1.3 billion were selected, and private cost-sharing amounted to more than 60 percent. This competition brought to 29 the number of clean coal demonstration projects selected for which contracts have been signed or are being negotiated. The total value of the 29 projects is $2.3 billion; the federal share is approximately 40 percent.

In 1989, a third call for clean coal demonstration proposals was issued. Companies competed for more than $350 million in federal matching funds. By the time the program is over, and rounds four and five have been completed, there could be as many as 75 to 100 commercial-scale demonstration projects underway in the U.S. All will be showcasing new, more efficient, and more effective ways to use coal.

INTERNATIONAL COLLABORATION IN THE RESEARCH, DEVELOPMENT, DEMONSTRATION, AND UTILIZATION OF CLEAN COAL TECHNOLOGIES

As mentioned before, the United States promotes the expanded use of clean coal technologies worldwide. For many countries coal offers an alternate fuel source based on secure domestic and/or international supplies. Worldwide concerns with health and environmental impacts of elements in emissions and effluents of fossil fuel utilization projects call for improved efficiency and environmental performance of utilization technologies. In addition, the projected shortfall in electric generation capacity in many countries for the next several decades highlights the need for easily deployable, modular, prefabricated units to make the utilization of coal environmentally acceptable.

The DOE has entered into collaborative research and development agreements with several countries for the expanded use of clean coal technologies. An example is the collaborative agreement with the Italian government which has been very active and successful. It has involved extensive participation by Italian and U.S. companies. Examples of projects undertaken in this program include the following:

1. Coal water mixtures involving a 75 MW oil retrofit including slurry transport and stability combustion performance and electrostatic precipitator performance; models for predicting performance; 1,000 hour testing using slurry prepared from U.S. coal; testing of U.S. prepared slurry; and construction of two 250 MW thermal power plants utilizing coal water slurry.

2. Combustion studies involving four coal-fired units, with six more units planned, to assess the economic impact of coal of varying quality on power generation.

3. Testing of 10,000 tons of super-clean slurry at both U.S. and Italian sites.

4. Retrofit and repowering of Italian plants already in operation now being evaluated.

5. Use of hospital waste cofired with coal.

Because many countries utilizing domestic or imported coal have not considered the application of clean coal technologies, a new format for collaborative agreements has been developed by DOE which is oriented in that direction. The relationship between the countries is focused on commitments to use clean coal technologies. The DOE provides technical expertise to the partner country on methods for evaluating the clean coal technologies most suitable to its situation. DOE also facilitates opportunities for the partner country to visit demonstration and operational sites of clean coal technologies in the United States and to discuss the projects with DOE and with the private sector sponsors. A collaborative agreement is now under consideration between the Costa Rican government and the DOE.

In summary, the DOE is working toward solution of some problems involving use of coal by development of clean coal technologies. Because joint efforts generally result in more progressive and practical technology development and use, the DOE welcomes additional opportunities for international collaboration.

Circum-Pacific Council for Energy and Mineral Resources
Earth Science Series, Vol. 16
R. L. Miller, G. Escalante, J. A. Reinemund, and M. J. Bergin (Eds.)
Energy and Mineral Potential of the Central American-Caribbean Regions
Springer-Verlag, Berlin Heidelberg 1995

PEAT DEPOSITS OF CENTRAL AMERICA AND THE CARIBBEAN REGION

Arthur D. Cohen
Department of Geological Sciences, University of South Carolina, Columbia SC 29208 USA

Oldemar Ramirez, Luis Obando, Luis Malavassi
Refinadora Costarricense de Petróleo, San José, Costa Rica

Arturo Ramirez
Instituto de Recursos Hidraulicos y Electrificación, Panama

ABSTRACT

Peat deposits are found extensively throughout Central America and the Caribbean region, with large deposits occurring in Jamaica, Costa Rica, and Panama, and smaller ones in the Dominican Republic, Puerto Rico, Belize, Honduras, Cuba, Nicaragua, and Guatemala. It is likely that many others will be discovered as more exploration takes place. This exploration could be greatly aided by an understanding of the geologic and ecologic conditions under which these deposits are formed. This paper describes come of these conditions.

All of the deposits investigated to date can be grouped into three geomorphic types: (1) back-barrier (shoreline-related), (2) river flood plain, or (3) high altitude (mountain-top). Some examples of these types are the Changuinola deposit of Panama (back-barrier type), the Rio Medio Queso deposit of Costa Rica (flood plain), and the Talamanca Mountain blanket deposits of Costa Rica (mountain-top).

Some factors that control the locations of these deposits include subpeat and surrounding sediment or rock type, hydrologic conditions, and tectonic activity (such as faulting). The composition of the peats formed in these settings is controlled not only by the above factors but also by water chemistry and nearness to sources of detrital minerals (such as volcanos, rivers, open ocean).

INTRODUCTION

Peat deposits were once thought to be rare in tropical regions due to the rate of decomposition that can occur. However, numerous studies in the past decade (e.g., International Peat Society, Tropical Peat Resources Symposium, 1985) have shown this conventional wisdom to be in error. Not only are peat deposits very common in the tropics, they are often widespread and very thick. In Panama, for example, we recently discovered a deposit that is over 82 km^2 in area and averages over 8 m in thickness (Cohen et al., 1989). Even larger deposits have been found in Indonesia (e.g., Esterle and Ferm, 1989).

In the Central American and the Caribbean region (the subject of this paper), peat deposits have been found in Costa Rica, Panama, Guatemala, Honduras, Mexico, Cuba, Jamaica, Puerto Rico, and the Dominican Republic. All deposits discovered to date can be placed in one of three geomorphic types: back-barrier, flood plain, or mountain. Characteristics of these three types are described below using examples from Costa Rica and Panama.

THE BACK-BARRIER DEPOSIT

Back-barrier peat deposits are ones that occur behind (i.e., inland) of shoreline barrier features. By this definition we include not only deposits that occur behind active, quartz sand dominated beach-barriers (the more standard definition) but also deposits that occur behind stranded, inactive (former)

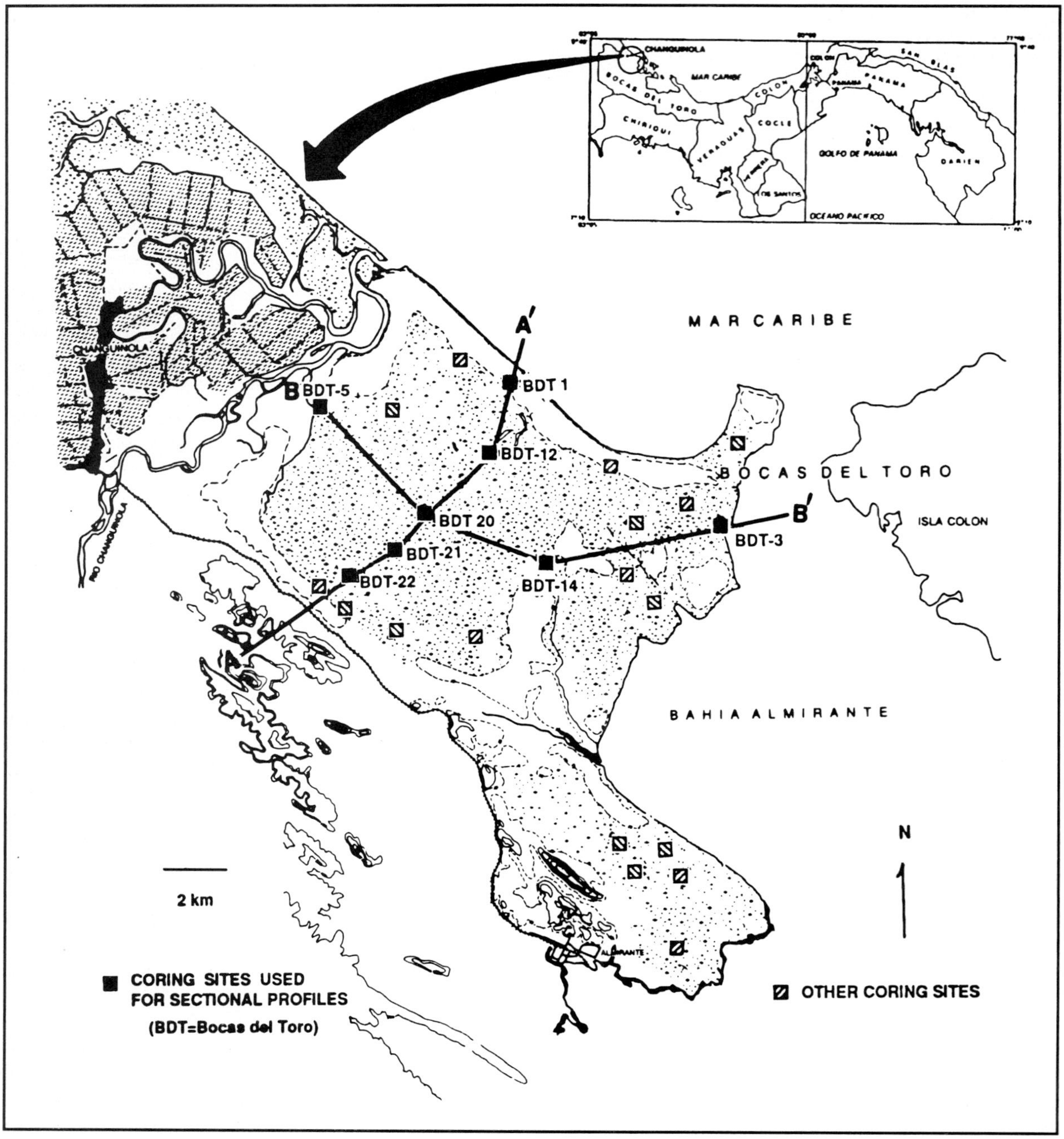

Figure 1. Location of coring sites and sectional profiles, Changuinola, Panama.

beach-barriers and also barriers that are dominated by clastics other than quartz sand, such as clastic carbonates.

All of these deposits, besides occurring behind some sort of beach-barrier feature, also often have certain other characteristics in common. First, they often occur on top of marine or brackish sediments of salt marsh, mangrove swamp, or lagoonal origin. Second, these deposits are almost always elongated roughly parallel to the long axis of the bordering shoreline feature. On the other hand, these deposits may vary greatly in such characteristics as botanical composition, salinity (i.e., may contain varying amounts of marine to fresh-water peat types), mineral matter content, and three-dimensional geometry such as having a domed or planar upper surface.

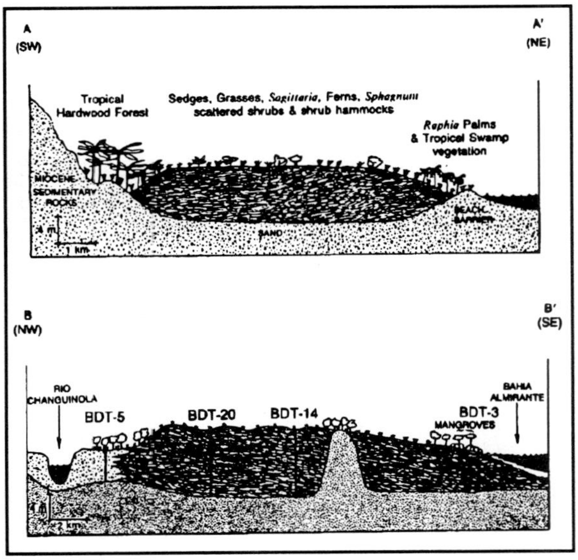

Figure 2. Schematic diagram illustrating geometry and vegetation types, Changuinola, Panama.

The Changuinola peat deposit of Panama is an example of a back-barrier deposit (Cohen et al., 1989). It is located on the northwestern coast of Panama near its border with Costa Rica, just southeast of the Rio Changuinola (Figure 1). The peat has formed over unconsolidated sediments behind, and roughly parallel to, a beach-barrier feature that extends continuously to the southeast for about 10 km. Bordering the deposit on the southwest are the foothills of a mountain range composed of Upper Cretaceous to Eocene sandstones, shales, and volcanics and Miocene rocks of the Gatun Formation consisting of shales, sandstones, coals, conglomerates, and pyroclastics. Bahia Almirante, a brackish-water bay, borders the deposit to the southeast.

The geometry, facies relationships, and surficial vegetation of this deposit are illustrated by Figure 2. Core locations and transect lines are shown in Figure 1. In cross section (Figure 2A), this deposit has the appearance of a flattened dome, colonized by tropical hardwoods and palms at its margins and by sparse herbaceous and shrub vegetation toward its center. In longitudinal section (Figure 2B), the peat interfingers at its northwestern boundary with freshwater sands and clays from the Rio Changuinola. Sediments from Bahia Almirante cover an undetermined portion of the peat deposit to the southeast. The peats composing the deposit are all of freshwater type, except for a few lenses of mangrove peat that interfinger with the other types near Bahia Almirante (Cohen et al., 1989).

This deposit averages over 8 m in thickness and covers an area of over 82 km^2. It is low in sulfur (< 2 percent, dry wt.) and ash (< 2.5 percent) and has high potential to be used for many purposes (energy, agriculture, or industrial chemicals).

Some other typical back-barrier peat deposits in the Central American-Caribbean region include the Moin deposit in Costa Rica (a planar, high-ash, high-sulfur deposit), and the Negril morass deposit in Jamaica (a very thick, carbonate-bounded, moderate-ash, high-sulfur deposit) (Blackwood and Robinson, 1985).

THE FLOOD PLAIN DEPOSIT

Peat deposits developed on flood plains of rivers are fairly common in this region. These deposits tend to form in cut-off meanders or old channels, but may also develop on the flood plain surface, if high rainfall and/or springs from bordering uplands allow organic sedimentation to exceed normal flood plain clastic sedimentation. In the latter case, these deposits tend to be near to the edge of the flood plain.

The Rio Medio Queso deposit, located at the Costa Rican border with Nicaragua to the east of the town of Los Chiles (Figure 3), will be used to illustrate the characteristics of this type of deposit. This deposit has formed on the flood plain of the Rio Medio Queso, primarily to the west of the river. An estimate of the potential size of this deposit in Costa Rica is about 70 km^2, but the actual size of the deposit is undetermined at this time because it appears to thicken and widen toward the north and may extend some unknown distance into Nicaragua. The deposit is covered with grasses, sedges, and occasional small shrubs.

Figure 4 illustrates the relationship of the peat to the other flood plain sediments. The economic peat layers (i.e., those with less than 25 percent ash content) tend to be relatively thin and to interfinger with layers with higher ash contents (clastic flood plain clays and silts derived from the river). The thickest peat occurs in a trough that roughly parallels the present river and probably represents the filling of a former channel.

These types of peats tend to be highly decomposed (low in fiber content), high in ash (e.g., the economic peat layers in the Rio Medio Queso deposit have an approximate ash content of 22 percent), and low in sulfur content (this one averaging about 0.3 wt. percent). The greatest economic potential for deposits of this type would probably be agricultural use. This deposit is further described in Cohen et al. (1985) and Obando (1988).

THE HIGH ALTITUDE (MOUNTAIN) PEAT DEPOSIT

The paramount peat deposits of the higher elevations of the Talamanca Mountains in Costa Rica are used to illustrate this type of deposit (Figure 5A). The soils of this region are referred to as paramount soils (Otarola, 1976) because they occur at the tops of mountains above about 2,500 m. Since clouds cover the mountains at this altitude for most of the year, the soils remain moist and cool. This

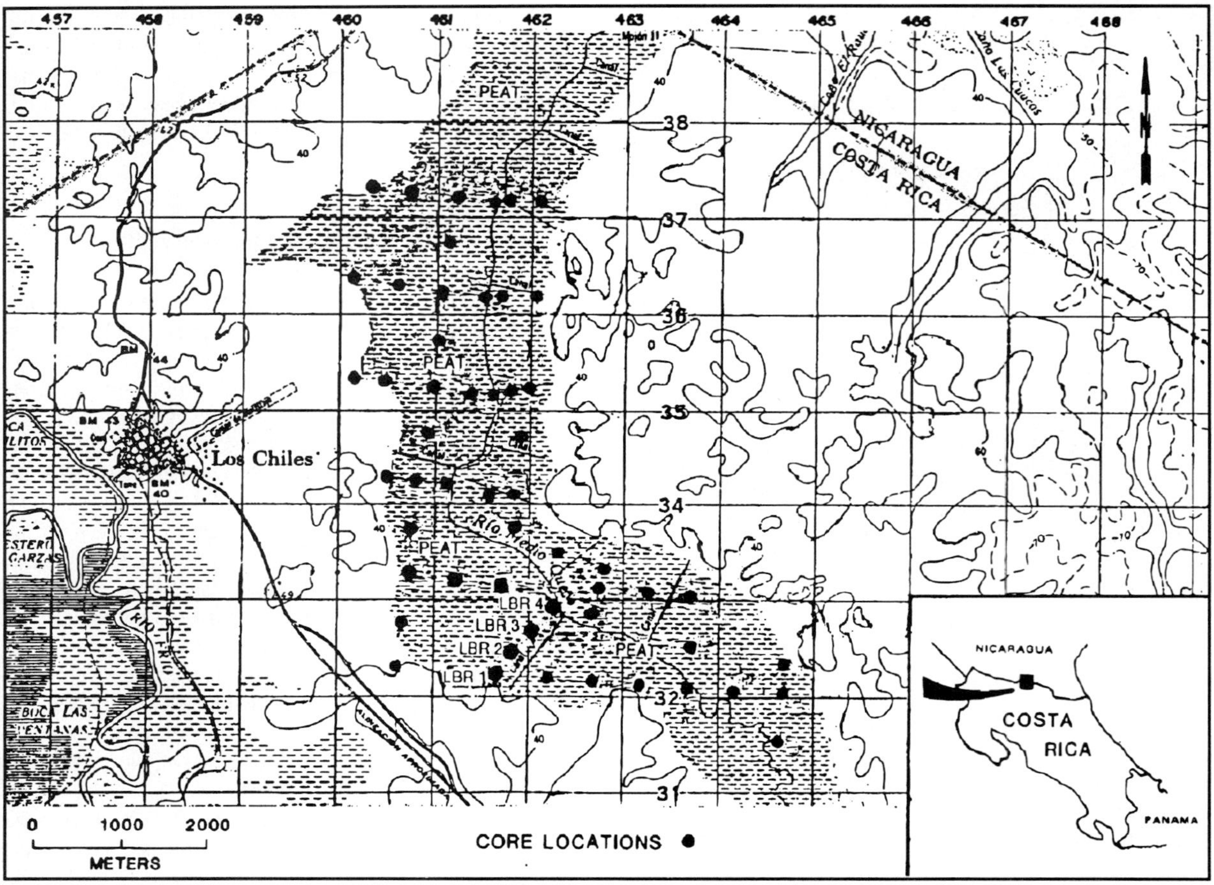

Figure 3. Location of coring sites at Rio Medio Queso peatland (topographic contours in meters).

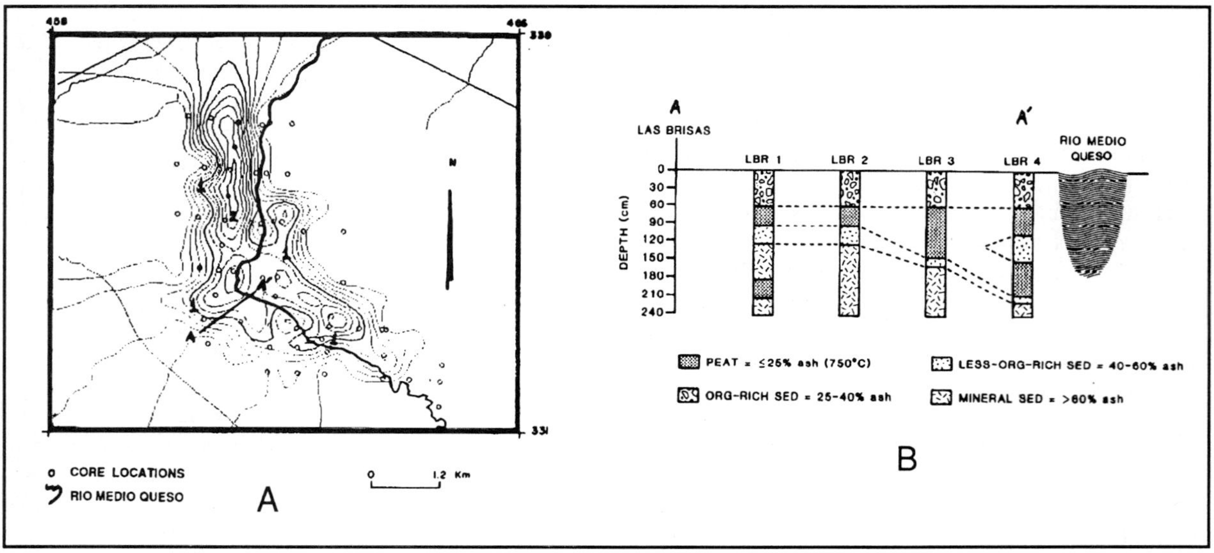

Figure 4A. Base of peat sequence (meters) and location of transect A-A'. **Figure 4B.** Transect of Rio Medio Queso deposit.

accounts for the ability of peats to develop in blanket-like deposits over much of the area. The vegetation of this region consists of both tropical cloud forests and extensive open areas covered by sedges, grasses, ferns, peat moss (*Sphagnum*), and heaths.

In the areas we sampled, the peats tend to form a surface blanket up to 1 m thick. This blanket extends up the sides of hills but tends to be thickest in depressions. Some typical cores from this region are shown in Figure 5B. The peats from this area tend to be highly decomposed and reworked (containing abundant fecal pellets and fungal remains). The dominant source plants that produced these peats are grasses, sedges, ferns, mosses and, occasionally, trees. The potentially economic peats tend to be high in ash (averaging about 21 percent dry wt.) but low in sulfur (0.2 percent dry wt.).

Although these peats might have some horticultural/agricultural use, the thinness of these deposits (even though they may cover large areas) and the ecological fragility of the cloud forest environment would argue against ever exploiting them.

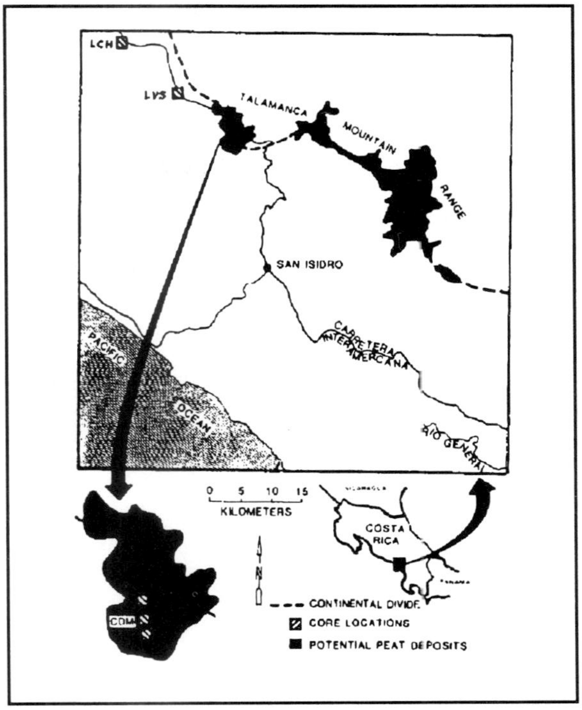

Figure 5A. Location of blanket peat deposits and coring sites in Talamanca Mountains region, Costa Rica.

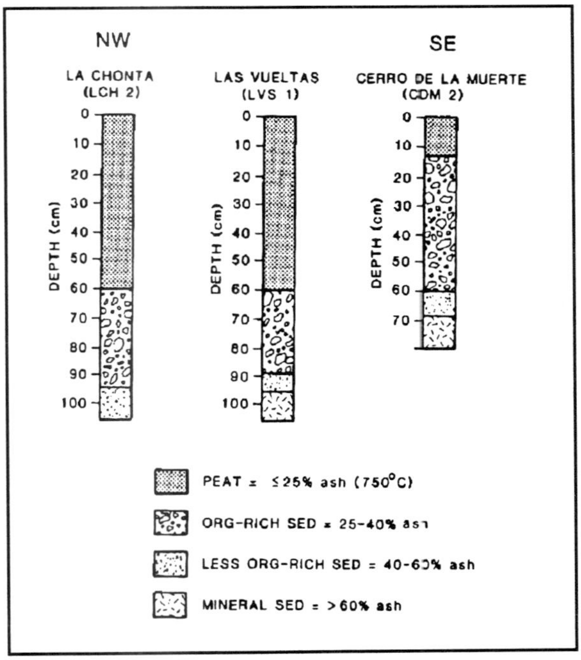

Figure 5B. Typical stratigraphic sections in Talamanca Mountains area.

ACKNOWLEDGMENTS

This work has been funded in part by the National Science Foundation (Grant #EAR-8904208), the Inter-American Development Bank, the U.S. Agency for International Development, Refinadora Costarricense de Petróleo (Costa Rica), and Instituto de Recursos Hidraulicos y Electrificación (Panama).

REFERENCES CITED

Blackwood, M., and E. Robinson, 1985, Peat resources of Jamaica—An overview, *in* Tropical peat resources—Prospects and potential: International Peat Society, p. 121-135.

Cohen, A. D., R. Raymond, S. Mora, A. Alverado, and L. Malavassi, 1985, Economic characteristics of the peat deposits of Costa Rica: Preliminary study, *in* Tropical peat resources—Prospects and potential: International Peat Society, 480p.

Cohen, A. D., R. Raymond, A. Ramirez, Z. Morales, and F. Ponce, 1989, The Changuinola peat deposit of northwestern Panama: A tropical, back-barrier, peat (coal)-forming environment: International Journal of Coal Geology, v. 12, p. 157-192.

Esterle, J. S., and J. C. Ferm, 1989, On the use of tropical domed peats as analogues for petrographic variation in carboniferous coal beds: Abstracts, 28th International Geological Congress, v. 1., p. 463-464.

International Peat Society, 1985, Tropical peat resources—Prospects and potential: International Peat Society 480p.

Obando, L., 1983, Evaluación de la turbera del Rio Medio Queso: Refinadora Costarricense de Petróleo, Division de Recursos Carboniferos, 28p.

Otarola, C. E., 1976, Characterización y classificación de algunos suelos de la Cordillera de Talamanca: University of Costa Rica (Thesis, Grado de Ingeniero Agronomo), 57p.

Circum-Pacific Council for Energy and Mineral Resources
Earth Science Series, Vol. 16
R. L. Miller, G. Escalante, J. A. Reinemund, and M. J. Bergin (Eds.)
Energy and Mineral Potential of the Central American-Caribbean Regions
Springer-Verlag, Berlin Heidelberg 1995

DEPOSITS OF PEAT IN COSTA RICA

Luis Gmo. Obando
Refinadora Costarricense de Petroleo, S.A. (RECOPE), P. O. Box 4351, San Jose 1000, Costa Rica
Escuela Centroamericana de Geología, Universidad de Costa Rica, P. O. Box 35,
Ciudad Universitaria Rodrigo Facio, San Jose, Costa Rica

Luis R. Malavassi and Rodrigo Estrada
Refinadora Costarricense de Petroleo, S.A. (RECOPE), P. O. Box 4351, San Jose 1000, Costa Rica

ABSTRACT

Peat deposits of Costa Rica developed in major alluvial plains of the Atlantic coast (back arc basin) and in the mountains in small depressions of the magmatic arc.

The peat swamps of the Caribbean coastal plain display both elongate morphology typical of a back-barrier environment, or irregular forms associated with inundated plains of meandering rivers. The peat deposits range in thickness from 0.5 m to 15 m. Calorific values vary between 2975 and 4695 Kcal/kg (dry base) and display low sulfur values. The organic content is mainly fragments of Yolillo (*Raphia*), a palm which is common in peat bogs in the Caribbean zone. The inorganic components were derived from the magmatic-arc mountains located to the south, which provided sediments principally of volcanic origin.

The peats in the intermontane depressions are found in the higher zones of Costa Rica (magmatic-arc of the Cordillera de Talamanca). These types of peat are poorly developed and have irregular shapes. The cold climate, high precipitation, poor drainage, and the lack of inorganic sedimentation produce a peat low in sulfur with a relatively high calorific value. The most important organic components here are sphagnum, grasses, ferns, and sedges.

PEAT DEPOSITS OF COSTA RICA

Introduction

In Costa Rica, the first effort to define the areas of peat potential occurred during the preparation of soils maps of the nation by the Association of Soils of Costa Rica. The maps were compiled at a scale of 1:200,000, using base maps from the National Geographic Institute.

After this, the Ministry of Industry, Energy and Mines assigned RECOPE the task of carrying out a program to investigate and develop the peat resources of the country on a national level. With the soils maps as a base, personnel from the coal resources division identified the areas with peat potential within the country.

In March 1984, Costa Rican personnel and Dr. Arthur Cohen of the University of South Carolina and R. Raymond of the Los Alamos National Laboratory in New Mexico collected the first peat samples from the areas of San Isidro de El General, Medio Queso de San Carlos, and from the coastal plain between Limon and Parismina. The project acquired preliminary field data and made preliminary evaluations of the peat resources of these swampy sites. They also defined the logistical requirements necessary for development of the peat resources.

Since 1986, the principal effort of evaluating peat resources has centered on the northern swamps near Medio Queso, and the swamps of Moin, the Airport, and El Cairo in the Atlantic coastal plain, but has included some work in other areas of Costa Rica.

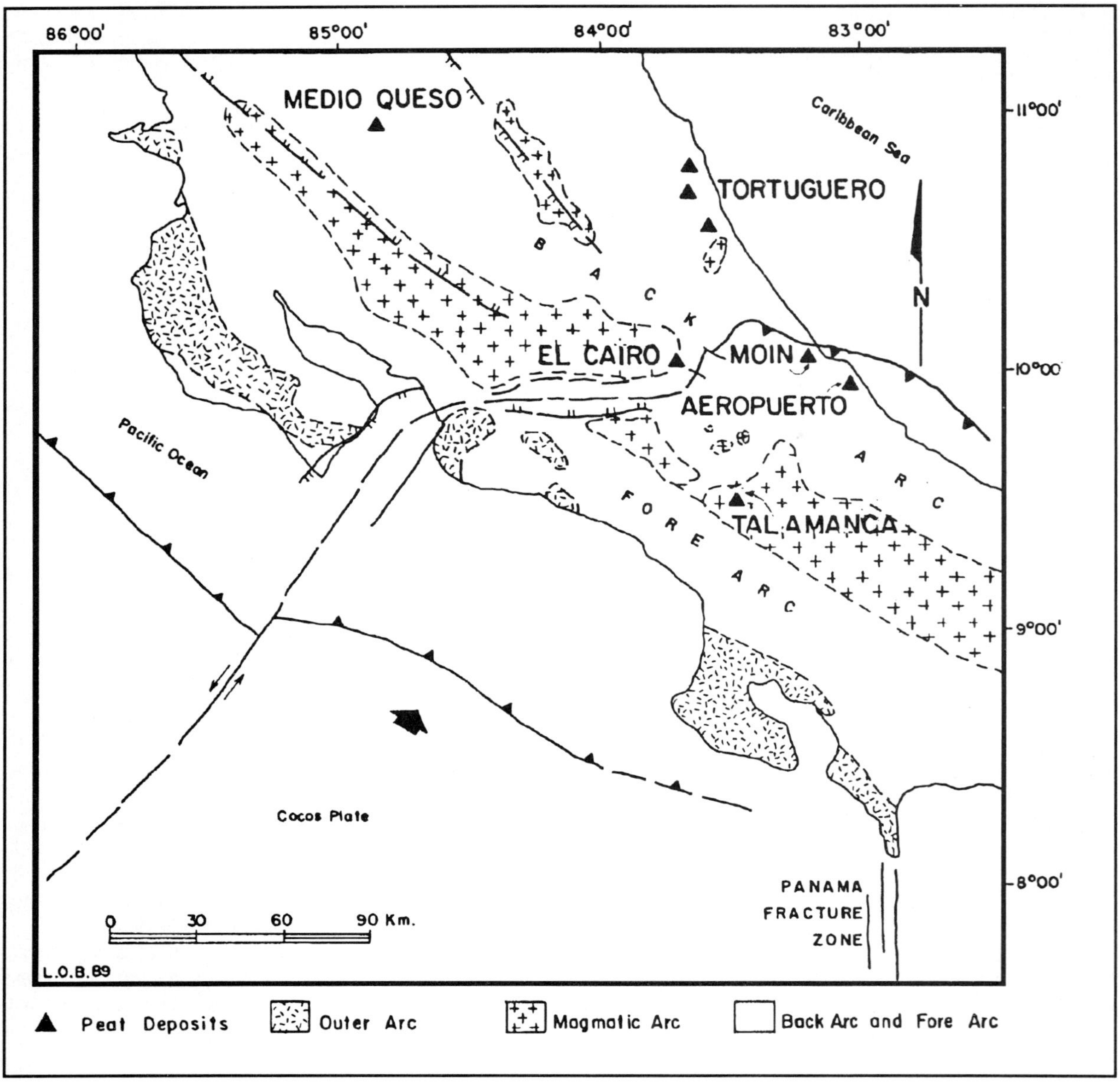

Figure 1. Location map of peat deposits in Costa Rica. Modified from Astorga and others, 1989.

Location of the Deposits

The swamps of the nation are found in different tectonic basins (Figure 1). The majority of them are located in the back arc basin where they formed in flat areas with poor drainage (for example, the peat deposits of Tortuguero, Moin, and the Airport, which formed behind barrier beaches), in drowned alluvial plains (peats of Medio Queso) or lateral to meandering rivers of the region (peats of El Cairo and Tortuguero).

Other deposits are found in small intermontane basins in the Costa Rican magmatic arc (mainly the Cordillera de Talamanca). In these mountains, limited drainage and cold and humid climate offered favorable conditions for the development of the flora and the accumulation of peat in deposits of variable size and shape.

The largest swamp areas of the country are found on the Caribbean coastal plain. Areas such as Barra del Colorado, Tortuguero, Parismina, and Moin have swamps with important potential and deserve to be intensively investigated and evaluated (L. G. Obando, internal report, Peat project, preliminary report, Division of Coal Resources, RECOPE, 1986). Other areas of interest are the northern plains of Costa Rica: Rio Medio Queso (Figure 1) and Caño Negro (not shown). In

Table 1. Physical characteristics of some peat deposits in Costa Rica.

Peat Deposits	Tortuguero [1]	Moin [2]	El Cairo [2]	Airport [2]	Medio Queso [3]	Talamanca [4]
Topographic Map (1:50,000)	Barra Colorado	Moin	Guacimo	Rio Banano	Los	Cuerici-Vueltas
Location		Provincia Limon			Alejuela	San Jose
Investigation Stage	Exploration	Exploration	Exploration	Exploration	Prefeasibility	Exploration
Boreholes	2.0	17.0	5.0	2.0	53.0	5.0
Samples	12.0	85.0	187.0	14.0	264.0	29.0
Thickness (max. m)	1.5	2.0	15.0	1.5	2.0	1.3
Thickness (avg. m)	1.2	1.6	10.0	1.0	1.6	1.0
Area (km^2)	+70.0	34.0	2.0	3.0	5.5	175.0
Peat Resources (dry MT)	[5]ND	12,852,000	[4]1,300,000	5,400,000	1,024,264	38,500,000
Quality (dry basis, avg.):						
Calorific Value (Kcal/kg)	3929	3862	2468	4695	2975	4277
End Moisture (wt%)	17.50	12.50	14.60	11.33	11.16	83.60
Ash (wt%)	22.93	27.26	49.00	21.60	40.14	26.74
Volt. Matter (wt%)	29.14	43.84	30.14	24.79	30.40	50.00
Fix. Carbon (wt%)	37.16	17.00	16.42	46.80	17.22	21.27
Sulfur (wt%)	0.43	1.50	0.20	0.64	0.33	0.23

[1]L.G. Obando, internal report, Peat project, preliminary report, Division of Coal Resources, RECOPE, 1986.
[2]L.G. Obando and L. Malavassi, internal report, Peat in Costa Rica, potential and uses, Division of Coal Resources, RECOPE, 1987.
[3]L.G. Obando, internal report, Evaluación de la turbera del Rio Medio Queso, Pantano Medio Queso, Los Chiles, Division of Coal Resources, RECOPE, 1988. Note that Cohen and Raymond's 1984 estimate (cited in text) was much larger for the Medio Queso deposit.
[4]Cohen and Raymond (1984).
[5]No data.

areas along the Pacific coast, Rio Serpe, Rio Naranjo, and Rio Tempisque may merit investigation, but given the geologic-structural characteristics of these Pacific areas, the peat potential is probably minimal. Finally, important deposits have been observed in the Cordillera de Talamanca, near la Chonta en San Isidro de El General (Cohen and Raymond, 1984) (Figure 1). Preliminary studies there show peat thicknesses which vary between 0.5 m to 15 m (El Cairo, for example), with calorific values varying between 2468 Kcal/kg and 4695 Kcal/kg on the dry basis (Table 1).

Total resources of peat at the following deposits were estimated by Cohen and Raymond (1984), and are as follows:

Medio Queso	8,400,000 dry MT (0.6 m thick)
El Cairo	1,300,000 dry MT (4 m thick)
Talamanca	38,500,000 dry MT 1 m thick)
TOTAL	48,200,000 dry MT

In addition, L. G. Obando and L. Malavassi recently investigated the Moin and Airport deposits, with preliminary results showing 12,852,000 and 5,400,000 MT (dry base), respectively. Thus, at the national level, the preliminary estimates of the five peat deposits that have been evaluated total 66,452,000 MT (dry base) (L. G. Obando and L. Malavassi, internal report, Peat in Costa Rica, potential and uses, Division of Coal Resources, RECOPE, 1987).

Deposits

Table 1 summarizes the peats studied and explored to date. In Costa Rica, a great peat potential exists in the swamps along the Caribbean coast. These are characterized by the genus *Raphia*, a palm which is locally called Yolillo. It is abundant in the fresh water environment along the coast (Cohen and Raymond, 1984).

Rio Medio Queso Peat

The Rio Medio Queso peat deposit is on the floodplain of that river on the back arc basin (Figure 1). The river is located east of the Alajuela Province town of Los Chiles (Figure 2). A paved road from San Jose to Los Chiles facilitates access to the region. Traveling within the region is possible via public road or private roads; during the winter a four-wheel drive vehicle is necessary, whereas during the summer a normal passenger vehicle can be used. Toward the end of March, one can cross the Medio Queso River plain with a four-wheel drive vehicle.

The Medio Queso floodplain is occasionally totally covered by 1 to 2 m of water during the winter. The plain is well defined with the topographic change at the outer edges of the plain abrupt and clear. It is devoid of trees except along the river's edge. The vegetation is basically characterized by sedges, reeds, shrubs, a few under-developed palms, and other small plants.

The land on the Medio Queso River plain has been used for pasture and for agriculture—especially the growing of beans and various tubers, some of which are unique to the country. The inhabitants have constructed numerous canals for the drainage of the plain. During the summer, controlled burns are frequent and this systematic practice has caused oxidation and strong carbonization of the upper layers of the peat, which explains its characteristic black color.

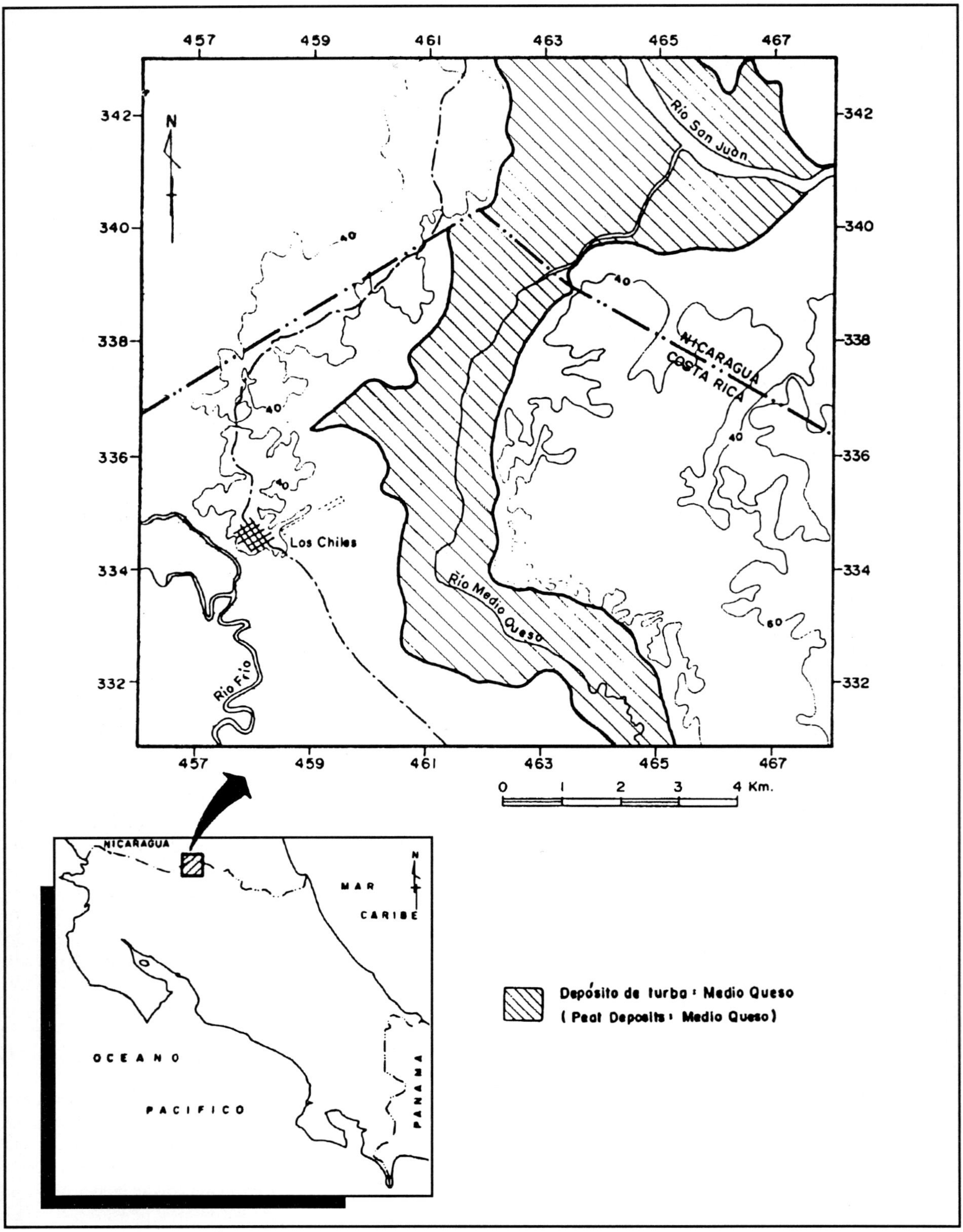

Figure 2. Location map of peat deposit: Medio Queso.

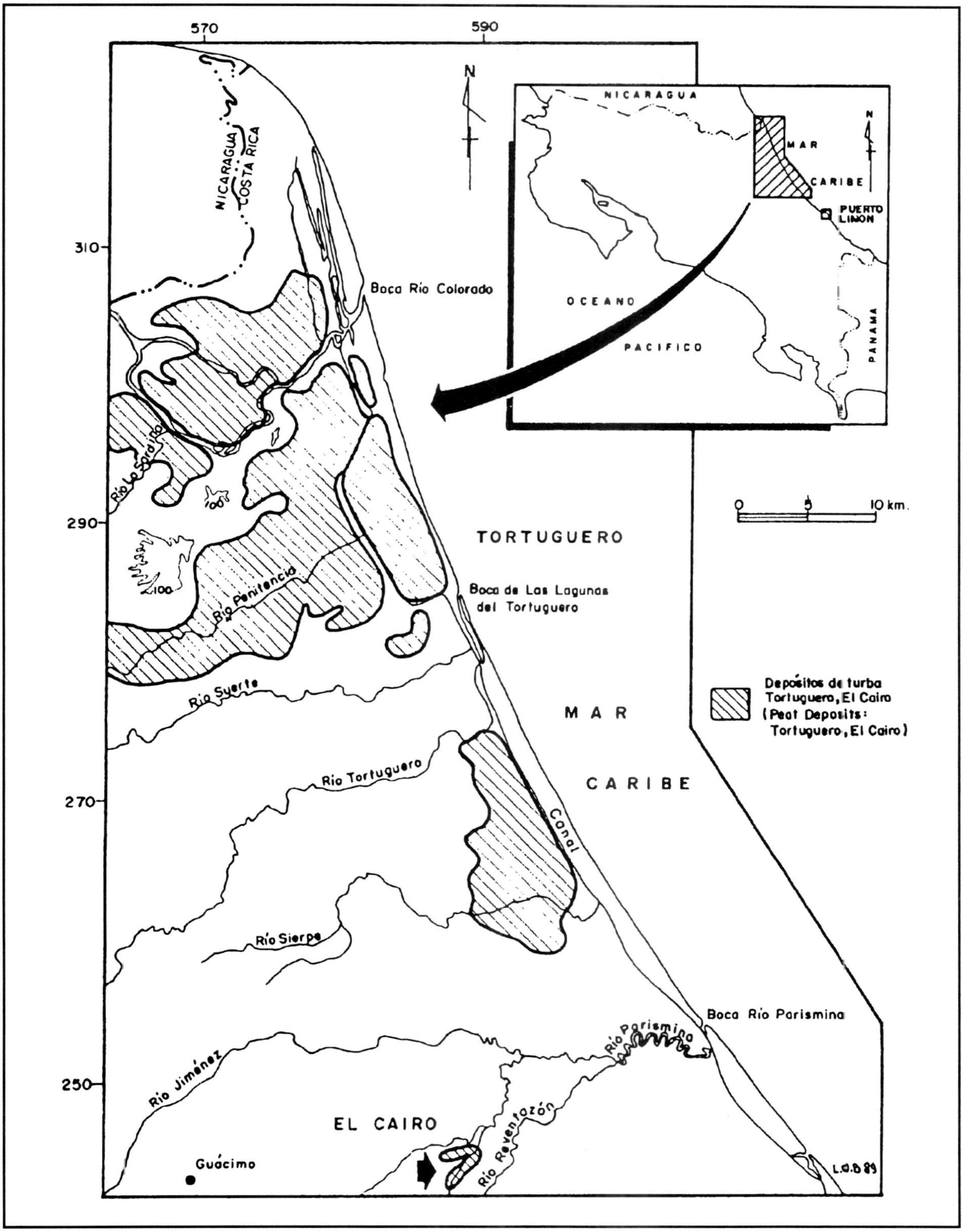

Figure 3. Location map of peat deposits: Tortuguero and El Cairo (Rio Silencio). Boca = mouth, lagunas = lagoons.

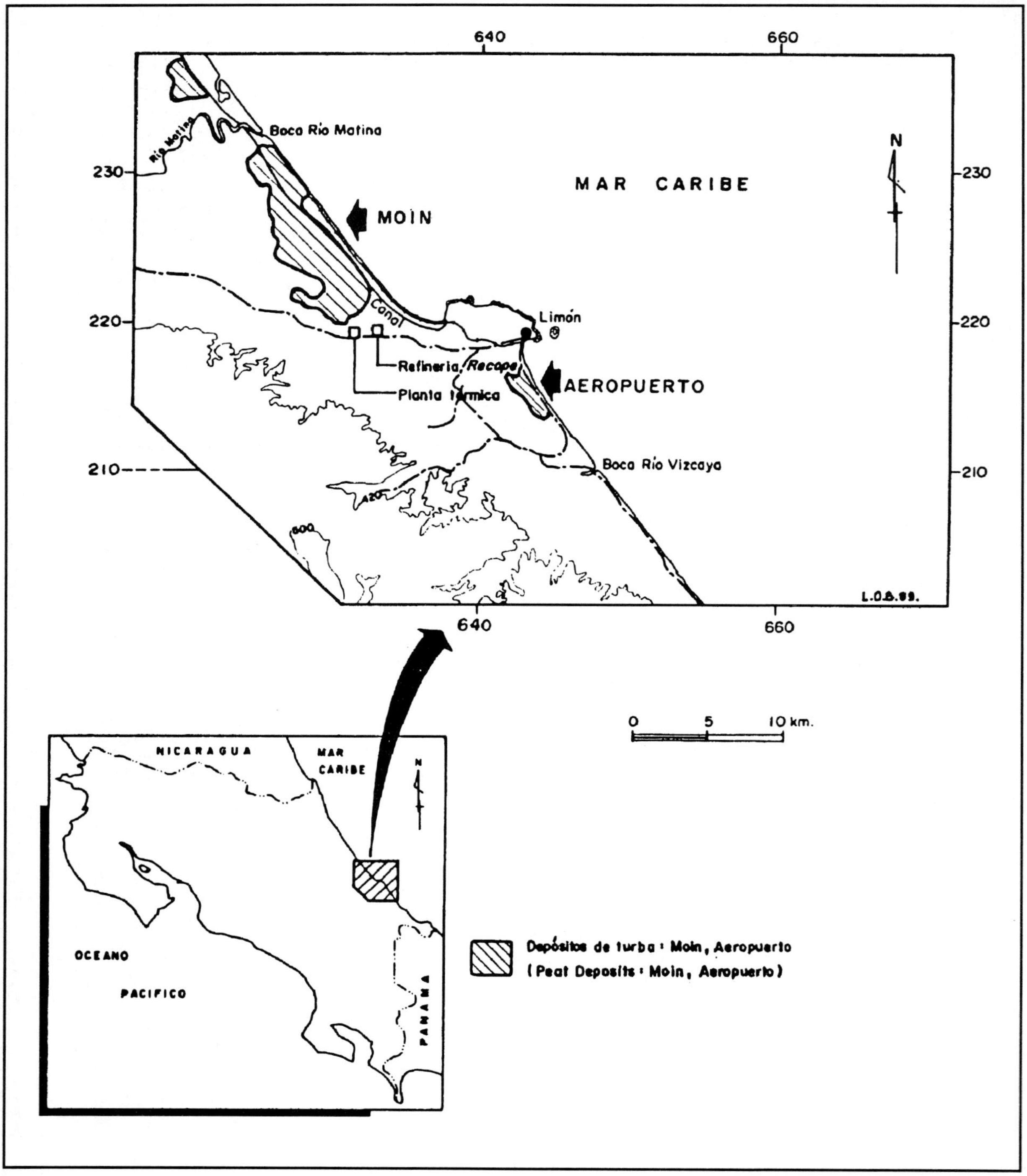

Figure 4. Location of peat deposits: Moin and Airport. Boca = mouth.

The Medio Queso peat deposit spans the Costa Rican/ Nicaraguan border (Figure 2). The high areas along the floodplain's outer edge have caused influx of inorganic material within the peat deposit. For this reason, the ash content of the peat is always very high (Table 1). Also, most of the non-peat samples analyzed chemically have indicated that those sediments are also organically rich (L. G. Obando, internal report, Evaluación de la turbera del Rio Medio

Queso, Pantano Medio Queso, Los Chiles, Division of Coal Resources, RECOPE, 1988). In Medio Queso the *Raphia* palm, which is the predominant plant of most of the peat deposits along the Atlantic coast, here is absent. Botanical studies indicate that ferns, water lilies, sedges, heaths, and the aquatic herb *Sagittaria* dominate the Medio Queso peat deposit (Cohen and Raymond, 1984).

Cohen and Raymond (1984) made the first quantitative studies of this deposit. They measured 8,400,000 MT (dry base) of peat within an area of 70 km^2. More recently, L. G. Obando (internal report, Evaluación de la turbera del Rio Medio Queso, Pantano Medio Queso, Los Chiles, Division of Coal Resources, RECOPE, 1988) performed detailed studies on 264 samples (Table 1), all acquired from the Costa Rican side of the border with Nicaragua. His studies showed that peat suitable for the production of energy is confined to an area of barely 5.48 km^2. He identified 181,989 MT of dry peat with less than 25 percent ash and 791,116 MT of dry organic sediment with 25 to 50 percent ash, and an additional 51,159 MT of organic material with more than 50 percent ash.

Cohen and Raymond (1984) have stated that this peat has excellent potential for agricultural and horticultural usage. L. G. Obando (internal report, Evaluación de la turbera del Rio Medio Queso, Pantano Medio Queso, Los Chiles, Division of Coal Resources, RECOPE, 1988) suggested the use of the peat for biogas (Ch_4) production.

Tortuguero Peat Deposits

The Tortuguero peat deposits are located in the northern part of the country, along the Caribbean coast (Figure 3) in the Limon Province. They extend from the Nicaraguan border almost to the Parismina River and cover an area of approximately 830 km^2. The deposits are associated with both meandering rivers and back-barriers. They are drained by the Sierpe, Tortuguero, Penitencia, and La Sardina Rivers. Access to the villages of the area is possible only through a canal (Figure 3) that links the Caribbean coast with the town of Limon to the southeast.

Currently, these peat deposits are in an early stage of exploration (L. G. Obando, internal report, Peat project, preliminary report, Division of Coal Resources, RECOPE, 1986). Only 12 samples have been obtained, the chemical analyses of which are shown in Table 1.

El Cairo Peat Deposit (Rio Silencio)

This deposit is located east of the town of Guacimo (Figure 3), 11 km north of Herediana in Limon Province. The bordering villages are Josefina Peje, Silencio, and Herdiana. The El Cairo peat deposit, covering 2 km^2, is on a small alluvial plain of the back arc basin that is drained by tributaries of Rio Jiménez. The vegetation is characterized by palms of the genus *Raphia* and by other small, under-developed trees.

In this peat deposit, systematic sampling has been difficult because the holes need to be as much as 15 m in depth, which is unusually thick for the Costa Rican peats (L. G. Obando and L. Malavassi, internal report, Peat in Costa Rica, potential and uses, Division of Coal Resources, RECOPE, 1987). Exploration of the deposit has begun, and 187 samples have so far been obtained (Table 1). Because of its high ash content (49 percent, Table 1), possible uses of the peat have been directed toward the agricultural sector and also toward the horticultural sector in furthering the growth, sale, and export of ornamental plants.

Moin Peat Deposit

This peat deposit is located north of the town of Limon, along the Caribbean coast in Limon Province (Figure 4). The deposit is present over an area of 34 km^2. The villages which border the Moin peat deposit are Liverpool, Bufalo, Maravilla, Cedro, Larga Distancia, Catorce Millas, and Cocal. Within the Limon Province, this deposit is strategically located close to the RECOPE oil refinery (4 km to the southeast), the Moin thermal plant, and the port of Moin. To the southeast, 6 km distant, are the port of Limon and the airport.

This is a typical back-barrier deposit within the back arc basin. It is drained by the Vueltas, Toro, and Pascual Rivers. These small rivers are not shown on Figure 4. The deposit is in part bordered and in part traversed by the coastal canal, also shown in Figure 3. Most of the vegetation along the shoreline of the canal is *Raphia* palms, with a few hardwood trees, bryophytes, and various vascular plants (L. G. Obando and L. Malavassi, internal report, Peat in Costa Rica, potential and uses, Division of Coal Resources, RECOPE, 1987).

Up to now, 17 holes have been drilled. They showed an average thickness of the peat of 1.6 m. Analyses of 85 samples (Table 1) indicate a calorific value of 3862 Kcal/kg. Ash content is 27 percent and the sulfur content is 1.5 percent, the highest of all the analyzed peat in Costa Rica.

Considering that RECOPE's oil refinery is as close as it is to the Moin peat deposit, current thought is to use the peat to generate electricity for the refinery and to substitute in part for the bunker fuel used.

Airport Peat Deposit

This deposit is situated in the back arc basin of the country, south of the town of Limon. This location is directly opposite the airport and is next to an asphalt road which connects Limon with the southeastern part of the country (Figure 4). The villages of Cieneguita, Westfalia, Beverly, and La Bomba border the deposit.

It is elongate in shape parallel to the coast, a typical back-barrier deposit. It covers approximately 3 km^2, but its uncultivated areal distribution has been diminished by cattle

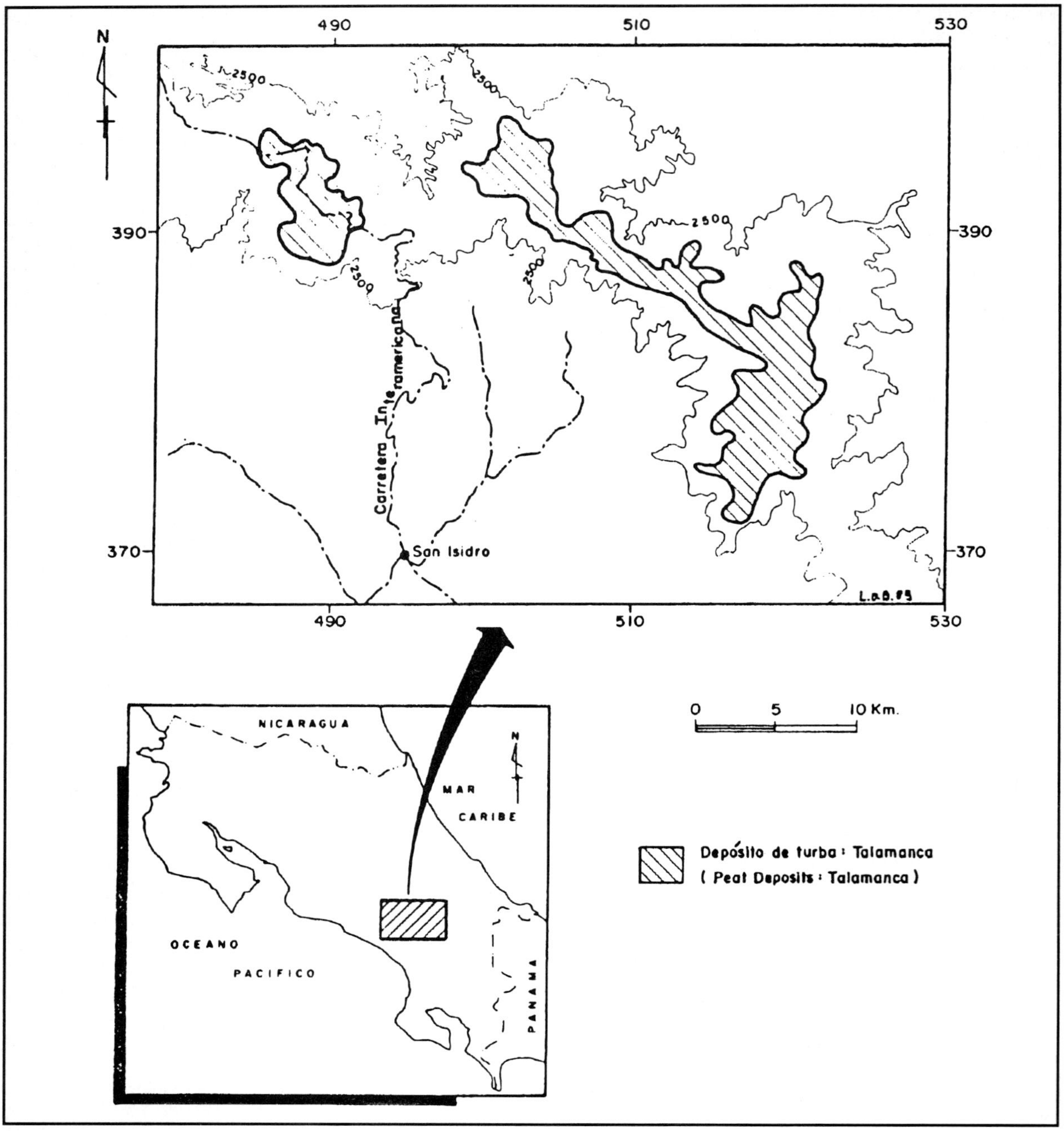

Figure 5. Location map of peat deposits: Talamanca Range.

grazing. The deposit is drained by Westfalia stream. Vegetation consists primarily of the *Raphia* palm.

Talamanca Deposits

Two large peat deposits are located in the mountainous area of Talamanca (Figure 1). They lie about 20 km north and northeast, respectively, of San Isidro. They form moors in intermontane basins. The sizes and shapes of the deposits are shown in Figure 5. Results of preliminary evaluation and analyses of the quality of the peat are shown in Table 1. Beds are thin but the large areal extent (about 175 km^2) results in very large estimated tonnages. The cold weather, high humidity, and heavy precipitation, as well as ineffective drainage, or no drainage at all, prevent the decomposition of organic matter resulting in a low sulfur peat with relatively high calorific value. The most important organic components here are moss (sphagnum), ferns, and sedges, as well as pasture (Cohen and Raymond, 1984). Because of their remoteness from large centers of industry and population, these deposits have so far received little exploration, but they may constitute a large resource for the future.

USAGE

Thayer et al. (1985) have estimated 13 possible uses for the peat in Costa Rica, taking into account cost, technical problems, needs, and domestic priorities. These uses are classified as follows: fluidized beds, gasification (low BTU), substitute for natural gas, ammonia, electricity production, refrigeration, and agricultural and horticultural uses.

ACKNOWLEDGMENTS

We wish to thank RECOPE S.A. for supporting this study and Emilia Morales for her help.

REFERENCES

Astorga, A., J. Fernandez, G. Barboza, L. Campos, J. Obando, A. Aguilar, and L. Obando, 1989, Sedimentary basins of Costa Rica: Upper Cretaceous-Cenozoic evolution and hydrocarbon potential: San José, Costa Rica, Refinadora Costarricense de Petroleo (RECOPE S.A.), Management of Primary Production, this volume.

Cohen, A. D., and R. Raymond, Jr., 1984, Final report of short-term mission for preliminary evaluation of the peat resources of Costa Rica: InterAmerican Development Bank, Los Alamos National Laboratory Report LA-UR-84-3436, p.31.

Thayer, G. R., K. D. Williamson, Jr., A. D. Cohen, and O. Ramirez, 1985, Costa Rica peat program: Los Alamos National Laboratory Report LA-UR-3456, p. 140.

Circum-Pacific Council for Energy and Mineral Resources
Earth Science Series, Vol. 16
R. L. Miller, G. Escalante, J. A. Reinemund, and M. J. Bergin (Eds.)
Energy and Mineral Potential of the Central American-Caribbean Regions
Springer-Verlag, Berlin Heidelberg 1995

ECONOMICS OF SELECTED ENERGY APPLICATIONS OF PEAT IN PANAMA AND COSTA RICA

Gary Thayer
Los Alamos National Laboratory, Los Alamos, New Mexico 87545, USA

Oldemar Ramirez E.
RECOPE, Apartado Postal 4351 Zona 1000, San Jose, Costa Rica

Arturo Ramirez
IRHE, Apartado Postal 5285, Panama 5, Republic of Panama

ABSTRACT

Studies were performed to determine the economic competitiveness of peat in Costa Rica and Panama. The options examined were (1) electrical production in Panama, and (2) industrial boilers and cement plants in Costa Rica. Using estimates of peat mining costs and the end-use costs, we calculated for each application the price of coal and oil at which the representative annual cost of energy using peat was the same as that when coal or oil was used. We found that a peat-fueled power plant in Panama would be economic if the price of fuel oil was above $0.10 per liter ($15.90 per barrel) and the cost of coal was above $40.00 per metric ton ($36.30 per short ton), delivered. In Costa Rica, peat was competitive with fuel oil for large boilers (34,000 kg of steam per hour) when the cost of oil was above $0.10 per liter ($15.90 per barrel). For smaller boilers (5,000 kg of steam per hour) peat was cheaper than fuel oil when oil was above $0.08 per liter ($12.70 per barrel). Peat would be competitive in a cement plant when fuel oil prices were above $0.075 per liter ($11.90 per barrel).

INTRODUCTION

Since 1986, Los Alamos National Laboratory has been involved in a project, funded by the Agency for International Development (AID), to provide technical assistance on energy matters to Central American countries. One portion of this project was to identify peat resources and examine their possible uses. The identification of the peat resources was led by Arthur D. Cohen, University of South Carolina, and is reported elsewhere in this volume. This paper reports on the examination of the economics of using peat in selected applications: a 30 MW power plant in Panama, two sizes of boilers in Costa Rica, and a substitute for oil in cement plants in Costa Rica.

The method used to compare the costs of the different technologies was to calculate the levelized cost for each application. The representative annual cost is the price charged for the output of the facility (i.e., electricity and steam) that would cause the total revenues over the lifetime of the facility to equal the total charges for the facility including capital, interest, return on investment, operating costs, and maintenance. To find the break-even costs, the levelized cost for the conventional facility was calculated without the fuel cost. The fuel price that would make the levelized cost of the peat-run facility equal to that of a conventional facility was calculated. This fuel cost is the break-even cost of fuel for the application or it is the fuel price where the cost of the conventional facility and that of the peat-fueled facility are the same. The assumption is that the real cost of the fuel is constant over the lifetime of the plant. Examining the economics in this manner removes assumptions about the future prices of coal or oil from the analysis.

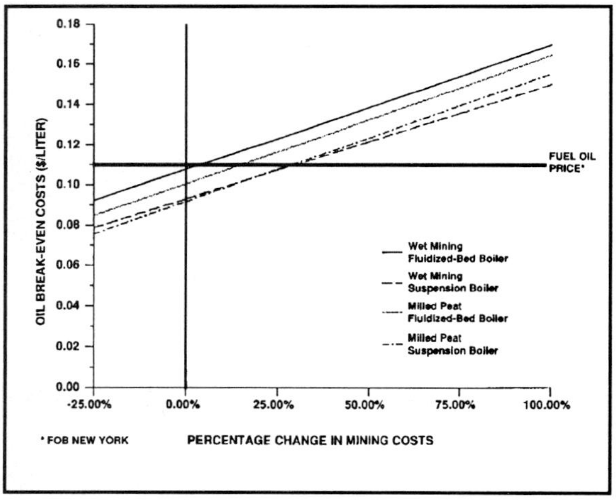

Figure 1 Break-even oil cost for a peat-fueled 30 MW electric power plant. Computer-drawn straight lines in the explanation are represented by jagged lines in the graph.

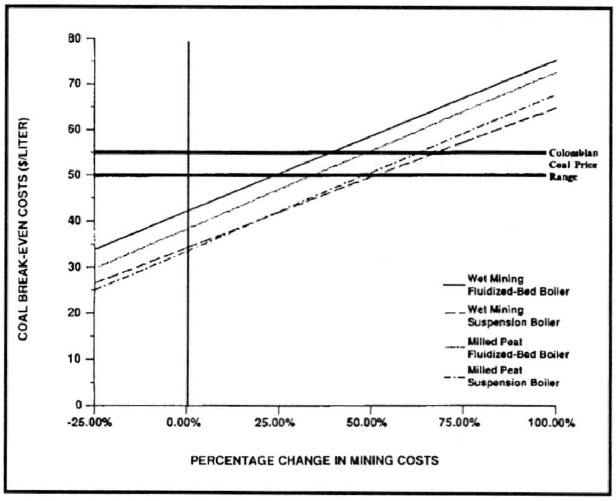

Figure 2. Break-even coal cost for a peat-fueled 30 MW electric power plant.

The costs of using peat in the selected applications were compared with the costs of using conventional fuels: coal and oil for the power plant, oil for the boilers, and oil for the cement plant. A break-even cost was calculated for the prices of the conventional commodities as described above.

As of 1986 about 6,000 megawatts of peat-fueled electrical capacity were on line world-wide; peat was being used as a boiler fuel for community space heating and for other applications in Finland, Ireland, Sweden, and the Soviet Union. Hence, the technology for burning peat is well developed. However, most of the experience in mining peat has been in northern Europe and Canada. Thus uncertainties exist as to the cost of mining peat in the tropics, where weather and drying conditions are considerably different from those in northern latitudes. Estimates of the mining costs were made for the applications examined here, but because of the uncertainty of these costs, the break-even costs will be presented as a function of the mining costs.

PEAT APPLICATIONS FOR ELECTRICAL PRODUCTION IN PANAMA

The application considered for the Changuinola peat area in Panama was a 30 MW electric power plant located next to the peat area. Two mining methods were examined: milled peat and wet-mined peat. Also, two boiler types were examined: a conventional suspension boiler and a fluidized-bed boiler.

Mining Methods

Milled peat is the most common peat mining method used today to produce fuel peat. In this method the peat bog is drained and leveled. The top few centimeters of peat are then removed by grinding the surface. This process produces peat of the consistency of sawdust. The peat is allowed to dry, which will take one to two days, and then is raked into rows. The dried peat is later moved to large storage piles. Peat from these piles is subsequently trucked to the power plant. Milled peat can be used in both suspension-fired and fluidized-bed boilers with no further drying.

Wet mining is a method that has been examined for mining the peat areas in Jamaica. The first step in wet mining peat is to remove the peat using a clamshell dredge. The peat is ground into a slurry and pumped in a pipeline to the power plant where it is put through presses similar to those used in paper manufacturing. This step reduces the water content of the peat to approximately 70 percent. It must be dried further, using flue gases and steam from the boilers, before being sent either to a suspension boiler where the water content must not exceed 45 percent or to a fluidized-bed boiler where the water content is expected to be about 55 percent.

Boiler Types

The first boiler type is a suspension boiler, the most common boiler type used for coal. In a suspension boiler, the peat is ground to a fine dust and blown into the boiler where it is ignited.

The second boiler type is a fluidized-bed boiler. Because of its versatility and ease of pollution control, this type of boiler is being installed to utilize lower grade fuels such as coal and wood wastes. In a fluidized-bed boiler, air is blown upward in a bed of fine-grained material such as sand. The upward flowing causes the bed to "fluidize" or act as a viscous fluid. The peat is added to the bed, and as it works its way through the bed, it is burned. The fluidized-bed boiler product has the advantages that it can be used as milled peat or it can be used directly from the dryers without further grinding. Also the fluidized-bed boiler can burn higher moisture peat.

Cost Calculations and Results

In the economic calculations, a plant lifetime of 30 years for both the conventional and the peat-fueled systems was assumed. A real (actual minus inflation) interest rate of 6 percent was used in the calculations of the levelized cost. The plant was assumed to be a government-owned operation so that there were no taxes and no return on investment. The plant was assumed to be financed by bonds.

The fuel break-even costs for an oil-fired electrical plant and a coal-fired electrical plant are given in Figures 1 and 2. The break-even costs are plotted against the mining costs. The best base-case values (milled peat in a suspension boiler) for the break-even costs were \$0.09 per liter (\$14.30 per barrel) for oil and \$35 per metric ton (\$31.80 per short ton) for coal. The present day prices shown on the graphs are estimates of the price for Bunker C f.o.b. New York, and of the cost of imported Colombian coal delivered to the electrical facility.

PEAT APPLICATIONS IN COSTA RICA

In Costa Rica, the applications considered for peat were for use as a fuel in boilers and for use as a partial substitute for oil in a cement plant. The peat was assumed to come from the El Cairo peat area and the boilers and cement plant were assumed to be located in the central valley region of Costa Rica. Transportation costs for the peat from El Cairo to the central valley are included in the cost estimates.

Sod Peat Process

In addition to the two mining methods examined above for the Changuinola peat deposit of Panama, the cost of producing sod peat was examined for the applications in Costa Rica. Sod peat is produced by first draining and leveling the bog as in milled peat production. Then the peat is cut from the bog, masticated, and extruded in the form of cylinders. These cylinders have the consistency of clay and are left on the fields to dry. Two to three weeks are typically required for the sods to dry sufficiently for transport. After a day or so the sods will produce a waxy layer on their surfaces that will repel water, so the sods will not rewet if rain occurs while they are drying on the fields. Sod peat has been the favored peat form for the present tests done in the mining of tropical peat. It is also the easiest form of peat to transport.

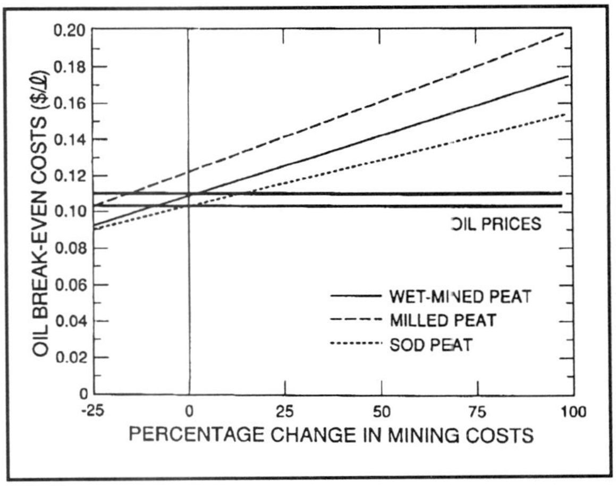

Figure 3. Break-even oil cost for a peat-fueled 34,000 kg of steam per hour boiler.

Boiler Types

Two sizes of boilers were considered for Costa Rica: a larger boiler producing 34,000 kg of steam per hour and a smaller boiler producing 5,000 kg of steam per hour. The cost of the larger boiler was based on boilers designed for the United States and included automatic controls and pollution control. The smaller boilers were based on designs for wood-burning boilers and did not include as many automatic controls and pollution control devices. Thus the relative capital cost of the larger boilers was greater than for the smaller boilers. Because the differential costs between a peat boiler and an oil boiler are large, the initial cost of the boiler has a significant effect on the competitiveness of the peat-fueled system. This competitiveness is reflected in the break-even costs which show that peat use in smaller boilers will compete with a lower oil price than would peat use in the larger boilers.

Cost Calculations and Results

Figure 3 shows the break-even costs for peat compared with oil when used in the larger boilers. Again, the break-even costs were plotted against the mining costs because the mining costs are uncertain. The costs for three different mining methods are shown in Figure 3. The base-case break-even cost for oil is \$0.10 per liter (\$15.90 per barrel). The peat mining operation assumes that the peat is being mined exclusively for the boiler. Mining of peat for additional applications would probably lower the cost of peat for this application.

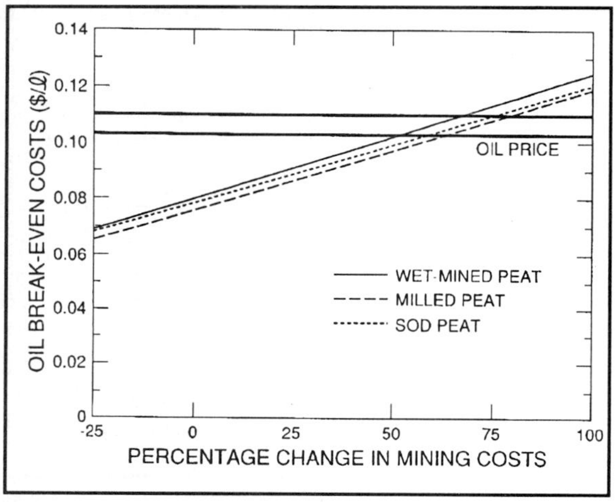

Figure 4. Break-even oil cost for a peat-fueled 5,000 kg of steam per hour boiler.

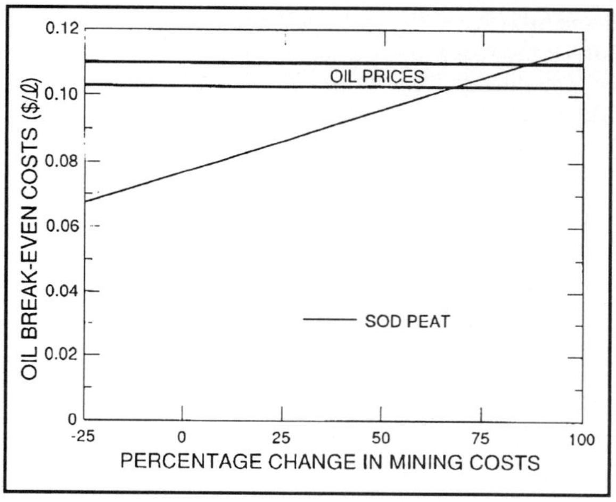

Figure 5. Break-even oil cost for peat use in a cement plant.

Figure 4 shows the break-even costs for oil for the smaller boilers plotted against the peat mining costs. As was mentioned above, the smaller peat-fired boilers are more competitive because they are not as sophisticated units as are the larger boilers. The base-case break-even cost for oil is $0.08 per liter ($12.70 per barrel). The mining costs presented here assume that the peat mining operation is larger than is needed for one boiler: mining costs for only one small boiler would be considerably higher.

Peat can be a partial substitute for oil in cement production. Depending on the design of the cement kiln, peat can substitute for 10 to 60 percent of the oil usage in the plant. This study assumed that 50,000 metric tons of peat were provided to the cement plant per year and that it replaced an equivalent energy value of oil. Costs of providing special fuel handling equipment for a drying facility for the peat were included in the cost estimates. Figure 5 gives the break-even costs for oil and peat in a cement plant. The base-case break-even cost was $0.075 per liter ($11.90 per barrel) of oil.

SUMMARY

These cost studies have shown that peat can be competitive with oil and coal in energy applications. Peat represents a domestic source of energy, and if it were used, it could reduce oil imports. The major obstacles facing peat use are the initial cost of developing the peat area for mining and the cost of demonstrating that peat can work. Mining studies and tests need to be made to determine the most economical method to produce peat for energy. However, these studies will be expensive and time consuming. Costa Rica's approach of establishing peat mining by first using the peat in horticultural applications, where it has a higher value than for energy uses, may help to overcome the initial cost obstacle and may provide the impetus to examine the possibility of using peat in energy applications.

Circum-Pacific Council for Energy and Mineral Resources
Earth Science Series, Vol. 16
R. L. Miller, G. Escalante, J. A. Reinemund, and M. J. Bergin (Eds.)
Energy and Mineral Potential of the Central American-Caribbean Regions
Springer-Verlag, Berlin Heidelberg 1995

RECOMMENDATIONS OF THE WORKING GROUP ON COAL AND PEAT

Jean Weaver
U.S. Geological Survey

The Working Group on coal and peat discussed many of the same problems and concepts that were identified by other working groups, especially those on petroleum and mineral resources. Particularly, the panel strongly agreed that a vastly increased amount of basic geologic and related engineering studies are needed before we can properly evaluate the fuel and mineral potential of the Central American-Caribbean region. Members of the coal panel agreed that a great potential for coal and peat does exist within the region and they emphasized that a greater effort should be made in their geologic and economic assessment. Successful coal and peat programs are currently underway in Costa Rica, Colombia, and Venezuela.

From the perspective of solid fuels, a more thorough study of the sedimentary basins, large and small, is vital to the improved understanding of the energy potential of the region. The sedimentary basins and their fringes are the habitat of fuels and related mineral resources, such as phosphate. The geologic factors that control occurrence and recoverability, the quality factors that control utilization, are poorly understood in most of the basinal areas of the Central American-Caribbean region.

The panel recognized three facets of the basic and applied information problems:

1. existing relevant information on solid fuels should be gathered, integrated, and summarized for regional use;
2. identified data needs should be satisfied by basic and applied research; and
3. a mechanism or an organism for transfer and dissemination of information among individuals and groups with interests and responsibilities in solid fuels should be created.

The subject of educational opportunities was discussed and general support for an increase in scholarships was recognized. However, the panel specifically recognized a lack of opportunity to study solid fuels at both the undergraduate and graduate levels in existing educational facilities. They strongly encouraged establishment of study courses in solid fuels at institutions such as the Central American School of Geology in San Jose, Costa Rica.

GEOTHERMAL RESOURCES

Circum-Pacific Council for Energy and Mineral Resources
Earth Science Series, Vol. 16
R. L. Miller, G. Escalante, J. A. Reinemund, and M. J. Bergin (Eds.)
Energy and Mineral Potential of the Central American-Caribbean Regions
Springer-Verlag, Berlin Heidelberg 1995

STATUS AND GEOLOGIC SETTING OF GEOTHERMAL FIELDS IN CENTRAL AMERICA, MEXICO, AND THE CARIBBEAN

David Sussman
Unocal Geothermal Division, 3576 Unocal Place, P. O. Box 6854, Santa Rosa, California 95406, USA

ABSTRACT

The current installed geothermal capacity in Mexico, Central America, and the Caribbean is 869 MWe (Megawatts of electricity), of which 71 percent is in Cerro Prieto field, Mexico. The projected installed geothermal capacity of the region is 1,020 MWe in 1992, reflecting power plants to be brought on line in Costa Rica, El Salvador, Guatemala, St. Lucia, and Mexico.

With few exceptions, high enthalpy (heat content) geothermal fields in Mexico, Central America, and the Caribbean region are associated with active volcanic belts at plate boundaries. In Mexico, three high enthalpy fields occur within the 1,200 km-long Trans-Mexican Volcanic Belt (TMVB) along the western margin of the North American Plate. The 1,100 km-long Central American Volcanic Belt (CAVB) and TMVB result from subduction of the Cocos Plate beneath the western margins of the Caribbean and North American Plates, respectively. Several of the Central American and Mexican geothermal systems are related to young silicic volcanism, e.g., Los Humeros and La Primavera, Mexico, and Miravalles, Costa Rica. However, Momotombo (Nicaragua) and Ahuachapán (El Salvador) geothermal fields appear to be associated with mafic to intermediate composition volcanic centers.

In addition to magmatic heat sources, Central America hosts an extensive system of grabens in and behind the volcanic arc. The intersection of tensional structures and young volcanic centers yields highly favorable settings for exploitable geothermal systems. Several geothermal fields are being explored in Honduras, all of which are well to the east of young arc volcanism. These fields are related to graben faults and are thus similar to geothermal areas under development in the Basin and Range Province of Nevada and Utah in the United States.

In the eastern Caribbean, the density and volume of Quaternary volcanos is lower than in Central America and Mexico. The relative plate velocity at the convergent boundary between the Caribbean and the North and South American Plates is low. To date, one geothermal field is operating at La Bouillante, Guadeloupe (4 MWe), and a discovery well was drilled at Soufriere, St. Lucia.

INTRODUCTION

Geothermal electricity development in the Central American-Caribbean region began at Pathe, Mexico, in 1959, with the installation of a small experimental power plant, and has accelerated rapidly since then. This paper briefly reviews the tectonic and volcanic settings that influence the geographic distribution of the region's geothermal resources and presents the current status of geothermal development. The main focus in this paper is on high-temperature geothermal resources capable of generating electricity (generally >150°C).

GEOLOGIC SETTING

Most known high-temperature geothermal fields in Central America, Mexico, and the Caribbean are associated with Quaternary volcanos in three volcanic areas marking the edges of the North American and Caribbean Plates (Figure 1).

The Trans-Mexican Volcanic Belt (TMVB) and Central American Volcanic Belt (CAVB) extend 1,200 km and 1,100 km, respectively, along the western margins of North and Central America and result from the subduction of the Cocos Plate beneath the North American and Caribbean

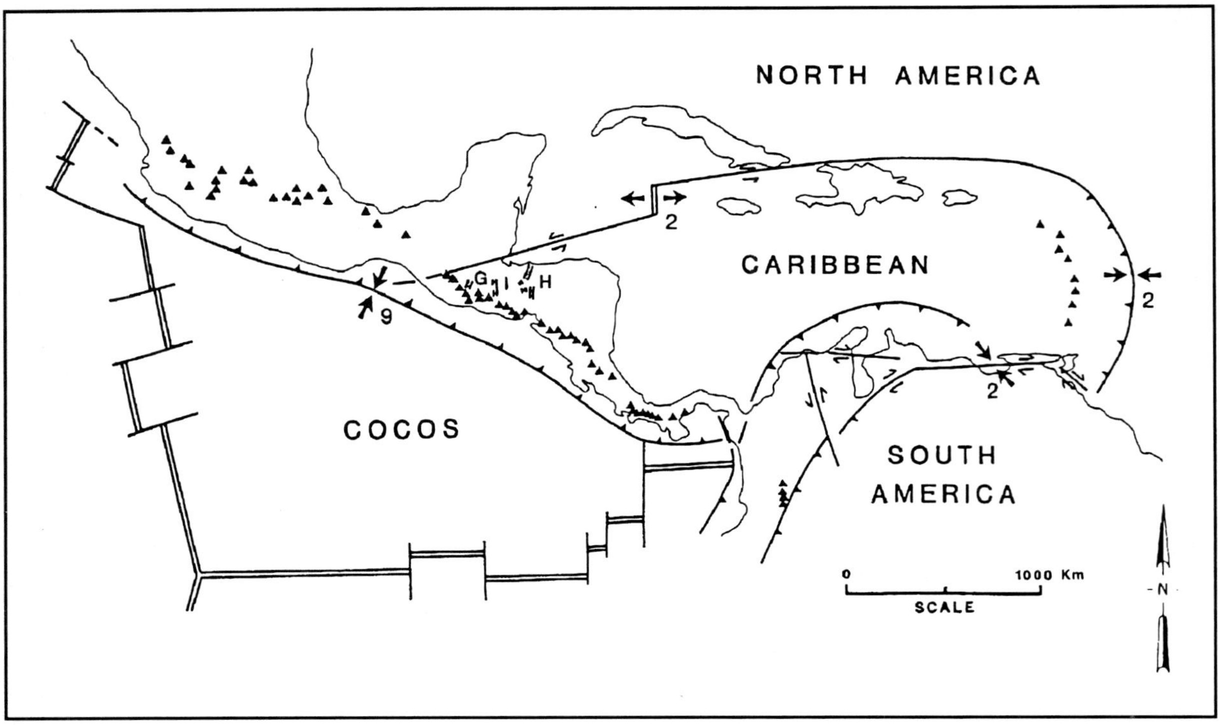

Figure 1. Generalized tectonics of Central America and Caribbean region. G = Guatemala City graben, I = Ipala graben, H = Honduras depression.

Plates. The 750 km-long volcanic arc of the Lesser Antilles results from underthrusting of the North and South American Plates beneath the eastern end of the Caribbean Plate.

Quaternary volcanos in the TMVB and CAVB are spaced 25-30 km apart, comparable to the high density of volcanos on Java, Indonesia. The Lesser Antilles arc is a less active volcanic front than the TMVB and CAVB, and the average spacing between volcanos is about 60 km. The CAVB is usually thought to extend from the Mexico-Guatemala border to central Costa Rica (Carr and Stoiber, 1988). However, at least ten major Quaternary volcanic centers have recently been identified in western Panama (de Boer et al., 1988). Active subduction beneath Panama is suggested by calc-alkaline volcanism, subdued but active seismicity, and folded, thrusted young sediments along the Panama Trench (de Boer et al., 1988).

Most of the high-temperature fields discovered to date in the three volcanic belts are associated with late Quaternary silicic volcanic centers whose magma chambers provide a heat source for hydrothermal convection. In general, young silicic centers are fewer and more widely spaced along these arcs than intermediate to mafic volcanos (Carr and Stoiber, 1988). At least two high-temperature geothermal systems appear to be associated with mafic to intermediate composition volcanic centers. Momotombo field, Nicaragua, for example, is located on the flanks of a fumarolic basaltic andesitic volcano, and the resource at Ahuachapán field, El Salvador, is probably related to a nearby andesitic volcano.

In the northwest corner of the Caribbean Plate the dominant tectonic features behind the arc are the Guatemala and Ipala grabens, trending north-south, and a series of grabens together called the Honduras Depression (Figure 1). Several deep-circulation type geothermal systems unrelated to Quaternary volcanism have been identified within the grabens in western and central Honduras.

The coincidence of tensional structures (implying subsurface permeability) and young volcanic centers can yield highly favorable settings for high-temperature geothermal systems. Miravalles (Costa Rica) and Amatitlan (Guatemala) geothermal fields occur within young calderas cut by graben faults oriented transverse to the volcanic arc.

ELECTRICAL POWER OVERVIEW

Table 1 summarizes the most recent electric power data for those countries in the region which produce (or have potential to produce) geothermal electricity. No data were available for Guadeloupe.

The installed power capacity in Mexico is more than five times the total installed power capacity in Central America. Hydropower and thermal-fired power plants generate most of

Table 1. Electric power data for Mexico, Central America, and Caribbean

Country	Population Million	Installed Capacity MW (Year)	Installed Geothermal Capacity MW (1989)	Projected MW (1992)	Annual Generation GWh (Year)	Geothermal Generation GWh
Mexico	78.5	25,000 (1988)	700	750	79,630 (1985)	3160 (1985)
Guatemala	8.2	791 (1985)	0	15	1,775 (1986)	0
Honduras	5.0	558 (1985)	0	0	1,450 (1986)	0
El Salvador	4.4[a]	651 (1989)	95	115	1,670 (1986)	367
Nicaragua	3.4	360 (1989)	70	70	1,118 (1988)	212[c]
Costa Rica	2.6[b]	863 (1988)	0	55	3,243 (1987)	0
Panama	2.2[b]	854 (1985)	0	0	2,585 (1986)	0
Guadeloupe	0.3	d	4.2	4.2	d	d
St. Lucia	0.1	18 (1988)	0	10	d	O
TOTAL		29,077 MWe	869.2	1019.2		

a 1987
b 1986
c Estimated to be 445 GWh (40% of total generation) during 1989 following addition of Momotombo Unit II.
d Data unavailable

Sources:

Brooks, 1987
Dickson and Fanelli, 1988
Trocki and Hanold, 1988
Gustavo Cuellar, pers. comm., 1989 (El Salvador)
Manuel Corrales, pers. comm., 1988 (Costa Rica)
Hernandez-Galan, pers. comm., 1989 (Mexico)
Aloysius Barthelmy. pers. comm., 1988 (St. Lucia)

the electricity in that region. More than 92 percent of the power in Costa Rica, Guatemala, and Honduras is generated by hydro plants (Trocki and Hanold, 1988). Of the three countries producing geothermal electricity in the region for which data are available, geothermal accounts for about 4-40 percent of baseload generation.

GEOTHERMAL DEVELOPMENT STATUS—MARCH, 1989

The current geothermal activities in Central America have been most recently reviewed by DiPippo (1986) and Dickson and Fanelli (1988). The purpose of this section is to briefly summarize geothermal developments in the broader region and provide new data on 1987-89 activities.

Mexico

Mexico is the largest producer of geothermal electricity in the Central American-Caribbean region, and is the third largest geothermal electric producer in the world. The current installed geothermal capacity of 700 MWe is divided between Cerro Prieto and Los Azufres fields (Figure 2).

Cerro Prieto

Cerro Prieto is located in northern Baja California, approximately 30 km south of the U.S. border. The field lies in the southern extension of the Salton Trough, an active rift zone along the boundary of the Pacific and North American Plates. The installed capacity of 620 MWe at Cerro Prieto is generated by nine units which came on line between 1973 and 1986. An additional 110 MWe are scheduled for 1993 (J. L. Hernandez-Galan, personal communication). Cerro Prieto has been intensively studied since the late 1950s and much of this work is summarized in a review paper by Lippman and Manon (1987).

Los Azufres

The second largest developed field in Mexico is Los Azufres, which is located in the TMVB abut 150 km northwest of Mexico City. Production occurs in two widely-spaced reservoirs that are spatially related to a Pleistocene silicic dome complex (Dobson and Mahood, 1985) and within a structurally dissected caldera (Teyssier, 1988).

Los Azufres presently has six 5 MWe units installed, most connected to single wells. A 50 MWe plant has been delivered and will be operational in 1989, bringing the total installed capacity at Los Azufres to 80 MWe. Five additional wellhead generators will be installed during 1989-1990, and longer range plans include the addition of a 37.5 MWe unit in 1993 (Hernandez-Galan, 1988).

Los Humeros

Los Humeros geothermal field is located 200 km east of Mexico City within a multi-collapse caldera. Fifteen wells (of 26 drilled) delineate a reservoir in the central collapse zone (Kruger et al., 1987). Four 5 MWe wellhead generators of the type operating at Los Azufres are scheduled for operation by 1990 (Hernandez-Galan, 1988).

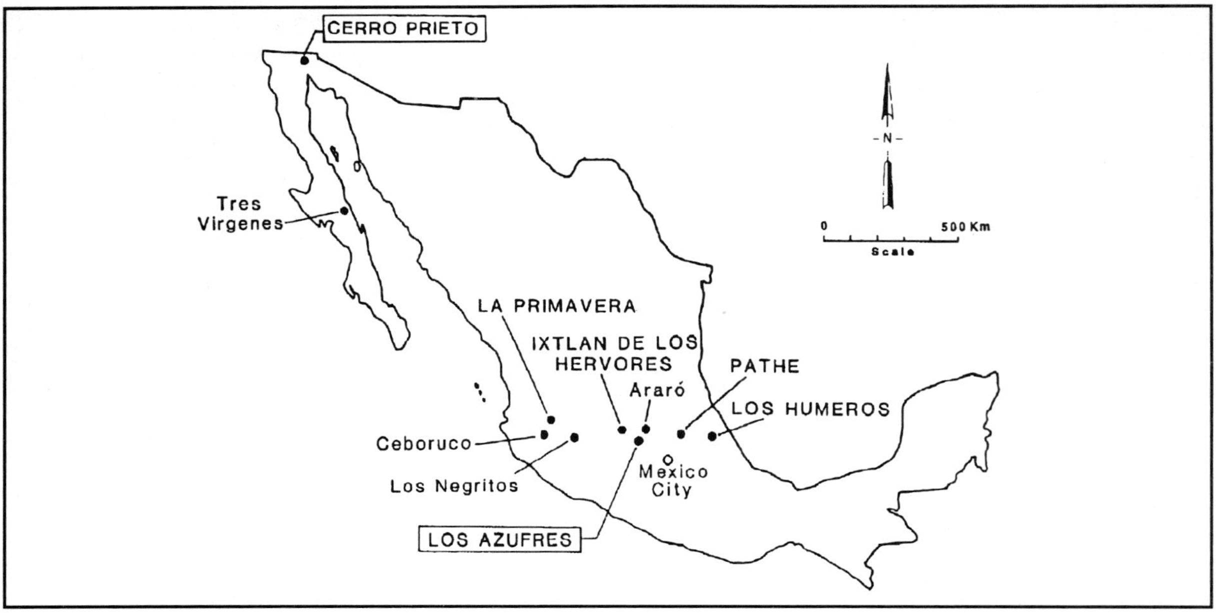

Figure 2. Geothermal plant sites (in boxes), drilled fields with no production (no boxes), and prospects (lower case) in Mexico, modified from DiPippo (1988).

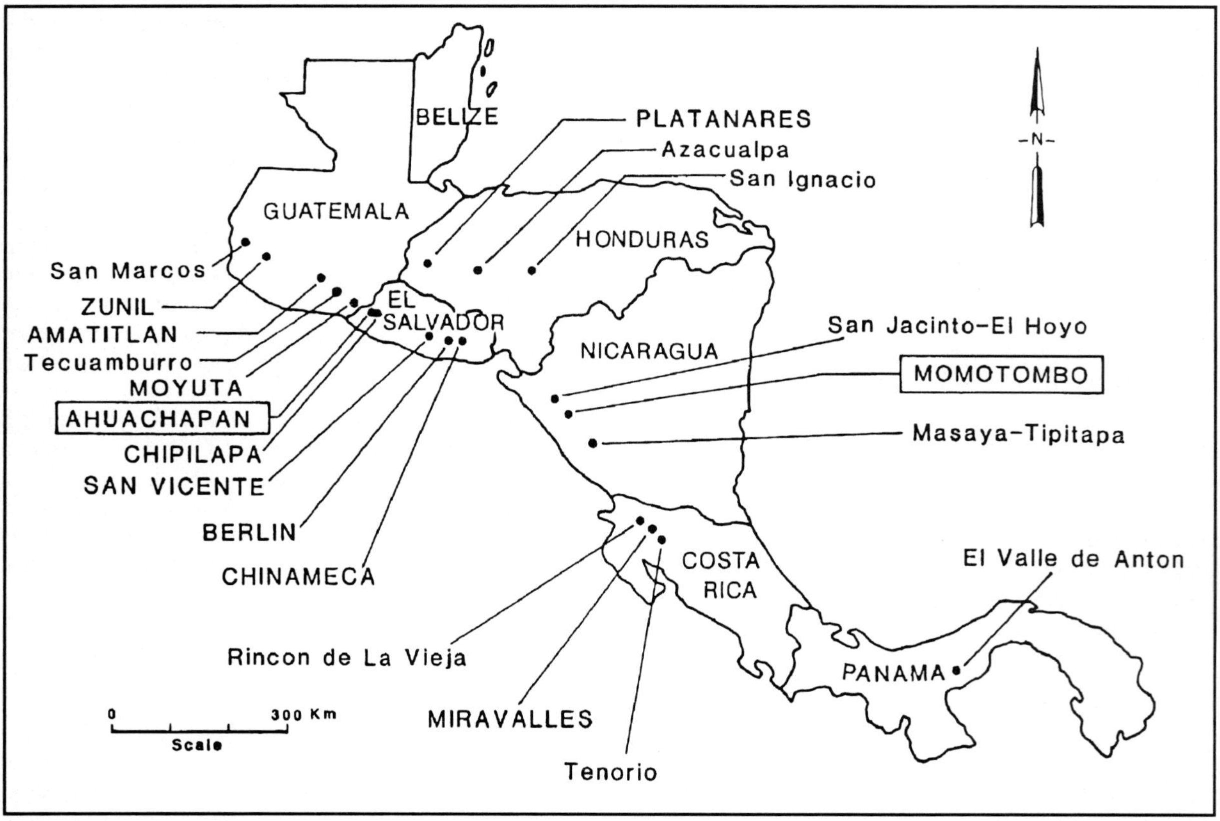

Figure 3. Geothermal plant sites (in boxes), drilled fields with no production (no boxes), and prospects (lower case) in Central America, modified from DiPippo (1988).

La Primavera

La Primavera field, located on the outskirts of Guadalajara, also occurs within a Pleistocene caldera. Eight deep wells have been drilled, reaching temperatures up to 305°C (Hernandez-Galan, 1988). A single 5 MWe wellhead generator is scheduled for 1990 (J. L. Hernandez-Galan, personal communication).

Other Areas

Reconnaissance and prefeasibility studies were carried out in Tres Virgenes (Baja, California), Los Negritos, Ixtlan, Ceboruco, and Araró geothermal fields (Figure 2) (Hernandez-Galan, 1988), but no specific development plans have been published for these areas.

Guatemala

Geothermal activities in Guatemala are described by DiPippo (1986) and Caicedo (this volume). Successful production wells have been drilled at Zunil (Figure 3), but no other high-temperature reservoirs have been confirmed. Temperature gradient drilling is underway at Amatitlan field, and reconnaissance level studies are in progress at Tecuamburro and San Marcos.

Zunil

Zunil is located in western Guatemala. The geothermal field occurs on the southeastern flank of the Cerro Quemado volcanic dome complex and about 4 km northeast of the active Santa Maria volcano. Zunil has five productive wells in a small area (<1 km^2) with a maximum measured temperature of 287°C and a power potential in existing wells of 16 MWe (DiPippo, 1986). The presence of impermeable granite at an average depth of 1,000 m may limit the ultimate potential of the Zunil reservoir (Bethancourt and Dominco, 1982). The Guatemalan electric authority (INDE) plans to install a 15 MWe power plant, and is continuing to study the surrounding area (called Zunil II).

Amatitlan

The Amatitlan geothermal area is located 15 km south of Guatemala City (Figure 3), in the southern part of the Amatitlan Caldera (Wunderman, 1982). The caldera produced numerous silicic domes and tuffs over the last 0.3 m.y. The active, more mafic Pacaya volcano lies astride the southern rim of the caldera. A broad area of hot springs and fumaroles south of Lake Amatitlan has been the focus of detailed investigation. Nine temperature gradient wells have been drilled from 85-550 m deep in a 20 km^2 area southwest of Lake Amatitlan. At least three wells encountered temperature reversals, and the maximum temperature has been 174°C at 500 m (DiPippo, 1986). A tenth gradient hole planned for 700-m depth was started in April, 1989 (Andres Caicedo, personal communication).

Other Areas

Early exploration efforts focused on the Moyuta field near the border with El Salvador. The latest reconnaissance studies have been carried out at Tecuamburro volcano, which is located between Guatemala City and Moyuta. The San Marcos area lies near the western end of the Central American volcanic arc 50 km east of the Mexican border. Fluid geochemistry suggests that a high-temperature reservoir (260°C) may be present in this area (Fournier and Hanshaw, 1986).

Honduras

Geothermal fields identified in Honduras are deep-circulation systems related to extensional faulting (and not to young volcanism), similar to geothermal fields in the Basin and Range Province of the United States. Three sites, Platanares, Azacualpa, and San Ignacio (Figure 3), were considered to have the best potential for high-temperature reservoirs (Flores, this volume). Temperature gradient holes were drilled in all three areas in 1988 (Goff et al., 1987; F. Goff, personal communication), with the most favorable results achieved at Platanares.

Platanares

Platanares geothermal field is located in western Honduras near the Guatemalan border. Three small diameter holes were drilled to a maximum depth of 680 m. One of the wells reached a maximum temperature of 165°C and flowed under artesian pressure (Goff et al., 1987). Higher estimated subsurface temperatures of 225°C may require wells more than 1,200 m deep.

El Salvador

El Salvador is richly endowed with high temperature geothermal fields and is the largest producer of geothermal electricity in Central America. Three power plants with an installed capacity of 95 MWe are presently operating at Ahuachapán (Figure 3), and an additional 20 MWe are scheduled to begin operations in the Berlin and Chipilapa areas by 1990.

Ahuachapán

Ahuachapán is a densely drilled field located in western El Salvador near the Guatemalan border. Over-production and poorly-sited injection wells (Vides-Ramos, 1985) have resulted in pressure and temperature declines and diminished generating capacity. Reinjection was abandoned in 1983. The plant has been running at about 55 MWe (Dickson and Fanelli, 1988).

Chipilapa

Chipilapa field is 4 km east of Ahuachapán, but geophysical evidence suggests the two fields are not connected. CEL, the Salvadoran government utility, plans to begin commercial operation in 1989 with two 5 MWe wellhead generators.

Berlin

The Berlin field is in east-central El Salvador. Five wells have been drilled, and two of these wells will supply steam to two 5 MWe wellhead generators to be installed in 1989. The ultimate size of this field is still unknown, as existing wells have not defined the limits of the productive reservoir.

Other Areas

Exploratory drilling has begun in the Coatepeque area just east of Ahuachapán. Exploratory wells have been drilled at San Vicente and Chinameca fields (Figure 3). Encouraging results were obtained at San Vicente (230°C at 1,300 m depth) but no further work is planned for the near future.

Nicaragua

With the addition of a second 35 MWe power plant at Momotombo field, the geothermal electric capacity of Nicaragua increased to 70 MWe in early 1989. About 40 percent of Nicaragua's baseload power is generated by these plants. Momotombo and several other high-quality geothermal prospects lie along the active volcanic arc, coincident with the Nicaraguan Depression.

Momotombo

Momotombo field (Figure 3) is adjacent to Lake Managua on the southern flank of Momotombo volcano, which last erupted in 1905 (Mooser, Meyer-Abich, and McBirnery, 1958). The first 35 MWe power plant was placed in service in 1983, and the second 35 MWe plant was being tested in February, 1989. Four wells are required for each plant, with two wells on standby, and two injection wells are presently accepting 65 percent of the produced fluid (R. Arcia, personal communication). Production wells feeding the first plant utilize steam from a shallow steam cap, while wells for the second plant tap a deeper two-phase zone with temperatures reaching 350°C. Pressure decline in the Unit 1 wells has been quite low (5 percent over five years; Martinez and Arcia, this volume).

Other Areas

Several high-quality geothermal prospects have been identified in Nicaragua, and reconnaissance work has been carried out at El Hoyo-Monte Galan, San Jacinto-Tizate, and Masaya-Nandaime areas (Martinez and Arcia, this volume).

Costa Rica

Miravalles field is the only geothermal area in Costa Rica under development at this time. Reconnaissance studies have been carried out in the two main volcanic belts, Guanacaste and Central Cordillera, and at the active Arenal volcano, situated between them. Geologic and geophysical surveys are currently underway at Tenorio and Rincon de la Vieja, two Pleistocene volcanic centers which are, respectively, southeast and northwest of Miravalles (Figure 3).

Miravalles

A total of nine deep exploration and development wells have been drilled at Miravalles since 1978. Existing wells are capable of producing 37 MWe (Corrales, 1986). The first 55 MWe power plant, for which construction should begin this year, will require the drilling of an additional five production and four injection wells (A. Mainieri, personal communication). The Costa Rican government utility, the Instituto Costarricense de Electricidad (ICE), has studied the feasibility of initially installing three wellhead units of 5 MWe each, but has not decided to proceed (A. Mainieri, personal communication).

Panama

No high-temperature resource has been discovered to date in Panama. A systematic evaluation of the main hot spring areas was carried out in the late 1970s and early 1980s under the auspices of the Instituto de Recursos Hidraulicos y Electrificacion (IRHE), the government utility. The El Valle de Anton area, located 100 km west of Panama City, was identified in 1985 as the most likely prospect to host a high-temperature reservoir, and a 600 m-deep slim hole has been proposed (Ramirez, 1988). Reconnaissance work conducted

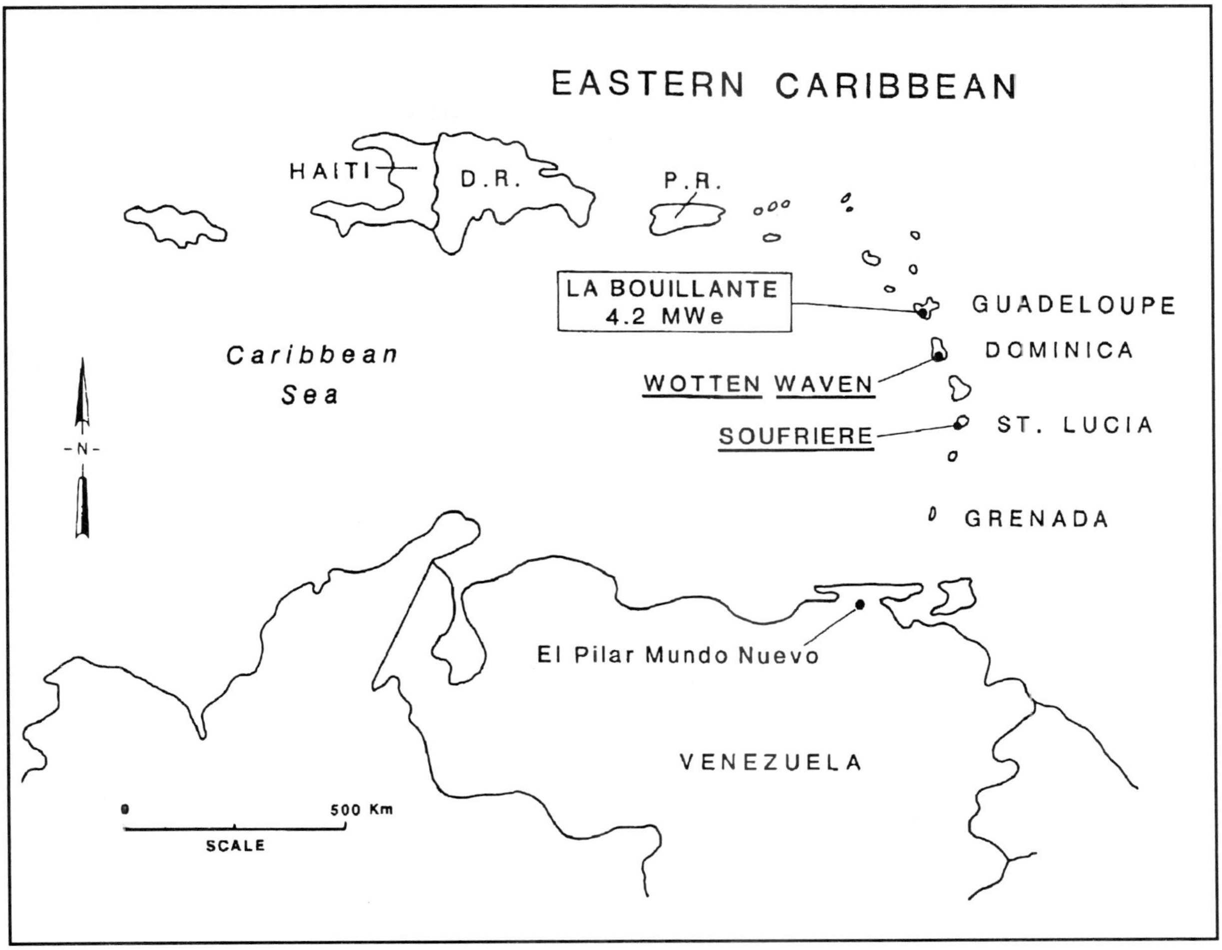

Figure 4. Geothermal plant sites (in box), drilled fields with no production (underlined), and prospects (lower case) in eastern Caribbean and Venezuela.

in four other areas suggests the presence of low-to moderate-temperature reservoirs (Ramirez, 1988).

Eastern Caribbean

Geothermal development in the eastern Caribbean is limited to a small power plant on the island of Guadeloupe (France) and exploration drilling in St. Lucia and Dominica (Figure 4). Reconnaissance studies have been carried out in Grenada, Haiti, the Dominican Republic, and Venezuela.

Guadeloupe

A 4.2 MWe power plant has been operating at La Bouillante field on Guadeloupe since 1984 (DiPippo, 1986). No publications were found describing the status of this plant or future geothermal development plans on Guadeloupe.

St. Lucia

St. Lucia is a small volcanic island in the Lesser Antilles south of Guadeloupe (Figure 4). Geothermal exploration has been concentrated on the Soufriere thermal area within a Pleistocene caldera on the southwest side of the island. Eight shallow wells were drilled in 1975-1976, and three of them produced minor quantities of steam. Following additional geologic and geophysical surveys, two deep wells were drilled in 1987-1988.

The first well was drilled to 2,214 m and was not productive. The second well was drilled to 1,413 m and produced about 32 T/hr of superheated steam at the end of a 10-day flow test (A. Barthelmy, personal communication). Non-condensible gases accounted for about 20 percent by weight of steam. The St. Lucian Government Energy Ministry plans to install a 10 MWe power plant within the next few years.

Dominica

Exploration work has been carried out at Soufriere, Boiling Lake, and Wotten Waven prospects on the island of Dominica. Moderately high temperatures (200-230°C) were calculated for a deep reservoir at Wotten Waven (DeMange, Iundt, and Puvilland, 1985). Ten to 15 wells averaging 500 m deep were drilled at Soufriere and Wotten Waven with encouraging results (DiPippo, 1985).

Venezuela

The El Pilar-Mundo Nuevo prospect in northeastern Venezuela contains hot springs and fumaroles discharging along a major transform fault (Almandoz and Rojas, 1988). The absence of young volcanos in the area suggests that these may represent a deep-circulation system. No specific development plans have been published.

CONCLUSIONS

Geothermal electric production in the Central American-Caribbean region (including Mexico) represents about 15 percent of the installed world geothermal capacity, and is growing at around 6 percent per year. Significant geothermal capacity coming on-line in the next few years will contribute to diversification of the region's electricity base and will continue to save millions of dollars in reduced petroleum imports.

REFERENCES

Almandoz, A., and J. Rojas, 1988, Geothermal prospects in the central region of Sucre State, Venezuela: Geothermics, v. 17, p. 369-375.

Bethancourt, H. R., and E. Dominco, 1982, Characteristics of the Zunil geothermal field (Western Guatemala): Geothermal Resources Council Transactions, v. 6, p. 241-244.

Brooks, J., ed., 1987, 1988 South American handbook: Bath, England, Rand McNally, Trade and Travel Publications Ltd., 1340p.

Carr, M. J., and R. E. Stoiber, 1988, Volcanism, *in* G. Dengo and J. E. Case, eds., The Caribbean region: Boulder, Colorado, Geological Society of America, The Geology of North America, v. H, in press.

Caicedo, A. A., this volume, Current status of geothermal activities in Guatemala.

Corrales, M. F., 1986, Costa Rica, country update report, *in* U.N. workshop on the development and exploitation of geothermal energy in developing countries: Reykjavik, TDCD/NRED/CTP/7, p. 6.

de Boer, J. Z., M. J. Defant, R. H. Stewart, J. F. Restropo, L. F. Clark, and A. H. Ramirez, 1988, Quaternary calc-alkaline volcanism in western Panama: Regional variation and implication for the plate tectonic framework: Journal of South American Earth Science, v. 1, no. 3, p. 275-293.

DeMange, J., F. Iundt, and P. Puvilland, 1985, Geothermal field model of Wotten Waven Island of Dominica Lesser Antilles: Geothermal Resources Council Transactions: v. 9, p. 409-415.

Dickson, M. H., and M. Fanelli, 1988, Geothermal R&D in developing countries: Africa, Asia and the Americas: Geothermics, v. 17, p. 815-877.

DiPippo, R., 1985, Geothermal electric power, the state of the world—1985, *in* 1985 International symposium on geothermal energy: Sacramento, Cal Central Press, Geothermal Resources Council, p. 3-18.

DiPippo, R., 1986, Geothermal energy, developments in Central America: Geothermal Resources Council Bulletin, v. 15, no. 10, p.3-14.

DiPippo, R., 1988, International developments in geothermal power production: Geothermal Resources Council Bulletin, v. 17, no. 5, p.8-19.

Dobson, P. F., and G. A. Mahood, 1985, Volcanic stratigraphy of the Los Azufres geothermal area, Mexico: Journal of Volcanology and Geothermal Research, v. 25, p. 273-287.

Flores, W., this volume, Geothermal prefeasibility studies in Honduras.

Fournier, R. D., and B. Hanshaw, 1986, Geothermical evolution of the geothermal resources in thermal waters in the San Marcos region, Guatemala: Applied Geochemistry, v. 1, p.189-197.

Goff, R., L. Shevenell, S. Kelkar, D. Smith, J. Meert, et al., 1987, Stratigraphy, temperature profiles, and flow test data from PLTG-1 and PLTG-2 coreholes, Platanares geothermal system, Honduras: Geothermal Resources Council Transactions, v. 11, p.253-259.

Hernandez-Galan, J. L., 1988, Small geothermal schemes: The Mexican experience: Geothermics, v. 17, p. 303-309.

Kruger, P., J. Ortiz, C. Miranda, and M. Gallardo, 1987, Response of the Los Azufres geothermal field to four years of 25 MW wellhead generation: Stanford, Stanford University, Proceedings, 12th Workshop on Geothermal Reservoir Engineering, p. 181-187.

Lippman, M. J., and A. Manon, 1987, The Cerro Prieto geothermal field: Geothermal Science and Technology, v. 1, no. 1, p. 1-38.

Martinez T., and R. Arcia L., this volume, Momotombo geothermal field.

Mooser, F., H. Meyer-Abich, and A. R. McBirnery, 1958, Catalogue of the active volcanoes of the world including solfatara fields; Part VI Central America: Naples, Italy, International Volcanological Association, p. 123-126.

Ramirez, A. H., 1988, Present status of geothermal reconnaissance studies in the Republic of Panama: Geothermics, v. 17, p. 355-367.

Teyssier, C., 1988, Los Azufres (Mexico) half caldera: A model of interaction between caldera and graben structures (abs.): Denver, Colorado, Geological Society of America Annual Meeting Transactions, p. A271.

Trocki, L. K., and R. J. Hanold, 1988, Electricity production in Central America—Geothermal versus other options: Geothermal Resources Council Bulletin, v. 17, no. 9, p. 3-10.

Vides-Ramos, A., 1985, Ahuachapán, El Salvador, field management: Geothermal Resources Council Transactions, International Volume, p. 397-404.

Wunderman, R. L., 1982, Amatitlan, an active resurgent caldera immediately south of Guatemala City, Guatemala: Masters Thesis, Michigan Tech. University, 192p.

Circum-Pacific Council for Energy and Mineral Resources
Earth Science Series, Vol. 16
R. L. Miller, G. Escalante, J. A. Reinemund, and M. J. Bergin (Eds.)
Energy and Mineral Potential of the Central American-Caribbean Regions
Springer-Verlag, Berlin Heidelberg 1995

SEISMOLOGICAL STUDIES AT THE MIRAVALLES GEOTHERMAL PROJECT

Rafael Barquero
Sección de Sismología e Ing. Sismica, Departamento de Geología, Instituto Costarricense de Electricidad
Apdo. 10032, San José, Costa Rica

ABSTRACT

The Miravalles Geothermal Project (MPG) is located in the Guanacaste Mountain Range within the Guayabo Caldera in northwest Costa Rica. The Guanacaste Range is a chain of andesitic Quaternary stratovolcanos aligned northwest, composed of pyroclastic rocks, lava, and fluvio-lacustrine deposits. Glowing avalanche deposits (ignimbrites) formed gently-sloping plateaus on both sides of the mountain range. This area is under constant regional stress derived from the subduction of the Cocos Plate under the Caribbean Plate and the regional uplift of the volcanic arc, resulting in a complex system of faults with main orientations northwest, northeast, and north.

The Costarrican Institute of Electricity (ICE) through its Department of Geology has undertaken seismological and volcanological studies in the northwest part of the country since 1974. From 1974 to 1978 a ten-station seismic network was in operation in the Guanacaste region. A local network was set up during 1977-1978 at the MGP in order to study, in detail, the microseismicity in this area.

Microseismic observations at MGP have been successful in determining the seismicity, stress fields, seismic alignments, and active or potentially active faults within the area of the geothermal project.

INTRODUCTION

Given the great importance of micro-earthquakes as a fundamental element in determining seismic risks associated with the generation of electrical energy, ICE's Department of Geology has completed a series of seismological studies in the area of the Miravalles Geothermal Project (MGP) from 1977 to 1988. The principal objectives have been as follows:

1. To offer seismologic information for future seismic studies in the region;

2. To furnish information relating to the location and levels of activity of the microearthquakes in the region; and

3. To determine the alignment and activities or potential activity of faults within the geothermal project zone.

The study area is located in the northwest part of Costa Rica, within the coordinates of latitude 10°30'N and 10°55'N and longitude 85°00'W and 85°20'W (Figure 1).

GEOLOGIC-TECTONIC FRAMEWORK

The Miravalles Geothermal Project is located in the Guanacaste Range, on the flanks of the Miravalles volcano. This volcanic chain forms part of the Interior Arc, resulting from active subduction of the Cocos Plate beneath the Caribbean Plate. This phenomenon also causes a seismic zone which descends to the northeast from the oceanic trench in the Pacific.

The Guanacaste Range is a chain of andesitic Quaternary stratovolcanos that are aligned in a northwest-southeast direction, and are composed primarily of pyroclastic rocks, lavas, and fluvio-lacustrine deposits. Glowing avalanche deposits (ignimbrites) form plateaus with gently dipping slopes. These extend outward on both sides of the range (Alvarado, 1984). The whole region is severely stressed due to the subduction process, and the rise of the magmatic arc, in

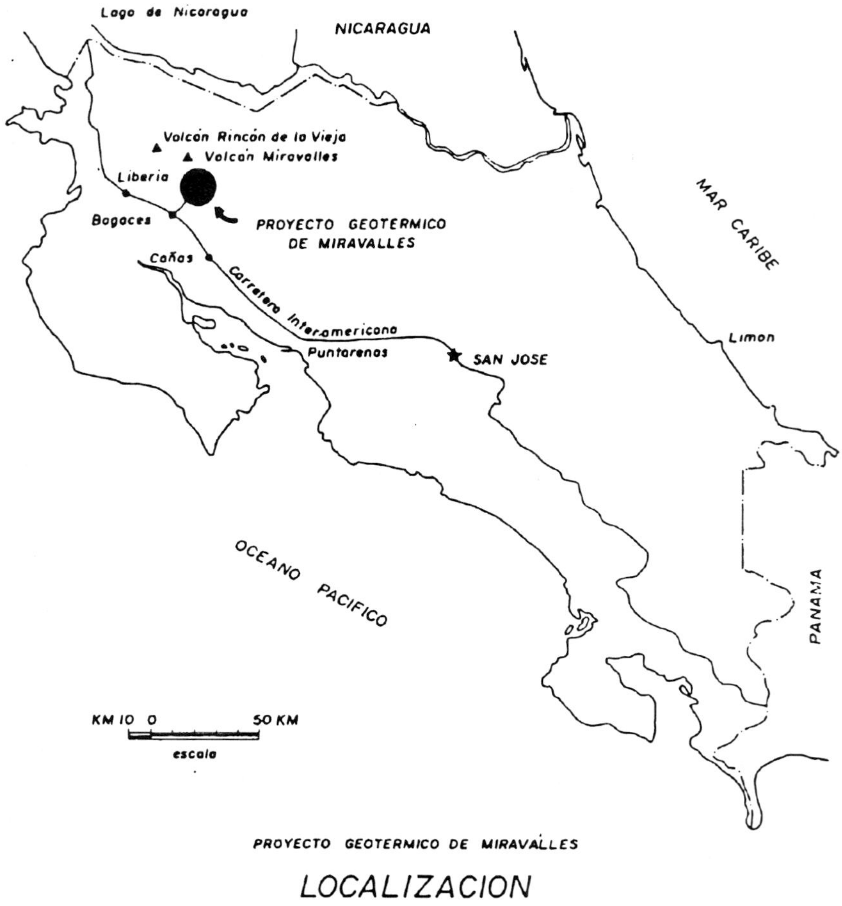

Figure 1. Map showing location of the Miravalles geothermal project area, nearby volcanos, the Interamerican Highway, and selected population centers.

particular, gave rise to a complex system of faults oriented mainly northwest-southeast, northeast-southwest, and north-south.

The Miravalles volcano is a complex stratovolcano in whose summit are at least five eruptive centers and a thick recent lava flow on the southwest flank. The volcano has not erupted in historic time, but solfatara activity and hot springs testify to its latent energy. The local tectonic structures are mainly due to the collapse of Guayabo Caldera and to the tectonic stresses with alignments northwest-southeast, north-east-southwest, and north-south.

BACKGROUND

ICE, concerned for the future stability of its energy projects, initiated seismologic and volcanic studies in the northwest region of the country beginning in 1974. The seismological network in the area of Arenal Volcano was installed in May 1974 (Figure 2), with signals from ten seismic stations collected at a central location. This network provided good coverage of the region. Around 3,500 seismic events were recorded from 1974 through 1978, some of which originated in the area currently being studied. Also, other studies were performed that permitted a pattern to be recognized for this region of the country (Matumoto et al., 1977). In October 1977, a local short-term seismic network was installed for the Miravalles Geothermal Project. Actually, the planning of a permanent local seismic network for the geothermal project is in progress. It will be in operation in 1991.

The seismologic studies in the Miravalles zones were resumed in July of 1983 with the installation of a portable seismological network of four stations. This functioned at various times (August-November, 1983; April-November, 1984; and December 1985-September, 1986).

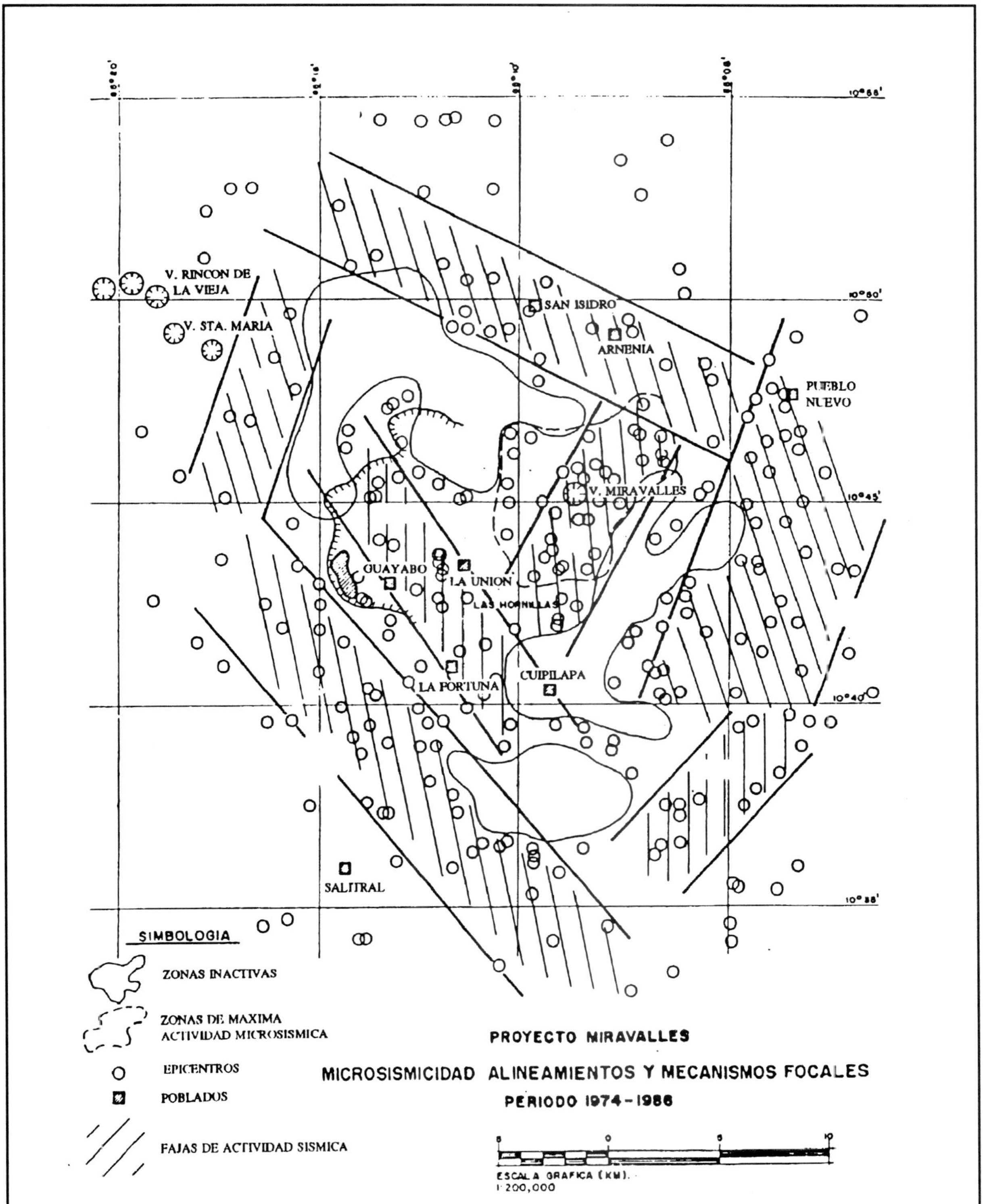

Figure 2. Map showing the locations of seismic stations of the Arenal seismic network of 1974-78.

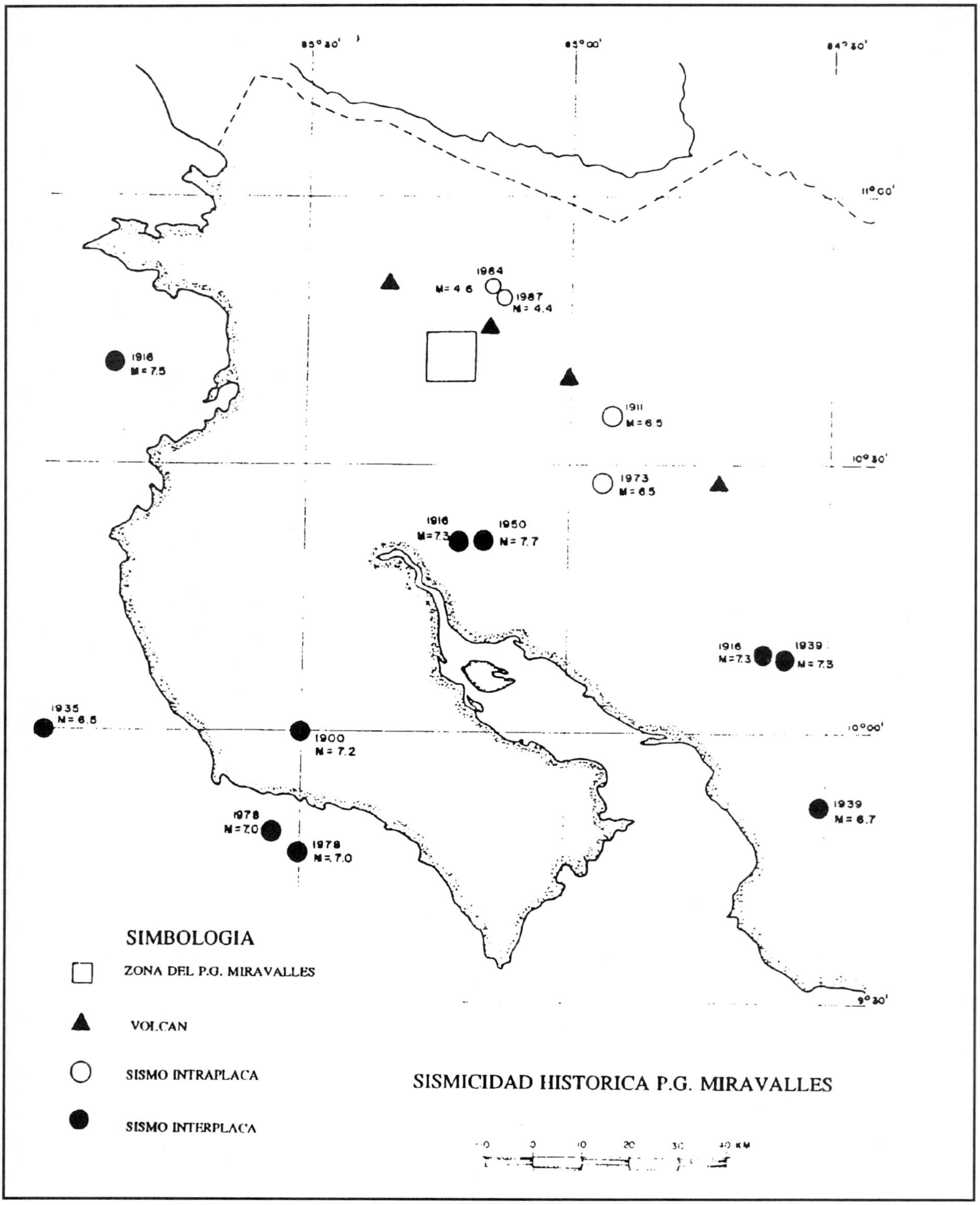

Figure 3. Map showing the location of the Miravalles project area (P.G.) and the epicenters of historlcal earthquakes, whose revival could threaten the Miravelles geothermal project. Distinction is made between earthquakes resulting from stresses originating within the Caribbean Plate and those from stresses originating outside the Plate. Date and magnitude (M-Richter Scale) of each quake is also shown.

SEISMIC HISTORY

After a thorough search of the published and unpublished information on seismicity of the region, it was decided to compile a list of the principal earthquakes that could have an effect on the Miravalles Geothermal Project (Figure 3), i.e., those which have epicenters within a 100 km radius of the geothermal area.

The seismic activity which has affected the northwest part of the country has two principal sources:

1. Seismic activity originating through the process of active subduction involving the Cocos and Caribbean Plates (inter-plate seismicity). Quakes with this origin reach magnitudes as high as 8 on the Richter Scale, with focal depths anywhere between 20 and 200 km. These are the most frequent quakes in the region. For strong earthquakes the probable period of recurrence has been calculated to be 21.6 $\pm$ 11.8 years.

2. Seismic activity directly related to local faults (intra-plate seismicity) is manifested by earthquakes of moderate magnitude, with a maximum of 6.5, and with focal depths less than 20 km. These present the greatest risk to the geothermal project.

SPACIAL SEISMICITY

The spacial distribution of the seismic activity within a specific region is, without doubt, one of the more important aspects of seismology. It involves the determination of (a) the geologic characteristics of the area, (b) the probable sources of seismic generation, and (c) zones of major and minor activity. All of this serves as a basis for further studies of seismic risks.

Using the distribution of the shallow quakes in the area of the Miravalles Geothermal Project as a base, various structural lineaments have been determined. They seem to have been inactive throughout the study period.

SEISMICITY IN TIME

The seismologic studies completed by ICE have been able to locate about 250 seismic foci in the Miravalles Geothermal Project area. In general, the level of microseismicity in the area for the entire period was relatively low, with an average of two to six daily events. Figure 4 shows the frequency by month over the recording periods. The seismological information obtained up to now is of great importance

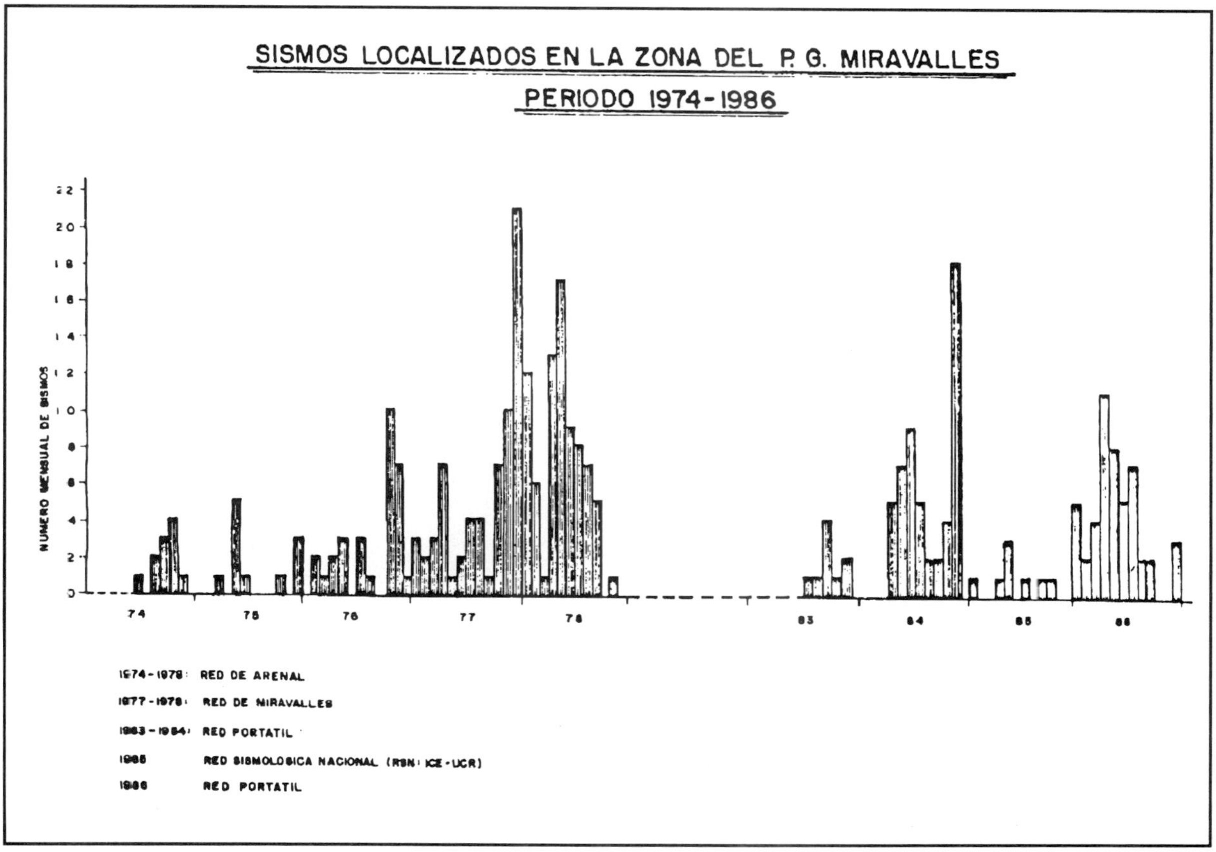

Figure 4. Chart showing numbers of mircoseisms within the area of the Miravalles project for the periods 1974-1978 and 1983-86.

because it will serve as a record for comparison with the seismicity after exploitation of the Miravalles geothermal field begins, including reinjection of the spent fluids into the reservoir. This could eventually cause a hydrodynamic imbalance and thus have repercussions on the seismic behavior of the region.

CONCLUSIONS

The principal seismic sources in the northwest region of Costa Rica are related to the subduction process of the Cocos Plate below the Caribbean Plate and the local faults. The seismic history for the northwest region includes at the very least 12 earthquakes in this century, with the ones originating from the local faults being the most destructive. Examining the seismic records completed from 1977 to 1986, about 250 shallow microquakes of tectonic origin were identified. Their distribution defines lineaments that coincide with known faults in the region. Measurements to date show that the level of microseismic activity is relatively low, with an average of two to six quakes per day.

REFERENCES

Alvarado, F., 1984, Geología estructural y tectonica al sus del Valle Central de Costa Rica (Tarbaca): Revista Geológica de América Central, v. 2, p. 96.

Matumoto, T., M. Ohtake, G. Latham, and J. Umaña, 1977, Crustal structure in southern Central America: Seismological Society of America Bulletin, v. 67, no. 1, p. 121-134.

Circum-Pacific Council for Energy and Mineral Resources
Earth Science Series, Vol. 16
R. L. Miller, G. Escalante, J. A. Reinemund, and M. J. Bergin (Eds.)
Energy and Mineral Potential of the Central American-Caribbean Regions
Springer-Verlag, Berlin Heidelberg 1995

GEOPHYSICAL EXPLORATION OF THE LAS PAILAS GEOTHERMAL FIELD, RINCON DE LA VIEJA, GUANACASTE PROVINCE, COSTA RICA

Arturo Quesada Muñoz
Escuela de Física, Universidad de El Salvador, San Salvador, El Salvador

Germán Leandro
Oficina de Geofísica Aplicada, Instituto Costarricense de Electricidad, P. O. Box 10032-1000, San José, Costa Rica
Escuela Centroamericana de Geología, Universidad de Costa Rica, San José, Costa Rica

Luis Diego Morales
Centro de Investigaciones Geofisicas y Escuela Centroamericana de Geología, Universidad de Costa Rica, San José, Costa Rica

ABSTRACT

Geological mapping and geophysical surveys (electrical, magnetometric, and gravimetric methods) were undertaken in the area of the Las Pailas geothermal field at the southern flank of the volcano Rincón de la Vieja, Guanacaste Volcanic Cordillera, in order to obtain a preliminary model of this field. Four rock units of Quaternary age were mapped, namely an ignimbrite unit, an andesitic lava unit, a lahar unit composed of andesitic blocks, and a unit of alluvium and colluvium. A detailed gravimetric survey indicates a positive increase of Bouguer anomalies to the northwest which may indicate a dip of the local basement to the southwest. The local magnetometric map shows a concentration of negative anomalies (up to -600 gammas) within the central region. This coincides with a higher concentration of surface hydrothermal features in this area. The rest of the mapped area is also dominated by negative anomalies which may correspond to a thick hydrothermally altered igneous body at depth.

The resulting model is composed of four layers: (1) a shallow layer, 100-300 m in thickness, composed of andesitic lavas, lahars, and pyroclastic flows with resistivities ranging between 20° and 200°, ±20% ohm-m; (2) a conductive cap-layer (1.5-15 ohm-m) that is probably composed of hydrothermally altered lavas and pyroclastic flows (the lower part of this layer at a depth of 400 m to 800 m has high temperatures and vertical convection channels); (3) a high resistivity layer (50-500 ohm-m) underlying the cap layer and believed to be the geothermal reservoir (this layer may consist of highly fractured volcanic and volcano-related sedimentary rocks); (4) a local gravimetric basement (ρ=2.5-2.7 g/cm^3) underlying the high resistivity layer (the top of this basement is at depths ranging from 1.4 to 2.0 km). Based on a comparison of this model with the Miravalles geothermal field (located 20 km to the east), the data obtained from Las Pailas indicates that this field has a high enthalpy (heat content) with temperatures probably ranging between 200°C and 250°C.

INTRODUCTION

Las Pailas geothermal field comprises an area of approximately 2 km^2, containing surface hydrothermal features such as thermal springs, mud volcanos, boiling mud pits, fumaroles, solfataric lagoons, and highly altered rocks. Surface measurements made in February 1988 at some of these hydrothermal features indicated temperatures between 37.8°C and 98.7°C and pH values ranging from 2.5 to 5.5 ±0.5. The field is located at the base of the south slope of the Rincón de

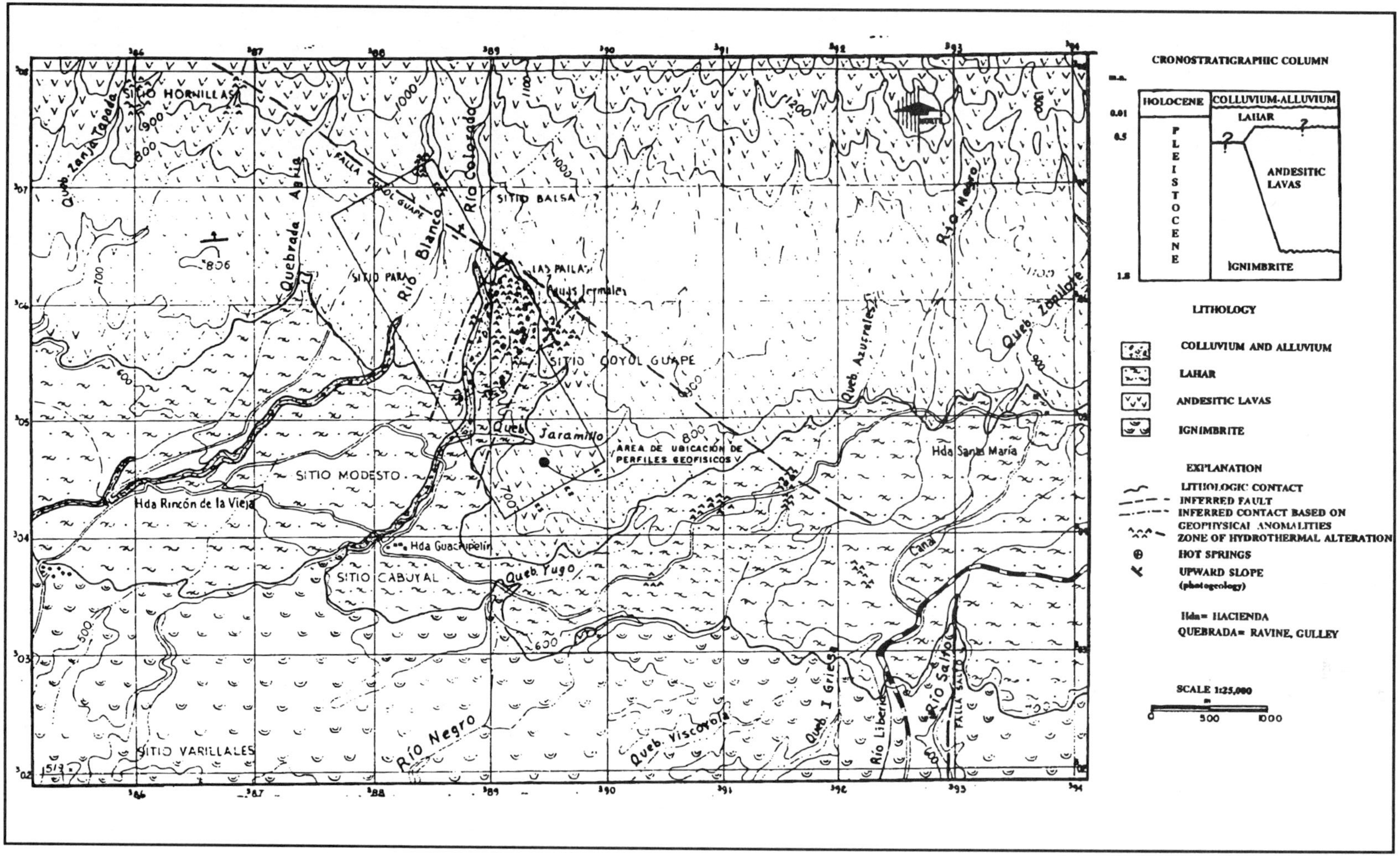

Figure 1. Geologic map of the Las Pailas geothermal area and surroundings. Hda = hacienda, Queb. = Quebrado = ravine or gully, ubicacion = location, sito = place, falla = fault.

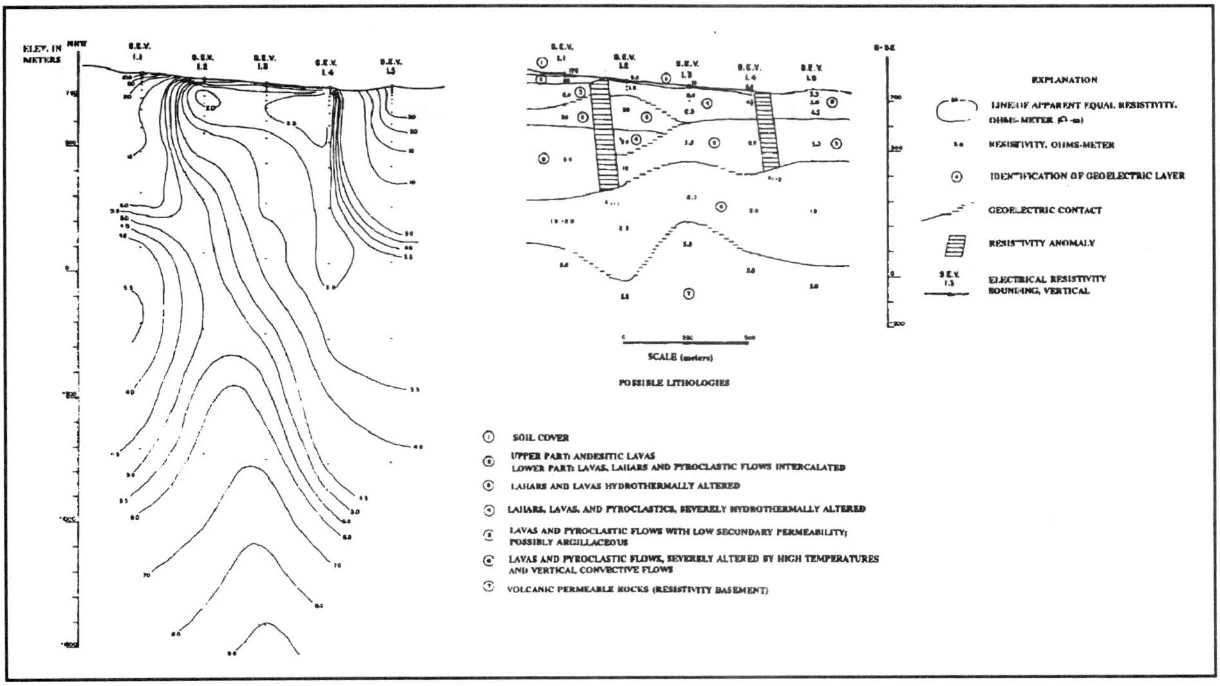

Figure 2. Local map of magnetic anomalies. FMl-FM2, fault inferred from magnetic data.

la Vieja volcano, the only active volcano in the Guanacaste Volcanic Range (Cordillera Volcánica de Guanacaste), at a distance of 6.5 km from the craters and 19 km north-northeast of Liberia (the capital city of the Guanacaste Province). It is within the 1:50,000 Curubandé topographic sheet published by the Instituto Geográfico Nacional (National Geographic Institute). The region is characterized by a humid and hot climate, with an annual mean rainfall of 2,300-2,800 mm and a temperature range of 18°C-24°C. The most important objectives of the present study are (a) to prepare a general geologic map of the Las Pailas area and its surrounding country and (b) to obtain a preliminary model of the geothermal field using geophysical prospecting methods including electrical resistivity, magnetometry, and gravimetry.

GEOLOGY

The Guanacaste Volcanic Range consists of a chain of volcanic cones, which were built on an ignimbrite field of Plio-Pleistocene age (Thorpe et al., 1985). These stratovolcanos, which began forming toward the end of the Pliocene, continue to be recently active. They are aligned along an axis (Carr, 1984), which parallels the Middle American Trench. Two important aspects that characterize this range are the surface hydrothermal features, and the presence of calderas along the southern flank of the range.

Hydrothermal activity is related to recent silicic volcanism. Magmatic intrusions that fed caldera-forming eruptions have provided heat for geothermal systems, and caldera eruptions have produced permeable fracture zones for hydrothermal convection (Granados, 1979; Chiesa, Corella, and Mora, 1987).

Four informal lithostratigraphic units of Quaternary age crop out in the area studied (Figure 1): (1) a unit of whitish ignimbrite, more than 100 m thick, slightly welded, with a matrix consisting of large amounts of biotite, quartz, and plagioclase and some pyroxene; (2) a unit of andesitic porphyritic lavas, gray, with plagioclase and pyroxene phenocrysts in an aphanitic matrix; (3) a unit of lahar composed of andesitic blocks imbedded in an argillaceous matrix; and (4) a colluvial and alluvial unit composed of subrounded volcaniclastic clasts. This last unit is of small areal extent and is mainly exposed in the beds of some of the rivers.

The Coyol-Guape fault (northwest-southeast) and the Salto fault (north-south) have been inferred. Possible faults F1 and F2, on the other hand, have been deduced from geophysical anomalies (Figure 1).

ELECTRIC RESISTIVITY

Fifteen vertical electric soundings (V.E.S.) were performed along the E1, E2, and E3 profiles using a Schlumberger instrument. Location of the lines is shown on Figure 2. The measurements were interpreted using the asymptote method and the computer program RHO INV (Gosh filter with 9 coefficients). The isoresistive pseudo-section and the interpreted geoelectric layers of the E1 profile are shown in Figure 3.

An analysis of the results of the three profiles leads to the postulation of a three-layer model: (1) a shallow layer, 100 m to 300 m in thickness, with a resistivity between 20 and 200 ohm-m, composed of soil, lahars, andesitic lavas, and pyroclastic flows; (2) an underlying caprock layer to the geothermal reservoir, with its base at depths of 400 m to 800 m (this layer is probably composed of lava and pyroclastic flows, both hydrothermally altered, and possibly with low permeability because of argillaceous composition; the lower part of this layer is electrically very conductive (1.5-2.5 ohm-m) and is probably composed of hydrothermally altered lavas and pyroclastic flows, which were transformed by high temperatures and by vertical convection flows); and (3) a high resistivity basement (50-500 ohm-m) which may be associated with the geothermal reservoir (as previously noted, the top of this unit is at depths between 400 m and 800 m; the layer is probably composed of volcanic and volcano-sedimentary rocks, which have good secondary permeability possibly due to intense fracturing).

This geoelectric model is similar to that of the Miravalles geothermal field, located 20 km to the east, which has a geothermal field of high enthalpy with temperatures ranging from 200°C to 250°C.

MAGNETICS

The magnetic prospecting was done by measuring the magnitude of the magnetic field along 10 profiles, which provided the data to construct a magnetic anomaly map (Figure 3). This map indicates the presence of a wide zone of negative anomalies (up to -600 gammas) in the central part of the area. This zone coincides with the highest concentrations of surface hydrothermal features. On the rest of the map other negative anomalies dominate, probably due to the presence of a thick sequence of hydrothermally altered volcanics. Elsewhere positive anomalies are associated with surficial andesitic lava flows composed of unaltered or slightly altered rocks. Two major faults (FM1 and FM2) with a north northeast-south southwest strike have been assumed. They delimit the zone of negative anomalies.

GRAVIMETRY

The regional map of Bouguer anomalies (ρ=2.4 g/cm^3) contains zones with negative values up to -6.0 mgal. These can be accounted for either by thick bodies of ignimbrites or by fractures and feeder dikes. The positive anomalies are associated with lava flows (Leandro, 1985). The most striking regional feature pointed out by Brown, Rymer, and Thorpe (1987) is an elongated negative anomaly (-15 mgal) extending northwest-southeast across the crest of the Rincón de la Vieja volcano.

The gravimetric measurements were made along lines E1, E2, E3, and LC from stations spaced every 50 m. The local gravimetric map (Figure 4) of Bouguer anomalies (ρ= 2.3 g/cm^3) shows positive values increasing in a northeast direction which may indicate a dip of the local basement to the southwest. On the other hand, this gravimetric map fits the topography and thus may be principally influenced by the surficial lava flows. In the central part, a strong variation of the isogal curves to the northeast may be accounted for by two postulated parallel faults (FG1 and FG2) which strike north northeast-south southwest. These presumed faults may be the same as the FM1 and FM2 faults inferred from the magnetometric survey. They have been interpreted and shown as the F1 and F2 faults on Figure 5.

The E1 profile was modeled using the method of Talwani (Talwani, Worzel, and Landisman, 1959) by initially taking into consideration the model from the geoelectric survey (Figure 3). The model for the gravity profile E1 is interpreted to reflect a local gravimetric basement, which has densities of around 2.5 and 2.7 g/cm^3 and whose top is located at depths between 1.4 and 2.0 km below the ground surface.

INTEGRATED INTERPRETATION

On the basis of the local magnetometric map, the Bouguer anomalies map, and the resistivity map the possible F1 and F2 faults are assumed. These trend north northeast-south southwest, and delineate the zone within which the surface geothermal manifestations at Las Pailas are concentrated (Figure 5).

The integrated model obtained on the basis of all the geophysical data points to the presence of four layers:

1. A shallow layer, 100-300 m thick and resistivity values of 20 to 200 ±20% ohm-m, composed of andesitic lavas, lahars, and pyroclastic flows.

2. A conductive cap layer (1.5-15 ohm-m in the upper part and 1.5 to 2.5 ohm-m in the lower part) composed of andesitic lavas and pyroclastic flows, highly altered, may have a low permeability. Its bottom is located at depths between 400 m and 800 m.

3. A resistive basement (50 to 500 ohm-m) probably composed of volcanic and volcano-sedimentary rocks with possible secondary permeability due to intense fracturing. This layer is associated with the presumed geothermal reservoir. Its top is located at a depth of between 400 m and 800 m.

4. A local gravimetric basement (ρ=2.5-2.7 g/cm^3) underlying the resistivity basement at depths between 1.4-2.0 km.

A comparison of this model with the Miravalles geothermal field, located 20 km to the east, suggests a high enthalpy field for Las Pailas with temperatures between 200°C and 250°C. This field represents a good prospect for the development of the use of geothermal energy in Costa Rica. Therefore, a more detailed study of the area should follow, including geochemical testing of surface waters to determine subsurface temperatures.

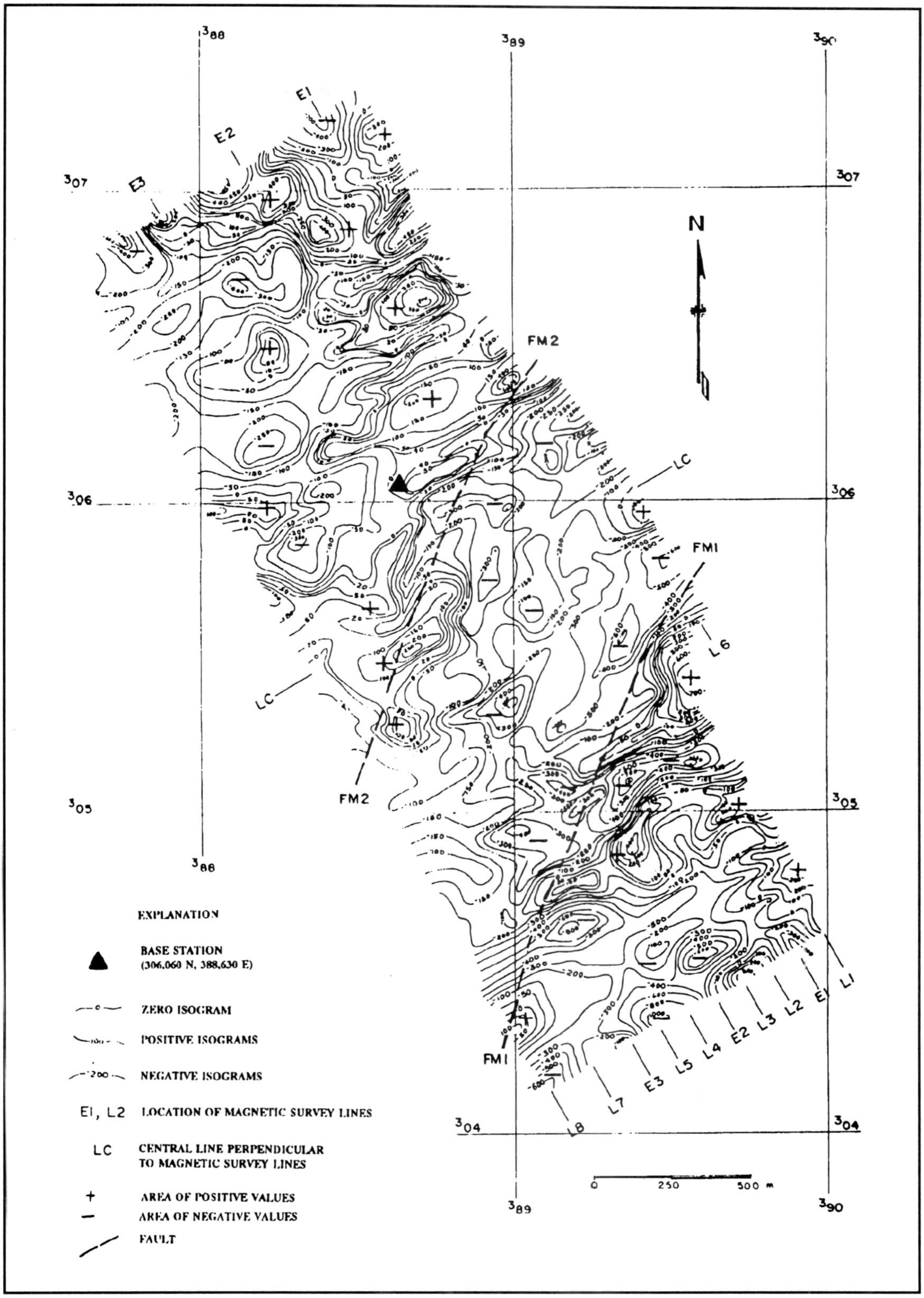

Figure 3. Geoelectric profile E-1.

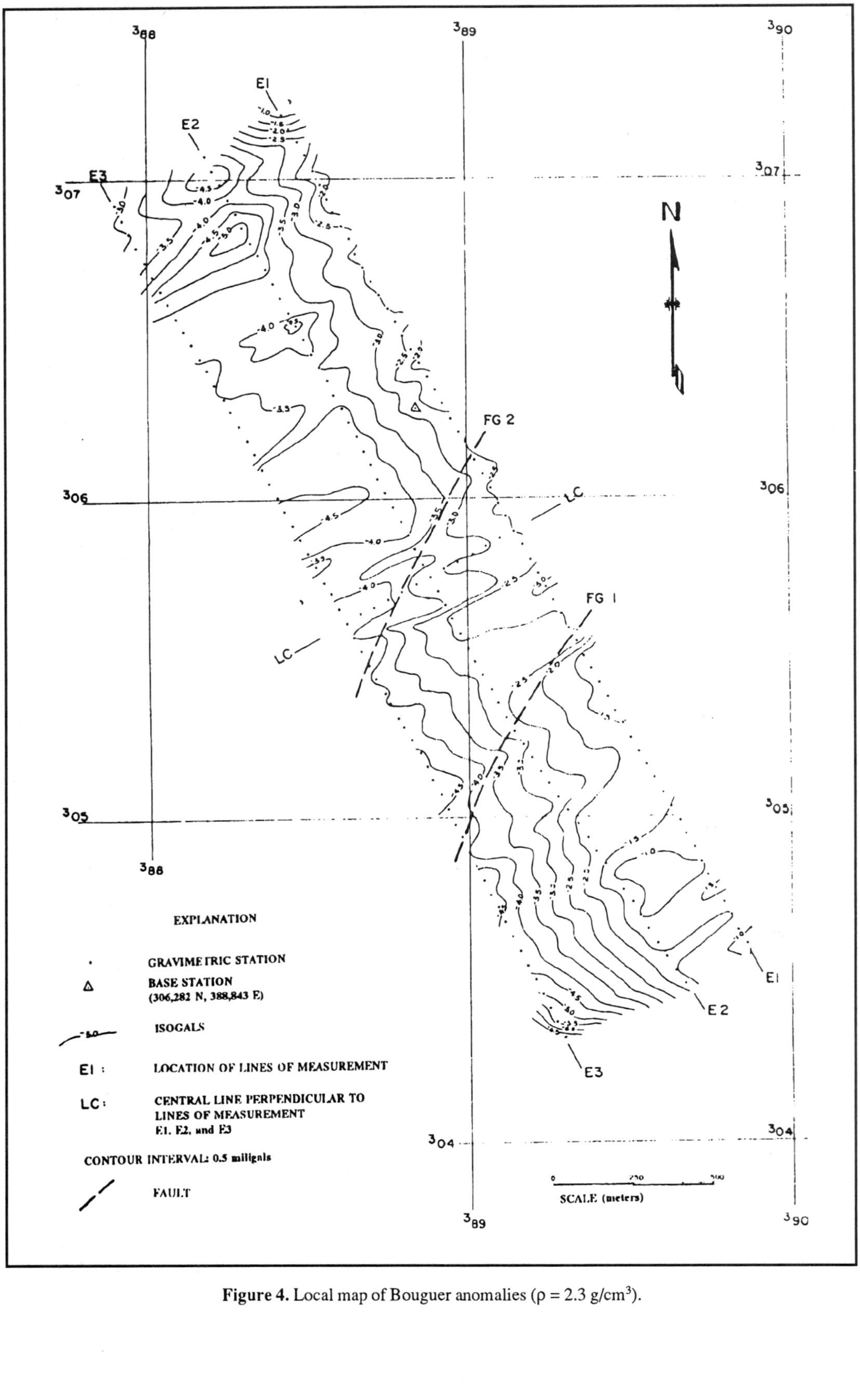

Figure 4. Local map of Bouguer anomalies ($\rho = 2.3$ g/cm^3).

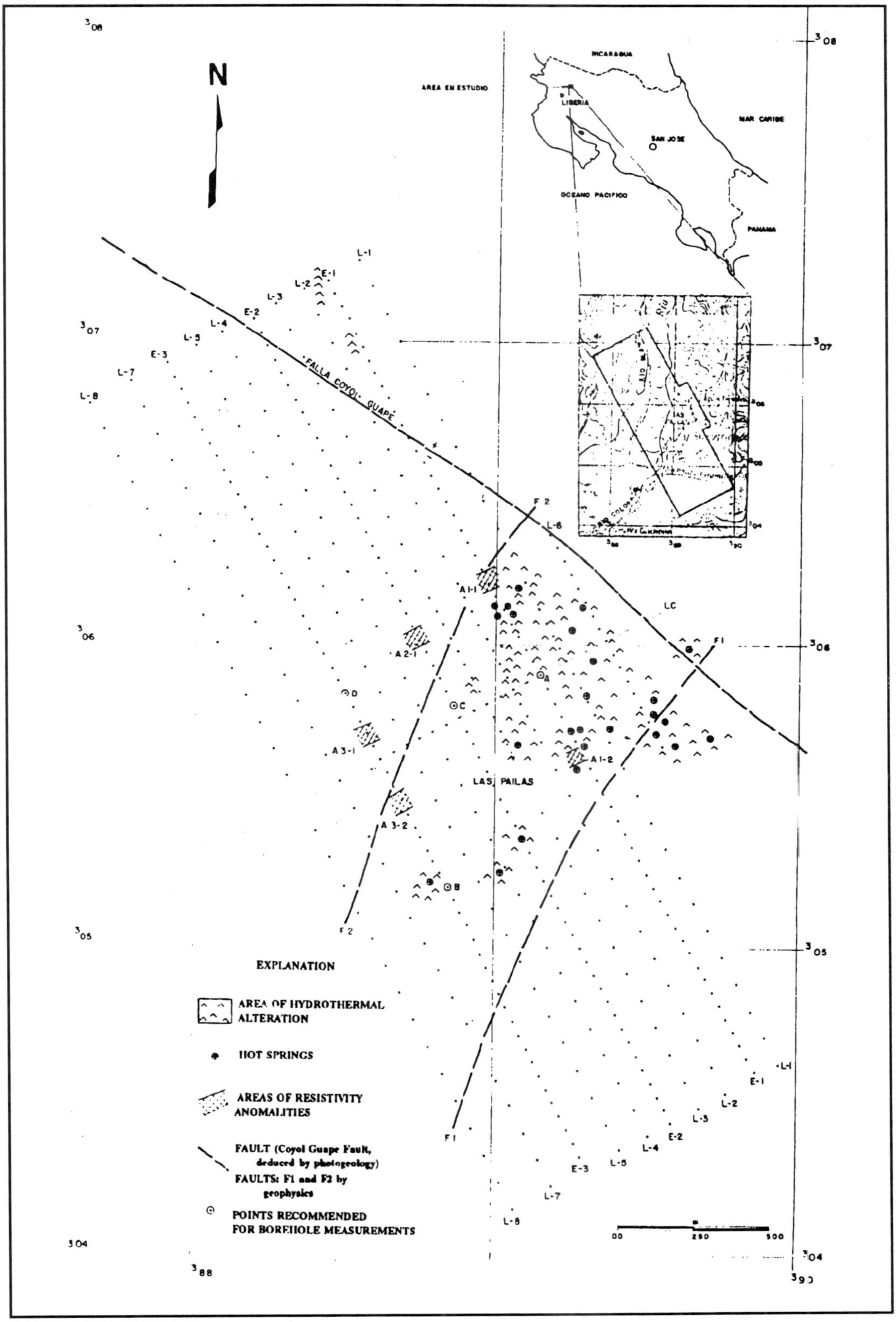

Figure 5. Location maps and plan of the principal geological-geophysical features of the Las Pailas area.

ACKNOWLEDGMENTS

We thank the Instituto Costarricense de Electricidad (ICE), the Consejo Superior Universitario Centroamericano (C.S.U.C.A.), the German Academic Exchange Program (DADD), and the Escuela Centroamericana de Geología (Central American School of Geology) of the Universidad de Costa Rica for the aid rendered.

REFERENCES CITED

Brown, G., H. Rymer, and R. Thorpe, 1987, The evolution of andesite volcano structures: New evidence from gravity studies in Costa Rica: Earth and Planetary Science Letters, v. 82, p. 323-334.

Carr, M. J., 1984, Symmetrical and segmented variation of physical and geochemical characteristics of the Central American volcanic front: Journal of Volcanology and Geothermal Research, v. 20, p. 231-252.

Chiesa, S., M. Corella, and O. Mora, 1987, Geología de la meseta ignimbrítica de Santa Rosa, Guanacaste, Costa Rica: San José, Informe Instituto Costarricense de Electricidad (ICE), unpublished report, 60p.

Granados, R., 1979, Investigación geológica en la zona de la caldera de Guayabo y alrededores: San José, Departamento de Geología, Instituto Costarricense de Electricidad (ICE), unpublished report, 43p.

Leandro, G., 1985, Mapa regional de anomalías de Bouguer, proyecto geotérmico Miravalles: San José, Informe Instituto Costarricense de Electricidad (ICE), unpublished map.

Talwani, M., J. Worzel, and M. Landisman, 1959, Rapid gravity computations for two-dimensional bodies with application to the Mendocino submarine fracture zone: Journal of Geophysical Research, v. 64, p. 49-59.

Thorpe, R., G. Brown, H. Rymer, S. Barritt, and M. Randal, 1985, Recent volcano monitoring in Costa Rica: Earthquake Information Bulletin, v. 17, no. 2, p. 44-49.

Circum-Pacific Council for Energy and Mineral Resources
Earth Science Series, Vol. 16
R. L. Miller, G. Escalante, J. A. Reinemund, and M. J. Bergin (Eds.)
Energy and Mineral Potential of the Central American-Caribbean Regions
Springer-Verlag, Berlin Heidelberg 1995

MOMOTOMBO GEOTHERMAL FIELD

Ernesto Martinez Tiffer and Roger Arcia Lacayo
Instituto Nicaragüense de Energía (INE)
Dirección General de Recursos Geotermicos, Apartado Postal 55, Managua, Nicaragua

ABSTRACT

With the beginning of operations of a 35-MW unit in Momotombo in August, 1983, Nicaragua joined the select group of countries that exploits geothermal resources for the purpose of producing electricity. Considering the old age of existing steam power plants, Nicaragua's dependence on electricity, and its economic hardships, geothermal energy has become a vitally important source of base load power and the government is rapidly accelerating geothermal development.

A second 35-MW unit has recently been placed in service in Momotombo field, increasing installed capacity to 70 MW. Current plans for Momotombo field include the drilling of four additional production wells for field development. These will be needed even though there is currently sufficient steam for both plants, because a reservoir pressure decline of five percent is anticipated during the next five years.

Several other geothermal areas in different regions of Nicaragua are being studied. These include the El Hoyo-Monte Galan, San Jacinto-Tizate, and Granada-Masaya-Nandaime prospect areas.

INTRODUCTION

Nicaragua, located in the Central American region, has an area of 135,000 km^2 and a population of a little more than three million. It is bordered by Honduras to the north, Costa Rica to the south, the Atlantic Ocean to the east, and the Pacific to the west. The capital, Managua, has one million inhabitants. Ninety percent of the population lives in the Pacific zone. This zone includes a chain of volcanos, most of them active recently or in historic time. The most promising geothermal zones are located in the vicinity of these volcanos (Figure 1).

Of all possible sites investigated since 1966, Momotombo was selected as the first geothermal area to explore and exploit because of its accessibility by highway. Cooperating agencies in this work included Electroconsult in 1966; Texas Instruments in 1969-1970; United Nations in 1971; Electroconsult-Foramines in 1973-1976; California Energy-Energeticos in 1975-1979; Electroconsult in 1980; Electroconsult-GIE-SPEG in 1981-1983 (construction and start of operations of the first unit); and DAL-INTESA-GIE in 1986-1989 (construction and start of operations of the second unit). Combined efforts of these agencies working with INE over the last 23 years have brought to reality the use of geothermal energy for generation of electricity in Nicaragua.

IMPORTANCE OF GEOTHERMAL ENERGY

Considering that Nicaragua is a small country with a small demand for energy, the relative importance of production of geothermal energy becomes very significant, as shown in Table 1.

In the table the geothermal electricity generated includes only the first unit, with a capacity of 35 MW, because the figures are for 1988. According to the program for 1989, two units will be in operation. Then electricity generated from geothermal production will be 40.9 percent of the total for the country. This will result in a probable reduction in the production of the steam thermal plants and, therefore, a saving of expensive Bunker C fuel.

THE MOMOTOMBO GEOTHERMAL FIELD

The Momotombo geothermal field, as presently developed, covers an area of about 12 km^2. It is located at the foot of the Momotombo volcano, which is part of the volcanic

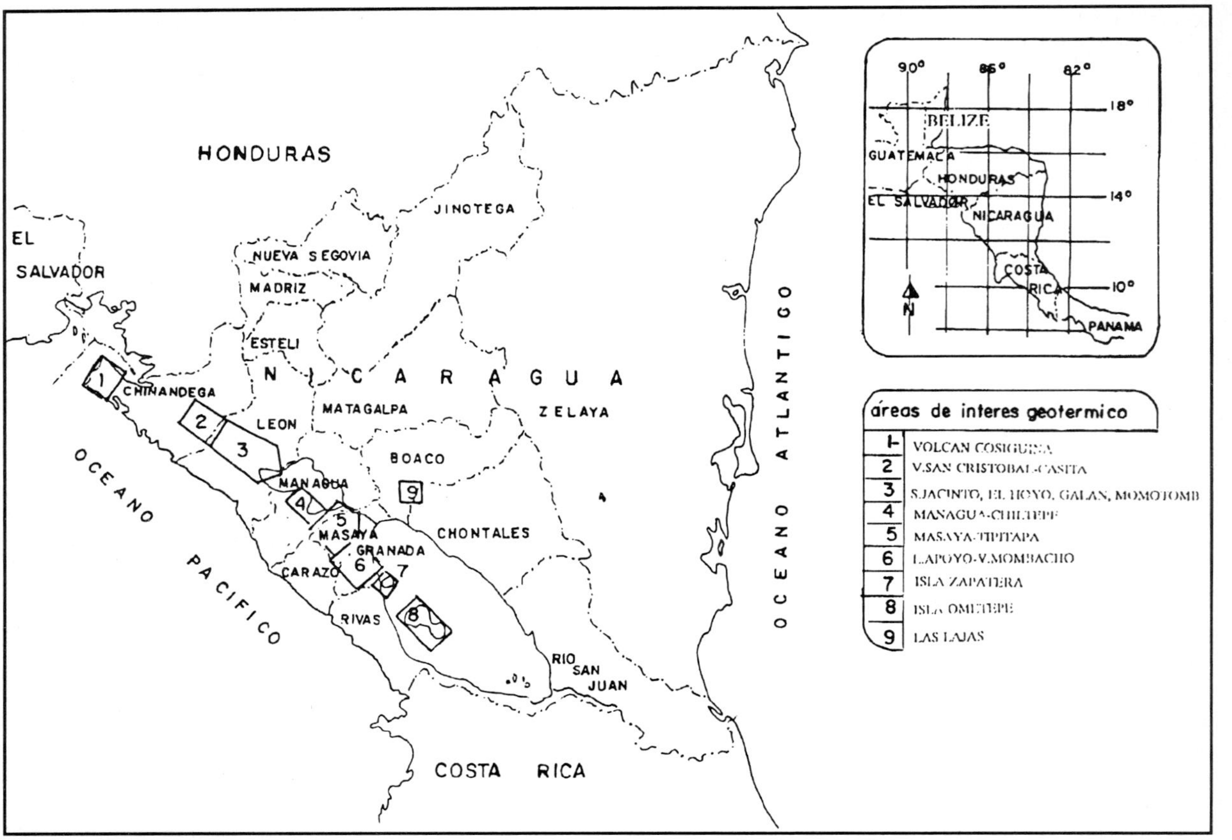

Figure 1. Location of areas of geothermal interest in Nicaragua. All but area 9 are along the volcanic chain.

chain that traverses the country in a northwest-southeast direction. The area under exploitation is part of a more complex hydrothermal system which extends to the northwest for an undefined distance. Presently, drilling operations in this direction have been suspended, but they have already identified temperatures of 220°C at a depth of 300 m through 350°C at a depth of 1,500 m. The surficial shows of the Momotombo hydrothermal system are characterized by fumaroles accompanied by large areas of strong hydrothermal alteration.

Table 1.

TYPE	INSTALLED CAPACITY IN MW (1,000 watts)	PERCENTAGE
Thermic plants	175	48.6
Hydroelectric plants	100	27.8
Geothermal plants	70	19.4
Turbo gas	15	4.2
	360	

TYPE	ANNUAL ELECTRICITY GENERATION IN GW (1 billion watts)	PERCENTAGE
Thermic plants	537 GW-hr	48
Hydroelectric plants	391 GW-hr	85
Geothermal plants	190 GW-hr	17
	1,118 GW-hr	

Up to now 37 wells have been drilled in the area, with depths ranging from 300 m to 2,250 m. Eleven of these wells are producers, five are reinjection wells, and 21 have different ranges of potential production but are not now tied into the system. The wells discussed in this paper are shown in Figure 2.

The producing wells discharge a mixture of liquid and steam in the proportion of 19 to 76 percent. The range of production of the mixture is from 200-400 MT/hr. Recommendations based on the results of the production tests led to installation of two 35-MW units, both of which are fed with steam from eight wells. Production of the first unit began in August, 1983, and the second unit in February, 1988. Production from the second unit is presently in an experimental stage.

The Momotombo geothermal field has two reservoirs, one shallow and one deep. The first geothermal unit is fed by the steam vapor from the shallow wells, the depths of which range from 310 m to 821 m. The flow from the deep reservoir, possibly originating below 1,500 m, may rise along an en echelon fault system to heat the shallow levels. The second unit is being fed by steam coming partly from the shallow wells and partly from those wells that have reached the deep reservoir. The engineering data on the reservoirs indicate that the production characteristics of the shallow reservoir are very good, but the capacity is not large enough to maintain continuous and stable production. The characteristics of the deep reservoir are equally as good or better, but its long-range capacity has not yet been determined. Because of the need to learn of the performance of the deep reservoir under intensive exploitation, a programmed control of testing consisting basically of measurements of the well-head pressure, water and steam flows, production monitoring, and temperature has been established. In addition, fluids will be sampled to detect and evaluate any changes in the physico-chemical conditions of the reservoir.

During the first five years of production of the first unit, the following phenomena have been observed:

1. A gradual change in the reservoir from the liquid to the steam phase. At the beginning of production of the reservoir, only the liquid phase existed, although very close to saturation.

2. As a result of this change, production has decreased by five percent.

3. The amount by volume of non-condensible gases has increased from 0.5 percent initially to a 0.75 percent average presently.

These three changes are being studied to ascertain their causes and effects.

PRODUCTIVE WELLS

The productive wells of the Momotombo geothermal field are separated into two groups: wells of the first unit and wells of the second unit. Location of the wells is shown in Figure 2, with the wells supplying the first unit represented by solid circles enclosed by boxes.

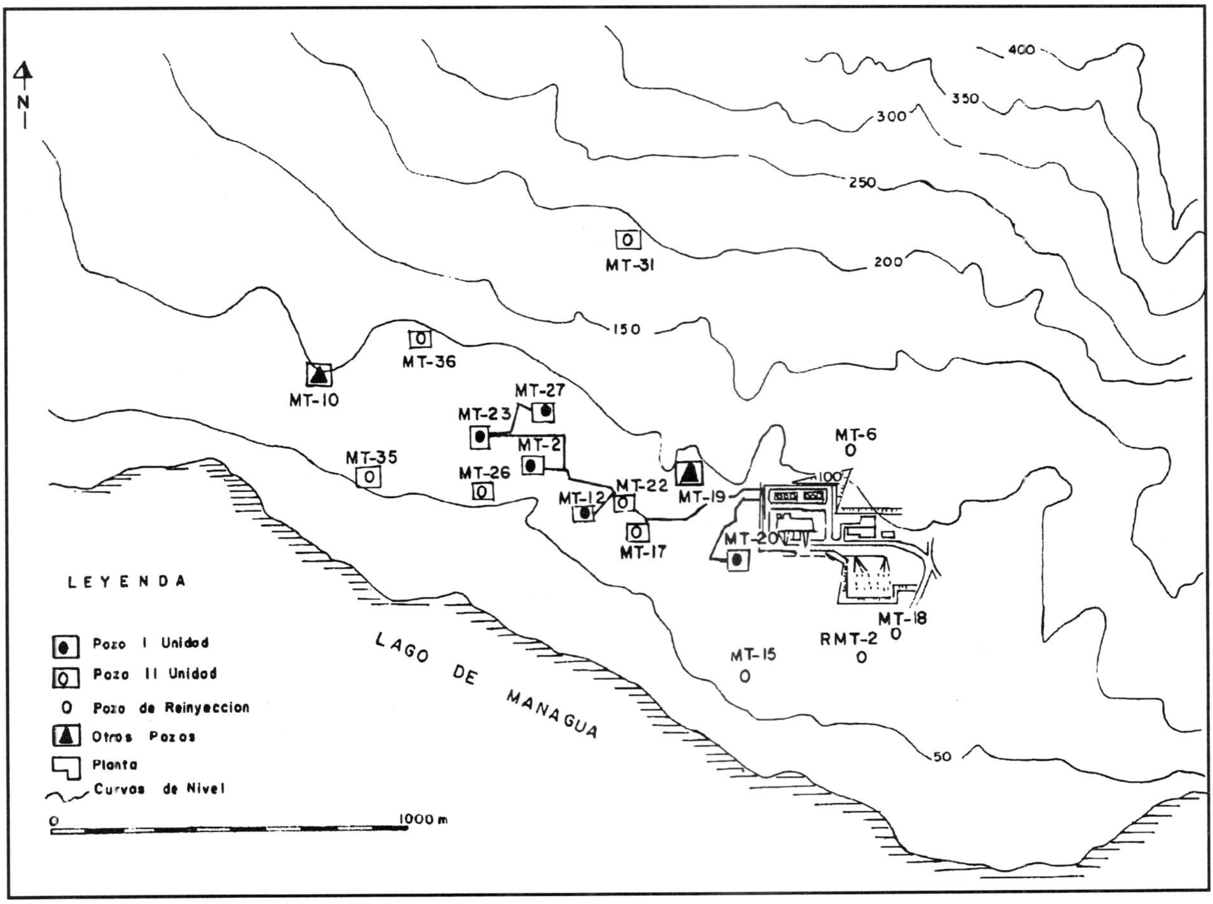

Figure 2. Location of wells drilled in the Momotombo area that are mentioned in the text. Pozo = well, curvas de nivel = contour lines, lago = lake. Wells of the first unit are shown by solid circles inside boxes. Wells of the second unit are shown by open circles inside boxes. Solid angled lines show connections from producing wells to the plant. Topographic contours are in meters.

Table 2. Steam Available for First Unit.

WELL	TONS/HOUR	MEGAWATTS	RESERVOIR
MT-2	60	7.5	Shallow
MT-12	70	8.7	Shallow
MT-20	70	8.7	Shallow
MT-23	40	5.0	Shallow
MT-27	80	10.0	Shallow
	320	39.9	

First Unit Wells

This group consists of five wells producing a steam-water mixture at depths from 304 m (the shallowest) to 823 m (the deepest) and with different rates of production. The wells MT-9, MT-12, MT-20, and MT-27 were planned as steady producers, and the MT-23 well was held as a reserve. The productive capacity of the wells is shown in Table 2 and, as can be observed, it is variable.

After three and a half years of production, the MT-19 well had to be abandoned due to many failures in the production casing; presently (1989), the completion of well MT-2 is under way to replace MT-19. Comparison of the initial production data and recent measurements show that a drop of five percent has been experienced. During the past year (1988), however, the rate of drop in production has slowed. The shallow hot water reservoir, dominant at the initial saturation conditions, has with continuous exploitation created a layer of steam in its upper part and has caused the water-steam relation to vary. Because of this the shallow wells MT-12 and MT-20 have not been equipped for reinjection.

Second Unit Wells

This group consists of six wells producing a mixture of steam and water from depths ranging from 328 m through 1,653 m. The wells are MT-17, MT-22, MT-26, MT-31, MT-35, and MT-36 (Figure 2, open circles in boxes). The productive capacity of the second unit wells is shown in Table 3. It has not yet been decided which four wells will be connected to the plant and which two wells will be held on a stand-by basis. This will be decided after completion of a 45-day testing period of the second unit wells. Observations of the behavior of the two units working together also will be made to define the optimum handling of the wells and, therefore, of the field.

Table 3. Steam Available for Second Unit.

WELL	TONS/HOUR	MEGAWATTS	RESERVOIR
MT-17	52	6.5	Shallow
MT-22	39	4.8	Shallow
MT-26	37	4.6	Shallow
MT-31	94	11.7	Deep
MT-35	136	17.0	Deep
MT-36	62	7.7	Deep
	420	52.3	

REINJECTION

Reinjection of the residual waters is in an experimental stage. Reinjection has been only partial. So far we have found that when the residual fluids are reinjected into the deeper part of the field (wells MT-15 and RMT-2) there is a slight response of the more shallow producing wells (MT-12 or MT-20), manifested by formation of a layer of steam during continuous exploitation. We are seeking the verification and evaluation of this effect in order to determine the optimum conditions for utilizing the residual waters and for achieving the hydraulic refeeding of the field. The results so far are promising because they seem to confirm the presence of a flow of high temperature heat from the lower to the higher parts of the reservoir. We have omitted MT-18 from the five reinjection wells, the locations of which are shown in Figure 2, because it has a negative effect on the shallow wells. We shall experiment exhaustively on MT-10 as a possible reinjection well because we think it may have favorable characteristics.

In tests so far the MT-6 well has accepted all the residual waters and an incremental increase in the water content of the producing wells has been observed. Perhaps because of this we have been able to keep the drop in production of the field to only five percent in an operating period of five years. Based on all information now in hand, a conceptual model of Momotombo geothermal field has been prepared, as shown in Figure 3.

THE GEOTHERMAL PLANT

The First Unit

Construction of the first unit of the plant began in 1981. It was supported by the Italian Government, the Investment Fund of Venezuela, and the Central American Bank for Economic Integration. Construction was completed in 1983, and commercial operation began in September of the same year. With four productive wells on line and one in reserve, up to December, 1988, the plant had produced a total of 1,325,437 MW-hr. During the first months of operation, the unit produced at a rate of 35 MW but later this increased to 40 MW.

In late 1986 problems in the cooling tower resulted in a loss of efficiency such that the plant, which in 1985 provided 25 percent of the total electrical output of the country, in 1988 was providing only 17 percent. Maintenance

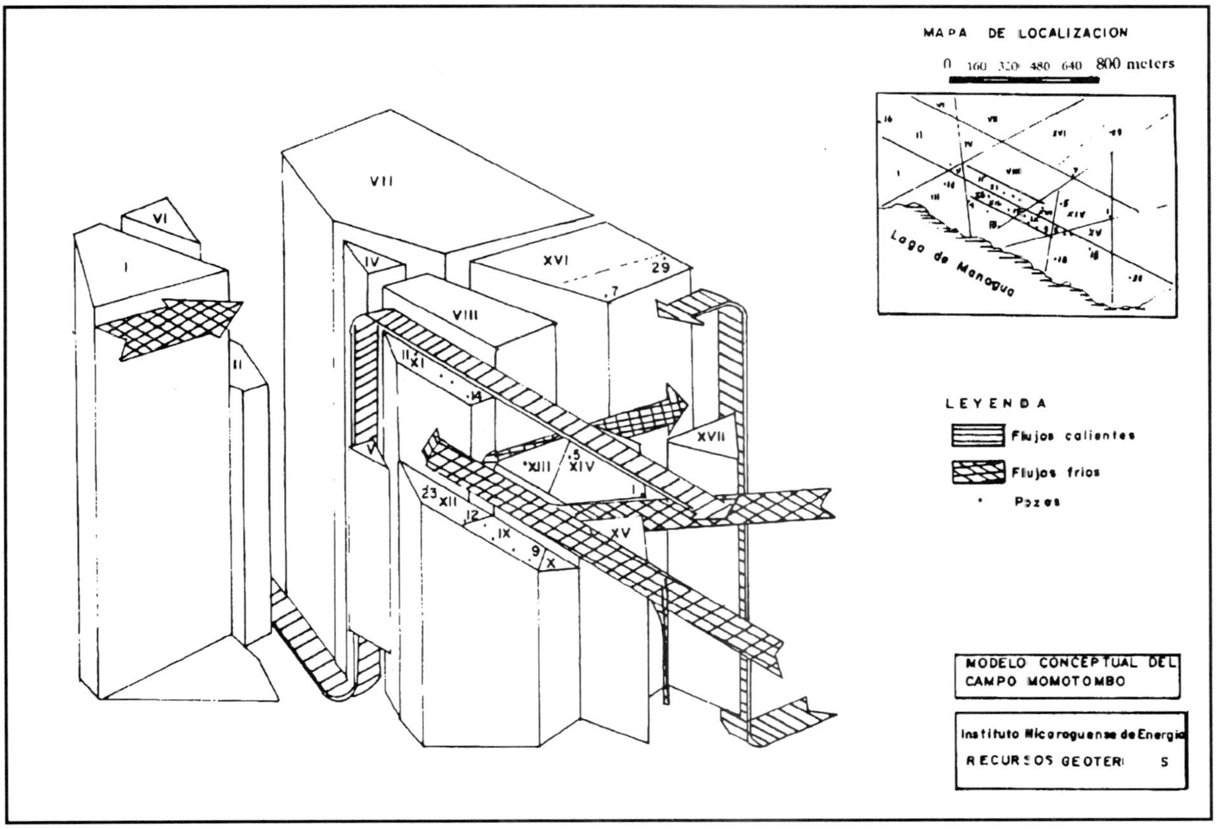

Figure 3. Conceptual model of the Momotombo geothermal field. Pozo = well, flujos = fluids, caliente = hot, frio = cold, lago = lake.

work conducted on the turbine that year did not discover any incrustations or deposits. Nonetheless, it was decided to continue maintenance service every three years. Problems also arose in the MT-19 well, so it was necessary to use the reserve well of smaller capacity. The MT-19 well had to be abandoned in 1988 due to collapsing of the production tubing. A replacement well, MT-2, with better qualities was added. At the present time (early 1989) the first unit cannot generate more than 29 MW.

The Second Unit

Construction on the second unit began in 1986 with contributions from and support of the Italian and Canadian Governments. It is at present operating on an experimental basis before being added to the commercial operation. Six producing wells have been assigned to this unit, four of them on line and the other two in reserve.

Its structure is similar to that of the first unit, with the innovation of mixing the inflow to the two collectors and in that way utilizing in each unit the steam from all the producing wells. This makes for easier handling of the reservoir. So far no relevant problems with the second unit have been encountered.

PROJECTS AT MOMOTOMBO

Plans of the Instituto Nicaragüense de Energía for the near future include two new projects of vital importance at Momotombo. The first project involves the rehabilitation of the first unit cooling tower. In order to recover the efficiency of the turbine, which presently has an average production of only 9.5 TV/MW, the cooling tower will be rehabilitated in steps, i.e., cell by cell, to avoid shutting down entirely the operation of the turbine. The cost of this project is around US $600,000.

The second project involves drilling four new production wells at Momotombo. Taking the described situation of the field into consideration, we have concluded that in order to keep production at the 70 MW constant for a long period the number of producing wells has to be increased. This will provide a margin of reserve steam, and thus give the system more flexibility, as well as compensate for the gradual reduction of production from the developed zone. To achieve this objective, INE has taken the necessary steps to drill four additional operating wells, two of which will be located in the shallow reservoir to replace damaged or unreliable wells. They will be connected to the plant immediately. In addition, two new wells will be drilled into the deep reservoir, thus assuring a more reliable source of steam. These wells will also be connected into the system at once. The cost of this project is around US $8 million.

OTHER PROJECTS

As a result of a reconnaissance program in the Pacific area, Nicaragua has identified and made prefeasibility studies of three other geothermal projects: El Hoyo-Monte Galan, San Jacinto-Tizate, and Granada-Masaya-Nandaime. Of these projects (see Figure 1 for approximate locations) the most promising is El Hoyo-Monte Galan. It is being explored with the help of the Italian Government and the United Nations. The purpose of the project is to evaluate the potential of the El Hoyo-Monte Galan area by drilling four deep wells and by installing two portable turbines of 5 MW each to make feasibility studies. These studies could lead to installation of a commercial first unit.

The El Hoyo-Monte Galan field is located to the northwest of Momotombo (Figure 4). It was investigated in a prefeasibility study by OLADE (Organización Latinoamericana de Energía) in 1982, which indicated that the field is a geothermal resource of high enthalpy (heat content). Interpretations of the findings, including (1) a strong superficial thermal anomaly, (2) temperatures of 300°C, based on geochemical data, (3) temperatures of 135°C at 80 m depth, (4) a wide anomaly of low resistivity identified by geoelectrical studies, and (5) geostructural and hydrogeological similarities to the Momotombo field, indicate the possibility of a geothermal reservoir at economic depths. The cost of this project would be about US$ 12 million for the first stage (four wells). We hope to begin work on this project soon.

Farther to the northwest is the San Jacinto-Tizate area (Figure 5) where preliminary investigations by Electroconsult have been brought to a prefeasibility level. This prospect is very similar to El Hoyo-Monte Galan. No financing for further work is yet in sight.

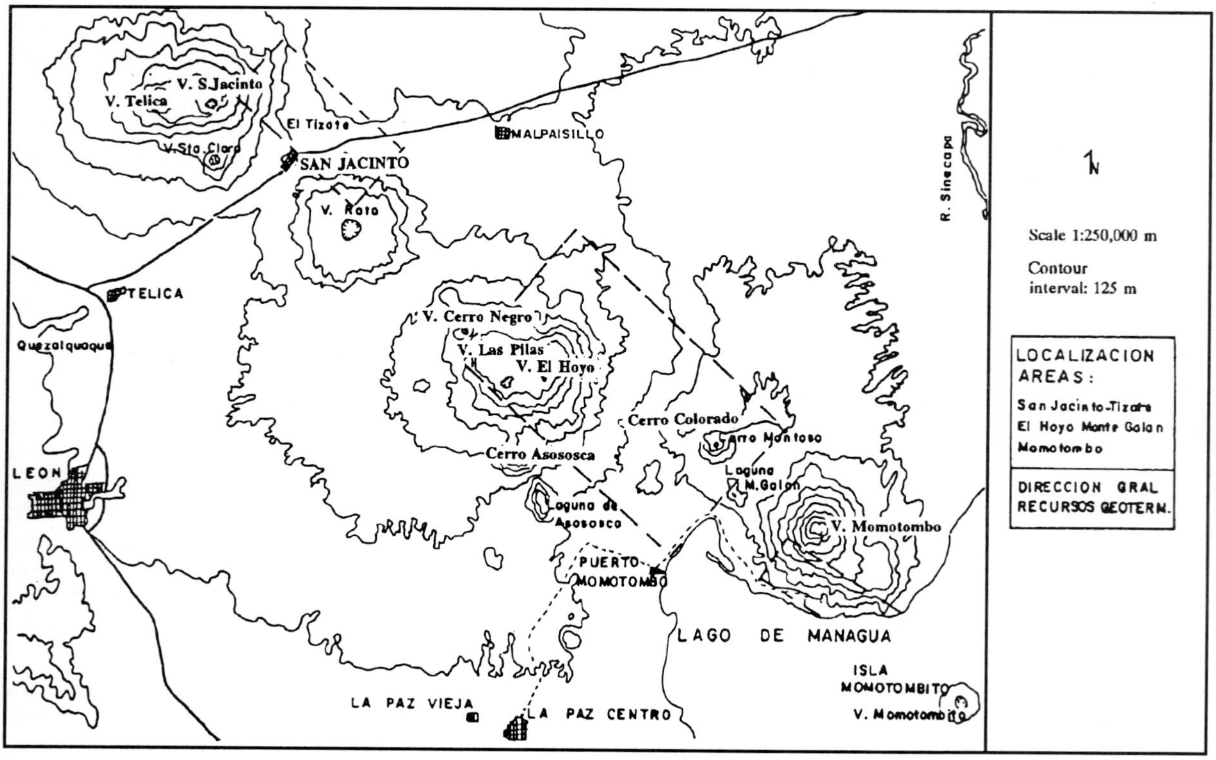

Figure 4. Location of principal volcanos and towns in the El Hoyo-Monte Galan and San Jacinto-Monte Galan geothermal areas. Scale is approximately 1:250,000, contour approximately 125 m. V = volcano, cerro = hill (mountain), puerto = port, laguna = small lake (pond).

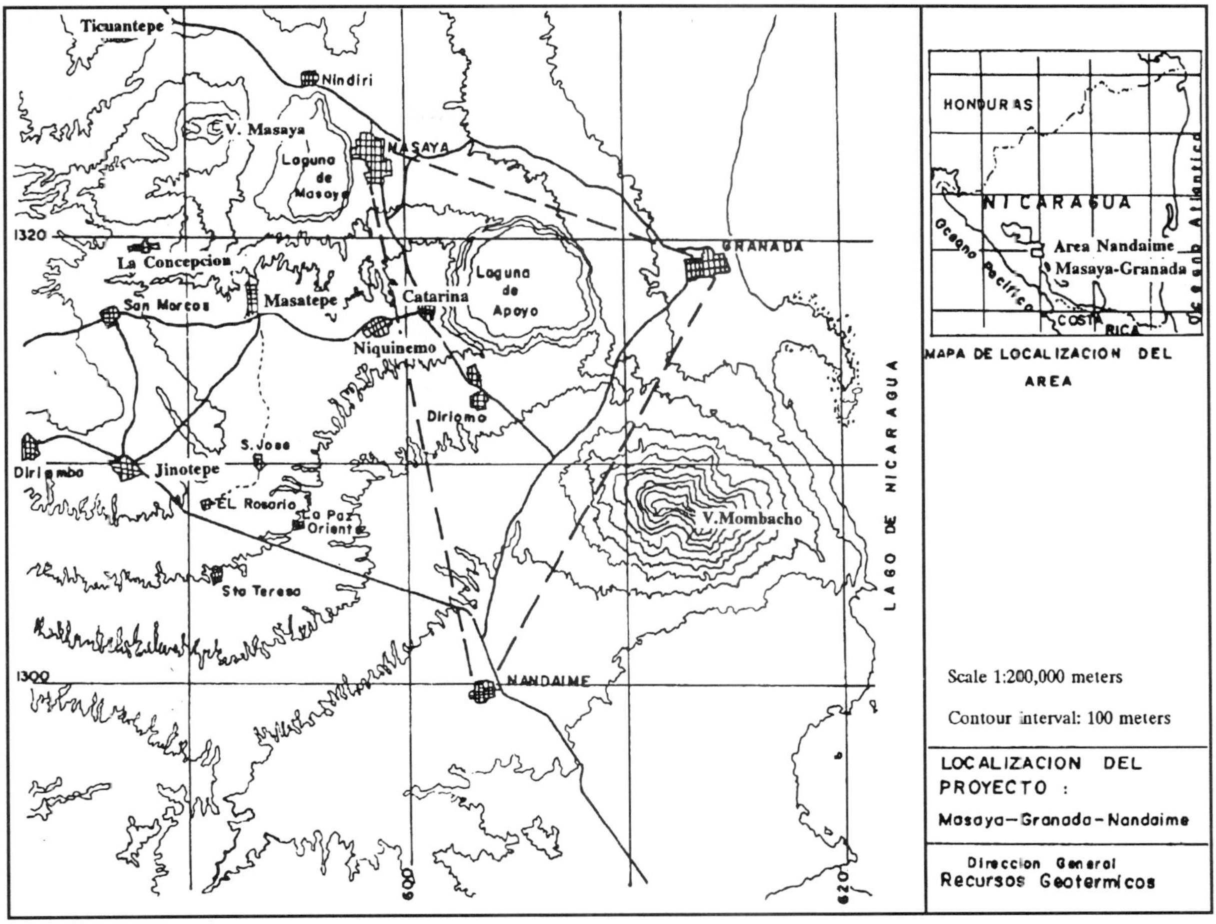

Figure 5. Map showing volcanos and towns of the potential geothermal area Masaya-Granada-Nandaime. No exploration of this prospect area has started, but is planned. Elevation of Lake Nicaragua 29 m (95 ft).

To the southeast of the previously described projects is the Granada-Masaya-Nandaime geothermal area (Figure 5). According to the studies made so far, it could be the largest geothermal field in Nicaragua. The main problem, however, is the depth to the prospect (2,500-3,000 m). Exploratory drilling will be very costly, but financing has been secured.

Circum-Pacific Council for Energy and Mineral Resources
Earth Science Series, Vol. 16
R. L. Miller, G. Escalante, J. A. Reinemund, and M. J. Bergin (Eds.)
Energy and Mineral Potential of the Central American-Caribbean Regions
Springer-Verlag, Berlin Heidelberg 1995

CURRENT STATUS OF GEOTHERMAL ACTIVITIES IN GUATEMALA

A. A. Caicedo
Unidad de Desarrollo Geotermico, Instituto Nacional de Electrificación (INDE)
Torre Profesional I, 5o nivel, Centro Comercial Zona 4, Guatemala City, Guatemala

ABSTRACT

Recent geothermal investigations in Guatemala by the Unit of Geothermal Developments of the National Electrical Institute (INDE) are the subject of this paper. Some of the studies were made by or in collaboration with foreign organizations.

The volcanic belt of Guatemala is the area favorable for geothermal phenomena. A reconnaissance of the belt identified 23 localities showing some geothermal manifestations. From these, three most favorable sites were selected for early additional consideration. Prefeasibility studies have begun and are described for these three, but principal emphasis has been on one named Zunil, where prefeasibility work is completed and feasibility investigation well advanced. Plans for installing an electricity generating plant in this field are moving forward, with the hope that the first geothermal electricity in Guatemala will be on line in 1995.

INTRODUCTION

Exploration and geothermal development studies have been carried out in Guatemala by the Instituto Nacional de Electrificación (INDE) through the Unit of Geothermal Developments, for the purpose of using this resource to generate electricity.

Beginning in 1972, the Unit of Geothermal Developments of INDE has carried on reconnaissance, prefeasibility, and feasibility studies of a geothermal prospect in the municipality of Zunil, Department of Quetzaltenango, looking forward to installation in 1995 of the country's first geothermal electric plants. At the same time they are completing feasibility studies for a second geothermal plant at either Zunil II or at Amatitlan.

RECONNAISSANCE STUDIES OF THE GEOTHERMAL RESOURCES

The part of Guatemala where geothermal manifestations are present is the volcanic belt (Figure 1) which extends from the Mexican border on the west to the borders of El Salvador and Honduras on the east. The belt is bounded on the north by the Montagua Fault and on the south by the Coastal Plain. Numerous volcanos, some active in historic time, are included. Reconnaissance studies of the entire volcanic belt identified 23 sites of hot rocks where potential for developing geothermal energy might be present. Additional work, including some geological mapping, volcanology, hydrology, and geochemistry, eliminated some of these prospects, at least for early investigation. First priority was assigned to Zunil in the Department of Quetzaltenango, Amatitlan in the Department of Guatemala, and Moyuta in the Department of Jutiapa. A second category of prospects believed to merit additional study because of hot springs or high temperatures in shallow wells includes San Marcos in the Department of that name, Tecuamburro in the Department of Santa Rosa, Los Achiotes also in Santa Rosa, and Ixtepeque Ipala in the Department of Chiquimula. A third group of geothermal prospects includes Palencia, Retana, Ayarza, Atitlan, and Motagua, but the potential of these areas could not be adequately appraised because of the absence or scarcity of hot springs.

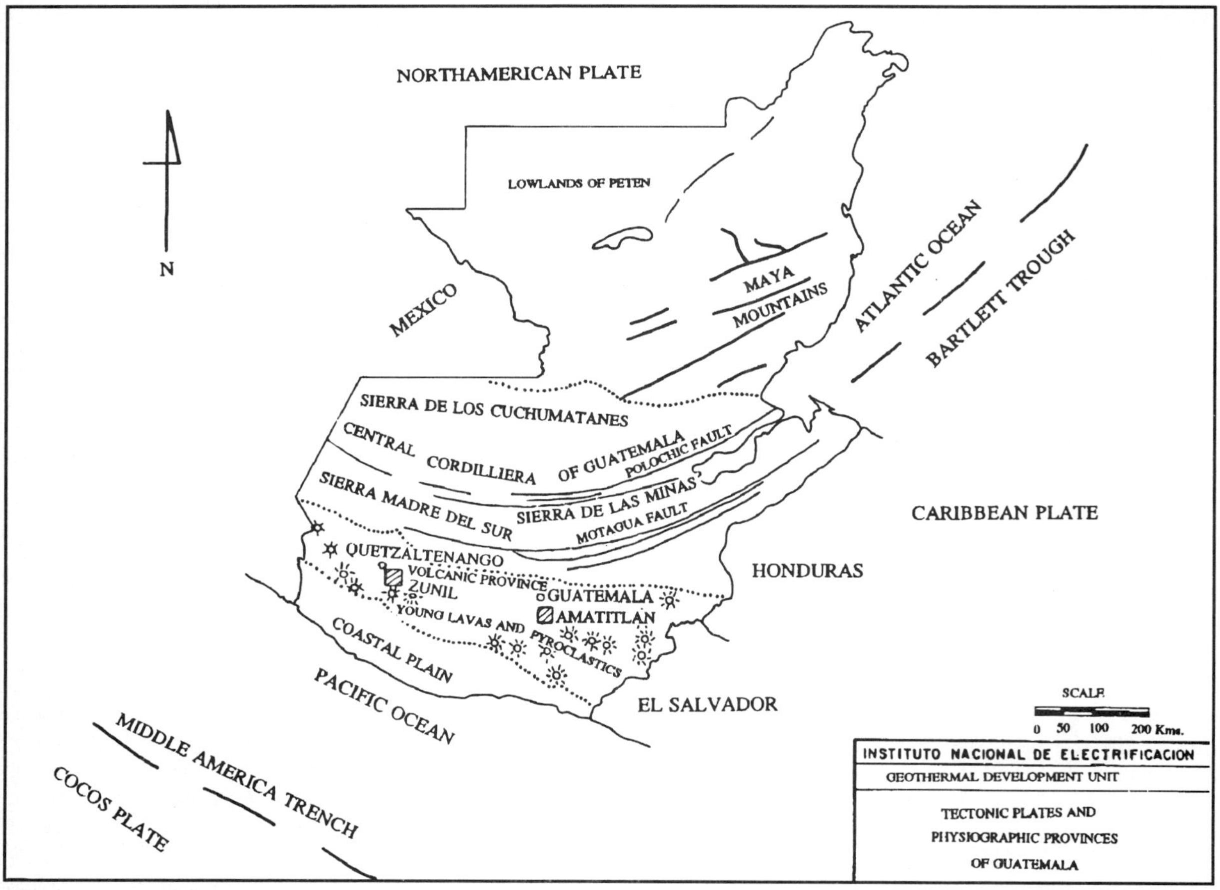

Figure 1. Tectonic plates and physiographic provinces of Guatemala. The area between the dotted lines is the volcanic province. Stars are volcanoes.

PREFEASIBILITY STUDIES FOR THE GEOTHERMAL RESOURCES OF HIGH ENTHALPY

Prefeasibility Studies—General

Investigations in the geothermal field of Zunil have established the feasibility to install the first geothermoelectric plant in the country. Its final output capacity will be determined when the re-evaluation studies of the reservoir are completed in February 1989. In addition, prefeasibility exploratory studies have been done in the Moyuta, Amatitlan, and Zunil areas, and are under way in Tecuamburro.

Moyuta Geothermal Area

The geothermal area of Moyuta is located within the volcanic province in the eastern part of the country. Reconnaissance studies by INDE with local technical personnel, particularly geologists, geochemists, and geophysicists, began in 1972. They covered an area of about 1,000 km^2.

In 1974 an area of 330 km^2 was selected for prefeasibility investigations, and for this purpose the consulting firm ELC-Electroconsult was hired in November 1975. Their task was to evaluate the studies made so far and complete those which according to their judgement were necessary to accomplish the evaluation. The interpretation of the geochemical data from the hot springs did not indicate the presence of a deep aquifer; therefore, its temperature could not be estimated. However, the sampling of gases from the main fumaroles gave qualitative information suggesting temperatures around 200°C. The results of the geoscientific investigations, together with the drilling of 12 wells of small diameter, showed the existence of three areas with high conductance related to low electrical resistivity anomalies (5 ohm-m or less). These data were interpreted as indicating a possible geothermal reservoir with thicknesses ranging between 300 and 600 m. A formation of basic andesites at depth is postulated.

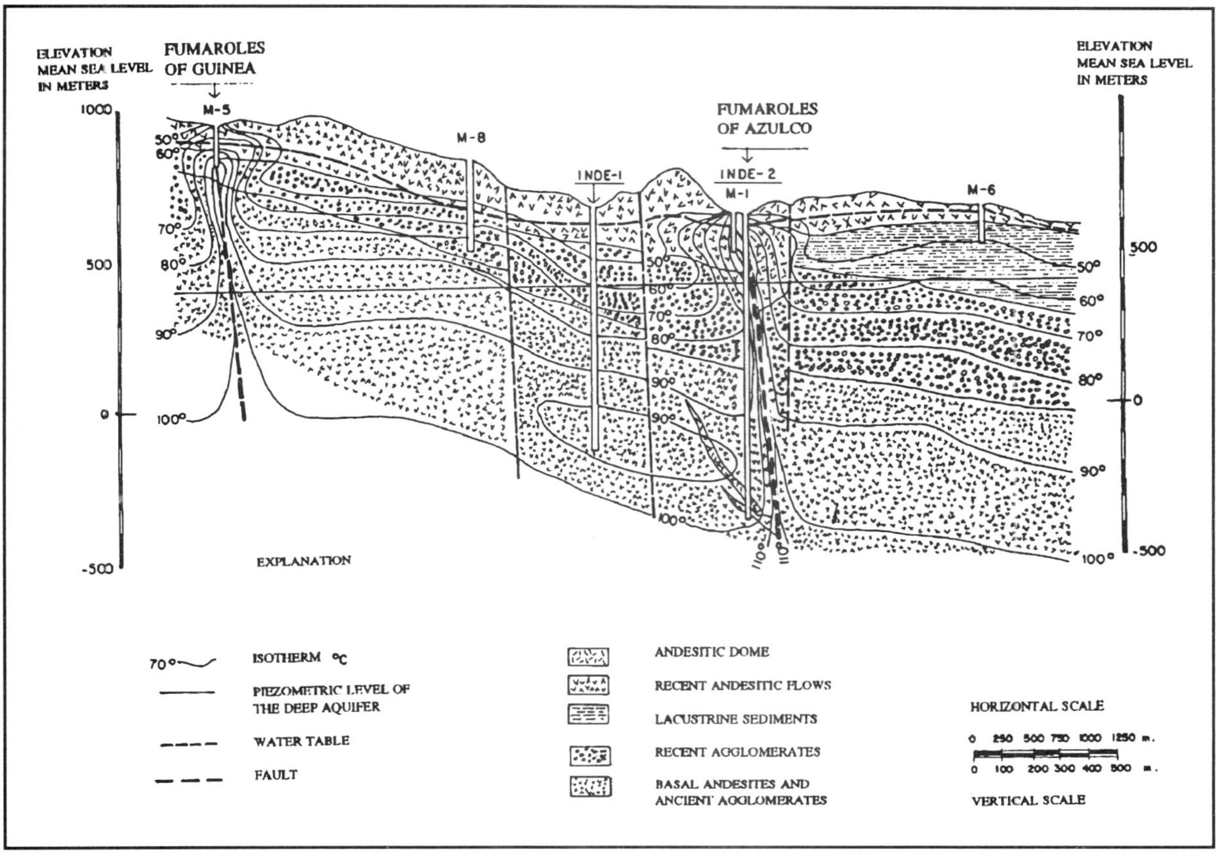

Figure 2. Cross section showing interpreted thermal profile at the Moyuta geothermal prospect. Based on prefeasibility studies by ELC-Electroconsult.

On the other hand, the 12 gradient wells drilled showed an anomaly over an area of approximately 10 km^2 with a thermal gradient of 2.5°C/10 m. This anomaly, which comprises the zone of maximum hydrothermal activity, is located along the slopes of Cerro Caliente. It overwhelmed the geoelectrical anomalies except in its central portion that is characterized by a strip of high resistivity. These results permitted the making of a first model of the hydrothermal system in which the main zone of hydrothermal activity was interpreted on the basis of surficial expression of the feeding channels as a possible geothermal reservoir located within the basic andesitic formation and below the recent volcanic rocks (Figure 2). Based on the model, two exploratory wells of commercial diameter were drilled, having as their objective the investigation of the beds of low electrical resistivity. The records of the two wells, drilled down to 797 m and 1,000 m, indicated that the maximum temperatures are around 114°C. This indicated that the interpreted zones of low resistivity were in part due only to hydrothermal alteration caused by the thermal fluids which at some time existed within the fractures or cavities of the rocks. These, when cooled, lowered the piezometric level of the area and caused the precipitation of secondary minerals that filled the original fissures. The conclusion is that the temperature and the superficial thermal gradient are due to lateral flow only. The exploration activity was suspended in March 1976 in order to go on studying the more promising geothermal field of Zunil.

The results at Moyuta are not conclusive enough to abandon the studies completely. Additional investigations should be undertaken to evaluate the possibility of using the low enthalpy fluids for binary cycle electrical generation or non-electrical uses such as heating for industrial and agricultural applications. Additional studies should also be done in the Rio San Pedro area, located on the south slope of the Moyuta Cordillera, where the geothermometry of the geochemical sampling indicates temperatures around 180°C.

Geothermal Area of Amatitlan

The Amatitlan geothermal area is located 24 km south of Guatemala City, within the municipalities of Amatitlan,

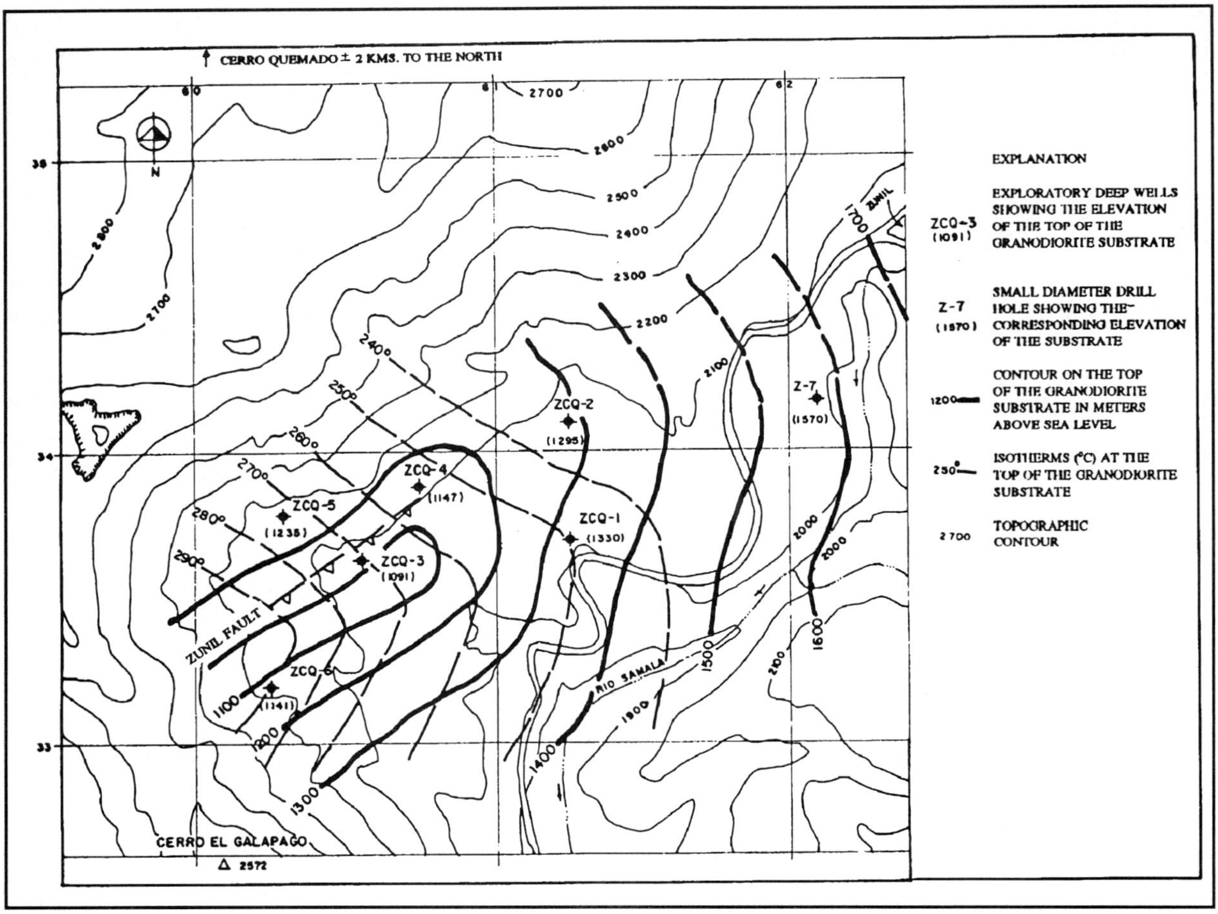

Figure 3. Map showing the locations of the deep exploratory drill holes and small diameter shallow exploratory drill holes in a part of the Zunil I geothermal field. Map also shows the elevation of the top of the granodiorite substrata and the measured temperature (°C) at that horizon.

San Vicente Pacaya, and Villa Canales. The first geovolcanological, geochemical, and geophysical investigations of this geothermal area were done by the Unit of Geothermal Developments of INDE beginning in 1977 with local technical personnel. Work was suspended temporarily until 1979 in order to continue the studies in Zunil. Systematic studies began again in 1980 and were carried to a prefeasibility level. The geothermal prospect covers an area of about 170 km^2.

Structurally the area is located on the southern end of the Guatemala City Graben within a volcanotectonic depression classified as a caldera. Within the caldera a lacustrine basin was developed which has been reduced in size to end up as Lake Amatitlan.

In the final stage of the subsidence movement, acid volcanic activity took place, which resulted in the formation of lava dome structures in the west central part of the depression. Geological studies of other recent acid domes indicate that a shallow magmatic chamber can exist, and that it could radiate enough heat downward to heat geothermal fluids that may make up a deep reservoir.

The main shows of hot springs are located in Lake Amatitlan near its south shore, whereas the areas of fumaroles and hydrothermally altered rocks are distributed from the southern edge of the lake through the Laguna de Calderas region. Isotopic and chemical analyses were made of two groups of springs. Those on the shore of the lake are of the sodic chloride type and those located in the Rio Michatoya Valley are of the carbonate type. The measurements made with geothermometers indicate a temperature of the reservoir higher than 230°C.

Interpretation of the electrical resistivity studies indicated that the area of greatest geothermal interest is where the beds of less resistivity lie within the volcanotectonic depression of Amatitlan. This is limited structurally to the southeast and west by the fault systems of Mixco and Santa Catarina

Pinula and to the north by the Amatitlan fault, which has been interpreted as an extension of the Jalpatagua fault.

Eight wells of small diameter have been drilled within the area for the purpose of correlating the results of the surficial studies with the stratigraphic data and the temperature and nature of the geothermal fluids at depth. Ninth and tenth wells are being drilled. The maximum temperatures measured at the bottom of those wells have been obtained in the wells AM-2 a reading of 174°C at a depth of 490 m; in a well drilled by the firm Bloteca a reading of 140°C at a depth of 80 m; and in the well AM-7 a reading was 180°C at a depth of 463 m.

With the loans to the Guatemalan government from the Banco Inter Americano de Desarrollo (BID) and the Organización de Paises Exportadores de Petróleo (OPEP), this last one through OLADE, the prefeasibility studies of Amatitlan will be completed. The consulting services of the Italian firm ELC-Electroconsult have been obtained to evaluate the work done so far by INDE and to complete the studies. The Mexican firm Perforadora Latina will drill the small diameter wells. The work should be done over a period of ten months, beginning in December 1988.

Zunil Geothermal Area

The geothermal area of Zunil is located 200 km west of Guatemala City in the municipality of Zunil, Department of Quetzaltenango. It lies near the southern edge of the Zunil depression which consists of a granite basement of Cretaceous age covered by Tertiary and Quaternary formations. The geothermal field is associated in the volcanic activity of the southern part of the Quetzaltenango basin and belongs to the hydrothermal system related to the active Santa Maria volcano and the plug and lava domes of Cerro Quemado. Preliminary reconnaissance studies were made between 1973 and 1977, with technical assistance from the Japanese government, which sent three missions. From 1977 on, the Unit of Geothermal Developments of INDE, with Guatemalan scientific personnel, carried out geovolcanologic, geochemical, and geophysical studies in an area of about 310 km^2. In order to bring the project to a prefeasibility level, 18 small diameter wells were drilled in the areas considered most favorable. The maximum temperature reading was in well Z-2, where at 646 m the temperature was 257°C.

With the results of the prefeasibility studies completed, an area of 4 km^2 was selected as being the most promising for feasibility studies, the results of which are described in the section on feasibility studies. The prefeasibility studies at Zunil should be completed by April 1990.

San Marcos and Tecuamburro Geothermal Area

Using the results of the regional reconnaissance studies, the Unit of Geothermal Developments of INDE has made some exploratory investigations in the geothermal areas of San Marcos, located in the western part of the country in the department of the same name, and in Tecuamburro to the southeast in the Department of Santa Rosa. The geovolcanologic, geochemical, and geophysical studies of these areas in a preliminary or semi-detailed form involve an area of 90 km^2 at San Marcos and of 273 km^2 at Tecuamburro. Beginning in July 1989, the personnel of the Unit of Geothermal Developments of INDE, together with technical personnel of the National Laboratory at Los Alamos, New Mexico, will conduct a program to bring the Tecuamburro geothermal area up to prefeasibility level by carrying out geovolcanologic, geochemical, and geophysical investigations. This program is being financed by ROCAP (U. S. Regional Organization for Central America and Panama).

FEASIBILITY STUDIES

Zunil Geothermal Area

From the results of the prefeasibility studies at Zunil a promising area was selected to implement an exploratory program of deep commercial diameter drilling. From March 1980 through October 1981, six wells were drilled in the Zunil field within an area of about 1 km^2 (Figure 3). This program was funded by INDE. The evaluation of the field and, in case of positive results, feasibility study for a geothermoelectric plant, are being performed by the consulting firm ELC-Electroconsult, which is under contract with INDE for that purpose.

The data provided by the deep wells has been used for geologic and mineralogic studies, for static and dynamic records of temperature and pressure, for pressure recovery tests in the reservoir, and for production measurements. The feasibility report incorporating this data and providing an evaluation of the field was completed in March 1982.

The geological and volcanological data gathered by the surface studies and the information provided by the wells may be summarized as follows (Figure 4). The Zunil geothermal field, here designated as "exploitation block," gets its hydrothermal fluids by lateral migration from the Cerro Quemado zone, here designated as "recharge block." Circulation takes place mainly along the contact of Tertiary volcanics with the underlying granodioritic basement. A permeable zone is generated there by the presence of a basal conglomerate and an erosional and fractured layer of the granodiorite. The Zunil fault helps to channel fluids from other structures into this permeable zone, and other similar and parallel faults may exist.

In the area drilled, approximately 1 km^2, the temperature increases from east to west, reaching maximum value in the ZCQ-6 well of 288°C. The cover of the geothermal reservoir consists of a lava complex, the thickness of which in the drilled zone ranges between 500 and 600 m. The underlying unit has a variable but locally well developed

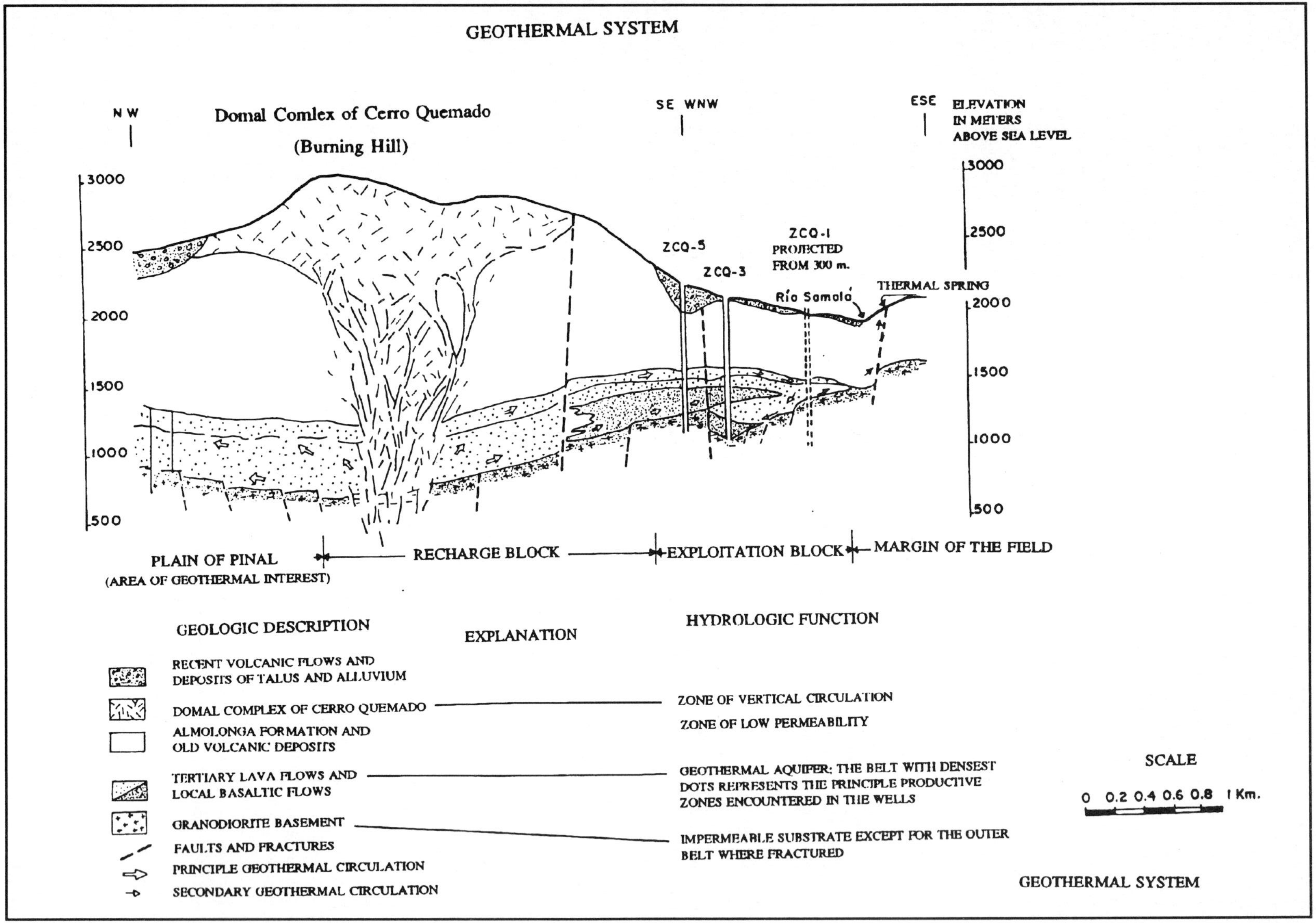

Figure 4. Interpretive cross section of Zunil I showing the structure of the rock formations, locations of drilled wells, and the circulation of fluids. Scale of the section is approximately one-half scale of Figure 3.

permeability in certain zones; it reaches a maximum thickness of 400 m in the ZCQ-3 well, but thins eastward and northeastward as the granodioritic basement rises. In the ZCQ-1 well (Figure 3) it is about 100 m thick.

Thermohydraulic and Production Characteristics

The thermohydraulic characteristics of the field have been determined from the static profiles, dynamic profiles of temperature and pressure, injection tests, production tests, and pressure recovery tests. Figure 4 shows an interpreted cross section of a part of the Zunil field, as now known. Horizontal scale is approximately one-half of Figure 3.

The records and tests have shown that:

1. The source of heat (288°C) is located to the west of wells ZCQ-5 and ZCQ-6 (Figures 3 and 4). Figure 5 shows detailed geologic columns for three wells and the temperatures measured in these wells.

2. In the drilled area, a circulation of the convective type exists in permeable levels between 1,400-1,600 m and 1,000-1,200 m above sea level. The fluid is in a state of pressurized fluid in all the wells.

3. When the wells are open a decrease in the bottom pressure ensues, measurable in tens of bars.

4. The four productive wells are divided into two groups based on their heat production: wells ZCQ-3 and ZCQ-6 have a heat production of 1,000 kj/kg, while the ZCQ-4 and ZCQ-5 produce dry steam with a heat value of 2,700 kj/kg.

5. The flow of steam separated at 8 bar is between 5 and 10 kg/sec for the four productive wells drilled. From these a production of electricity of 11.5 MW can be achieved.

Chemical Characteristics of the Fluids

The chemical character of the geothermal waters produced by the wells is of the sodium chloride type. The concentration (1,600 ppm) is relatively low compared with most of the geothermal fields associated with volcanic rocks.

The non-condensable gas content measured in the productive wells ranges between 0.4 and 0.85 percent, with an average of about 0.6 percent by weight of the vapor flow. The average chemical composition of these gases is as follows:

CO_2	87.5%
H_2S	7.0%
H_2	0.5%
Residual gases	5.0%

Field Modelling

Using the geological, hydraulic, and thermodynamic data provided by the exploratory wells in simulation mathematical models, it has been possible to determine that the area of Zunil has a potential of 15 MW over a useful life period of the plant, estimated to be 20 years. In the rectangular area utilized for the numerical simulation of the potential of the reservoir three more wells in the interior of the block were included. These will have to be drilled to complete the production of the field to 15 MW. Two will be used as producers and the third will be held as a reserve.

The mathematical model verified that the size of the recharge area has no influence on the simulated evaluation of the field. Its influence will be secondary with respect to recharge, indicating reinjection will be a useful factor for an efficient exploitation of the field.

Therefore, the location of the new wells has to be selected carefully to restrict to a minimum interference between the fluids that will feed the field and to keep in equilibrium the pressure drop in the zone of extraction.

Feasibility of Zunil II or Amatitlan

The Unit of Geothermal Developments of INDE will make complementary prefeasibility studies of the geothermal areas of Amatitlan and Zunil II, involving photogeological, geovolcanological, hydrogeologic, geochemical, and geophysical investigations and small diameter drilling for the purpose of obtaining a conceptual model of the reservoirs in both fields. From the information gained from these studies, the most promising areas will be selected within which to continue work up to the feasibility stage. This program of work should be completed by 1992. If the investigative results are favorable, a second geothermoelectric plant will be installed at Amatitlan or Zunil II.

DEVELOPMENT STAGE

Geothermal Field Zunil I

The Banco Inter Americano de Desarrollo (BID) provided a loan to the Guatemalan government to finance the design and construction of a geothermoelectric plant in Zunil I and to complete the prefeasibility studies of Zunil II and Amatitlan and the feasibility studies of the more promising field of these two. Part of the studies for Amatitlan will be financed by the OPEP Fund for International Development, a loan that was requested through OLADE.

Design and Construction of the Geothermoelectric Plant Zunil I

Additional reservoir engineering studies will be made at Zunil I to re-evaluate the thermohydraulic characteristics of production of the fluids to confirm the practicability of the 15 MW capacity as determined by the studies of ELC-Electroconsult and to select the optimum design for the plant.

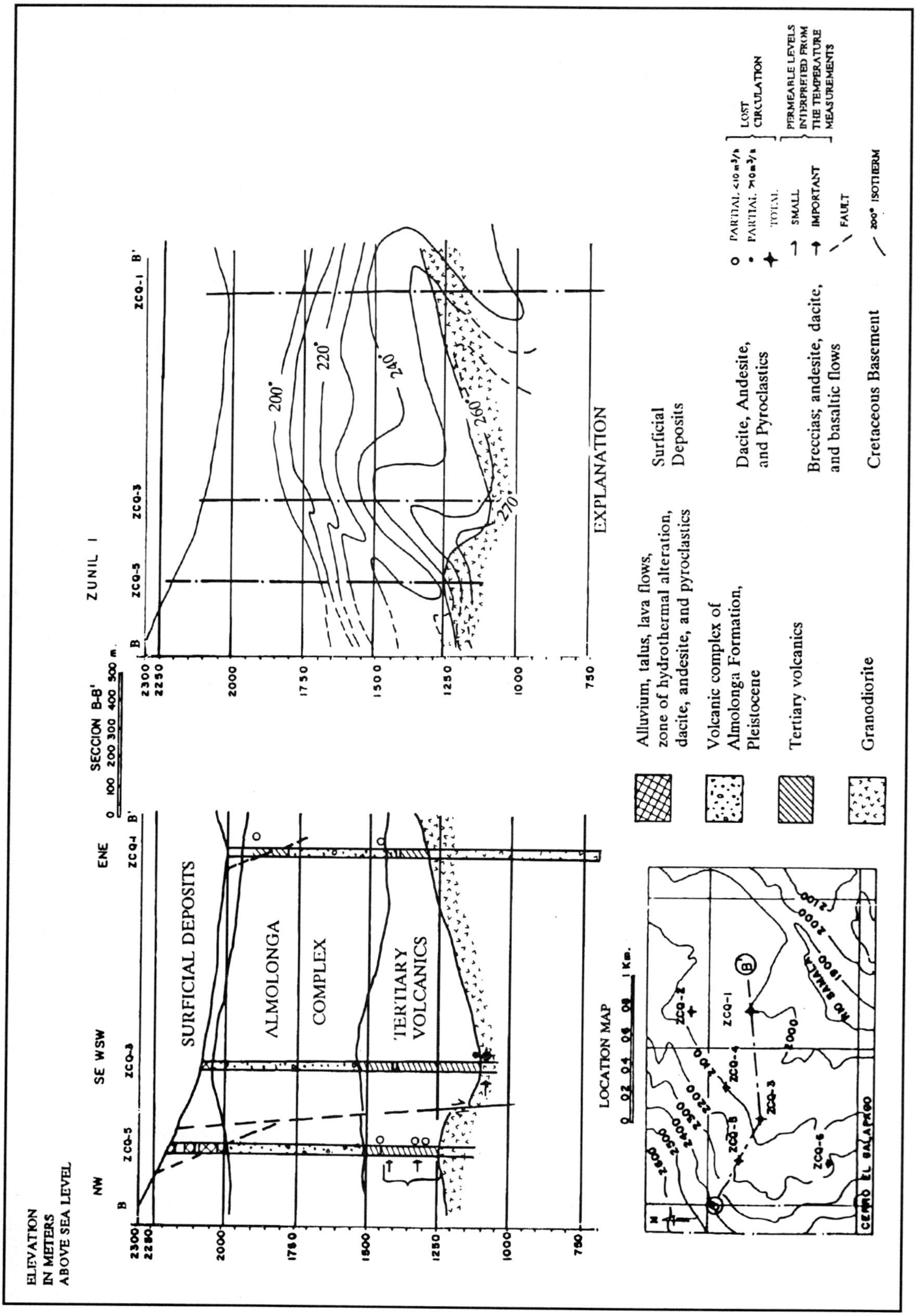

Figure 5. Geologic logs, permeable zones, and heat regime in a part of Zunil I geothermal field, based on measurements in three deep wells.

The civil works, substation, transmission lines, and installation of the 15 MW plant are scheduled to be completed no later than April 1995. The additional geoscientific studies including volcanologic, geochemical, and geophysical, to confirm or change the conceptual model of the Zunil I field, to re-evaluate the deposit, to locate the development wells, to provide supervision during the drilling, evaluation of the wells, reservoir engineering, and design and install the plant will be accomplished under contracts with the consulting firms MK-Ferguson, Morrison-Knudsen, and Cordon and Merida.

EXPLOITATION

Exploitation of the geothermal resources of Guatemala to generate electricity has not yet begun, but the geothermal field Zunil I is in the development stage.

EXPLORATION OF LOW AND MEDIUM ENTHALPY GEOTHERMAL RESOURCES

The Unit of Geothermal Developments of INDE has not made studies for the use of low and medium enthalpy fluids to generate electricity using a binary cycle or to be used as direct heat in agroindustrial processes. Direct-heat applications can result in a substantial saving of electrical energy.

Exploration, development, and exploitation of geothermal fluids of low and medium enthalpy for purposes other than the generation of electricity are being handled by the Dirección General de Fuentes Nuevas y Renovables de Energía of the Ministerio de Energía y Minas. However, in order to take advantage of the studies already made by the Unit of Geothermal Developments in the Zunil I geothermal field, INDE, through the Ministerio de Energía y Minas, has signed an agreement of technical cooperation with the National Laboratory of Los Alamos, New Mexico, for the development of a cooperation program of projects for the use of direct heat. These projects are planned to explore the use of geothermal fluids of high, low, and medium enthalpy in agroindustrial processes. One of these projects involves the construction of an experimental pilot plant in the Zunil I field to test the practicability of using these fluids for the dehydration of the agricultural produce of the region. A mixture of water-steam from the small diameter well Z-11 will provide the source of heat.

Circum-Pacific Council for Energy and Mineral Resources
Earth Science Series, Vol. 16
R. L. Miller, G. Escalante, J. A. Reinemund, and M. J. Bergin (Eds.)
Energy and Mineral Potential of the Central American-Caribbean Regions
Springer-Verlag, Berlin Heidelberg 1995

GEOTHERMAL PREFEASIBILITY STUDIES IN HONDURAS

Wilmer Flores
National Company of Electrical Energy, Tegucigalpa, Honduras

ABSTRACT

From 1985 to 1987, the National Company of Electrical Energy developed simultaneously two evaluation projects of geothermal resources in Honduras. One of these was with the assistance of the U.S. Agency for International Development and the other of the United Nations Development Program; the first project rapidly evaluated geothermal areas previously identified, then concentrated on the Platanares area; the second evaluated the central region of Honduras, undertaking detailed work in the Azacualpa and San Ignacio areas as well as in the Comayagua and Sula grabens.

The nature of the exploration studies in the Platanares, Azacualpa and San Ignacio areas were similar: essentially they included geologic, geochemical, and gravity/magnetics studies and shallow drill holes to a maximum depth of 680 m. Studies of this type, with the exception of the drilling, also were made in the Comayagua and Sula grabens.

In the thermal systems that have been studied, no evidence exists of a magmatic type of heat source. The thermal fluids that do exist are probably heated through the deep circulation of meteoric waters along faults. I infer that this circulation has created shallow thermal deposits with temperatures approaching 170°C.

Results of this work indicate that Platanares has first priority because temperatures measuring over 160°C were found at only 250 m in depth. The well tests indicate that Platanares may have important potential for energy production from plants adapted for binary cycles.

BACKGROUND

The Los Alamos National Laboratory (LANL), the U.S. Geological Survey (USGS), and the National Company of Electrical Energy (ENEE), with support from the U.S. Agency for International Development, conducted geothermal reconnaissance studies and examined six previously identified areas. Following these studies by LANL/USGS, another project was implemented with the assistance of the United Nations Development Program (UNDP) in the central region of Honduras. The objective of the studies was to assist ENEE in deciding whether to undertake more intensive geothermal investigations in Honduras. Results of the reconnaissance studies clearly indicated that Platanares has the greatest potential. Hence it was selected for the first prefeasibility study. In the central region, Azacualpa and San Ignacio were the most promising.

PLATANARES PROSPECT

Location

The geothermal prospect at Platanares is in western Honduras, some 16 km from the city of Santa Rosa de Copan.

Previous Work

Los Alamos, USGS, and ENEE have completed studies amounting to a prefeasibility study at Platanares. These

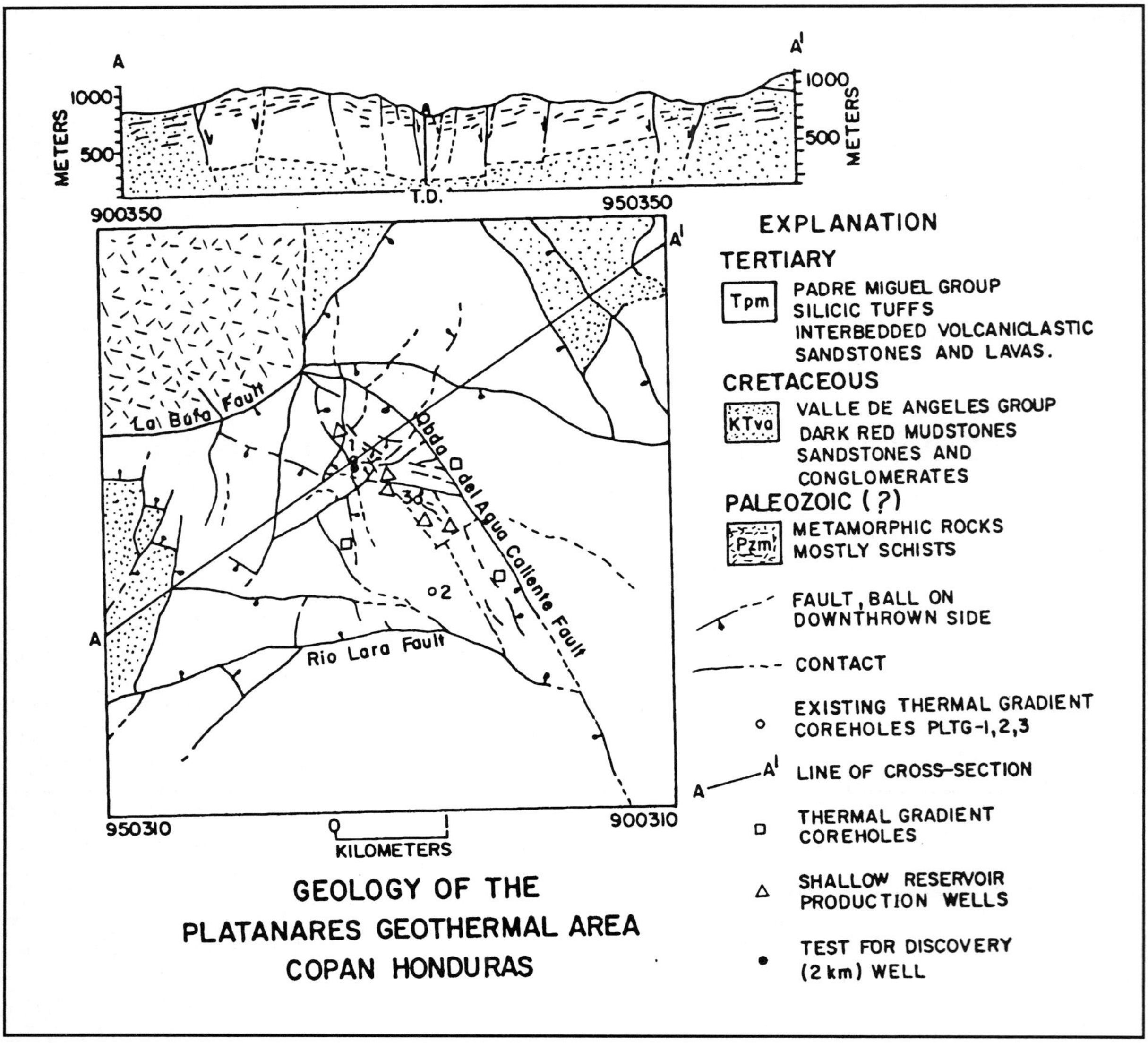

Figure 1. Geologic map of the Platanares geothermal site.

included detailed mapping, both geologic and structural; geochemical and geothermometric testing of the water; detailed geophysical readings; slope studies; and an economic evaluation.

Results

Results of the prefeasibility study were quite promising. The geothermal field is located in a region of Quaternary tectonism in western Honduras. The boiling water areas are most numerous along a northwest structural trend that parallels the Agua Caliente (Hot Water) stream. The thermal activity is strongest along the Agua Caliente stream fault (Figure 1), whose natural discharge reaches 3,300 liters/min. Other numerous faults also act as conductors for the rise of the hot water. Temperatures at the surface range between 35°C to 100°C, the pH ranges from 7 to 9, and dissolved solids are minimal (1,100 mg/kg). Chemical geothermometry analyses showed a reservoir temperature from 225°C to 240°C. Three temperature gradient wells were drilled, the deepest to 680 m. Two of the wells produced superheated water under artesian pressure of 5 atmospheres at rates of 515 liters/min from the red siliciclastic beds of the Valle de Angeles Group (Cretaceous-Eocene); the third, a dry hole, measured an apparent temperature gradient of 139°C/km at a depth of 400 m. The maximum thermal capacity measured at one of the producing wells into the shallow reservoir is 4.5 MWt. The freed CO_2

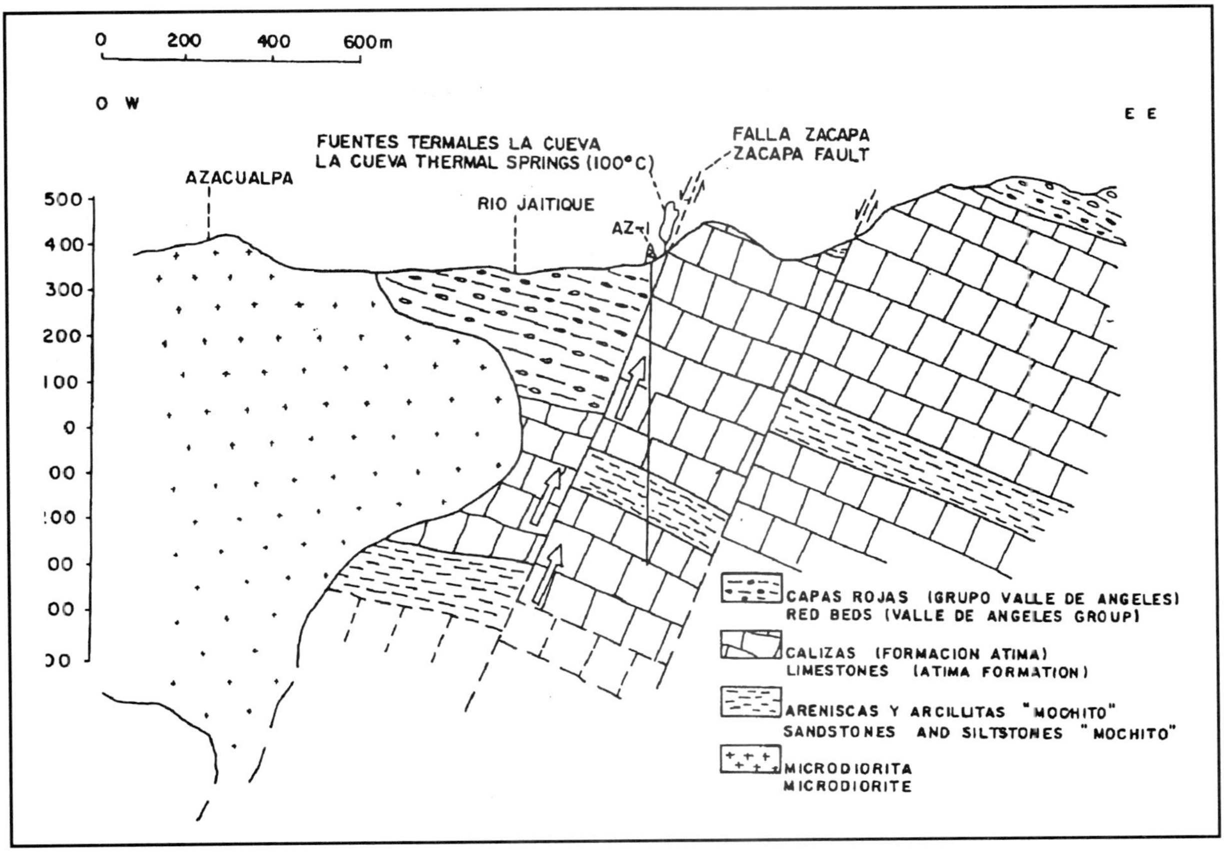

Figure 2. Interpretive geologic cross section of the area of Azacualpa.

during the "flashing to steam" created aragonite buildup on the casing of one of the holes.

Geothermal Model

The results indicate that Platanares consists of a geothermal system of chloro-alkaline water in two reservoirs. The thermal water in the shallow reservoir has an average temperature of 165°C. In the deeper reservoir at 1.2 to 1.5 km, temperatures of 225°C - 240°C were estimated through geothermometry. The source of this heat is not volcanic. Permeability is controlled by the numerous faults and fractures.

CENTRAL REGION OF HONDURAS

Location

The central region contains the Azacualpa and San Ignacio prospects. The first is located 10 km south-southwest of the southern edge of Lago de Yojoa and the second is 65 km north-northeast of the capital city, Tegucigalpa.

Previous Work

DAL/Geotérmica Italiana (1988) and ENEE made the same studies in the Azacualpa and San Ignacio areas as did LANL/USGS and ENEE at Platanares, with the exception of the economic evaluation.

Results

The Azacualpa zone is located in the extreme southwest of the Santa Barbara Depression. The main thermal springs are controlled by the Zacapa Fault (Figure 2), which forms the southeast border of the Santa Barbara Depression. Other small faults control other less important hot water sources, that surface along the Jaitique River. Above the La Cueva site and above the Jaitique River, a hot spring emerges in the form of jets of superheated water (115°C). The thermal activity is most evident in the group of occurrences that is controlled by the Zacapa Fault at the La Cueva site. The natural discharge of all the sources in the area reaches 3,340 liters/min. Temperatures range from 5°C up to 115°C and the pH from 7 to 9. Geothermometric chemistry indicates a conservative temperature in the reservoir at 185°C and pos-

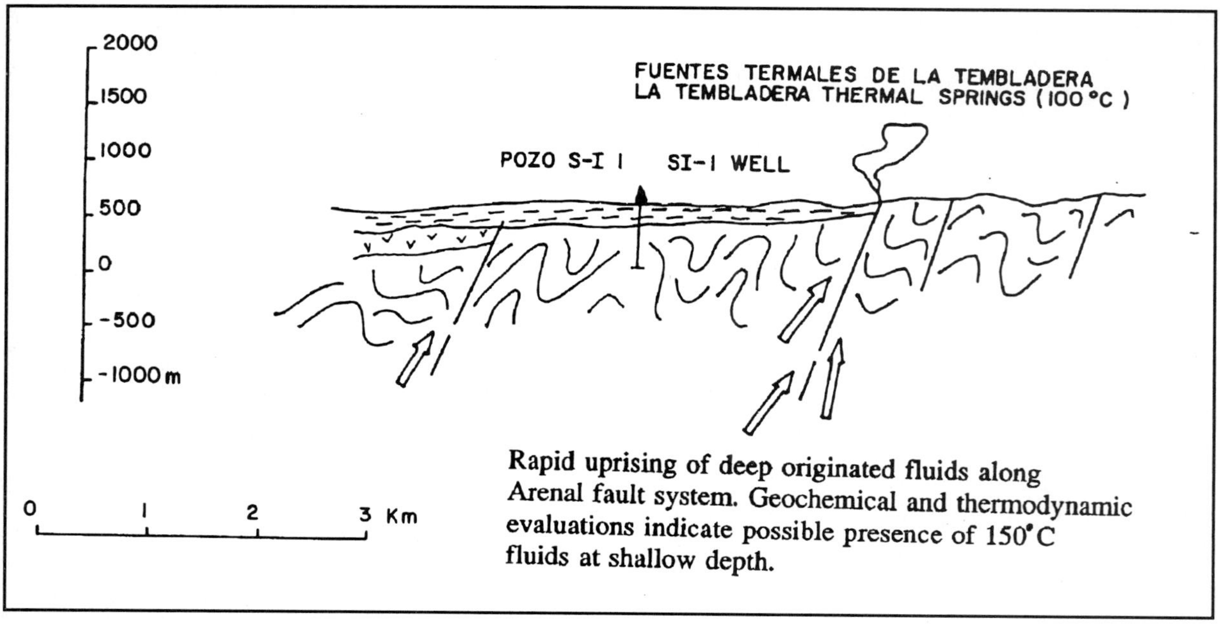

Figure 3. Interpretive geologic cross section of the valley of San Ignacio.

sibly as high as 200°C. Two temperature gradient slim hole wells were drilled to a maximum depth of 650 m. One of the wells (Az No. 1) is producing superheated water under artesian pressure at the rate of 200 liters/min. The producing zone is between 100 m and 350 m depths and is composed of massive micritic limestone of the Cretaceous Atima Superior Formation. In the second drill hole (Az No. 2), fluids did not circulate, even though the average temperature of 121°C was comparable with that in drill hole Az No. 1, which was 139°C.

Geothermal Model

It is evident that the deep hot fluids rise along faults and fractures. The limestones of the Atima Formation have adequate thickness and permeability to act as a reservoir. The clastic rocks of the overlying Valle de Angeles Group would serve as a seal in this area. Extrapolating the gradients observed in the drill holes Az No. 1 (5°C/100 m) and Az No. 2 (15°C/100 m), one would find the temperatures estimated by geochemical analysis (185°C) at 1,750 m and 1,050 m, respectively. In this area the possibility exists that a second shallow reservoir may be present which would be fed by the rise of the deeper superheated fluids (Figure 2).

SAN IGNACIO

The San Ignacio area is located in central Honduras north of Tegucigalpa. It is in the northern part of the small structural valley of Siria. The Siria Valley is an "asymmetric graben" oriented northwest-southeast, with the northeast wall being higher. The graben is a part of a system of normal faults, with the Arenal Fault being the most prominent. A second fault and fracture system has a transverse direction to the axis of the graben (Figure 3).

The Arenal Fault controls La Tembladera, an important group of boiling thermal springs, probably associated with the rise of a rhyolitic dome. Metamorphic basement rocks are the principal lithologic components. The natural discharge of the springs reaches a maximum of 1,200 liters/min, temperatures range from 30°C up to boiling, the pH is between 7.8 and 9.8, and the dissolved solids range from 900 to 1,000 mg/l. The hydrogeochemical analyses indicate that temperature at depth ranges from 200°C to 210°C. A drill hole was drilled 1,500 m south-southwest of the thermal springs of La Tembladera. It reached a depth of 500 m (Figure 3) at which fluids were not circulating. The conductive gradient was approximately 5.5°C/100 m. From the results of this hole, one can deduce that the Siria Valley area does not have regional geothermal potential. The hydrogeologic system of the La Tembladera hot springs area is developed outside the graben. The springs are fed by circulation from depth upward along the faults that affect the northwestern elevated areas.

Geothermal Model

The geochemical data indicate that the thermal waters of La Tembladera are related to a deep hydrothermal system originating in the metamorphic basement where temperatures range from 200°C to 210°C. One may deduce that the thermal activity occurs at the crossing of two important lineaments in a highly fractured sector with the deep faults providing channels for the hot fluids to rise from lower levels.

REFERENCES

DAL/Geotérmica Italiana, 1988, Estudio de prefactibilidad geotérmica en la región central de Honduras: Proyecto HON/85/001-PNUD.

Circum-Pacific Council for Energy and Mineral Resources
Earth Science Series, Vol. 16
R. L. Miller, G. Escalante, J. A. Reinemund, and M. J. Bergin (Eds.)
Energy and Mineral Potential of the Central American-Caribbean Regions
Springer-Verlag, Berlin Heidelberg 1995

PREFEASIBILITY GEOTHERMAL ASSESSMENT OF PLATANARES, DEPARTMENT OF COPAN, HONDURAS

Sue Goff, Fraser Goff, Grant Heiken, and A. W. Laughlin
Los Alamos National Laboratory, Los Alamos, NM 87545 USA

Wendell A. Duffield
U.S. Geological Survey, Flagstaff, AZ 86001 USA

Alfred H. Truesdell
U.S. Geological Survey, Menlo Park, CA 94025 USA

Wilmer Flores
Unidad Proyecto Geotérmico, Empresa Nacional de Energía Eléctrica, Tegucigalpa, Honduras

ABSTRACT

The Platanares geothermal system is located in a region of active Quaternary tectonism in western Honduras. Although the geothermal area is partially blanketed by Miocene ignimbrites (14.5 m.y.), there are no nearby Quaternary volcanic rocks to suggest a young, subsurface, magmatic heat source. No acid-sulfate waters, indicative of vapor-dominated conditions, exist in the area. Hot spring activity is most vigorous along a 2 km stretch of the Quebrada del Agua Caliente fault zone. Natural discharge is high (3300 l/min), spring temperatures range from 35 to 100°C, pH ranges from 7 to 9, and total dissolved solids are low (1100 mg/kg). Chemical geothermometers indicate a subsurface reservoir temperature of about 225°C. Three exploration core holes (7.8 cm diameter) have been drilled to a maximum depth of 680 m and maximum temperature of 165°C. Two holes produce copious amounts of water under artesian conditions (500 l/min maximum; 5 bars flowing) from fractured red beds of Cretaceous to Eocene age (Valle de Angeles Group). Maximum power output is 4.5 MW (thermal) but CO_2 released during flashing formed some aragonite scale in one hole. The third core hole has an apparent conductive gradient of 139°C/km at 400 m. Downward continuation of this gradient implies that the minimum depth to a possible geothermal resource (225°C) is about 1.5 km.

INTRODUCTION

Geothermal efforts in Honduras are part of a coordinated effort among the Empresa Nacional de Energía Eléctrica (ENEE), the U.S. Geological Survey (USGS), and the Los Alamos National Laboratory (LANL) and their consultants to locate, evaluate, and develop a geothermal resource for electrical power generation. Initial studies (1985 to 1986) focused on ten areas: Platanares, San Ignacio, Azacualpa, Choluteca region (Pavana), La Ceiba region (Sambo Creek), El Olivar, Isla de Aguas Calientes, San Francisco de Ojuera, El Cajon Dam, and Agua Caliente (west of San Ignacio) (Figure 1). Further evaluation during 1986 concentrated on the three sites having the best apparent potential: Platanares, San Ignacio, and Azacualpa. From these studies Platanares was chosen as the best site for geothermal development because of relatively high estimated subsurface reservoir temperatures (225°C), large natural discharge of the hot spring system (3300 l/min), and well-defined structural setting. In late 1986 to middle 1987, three slim core holes were drilled at Platanares to obtain information on the geothermal gradient, hydrothermal alterations, fracturing, and possible inflows of hydrothermal fluids. The object of this paper is to briefly outline the results of our studies at Platanares. Reports

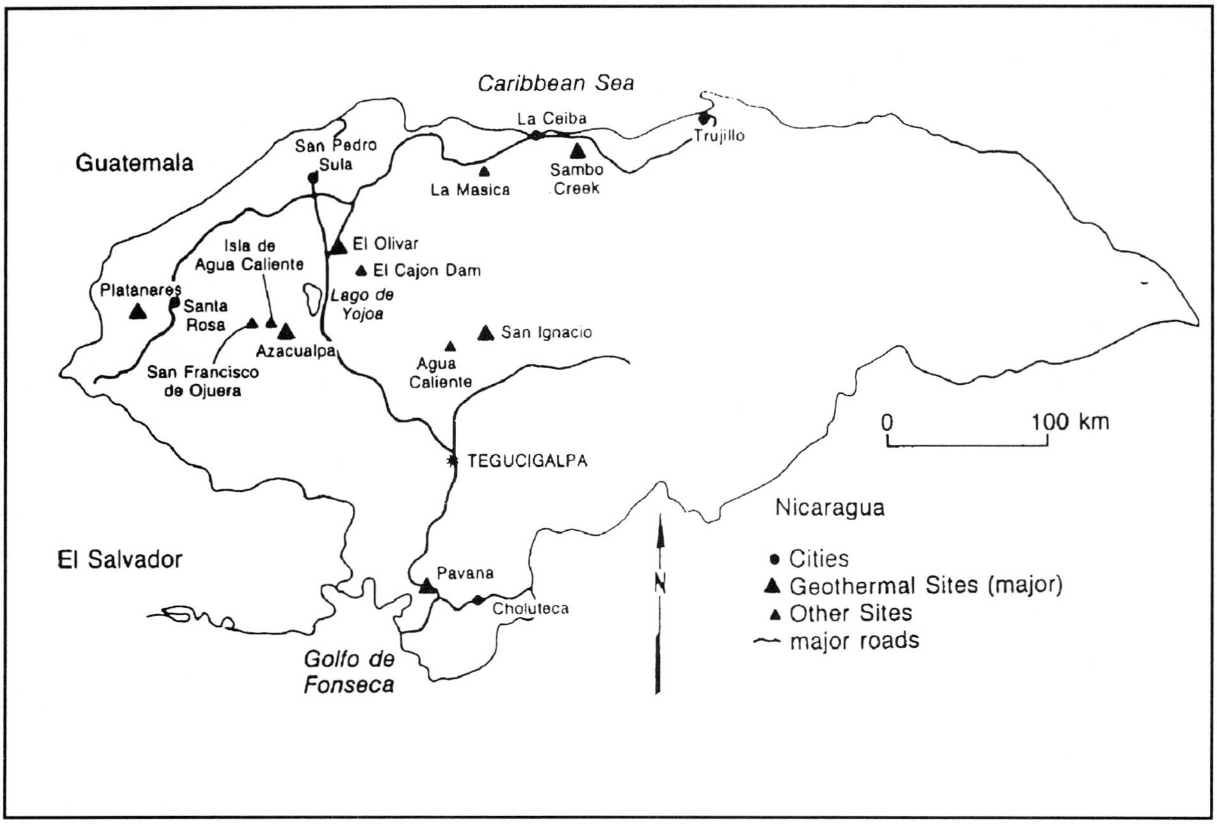

Figure 1. Map of Honduras showing locations of major geothermal sites (large triangles) and other hot spring areas (small triangles) sampled during our investigations from 1985 to 1987.

on the area have been presented by Heiken et al. (1986; 1987), F. Goff et al. (1986; 1987a; 1987b; 1988), S. Goff et al. (1987; 1988), Laughlin (1988), Truesdell et al. (1987a; 1987b), Hoover and Pierce (1988), and Barger (in press).

GEOLOGIC BACKGROUND

The Platanares geothermal site is located along the Quebrada del Agua Caliente about 16 km west of Santa Rosa de Copán (Figure 1). The region around Platanares is mountainous and deeply dissected. Elevations of nearby ridges exceed 1,400 m, whereas the hot springs in the quebrada issue at 700 to 800 m. A gold mine is actively being worked in the village of San Andres and an antimony mine (El Quetzal mine), which is no longer being worked, lies about 2 km north of San Andres (Figure 2).

Rocks in the Platanares area consist of Paleozoic schists and gneiss in the north, which are juxtaposed along La Bufa Fault against Oligocene to Miocene andesites (Matagalpa Formation), basin-fill sedimentary rocks (Subinal Formation) and tuffs, and tuffaceous sedimentary rocks (Padre Miguel Formation) to the south. Late Cretaceous to Eocene red beds and conglomerates of the Valle de Angeles Group lie in fault contact with both Paleozoic and Tertiary rocks west, east, and northeast of the geothermal area and unconformably underlie tuffaceous rocks in the Quebrada de las Juniapas 1 km east of the area.

The geothermal area is highly faulted and fractured and occupies a structural graben trending in a northwest direction. Of primary importance are La Bufa Fault, which marks the northern limit of hot spring activity, and the Quebrada del Agua Caliente fault zone, which controls the discharge of 95 percent of the hot springs in the area. Hydrothermal alteration in the area varies from intense to nonexistent. Intense alteration (argillic to phyllic grade) occurs in the San Andres mine, the El Quetzal mine, along the Quebrada del Agua Caliente fault zone, along the Rio Lara and other faults within the structural graben, and within the PLTG-1 and PLTG-3 core holes. Alteration in the PLTG-2 hole is less intense. At least two distinct hydrothermal events have affected the area: an early (Oligocene to Miocene?) event associated with gold, mercury, antimony, arsenic, and copper mineralization and quartz veins visible in the mines and in the creeks, and the present hydrothermal event.

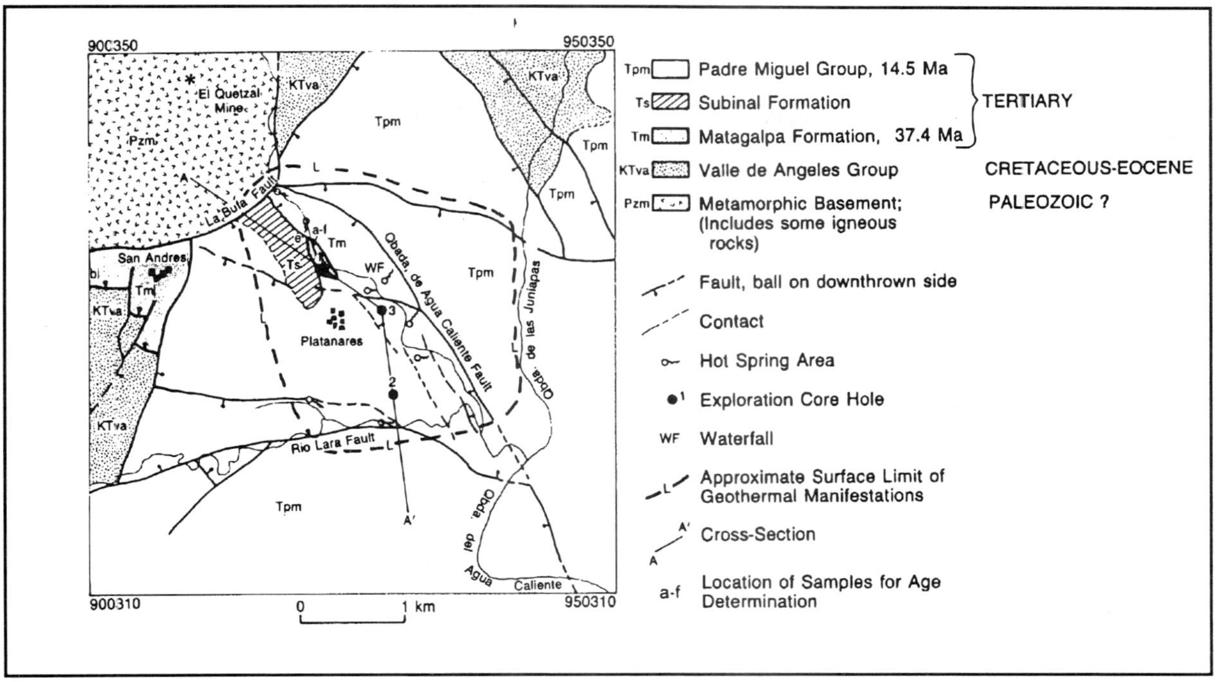

Figure 2. Simplified geologic map of the Platanares geothermal site showing locations of major hot spring areas and three exploration core holes in relation to rock units and structure. "Quebrada" as used in this report means ravine.

GEOPHYSICAL INVESTIGATIONS

Detailed gravity survey results generally reflect the location of major faults and a northwest-trending graben. Steep gravity gradients clearly define the Agua Caliente fault zone on the northeast side of the graben. A positive gravity anomaly about 300 m long trends northeast, intersecting the Quebrada del Agua Caliente between the PLTG-1 and PLTG-3 boreholes and is interpreted as being the result of an andesitic intrusion or stack of andesitic flows. The contact between andesites and overlying rocks is exposed in the quebrada upstream of PLTG-1. The natural-source electrical surveys (audiomagnetotelluric sounding and multifrequency telluric traversing) indicate very low resistivities (<10 ohm-m) throughout the Platanares survey area. The depth of penetration, however, is shallow, i.e., less than 300 m, thus identifying the fractured, more permeable areas where geothermal fluids may have penetrated or where altered areas may be present. The low resistivity zone is a large area coincident with the Quebrada del Agua Caliente fault zone where geochemical evidence indicates that relatively unmixed thermal waters ascend. The lowest resistivities lie east of the fault, but because of the scarcity of data in this direction, the shape of this low-resistivity zone is not constrained. A second area (0.3 km^2) of very low resistivity occurs about 500 m west of the PLTG-2 core hole. This anomaly appears to coincide with several faults and fault intersections along which fluids could be ascending although there are no surface thermal manifestations visible in this area.

THERMAL FEATURES AND HYDROGEOCHEMISTRY

Most of over 100 thermal springs and seeps, with temperatures ranging from 35° to 100°C, discharge near river level along both the Quebrada del Agua Caliente and Rio Lara with the greatest number along a 1 km stretch downstream from the waterfall (Figure 2). All boiling springs are depositing sinter. No single hot spring has a flow exceeding 150 l/min but the combined discharge along the quebrada averages 3,300 l/min, which during dry seasons raises the temperature of the river to about 40°C.

Thermal waters are dominated by Na, K, HCO_3, SO_4, and Cl. They contain significant concentrations of As, B, F, Li, and NH_4 but low concentrations of Ca and Mg. Total dissolved solids are only 1,100 mg/kg. Stable isotope data indicate that recharge to the geothermal reservoir comes from local precipitation, and very anomalous concentrations of B, SO_4, and NH_4 and a ^{13}C-CO_2 signature of 11 percent indicate that the water in the geothermal reservoir is in chemical equilibrium with red beds of the Valle de Angeles Group. Good agreement exists among Na-K-Ca, Na-K, ^{18}O-SO_4, and gas geothermometers, suggesting a reservoir temperature of 225°C, possibly as high as 240°C. (For a discussion of

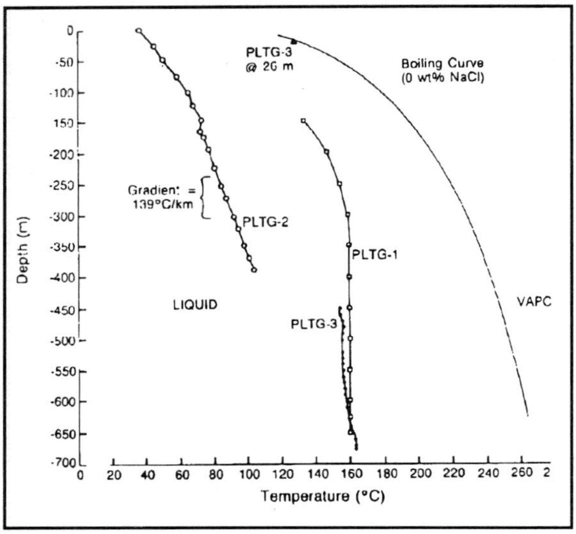

Figure 3. Plot of temperature versus depth for the temperature logs compared to the boiling curve for pure water.

the theory behind "chemical" geothermometers, consult Fournier, 1981. For a discussion of the application of chemical geothermometers to the Platanares system, see Goff et al., 1987b.)

EXPLORATION CORE HOLES

Three slim exploration diamond core holes (PLTG-1 to PLTG-3) were drilled to depths of 650, 428, and 676 m, respectively, at the Platanares site during the last stages of the prefeasibility study. PLTG-1, located along a part of the quebrada with very few hot springs, and PLTG-3, located within the area of most vigorous hot spring activity, bottomed in hydrothermally-altered shales and conglomerates of the Valle de Angeles Group (Figure 2). Fractured zones in PLTG-1 at 252 m, 574 m, and 625 to 644 m individually produce as much as 300 l/min of hydrothermal fluids. A large water entry (>200 l/min) was encountered at 26 m in PLTG-3. Other large water entries in this hole were encountered at depths of 459 m and 622 to 635 m. PLTG-2 was drilled south of the main hot spring areas in order to avoid large water entries so as to obtain conductive thermal gradients.

TEMPERATURE LOGS

The temperature logs for the three boreholes are shown in Figure 3. The bottomhole temperature of 160°C in PLTG-1 measured after 10 days of flow, which persists from a depth of 350 to 650 m, is caused by heating during flow in the wellbore. In PLTG-2, which produces only very small quantities of water, the log displays two zones of linear (conductive) gradients separated by a perturbed zone at about 170 m, which is assumed to be caused by lateral flow of water around the casing at this horizon. The deeper gradient is 139°C. The temperature of 163.9°C at total depth in PLTG-3 (measured 24 hours after completion) is considered to be slightly low because of insufficient time to let the well bore equilibrate.

Comparing the temperature gradients with the boiling point curve for pure water shows that the subsurface temperatures lie well within the field for liquid water. There is no evidence for a well-defined cap rock to the system or for a sudden increase in temperature to occur beneath PLTG-1 and PLTG-3. If a deeper reservoir at 225°C to 240°C underlies the Platanares area, as suggested by chemical geothermometry, downward continuation of the PLTG-2 gradient would suggest that the minimum depth to the source geothermal reservoir is at least 1.5 km.

FLOW TESTING PLTG-1 AND PLTG-3

Although the exploration core holes were never intended to be production wells, they do produce copious amounts of water. PLTG-1 will produce as much as 400 l/min and PLTG-3 as much as 550 l/min of water from their open hole (uncased) sections (3-in.-diameter holes). In February and June of 1987 LANL, ENEE, and USGS staff were on site at Platanares to collect fluid samples and perform flow tests on PLTG-1 and PLTG-3 to obtain information on aquifer chemistry and flow characteristics. The eight-day flow test of PLTG-1 showed a gradual decrease in flow rate up to about 100 hours and then a sharp decrease in flow rate to about 190 hours. When we dismantled the surface pipes after this test, we found aragonite scale lining the upper casing and wellhead. A similar test performed on PLTG-3 did not produce scale after ten days of flow. We conclude that higher flowing pressure in PLTG-3 (70 psia as opposed to 30 psia in PLTG-1) prevented the scaling problem. Although PLTG-1 and PLTG-3 are exploration core holes, we did estimate the maximum thermal power from our flow data to be 3.1 MW (thermal) and 5.1 MW (thermal) for PLTG-1 and PLTG-3, respectively.

RESERVOIR MODEL, CONCLUSIONS, AND RECOMMENDATIONS

Figure 4 presents a reservoir model of the Platanares system. Our data indicate that geothermal water is probably in chemical equilibrium with rocks of the Valle de Angeles Group of Cretaceous to Eocene age. Geochemical data indicate that the source reservoir has an equilibrium temperature of 210° to 240°C. More than 95 percent of reservoir fluids discharge along the Quebrada del Agua Caliente fault zone. The northwestern limit to the hydrothermal system is La Bufa Fault, which juxtaposes Paleozoic schist (north) against Tertiary volcanic rocks and Cretaceous to Eocene red beds. Other limits to the reservoir are not as well defined, but, as shown

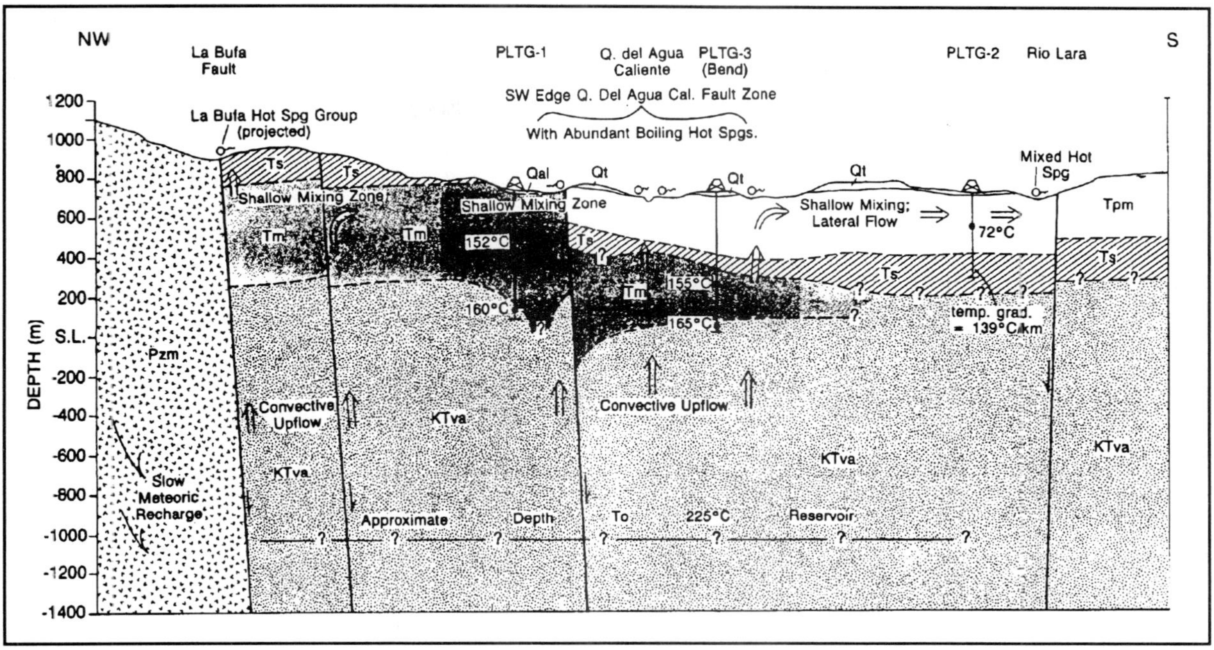

Figure 4. Schematic northwest-south cross-section of Platanares geothermal site; rock symbols same as on Figure 2 except Qal = Quaternary alluvium and QT = Quaternary terrace gravels. Horizontal scale: 7 cm = 1 km. A postulated reservoir model is shown as are the temperatures of major fluid entries in the core holes.

in Figure 2, the reservoir appears to underlie an area of at least 3.5 km^2. The thermal gradient data suggest that the minimum depth to a possible reservoir of 225°C is about 1.5 km. More likely, temperatures as great as 225°C occur at depths of 2 km and greater.

Flow test data from PLTG-1 and PLTG-2 indicate that fluids of 160° to 165°C are produced without stimulation from fractured andesite and metaconglomerate along the Quebrada del Agua Caliente fault zone. Scaling by aragonite in the upper casing and wellhead of PLTG-1 resulted from flashing and release of CO_2 because of lower flow rates and flowing pressures compared to PLTG-3. Downhole pumps with binary-cycle heat extraction technology and/or injection of scale inhibitors could probably prevent scaling if the relatively shallow 160°C resource is developed.

The LANL/USGS team submitted a recommendations report to ENEE in early 1988 (Laughlin, 1988). This report recommends that feasibility stage assessment begin at Platanares and that the identified 160°C resource be developed using binary generators. Results from this production drilling would be integrated with existing data to site a 2 km "test-for-discovery" well to locate the predicted 225° to 240°C resource.

ACKNOWLEDGMENTS

We thank the staffs of ENEE, Tegucigalpa, Honduras; the San Andres mine; Swissboring, Ltd., Guatemala City; and the people of Platanares. This project was funded by the U.S. State Department, Agency for International Development.

REFERENCES CITED

Bargar, K., in press, Fluid inclusions and preliminary studies of hydrothermal alteration in core hole PLTG-1, Platanares geothermal area, Honduras: Journal of Volcanology and Geothermal Research.

Fournier, R. O., 1981, Application of water geochemistry to geothermal exploration and reservoir engineering, in L. Rybach and L. J. P. Muffler, eds., Geothermal systems: Principles and case histories: New York, J. Wiley and Sons, p. 109-143.

Goff, F., L. Shevenell, C. Janik, A. Truesdell, C. Grigsby, and R. Parades, 1986, Hydrogeochemistry and preliminary reservoir model of the Platanares geothermal system, Honduras, Central America: Geothermal Research Council Transactions, v. 10, p. 125-130.

Goff, F., L. Shevenell, S. Kelkar, D. Smith, J. Meert, G. Heiken, K. Bargar, N. Ramos, A. Truesdell, M. Stallard, and J. Musgrave, 1987a, Stratigraphy, temperature profiles, and flow test data from the PLTG-1 and PLTG-2 core holes, Platanares geothermal system, Honduras: Geothermal Research Council Transactions, v. 11, p. 253-259.

Goff, F., A. Truesdell, C. Grigsby, C. Janik, L. Shevenell, R. Parades, J. Guiterrez, P. Trujillo, and D. Counce, 1987b, Hydrogeothermal investigation of six geothermal sites in Honduras, Central America: Los Alamos National Laboratory Report LA-10785-MS, 170p.

Goff, F., A. Truesdell, L. Shevenell, C. Janik, C. Grigsby, R.

Parades, P. Trujillo, D. Counce, J. Gutierrez, A. Adams, F. Urbani, and R. Perdomo, 1988, Hydrogeochemical report of the second Honduras sampling trip, January-February, 1986: Los Alamos National Laboratory Report, 105 p.

Goff, S., H. Rüfenacht, A. Laughlin, A. Adams, H. Planner, and N. Ramos, 1987, Geothermal core hole drilling and operations, Platanares, Honduras, Central America: Geothermal Research Council Transactions, v. 11, p. 37.

Goff, S., A. Laughlin, H. Rüfenacht, F. Goff, G. Heiken, A. Adams, J. Musgrave, H. Planner, and N. Ramos, 1988, Exploration geothermal gradient drilling, Platanares, Honduras, Central America: Los Alamos National Laboratory Report LA-11349-MS, 29 p.

Heiken, G., D. Eppler, K. Wohletz, W. Flores, N. Ramos, and A. Ritchie, 1986, Geology of the Platanares geothermal site, Departmento de Copán, Honduras, Central America: Los Alamos National Laboratory Report LA-10634-MS, 24p.

Heiken, G., W. Duffield, K. Wohletz, S. Priest, N. Ramos, W. Flores, D. Eppler, A. Ritchie, and C. Escobar, 1987, Geology of the Platanares geothermal area, Copán, Honduras: Geothermal Research Council Transactions, v. 11, p. 263-266.

Hoover, D., and H. Pierce, 1988, Electrical geophysical studies of the Platanares geothermal area, Honduras: U.S. Geological Survey Report, 24p.

Laughlin, A. W., 1988, Recommendations report for the Platanares geothermal site, Department of Copán, Honduras: Los Alamos National Laboratory Report LA-11410-MS, 73p.

Truesdell, A., C. Janik, F. Goff, L. Shevenell, P. Trujillo, D. Counce, B. Kennedy, and R. Parades, 1987a, The origin of thermal water of Honduras and puzzling variations in spring chemistries, in Proceedings of the ninth New Zealand geothermal workshop: University of Auckland, Geothermal Institute, p. 79-88.

Truesdell, A., M. Stallard, P. Trujillo, D. Counce, C. Janik, T. Winnett, F. Goff, and L. Shevenell, 1987b, Interpretation of fluid chemistry from the PLTG-1 exploratory drill hole, Platanares, Honduras: Geothermal Research Council Transactions, v. 11, p. 217-222.

Circum-Pacific Council for Energy and Mineral Resources
Earth Science Series, Vol. 16
R. L. Miller, G. Escalante, J. A. Reinemund, and M. J. Bergin (Eds.)
Energy and Mineral Potential of the Central American-Caribbean Regions
Springer-Verlag, Berlin Heidelberg 1995

AN ECONOMIC PREFEASIBILITY STUDY OF GEOTHERMAL ENERGY DEVELOPMENT AT PLATANARES, HONDURAS

Linda K. Trocki
Los Alamos National Laboratory, Los Alamos, New Mexico 87545 USA

ABSTRACT

The expected economic benefits from development of a geothermal power plant at Platanares in the Department of Copán, Honduras, are evaluated in this report. The economic benefits of geothermal plants ranging in size from a 10-MW plant in the shallow reservoir to a 20-, 30-, 55-, or 110-MW plant in the assumed deeper reservoir were measured by computing optimal expansion plans for each size of geothermal plant. Savings are computed as the difference in present value cost between a plan that must rely on other alternatives and one that contains a geothermal plant. Present value savings in 1987 dollars range from $25 million for the 10-MW plant to $110 million for the 110-MW plant—savings of 6 percent to 25 percent over the time period 1988 through 2008.

The existence of the shallow reservoir is relatively well-characterized, and much indirect scientific evidence indicates the existence of the deeper reservoir. Based on probability distributions estimated by geologists of temperature, areal extent, depth, and porosity, the expected size of a power plant that the deep reservoir can support was calculated with the following results: 0-MW—16 percent (i.e., there is a 16 percent chance that the deep reservoir cannot support a power plant); 20-MW—38 percent; 30-MW—25 percent; 55-MW—19 percent; and 110-MW—2 percent. When the cost savings from each size of plant are weighted by the probability that the reservoir will support a plant of that size, the expected monetary value of the deep reservoir can be computed. It is $42 million in present value 1987 dollars—a cost savings of 10 percent. The expected savings from the 10-MW plant in the shallow reservoir are expected to be close to the computed value of $25 million, i.e., the probability that the shallow reservoir can support the plant is high.

INTRODUCTION

In the course of a geothermal resource assessment of Honduras, geoscientists from the Empresa Nacional de Energía Electrica (ENEE), Los Alamos National Laboratory (Los Alamos), and the U.S. Geological Survey (USGS) confirmed the potential for development of geothermal resources at the Platanares site in the Department of Copán, Honduras. The scientific recommendations appear in a report by the Los Alamos/USGS Assessment Team (1988).

During the exploration of the Platanares site, existence of a shallow reservoir was proven by drilling. Two of the three exploration core holes intersected a shallow reservoir. Temperatures of 160°C or more were observed at depths of 650 m or more (Los Alamos/USGS Assessment Team, 1988).

The main purpose of this study is to determine the economic prefeasibility of developing the deeper, 225-240°C reservoir indicated to exist at depths of 1.2 to 1.5 km and to determine the benefits of early development of a 10-MW plant in the shallow reservoir. A prefeasibility study is performed to assess whether a project is favorable enough to warrant a feasibility study. "Prefeasibility" is a term recommended by the InterAmerican Development Bank.

Because the size of the deeper reservoir cannot be measured until the feasibility study of the Platanares geothermal field is completed, the economic benefits of geothermal energy development were determined parametrically; that is, first, the geoscientists involved in the field evaluation estimated the probability of the temperature, surface extent, thickness, and porosity of the host for the deeper reservoir, which would affect the size of the power plant that could be installed at Platanares. Second, optimal expansion plans were computed based on the assumptions that the deeper reservoir may support a 20-, 30-, 55-, or 110-MW power plant. Third, when the cost savings from geothermal power plants of various sizes were weighted by the probability that the deeper reservoir can support a given size power plant and these weighted benefits were summed, an expected monetary value of the feasibility stage and geothermal energy development at Platanares was obtained.

ENERGY SITUATION IN HONDURAS

Less than one-fifth of the energy consumed in Honduras is in the form of electricity. Almost two-thirds of the country's energy derives from fuelwood, which is used for residential cooking and for heating in some industrial processes. The remaining quarter of the country's energy comes from petroleum products, consumed mainly in the transport sector and, to a lesser extent, in the industrial sector. The share of electricity in total energy consumption will increase significantly, however, as more consumers are connected to the electrical grid. Although Honduras' per capita consumption of electricity approximates only one-third the level in Costa Rica and Panama, the most developed Central American countries, consumption is growing at a faster rate than in any other Central American country. Electricity use increased at a rate of 8 percent per year in the 1970s through the mid-1980s (Trocki et al., 1987). In this study, the assumed growth rate in electricity demand was approximately 8 percent per year from 1987 through 1992, leveling off at 5 percent per year through the end of the period considered—2008.

Honduras' electric generation capacity comprised 558 MW in 1987. Like other Central American countries, Honduras aggressively developed its abundant hydroelectric resources after the oil price shocks of the 1970s, so that more than three quarters of their present capacity is hydroelectric (Trocki et al., 1988). The remainder consists of aging diesel generators and gas turbines. One hydroelectric plant, El Cajon, which commenced operation in 1985, has 300 MW of capacity—more than half the country's total capacity. In 1987, peak demand reached approximately 260 MW. Although it appears that Honduras possesses a large amount of excess capacity, oil- and gas-fired plants will require replacement; and geothermal and thermal capacity is necessary to complement hydroelectric capacity, especially during the dry climatic season. Planners in the national utility expressed the need for at least 20 MW of new capacity by the year 1992. The purpose of this study was to determine if geothermal plants could serve this need. Feasibility studies are underway for hydroelectric projects, and thermal power plants are also alternatives.

Geothermal plants are considered a good option for Honduras for at least three reasons. First, abundant geothermal resources are being discovered and neighboring countries of Nicaragua and El Salvador have successfully exploited these resources; Costa Rica and Guatemala are also installing geothermal plants. Second, geothermal plants present a diversification away from thermal and hydroelectric capacity, and provide a steady source of electricity generation, useful to complement hydroelectric generation during the dry season. Third, although more costly to install, geothermal plants have an advantage over thermal plants in that imported fuel, subject to price increases, is not required for geothermal plant operation.

GEOTHERMAL RESERVOIR DATA AND ASSUMPTIONS

The standard procedure for determining the benefits of a new energy plant is to compute an optimal capacity expansion plan for the next 20 years, considering all of the options available to Honduras (e.g., geothermal plants, combustion turbines, diesel generators, and several hydroelectric plants). According to the results, described later, almost 800 MW of additional capacity will be required between the present and 2008. Of this amount, four hydroelectric projects totaling 323 MW are included in the plan; the remaining amount consists of mixtures of thermal and geothermal plants. The plan assumes that the reservoir at Platanares could support from 10 MW in the shallow reservoir and up to 110 MW of geothermal capacity in the deeper reservoir. The rationale supporting these assumptions is described in this section.

The cost of geothermal development, in the absence of a feasibility study, was assumed to be the same as the cost of development at the Miravalles field in Costa Rica. Those costs are based upon figures provided by the InterAmerican Development Bank for Miravalles II and include quotes from Ormat, Inc., who built the 10 MW plant. Cost estimates (per kw) at Platanares ranged from $1,715 for a 55-MW plant to $2,350 for a 10-MW plant, including costs of wells, engineering and administration, turbines, pipes, land, and transmission lines. The cost for the 10-MW plant is for a binary-cycle, transportable power plant that can be ordered on a turn-key basis. Costs do not include the test-for-discovery well ($1.5 million) and the feasibility study for the deeper reservoir development ($12 million).

The size of the deeper reservoir was parametrically studied, based on expected reservoir characteristics shown in Table 1. These parameters were established by the geologists who performed the prefeasibility study (Goff et al., this volume).

It is known with certainty that the shallow geothermal reservoir exists. Flow tests on new wells that can later be used in production will prove whether the shallow field can support a 10-MW plant. The existence of the deeper reservoir, which would need to be tapped to support a 20-, 30-, 55-, or 110-MW field, is based upon indirect geologic evidence (Goff et al., this volume). The probable temperature range of the deeper reservoir was determined using silica, cation,

Table 1. Expected reservoir characteristics—Deeper Reservoir

	Low		Medium		High	
		Probability [a]		Probability [a]		Probability [a]
	Value	%	Value	%	Value	%
Temperature (°C)	205	5	225	80	240	15
Areal Extent of Reservoir (km^2)	3.5	50	7.0	45	10	5
Thickness of Reservoir (km)	0.5	45	1.0	50	2.5	5
Porosity (%)	5	33	7.5	33	10	33

a"Probability" refers to geoscientists' estimate of probability of occurrence.

isotope, and gas geothermometers (Los Alamos/USGS Assessment Team, 1988). The results of the geothermometers indicate reservoir temperatures of 205 to 240°C.

The exergy of the reservoir was calculated for the low, medium, and high temperatures (205, 225, and 240°C). Based on the exergy calculations, the number of wells required to support various sizes of plants and the required fluid volumes was determined. An efficiency factor of 35 percent was applied to the exergy figures to represent the amount of electrical energy that can be extracted from the exergy (Ronald DiPippo, Southeastern Massachusetts University, North Dartmouth, Massachusetts, May, 1988, personal communication). Varying fluid flow rates were assumed to determine how many wells would be required to support the sizes of geothermal power plants under consideration in this study. The required volume of fluids to support a given power plant for 25 years at an 80 percent capacity factor was also determined. The probable volume of reservoir fluids was estimated using the expected surface area of the reservoir times the expected thickness and effective porosity.

The top of the deep reservoir is expected to occur at a depth of 1,200 to 1,500 m. Shallow test wells drilled at Platanares have intersected the Valle de Angeles Group at depths of approximately 600 m. If the Valle de Angeles Group is greater than 3,000 m thick in the region, if the top of the Group is 600 m below the surface and if the top of the deep reservoir is at a 1,200-m depth, then the maximum thickness of the reservoir could be 2,400 m.

Effective porosity is expected to range from 5 percent to 10 percent, and for the purpose of this approximation, it is assumed that all of the fluids and exergy are recoverable. Artesian flow rates observed in 600-m-deep wells also led the geoscientists who conducted the prefeasibility study to believe that a large percentage of exergy in the fluids can be captured to produce electricity. Based on advice from R. DiPippo, we too calculated exergy levels from standard 9-5/8 in. slim holes assuming flow rates varying from 50 to 100 kg/s. The drilling of deeper wells during the feasibility stage will provide more data for optimization of design of the system.

The synthesis of the probability estimates of reservoir characteristics and required energy for various size power plants result in estimates of the probability of occurrence of 20-, 30-, 55-, and 110-MW-size fields at Platanares as shown in Figure 1. The required fluid volume necessary to support various size power plants for reservoirs at possible exergy levels corresponding to temperatures between 205 and 240°C was applied to derive the probabilities that Platanares will support 20-, 30-, 55-, or 110-MW power plants. This treatment implicitly assumes that the geothermal resource is exhaustible; i.e., there is no replenishment of the fluids from recharge or reinjection. This assumption is a conservative one.

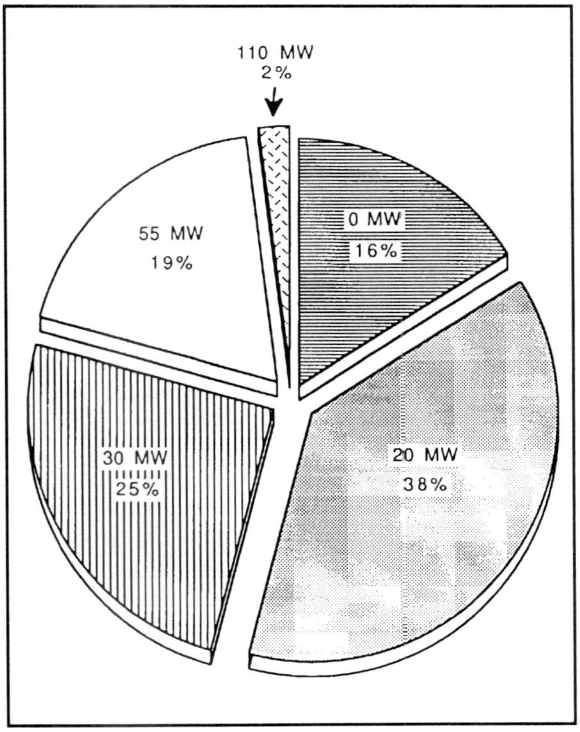

Figure 1. Probabilities of the deep reservoir at Platanares being able to support geothermal plants of assumed sizes from 0 to 110 MW.

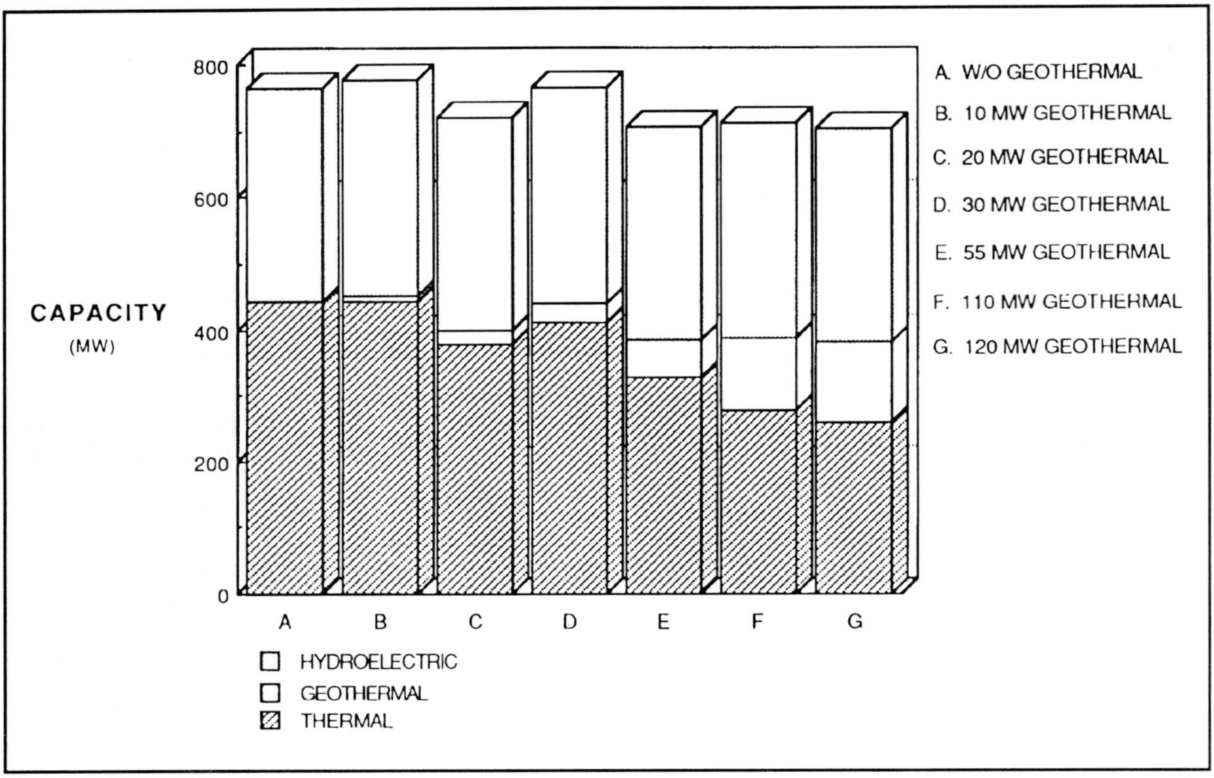

Figure 2. Optimal expansion plans with varying amounts of geothermal power.

RESULTS

Optimal expansion plant studies were conducted using the Wien Automatic System Planning package (WASP) Version III (International Atomic Energy Agency, 1980). WASP is a dynamic optimization code that computes an optimal electric generating expansion plan, i.e., in the present study, the schedule and type of plants that should be installed by Empresa Nacional de Energía (ENEE) to minimize the present value cost of electricity generation for Honduras. The benefits of geothermal energy development at Platanares are measured as the difference in the present cost of electricity and cost following a plan that includes geothermal plants.

Required data inputs to WASP include an estimate of growth in peak demand; characteristics of the load duration curve; operating characteristics of existing thermal and hydroelectric plants in similar environments; characteristics of possible new hydroelectric, thermal, and geothermal plants; cost figures for fuel, operation and maintenance, and capital for new plants; constraints on the expansion schedule; and other data, such as the maximum acceptable probability of loss of load.

Five possible sizes of geothermal fields were evaluated—10 MW in the shallow reservoir, plus 20, 30, 55, and 110 MW in the deeper reservoir. The evaluation period was 1988 through 2008. The earliest year in which it would be possible to install the 10-MW plant is 1991. For the larger plants, the earliest possible installation date is 1995.

The optimal expansion plans determined by WASP appear in Figure 2 and the costs of the plans appear in Figure 3. The cost of the plans includes construction costs of any new plants selected in the plan, salvage values, operation and maintenance costs for existing and new plants, and the cost of unserved energy. All costs are expressed as present value costs in 1987 US dollars. Costs incurred in the future have been discounted to 1987 dollars using a 12 percent discount rate.

The effect of introducing geothermal plants into the energy scenario would be a sizable reduction in the present value cost of electricity production in Honduras. Because geothermal plants can be programmed to operate at a high capacity during the dry season, their operation can substitute for some thermal generation, which is much more expensive. The geothermal plants substitute for thermal capacity, demonstrated by Figure 2. Plan A, with no geothermal alternatives, calls for the installation of almost 450 MW of gas turbines between 1988 and 2008; Plan G, with the maximum amount of geothermal energy, calls for 260 MW of gas turbines. The present value operating costs decrease as the amount of geothermal capacity increases and the thermal capacity decreases, because the thermal plants are more expensive to operate.

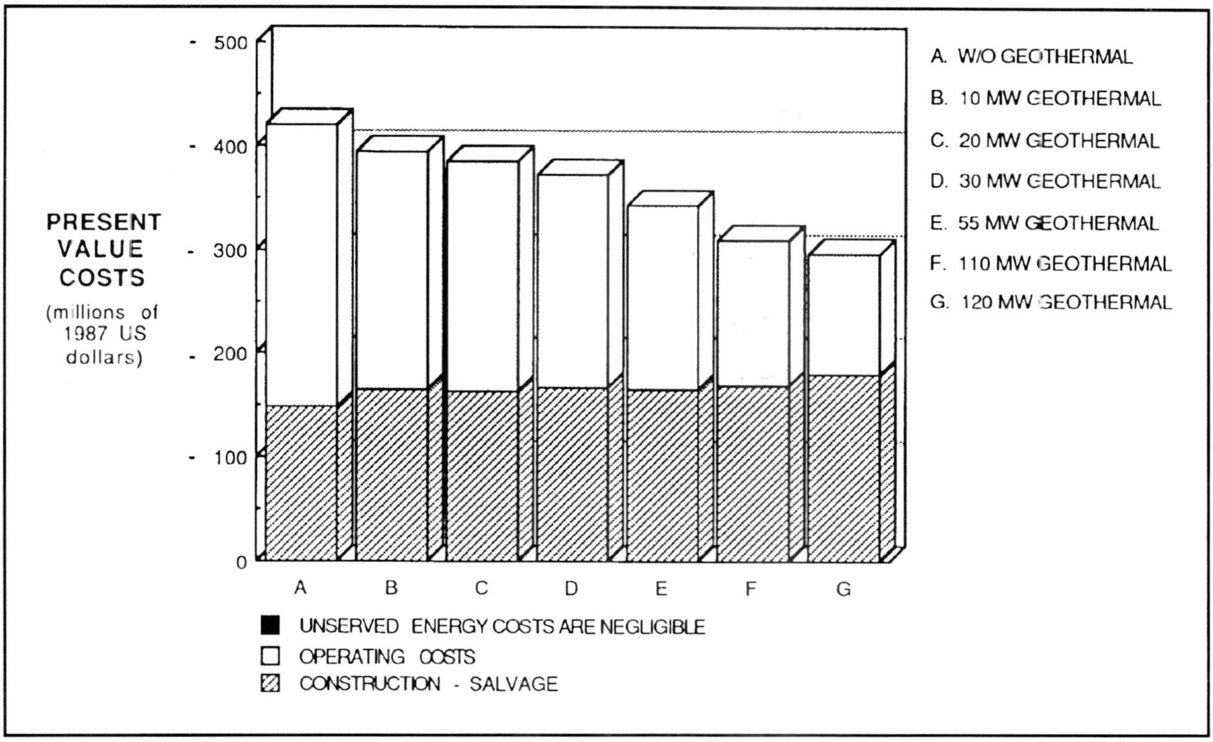

Figure 3. Present value cost of optimal expansion plans with varying amounts of geothermal power.

From an economic point of view, construction of the 10-MW plant, which would operate at lower temperatures from fluids in the shallow reservoir, affords significant economic benefits. In the cases where the 10-MW plant is included as an option (Plans B and G), it is selected in the least cost solution in the first or second year in which it would be possible to install it—1991. Cost savings resulting from the 10-MW plant were determined by subtracting the cost of the plan that contains the 10-MW geothermal plant from the comparable plan that does not contain it—the cost of Plan A minus Plan B equals \$25 million, and the cost of Plan F minus Plan G equals \$14 million.

The expected value of the feasibility stage is approximated as the benefit of geothermal power to Honduras times the probability that its feasibility will be proven and developed, or

$$EMV = p(10\ MW)xB(10\ MW) + p(20\ MW)xB(20\ MW) + p(30\ MW)xB(30\ MW) + p(55\ MW)xB(55\ MW) + p(110\ MW)xB(110\ MW),$$

where

EMV = expected monetary value,

p = marginal probability of occurrence of the given sized field, and

B = monetary benefits from the development of the given size field (Table 2).

According to IDB guidelines, if the EMV of the feasibility study exceeds its cost, the feasibility study is warranted. The breakdown for expected monetary value appears in Table 2. The benefits of each size of geothermal power plant equal the present value cost of the optimal plan without geothermal power minus the present value cost of the plan with the geothermal power plant of the given size. For example, the benefit of the 20-MW plant (Table 2) equals \$419.7 million (Plan A, Figure 3) minus \$385.0 million (Plan C, Figure 3), or \$34.7 million. The expected monetary value of the feasibility study, assuming that construction of the power plant follows in time for the plant to come onstream when needed in the WASP simulation, is \$42.4 million. The cost of the feasibility stage is expected to be approximately \$12 million.

SUMMARY

The purpose of this study was to ascertain if geothermal plants should be included in ENEE's optimal expansion plan, and if so, what the cost savings are from installation of a geothermal plant.

In this analysis, geothermal plant alternatives were always chosen for inclusion in the optimal expansion plans,

Table 2. Expected monetary value of geothermal energy development in Honduras

Capacity of Geothermal Reservoir (MW)	Cost of Expansion Plan (1987 US$ million)	Benefits (1987 US$ million)	Probability of Occurrence %	EMV of Deep Reservoir (1987 US$ million)
W/O Geothermal	419.7	0	16	0
10	394.3	25.4	—	—[a]
20	386.0	34.7	38	13.2
30	371.3	48.4	25	12.1
55	341.4	78.3	19	14.9
110	309.0	110.7	2	2.2
120	294.7	125.0		—[b]
				42.4

[a]Benefit of shallow reservoir development is $26.4 million times a probability much greater than 50 percent that the shallow reservoir will support development.
[b]This plan is a combination of the installation of a 110-MW plant in the deep reservoir and a 10-MW plant in the shallow reservoir. Its benefits are included in these above two cases.

often in the first year in which it was possible to install them. Between 1991 and 2008, 323.1 MW of hydroelectric plants were selected from the list of alternatives for installation as well as 278 to 442 MW of thermal plants, depending on the capacity of geothermal plants included in the plan.

The expected value benefits from cost savings of development of the 10-MW plant alone in the shallow reservoir are $25.4 million. In order, however, to prove the feasibility of the 10-MW plant, it will be necessary to test drill the shallow reservoir. Test drilling and flow tests could be completed within one year. The feasibility study for the shallow reservoir is not expected to add significantly to investment costs for the 10-MW plant. The outcome of this short-term feasibility study is expected to confirm the potential for 10 MW or more of geothermal capacity with a probability assigned by an expert of greater than 50 percent. Binary-cycle geothermal power plants with sufficient energy to total 10 MW can be ordered on a turn-key basis from companies that manufacture such plants, and can be installed rapidly. The binary wellhead generators are transportable and could be moved to another field if ENEE desires.

The expected monetary value of development of the deeper reservoir at Platanares is $42.4 million (1987 present value) of cost savings resulting from installation of the geothermal plant. Cost savings arise mainly from a reduction in operation costs in plans that contain geothermal power plants because they substitute for thermal plants. The cost savings do not include drilling costs to confirm the presence of a deep reservoir or a feasibility study at Platanares. A discovery well is expected to cost approximately $1.5 million including mobilization costs. If the results are favorable, the following feasibility study would cost $12 million. As with the 10-MW plant, some of the wells drilled during the feasibility study could also be used as production wells. If the test-for-discovery well drilling occurred in 1989 and the feasibility study followed immediately in 1990 and 1991, the present value cost of this work would be approximately $10 million, applying a discount rate of 12 percent. The net expected monetary benefits would thus be $42.4 million minus approximately $10 million or $32.4 million.

ACKNOWLEDGMENTS

The author wishes to gratefully acknowledge the cooperation of Lic. Mauricio Mossi and Ings. Danilo Esquivel, Gerardo Salgado, and Wilmer Flores of ENEE. Without their assistance, this study could not have been performed. The author is also appreciative of data provided by other members of ENEE, including Ings. Marco Waimin, Edgardo Ulloa, and Roberto Avalos.

Ing. Gustavo Calderon of the InterAmerican Development Bank has been extremely helpful throughout the study, providing both data and guidance. Ronald DiPippo of Southeastern Massachusetts University has also kindly provided advice.

The author had the advantage of using the ENPEP version of WASP-III adapted by the Argonne National Laboratory. She is especially grateful to William Buehring, Karen Guziel, and Bruce Hamilton of Argonne National Laboratory for demonstrating ENPEP's use.

The paper benefited from a review by David Sussman.

Many thanks are due to the managers R. Wayne Hardie and Edward Van Eeckhout of Los Alamos National Laboratory for their support and to the scientists at the Los Alamos National Laboratory who contributed to the study, especially William Laughlin, Sue Goff, Fraser Goff, and Richard Beckman. In addition, the report has benefited from internal review by several scientists and engineers at Los Alamos. Kay Grady and Hazel Kutac masterfully typed the report.

REFERENCES

Goff, S., F. Goff, G. Heiken, W. Duffield, A. Truesdell, A. W. Laughlin, and W. Flores, Prefeasibility geothermal assessment, Platanares, Department of Copán, Honduras, this volume.

International Atomic Energy Agency, 1980, Wien automatic system planning package (WASP)—A computer code for power generating expansion planning: Vienna, IAEA, 250p.

Los Alamos/USGS Assessment Team, 1988, Recommendations report for the Platanares geothermal site, Department of Copán, Honduras, 59p.

Trocki, L. K., S. R. Booth, A. Umana Q., P. C. Aragon, A. Youngblood-Turpin, R. H. Drake, D. W. Morris, F. Roach, and T. A. Selva, 1987, The energy situation in five Central American countries: Los Alamos National Laboratory Report No. LA-10988-MS, UC-98, 365p.

Trocki, L., C. K. Newman, F. Gurule, P. C. Aragon, and C. Peck, 1988, Un atlas energetico de cinco paises Centroamericanos: Los Alamos National Laboratory Report No. LA-11205-MS, UC-98, 126p.

Circum-Pacific Council for Energy and Mineral Resources
Earth Science Series, Vol. 16
R. L. Miller, G. Escalante, J. A. Reinemund, and M. J. Bergin (Eds.)
Energy and Mineral Potential of the Central American-Caribbean Regions
Springer-Verlag, Berlin Heidelberg 1995

APPLICATION OF GEOTHERMAL ENERGY TO MINERAL PROCESSING: CYANIDE HEAP-LEACHING OF LOW-GRADE GOLD ORE

Thomas Flynn and Dennis Trexler
University of Nevada, Las Vegas, Environmental Research Center
Division of Earth Sciences, 100 Washington Street, Suite 201, Reno, NV 89503 USA

James L. Hendrix
Mackay School of Mines, University of Nevada, Reno, Reno NV 89557 USA

ABSTRACT

The advent of cyanide heap-leaching for low-grade gold and silver ores has resulted in an unprecedented increase in the output of mines throughout the world. This simple and economical process entails open-pit mining, milling, and stacking ore, which is then soaked with a dilute solution of sodium cyanide. The cyanide solution dissolves both gold and silver, which is subsequently recovered by additional processing. Many of these mines are located in remote, inaccessible areas and are subject to severe climatic conditions. Some operations are terminated during extreme cold or rainy seasons. As a result of United States Department of Energy sponsored investigations conducted by the Division of Earth Sciences, University of Nevada, Las Vegas, and the Mackay School of Mines, University of Nevada, Reno, some operating mines in Nevada were found to be co-located with geothermal resources suitable for use in cyanide heap-leaching operations. Low-to moderate-temperature geothermal resources may be used to heat the cyanide circuit, providing a low-cost method of maintaining the operation during freezing conditions. This paper describes the results of bench-scale heap-leaching experiments using a simulated geothermal source to increase metal dissolution rate. In addition, examples from two active mines that utilize geothermal energy in commercial-scale operations are included. The implications for co-development of gold and geothermal resources in Central America are discussed.

INTRODUCTION

Mining and mineral processing activities throughout the world are often located in remote, poorly accessible areas. A reliable and affordable energy source is an essential factor in profitable mine management. Geothermal energy resources, which appear at the surface as fumaroles and hot springs, are often co-located with hydrothermal mineral deposits (Trexler, Flynn, and Hendrix, 1987). Recent technological advances permit utilization of moderate temperature resources for process heating and electric power generation. In addition, large-scale gold and silver mines, which exploit low-grade, disseminated ore bodies, may operate in one location for 20 to 30 years. The critical need for low-cost energy and close proximity of geothermal resources combine favorably for the mining and mineral processing industry. This paper describes the results of ongoing investigations on geothermal energy applications in the mining industry. Bench-scale experimental results, examples of present-day use, and suggestions for geothermal energy utilization at selected mine sites are included.

The original purpose of this research was to demonstrate a practical application of geothermal energy for active mines in Nevada that use cyanide heap-leaching to recover gold and silver from low-grade ore. During the course of the investigation, two other applications of geothermal energy to

Table 1. Proximity of Geothermal Resources to Active Mines in Nevada.

Name of Mine	Type of Ore	Geothermal Proximity	Fluid Temperature
Pegasus Gold Corp.	Gold/Silver	On Site	100°C
Round Mountain Gold	Gold/Silver	On Site	86°C
West Northumberland	Gold	13 km	68°C
Jerritt Canyon	Gold/Silver	6 km	90°C
Boot Strap Mine	Gold	On Site	50°C
Rawhide Mine	Gold/Silver	10 km	62°C
16 to 1 Mine	Silver	1 km	48°C
Maggie Creek Mine	Gold	On Site	warm
Gold Quarry Mine	Gold	On Site	warm
Tonopah Divide	Gold/Silver	On Site	warm

the mining industry were identified. One process is an experimental pre-treatment of sulfide-rich gold and silver ores using bacteria. The other process is the use of moderate- to high-temperature geothermal fluids to produce electricity using binary fluid technology. This paper focuses on the application of geothermal energy to cyanide heap-leaching operations. The principal effort was to illustrate the benefits of low-temperature geothermal fluids as an economical alternative to fossil fuels used to maintain temperatures in aqueous cyanide solutions during cold winter months.

CO-LOCATION OF PRECIOUS METAL MINES AND GEOTHERMAL RESOURCES

The benefits of using geothermal fluids are directly related to the proximity of the resource to the mine site. In Nevada, for example, 32 mines that produce gold, silver, or both, were identified, but only ten were found to be closely associated with known geothermal areas (Table 1).

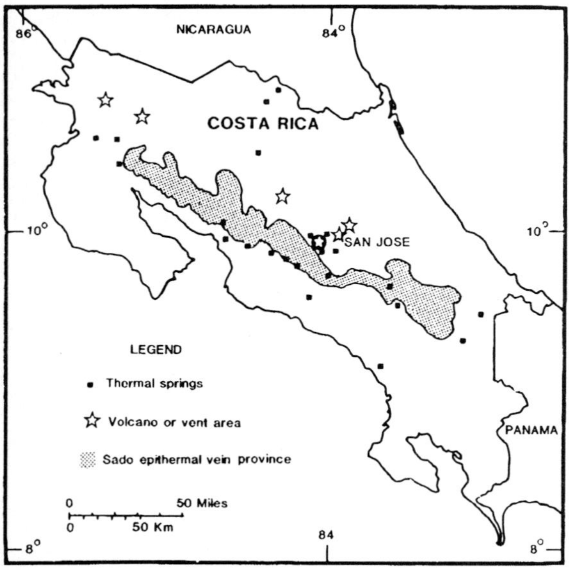

Figure 1. Location of thermal manifestations in relation to the Sado epithermal vein metallogenic province of Costa Rica.

Geothermal springs are widespread throughout Central America and geothermal resource exploration and development are underway from Guatemala to Panama. Gold mining, especially of low-grade ores, has become increasingly more important to the economic development of all of the countries in Central America. Because mineral and geothermal resources are both site-specific activities, a reasonable degree of co-location must be achieved to maximize the potential economic benefits of joint exploitation. Figure 1, for example, shows the location of geothermal activity in relation to the principal metallogenic province in Costa Rica. Potential sites for co-development appear along or close to the Sado epithermal vein province.

The principal benefit of using geothermal energy in cyanide heap-leaching operations is the increase in productivity that results from chemical processing conducted at elevated temperatures. The use of geothermal fluids in cyanide heap-leaching operations should include a careful consideration of the following parameters.

Effects of Temperature

Julian and Smart (1903) were the first to demonstrate the relationship between temperature and rate of dissolution of gold by cyanide solutions and established 85°C as the maximum temperature for cyanide dissolution of gold. Several years later Meyer (1931) determined a maximum dissolution temperature of 80°C.

Solution temperature is critical because it is difficult to control and because external adjustments will alter chemical reaction rates. Temperature is difficult to control because in Nevada, for example, heap-leaching operations are performed outdoors where diurnal temperature variations of 10°C to 21°C are not uncommon. Operational difficulties in freezing temperatures can be reduced or eliminated by increasing the cyanide solution temperature.

Effect of Oxygen

Temperature is not the only factor that controls the rate of dissolution of gold in cyanide solutions. Elsner (1846) first recognized the essential need for atmospheric oxygen in the

Table 2. Ions that affect cyanidation.

Accelerating	Neutral	Retarding
Pb, Hg, Bi, Tl	Na, K, Cl, NO_3, SO_4	Fe, Cu, Zn, Ni, Ca, Ba, C, (Pb), Mn, S_2^-

dissolution of gold in cyanide solutions. Lund (1951) pointed out that in cyanide dissolution experiments with silver, oxygen is continuously consumed during the reaction. Habashi (1967) noted that atmospheric oxygen (20 percent of a volume of air) is the oxidizing agent universally used in cyanide gold mills.

The solubility of oxygen is inversely proportional to the temperature of aqueous solutions. It is essential that dissolved oxygen levels not decline as a result of adding heat to the aqueous solution. Habashi (1967) indicated that the important consideration is not the absolute amount of cyanide ion and oxygen, but the ratio of the molar concentrations. He determined that the maximum dissolution rate occurs when the molar ratio $[CN^-]/[Q]=6$.

Effect of pH

Another critical factor is the pH of the aqueous solution. In near-neutral pH or acidic solutions, the cyanide ion, $[CN^-]$, is subject to hydrolysis by water or decomposition by atmospheric carbon dioxide. In cyanide practice, the pH of mill solutions usually ranges from 11 to 12.

Effect of Foreign Ions

Habashi (1967) summarized the impact of foreign ions on the cyanide dissolution of gold and silver. He found that ions can accelerate, retard, or have no effect on the dissolution rate. Table 2 lists the ions most likely to be encountered in gold ores and their associated effects.

EXPERIMENTAL RESULTS

A series of bench-scale experiments were performed to determine the effects of elevated temperatures on typical Nevada gold and silver ores. All tests were conducted using a flask and shaker bath apparatus. One hundred grams of ore were placed in the 500 ml Erlenmeyer flask and 200 ml of 0.5 g/l NaCN solution with a pH of 10.5 was added to the flask. The flask was then attached to a shaker arm and immersed in a thermostatically controlled temperature bath. The flask was oscillated in the bath for a prescribed period of time at a constant temperature. The independent variable in these experiments was temperature.

The gold ore was obtained from the Freeport Jerritt Canyon mine, in Elko County, Nevada, and contained 0.25 oz/T gold. The silver ore was obtained from the Gooseberry mine in Washoe County, Nevada; it contained 6.30 oz/T. These data, reported as Troy ounces, were obtained by fire assay technique.

Gold Ore Results

Figure 2 shows that the extraction of gold from the Freeport ore is significantly increased at higher temperatures. Extraction is measured as a percentage of the total amount of gold in the ore. Maintaining the cyanide solution at 20°C provides an increase in the extraction rate of eight to twenty percent relative to 5°C. In addition to demonstrating the enhanced extraction rates at higher temperatures, these data show the influence of carbonaceous material in the ore on the pregnant cyanide fluids. In all three cases, but especially at 10°C, the data suggest that gold recovery actually decreases after two hours. This phenomena is caused by allowing pregnant solutions to remain in direct contact with carbonaceous material in the ore. In actual cyanide practice, the pregnant solutions are piped from the ore body to a series of columns that contain carbon. The gold is stripped from solution by the carbon and returns to the heaps depleted of gold. The gold is removed from the carbon columns in another processing step.

Silver Ore Results

The silver ore experiments were nearly identical to the gold ore tests. Results of the silver extraction from the Gooseberry ore for four different temperatures are shown in Figure 3. The effect of increasing temperature on the extraction rate for silver is also evident. In 24 hours, at 35°C, more than 90 percent of the available silver was extracted. At 5°C,

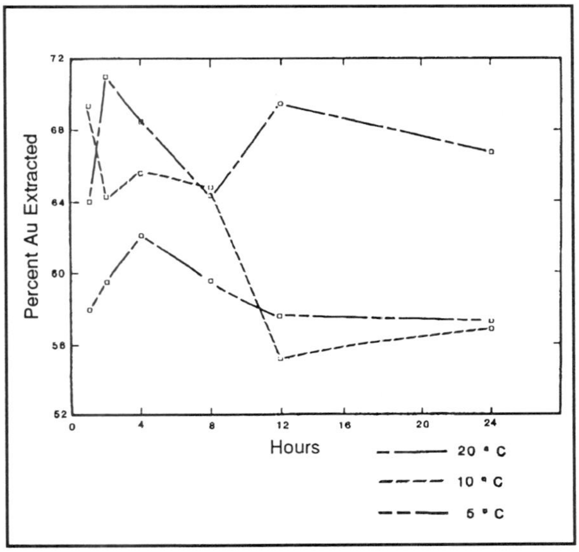

Figure 2. Gold extraction from Freeport ore when leached at various temperatures.

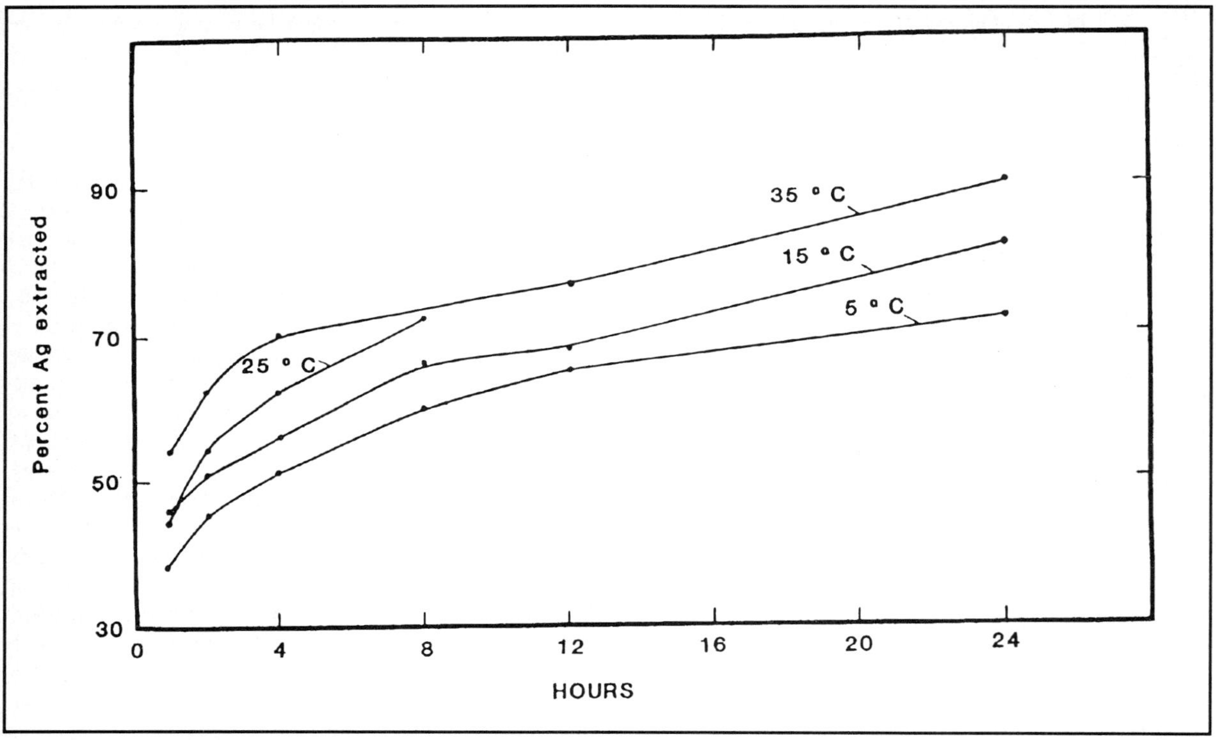

Figure 3. Silver extraction from Gooseberry ore when leached at various temperatures .

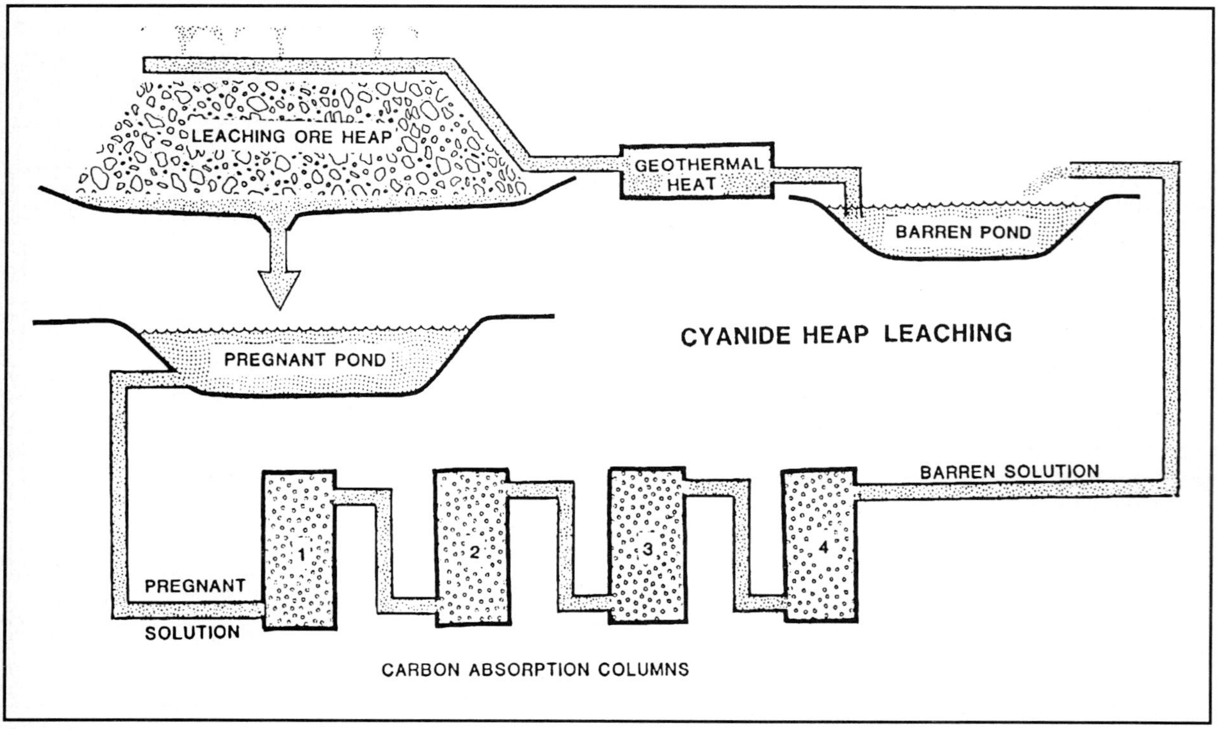

Figure 4. Diagram showing the major components of a cyanide heap-leaching operation using geothermal heat.

only 72 percent was extracted. Increases in recovery of silver with cyanide solutions at 35°C range from 17 to 39 percent for these short duration laboratory tests. These experiments demonstrate that higher temperatures can effectively increase the rate of gold and silver ore extraction from typical Nevada ores.

EXAMPLES OF GEOTHERMAL APPLICATIONS AT TWO NEVADA MINES

Figure 4 shows the configuration of a typical cyanide heap-leaching operation heated by geothermal fluids. The ore heap is soaked with a dilute solution of sodium cyanide, maintained at a pH of 11. Gold-rich solutions are pumped from the pregnant pond to carbon-filled columns that strip the gold and return the solution to a barren pond. From there, the cyanide solution is pumped back to the heap and the process is repeated. Many mines use sprinkler heads because the fluids are aerated in the process. Heating is usually applied to the solution immediately before it is sprayed on the heap. Figure 5 is a flow chart that illustrates the use of thermal fluids as a water resource, and the use of injection wells for geothermal fluid disposal.

Pegasus Gold Corporation - Florida Canyon Mine

The Florida Canyon gold deposit is located on the west flank of the Humboldt Range, within the Imlay mining district of central Nevada. The ore body consists of hydrothermally altered argillaceous sediments of the Late Triassic Grass Valley formation. The age of the hydrothermal alteration is uncertain, but K-Ar data from nearby quartz-adularia assemblages suggest an age of 70 to 75 m.y. (Silberman et al., 1973). The deposit is adjacent to the juncture of northeast-trending lineament and north-south-trending Basin and Range normal faults. The present open-pit minable reserves are 17.8 million t with an average grade of 0.025 oz/t. The waste to ore mass ratio is 1.09:1. The annual production for 1988 was 60,000 oz.

Geothermal fluids are pumped from a well, 180 m deep, which is located on the property. The mine is adjacent to the Rye Patch KGRA (Known Geothermal Resource Area). The maximum temperature in the well is 115°C and fluids are produced at 98°C. Geothermal fluids are high in total dissolved solids (TDS) and are the only source of groundwater at this mine. Pumped geothermal fluids are not immediately used in this operation, but are stored in an open pond where they are allowed to cool and absorb oxygen from the atmosphere (Figure 5). Once cooled, they are chemically treated with lime, sodium cyanide and an anti-scaling compound. The cyanide rich solution is then pumped through a geothermal heat exchanger to the heap. The heat exchanger receives fluids from the geothermal well and those fluids are piped to the cooling pond. The gold-laden pregnant cyanide solution is processed, the gold is removed, and the barren cyanide fluid is returned to the heap. Detailed geothermal fluid flow rates and temperature data are not available, but it has been estimated that the annual energy budget for this system is 40 billion BTU/yr.

Round Mountain Gold Corporation

Round Mountain Gold is currently the world's largest open-pit, heap-leach, gold mine. The mine is located in Big Smoky Valley, in central Nevada, at an elevation of 2 km above sea level. The ore body occurs within hydrothermally altered rhyolitic tuffs and ignimbrites. Gold is disseminated throughout the volcanic rocks and individual particles range in size from 6 mm to 2 microns. Only the smallest gold particles are leached.

In 1988, 234,000 oz of gold were produced. Ore production was recently increased from 18,000 t/d to 40,000 t/d with an anticipated recovery of 320,000 oz of gold in 1989 (production increase of 37 percent).

During the cold winter months, geothermal energy is used to heat the cyanide circuit from a temperature of 5°C to 22°C. This prevents the solution from freezing on the top of the heap, which would effectively terminate the operation. Geothermal fluids at a temperature of 82°C are pumped from three wells, located on the property, at a combined flow of 158 l/s to two flat-plate heat exchangers. The separate cyanide circuit has a flow rate of approximately 632 l/s, and this solution is also recycled (Figure 5).

Make-up water, i.e., water lost to evaporation, is piped from a non-thermal well, located approximately eight miles from the mine. Geothermal fluids, therefore, are not required as a fluid source and are injected back into the thermal aquifer.

Prior to the installation of the geothermal heating system, two oil emersion burners were used to heat the cyanide circuit during the cold winter months at an estimated

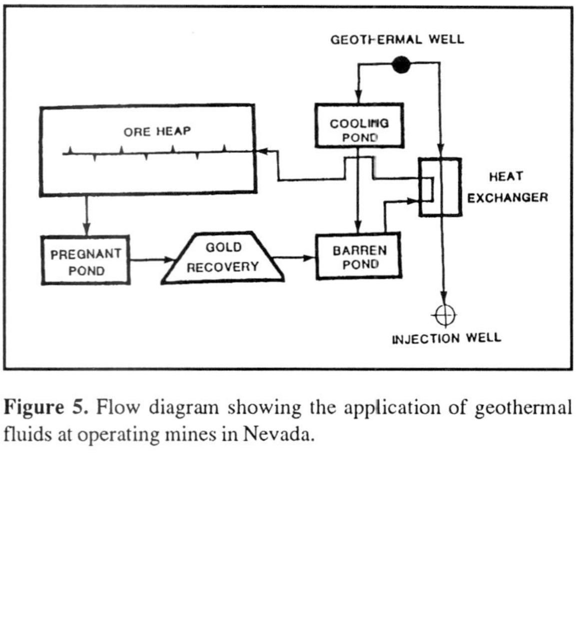

Figure 5. Flow diagram showing the application of geothermal fluids at operating mines in Nevada.

cost of $3,000/d. At current production rates, approximately 58 billion BTU per month are supplied by the geothermal resource to maintain leaching operations during the winter.

IMPLICATIONS FOR CENTRAL AMERICAN GOLD MINING OPERATIONS

Precious metal deposits and geothermal resources have been shown to be co-located in Nevada and Costa Rica. It is likely that this relationship is repeated throughout Central America. Unlike Nevada, however, Central America is not subject to freezing conditions, but climate-related problems may be minimized by using geothermal energy in cyanide heap-leaching operations.

Average daily temperatures, for example, vary with both latitude and elevation. Although nighttime temperatures may approach freezing at some mine sites in Central America, substantial damage to the heap is unlikely. Instead, cyanide dissolution of gold and silver simply slows down, increasing residence time on the heap and decreasing productivity. The application of geothermal heat increases the dissolution rate and can maintain high levels of productivity.

Another climate-related problem, which is characteristic of tropical regions, is a prolonged rainy season. Clay-rich hydrothermally altered volcanic and sedimentary rocks, which are the principal hosts for low-grade disseminated ore, present special logistical problems for heavy equipment operation. The combination of mud and near-constant rainfall disrupts production. Optimum productivity will occur by increasing production rates during the dry season. Using geothermal fluids can significantly increase productivity by reducing the time required for leaching. The end result is a more efficient use of time available.

CONCLUSIONS

The economic recovery of gold from low-grade, disseminated ore bodies can be enhanced by the application of low- to moderate-temperature geothermal fluids in cyanide heap-leaching operations. Large-scale, long-term mining activities should consider geothermal applications to prolong leaching operations in cold climates and to accelerate leaching in tropical climates that may have a rain-shortened mining season. Gold mines co-located with geothermal resources are common throughout the Pacific Basin and co-development of these resources is now practiced in Nevada. The identification of co-located resources may be important for leasing or concessions. In the western United States, for example, mining companies are encouraged to lease both mineral and geothermal rights to capitalize on the low cost energy resource and to eliminate potential conflicts with other organizations. This technology, which has been proven and demonstrated at two mines in Nevada, may find future application to precious metal mining in Central America.

ACKNOWLEDGMENTS

The authors gratefully acknowledge the symposium conveners and participants for their encouraging remarks. David Sussman of Unocal and Ralph Miller of the U.S. Geological survey provided invaluable technical and editorial support. Thanks also to the Nevada mining industry and the mines described in this report for their support.

This work was funded by the U.S. Department of Energy, San Francisco Operations Office, under grant number FG03-85SF15603. Detailed research, funded by the U.S. Bureau of Mines, is continuing at the Pegasus and Round Mountain gold mines in Nevada.

REFERENCES

Elsner, L., 1846, Uber das verhalten verschiedener metalle in einer wassrigen losung von cyankalium: J. Prak. Chem., v. 37, p. 441-446.

Habashi, F., 1967, Kinetics and mechanism of gold and silver dissolution in cyanide solutions: State of Montana Bureau of Mines and Geology, Bulletin 59.

Julian, H. F., and E. Smart, 1903, Cyaniding gold and silver ores: London, Griffen, 3rd ed., 1922.

Lund, V., 1951, The corrosion of silver by potassium cyanide solutions and oxygen: Acta Chim. Scand., v. 5, p. 555-567.

Meyer, H., 1931, Uber die wirksamkeit der arbeitslosungen in cyanidprozess: Metall. U. Erz., v. 28, p. 261-280.

Silberman, M. W., M. G. Johnson, R. A. Koski, and R. J. Roberts, 1973, K-Ar ages of mineral deposits at Wonder, Seven Troughs, Imlay, Ten Mile, and Adelaide mining districts in central Nevada: Isochron/West, no. 8, p. 31-35.

Trexler, D. T., T. Flynn, and J. L. Hendrix, 1987, Enhancement of precious metal recovery by geothermal heat: Geothermal Resources Council, Transactions, v. 11, p.15-22.

Circum-Pacific Council for Energy and Mineral Resources
Earth Science Series, Vol. 16
R. L. Miller, G. Escalante, J. A. Reinemund, and M. J. Bergin (Eds.)
Energy and Mineral Potential of the Central American-Caribbean Regions
Springer-Verlag, Berlin Heidelberg 1995

ENVIRONMENTAL IMPACTS ASSOCIATED WITH GEOTHERMAL EXPLORATION, DEVELOPMENT, AND POWER GENERATION

Agustin L. Viesca
Unocal Geothermal Division, P. O. Box 6854, Santa Rosa, California 95406 USA

ABSTRACT

The environmental impacts associated with geothermal exploration, development, and power generation include air, land, noise, and water pollution, visual impacts, waste generation, subsidence, and seismicity. These impacts become more important as geothermal activity advances from the exploration phase into the development and production phases. During exploration, none to a few of these impacts need to be addressed, depending on the project specifics (location, size, etc.); however, most of them eventually come into play in a successful project. Appropriate mitigation measures are also discussed. Emphasis is placed on noise, air and water pollution, and waste generation because these environmental impacts have been of special concern. The geothermal industry is relatively new in the power generation field and its impacts are not as severe as those of other electric power generating facilities. Nevertheless, progress in the development and implementation of geothermal mitigation technology has been substantial, and effective abatement equipment is available.

INTRODUCTION

The environmental impacts associated with geothermal power generation occur in three basic stages. These stages reflect the three phases of geothermal activity, which are exploration, development, and power generation (County of Lake, California, 1989). None of these stages involve the more difficult fuel-related problems faced by nuclear and fossil fuel power generation; however, they do present environmental impacts that must be properly mitigated.

This paper addresses the environmental impacts in each of these stages. The objective is to point out that the geothermal industry must also contend with environmental concerns, and that to mitigate these concerns one does not have to re-invent the wheel because there are effective mitigation measures for most situations. An overview of these impacts is included, placing emphasis on those problems that have been of special concern—noise, air pollution, water pollution, and waste generation.

EXPLORATION

In the exploration phase, the environmental impacts can range from none to a few. Table 1 lists these impacts along with suggested mitigation measures. Initial geological, geochemical, and geophysical exploration work has essentially no impacts associated with it. It may become necessary to conduct advanced exploration work to learn more about the geothermal potential in a given area. This additional work consists of drilling temperature gradient holes and exploration wells. The environmental impacts from these activities are of low magnitude and are short lived.

DEVELOPMENT

When an exploratory project results in the discovery of a resource large enough to offer commercial benefits, the project can then advance from the exploration into the devel-

Table 1. Environmental Impacts/Mitigation. Geothermal Exploration Phase.

ACTIVITY	IMPACT	MITIGATION
Geologic, Geochemical and Geophysical Surveys	None	--
Temperature Gradient Hole	Surface Damage	Use Berms and Drains. Revegetate.
	Visual	Use Natural Topography and Vegetation.
	Noise	Use Mufflers and Natural Topography.
Exploratory Drilling	Surface Damage	Minimize Pad Size and Road Width and Length. Use Berms and Drains. Revegetate.
	Visual	Use Natural Topography and Vegetation.
	Noise	Use Mufflers and Natural Topography. Conduct Noisy Operations During Least Sensitive Time. Maintain Neighbors Informed.
	Air Pollution	Control H_2S with Chemicals and Dust with Water.
	Water Pollution	Use Berms, Drains and Lined Sumps.
	Waste Generation	Evaporate and Bury on Site in Lined Sump.

opment phase. The development phase consists of production well drilling, construction of pipelines and other production facilities, and power plant construction. The environmental impacts associated with this phase depend on the following:

1. The project location. The closer the project is to sensitive areas, the higher the impact potential.

2. The project size. As the project increases in size, so does the impact potential.

3. The project life. If the project life is limited to a few years, as is the life of demonstration plants, then the impacts will be less than those associated with a long-life project.

4. The resource type. A dry resource may have fewer problems with waste generation than a hot-water resource, but more problems with air pollution and noise.

5. The degree of geothermal activity nearby. In highly developed areas, such as The Geysers in northern California, the incremental environmental change brought about by a new project may be of greater concern than in less developed or undeveloped areas.

6. The weather during construction. Construction during rainy seasons, for example, can result in more erosion and stream siltation and contamination problems than dry season construction.

The basic types of environmental impacts for development are the same as those for exploration. However, the environmental planning needed for development is more important than for exploration. In a development scenario, long-term needs and impacts must be considered.

Surface Damage

Geothermal wells can be drilled directionally, so that two wells may be located close to one another on the surface with their bottoms far apart. Therefore, where practical, drilling pads should be built to accommodate several wells to minimize surface damage. Grouping other production equipment, such as steam separators, will also reduce land disturbance. Steep terrain should be avoided, where possible, to reduce the number of cuts and fills and landslide problems. For example, in mountainous fields of 250 hectares dedicated to a 50 MW project, about 7 percent of the field, or 18 hectares may be disturbed (Eric Steger, personal communication). On flat land the disturbed area is, of course, less.

Visual

If possible, natural topography and vegetation should be used to reduce visual problems. Tree planting may be necessary to reduce equipment exposure. Facilities should be painted to blend with the environment.

Noise

To reduce noise, the use of mufflers, temporary barriers, and natural topography may be necessary. Furthermore, because noise has been found to be a particularly irritating problem, it is extremely important to conduct noisy drilling and construction operations during the least sensitive times of the day and to keep local regulators and neighbors informed about project activities.

Air Pollution

Although neither hydrogen sulfide (H_2S) nor dust emissions have been a major concern when drilling in hot-water fields, they can be a problem in dry steam fields. If air pollution is found to be a problem during drilling operations, emissions can usually be abated with existing technology. The sodium hydroxide/hydrogen peroxide abatement system, for example, has been used for many years to reduce to acceptable levels the emissions of H_2S, the air pollutant of greatest concern to the industry because of its bad odor (Hampshire, Woertz, and Castrantas, 1977, Unocal internal report). Abatement levels of 90 percent or greater are common with this system so that if, for example, a typical Geysers well emits 23 kg/hr of H_2S before treatment, it will only emit 2.3 kg/hr after treatment.

Water Pollution

To prevent the contamination of nearby sources of water by well testing and other activities, injection wells, berms, drains, and lined drilling sumps may be required. When possible, construction activities should be conducted during dry weather.

Waste Generation

Finally, a development project will generate drilling waste. The water in this waste can be allowed to evaporate, and the remaining solids can be buried in individual or centrally located sumps. The sumps may require clay or synthetic liners if, for example, contamination of local aquifers is a possibility. Waste dewatering processes expedite this waste minimization activity and reduce waste volumes by as much as 60 percent (Eric Steger, personal communication). The separated water can be disposed in injection wells.

POWER GENERATION

The power generation phase can be conducted with a minimum of environmental impacts, provided the proper environmental equipment is incorporated during project construction. All of the basic types of environmental impacts mentioned earlier also apply to operations. A further discussion of noise, air and water pollution, and waste generation, the impacts of special concern, follows a brief discussion of two additional impacts, not mentioned earlier, which are products of operations—seismicity and subsidence.

Seismicity

Seismicity can occur when the resource is produced and, more frequently, during reinjection. The Geysers is believed to be the area of greatest activity. There, about 30 micro earthquakes per day are recorded which are of magnitudes too low to be felt by people (Mitchel Stark, personal communication). A few earthquakes per year, ranging from three to four on the Richter scale, also occur (Mitchel Stark, personal communication). No serious incidents resulting from production or reinjection-related earthquakes have been reported. In places such as The Geysers and the Imperial Valley, California, and Larderello and Latera, Italy, monitoring networks track earthquake activity. Data from these networks can be useful in the event of a problem.

Subsidence

Resource production over a period of years can result in substantial subsidence. In Wairakei, New Zealand, for example, where production began in 1950, the maximum amount of vertical movement exceeds 10 m (Dench, 1986). In most other places, however, it is not nearly as high. Because subsidence occurs gradually, measures to prevent equipment damage can be implemented in a timely manner. Furthermore, through reinjection and deep production, subsidence can be reduced or prevented. Reinjection also provides support for the reservoir and prevents water contamination by providing for the disposal of condensate and produced fluids.

Problems of noise, air and water pollution, and waste generation occur in both field and power plant operations. In general, their impact can be reduced by properly coordinating operations. This observation is particularly important in areas where the field may be operated by one party and the power plant by another one, as is generally the case in California.

Noise

Although power plants can be noisy, field operations are the main source of noise problems (Table 2) (Long and Stern, 1979-82, Department of Energy internal report). Wells and steam lines are normally the sources of field noise. Well-related noise occurs during drilling, cleaning, and venting operations. Pipeline noise occurs at small pipeline vents and during stacking. It is important to note that the noise readings in Table 2 were taken just 30 m from the source. These readings probably would be in the 40s or lower at a distance of one kilometer from the source. Briefly, natural noise levels

Table 2. Noise Impacts—Generation.

NOISE SOURCE	UNMITIGATED (dBA @ 30M.)	MITIGATED (dBA @ 30 M.)
Well:		
Drilling	103	88
Cleaning	108	83
Venting	65	22
Pipeline:		
Venting	92	55
Stacking	100-110*	68

*Approximate Levels

in rural areas are usually 40 dBA or lower, and an increase of 10 dBA sounds twice as loud. The use of natural or man-made barriers and mufflers reduces noise to acceptable levels. Rock mufflers have been particularly effective in reducing noise.

Air Pollution

Air pollution can also be a problem if it is not properly mitigated. The geothermal resource produces air pollutants such as hydrogen sulfide, mercury, arsenic, and boron. Of these pollutants, only H_2S will be discussed; this pollutant has presented the most concerns due to its rotten egg smell.

During operations, the main sources of H_2S emissions are makeup well drilling, stacking facilities, and power plants. The following discussion addresses stacking and power plant emissions. Drilling emissions have already been discussed.

The H_2S emissions from stacking episodes can be abated chemically, using the previously mentioned sodium hydroxide/hydrogen peroxide system, or mechanically. Stacking emissions occur primarily when a power plant is shut down and steam is diverted to the atmosphere, via a muffler, until the steam wells are shut in or the power plant resumes operations. The preferred method for stacking abatement is the mechanical one because it is effective, it does not require chemicals nor does it generate waste, and it also abates any other pollutant in the steam. Using the mechanical system, when steam diversion begins, the wells are quickly shut in remotely from a control center, thereby eliminating emissions usually in a matter of minutes and minimizing the amount of untreated steam released into the atmosphere. In regions where the H_2S flow rate is so high that even a few minutes of stacking will create a problem, a portion of the steam flow can be directed to air-cooled heat exchangers where it is condensed and then reinjected into the reservoir.

Power plant emissions can also be mitigated so the steam emitted is very low in H_2S. Unlike stacking emissions, which are brief, power plant emissions occur whenever the plant is operating. These emissions can be reduced by more than 90 percent. Two processes that will achieve this reduction are the Stretford and the incineration processes. The Stretford process converts H_2S into element sulfur. In the incineration process, the final product is soluble and is reinjected into the reservoir.

California has the reputation of having one of the most, if not the most, stringent sets of environmental regulations in the world. California's H_2S ambient air limitation is 30 ppb (County of Lake, California, 1989). The Geysers, the largest geothermal development in the world, has an installed power generating capacity of nearly 2,000 MW (County of Lake, California, 1989), for which about 16 million kg/hr of steam is needed. There, by using proper technology and operations, California's stringent limitation is being met.

Water Pollution

The best way to combat water pollution is through reinjection, which, as mentioned earlier, also has other benefits. Spent fluids from power plants, pipelines, drilling rigs, and other locations can be directed to centralized reinjection wells.

Waste Generation

The power production phase generates substantial amounts of waste. More waste is generated in hot-water fields than in dry-steam fields. In the Imperial Valley, California, a hot-water field, about 36 MT/d of waste (sludge) can be generated from a 50 MW plant (Tim Garison, personal communication). To mitigate this problem, the sludge can be dewatered and turned into concrete blocks.

CONCLUSIONS

To conclude, two brief observations are made. First, electrical power is largely generated using steam to turn a turbine which, in turn, turns an electric generator. The geothermal industry is indeed fortunate because it does not have to contend with the environmental problems associated with heating water with fuel in order to produce the required steam, as other power generating industries must do. Second, although the geothermal industry is relatively young, it has already made great progress regarding environmental matters. Its pioneers have faced the fundamental problems, conducted the necessary research, and implemented what are now common solutions to many environmental problems. Also fortunate, then, are those entities new in geothermal energy development that can benefit from past efforts by calling upon these known solutions as they seek to balance energy needs and environmental concerns.

REFERENCES

County of Lake, California, 1989, Geothermal resource and transmission element of the Lake County General Plan, v. 1, 2: p. 16-29, 34; v. 2, 1: p. 44.

Dench, N. D., 1986, Hazards and safety measures in geothermal projects: Proceedings, 10th New Zealand Geothermal Workshop 1988, p. 200.

Circum-Pacific Council for Energy and Mineral Resources
Earth Science Series, Vol. 16
R. L. Miller, G. Escalante, J. A. Reinemund, and M. J. Bergin (Eds.)
Energy and Mineral Potential of the Central American-Caribbean Regions
Springer-Verlag, Berlin Heidelberg 1995

SUMMARY AND RECOMMENDATIONS OF THE WORKING GROUP ON GEOTHERMAL RESOURCES

David Sussman
Unocal

The 1989 installed electric capacity of Central America, Mexico, and the Caribbean is 869 MW, representing 15 percent of the world's geothermal electric power plant capacity. Geothermal plants are operating in Mexico, El Salvador, Nicaragua, and French Guadeloupe, and will soon be operating in Costa Rica and Guatemala.

Between 75 and 98 percent of the electricity produced in the Central American countries is generated by hydropower plants, so these countries are vulnerable to periods of drought. Costa Rica, for example, has been importing gas and oil to fuel thermal plants at very high cost to the country due to lower hydro reserves caused by the recent drought.

Geothermal installations in Nicaragua and El Salvador generate 40 percent and 22 percent, respectively, of base load electricity in those countries. Geothermal plants are being developed at Miravalles in Costa Rica and Zunil in Guatemala. Continued development in these and neighboring countries will contribute to the diversification of the region's energy base, saving millions of dollars by reducing the need to import oil and gas for electricity generation.

In Central America, the system of interconnected transmission lines could provide a larger market for geothermal development, especially in those countries not yet affected by electricity deficits. In turn, geothermal power can provide greater regional stability through a diversification of power supply, employment and international cooperation.

RECOMMENDATIONS

1. From identifying a geothermal resource to bringing a power plant into operation has taken 15 to over 20 years in the Central American countries. Stringent requirements by the international banking community for extensive or repeated feasibility studies prior to committing funds for actual development have resulted in long time delays and added project expense. This process needs to be improved and streamlined in order to allow expedient development of geothermal resources.

2. Geothermal development should be treated in the same way as oil and gas development, allowing countries to negotiate contracts with private domestic or foreign companies. Encourage the private sector, both domestic and foreign, to invest in geothermal energy projects.

3. Encourage those countries with geothermal exploration and development expertise to help train professional personnel in countries lacking this expertise.

4. Explore the economics of developing geothermal fields in incremental stages, initially using small wellhead generators connected to the first producing wells. Power can be brought on-line very early during field development. At the same time, this strategy allows detailed reservoir studies to define the most appropriate size and type of electric generating equipment to match the characteristics of the resource.

5. Environmental impacts resulting from the various phases of geothermal development need to be adequately understood prior to development. Effective mitigation equipment and technology are available, and should be disseminated to technical and governmental agencies in the region.

INDUSTRIAL MINERALS AND METALLIC MINERAL RESOURCES

Circum-Pacific Council for Energy and Mineral Resources
Earth Science Series, Vol. 16
R. L. Miller, G. Escalante, J. A. Reinemund, and M. J. Bergin (Eds.)
Energy and Mineral Potential of the Central American-Caribbean Regions
Springer-Verlag, Berlin Heidelberg 1995

THE OUTLOOK FOR VOLCANIC-HOSTED GOLD DEPOSITS IN THE REPUBLIC OF COSTA RICA

Steve Ludington
U.S. Geological Survey, MS 984, 345 Middlefield Road, Menlo Park, California 94025, USA

William C. Bagby
U.S. Geological Survey, National Center, MS 913, Reston, Virginia 22092, USA

ABSTRACT

Gold deposits in Costa Rica are found in lavas and pyroclastic rocks in volcanic arcs ranging in age from Miocene to the present. Although the arcs and some individual volcanos vary widely in alkalinity and other compositional parameters, there seems to be no close relationship between composition and associated type of epithermal precious-metal deposits.

Two deposit types are known: (1) Sado-type epithermal veins (Bellavista-Montezuma, Esperanza de Líbano, Sacrafamilia district, Juntas district), similar to deposits found in volcanic arcs in the Cascades and in Japan, and (2) hot-spring type deposits (Cañamazo, Beta Vargas, Santa Clara?), similar to large, low-grade deposits found in near-surface environments such as those in western North America. The Sado-type deposits are epithermal quartz veins and related stockworks that contain gold closely associated with pyrite. The hot-spring deposits are near-paleosurface deposits that consist primarily of stockworks of chalcedonic quartz and pyrite. The veins commonly contain minor galena, sphalerite, and chalcopyrite. Both deposit types contain anomalous amounts of arsenic, antimony, and, in some cases, mercury. In addition, auriferous stockworks commonly surround subvolcanic rhyolite intrusions, but none have been shown to contain substantial ore-grade material.

Conceptual models for undiscovered deposits include the following:

1. Hot-spring deposits may be present in a geochemically anomalous area near Ciudad Quesada.

2. Deposits associated with alkaline volcanic rocks, similar to those at Lihir Island, Papua New Guinea, may be present on the Atlantic Slope.

3. Large areas of Tertiary and Quaternary volcanic rocks to the southeast of the traditional gold belt are permissive for Sado-type veins. Also in the southeast, gold-bearing skarns are possibly present where Tertiary plutons intrude carbonate strata.

4. Recent studies suggest the possibility for gold deposition related to acid, summit-caldera lakes in stratovolcanos. Independent regional geochemical studies are consistent with this possibility.

GEOLOGIC SETTING

The geologic framework of Costa Rica is that of a mature island-arc in which the islands have coalesced into a peninsula and have been surrounded by an apron of sedimentary rocks. Several blocks of pre-Tertiary basaltic rock that formed on the sea floor were accreted to the Pacific margin of Costa Rica by plate tectonic activity. These are (1) the ultramafic rocks of the Santa Elena Peninsula and (2) mafic volcanic rocks of the Nicoya Group, found on the mainland northwest of Quepos, and on the Osa Peninsula (Figure 1). They do not host lode gold deposits, and they will not be considered further, though they do host important placer gold deposits on the Osa Peninsula (Berrangé, this volume). Place

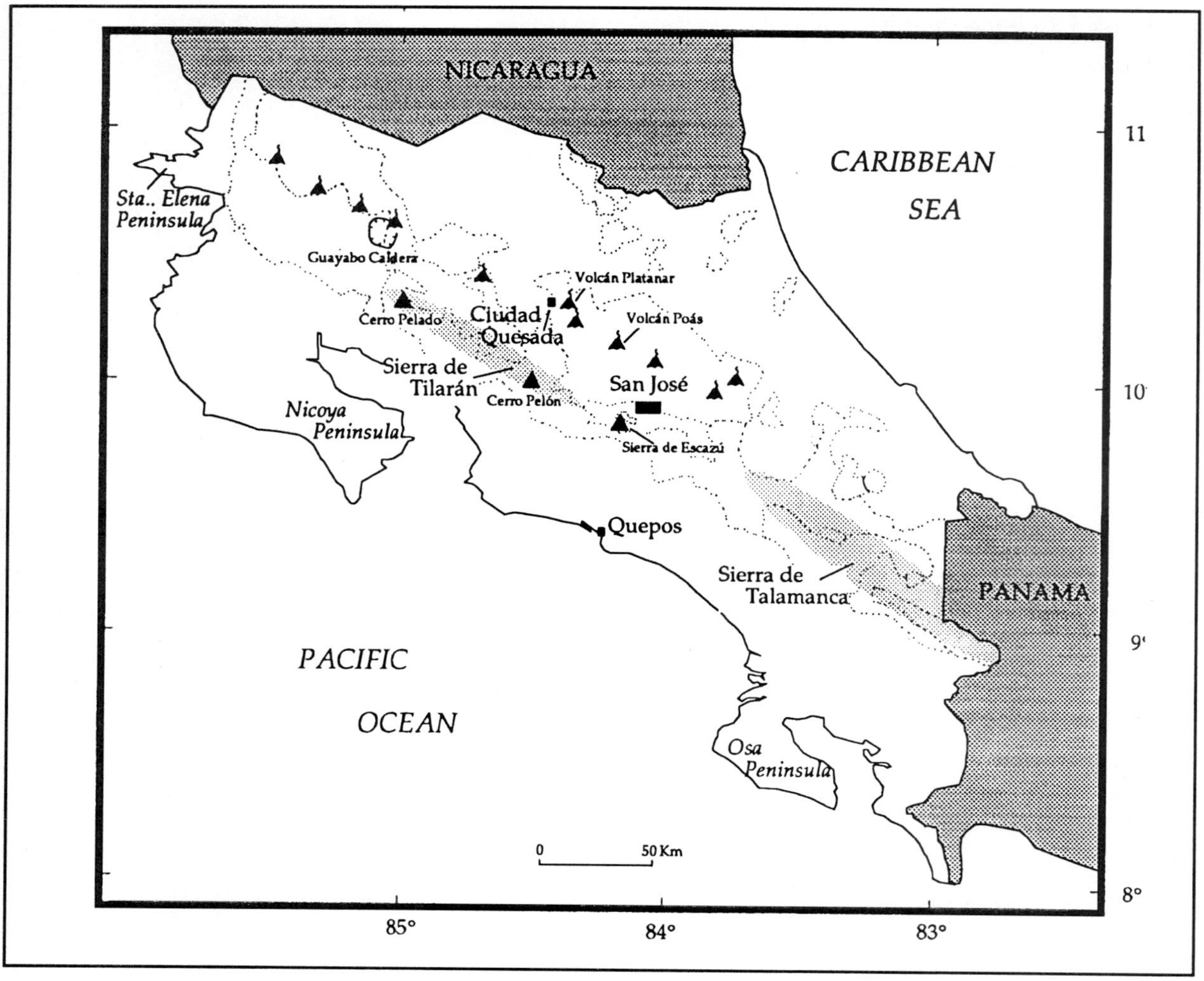

Figure 1. Map showing many of the geographic features of Costa Rica that are referred to in the text. Outcrop patterns (dotted) are the same as those shown in Figure 2.

names not on figures accompanying this paper may be found in Ludington et al. (1987) or Schultz et al. (1987).

Volcanism related to subduction along the Middle American Trench apparently began in the Oligocene; however, most of the activity has probably occurred after middle Miocene time. Most of the volcanic rocks are near andesite in composition, and are the products of a nearly continuous chain of stratovolcanos that stretches throughout the country from the northwest to the southeast (Ludington, Castillo, and Azuola, 1987). In the Cordillera de Tilarán, where many of the precious-metal mines in the country are found, most of these rocks are included in a formal stratigraphic unit named the Aguacate Group (Sprechmann, 1984), that has an age of greater than 5 Ma (Schultz et al., 1987) (see Figure 2).

Volcaniclastic sedimentary rocks have formed laterally to the arc since volcanism began. These sedimentary rocks are in many places complexly interlayered with volcanic rocks and they contain some carbonate strata.

The andesitic Monteverde Formation and the products of the presently active stratovolcanos, which are similar in composition to the Miocene and Pliocene rocks, have been grouped together by Schultz et al. (1987). All these younger volcanic rocks appear to be less than 2 Ma and to lie unconformably on the Aguacate Group. In general, they are not mineralized, though there appear to be exceptions.

In the northern part of the country, there is a large field of ignimbrites and ash-flow tuffs (Figure 1) that is dacitic to rhyolitic in composition. These rocks range in age from 7.5 Ma to less than 1 Ma (Mainieri, 1976). One caldera, the Guayabo (Figure 1), is known, but sources for most of the ash-flow tuffs have not been found. A few scattered exposures of silicic ash-flow tuff are also found in the Cordillera de Tilarán

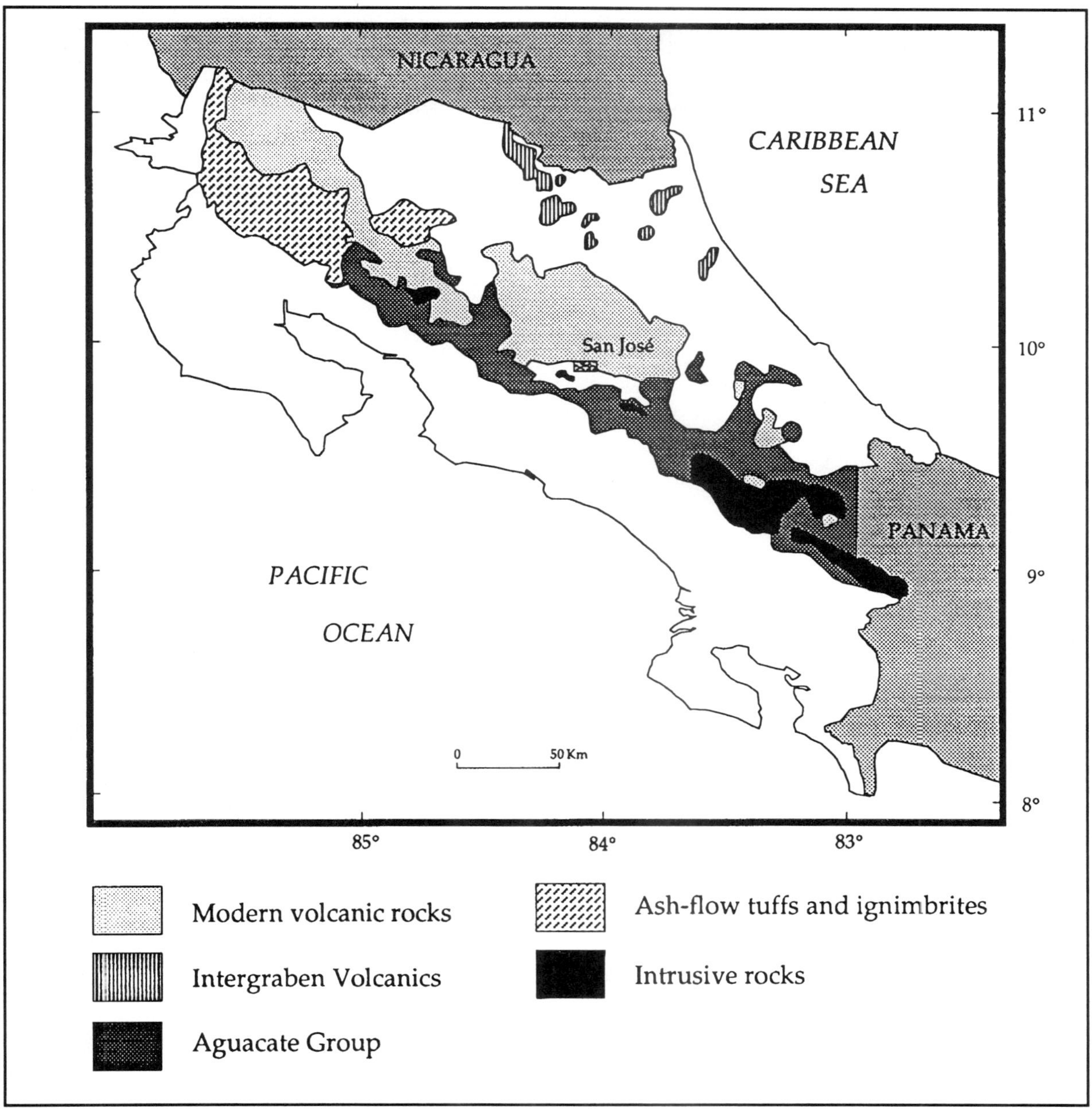

Figure 2. Map showing Tertiary and Quaternary igneous rocks of Costa Rica. Unpatterned areas are pre-Tertiary rocks and Tertiary sedimentary rocks.

(Schultz et al., 1987). In addition, on the Caribbean slope, there is a group of poorly known rocks that are alkalic in composition and that have been termed the Intergraben Volcanics (Sprechmann, 1984). Their composition may be influenced by Neogene extension or possibly by their formation above very deep parts of the subduction zone.

All these volcanic rocks are intruded by a wide variety of plutonic and subvolcanic rocks, ranging in composition from less than 50 percent to more than 70 percent silica. Scattered radiometric age data suggest they are progressively younger northwestward; the intrusions are 12 to 8 Ma in the Cordillera de Talamanca, where they consist of granodiorite to granite plutons that together form a batholith, 7 to 5 Ma in the dioritic stock at Sierra de Escazú, 10 km south of San José, and 5 to 4 Ma at Guacimál, a large quartz monzonite pluton in the Cordillera de Tilarán. Subvolcanic intrusions and flow-domes, some of them rhyolitic in composition, are also scattered throughout the volcanic terrane.

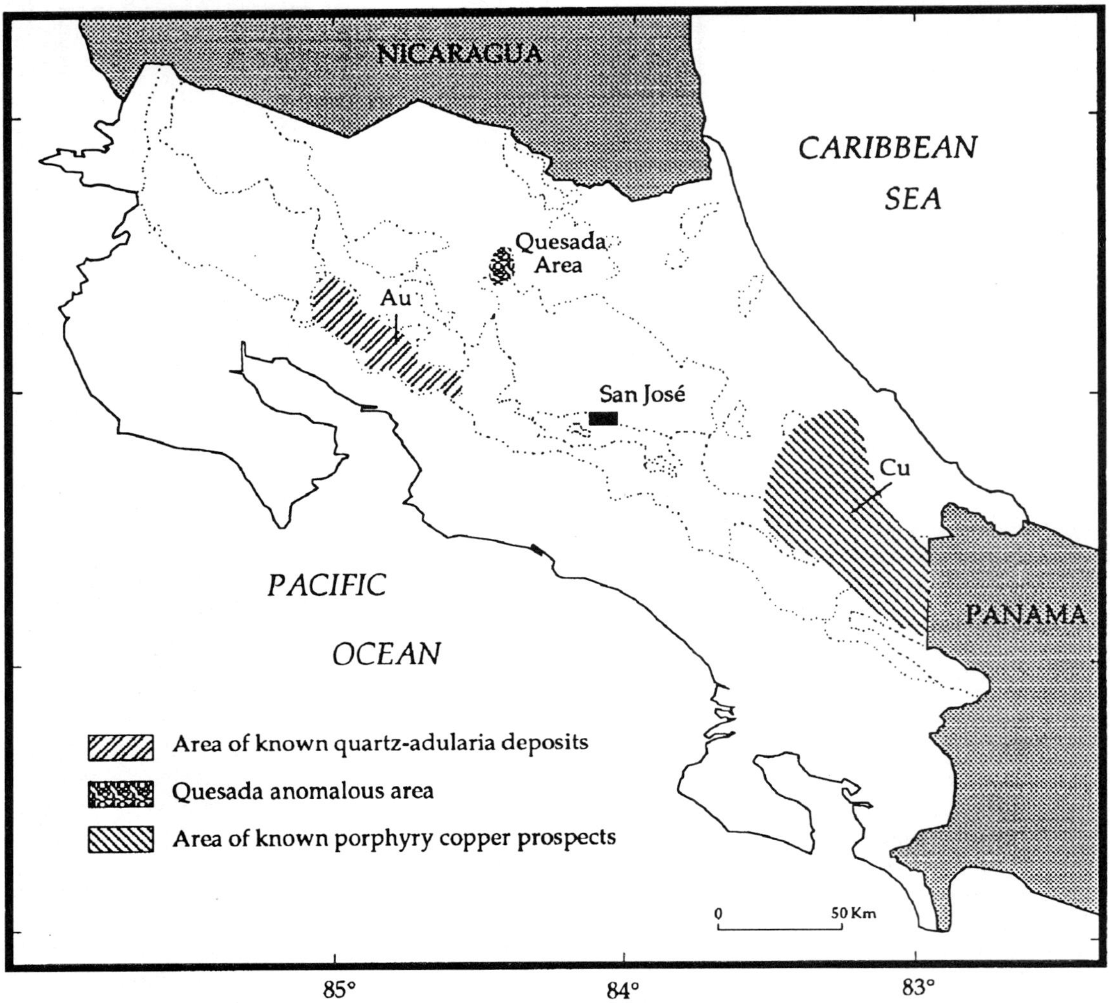

Figure 3. Map showing the general locations of some of the deposits and prospects described in the text. Outcrop patterns (dotted) are the same as in Figure 2.

PRECIOUS-METAL DEPOSITS

Precious-metal deposits in Costa Rica can be divided into two categories, those known to exist and those whose existence is only suspected or postulated. Known to exist are Sado-type quartz-adularia veins, hot-spring gold-silver deposits, and quartz stockworks around rhyolite plugs and flow-domes. More speculative are possible deposits related to a large geochemical anomaly in the Quesada area (Figure 3). There may be precious-metal deposits related to known porphyry copper prospects and any associated skarns. Finally, and most speculative, are deposits related to alkaline rocks in the northern part of the country and metal-rich brines associated with summit crater lakes.

Sado-Type Deposits

Sado-type deposits are epithermal volcanic-hosted quartz-adularia precious-metal deposits (Mosier, Berger, and Singer, 1986). They occur in areas underlain by thick deposits of volcanic rocks. In Costa Rica, this criterion applies to the major portion of the country; the post-Mesozoic part of the country is built almost entirely on a volcanic substructure.

Almost all of the known vein deposits are found in the northwestern part of the country in the Aguacate Group, the mid- to late-Tertiary belt of volcanic rocks that lies oceanward of the modern volcanos (Figure 3). The great majority of these deposits have been known at least since the nineteenth century and have been exploited periodically since that time.

There has also been recent exploration activity, and several deposits have been discovered in the 1980s.

The veins are as much as several kilometers long and as much as a few meters wide. Some high-grade ore occurs within distinct quartz veins, but many of the deposits have associated envelopes of quartz stockwork veins. In some cases, such as at the Bellavista deposit, currently under development, these stockworks constitute the significantly valuable part of the ore deposit. The median tonnage of Sado-type deposits worldwide is 300,000 MT; the median gold grade is 6 g/MT (Mosier and Sato, 1986). The Bellavista deposit in the central part of the Sierra de Tilarán, the only one in Costa Rica for which there is reliable information because it has been drilled out, contains in excess of 16 million grams of gold (Mora and Coates, 1987), and an unknown amount was removed during previous mining.

No studies have been made specifically on the mineralogical residence of gold in these epithermal veins, but it is clearly associated with pyrite and other sulfides. Minor amounts of galena, sphalerite, and chalcopyrite are present in some, but not all, veins.

Hot-Spring Deposits

It has recently become clear that some of the identified precious-metal deposits in Costa Rica are of the hot-spring type. These are deposits that formed at or very near the surface in an environment associated with an active thermal spring. Costa Rican examples include Beta Vargas and Cañamazo, both of which are prospects. Santa Clara, a producing deposit, is very probably also of the hot-spring type. Samples collected in 1986 and 1987 from float boulders peripheral to the Santa Clara property appear to be sinter breccia, similar to rocks observed at Río Cañamazo. Specific mine locations can be found in Bagby, Marsh, and Gamboa (1987); the three areas mentioned here are in the central and eastern parts of the Sierra de Tilarán, within the area shown on Figure 3 as containing known quartz-adularia deposits.

Two samples of the sinter breccia from near Santa Clara contain 1.72 and 8.45 ppm gold. The residence of the gold has not been determined, but it probably is included in the very fine-grained arsenian pyrite distributed throughout the sample, as determined by observation with a scanning electron microscope equipped with an energy-dispersive analyzer. The mineral assemblage in these samples includes quartz, arsenian pyrite, hydroxyapatite, and a rare potassium-bearing phase that may be adularia.

Auriferous Stockworks Related to Rhyolites

We believe it is important to note the common occurrence of pyrite-bearing quartz stockworks that are not related to any known productive veins. These stockworks are almost always associated with subvolcanic rhyolite intrusions. Examples of these rhyolite intrusions include Cerro Pelón, Cerro Pelado (Figure 1), and the area near Río Jesús, a few kilometers north of Cerro Pelón. Many of the rhyolites from these dome complexes clearly contain resorbed quartz phenocrysts. The rocks at these localities are more fully described, and chemical analyses presented, in Schultz et al. (1987). The association of rhyolites with gold mineralization was also noted by DeRuyter (1985), who emphasized the intimate involvement of rhyolite with ore at the Esperanza de Líbano and Río Chiquito mines.

From the little sampling we have done (Bagby, Marsh, and Gamboa, 1987), the stockworks appear to be auriferous, but all are deeply oxidized. The fresh material needs to be sampled to determine whether these areas are related to economic mineralization and, if so, of what type.

Quesada Area

Stream-sediment samples from an area northeast of Ciudad Quesada were reported by Los Alamos National Laboratories (1987) to be anomalous in a number of elements indicative of epithermal gold deposits, especially gold, silver, arsenic, antimony, and mercury. Because the area is one of young volcanic rocks that are little-eroded, we interpreted the anomalies to represent a hot-spring gold deposit.

The anomalous area is at the base of the northern flank of Volcán Platanar, between the towns of Ciudad Quesada and Aguas Zarcas, in Alajuela Province. The locality is shown in Figure 1. The area, which covers more than 100 km^2, is largely devoted to grazing and farming, though the steeper slopes of the volcano are mostly forested. Outcrops are scarce, and during our two-day visit we saw no obviously mineralized rock in outcrop. There are some native sulfur prospects to the south, higher on the slopes of the volcano.

An active hot spring in the area, at La Marina, is being developed as a bathing resort. A sample of siliceous sinter from this spring shows no unexpected geochemical concentrations other than an elevated As value of 76 ppm.

Sampling in 1987 is described in detail in Ludington and Bagby (1988). The most significant results are from a site at 494700 East, 263600 North (UTM coordinates), on the Río la Ceiba, where a significant proportion of the stream boulders (>1%) are composed of silica- and sulfide-rich material. The textures shown by these samples appear to be consistent with formation in a hot-spring environment. Much of the silica is opaline, and much is dark gray, owing to very fine-grained sulfide materials. Eleven samples were taken of this material; some of the results are shown in Table 1.

Precious-metal contents are not ore grade but are anomalously high; the highest Au value is 0.053 ppm in sample 87CRB01 and the highest Ag value is 0.8 ppm in sample 87CRB10. Enrichment in As and Sb is moderate; maximum values are 131 and 12 ppm, respectively. Unexpected in this probable hot-spring environment is the base-metal enrichment shown by these samples. Sample 87CRB03

Table 1. Analytical results for samples from the Quesada anomalous area.

ID	Description	Au	Ag	As	Sb	Mo	Cu	Pb	Zn
87CRB15	La Marina Hot Spring, sinter	<.001	<.1	76	6	<1	6	<2	5
87CRL05	siliceous chips	<.001	<.1	47	<2	3	39	<2	12
87CRB01	opaline sinter	.053	.2	<2	<2	<1	5	3	3
87CRB02	altered basalt?, trace native sulfur	.019	<.1	10	<2	4	99	9	3
87CRB03	opaline silica breccia	.011	<.1	50	<2	264	26	120	10
87CRB04	silicified rock with pyrite	.014	<.1	99	3	3	467	11	29
87CRB05	altered rock with pyrite	.002	<.1	7	<2	8	41	19	11
87CRB06	sulfide-silica breccia	<.001	<.1	5	<2	2	156	<2	10
87CRB07	sulfide and silica rich rock	.002	<.1	23	<2	2	111	<2	6
87CRB08	sulfide-silica breccia	<.001	<.1	3	<2	2	114	4	9
87CRB09	altered rock with pyrite	<.001	<.1	4	<2	<1	122	6	11
87CRB10	silica rock with massive pyrite	<.001	.8	131	12	55	910	93	92
87CRB11	sulfide-silica breccia	<.001	<.1	41	4	90	363	19	19

All results are reported in parts per million (ppm).
Analyses are by inductively coupled plasma (ICP) method, except gold analysis, which is by atomic absorption (AA) method.
C. Leong (Acme Analytical Laboratories, Vancouver, B.C.) performed the analyses.

contains 264 ppm Mo, and two other samples contain more than 10 ppm. Sample 87CRB10 contains 910 ppm Cu, and seven of the 11 samples contain more than 100 ppm. Maximum Pb and Zn values are 120 and 92 ppm, respectively. The stream-sediment sample from this locality does not contain abnormal amounts of base metals. However, other stream-sediment samples within the Quesada area are anomalous in Zn and Mo (Los Alamos National Laboratories, 1987).

Using the scanning electron microscope, we observed molybdenite in sample 87CRB03, which contains 264 ppm Mo. The molybdenite appears to be coeval with fine-grained quartz and pyrite and to precede later veining by coarse pyrite.

These high base-metal contents are not typical of identified hot-spring gold deposits, although high molybdenum values are known from at least one other hot-spring environment, at Sulphur Springs in the Valles Caldera, New Mexico, where recent drilling has encountered a very fine-grained molybdenite in rhyolitic tuffs in a modern hot-spring environment (Hulen et al., 1987). The Quesada area results remain enigmatic.

Porphyry-Related Deposits

Prior to 1980, there was a significant amount of exploration for porphyry copper deposits associated with Miocene intrusive rocks in the Cordillera de Talamanca in the southeastern part of Costa Rica. A number of prospects were discovered and partially developed at that time (Singer et al., 1987). Their general location is shown in Figure 3. With the recent rise in copper prices, some of these porphyry copper prospects may attract interest again, particularly if they contain recoverable amounts of precious metals.

In addition, several other deposit types that are associated with porphyry systems may contain precious metals as a primary product. First, significant gold deposits have been developed in skarns related to circum-Pacific porphyry copper systems in many parts of the world, and Costa Rica should be no exception. Skarns are possible because limestones, as well as limy siliciclastic rocks, form part of the Tertiary sedimentary rocks that interfinger with distal volcanics in this part of Costa Rica. Second, polymetallic veins related to porphyry systems have been important sources of gold and silver in the circum-Pacific region. These may occur in either volcanic or sedimentary host rocks. Finally, volcanic-hosted Cu-As-Sb deposits (Cox, 1986), which may be important sources of precious metals, are found near porphyry systems where they intrude volcanic sequences.

Alkaline Rock-Related Deposits

A special group of rocks, known as the Intergraben Volcanics, occurs on the Caribbean slope of Costa Rica, in the northern part of the country (Figure 2). These rocks are strongly alkalic and their composition may be influenced by Neogene extension or possibly by their formation above very deep parts of the subduction zone (Ludington, Castillo, and Azuola, 1987); Malavassi and Madrigal, 1970). The age is poorly documented, but many are probably Quaternary.

The geology of the area is not well known and no precious-metal deposits are definitely known to occur here, but there have been reports of sulfide veins. If the nature of the gold deposit at Lihir Island, Papua New Guinea, is indeed related to the alkalinity of the enclosing volcanic rocks, as many believe, then any deposits discovered in this part of Costa Rica could be particularly significant (Cox and Rytuba, 1987). Gold-telluride veins, such as at the Emperor mine in Fiji (Cox and Bagby, 1986), are also found in similar alkalic environments.

Crater Lakes

Brantley et al. (1987) have reported on the very acid, metal-, sulfate-, and chloride-rich warm waters of Laguna Caliente, the lake in the summit crater of Poás volcano, to the north of San José (Figure 1). The pH of this lake approaches 0, the chloride content exceeds 30,000 ppm, the sulfate content exceeds 60,000 ppm, and the temperature is near 60°C. Elevated levels of Cr, Co, Ni, Cu, Zn, Cd, and Pb, as high as 3,000 ppm, were reported. The elements in this lake water are probably of volcanic origin and were introduced into the water either directly by volcanic gases or indirectly by the leaching of the surrounding volcanic rocks.

Henley (1985) has suggested that such acid, sulfate-, and chloride-rich lakes, and their accompanying advanced argillic alteration, may represent the upper portion of quartz-alunite-type gold deposits. Whatever its origin, the lake water may represent an ore-forming fluid in search of a depositing mechanism.

Mass balance considerations suggest that there must be a net outflow of lake water downward into the mass of the volcano. During 1988, phreatic explosions accompanied a drop in the level of the crater lake, and color changes in the water signalled changes in the chemistry. Perhaps the fluids are draining to another location in the superstructure of the volcano, where the conditions may be amenable to metal deposition.

SUMMARY AND CONCLUSIONS

Precious-metal deposits in Costa Rica include both known and suspected deposits. We know of numerous quartz-adularia vein deposits, and several hot-spring deposits, as well as ryholite-related quartz stockworks of unknown economic potential. We believe there is significant potential for the discovery of hot-spring deposits in the Quesada area and for precious-metal-bearing deposits in the porphyry copper province. We also speculate that there may be mineralized systems associated with the alkaline Intergraben Volcanics and with several summit crater lakes in active volcanos.

ACKNOWLEDGMENTS

We gratefully acknowledge the cooperation and help in the field that we have received form numerous representatives of the Universidad de Costa Rica, Minera Nacional, S. A., and the Dirección de Geología, Minas y Hidrocarburos, as well as the financial support provided by the United States Agency for International Development.

REFERENCES

Bagby, W. C., S. P. Marsh, and E. Gamboa, 1987, An analysis of the epithermal vein deposit type in the Cordillera de Tilarán and Montes del Aguacate, Costa Rica, *in* U.S. Geological Survey, Dirección de Geología, Minas y Hidrocarburos, Universidad de Costa Rica, Mineral resource assessment of the Republic of Costa Rica: U.S. Geological Survey Miscellaneous Investigations Series Map I-1865, p. 44-55.

Berrangé, J., this volume, Origin of gold from the Golfo Dulce Province, southern Costa Rica.

Brantley, S. L., A. Borgia, G. Rowe, J. F. Fernandez, and J. R. Reynolds, 1987, Poás volcano crater lake acts as a condenser for acid metal-rich brine: Nature, v. 330, p. 470-472.

Cox, D. P., 1986, Descriptive model of volcanic-hosted Cu-As-Sb, in D. P. Cox and D. A. Singer, eds., Mineral deposit models: U.S. Geological Survey Bulletin 1693, p. 123.

Cox, D. P., and W. C. Bagby, 1986, Descriptive model of Au-Ag-Te veins, *in* D. P. Cox and D. A. Singer, eds., Mineral deposit models: U.S. Geological Survey Bulletin 1693, p. 124.

Cox, D. P., and J. J. Rytuba, 1987, Lihir Island gold: A supplement to U.S. Geological Survey Bulletin 1693: U.S. Geological Survey Open-File Report 87-0272a, 4p.

DeRuyter, V. D., 1985, Tectonic setting of late Cenozoic gold mineralization in the gold belt of Costa Rica: Geological Society of America Abstracts with Programs, v. 17, no. 7, p. 561.

Henley, R. W., 1985, The geothermal framework of epithermal deposits: Chapter 1, *in* B. R. Berger and P. M. Bethke, eds., Geology and geochemistry of epithermal systems, reviews in economic geology, v. 2, p. 1-24.

Hulen, J. B., D. L. Nielson, F. Goff, J. N. Gardner, and R. W. Charles, 1987, Molybdenum mineralization in an active geothermal system, Valles Caldera, New Mexico: Geology, v. 15, p. 748-752.

Los Alamos National Laboratories, 1987, Geochemical atlas of the San José and Golfito quadrangles, Costa Rica: U.S. Government Printing Office, Report LA-1065-MS, UC-11, July 1987, 26p.

Ludington, S., R. Castillo M., and H. Azuola, 1987, Geology of Costa Rica, *in* U.S. Geological Survey, Dirección de Geología, Minas y Hidrocarburos, Universidad de Costa Rica, Mineral resource assessment of the Republic of Costa Rica: U.S. Geological Survey Miscellaneous Investigations Series Map I-1865, p. 4-7.

Ludington, S., and W. C. Bagby, 1988, Follow-up examination of field areas with gold potential in Costa Rica—Analytical results: U.S. Geological Survey Open-File Report 88-0522, 8p.

Mainieri, P. A., 1976, Proyecto geotermal de Guanacaste: San José, Instituto Costarricense de Electricidad, Informe de previabilidad técnica, 97p.

Malavassi, V. E., and R. Madrigal, 1970, Reconocimiento geológico de la zona Norte de Costa Rica: Costa Rica, Dirección de Geología, Minas, y Petróleo, Informes Técnicos y Notas Geológicas, no. 38, 12p.

Mora, M. A., and M. E. T. Coates, 1987, Geology of the Bellavista-Montezuma deposit: A case study, *in* S. Ludington, W. C. Bagby, and S. L. Bolivar, eds., Program, abstracts, and field trip guide for the International Gold Conference, San José, Costa Rica, October 20-22, 1987: U.S. Geological Survey Open-File Report 87-602.

Mosier, D. L., B. R. Berger, and D. A. Singer, 1986, Descriptive model of Sado epithermal veins, *in* D. P. Cox and D. A. Singer, eds., Mineral deposit models: U.S. Geological Survey Bulletin 1693, p. 154.

Mosier, D. L., and T. Sato, 1986, Grade and tonnage model of Sado epithermal veins, *in* D. P. Cox and D. A. Singer, eds., Mineral deposit models: U.S. Geological Survey Bulletin 1693, p. 155-157.

Schultz, K., R. Koeppen, S. Ludington, S. Kussmaul, and K. M. Gray, 1987, Volcanological framework for the gold deposits in the Cordillera de Tilarán and Montes del Aguacate, Costa Rica, *in* U.S. Geological Survey, Dirección de Geología, Minas y Hidrocarburos, Universidad de Costa Rica, Mineral resource assessment of the Republic of Costa Rica: U.S. Geological Survey Miscellaneous Investigations Series Map I-1865, p. 34-43.

Singer, D. A., N. J. Page, W. C. Bagby, D. P. Cox, and S. Ludington, 1987, Assessment of metallic mineral resources of Costa Rica, in U.S. Geological Survey, Dirección de Geología, Minas y Hidrocarburos, Universidad de Costa Rica, Mineral resource assessment of the Republic of Costa Rica: U.S. Geological Survey Miscellaneous Investigations Series Map I-1865, p. 58-64.

Sprechmann, P., ed., 1984, Manual de geología de Costa Rica, v. 1, Estratigrafía: San José, Costa Rica, Editorial de la Universidad de Costa Rica, 320p.

Circum-Pacific Council for Energy and Mineral Resources
Earth Science Series, Vol. 16
R. L. Miller, G. Escalante, J. A. Reinemund, and M. J. Bergin (Eds.)
Energy and Mineral Potential of the Central American-Caribbean Regions
Springer-Verlag, Berlin Heidelberg 1995

ORIGIN OF GOLD FROM THE GOLFO DULCE PLACER PROVINCE, SOUTHERN COSTA RICA

Jevan P. Berrangé
British Geological Survey, Overseas Directorate, Keyworth, Nottinghamshire NG12 5GG, United Kingdom

ABSTRACT

The Golfo Dulce goldfield has been mined since pre-Columbian times and currently produces at least twice as much gold as the entire Tilarán-Aguacate Gold Province in northern Costa Rica. The gold is mainly concentrated in eluvial, colluvial, and alluvial placers in the Osa Group (late Pliocene) and the Puerto Jiménez Group (late Pleistocene-Holocene) that unconformably overlie the Nicoya Complex basement (Late Cretaceous-early Tertiary). This is a segment of oceanic crust characterized by E-MORB-type basalts typical of a back arc basin such as the Mariana Trough.

Studies of the geographic and stratigraphic distribution of the placer gold, its habit and composition, and the associated heavy mineral suite, demonstrate conclusively that it was locally derived from the nearby Nicoya Complex. The gold was initially concentrated in epithermal auriferous quartz lodes in the basalts, and in interlayered pelagic sediments, by hot circulating fluids (seawater and magmatic fluids) related to one or more of three submarine volcanic events. Translative plate movement and subsequent obduction and isostatic uplift of this segment of oceanic crust during development of the Southern Central America Orogen allowed weathering and erosion of the ophiolite complex. This produced native gold and gold in solution that was concentrated during several cycles as placers in the associated sediments.

The primary gold deposits of the Southern Central America Orogen, in Costa Rica and Panamá, appear to include two distinct types and occur as (1) epithermal Au-Ag+(Pb-Zn-Cu) deposits of late Tertiary age associated with calc-alkaline acid-intermediate intrusives and/or volcanics in the inner arc, or (2) as epithermal low-Ag deposits in ophiolite terrains of the outer arc. These probably originated by hydrothermal activity related to submarine basaltic volcanism in an extensional environment such as a back arc basin, and are significantly older—Late Cretaceous to early Tertiary. The Golfo Dulce goldfield is the most westerly occurrence of an "ophiolite gold province." The recognition that virtually uncratonized oceanic crust in this region forms the "source beds" for gold deposits suggests that similar ophiolitic terrains in the circum-Pacific and Caribbean should be regarded as gold exploration targets.

INTRODUCTION

Gold mining in the Golfo Dulce Placer Province, that includes the Osa Peninsula, dates back to pre-Columbian time and it has been estimated (Ivosevic, 1979) that the peninsula had produced about 100,000 Troy ounces of gold during and since prehistoric times up to 1974. At present the Golfo Dulce Province produces at least 16,000 Troy ounces of 24 carat gold per year, which is at least twice as much as is produced in the entire Tilarán-Aguacate Gold Province in northern Costa Rica (Berrangé, 1987).

The Golfo Dulce goldfield occurs in the outer arc of the Southern Central American Orogen (Weyl, 1980; Dengo, 1985). Here, obducted oceanic crust of the Nicoya Complex forms the basement comprising a truncated ophiolite suite of basaltic lavas with associated intrusive dolerite dikes and sills, and gabbro plutons, with subordinate interstratified pelagic deposits including micritic limestone, radiolarian cherts, and argillites. The complex as a whole is variably deformed and altered with the resultant local development of shear and fracture zones, tectonic melanges, and vein networks filled with secondary minerals including hematite,

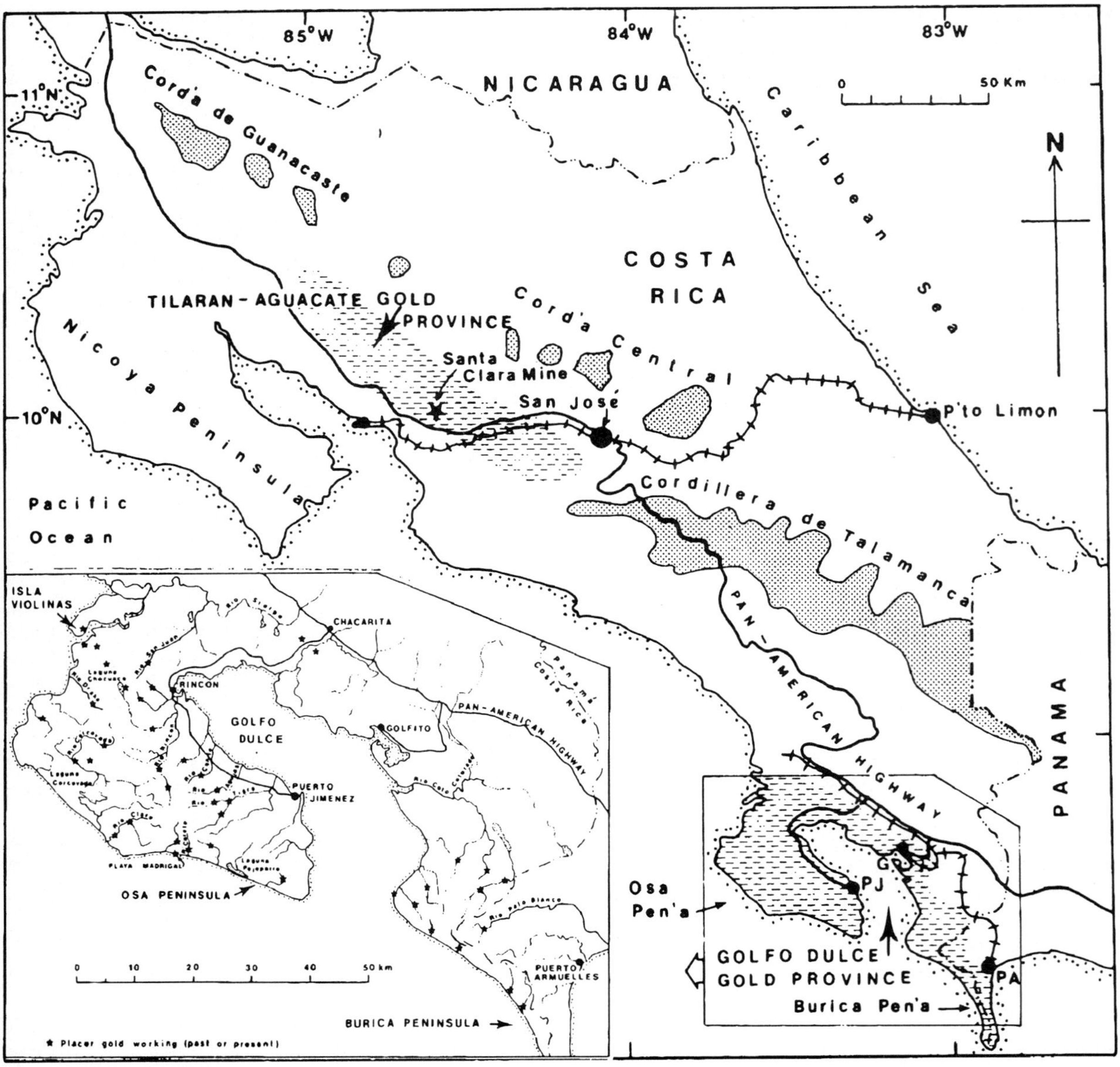

Figure 1. Geographic setting of Costa Rican gold provinces. Inset: Main placer occurrences of the Gulfo Dulce goldfield.

flaky and crystalline pyrite, calcite, barite, celsian, zeolites, chlorite, talc, epidote, and quartz. The basaltic rocks have affinities to large-ion-lithophile (LIL) enriched mid-ocean ridge basalts (E-MORB) of the sort formed in an extensional back arc basin such as the present-day Mariana Trough, the Lau basin, and/or in a setting comparable to that of the Gulf of California (Berrangé and Thorpe, 1988). Most of the igneous rocks of the Nicoya Complex were formed episodically during three volcanic events in the Santonian-Campanian (78.0 ±2 Ma), the Paleocene (60.2 ±7.6 Ma), and the middle Eocene (44.1 ±4.4 Ma) (Berrangé, Bradley, and Snelling, 1989).

In the southern two-thirds of the Osa Peninsula the basement complex is unconformably overlain by the middle to late Pliocene Osa Group comprising slightly lithified, well bedded, graded, greywacke-type conglomerates, sandstones, siltstones, and claystones derived from the underlying Nicoya Complex. The Osa Group is up to at least 800 m thick and was deposited in a wide range of environments from fresh-water fluviatile; brackish deltas, lagoons, and estuaries; and/or shallow shelf down to deep marine basins (down to 250 m) (Berrangé, 1990).

Unconsolidated gravels, sands, silts, and clays of late Pleistocene to Holocene age form the Puerto Jiménez Group

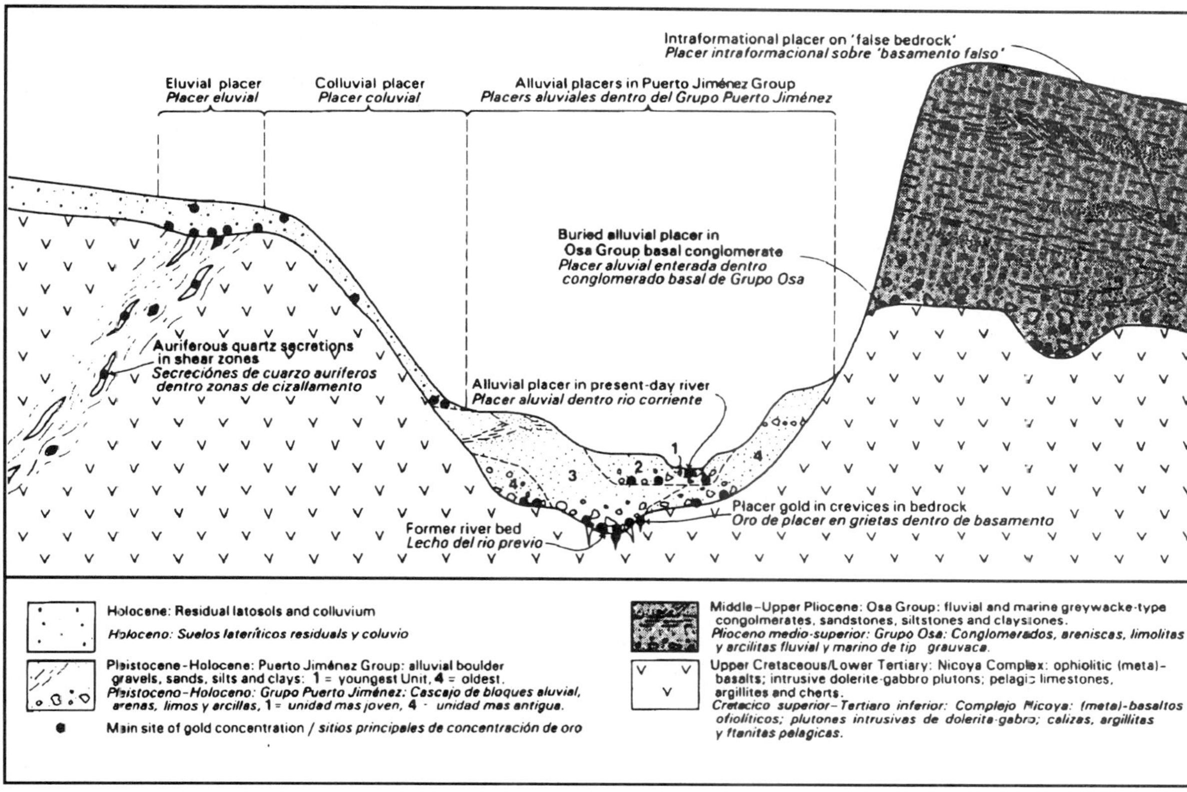

Figure 2. Schematic representation of the main types of gold placers found in the Osa Peninsula.

that unconformably overlies the older units (Berrangé, 1988). Where not covered by the Osa Group or alluvium, the rocks of the Nicoya Complex are mantled by up to 20 m of reddish-brown laterite.

THE GOLFO DULCE PLACER PROVINCE

The Golfo Dulce Gold Province in southern Costa Rica includes the Osa and the Burica Peninsulas and the Fila Golfito on the east side of Golfo Dulce (Figure 1). It has been suggested that the Golfo Dulce is a pull-apart basin that in pre-Pliocene time was filled by the landmass now forming the Burica Peninsula (Berrangé, 1990). If this is true, then the gold occurrences originally formed a smaller and more unified province. The gold deposits of the Osa Peninsula occur (1) as eluvial-colluvial placers in the weathered latosols derived from the Late Cretaceous to early Tertiary Nicoya Complex; (2) as buried fossil placers of Pliocene age in the conglomerates at the base of the Osa Group; (3) as alluvial placers in the Pleistocene and Holocene gravels along the courses of present-day rivers where they occur in a series of raised terraces and in the active stream sediments; and (4) as beach and offshore marine placers (Figure 2). The gold occurs in all sizes and shapes from flour gold barely visible to the naked eye, to flakes, grains, and nuggets weighing up to 7.7 kg. The Osa placer gold averages 93.8 percent Au and 5.4 percent Ag; this is equivalent to 22.5 carats, a fineness of 946 and an Au:Ag ratio of 1:17 (Berrangé, 1987).

ORIGIN OF THE PLACER GOLD

The occurrence of placer gold deposits in the outer arc of the Southern Central America Orogen, apparently related to an ophiolite complex but unrelated to acid-intermediate igneous activity, may be considered to be somewhat enigmatic. A number of theories have been advocated to explain the origin of the alluvial gold. These may be classified into two broad groups as follows:

1. It has been suggested (Levy, 1970; Anonymous, 1978; Lew, 1983) that the gold is derived from a distant source in the Talamanca Cordillera (inner arc) and is related to either Cu-Mo porphyry deposits or to epithermal gold-quartz veins in a felsic volcanic-subvolcanic sequence.

2. Alternatively, the gold is locally derived from the Nicoya Complex basement in which it may be derived from highly altered peridotites; or from massive exhalative sulphide

deposits; or from metalliferous pelagic sediments; and/or from quartz veins, stockworks, space fillings and replacement masses in the bedrock.

The evidence for and against these competing theories is examined below.

(a) Gold from the Osa Peninsula has an average Au:Ag ratio of 1:17, whereas that from the Tilarán-Aguacate Gold Province (Figure 1) generally ranges from 1:1 to 1:10 (U.S. Geological Survey, Dirección General de Geología e Hidrocarburos, and Universidad de Costa Rica, 1987). This shows that the gold from the two provinces is fundamentally different and suggests that it has different origins, even if allowance is made for depth-related differences or modification of Au:Ag ratios due to solution and reprecipitation in the exogenic cycle.

(b) The heavy mineral concentrates associated with the alluvial gold are dominated by a suite derived from basic igneous rocks and contain no significant occurrence of minerals indicative of a granitoid or rhyolitic provenance, such as rutile, monazite, sphene, tourmaline, and micas.

(c) A thermometric analysis of the fluid inclusions in gold-quartz samples gave a temperature range of 198-207°C and a salinity of 0.8-1.5 wt% NaCl equivalents. This is consistent with typical epithermal Au-Ag fluids and conforms to circulation of sea water through hot oceanic crust, but is incompatible with gold deposition in a Cu-Mo porphyry environment.

(d) The nearest known porphyry-copper deposits are in the Talamanca Cordillera some 90 km distant, and the nearest outliers of felsic volcanics are 40 km away. Furthermore, there are two major tectonic basins—the Valle del General and the Diquís/Golfo Dulce basin—between these hypothetical source rocks and the Osa Peninsula; and there are no known gulch or river placers on the western slopes of the Talamancas.

(e) Ivosevic (1979) has described "hydrothermal alteration" of the "mudstone" and "basalt" of the Nicoya Complex in the Río Carate sector where he recognized calcitization, silicification and metallization including "native gold filling fractures in vein quartz and in jasperized mudstone and quartz veined basalt containing 1-mm-thick pyrite grains and fracture fillings." He concluded that "the pre-detrital source of the placer gold apparently was gold-pyrite(?)-quartz lodes" in the centers of zones of silicified mudstone and maybe silicified basalt of the Nicoya Complex, although he found no lode gold in place.

(f) The present author has similarly recorded the presence of zones of brecciation, shearing and alteration in the Nicoya Complex including carbonitization, saussuritization, silicification, and pyritization. Fire assays of samples from different quartz veins have given up to 510 ppb Au, compared with a background value of 5-10 ppb Au for barren veins and country rock (Berrangé, 1987; Berrangé and Thorpe, 1988). Narrow quartz lenses cutting basalt on the west side of Playa Blanca near Rio Sierpe mouth have yielded free gold on crushing and panning (Collin Little, personal communication, 1989). At one locality on the upper Río Agujas "oreros" were seen winning gold by hand-mining a pocket of highly pyritized pelagic argillite interbedded with chert. (An apparently analogous gold occurrence has been described from the Palo Seco placer field of the Azuero Peninsula, Panamá, where "oreros" are reported to have been seen winning gold by grinding up limestone from the ophiolite sequence that outcrops in that area [M. Ward, personal communication, 1989]).

(g) Gold is found in eluvial-colluvial latosols and gravels developed directly from the underlying Nicoya Complex. Many of the large and super-large nuggets appear to derive from this type of material including the largest nugget (7.7 kg) found in the region.

(h) There is a general decrease in size downstream of the gold found in Holocene river gravels of the rivers draining the peninsula, indicating that the source is the Nicoya Complex that forms the basement in the central mountainous region (Azuola, 1985).

(i) The relative abundance of nuggets, including composite gold-quartz nuggets and gold flakes with attached quartz, the delicate or angular shape of some of the alluvial gold, are all features indicative of recent release and a nearby source—probably less than 10 km (cf. Boyle, 1979).

(j) The habit and luster of the gold from different rivers is different and distinctive which is not what would be expected if the gold had been derived from distant sources and spread over the area.

There can therefore be no doubt that the placer gold is locally derived from the Nicoya Complex. Some of it must come from gold-quartz veins and lenses, and some is derived from pyritiferous pelagic argillites. Exactly how and when the gold was initially concentrated in this segment of oceanic crust is less certain.

In recent years extensive metalliferous deposits have been found actively forming on the deep seabed at numerous localities in many of the seas and oceans around the world. These deposits are all genetically related to hydrothermal centers located along spreading ridges and island arc systems. Elements concentrated include Fe, Mn, Cu, Zn, Pb, Ag, and Au. Anomalously high gold values are common: samples containing up to 6700 ppb Au and averaging 4900 ppb Au have been recorded from the eastern Pacific (Gross, 1987; Gross and McLeod, 1987).

Back arc basins related to subduction zones can be expected to be particularly favorable sites for hydrothermal metallization, including gold, because (a) they provide large extensional tectonic settings; (b) they are characterized by episodic submarine volcanism active over long periods of time; (c) they contain basaltic rocks derived by partial melting and one can therefore expect enrichment in metalliferous elements contributed from contemporaneous or earlier subduction of oceanic lithosphere below the back arc basin; and (d) they are characterized by high heat flow.

The Nicoya Complex of the Osa Peninsula has been interpreted as an obducted segment of back arc basin-type oceanic crust that formed episodically during three distinct igneous events (Berrangé and Thorpe, 1988; Berrangé, Bradley, and Snelling, 1989). In other words there have been at least three "golden opportunities" for hydrothermal mineralization. It is postulated that the gold mineralization probably formed in Late Cretaceous to early Tertiary time in a volcanically active submarine back arc basin by circulation of heated seawater and magmatic fluids which leached the gold from the volcanic-sedimentary sequence and deposited it where physico-chemical conditions were favorable, in a reducing environment in pelagic argillites and/or in gold-quartz veins formed in low-pressure space fillings.

Elsewhere in the world, many of the gold occurrences associated with ophiolites appear to have formed by the alteration of peridotites to serpentinite, talc schists, and listwaenites (carbonitized ultramafic rocks) (Buisson and Leblanc, 1986). A gravity survey of the Osa Peninsula and environs (Barritt and Berrangé, 1987) has revealed the probable presence of deep-seated wedges of mantle peridotite within the ophiolite sequence; but only two boulders of ultramafic rock have been found to date. It is concluded that although a peridotitic source for the Osa gold cannot be ruled out, there is little or no evidence to support this mode of origin.

The Golfo Dulce placer gold could conceivably be related to polymetallic massive sulphide deposits, that are generally considered to be of submarine volcanogenic origin, and in many parts of the world contain up to 2 oz Au/t or more (cf. Boyle, 1979). However, massive sulphide mineralization has never been found in the Golfo Dulce region; and pyrite is the only essential sulphide mineral found associated with the gold. There is therefore no evidence to support this theory.

Another possible hypothesis is that the gold-quartz veins were deposited from gold-bearing siliceous hydrothermal solutions derived directly from a felsic calc-alkaline pluton derived by partial melting at depth along the subduction zone plunging beneath the Southern Central America Orogen; or that hot hydrothermal fluids from such a pluton scavenged gold from out of the ophiolitic country-rocks. However, no granitic or dioritic outcrop or float has ever been found in the Osa Peninsula and stream sediment heavy-mineral surveys have given no indication of buried granitoid rocks in the region. A gravity survey has also failed to reveal the presence at depth of a granitoid pluton to which the primary gold mineralization may be related. There is, therefore, no evidence to support this hypothesis.

MODEL FOR COSTA RICAN AND PANAMANIAN GOLD DEPOSITS

The gold deposits of the Southern Central America Orogen appear to occur in two distinct settings (Berrangé, in preparation):

1. The deposits of the inner arc are generally epithermal Au-Ag±(Pb-Zn-Cu) quartz veins, stockworks and disseminated(?) deposits, of late Tertiary age, associated with subduction-related calc-alkaline acid-intermediate intrusives and/or volcanics. These quasi-continental crustal rocks and their deposits show an intermediate level of metallogenetic maturity in the sense of Kesler (1978) and are characterized by a high Ag:Au ratio relative to those of the outer arc deposits.

2. The deposits of the outer arc occur in uplifted segments of virtually uncratonized oceanic crust—ophiolite complexes that are both the source and host rocks. They generally show a low level of metallogenetic maturity and are characterized by a relatively low Ag:Au ratio. The gold was initially concentrated by submarine hydrothermal circulatory systems in hot rock along spreading lineaments or near submarine volcanos. It occurs in gold-quartz lodes and disseminated in pelagic argillites or limestones. Native gold is common in these deposits which, therefore, tend to give rise to rich placers. The primary deposits as a whole can be expected to be relatively old and have ages similar to those of their ophiolite host rocks, namely Late Cretaceous to early Tertiary. Some of these ophiolite gold deposits, such as those in the Cordillera San Blas and the Azuero Peninsula, appear to cluster around, and be related to, younger subduction zone-related acid-intermediate plutons. The outcrop patterns are reminiscent of those typical of Precambrian granite-greenstone terrains. It is suggested that in these cases later hydrothermal activity connected to granitoid magmatism probably remobilized and reconcentrated the gold that was originally enriched in ophiolitic source rocks.

We should direct attention to the geologically young analogues of the Precambrian greenstone belts that have yielded a very large proportion of the world's gold; and ophiolite complexes of the circum-Pacific and Caribbean should constitute a target for gold exploration. Special attention should be directed toward ultramafic bodies within the complexes, and to ophiolites in the vicinity of younger felsic plutons. Further research is necessary to determine the geochemical parameters and the original tectonic setting of those ophiolitic rocks most likely to be gold source rocks and host gold deposits; e.g., back arc basins, volcanic arcs, mid-ocean ridges or within-plate volcanos. Ophiolite complexes are especially worthwhile targets because they can be expected to host a galaxy of other types of metalliferous deposits including platinum group elements, chrome, manganese, cobalt, nickel, copper, and zinc; and non-metallic deposits of barite, talc, asbestos, phlogopite, and magnesite.

ACKNOWLEDGMENTS

This study formed part of a research project that investigated the geology and gold mineralization of the Osa Peninsula and environs. It was sponsored by the British Government's Overseas Development Administration and by the Costa Rican Government's state mining enterprise, Minera Nacional S. A. (MINASA) whose support is gratefully acknowledged. The author wishes to thank his colleagues in the British Geological Survey who contributed to this study: namely, D. Bland for electron microprobe and XRF analysis of gold samples, B. Beddoe-Stephens for electron microprobe studies of gold samples, I. Basham for mineralogical examination of heavy mineral concentrates, and T. Shepherd for fluid inclusion study of gold-quartz samples. This paper is published by permission of the Director of the British Geological Survey (Natural Environment Research Council).

REFERENCES

Anonymous (Republica de Costa Rica and Organización de los Estados Americanos), 1978, Diagnóstico del Sector Minero: San José, Imprenta Nacional, 91p.

Azuola, H., 1985, Occurrences and origin of heavy mineral placers in braided stream facies of the Agujas River, Osa Peninsula, Costa Rica: Pennsylvania State University, Department of Geosciences, M.Sc. Thesis, 164p.

Barritt, S., and J. P. Berrangé, 1987, An interpretation of a gravity survey of the Osa Peninsula and environs, southern Costa Rica: Overseas Geology and Mineral Resources, no. 64, 18p.

Berrangé, J. P., 1987, Gold in Costa Rica: Mining Magazine (May), p. 402-407.

Berrangé, J. P., 1988, Geological maps of the Osa Peninsula: sheets Sierpe, Rincón, Llorona, Golfo Dulce, Madrigal and Carate, scale 1:100,000, includes explanatory notes in English and Spanish: British Geological Survey, Overseas Directorate, unpublished colored photocopies.

Berrangé, J. P., 1990, The Osa Group: An auriferous Pliocene sedimentary unit from the Osa Peninsula, southern Costa Rica: Revista Geológica de América Central, no. 10, p. 67-93.

Berrangé, J. P., in preparation, Gold from the Golfo Dulce Placer Province, southern Costa Rica.

Berrangé, J. P., and R. S. Thorpe, 1988, The geology, geochemistry and emplacement of the Cretaceous/Tertiary ophiolitic Nicoya Complex of the Osa Peninsula, southern Costa Rica: Tectonophysics, v. 147, p. 193-220.

Berrangé, J. P., D. R. Bradley, and N. J. Snelling, 1989, K:Ar age dating of the ophiolitic Nicoya Complex of the Osa Peninsula, southern Costa Rica: Journal of South American Earth Sciences, v. 2, no. 1, p. 49-59.

Boyle, R. W., 1979, The geochemistry of gold and its deposits: Geological Survey of Canada, Bulletin 280, 584p.

Buisson, G., and M. Leblanc, 1986, Gold-bearing listwaenites (carbonatized ultramafic rocks) from ophiolite complexes, in M. J. Gallagher, R. A. Ixer, C. R. Neary, and H. M. Prichard, eds., Proceedings of the conference: Metallogeny of basic and ultrabasic rocks, Edinburgh, Scotland, 9-12 April 1985: Institution of Mining and Metallurgy, p. 121-131.

Dengo, G., 1985, Mid America: Tectonic setting for the Pacific margin from southern Mexico to northwestern Colombia, in A. E. M. Nairn, F. G. Stehli, and S. Uyeda, eds., The ocean basins and margins, v. 7A, The Pacific Ocean: New York: Plenum Press, p. 123-180.

Gross, G. A., 1987, Mineral deposits on the deep seabed: Marine Mining, v. 6, p. 109-119.

Gross, G. A., and C. R. McLeod, 1987, Metallic minerals on the deep seabed: Geological Survey of Canada, Paper 86-21, 65p.

Ivosevic, S. W., 1979, Geology and mineral deposits of Costa Rica's south-central Osa Peninsula placer gold district: Transactions, American Institute of Mining Metallurgical and Petroleum Engineers, v. 266, p. 1876-1882.

Kesler, S. E., 1978, Metallogenesis of the Caribbean region: Journal Geological Society of London, v. 135, p. 429-441.

Levy, E., 1970, La metalogenesis en America Central: Instituto Centroamericano de Investigación y Tecnología Industrial (ICAITI), Publicaciones Geológicas del ICAITI, no. 3, p. 17-57.

Lew, L. R., 1983, The geology of the Osa Peninsula, Costa Rica: Observations and speculations about the evolution of part of the outer arc of the Southern Central American Orogen: Pennsylvania State University, Department of Geosciences, M.Sc. Thesis, 128p.

U.S. Geological Survey, Dirección General de Geología e Hidrocarburos, and Universidad de Costa Rica, 1987, Mineral resource assessment of the Republic of Costa Rica: U.S. Geological Survey Miscellaneous Investigations Series Map I-1865.

Weyl, R., 1980, Geology of Central America: Berlin, Gebrüder Borntraeger, 371p.

Circum-Pacific Council for Energy and Mineral Resources
Earth Science Series, Vol. 16
R. L. Miller, G. Escalante, J. A. Reinemund, and M. J. Bergin (Eds.)
Energy and Mineral Potential of the Central American-Caribbean Regions
Springer-Verlag, Berlin Heidelberg 1995

NICKEL POTENTIAL OF THE CARIBBEAN PLATE AND ADJACENT REGIONS

Fred Barnard and Faustino Prado
NiCobalt International Corporation, Golden, Colorado and Lakeland, Florida

ABSTRACT

The Central American-Caribbean region hosts a number of identified nickel resources in nickeliferous laterites. These have been mined in Guatemala, Cuba, Dominican Republic, and Colombia, and currently provide about 12 percent of the world's annual nickel production.

All the active nickel projects, and most of the known lateritic deposits and potential resources, are on margins of the Caribbean Plate, where slabs of mantle-derived ultramafics have been emplaced above or within crustal rocks. Tropical weathering of the peridotites has created enriched nickel zones within the lateritic weathering profile. Unexploited nickeliferous laterite resources occur in this setting in a number of areas. Nickel sulfide prospects are sparse in the region.

Within the Antilles, Andes, and Central America, very modest possibilities exist for discovery of additional ultramafics and associated nickeliferous laterites. Review of nickel deposit types worldwide indicates that nickel sulfides could occur in some geological environments within the region, and that sulfide deposits have economic advantages over laterites. In the Guiana Shield a reasonable potential exists for discovery of nickeliferous sulfides in Proterozoic greenstone belts.

Many factors affect the potential for future new sources of nickel, including the region's proximity to ocean transport and markets, abundant water, and benign climate. Offsetting factors in many countries include scarcity of indigenous fuel, limited infrastructure, and variable foreign investment climates. The most promising potential for new nickel projects in the region is for as-yet undiscovered sulfide deposits in the Guiana Shield.

INTRODUCTION

The Caribbean region encompasses a number of identified and potential nickel resources. Nickel production from the region has had a relatively brief and erratic history, but appears likely to hold its share of production, which was 88,000 MT Ni (See Table 1) or about 12 percent of the world total in 1988.

During the past three years, the world market price of nickel has risen spectacularly, from less than $US 2.00/lb to over $US 7.00/lb. This rise has prompted renewed interest in potential new sources of nickel. However, metals prices are typically volatile and cyclical, and in the face of increased production, prices tend to drop rapidly. Many nickel mines around the world have been closed or have been producing at partial capacity for several years, but the price increase has tended to stimulate production from these existing sources.

The goal of this paper is to point out the diversity of under-explored geological environments in the Caribbean region which have the potential to host nickel deposits, while maintaining a realistic framework of economic constraints on development possibilities. It focuses on the potential for future production from known deposits which have not been exploited, and the potential for discovery of new deposits. Prediction of metal production or prices is not attempted. The area discussed is mainly the Caribbean Plate, with some discussion of adjacent portions of the Guiana Shield and Andean Cordillera which share similar economic and political environments. The potential nickel production from sea-floor nodules in the region is outside

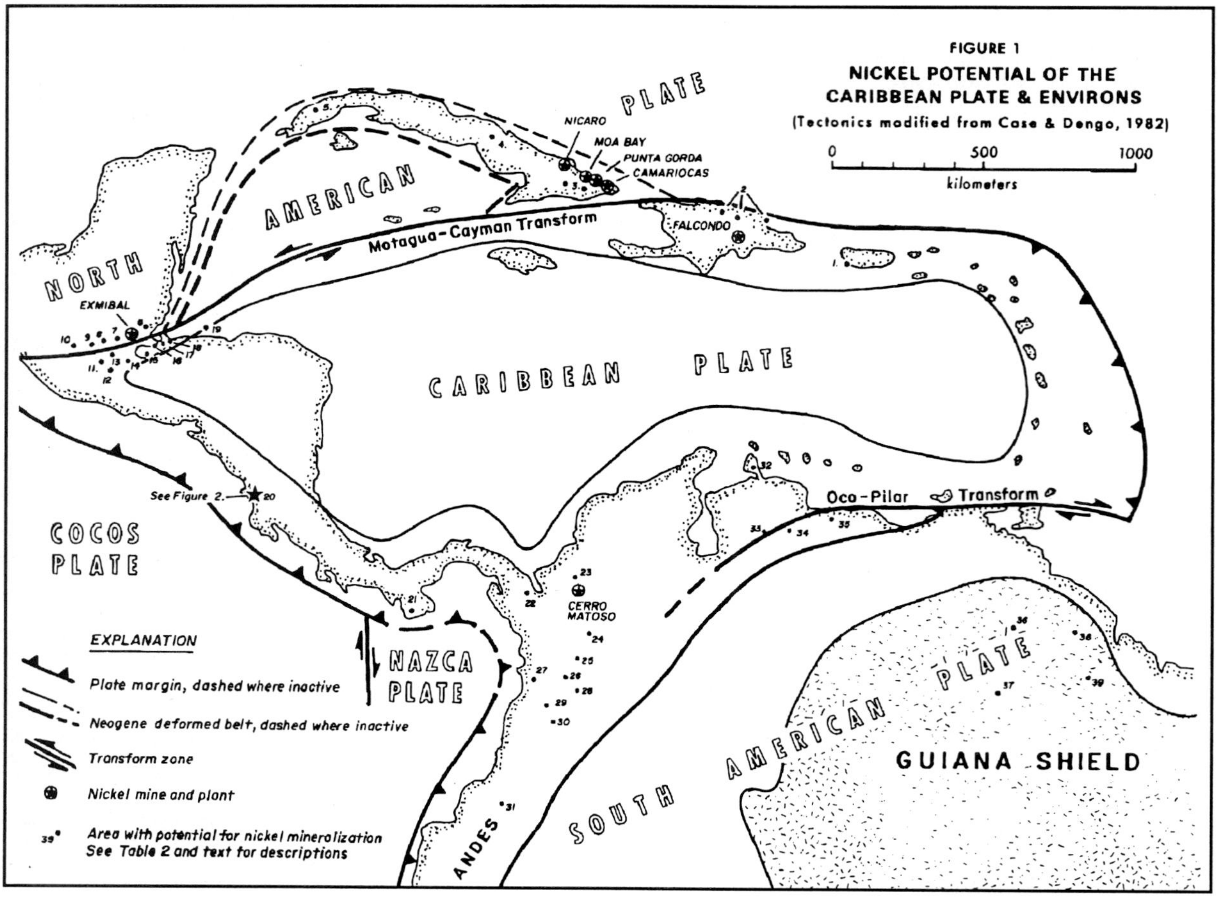

Figure 1. Nickel potential of the Caribbean Plate and environs.

the scope of this paper, but is not believed to be important compared to the land-based potential or the sea-floor potential in other oceanic basins.

The authors of this paper have worked at or visited all of the nickel projects listed in Table 1, as well as many of the prospective areas identified in Table 2.

GEOLOGIC SETTING

The Caribbean Plate, together with its marginal deformed belts, underlies about 5 million km^2, of which more than 75 percent is covered by ocean (Figure 1). As described by Case and Dengo (1982), the tectonic history of this plate has been complex and is not fully deciphered. It is apparent that a large slab of oceanic crust became detached from the adjacent oceanic plates during Mesozoic time, and has since behaved as a unit subject to interactions with the North American, South American, Nazca, and Cocos Plates.

As a relatively small plate buffeted by larger ones, the Caribbean Plate has undergone a large degree of deformation at the margins as well as internally. Undisturbed oceanic crust now underlies only about 25 percent of the plate. Subduction of oceanic crust from adjacent plates is currently active on the east (Antilles Arc) and on the west (Central American volcanic belt), while the northern and southern boundaries of the plate are undergoing lateral translation along the Montagua-Cayman and Oco-Pilar transforms.

With respect to genesis of nickel deposits, the most important tectonic process affecting the Caribbean Plate has been the emplacement at shallow levels of mafic and ultramafic rocks of deep crustal and mantle origin. This has occurred mainly at the northern, western, and southern plate margins, where ultramafic rocks (varieties of peridotite and their serpentinized equivalents), basalts, and gabbros occur within deformed belts. In some cases, such as in Cuba, Guatemala, and Costa Rica, large slabs of ophiolite, including gabbroic and ultramafic components, have been emplaced within and above Cretaceous and Tertiary rocks. In other portions of the plate margins, relatively small, highly-deformed pods and slices of deep-seated rocks lie within deformed and metamorphosed rocks of shallow crustal origin.

Table 1. Nickel production from the Caribbean Plate and environs.

DEPOSIT	LOCATION	YEAR STARTED	1988 Ni PRODUCTION tonnes	products	PRINCIPAL OPERATOR(s)	APPROX. INITIAL ORE GRADE, %Ni	REMARKS
Falcondo	Dominican Republic	1973	29,300	ferro-nickel	Falconbridge	1.6	estimated 5-10 years' reserves remaining as of 1989.
Camariocas	Cuba	1995?	0	oxide sinter	Cuban gov't.	?	ammonia-leach process; under construction since early 1980s.
Punta Gorda	Cuba	1986	9,000*	oxide sinter	Cuban gov't.	?	Che Guevara plant; ammonia-leach process, operating far below design capacity.
Moa Bay	Cuba	1959	17,000 *	sulfide#	Cuban gov't.	1.3	acid-leach process: inaugurated by Freeport Sulphur Co. in 1959.
Nicaro	Cuba	1952	16,000 *	oxide sinter	Cuban gov't.	1.4	ammonia-leach process; original plant built by U.S. government in 1944.
EXMIBAL	Guatemala	1977	0	Ni-S matte	INCO	2.1	mine and smelter closed in 1980 due to heavy financial losses during operation.
Cerro Matoso	Columbia	1983	16,600	ferro-nickel	Billiton-gov't.	2.7	mine and smelter now operating near capacity of 22,000 tonnes per year.

* = estimated, # = cobalt also produced in saleable form

NOTE: All production is based on lateritic ores. Production estimates in metric tonnes from U.S. Bureau of Mines. See Figure 1 for locations.

South of the Caribbean Plate, ultramafic and gabbroic rocks are found in the northern Andes in settings analogous to, but differing in age from, those in the Caribbean deformed belts. The northern portion of the Guiana Shield, south of the Caribbean Plate in Venezuela and Guyana, contains small areas of mafic and ultramafic intrusives. Mainly of Proterozoic age, these rocks are capable of hosting nickel deposits similar to those in the Caribbean Plate.

NICKEL DEPOSIT TYPES

Deposit Models

Nickel occurs in several geological environments and deposit types, discussed here insofar as they relate to this region. The relevant mineral deposit models, slightly modified from Cox and Singer (1986), are enumerated on Table 3. Several other models discussed by Cox and Singer are omitted, in the absence of any specific evidence that their favored environments occur in the Caribbean region. Another useful description of nickel sulfide deposit types is in Eckstrand (1984).

Nickel deposits are broadly divisible into sulfide types and laterite (silicate, oxide) types. Sulfide deposits are formed by magmatic (and locally hydrothermal) processes, and provide the bulk of current and historical worldwide nickel production. Laterite deposits are formed by supergene processes and constitute the majority of identified nickel resources worldwide, including all the important known deposits in the region under discussion. Many laterite deposits were produced during the 1960s and 1970s, but higher energy costs since the early 1970s have curtailed these operations.

Sulfide Deposits

Several types of nickel sulfide deposits are of economic importance, as shown in Table 3. These are usually hosted in ultramafic or mafic igneous rocks such as peridotites and gabbros, and most formed directly or indirectly through sulfide immiscibility in cooling silicate magmas. Most major producing deposits occur in Precambrian shield areas in Canada, Australia, the Kola Peninsula, and southern Africa, but examples are known in other geologic terranes as well. No important examples have been identified in the Caribbean region.

Nickel sulfide deposits normally carry other valuable metals, including copper, cobalt, platinum-group metals, silver, gold, and sometimes others. In most cases, concentrates carrying these metals can be prepared and treated in conjunction with treatment (usually smelting) of the nickel sulfides. Typically, a nickel mine based on sulfide ores will yield by-product metals with 25 to 75 percent of the value of nickel produced. The ore grade may vary from under 1 percent to about 3 percent Ni, depending on by-product credits, the mining method, and other factors.

Table 2. Areas with nickel potential in the Caribbean Plate and environs.

LOCATION	GEOLOGICAL	POTENTIAL DEPOSIT TYPES	STATUS	REFERENCE
Puerto Rico				
1. Mesillas, etc. SW corner	laterized serpentinite, gabbro	L G	20-50 million t @ 1.2% Ni development unlikely due to urbanization	Heidenreich & Reynolds (1959)
Dominican Republic				
2. three areas N. Coast	serpentinized peridotite	L	economic potential low	Haldemann and others (1982)
Cuba				Consejo de Redaccion (1978)
3. many areas Holguin & Guantanamo	serpentinite	L	extensive laterites	
4. Rio Caonao Camague	serpentinite	L	extensive laterites	
5. Cajalbana Pinar del Rio	serpentinite	L	extensive laterites	
Guatemala				Bonis, Bohnenberger and Dengo (1970)
6. Rio Sumache Izabal	peridotite in ophiolite	L	known laterite resource	
7. Oxec Izabal	basalt, peridotite in ophiolite	G LF	former copper mine	Peterson and Zantop (1980) this report
8. Rio Negro El Quiche & Baja Verapaz	peridotite	L	known laterite resource	
9. Cintac El Quiche	ultramafic	L	known laterite resource	
10. Cuchumatanes Huehuetenango	ultramafic	L G		
11. Rio Motague Baja Verapaz	ultramafic	L		
12. Morazan El Progreso	ultramafic	L		
13. Acasaguastlan El Progreso	ultramafic	L		
14. Rio Tambor Jalapa	ultramafic	L		
15. Cerro Colorado Zacapa	ultramafic	L	known laterite resource	
16. Montufar Izabal	peridotite	L	known laterite resource	
17. Morales Izabal	ultramafic	L	small area; low potential	
18. Rio Cacao Izabal	ultramafic	L	small area; low potential	
Honduras				
19. Bay Islands Islas Bahia	ophiolitic gabbro & serpentine	L D	small area; low potential	McBirney and Bass (1969)

Laterite Deposits

The only well-documented type of nickel deposit in the Caribbean region, and the only one of current economic importance there, is the lateritic type (38a on Table 3). The general geology and economics of lateritic nickel deposits are discussed by several authors in Evans, Shoemaker, and Veltman (1979). These deposits are formed by deep weathering of ultramafic rocks comprised primarily of olivine, or of serpentine derived from olivine. Olivine of mantle origin normally contains 80 percent or more of the magnesian forsterite molecule. Isomorphous substitution of Ni for Mg in forsterite normally yields a peridotite (or derived serpentinite) containing 0.2-0.3 percent Ni by weight.

If the rate of chemical weathering of a peridotite or serpentinite is greater than the rate of physical erosion, a deep laterite profile may be preserved, containing zones of leaching and of secondary deposition of nickel. Nickel enrichment occurs as poorly-crystalline nickeliferous hydrous iron oxides (goethite, etc.), and in nickeliferous serpentines (nepouite,

Table 2. Continued.

LOCATION	GEOLOGICAL	POTENTIAL DEPOSIT TYPES	STATUS	REFERENCE
Costa Rica				
20. Santa Elena Guanacaste	peridotite in ophiolite	L D LF	laterite remnants: partly in national park	Dengo (1962), this report
21. Azuero Los Santos	gabbroic ophiolite	D G		Del Guidice and Recchi (1969)
22. Obaldia Darien	ultramafic? pod	D		Instituto Geografico (1974)
23. Planeta Rica, Cordoba	laterized ultramafic	L	several Ni-Co deposits near Cerro Matoso	Wokittel (1960)
24. Yarumal Antioquia	serpentinite			Gonzales (1972)
25. Medellin Medellin	laterized ultramafic	L	Ni-Co prospect	Wokittel (1960) Gonzales (1972)
26. Anserma Risaralda	serpentinite			Gonzales (1972)
27. Condoto Choco	ultramafic	D	Ni-Co-Pt prospect	Wokittel (1960)
28. Manizales Caldas	serpentinite			Gonzales (1972)
29. Vijes Valle	ultramafic	L D	Ni prospect	Wokittel (1960)
30. Quilacho Cauca	laterized ultramafic	L	Ni-Co prospect	Wokittel (1960) Gonzales (1972)
31. Los Andes/Tambo Narino	ultramafic	L	Ni prospect	Wokittel (1960) Gonzales (1972)
Venezuela				
32. El Rodeo Falcon	gabbro in ophiolite	G		Bellizzia (1967)
33. La Bimba Lara	ultramafic			Bellizzia (1967)
34. Tinaquillo Cojedes	gabbro and peridotite	L	low Ni grades reported	Marval (1972)
35. Loma de Hierro Miranda	laterized ultramafic	L	drilled deposit; 75 million t @ 1.5% Ni	Rodriguez (1986)
36. Bochinche Bolivar	Precambrian buried dunite?	D G	near gold mine	Rodriguez (1986)
37. Piton de Uroy Bolivar	Precambrian ultramafic	D G	near gold mine	Rodriguez (1986)
Guyana				
38. Kauramembu N.W. district	ultramafic dikes in Precambrian greenstones	L D	recon shows to 1% Ni in soils	Kilpatrick (1969) Vallance (1970) this reprt
39. Wariri lower Cuyuni	Precambrian ultramafic pod	D G	recon shows Ni in soils; gold nearby	MacDonald (1968)

NOTE: Excludes producing deposits listed in Table 1. Second line of location indicates political subdivision (province, state, department, etc.). Deposit types: D = dunitic (6b), G = greenstone gabbro (7a), L = laterite (38a), LF = Limassol Forest (8c).

"pimelite," pecoraite, falcondoite, etc.). Although nickel-bearing sulfide minerals are often present in trace amounts in the parent ultramafic rock, their presence is not a factor in creation of a nickeliferous laterite deposit. The nickel content of ore-grade laterite is normally in the range 1.0-2.5 percent.

Nickel laterites in the Caribbean region and elsewhere often contain small amounts of cobalt, chromium, and manganese, in addition to large amounts of iron. Of these, only cobalt has been economically recovered in any nickel laterite operation.

Laterite ores are typified by low surface-mine operating costs, due to their occurrence as a soft surficial weathering crust. However, they also require high processing costs, because of their high free and combined moisture contents, and the absence of easily-separable discrete nickel minerals which can form a high-grade concentrate. Most nickel-laterite mining operations are, therefore, open-pit mines with nearby metallurgical plants. Metallurgical processes currently in use are pyrometallurgical (smelting), acid-leach, or ammonia-leach, all of which use more energy to process laterite ores than sulfide ores of comparable

Table 3. Nickel deposit models potentially applicable to the Caribbean Plate and environs.

DEPOSIT MODEL No.	Name (example).	TYPICAL HOST ENVIRONMENTS Lithology and tectonic setting	CARIBBEAN EXAMPLES Known	Potential	ECONOMIC POTENTIAL
5a.	Duluth gabbro (Duluth, USA)	layered mafic intrusions, external sulfur source; rift or craton	none	Guiana Shield	very low
6b.	dunitic Ni-Cu (Manibridge, Can.)	small dunite intrusives in greenstone belts; shield or plate margin ophiolite	none	Guiana Shield, Caribbean Plate margins	low
7a.	greenstone gabbroid (Selebi, Botswana)	small gabbroid intrusives in greenstone belts; shield or plate margin ophiolite	none	Guiana Shield, Andes; Carribbean Plate margins	low
8c.	Limassol Forest Co-Ni (Bou Azzer, Morocco)	serpentinized ultramafic; plate margin ophiolite	none	Caribbean Plate margins	very low
38a.	lateritic Ni (New Caledonia)	ultramafic (+60% olivine or serpentine); any setting in tropics	many (Table 1)	Caribbean Plate, Andes, Guiana Shield	high

NOTE: Models are slightly modified from Cox and Singer (1986)

grade. Thus the availability and cost of energy is a major factor in determining the economic feasibility of laterite ores (see P. T. O'Kane *in* Evans, Shoemaker, and Veltman, 1979).

Identification of Nickel Potential

The discussion above indicates that the existence of commercial nickel deposits depends upon the presence of mafic or ultramafic intrusive rocks. Furthermore, the formation of nickeliferous laterites depends preferably on tropical weathering of relatively flat areas of ultramafic rocks. In the Caribbean Plate and environs, favorable rock types and climatic conditions are widespread.

RECOGNIZED NICKEL RESOURCES IN THE CARIBBEAN REGION

Table 1 and Figure 1 identify the developed Caribbean nickel deposits, all but two of which are in current production. Most of these projects have sufficient mineral resources to remain in production for at least another decade, barring drastic economic or political developments.

In addition to the developed resources, a number of other nickel deposits, all laterites, have been identified and defined. These are shown on Table 2, along with some potential resource areas. The best-known of these are the deposits in southwest Puerto Rico, in several areas in Guatemala, and at Loma de Hierro in Venezuela.

The Puerto Rican laterites (1, Table 2) have been studied extensively over a period of several decades. A significant tonnage of serpentinite-derived ore of smeltable bulk composition and low- to moderate-grade (1.2 percent Ni, locally higher) has been defined. While the coastal location is a potential asset, the density of population in Puerto Rico and the failure of the Puerto Rican government to provide a suitable framework for mining investment, have resulted in urban development covering much of the surface of the deposits. Therefore, it is unlikely that these deposits will ever be mined.

Exploration in Guatemala since 1960 has revealed a number of nickel-enriched laterites in that country's extensive ultramafic terranes. None has as large a moderate-grade resource as the EXMIBAL property north of Lake Izabal (Figure 1). However, several have smaller resources of moderate- to high-grade which could conceivably feed a small local processing plant or the inactive EXMIBAL plant. These include Rio Sumache, Rio Negro, Cintac, Cerro Colorado, and Montufar (see Table 2 and Figure 1). Some of these are in isolated mountain localities. There is currently little activity at any of these deposits, and prospects for near-term development are poor.

Many nickeliferous laterites are known to exist in Cuba, in addition to those directly feeding the plants listed in Table 1. Whether additional deposits will be developed prior to commencement of full-scale production from the present projects (Table 1) is a matter for conjecture.

The Loma de Hierro laterite deposit in Venezuela (35, Table 2) contains a sizable nickel resource, well-located with respect to infrastructure. Given Venezuela's abundant energy supply and well-developed steel industry, this deposit stands a fair chance of development if attractive incentives for foreign investment are offered.

POTENTIAL NICKEL RESOURCES

General

In a broad sense, almost any ultramafic or gabbroic body can be considered as a potential host for laterite or sulfide nickel deposits, based on the models presented in Table 3. Figure 1 and Table 2 identify many areas containing these rocks, many of which have been investigated at least casually for nickel potential by government or company surveys. This list is not exhaustive,

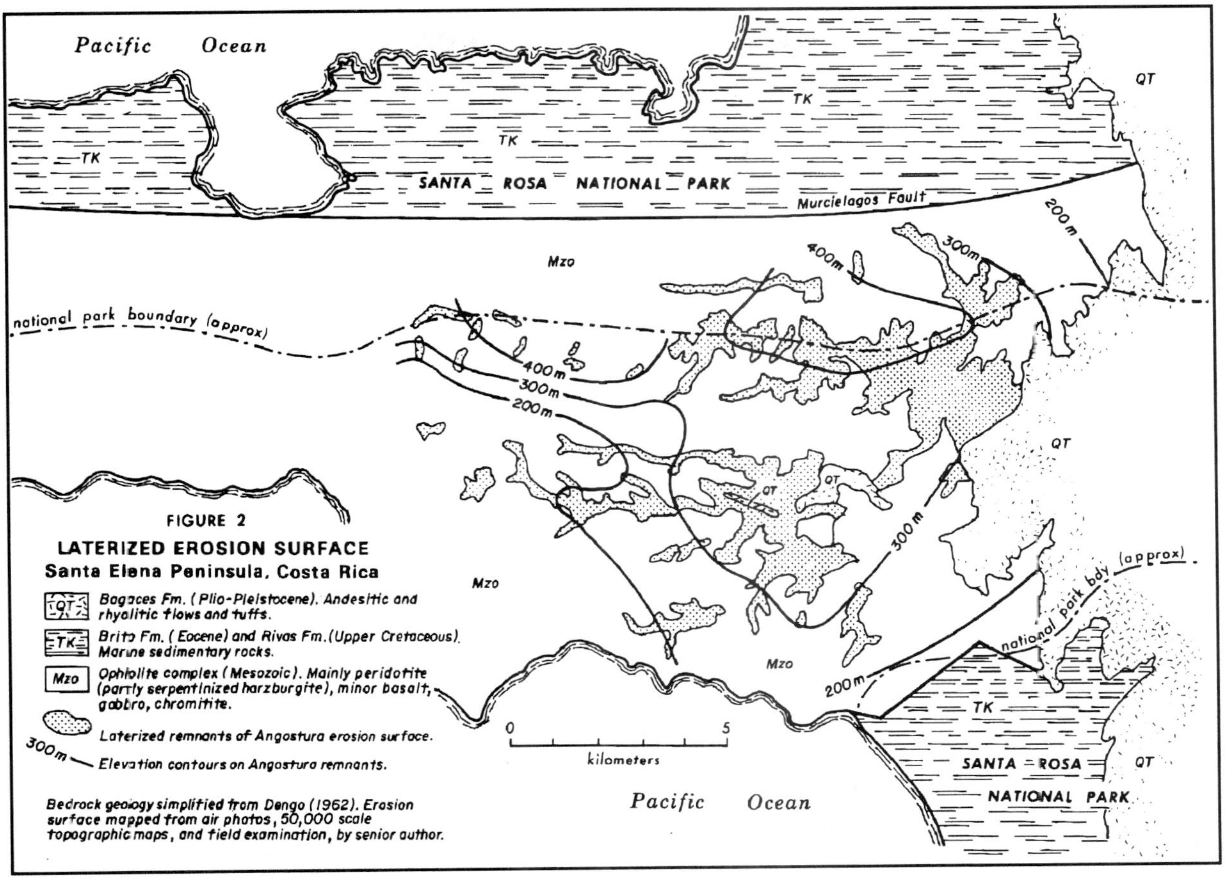

Figure 2. Laterized erosion surface, Santa Elena Peninsula, Costa Rica.

particularly with respect to small gabbroic intrusives. The discussion below relates to only a few of the areas, chosen to illustrate a variety of geologic settings rather than necessarily those with the best potential.

Oxec, Guatemala

The Oxec copper deposit is located mainly within pillow basalts in the upper part of an ophiolite sequence in eastern Guatemala (Petersen and Zantop, 1980). The deposit is similar to typical Cyprus copper deposits, and contains copper with traces of cobalt.

Although the Oxec deposit is not known to contain nickel mineralization, the relatively complete ophiolite sequence within which the copper mineralization occurs could host any of several types of nickel sulfide deposits: most likely are the Limassol Forest nickel-cobalt arsenide veins and greenstone-gabbroid type (8c and 7a, Table 3). Lesser probability exists for the dunitic and Duluth types (6b and 5a, Table 3). It should be emphasized that the potentials discussed are entirely hypothetical, as the writers are not aware of any specific evidence pointing toward the existence of nickel deposits (other than laterites, which are widespread in the area) at Oxec.

Santa Elena, Costa Rica

The Santa Elena Peninsula on Costa Rica's Pacific coast contains that country's only large ultramafic terrane. Serpentinized harzburgite with subsidiary basaltic rocks, covering some 200 km², has been emplaced within and probably above Cretaceous and Eocene sedimentary rocks (Dengo, 1962). Extensive weathering during late Tertiary time created a low-relief erosion surface, the Angostura surface (Dengo, 1962), with a lateritic weathering crust on the ultramafic rocks. Later this surface was covered by Plio-Pleistocene volcanics, which have now been nearly removed by erosion in the Santa Elena Peninsula (Figure 2).

The Angostura surface and related laterite crust exist in remnants on the Santa Elena Peninsula. A contour map on the erosion surface remnants (Figure 2) indicates the initial relief on the surface, as well as modification by Tertiary deformation. Many, but not all, of the erosion surface remnants shown on

Figure 2 host a truncated lateritic weathering profile which is enriched in nickel. Reconnaissance sampling of this crust by the senior author showed Ni grades averaging under 1 percent, but locally near 2 percent in small wedge-shaped root zones of low-iron saprolite. The total tonnage and average grade of this material was not determined.

In addition to the possibilities for nickel laterite orebodies, the area has an untested potential for nickel sulfides in the basaltic and gabbroic portions of the ophiolite sequence (see discussion of Oxec, above). Given the small area of such rocks on the peninsula, the potential is rated as low.

This area's coastal location might appear to enhance its potential for serious exploration and possible development. However, much of the Santa Elena Peninsula was recently incorporated into Santa Rosa National Park, in part due to its spectacular coastal setting. Most of the laterite remnants are sandwiched between two portions of the park (Figure 2), and further exploration and development of the laterite seem unlikely.

Kauramembu, Guyana

Several small areas of ultramafic rocks occur within the Guiana Shield in Guyana (Choudhuri, 1973). In the Kauramembu Mountains south of Kokerite (38, Table 2), Lower Proterozoic metasedimentary and metavolcanic rocks are intruded by peridotite dikes of probable Proterozoic age. Exploration work described by Kilpatrick (1969) and Vallance (1970) revealed some laterite located over the ultramafic dikes showed values up to 1 percent Ni. Due to the narrow width of the ultramafic intrusives belt (less than 200 m), no large areas of nickel-enriched laterite were found.

The Kauramembu occurrence is, however, of potential interest for nickel sulfide mineralization. While the remote location, dense jungle, and laterite cover precluded detailed mapping, it appears from Vallance's (1970) descriptions that the peridotite dikes could be either ultramafic flows within a volcanic-sedimentary sequence, or apophyses from a large ultramafic intrusive at depth. Vallance notes petrographic evidence for more than one ultramafic magmatic pulse. If there is a large intrusive at shallow depth, this area could have potential for nickel sulfide mineralization of the dunitic type (type 6b, Table 3), and possibly other types.

Other Areas

Some of the other areas shown on Table 2 are equally prospective for nickel as those discussed above. In addition, other areas with favorable geology exist in the region. For the most part, other small gabbroic or ultramafic massifs in the Antilles, the Andes, and Central America are too small and too dismembered by Cenozoic tectonic processes to host large laterite or sulfide nickel resources.

In the Guyana Shield, scope exists for undiscovered massifs of sufficient size and structural integrity to host sizable Precambrian nickel deposits. Any truly large ultramafic or mafic complexes probably have been already revealed by geophysical or geological surveys. However, smaller examples of dunitic, greenstone-gabbroid, and possibly even Duluth type, could remain in the lesser-known portions of the northern Guyana Shield.

ECONOMIC AND POLITICAL ENVIRONMENTS

The Caribbean region is favored by an abundance of water and indigenous labor, a relatively benign climate, and proximity to ocean shipping and to North American and European markets. However, most countries in the region suffer from scarcity of indigenous fuel, limited industrial infrastructure, and in some cases, from a foreign investment climate regarded as less than optimal by foreign investors.

A major constraint against development of additional Caribbean nickel resources is the economic risk for investors in the immense capital costs for installations necessary to process nickel laterites. This risk is composed of economic and political components, and to a lesser degree technical components. Economic risk, due to the cyclicity of nickel prices and the cost of installations, could be reduced in the case of nickel sulfide deposits (by preparation and shipping of a high-grade concentrate, obviating the need for a metallurgical plant at the site), but no sizable nickel sulfide resources are known in the region.

Most countries in the region have small economies, in which a large complex of the type needed to mine and process lateritic ores is highly visible economically and politically. In one country, for example, existing nickel plants were forcibly nationalized after they had achieved production. In two other countries, governments attempted to place additional taxes on established nickel plants, to raise revenue to meet the rising costs of imports, including petroleum. These incidents tend to accentuate the already-marginal economic environments of the project during periods of low nickel prices, and to deter further foreign investment.

CONCLUSIONS

The Caribbean region is well-situated geologically and with respect to markets, but faces a number of obstacles to significantly increasing the number of producing nickel deposits. The chief obstacles are the lack of known resources of nickel sulfides, with their lower energy consumption for processing and their attendant valuable by-products, and the widely-varying investment climate.

A comprehensive assessment of nickel sulfide potential in those countries with realistic investment climates could bring to light some areas with a serious potential for future production of nickel and by-product metals.

ACKNOWLEDGMENTS

The authors acknowledge helpful discussions with Mr. Gary Sidder of the U.S. Geological Survey, Caracas, and Mr. Gabriel Dengo of ICAITI, Guatemala City. Recent nickel production data were obtained from Ms. Ivette Torres of the U.S. Bureau of Mines, Washington, D.C. The Circum-Pacific Council for Energy and Mineral Resources provided partial financial support.

REFERENCES

Bellizzia G., A., 1967, Rocas ultrabasicas en el sistema montanoso del Caribe y yacimientos minerales, asociados: Caracas, Boletín de Geología, v. 8, no. 16, p. 159-198.

Bonis, S., O. Bohnenberger, and G. Dengo, compilers, 1970, Mapa geológica de la Republica de Guatemala, escala 1:100,000: Guatemala, Instituto Geografico Nacional.

Case, J. E., and G. Dengo, 1982, The Caribbean region, in A. R. Palmer, ed., Perspectives in regional geological synthesis: Geological Society of America, Decade of North American Geology Special Publication 1, p.163-170.

Choudhuri, A., 1973, Basic magmatism in relation to metamorphism and orogeny in northern Guyana: London, Geological Magazine, v. 110, p. 365-371.

Consejo de Redaccion del Atlas de Cuba, eds., 1978, Atlas de Cuba: Havana, Instituto Cubano de Geodesia y Cartografia, 168p.

Cox, D. P., and D. A. Singer, eds., 1986, Mineral deposit models: U.S. Geological Survey Bulletin 1693, 379p.

Del Guidice, D., and G. Recchi, 1969, Geología del area del proyecto minero del Azuero, mapa 5 (1:250,000): Panama, Instituto Geografico Nacional.

Dengo, G., 1962, Estudio geológico de la region de Guanacaste, Costa Rica: San Jose, Instituto Geografico de Costa Rica, 73p.

Eckstrand, O. R., 1984, Canadian mineral deposit types—A geologic synopsis: Geological Survey of Canada, Economic Geology Report 36, 86p.

Evans, D. J. I., R. S. Shoemaker, and H. Veltman, eds., 1979, International laterite symposium: New York, Society of Mining Engineers, 688p.

Gonzales I. H., 1972, Posibles aplicaciones industriales de la serpentinita en Colombia: Bogota, Boletín Geológico, v. 20, no. 3, p. 45-58.

Haldemann, E. G., R. Buchan, J. H. Blowes, and T. Chandler, 1982, Geology of lateritic nickel deposits, Dominican Republic: Society of Economic Geologists Fall Field Trip Guidebook, Economic geology of central Dominican Republic, October, 1972, p. 16-39.

Heidenreich, W. L., and B. M. Reynolds, 1959, Nickel-cobalt-iron-bearing deposits in Puerto Rico: U.S. Bureau of Mines Report of Investigations 5532, 68p.

Instituto Geografico Nacional Tommy Guardia, 1974, Mapa geológico de Panama, escala 1:1,000,000: Panama, Dirección General de Recursos Minerales.

Kilpatrick, B. E., 1969, Nickel, chromium and cobalt in tropical soils over serpentinites, northwest district, Guyana: Colorado School of Mines Quarterly, v. 64, part 1, p. 323-332.

Marval R. H., 1972, Estudio sobre la meteorización de la peridotita de Tinaquillo, Estado Cojedes: Caracas, Boletín de Geología, Publicación Especial no. 5, Memoria cuarto congreso geológico Venezolano, v. 4, p.2243-2244.

MacDonald, J. R., 1968, A guide to mineral exploration in Guyana: Guyana Geological Survey and Mines Department Bulletin 38, 91p.

McBirney, A. R., and M. N. Bass, 1969, Geology of the Bay Islands, Honduras, *in* Tectonic setting of northern Central America and the western Caribbean: American Association of Petroleum Geologists Memoir 11, p. 229-293.

Petersen, E. U., and H. Zantop, 1980, The Oxec deposit, Guatemala; An ophiolite copper occurrence: Economic Geology, v. 75, no. 7, p. 1053-1065.

Rodriguez M., S. E., 1986, Recursos minerales de Venezuela: Caracas, Boletín de Geología, v. 15, no. 27.

Vallance, G., 1970, Investigation of a nickeliferous "laterite" in the Kauramembu Mountains (Wanamu-Blue Mountains area), Barama River, North West District, Guyana: Contribution 14, *in* Proceedings of eighth Guiana geological conference: Georgetown, Department of Mines and Geology.

Wokittel, R., 1960, Recursos minerales de Colombia; Niquel y cobalto: Servicio Geológico Nacional, v. 10, p. 155-156.

Circum-Pacific Council for Energy and Mineral Resources
Earth Science Series, Vol. 16
R. L. Miller, G. Escalante, J. A. Reinemund, and M. J. Bergin (Eds.)
Energy and Mineral Potential of the Central American-Caribbean Regions
Springer-Verlag, Berlin Heidelberg 1995

THE JAMAICAN BAUXITE INDUSTRY: GLIMPSES INTO ITS PAST, PRESENT, AND FUTURE

Carlton E. Davis
Jamaica Bauxite Institute, Hope Gardens, P. O. Box 355, Kingston 6, Jamaica W.I.

ABSTRACT

The commercial possibilities of Jamaican bauxite were recognized in 1943 at a time when there was great need for aluminum for the Allied war effort, and when availability was being made difficult by the harassment by German U-boats of Allied bauxite ships plying from the sources of the ore in South America to the North American mainland. For technical reasons, however, it happened that Jamaican bauxite was not required for the war effort.

After the war, because of a number of factors, including the pre-eminence of the geographically-close United States as an economic and military power, and the emergence, at the instigation of the U.S. Government, of three companies (Alcan, Reynolds, and Kaiser), each of which needed its own independent source of bauxite, the Jamaican industry was rapidly developed to the point that the island became the number one world producer in 1957. Growth continued during the economically buoyant 1960s, and at the end of the decade six transnational companies—Alcan, Reynolds, Kaiser, Alcoa, Anaconda, and Revere—were well established, mining and/or processing bauxite in the island.

The industry faced enormous difficulties during most of the 1970s and 1980s as a result, among other things, of the oil price rises in 1973 and 1979, lower economic growth all around, the decline of the United States as a major alumina and aluminum producer and severe competition from new industries in Australia, Guinea, and Brazil where taxes were, by and large, lower.

With the lowering of oil prices since late 1985, a better supply/demand balance in the industry, a weaker U.S. dollar (to which the Jamaican currency is pegged) vis-a-vis other major currencies, and a new taxation regime for the industry with which the companies profess to be happier, the industry is poised to recover some of the ground lost during the troublesome 1970s and 1980s.

INTRODUCTION

As is well known, bauxite is the ore from which most of the world's aluminum is made. The ore was discovered in 1821 by the French chemist Pierre Berthier, near the village of Les Baux (from which it derived its name), Provence, France. It is a heterogenous, earthy, or rock-like, material comprising (apart from the dominant aluminum minerals) principally silica and iron minerals.

Among the conditions most favorable for the formation of bauxite are areas and geological conditions which provided warm wet climates. Thus most of the world's great deposits of bauxite occur in the tropical regions of the world such as Australia, Guinea, Brazil, Jamaica, Guyana, Suriname, and the warmer areas of temperate countries such as southern France and the southern United States.

Two celebrated inventions, roughly around the same period, were responsible for the founding of the aluminum industry that we know today. First was the Hall-H*roult process, named after the inventors Charles Martin Hall of the United States and Paul Louis Toussaint H*roult of France, for the electrolytic method of separating aluminum from its oxide. The other was the invention, in two stages, by the Austrian chemist Franz Josef Bayer in 1888 and 1894, respectively, of the process for extracting alumina from bauxite.

For most of the period up to the beginning of the second World War, most of the world's bauxite was derived from four countries: the United States, France, and what were

Table 1. Jamaica and World Bauxite Production ('000 tonnes). 1952 — 1959

YEAR	WORLD	USA	% of WORLD	EUROPE	% of WORLD	JAMAICA	% of WORLD	GUIANAS	% of WORLD
1952	12,769	1,694	13.3	1,208	9.5	346	2.7	5,504	43.1
1953	13,861	1,605	11.6	3,179	22.9	1,173	8.5	5,753	41.5
1954	16,253	2,038	12.5	3,924	24.1	2,077	12.8	5,800	35.7
1955	17,984	1,817	10.1	3,142	17 5	2,688	14.9	5,538	30.8
1956	19,000	1,772	9.3	3,350	17.6	3,192	16.8	6,005	31.6
1957	20,524	1,439	7.0	3,708	18.1	4,669*	22.7	5,615	27.4
1958	21,337	1,332	6.2	3,753	17.6	5,814	27.2	4,446	20.8
1959	23,064	1,727	7.5	3,821	16.6	5,208	22.6	5,131	22.3

*Replaced Suriname as the World's No.1 producer.

Table 2. Jamaica and World Bauxite Production ('000 tonnes). Selected Years 1960 — 1969

YEAR	WORLD	JAMAICA	% of WORLD	GUIANAS	% of WORLD	GUINEA	% of WORLD	AUSTRALI A	% of WORLD
1960	27,620	5,872	21.3	5,596	20.3	1,435	5.2	72	0.26
1962	31,220	7,706	24.7	5,164	16.5	1,660	5.3	30	0.10
1964	33,143	7,828	23.6	5,336	16.1	1,433	4.3	796	2.40
1966	41,057	9,062	22.1	8,921	21.7	1,609	3.9	1,827	4.40
1968	47,230	8,526	18.0	9,383	19.9	2,118	4.5	4,955	10.50
1969	55,458	10,499	18.9	10,542	19.0	2,459	8.3	7,921	14.30

then British Guiana (later Guyana) and Dutch Guiana (Suriname).

THE EARLY YEARS OF THE JAMAICAN INDUSTRY 1943-1959

The second World War resulted in, among other things, an enormous increase in demand for aluminum and the emergence of the United States as the world's major aluminum producer. These developments put tremendous stress on the available bauxite resources in the Western Hemisphere, principally the United States and the Guianas.

However, the supply routes from the Guianas to the North American mainland were harassed by German U-boats. The discovery of bauxite in Jamaica (ca. 1943), which is geographically much closer to the North American mainland than the Guianas are to the mainland, and the harassment by the U-boats focused interest in Jamaican ore as a source for the war effort. Because of technological difficulties, and perhaps because by that time the Allies were winning the war, Jamaican bauxite was not used for the war effort. The technological problems arose because the North Americans were used to the largely trihydrate and coarse-grained North and South American bauxite and not the mixed hydrate and fine-grained Jamaican bauxite.

However, after the war, because of a number of factors, including (1) the pre-eminence of the geographically close United States as an economic and military power; (2) increased demand for aluminum for military and peacetime uses (the Korean War, strategic stockpiling); (3) the emergence, at the instigation of the U.S. Government, of three independent companies (Alcan, Reynolds, and Kaiser), each of which needed its own independent source of bauxite, the local industry was soon rapidly developed so that by 1957 the island became the world's number one bauxite producer (Table 1).

THE JAMAICAN INDUSTRY DURING THE 1960S

Because of favorable economic conditions such as impressive growth rates in the developed world, low inflation, low oil prices, as a result, among other things, of the "guns and butter" policies pursued by the U.S. in providing for the needs of the Vietnam War as well as civilian needs, and very impressive market growth and penetration of other markets by the aluminum industry the Jamaican industry grew significantly during the 1960s.

As illustrated in Table 2, bauxite production grew from 5.9 million MT at the beginning of the decade to 10.5 million MT at the end. Some of this bauxite growth was in alumina (Al_2O_3) production. By the end of the decade, six transnational companies—Alcan, Reynolds, Kaiser, Alcoa, Anaconda, and Revere—were variously involved in the Jamaican Industry. Figure 1 shows the location of various mines and alumina plants around the end of the 1960s and the early 1970s.

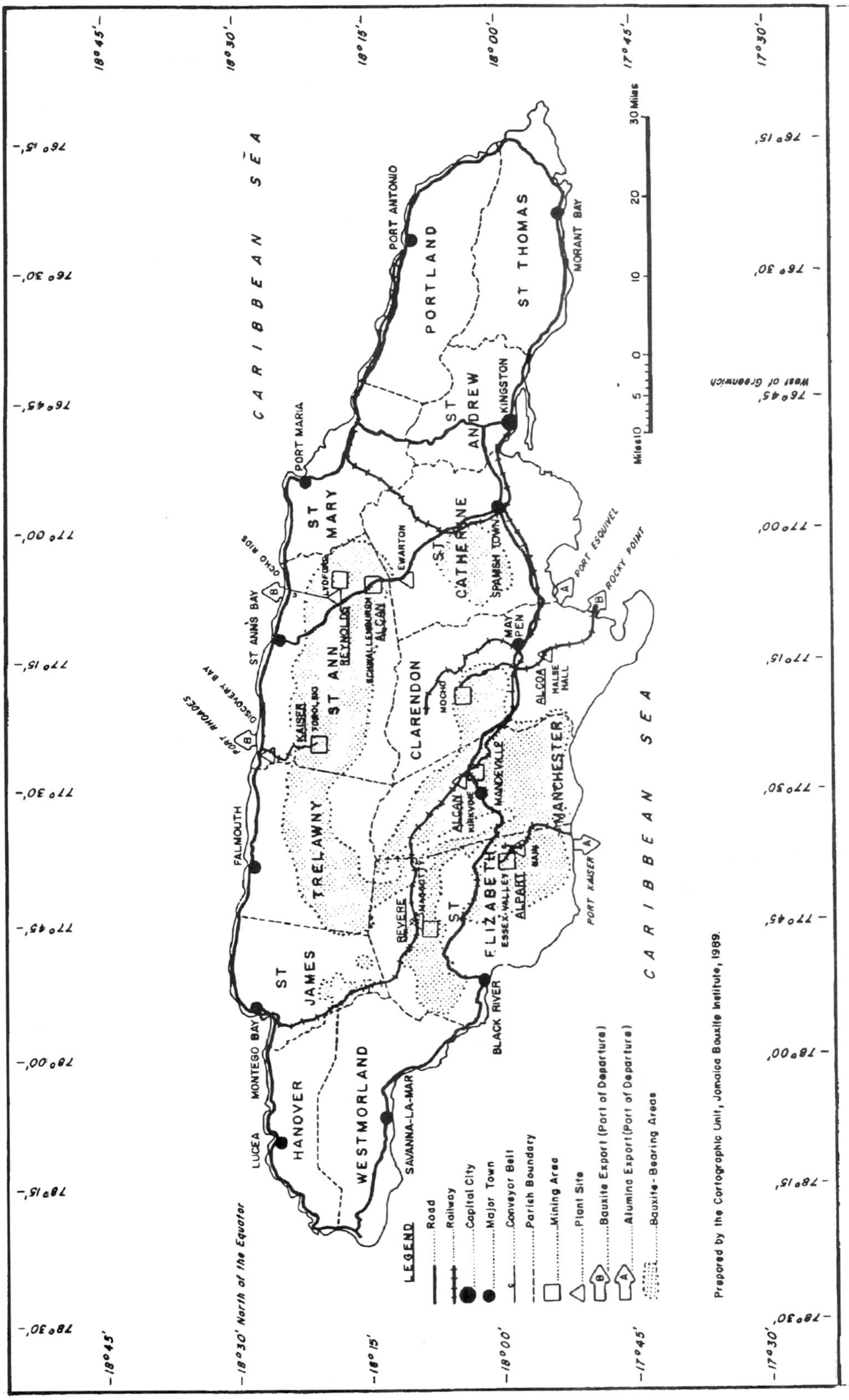

Figure 1. Industry location at about the end of the 1960s—early 1970s.

Table 3. Jamaica and World Bauxite Production ('000 tonnes). Selected Years 1970 — 1979

YEAR	WORLD	JAMAICA	% OF TOTAL	AUSTRALIA	% OF TOTAL	GUINEA	% OF TOTAL	BRAZIL	% OF TOTAL
1970	60,162	12,010	21.0	9,256	15.4	2,490	4.1	510	0.8
1972	70,806	12,539	17.7	14,437	20.4	2,600	3.7	785	1.1
1974	83,935	15,166	18.1	19,994	23.8	7,600	9.0	859	1.0
1976	80,492	10,296	12.8	24,085	29.9	11,316	14.1	998	1.2
1978	84,369	11,736	13.9	24,293	28.8	11,641	13.8	1,131	1.3
1979	88,009	11,505	13.1	27,583	31.3	12,199	13.9	1,642	1.9

Table 4. Jamaica and World Bauxite Production ('000 tonnes). Selected Years 1980 — 1988

YEAR	WORLD	JAMAICA	% OF TOTAL	AUSTRALIA	% OF TOTAL	GUINEA	% OF TOTAL	BRAZIL	% OF TOTAL
1980	91,338	11,978	13.1	27,178	29.8	13,311	14.6	4,152	4.5
1982	76,905	8,334	10.8	23,623	30.7	11,479	14.9	4,187	5.4
1984	90,323	8,735	10.0	32,121	36.0	12,745	14.0	6,271	7.0
1986	91,205	6,964	7.6	32,384	35.5	14,835	16.3	6,433	7.0
1987	94,670	7,660	8.1	34,000	35.9	16,282	17.2	6,565	6.9

THE JAMAICAN INDUSTRY DURING THE 1970S

The industry faced enormous difficulties during the 1970s. An important cause was the less buoyant international economic climate generated by massive inflation in commodity prices, especially oil, currency instability, and conflicts between bauxite-rich nations including Jamaica and transnational companies in the matter of resources taxation.

The Jamaican industry faced particular pressure from the consequential decline of the United States as a major alumina producer, severe competition from Australia, Guinea and Brazil in the industry, and the direct and indirect adverse impact of the oil price rises on Jamaica.

The strong competition from Australia especially arose from its much lower taxation rate compared to the Jamaican levy and its lesser dependence on oil.

With respect to Guinea, the competition arose mainly because the contracts for the supply of their high-quality ore were take-and-pay ones. The net effect was that Jamaica lost ground to its competitors during the decade of the 1970s (Table 3).

THE JAMAICAN INDUSTRY DURING THE 1980S

The situation continued during the 1980s (Table 4), punctuated by severe recessions in the industry in 1982 and 1985. The situation would have been worse had not the U.S. Government, as it did in an earlier period as a foreign aid gesture, purchased some 3.6 million MT of bauxite from Jamaica during 1983 and 1984.

A number of adverse developments characterized the period. The Reynolds bauxite mining operations, of a capacity of 2.8 million MT/yr, were closed in 1984. The Jamalco (Alcoa, 93 percent; Government of Jamaica, 7 percent) facilities were closed in February 1985. They were later re-opened by the Government in August 1985. The Alpart alumina plant, the country's single largest, was closed in September 1985. It was re-opened in 1988. In addition to these major developments, remaining companies cut back production during most of the 1980s.

PROSPECTS FOR THE JAMAICAN INDUSTRY IN THE 1990S

Recently a number of favorable factors have emerged, including (1) much lower oil prices since late 1985; (2) a better supply/demand balance in the industry; (3) a weaker U.S. dollar (to which the Jamaican dollar is pegged); (4) massive devaluation of the Jamaican dollar during the 1980s; and (5) a new taxation regime for the industry about which the companies appear to be happier. The industry, other things being equal, is poised for growth in the decade of the 1990s. For example, whereas we expect to produce some 8.5-9.0 million MT in 1989 (from a level of 7.4 million MT in 1988) we are looking at about 10.5 million MT in 1990 and more than 11.0 million MT in 1991. With luck, we could surpass the level of 12.0 million MT achieved in 1980 well before the middle of the 1990s.

Circum-Pacific Council for Energy and Mineral Resources
Earth Science Series, Vol. 16
R. L. Miller, G. Escalante, J. A. Reinemund, and M. J. Bergin (Eds.)
Energy and Mineral Potential of the Central American-Caribbean Regions
Springer-Verlag, Berlin Heidelberg 1995

MINERAL-ENERGY RESOURCES IN THE DOMINICAN REPUBLIC

Francisco J. Amézquita
Director General de Minería, Santo Domingo, D. F., Dominican Republic

ABSTRACT

Most of the geological exploration in the Dominican Republic in the present century has been by foreigners. Alcoa Exploration Company located an important deposit of bauxite in the southeastern part of the country and began production in 1958. However, production was terminated in 1982. Falconbridge Dominicana found important reserves of nickeliferous laterite in Pequero Hill near the town of Bonao in the central part of the country, and began production in 1972. Rosario Resources Corporation, after evaluating a previously-known mine of gold and silver in Pueblo Viejo near Cotui, began production in 1975.

The search for petroleum started early in the century, and has been concentrated in the southeast part of the country in the province of Azua. Some wells of 400 to 1,000 ft depth had modest production, but deep wells of 14,487 ft and 12,952 ft, named Charco Largo and Candelon, respectively, did not find commercial production.

The exploration for coal has found the first important deposit of coal in the country. Identified reserves total 40 million MT. The attempt to develop a source of geothermal energy has so far not resulted in the hoped-for success.

In the area of negotiations, the contract signed with Falconbridge in 1988 will permit the country to obtain substantial benefits in contrast with those obtained between 1972 and 1987.

BACKGROUND

The Dominican Republic has an area of 42,210 km^2. Its population of approximately six million is mostly of Spanish descent. Its principal exports are agricultural and mining, and its imports are those consumed by the people. It is not a country with a strong mining tradition, but extraction of gold and silver by the indians is known to have been carried on in pre-Columbian days. With the arrival of Columbus, production rose but then fell into a decline after 1530 because of the extinction of the indians and the discovery of the gold and silver in Mexico and Peru.

The investigation of the mining resources of the Dominican Republic, in a more or less systematic manner, dates from the beginning of this century. Geological reconnaissance was made by the U.S. Geological Survey in 1919. This same institution also studied the lateritic soils of the Sierra de Bahoruco in 1947 and the geology and the mineral resources of the Maimón-Hatillo mining district in 1950.

In 1945 the Alcoa Exploration Company started exploratory work in the southwestern part of the country and was successful in locating an important deposit of bauxite in Cabo Rojo, Pedernales. In 1958 the company began exploitation of this deposit.

In 1948, the Government of the Dominican Republic granted to the Compañía Minera y Beneficiadora Dominicana, CXA, exploration rights in the nickeliferous laterites of Loma Pequera in Bonao. These rights were transferred in 1955 to Falconbridge Dominicana which identified a promising deposit and initiated commercial exploitation in 1972. Production at this deposit has continued to the present.

During 1959, Dr. Renato Zoppis, consulting geologist, Managua, Nicaragua, carried out important geological investigations that extended the knowledge of our mineral and energy resources.

In 1968, Rosario Resources carried out an exploration program in Puerto Viejo, Cotui, which resulted in the discovery of a gold and silver deposit which began to be exploited in 1975. In 1979 the Dominican Government acquired all the shares of this company. The mine is currently being produced by the governmental company Rosario Dominicana, S. A.

PETROLEUM

The search for energy resources began early in this century and was concentrated mostly in the Azua basin. There, in 1904, a private firm drilled wells 400 ft (122 m) to 1,000 ft (305 m) deep obtaining production of 400 b/d from one of these wells for a short period of time. Subsequently, between the years 1919 and 1947, other companies continued drilling in this basin. A cumulative production of 19,000 bbl of 20° API was obtained from one of the wells.

In 1957 another private firm carried out drilling in the Cibao, Azua and Enriquillo basins and succeeded in drilling wells to depths of 10,000 ft (3,048 m).

In the 1960s other companies drilled a total of 27,934 ft (8,514 m) in the Bani basin in various wells which only gave some oil shows.

A seismic survey was carried out in 1969 on the submarine platform on the Ocoa, Neiba, and Samaná bays. In 1982, a seismic survey covering a total of 1,751 km of line was carried out within the free area of the submarine platform.

The deepest wells are those of Charco Largo No. 1 at 15,487 ft (4,720 m) in the Enriquillo basin and the Candelón No. 1 at 12,952 ft (3,948 m) in the San Juan basin. None of these wells encountered any commercial production.

In spite of the negative results obtained to the present, we cannot rule out the presence in the country of commercial accumulations of hydrocarbons. Further detailed investigations are required, but these cannot be undertaken at government expense due to economic limitations.

Because of the economic crisis provoked by the large increase in the price of petroleum during the early 1970s, the Dominican Republic, as well as many other countries which do not produce petroleum, decided to explore other sources of energy. This search began in 1975, and was directed toward exploration for possible deposits of coal. It has resulted in the identification of a deposit of lignite, which is of middle Miocene age. Recent exploration of this deposit, named the Sánchez deposit, located to the east of the town of Sánchez, has comprised 75 km^2 of geological mapping and the drilling of 223 boreholes. Geologic reserves are estimated to be about 72,300,000 MT and recoverable reserves to be about 43,800,000 MT. The deposit has a 30:1 cut ratio covering an area of 16 km^2. Based on the analysis of samples obtained from boreholes, the coal has been classified as lignite with a caloric value of 4,200 BTU/lb and with a high content of water, ash, and sulfur.

Various options as to the future use of this lignite were considered. Among these was the installation of a thermoelectric plant to be fired with a mixture of lignite from Sánchez and imported coal, in a ratio of 30 percent to 70 percent, respectively. This option was found to be more costly to operate than a plant using coal entirely imported.

Another of the alternatives that has been considered is installation of a plant to produce coal briquettes, with the hope of substituting the briquettes for charcoal, thus mitigating the destruction of the country's timber. This appears to be the best of the options and of highest national interest since such a project would contribute to the control of deforestation. However, no decision has yet been made on the eventual exploitation of the deposit.

Another of the country's attempts to find a solution to our energy problem that would contribute to the reduction of our petroleum bill was to explore the geothermal resources. In this respect, the Latin American Organization of Energy (OLADE) in the latter part of 1979 made a reconnaissance study of the zones with geothermal potential. This study resulted in a selection of four zones with geothermal possibilities, which in the order of their importance are as follows: (1) Yayas of Viajama-Constanza, (2) the thermal region of Azua, (3) the volcanic zone of San Juan, and (4) the region of Enriquillo Graben.

This selection was made taking into account the presence of heat related to recent volcanism, the presence of aquifers, and the presence of impermeable cover. On the basis of recommendations made by OLADE, a prefeasibility study was indicated in January of 1983 in an area of 3,800 km^2 that includes the volcanic axis Yayas de Viajama-Constanza. In the southwestern part of this area, earlier studies on the hydrology of the area and the drilling of two wells to obtain geothermal gradients had been undertaken.

On the basis of the prefeasibility study, the area of special interest was reduced to 140 km^2. Within this area a borehole was started for the purpose of measuring the geothermal gradient of the area and for investigation of the amount of hydrothermal alteration along the well. Unfortunately, the well had to be suspended due to problems penetrating the rocks and also with problems encountered with the drilling equipment.

THE INTERNATIONAL COOPERATION IN MINING DEVELOPMENT

The success enjoyed by the Dominican Republic in geological and mining exploration is due to a great extent to the cooperation of international organizations and friendly governments.

In this context I should mention the cooperative project begun in 1978 with the United Nations Development Program (UNDP), which resulted in the compilation of geological, structural, and mineralogical maps of the entire country, the set-up of a laboratory for chemical analysis, and a center for the documentation of information on geologic, mining, and energy matters.

Subsequently, another project with the same organization (UNDP), which was completed in 1985, involved the evaluation of gold resources in an area of approximately 700 km^2 in Mata Grande and 1,500 km^2 in the vicinity of the town of Miches in the northeastern part of the country.

Another program of exploration, started in 1984, was made possible with the aid of the Japanese International Cooperation Agency (JICA) and the Japanese Agency of Metallic Mining (MMJ). It involved geological exploration in the Mata Grande and Las Cañitas areas which include the sites of ancient metallic mines. The work entailed geological mapping, geochemical sampling, geophysical surveys, and drilling within designated areas of 700 km^2 in Mata Grande and 1,500 km^2 in Las Cañitas.

A training program on geophysics was carried out with the cooperation of the Swedish Government through the Swedish Geological Survey.

Another project, considered to be one of the most relevant in the last years in our country, is a Dominican-German cooperation program which is being carried out in collaboration with the Federal Institute for Geological Sciences and Raw Material of the Federal Republic of Germany and the Dirección General de Minería of the Dominican Republic. An important result of this program is a geological map of the entire country, which is to be published toward the end of 1989, and geological mapping at a scale of 1:50,000, which at this date has been completed in three quadrangles. In addition, thanks to the German cooperation, the country now has a modern laboratory for the analysis of clays and modern equipment for the analysis of rocks.

MINING EXPLOITATION

Mining operations in the Dominican Republic are currently limited to the exploitation of gypsum in Barahona, bauxite and limestone in Pedernales, marble in Samaná, small quarrying of clay and limestone in various places in the country, gold and silver by the state-owned Rosario Dominicana, S. A., and ferronickel by the transnational Falconbridge Dominicana, CXA. The last two of these exploitations are the most important to the national economy.

Circum-Pacific Council for Energy and Mineral Resources
Earth Science Series, Vol. 16
R. L. Miller, G. Escalante, J. A. Reinemund, and M. J. Bergin (Eds.)
Energy and Mineral Potential of the Central American-Caribbean Regions
Springer-Verlag, Berlin Heidelberg 1995

SELECTED PRECIOUS-METAL OCCURRENCES IN THE LESSER ANTILLES

Gregory E. McKelvey
U.S. Geological Survey, 210 E. 7th Street, Tucson, AZ 85701 USA

ABSTRACT

Significant quantities of gold and silver are being produced from epithermal deposits in the Greater Antilles and, although there are no known precious-metal ore deposits in the Lesser Antilles, there is potential for finding bulk minable deposits. The prospects and deposits in the Caribbean island arc can be roughly classified into five major groups: volcanic-hosted epithermal gold, sedimentary-rock-hosted skarns adjacent to granitic plutons, chemical precipitates in volcanic and sedimentary rock, gold-rich porphyry copper deposits, and placer deposits. Ages of mineralization span the geologic history of the island arc systems in which they occur. The hornfels-skarn gold prospects on St. Barthelemy may be as old as Eocene, the breccia-pipe controlled mineralization on Antigua is likely Pliocene, and the epithermal hot-spring systems on St. Lucia, Puerto Rico, Montserrat, Nevis, and other islands range in age from Cretaceous to Recent. Major mineralized areas in the Dominican Republic and St. Lucia appear to be coprecipitated with the reduced facies sedimentary and volcanic rocks; an occurrence on St. Johns and several of the deposits described in Haiti are reported to be volcanogenic massive sulfide deposits. Although placer deposits were of economic significance during the sixteenth and seventeenth centuries, they are less important today.

INTRODUCTION

The Greater and Lesser Antilles have produced significant quantities of gold. Although total production is not known, many believe it to be in excess of 10 million troy ounces (O. Martino, personal communication, 1989), mostly from lode and placer deposits on the larger islands. The most prominent deposits, such as Pueblo Viejo, have been found in the Greater Antilles.

Although there are no proven reserves of precious metals in the Lesser Antilles, more than 70 areas selected from the published literature have features commonly found in precious metal deposits. Field examination of 59 of these prospects has confirmed the presence of gold and indicates a potential for finding economic gold deposits in the rapidly urbanizing Lesser Antilles. On the basis of these preliminary field studies and by analogy with known deposits in the Greater Antilles and other gold-rich island-arc systems, these gold occurrences can be classified into five geologic environments: (1) volcanic-hosted epithermal gold; (2) skarn-tactite; (3) chemical precipitates coeval with sediments and volcanics; (4) gold-rich porphyry copper deposits; and (5) placer deposits. Selected examples from prospects in the Antilles illustrate these environments (Figure 1).

VOLCANIC-HOSTED EPITHERMAL GOLD

The best potential for minable precious metal deposits in the Lesser Antilles is probably in volcanic-hosted epithermal deposits. Examples are known on at least nine separate islands. These are Antigua, Dominica, Guadeloupe, Martinique, Montserrat, Nevis, Puerto Rico, St. Lucia, and St. Kitts. Three separate occurrences illustrate this type of deposit.

South Saddle Hill, Nevis

On the island of Nevis, south of St. Kitts, a low-silica pyroxene-dacite intrusion center called South Saddle Hill (Hutton, 1978) is dated at 1.8 Ma and is adjacent to a several-square-kilometer area of intensely altered rocks that include

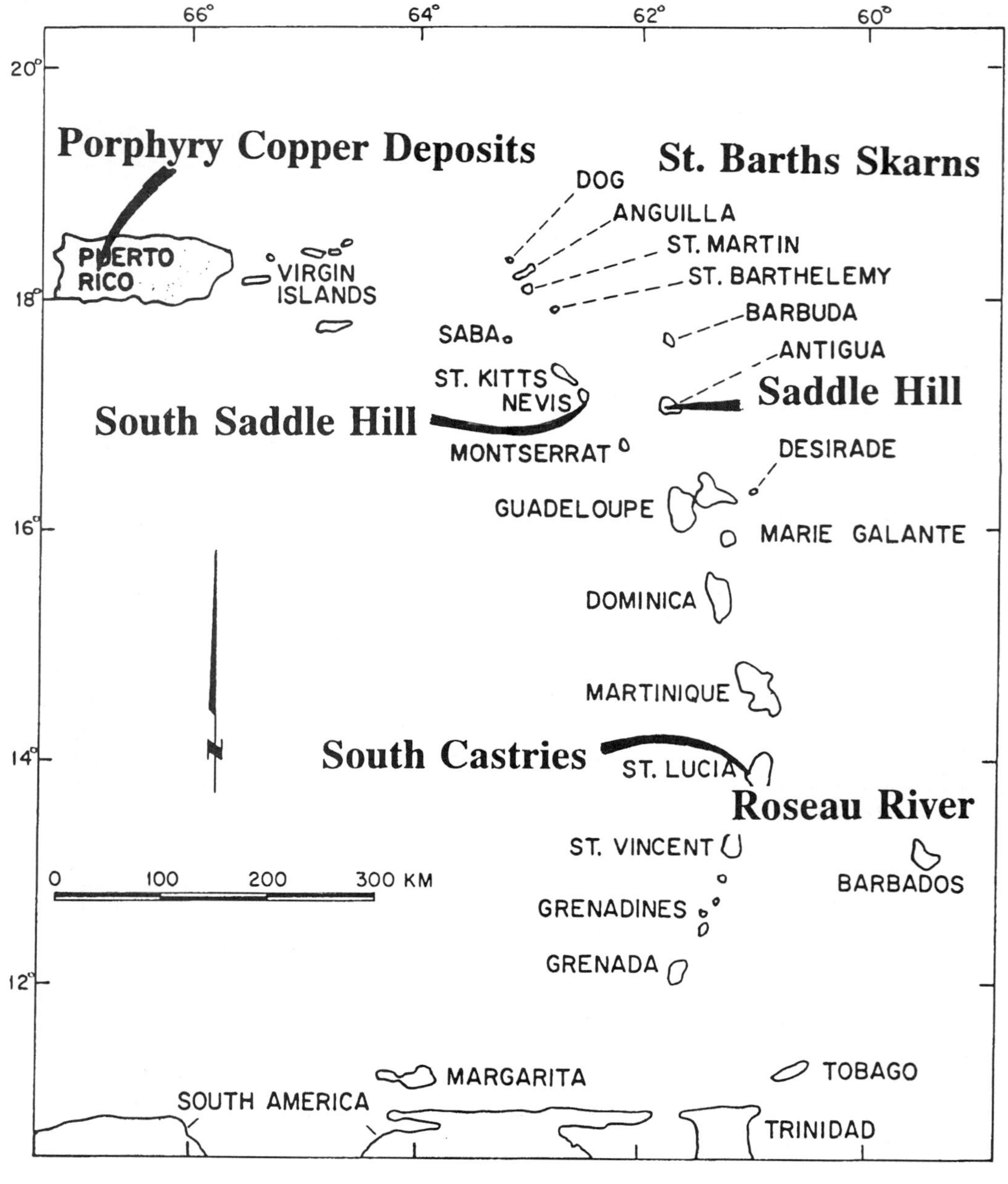

Figure 1. Location map of selected gold occurrences.

an intrusion breccia complex (Figure 2). An area of at least 0.6 km^2, partly covered by younger agglomerate and mud flows, contains rocks that are argillically altered and intensely silicified, in which the rock textures have been obliterated. This area of intense silicification, accompanied by stockwork veins, alunite (Hutton, 1978), jarosite, and geothite, is surrounded by several square kilometers of andesites altered to an undetermined clay assemblage. One area contains opal, alunite (dominantly sodic), hematite replacing plagioclase, and pyroxenes accompanying corroded quartz grains (Hutton, 1978). Reconnaissance field observations in 1985 confirmed the large alteration system and documented the occurrence of sinter and bedded opal. Halite, gypsum, kaolinite, halloysite, nontronite, and illite have also been recognized in the South Saddle Hill area.

The intrusion breccia is a mixture of dacite and several texturally different andesite clasts that range in size from fine rock flour to 10 cm in diameter. The breccia is clast supported, and the few open spaces contain coatings of opaline silica, alunite, and iron oxides resulting from the oxidation of pyrite. In several areas, the breccia is cut by a stockwork, 0.2 to 1 cm wide, of clear to cloudy, white quartz veins that are more abundant than 30 per meter.

Trace quantities of gold (140 ppb) have been detected. However, anomalous amounts of mercury (2,500 ppb), together with the presence of antimony (20 ppm) and arsenic (41 ppm) indicate that this geochemical environment is the type that has formed epithermal gold deposits elsewhere (Tooker, 1986) and suggest that South Saddle Hill may contain gold at shallow depths below this high-level hot-spring system.

South Castries, St. Lucia

A 1.5 x 2 km area along the southern residential areas of St. Lucia's capital city of Castries has exposed a vertically zoned sequence of altered rocks (Figure 3). Reconnaissance field mapping revealed the lowest exposures to be intensely stockwork-veined clay-altered andesites above which are clay-sericite, silica-clay-sericite, alunite, and a chalcedonic quartz capping. This cap contains more than 90 percent massive chalcedonic silica and sinter assemblage and is underlain by a several-square-kilometer zone of an advanced, argillically altered volcaniclastic sequence of andesites, rare tuffs, and interbedded medium-grained sediments. These flows are locally cut by a heterolithic intrusive breccia and a narrow felsite dike.

Veining is locally intense, and the original pyrite content exceeds 5 percent, as the abundance of jarosite indicates. Locally, pyrite is stable in surface exposures.

Rock samples supplemented by stream silt samples (Maassen, 1984) document the system's enrichment in precious metals. The highest quantity of gold found in reconnaissance rock sampling (1,430 ppb) is from a 5 m chip sample of andesite breccia flows altered to clay, silica, and sericite. Other rock samples in the quartz-sericite alteration zone contain up to 540 ppb Au. The silica cap, including the breccia (Figure 3), contains only 40 ppb Au. The system is anomalous in silver (78 ppm) as well as in arsenic, all samples below the silica cap containing at least 350 ppm As. Antimony was detected only in the silica cap (40 ppm), whereas mercury concentrations ranged up to 5,000 ppb. Anomalous arsenic concentrations in stream sediment samples (Maassen, 1984) outline the area of most intense alteration.

The system continues to the west, where it is covered by the Caribbean Sea, and to the south, where it is covered by gravel. The northern and eastern extensions are not well documented. One interpretation is that the system is slightly tilted to the west; thus, erosion has exposed a large area of alteration and coincident gold mineralization lying beneath the intensely altered silica zone.

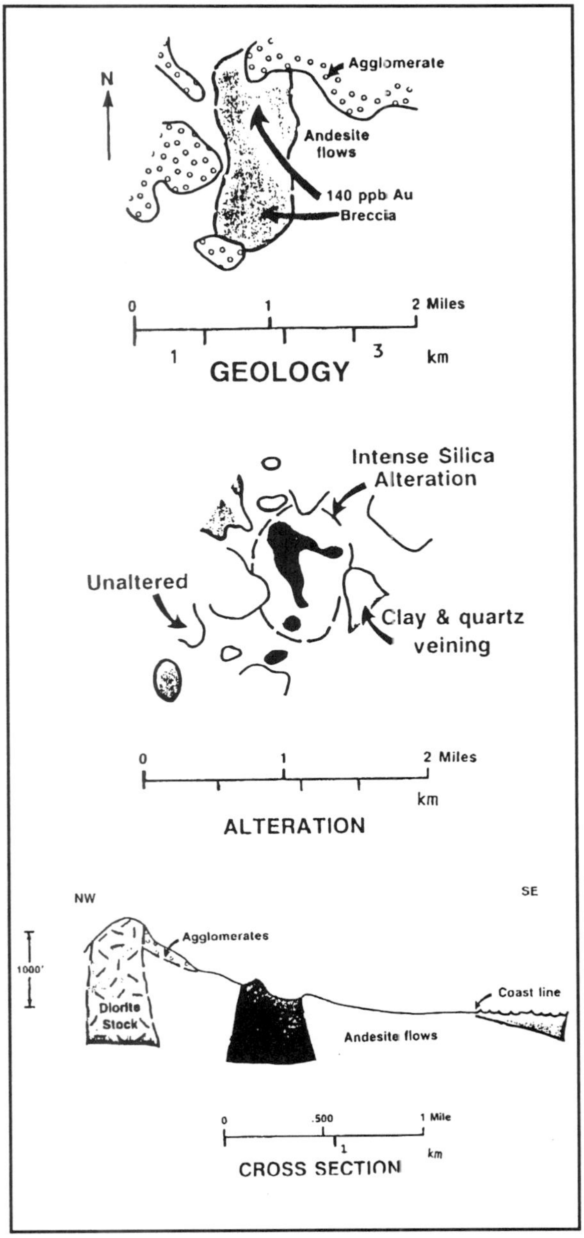

Figure 2. Geology and alteration, South Saddle Hill, Nevis.

Saddle Hill, Antigua

Described as "highly altered pale buff, hard, apparently silicified, in part brecciated rocks" (Martin-Kaye, 1955) cropping out along Saddle Hill (Figure 4), the prospect is reinterpreted as a breccia pipe-intrusion complex. One area consists of a zone 1,000 m in diameter of intensely silicified rocks displaying multicolored liesegang banding and cropping out along the beach. A second area, nearly 800 m in

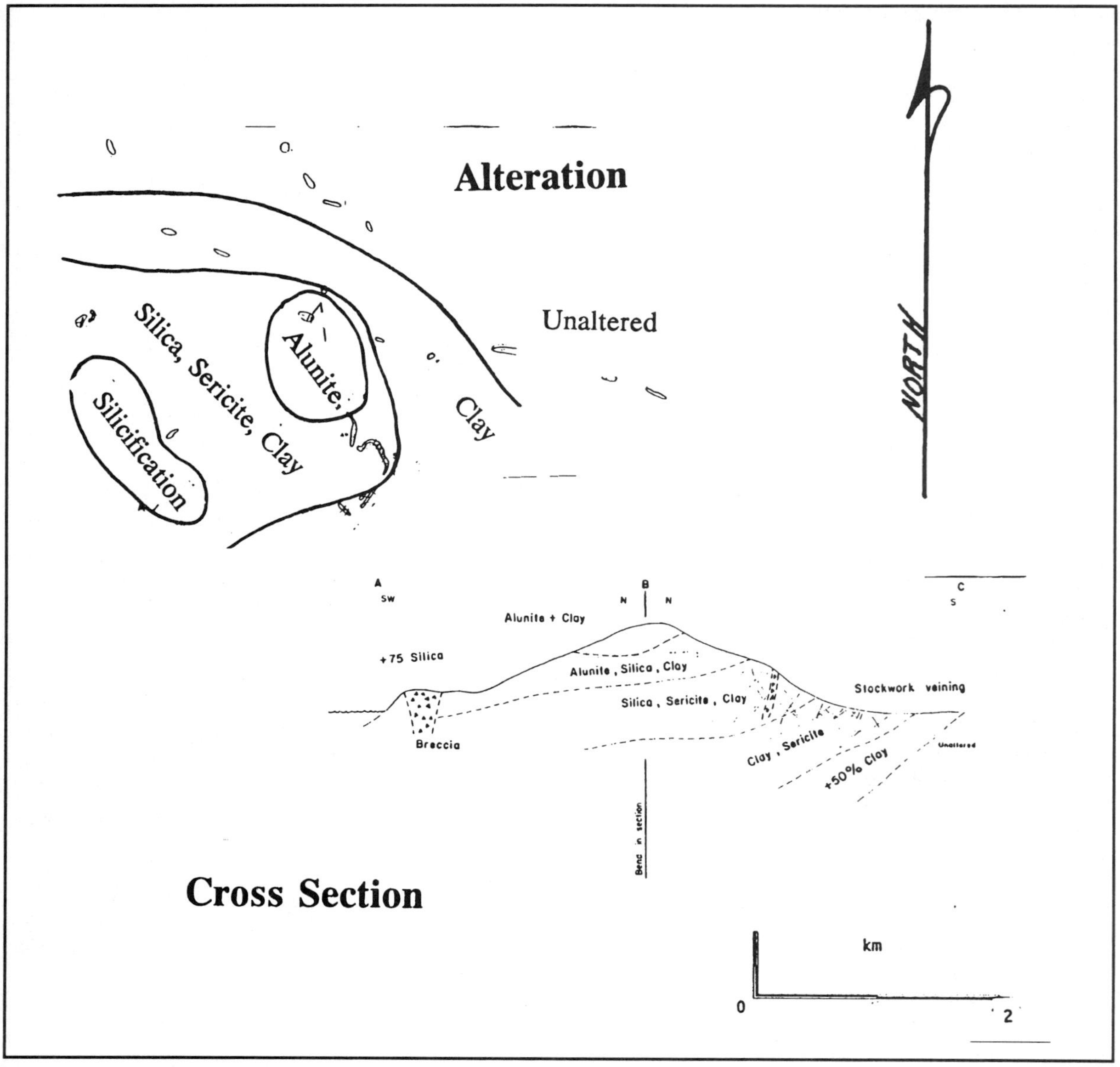

Figure 3. Distribution of alteration, South Castries, St. Lucia

diameter and locally known as "Sleeping Indian," consists of silicified and brecciated rocks cropping out along the crest of Saddle Hill. Few quartz veins are found anywhere in the area. These two altered areas are surrounded by volcanic rocks that appear to be unaltered. The fragment-supported breccia appears to be in a pipelike form; the space between the clast supported fragments is either open or filled by a rock flour matrix. In one place, the matrix appears as a fine-grained felsite. Martin-Kaye (1955) mapped these areas as dacites penetrating middle Oligocene volcanic rocks.

A series of rock chip samples each representing 10 m of outcrop from both the altered and the unaltered parts of the breccia-dacite contain gold. The highest assay is 3,150 ppb Au and 8.8 ppm Ag. Several additional samples confirm these values. Arsenic is enriched to a maximum of 105 ppm, but neither antimony nor copper is anomalous. Only one sample contained any anomalous mercury (120 ppb). The highest gold values are in the breccia, especially in the locally silicified areas; however, apparently unaltered rocks also contained up to 390 ppb Au.

SKARN-TACTITE

Gold-bearing skarns are documented on only two islands in the Lesser Antilles—St. Barthelemy and St. Martin. The dominant host rock assemblage consists of hornfels and rare tactite composed of chlorite-amphibole-biotite. The following brief field descriptions, without reconnaissance maps, illustrate that the skarn environment may be an important and perhaps somewhat under-prospected deposit type.

St. Barthelemy

The island of St. Barthelemy is composed of middle and late Eocene submarine volcanic and volcaniclastic rocks and, in part, of reef carbonates (Christman, 1953). These carbonate-rich units are intruded by latite and rhyolite intrusives and domes (Westercamp and Andreieff, 1983). Areas ranging from several square meters to 2 km^2 adjacent to weakly altered and mineralized intrusions have been altered to hornfels and epidote-chlorite-biotite skarns. Quartz veining is rare, and the total sulfide content of the most highly altered areas is less than 2 percent. Gold occurs in concentrations up to 1 ppm in selected samples from the most highly fractured outcrop. Silver (3.5 ppm), copper (200 ppm), lead (5,030 ppm), and zinc (174 ppm) are also present with gold in the samples.

Selected areas of altered sedimentary and volcanic rocks are enriched in metals, whereas areas of brecciated and silicified rocks within the rhyolite intrusions rarely contain significant quantities of precious metals. The only exception is a late, banded, 20-cm-wide milky-white quartz vein that cuts across the contact between the intrusion and the volcanic rock and contains 6.45 ppm Au, 4 ppm Ag, 1.1 percent Pb, 0.8 percent Zn, and 0.3 percent Cu.

CHEMICAL PRECIPITATES COEVAL WITH SEDIMENTS AND VOLCANICS

Gold in bedded cherts and pyritic black shales and within volcanogenic massive sulfide deposits is uncommon in the Lesser Antilles; however, several prospects may well represent examples of these important types of precious metal deposits. Geologic and geochemical characteristics of documented occurrences in the Dominican Republic (Amarante, Jiminez, and Lewis, 1988) and Puerto Rico of the Greater Antilles are similar to those in St. Lucia (Figure 5).

Roseau River, St. Lucia

Poorly-exposed bedded cherts and pyritic black shales layered within unaltered andesitic flows (Figure 5) are exposed along the Roseau River within the dense forest-jungle in the Government-controlled "nature preserve" of St. Lucia. These subhorizontal submarine flows and cherts appear to be cut, in several places, by a younger sequence of stockwork veining marked by clay, sericite, and silica alteration. Intrusion breccia is present in alluvium. Rock samples selected during this study and widely-spaced stream silt samples (Maassen, 1984) indicate that gold is present over a 6 km^2 area. The highest gold concentration found in rock samples is 378 ppb from an unveined, and apparently unaltered, finely-bedded chert. Samples from the younger altered rocks that crosscut andesitic flows contain up to 76 ppm Au and no detectable mercury, antimony, or silver. The pyritic black shales contain up to 184 ppm As and 562 ppm Cu and thus suggest that the Roseau River prospect is more complex than the high-level epithermal system exposed 35 km north of South Castries (described above).

There is not enough evidence to date to determine if this area has any potential for exhalite gold or volcanogenic massive sulfide deposits. In view of the sensitive location, there may never be enough information to determine if this

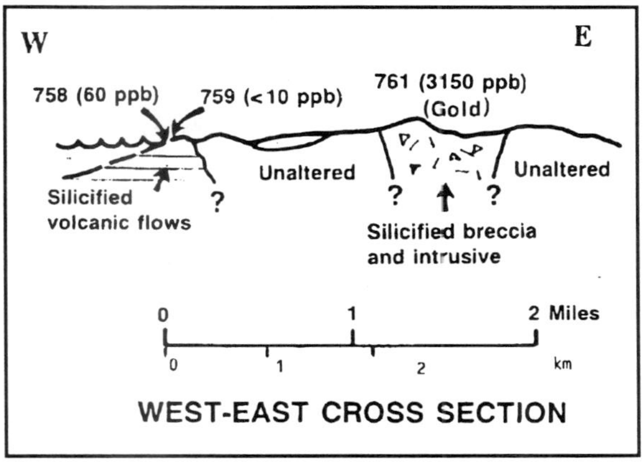

Figure 4. Diagrammatic geologic cross section, Saddle Hill, Antigua.

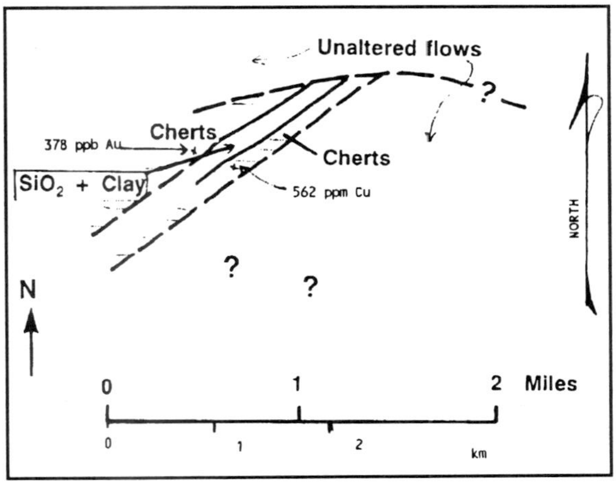

Figure 5. Reconnaissance geologic map, Roseau River, St. Lucia.

prospect has the potential for gold coeval with the sedimentary and volcanic rock hosts. This different deposit type may be present in other locations throughout the Caribbean.

GOLD-RICH PORPHYRY COPPER DEPOSITS

Gold is an important component of many island-arc porphyry copper deposits. Recently described by Cox (1985) in the Puerto Rico porphyry copper deposits and generalized as a specific deposit type by Cox and Singer (1986), these types of large ore deposits are becoming increasingly more important for both copper and gold. A zone of Cretaceous intrusions and associated porphyry copper deposits in the Utaudo region of west-central Puerto Rico was extensively drilled by AMAX at Rio Vivi and by Kennecott at Tanama in the 1960s; these deposits are known to contain gold. The gold concentration and spatial distribution are not known sufficiently to determine zoning or economic minable configurations; however, the deposits (Puerto Rico Department of Natural Resources private files) contain more than 100,000,000 MT at 0.8 percent Cu and 0.015 ounces gold per tonne for the Rio Vivi deposit and 127,000,000 tonnes at 0.54 percent Cu and 0.015 ounces per tonne gold for the Tanama porphyry copper deposits (Cox, 1985).

Virgin Gorda in the Virgin Islands and the French island of Desisrade may have some potential for gold-rich porphyry copper deposits. Prospects presently under exploration on the islands of Hispaniola, Jamaica, and likely Cuba, may offer better potential because of deeper erosion levels that expose the more metal-rich portions of these systems.

PLACER DEPOSITS

Placer deposits were the primary source of gold during Spanish mining in the Caribbean beginning in the 1600s. It is estimated that as much as one million troy ounces of gold were won from placers in Puerto Rico and perhaps more from Hispaniola, especially along the northern shore. The islands of the Greater Antilles have had longer periods of erosion and placer concentration than have the much younger island arcs in the Lesser Antilles. Placer deposits on the Leeward and Windward Antilles are not reasoned to be large enough or rich enough to contain major quantities of gold; their presence, however, will remain a direct guide to lode potential upstream.

SUMMARY

The best known gold prospects in the Lesser Antilles are high-level epithermal hot-spring systems. More than 35 areas on nine different islands contain either gold or trace elements and altered rocks commonly associated with volcanic-hosted epithermal gold deposits. Precious metals may also be present in skarns-tactites adjacent to alkali-calcic intrusions, within chemical precipitates coeval with submarine sedimentary and volcanic rocks, in porphyry copper deposits, and in some placer deposits. Several of the larger islands in the Lesser Antilles, which have longer erosion histories, have the potential for "laterite gold" occurrences similar to the Boddington deposit in western Australia. The islands offshore from the Precambrian shield rocks of South America may contain precious-metal deposits similar to those in Guyana, Venezuela, and Suriname. The Lesser Antilles have not been intensely explored, and urbanization is rapidly encroaching in several of the better prospects.

ACKNOWLEDGMENTS

I extend my thanks to Cominco Resources, Incorporated for permitting me to prospect in these islands while in their employ and for allowing me to present these data now. My gratitude is also due Charles G. Cunningham and Terry L. Klein for their constructive and critical review of this manuscript.

REFERENCES

Amarante, J. A., J. Jiminez, and J. F. Lewis, 1988, Geology, hydrothermal alteration and geochemistry of epithermal gold-silver mineralization in the Restauracion area, Dominican Republic: Transactions, 11th Caribbean Geological Conference, Barbados, 1986.

Christman, R. A., 1953, Geology of St. Bartholomew, St. Martin, and Anguilla, Lesser Antilles: Geological Society of America Bulletin, v. 64, p. 65-96.

Cox, D. P., 1985, Geology of the Tanama and Helcho porphyry copper deposits and vicinity, Puerto Rico: U.S. Geological Survey Professional Paper 1327, p. 59.

Cox, D. P., and D. A. Singer, eds., 1986, Mineral deposit models: U.S. Geological Survey Bulletin 1693, 379p.

Hutton, C. O., 1978, The petrology of Nevis, Leeward Islands, West Indies: Overseas Geology and Mineral Resources, no. 52, 31p.

Maassen, L. W., 1984, The geochemical atlas of St. Lucia, West Indies: Los Alamos National Laboratory report LA-UR-84-1747, 224p.

Martin-Kaye, P. H. A., 1955, Mineral prospects in the Leeward Islands: 1st Caribbean Geological Conference Report, December, 1955, p. 43-48.

Tooker, E. W., ed., 1986, Geologic characteristics of sediment-and volcanic-hosted disseminated gold deposits—Search for an occurrence model: U.S. Geological Survey Bulletin 1646, 150p.

Westercamp, D., and P. Andreieff, 1983, Saint-Barthelemy es ses itlets: Carte geologique a 1/20,000, avec notice explicative: Service Geologique National, Bureau de Recherches Geologiques et Minieres.

Circum-Pacific Council for Energy and Mineral Resources
Earth Science Series, Vol. 16
R. L. Miller, G. Escalante, J. A. Reinemund, and M. J. Bergin (Eds.)
Energy and Mineral Potential of the Central American-Caribbean Regions
Springer-Verlag, Berlin Heidelberg 1995

BIOGEOCHEMISTRY FOR FUTURE MINERAL RESOURCE EXPLORATION PROGRAMS IN THE CENTRAL AMERICAN-CARIBBEAN REGION

Frederick R. Siegel
Department of Geology, George Washington University, Washington, D.C. 20052 USA

Alfonso Segura Paguaga
Apartado No. 3, Esparza 5500, Puntarenas, Costa Rica

ABSTRACT

The variety of flora, its distribution density, and its range in the Central American-Caribbean region gives the area excellent potential for mineral resource exploration with biogeochemistry. When exploration by rock or stream sediment geochemistry targets an area, soil samples are used for detailed exploration. The typical soil sample represents a volume of 5,000 cm^3. Vegetation, however, samples a volume as large as a plant's root system. For example, a root volume of 1,000,000 cm^3 for a shrub has 200 times the volume of a soil sample, or a tree with roots reaching 20 m^3 of soil may take up trace elements from 4,000 times the volume of a soil sample. Vegetation geochemistry also avoids the "nugget effect" that may be associated with soil samples in gold exploration programs.

In Costa Rica, two samples of the small tree (3-6 m tall) *Curatella americana* L. were analyzed for gold by a mine assay technique using dissolution of 5 g of plant ash by a NaCN-NaOH solution and analysis by atomic absorption spectrometry. One sample contained 1.3 ppm Au and the other 0.92 ppm Au, extremely high values. The vegetation is from a disseminated gold region known since Spanish times. A third sample, collected during the flowering, fruiting season was sent to George Washington University where it was separated into open pods, seeds, and twigs before analysis by instrumental neutron activation analysis (INAA). For this dried sample (not ashed), the seeds contained 160 ppb Au, the twigs 430 ppb Au, and the open pods 1,200 ppb (1.2 ppm) Au, extraordinarily high concentrations. A fourth sample, also analyzed in Washington, D.C., contained 14 ppb Au in open pods, 16 ppb Au in closed pods, and 2.2 ppb Au in twigs + bark. Systematic sampling of *C. americana* L. in traverses across gold districts into barren terrain is necessary to determine if this species is an accumulator of gold so that it can be used in future exploration programs in Costa Rica and elsewhere in the Central American-South American-Caribbean region.

INTRODUCTION

Biogeochemistry of vegetation (the chemical element content of plants) is evaluated and used for mineral exploration in many countries (Soviet Union, New Zealand, Canada, Scandinavia, United States, and others). However, biogeochemical exploration has been little used in the Central American-Caribbean region or in South America, although there are scientists in the region who are well prepared to plan and carry out basic and applied research on the use of plant chemistry in mineral resource exploration. These include economic geologists to target areas of mineral potential, experienced botanists to identify and advise on the species that may be studied, chemists to analyze the samples, geochemists to relate the results and locations of chemical analyses to rock types, structure, and processes active in a sampled area, and statisticians to numerically evaluate the botanical-chemical-geological data base. All of these scientists would be part of the planning strategy for a biogeochemical exploration project as well as major players in carrying it to completion.

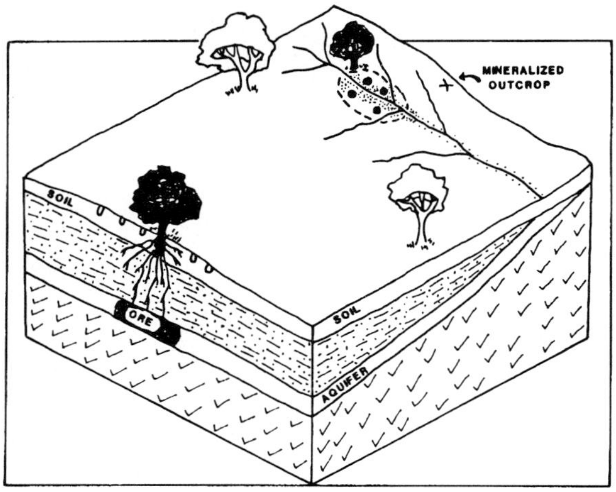

Figure 1. Different geochemical samples can indicate ore in different situations. Where tree roots tap ground water that carries ore metals from a hidden deposit, a tree may have high metals values (dark trees in this diagram), but the surrounding soil may not (open circles). Where hidden ore is shallow, soil may also have high values (dark circles). X marks a mineralized outcrop where a geologist would take a rock sample. Dots along the stream represent sediments with metal values that get lower downstream as sediment from non-mineralized zones washes in.

CONCEPT

The concept that is used in exploration with plants is quite straightforward, although the actual exploration may be complicated by factors to be considered later in this paper. First the premises are made that vegetation biogeochemistry often reflects soil geochemistry and that soil geochemistry represents subsurface (rock) geochemistry. Some plant species tend to accumulate a metal if it is present in the growth environment. These are called accumulator plants. Therefore, high chemical element concentrations in plants may reflect high metal concentrations in soils (Figure 1) that may indicate subsurface mineralization hidden beneath a vegetation-soil overburden or in a tropical rainforest environment beneath a significant thickness of humus, or through an alluvial fan (pediment) cover. Soil is the sample used for follow-up and detailed geochemical mineral prospecting after an area has been determined propitious by regional exploration using stream sediment geochemistry or rock geochemistry (lithogeochemistry). We find, however, that plant chemistry, as an adjunct to or alternative to soil geochemistry, can often be beneficial to an exploration program. Excellent reference material on biogeochemistry can be found in the works of Brooks (1972; 1983; 1988), Kovalevskii (1974; 1987), Malyuga (1964), Siegel (1974, Chapter 6; in press, Capitulo 4), Rose, Hawkes, and Webb (1979, Chapter 17), and Erdman and Olson (1985, specific for Au in plants).

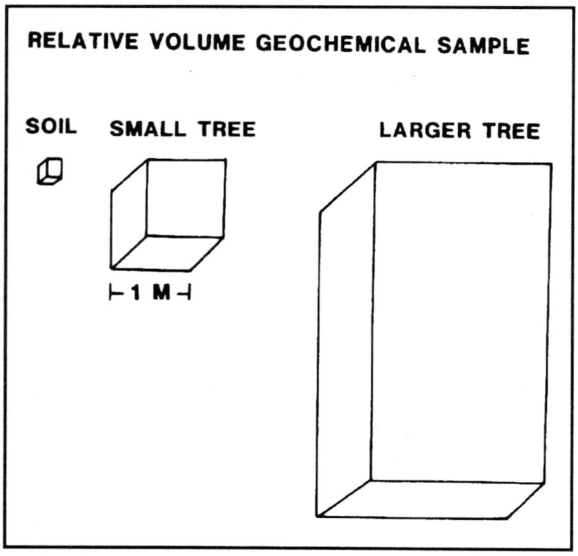

Figure 2. Relative volume of subsurface represented by soil and plant samples.

BENEFITS OF PLANTS IN EXPLORATION GEOCHEMISTRY

The most important benefit of using plant chemistry in mineral prospecting is that the plant represents a larger volume of the subsurface than the soil sample because of the volume reached by its root system. In the case of a small tree with a root system extending to 1 m^3, the volume of the subsurface sampled is about 200 times that of a soil sample. Similarly, a large tree with roots extending through 20 m^3 would give a sample representative of 4,000 times a greater volume than the corresponding soil sample (Figure 2). In addition, the plant gives a sampling of a deeper geochemical zone closer to hidden mineralization. In densely-vegetated, overgrown areas with a thick humus layer, it is easier to use plants rather than dig through the humus for a soil sample. The penetration of vegetation root systems allows a sampling of the subsurface in areas with poor soils, in areas with calcrete (caliche), or siliceous hardpan horizons where soil sampling is difficult. In addition to these benefits, the vegetation samples are relatively light weight. About 200 g of collected plant matter can be expected to yield 10 to 20 g of dried plants and <0.5 to 1 g of plant ash for geochemical analysis. Preparation of the vegetation sample for analysis is relatively simple and may include washing, drying, maceration, and perhaps ashing (high or cold temperature, or wet). In Au prospecting, the use of plants avoids the "nugget effect" inherent in the use of soils. This happens because an isolated small flake or tiny nugget of gold in a soil can give non-representative geochemical data for the true gold concentrations in the surroundings. In order to take advantage of these

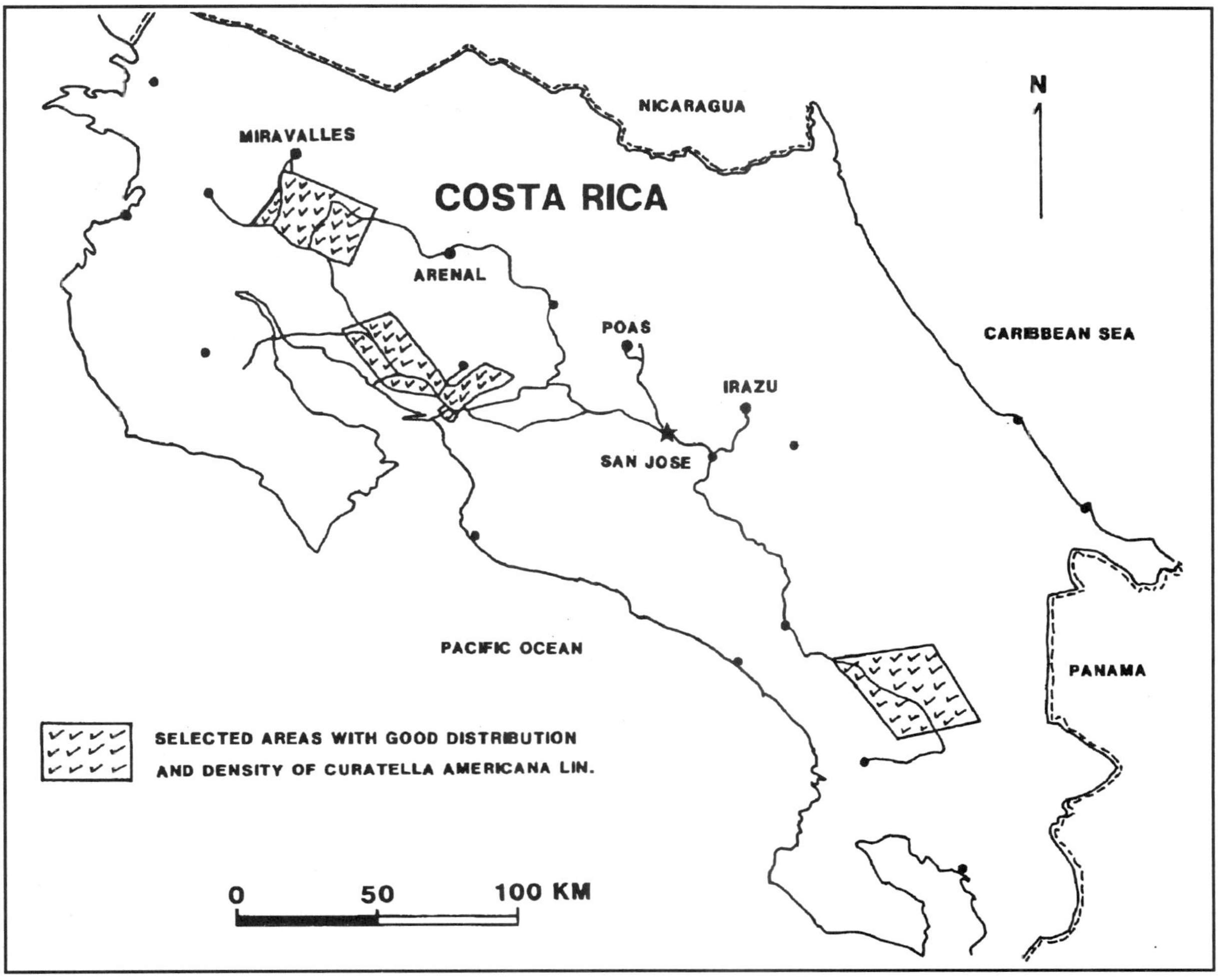

Figure 3. Selected areas in Costa Rica where *C. americana* L. ranges. Small dots represent population centers.

benefits of using plants in geochemical exploration, certain procedures should be followed.

CAUTIONS TO KEEP DATA COMPARABLE

Before carrying out a biogeochemical prospecting program in a little known area, a pilot study must be done to find plants that are accumulators of the target element(s) and/or pathfinder elements. Only one accumulator species (or other species statistically equivalent in their geochemistry) should be used in an exploration project. To keep the biogeochemical data comparable, only the same plant organ (leaves with leaves or twigs with twigs) and the same age plant organ (first year leaves with first year leaves or second year twigs with second year twigs) can be used. In addition, the organs should come from healthy plants at the same stage of maturity, collected during the same season (preferably during the vigorous growth season—spring), and with the same orientation with respect to the sun should be collected and analyzed. Certainly, the handling of the samples in the field and in the laboratory must be "clean" and constant operational procedures must be followed. Sample preparation techniques may be dictated by the element(s) sought (e.g., Au, Hg, As may volatilize and be lost during high temperature ashing). The same analytical technique should be used throughout a project in order to get truly comparable data. The geochemist-geologist doing interpretative work on the data has to know the geology-hydrology of the study area. This knowledge, together with an understanding of the topography and surficial processes active there, will allow the explorationist to recognize, among other things, the possibility that anomalies are displaced (offset) laterally at the surface from the position directly above the hidden mineralization.

DISADVANTAGES OF PLANTS IN EXPLORATION GEOCHEMISTRY

A major limitation to the application of biogeochemistry in many areas is that the density of an accumulator plant may be limited and its distribution may be irregular so as not to permit a relatively well-spaced sampling program. In addition, the desired organs to be sampled (leaves, needles, twigs) may be relatively inaccessible (too high). Interpretatively, the essential elements in plant nutrition may be difficult to use if they are the metal(s) being sought or pathfinder elements because their natural contents may mask concentrations originating from mineralization. As noted previously, samples should be handled carefully since they are subject to contamination.

PRELIMINARY RESULTS IN COSTA RICA TO IDENTIFY A GOLD ACCUMULATOR

Curatella americana L., also known as "raspa" in Central America, is a small tree found in several areas in Costa Rica (Figure 3) including the old Spanish gold mining area, Mina Santa Clara, in the Mondongo Zone. Four samples of *C. americana* L. were collected by Geologist Segura P. Two samples were analyzed for Au in Costa Rica by a mine assay technique using dissolution of 5 g of *plant ash* by a NaCN-NaOH solution and analysis by flameless atomic absorption spectrometry. These samples contained 1.3 ppm (1,300 ppb) and 0.92 ppm (920 ppb) of Au, values far greater than have been found in plants. Average values of Au in *plant ash* are between 1-10 ppb.

A third sample, collected during the flowering, fruiting season was sent to George Washington University where it was separated into open pods, seeds, and twigs before analysis of the *dried* organs for Au by neutron activation analysis. The seeds contained 160 ppb Au, the twigs had 430 ppb Au, and the open pods had 1,200 ppb Au (1.2 ppm), truly extraordinary values. The fourth sample, also prepared in Washington, D.C., gave 14 ppb Au in closed pods, 16 ppb Au in open pods, and 2.2 ppb in twigs + bark.

It is of special interest to note that ashed samples will contain about 20 times the metal content (by weight) than the same dry matter samples because more than 95 percent of the dry sample weight will be lost during ashing. Thus, in theory, if Au is not lost by volatilization during ashing (cyanogenic plants will lose Au during ashing), the ash of the open pods of the third sample should contain more than 24,000 ppb Au (24 ppm)!

There are three possible explanations for the high Au values obtained for samples of *C. americana* L.: first, the species is a Au accumulator; second, it is possible that fluids from the heap-leaching of the ores have mobilized gold to the sample plant locations via surface or groundwater flow and the plants then were able to take up the solubilized gold into their organs; and third, the extraordinarily high Au concentrations are due to contamination during the sampling or storage stage. Girling and Peterson (1980) found experimentally that the highest accumulation of gold was recorded in seedlings grown in a solution of gold potassium cyanide at pH = 5.5.

The preliminary data in this study are encouraging but additional systematic sampling must be done in different areas where *C. americana* L. occurs in both known Au areas and areas known to be barren of Au before this species can be designated as an accumulator plant. If *C. americana* L. is found to be an accumulator of Au, it can be used for exploration in areas where it has enough density and acceptable spatial distribution in the Central American-Caribbean-South American region.

ADVANTAGES FOR THE USE OF BIOGEOCHEMISTRY IN LATIN AMERICA AND THE CARIBBEAN

Biogeochemistry has not been used as an integrated part of mineral resource exploration programs in Latin America and the Caribbean, yet the area has extensive tropical forests (Figure 4) and other vegetation zones where the potential to find mineral resources is excellent. For example, ferns (helhechos) in the western cordillera in Colombia contained up to 30,000 ppm (3 percent) Cu (in plant ash) where there is chalcopyrite mineralization at the El Seis mine. In addition, the botany of many areas is well known and the science has a long tradition of excellence in the region. There are well established botanists to collaborate with geochemists in research applied to mineral resource exploration. The region also has many fine herbaria that can provide samples for preliminary evaluations of plants as accumulator species. Deforestation in the region, although locally severe, has probably not greatly affected the diversity of flora so that the potential to find accumulator species for ore metals is good.

PRIMARY NEEDS FOR THE APPLICATION OF BIOGEOCHEMISTRY

After geologists have targeted exploration areas on the basis of their geological characteristics (e.g., rock types, structural characteristics) and/or on the basis of regional geochemical data from rock or stream sediment analyses, a botanical study is necessary to determine which plants in the region have the desired density and spatial distribution. This is followed by pilot studies on the plants and their corresponding soils to find those whose chemistry reflects the chemistry of the soils (although this may not always be the case). The pilot studies will also determine if any of the plants with the desired densities and spatial distributions are accumulator species for Au and other metals, or perhaps for the pathfinder elements that may be even more useful in finding a potential deposit than the metal(s) being sought.

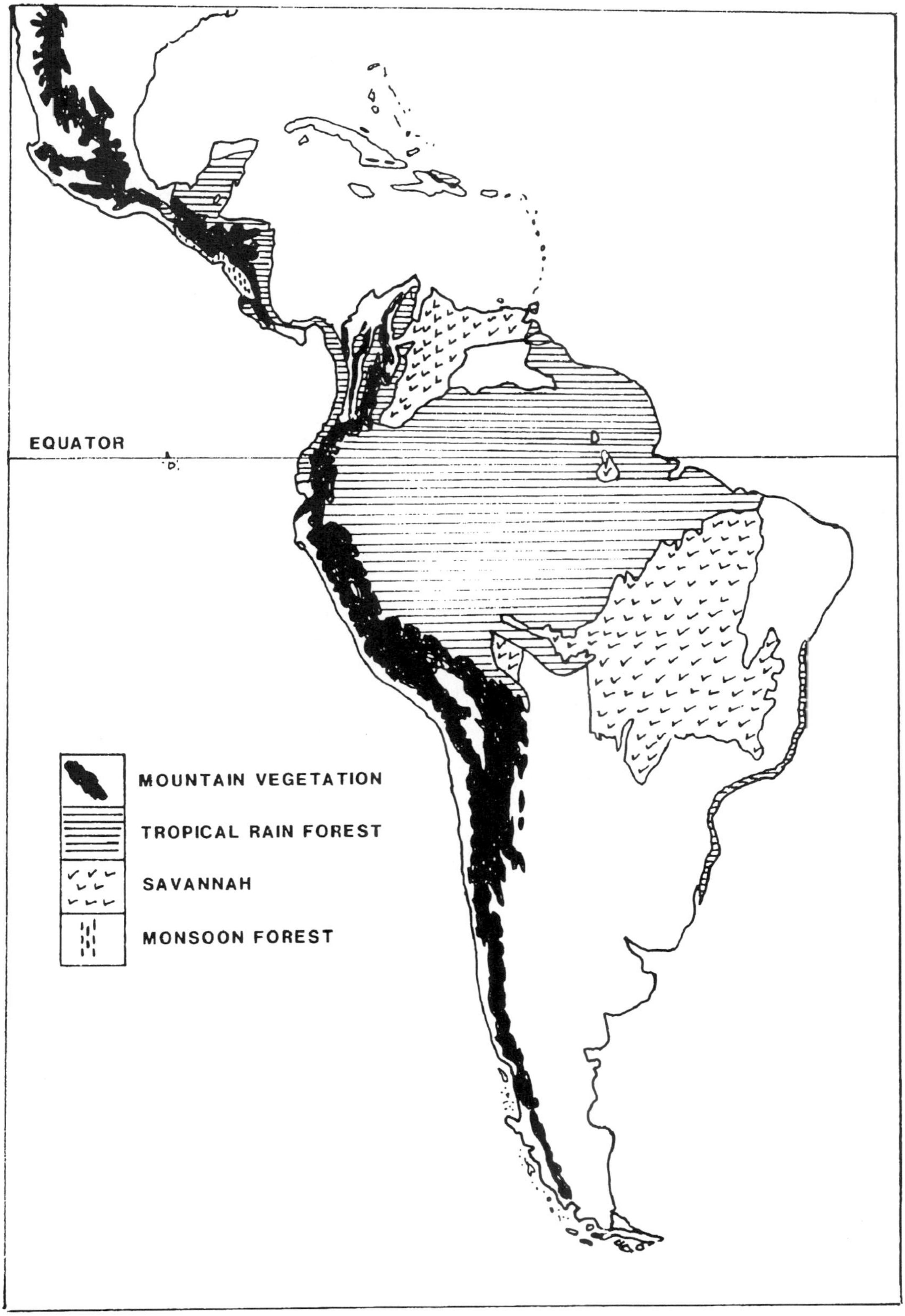

Figure 4. Selected vegetation zones of Latin America and the Caribbean which are propitious for biogeochemical exploration.

REFERENCES

Brooks, R. R., 1972, Geobotany and biogeochemistry in mineral exploration: New York, Harper & Row, 290p.

Brooks, R. R., 1983, Biological methods of prospecting for minerals: New York, Wiley Interscience, 322p.

Brooks, R. R., 1988, Biogeochemistry, in C. W. Finkl, ed., Encyclopedia of field and general geology: New York, Van Nostrand Reinhold, p. 30-35.

Erdman, J. A., and J. C. Olson, 1985, The use of plants in prospecting for gold: A brief overview with a selected bibliography and topic index: Journal of Geochemical Exploration, v. 24, p. 281-305.

Girling, C. A., and P. J. Peterson, 1980, Gold Bulletin, v. 13, p. 151-157.

Kovalevskii, A. L., 1974, Biogeochemical exploration for mineral deposits: New Delhi, Amerind. Publ. Co. Pvt. Ltd., 136p.

Kovalevskii, A. L., 1987, Biogeochemical exploration for mineral deposits: Utrecht, VNU Science Press, 220p.

Malyuga, D. P., 1964, Biogeochemical methods of prospecting: Consultants Bureau, 205p.

Rose, A. W., H. E. Hawks, and J. S. Webb, 1979, Geochemistry in mineral exploration, 2nd ed.: New York, Academic Press, 657p.

Siegel, F. R., 1974, Applied geochemistry: New York, Wiley Interscience, 353p.

Siegel, F. R., in press, Geoquimica aplicada, serie de quimica, monografia: Washington, Organización de los Estados Americanos.

Circum-Pacific Council for Energy and Mineral Resources
Earth Science Series, Vol. 16
R. L. Miller, G. Escalante, J. A. Reinemund, and M. J. Bergin (Eds.)
Energy and Mineral Potential of the Central American-Caribbean Regions
Springer-Verlag, Berlin Heidelberg 1995

INDUSTRIAL MINERALS—KEY TO ECONOMIC DEVELOPMENT

Haydn H. Murray
Department of Geology, Indiana University, Bloomington, Indiana 47405 USA

ABSTRACT

Industrial or non-metallic minerals are essential to economic development. The value of industrial mineral production in the United States is more than three times the value of metallic mineral production. In the developed countries of the world the value of non-metallic mineral production exceeds the value of metallic mineral production. The development of a modern industrialized society requires quality and reasonably priced industrial minerals in such industries as smelting of copper and iron, manufacturing cement, drilling oil wells, manufacturing ceramic materials, and a host of others. Because transportation costs are high most industrial minerals are not imported, so a country or region must have a good raw material source.

INTRODUCTION

A precise inclusive definition for industrial minerals or non-metallic minerals is difficult because it includes many unrelated minerals that range from low priced materials such as sand and gravel to high priced materials like industrial diamonds. The World Bank report (Noestaller, 1987) defines industrial minerals as comprising all non-metallic non-fuel minerals extracted and processed for industry end uses, some metallic minerals consumed in non-metallurgical applications, and consolidated and unconsolidated rock materials (e.g., sand, gravel, crushed rock, and dimension stone), and manufactured products (e.g., cement and refractories). The Mining Journal (1988) classifies industrial minerals into three economic groups: (1) low price, large volume commodities such as sand, gravel, and construction materials, (2) medium to high price, large volume commodities such as chemical and fertilizer minerals (e.g., salt, sulfur, phosphate, and potash), and (3) high price, small volume commodities such as fluorspar, talc, barite, and industrial diamonds.

The low value, high volume construction type materials are mainly used locally, and generally achieve their maximum use in the stage of population growth and the economic development of an area. The high value, low volume industrial materials are used in the more industrialized nations and are often traded internationally. A measure of industrial maturity is achieved when the value of industrial or non-metallic minerals and rocks exceeds that of metallic minerals. This point was reached in the United States early in this century, and today the value of the industrial minerals and rocks is about three times that of the metallic minerals. Australia just reached this point in the 1980s.

The standard of living in a particular country or area of the world correlates with the availability and production of industrial minerals. The United States, which has a very high standard of living, has an ample supply of most industrial minerals. Without their availability at a reasonable price, the U.S. could not have achieved this high standard of living nor its tremendous industrial growth. The same can be said of Western Europe.

As mentioned before, there is a great diversity of values for industrial minerals. These values range from a few dollars per ton for sand, gravel, and mineral aggregates such as limestone to several million dollars per ton for industrial diamonds. There is also a great diversity in the utilization of industrial minerals. Our building and process industries require quality and reasonably priced industrial minerals. These minerals are an integral and functional part of many, many products such as cement, ceramics, paper, paint, plastics, and a host of other manufactured products. The growth of the non-metallic or industrial minerals correlates to a large extent with the increase of our Gross National Product (GNP).

INDUSTRIAL UTILIZATION OF NON-METALLIC MINERALS

Industrial minerals are necessary in many industries, as mentioned above. Some examples of the minerals that are used and what they contribute to the manufactured products are given as follows. This brief account is by no means complete but does indicate the diversity and functionality of certain important industrial minerals.

The building industry could not function without industrial minerals. Cement, which is a necessary component of most of our buildings, is made from industrial minerals that include limestone and shale which are pulverized, mixed together in specified proportions, and heated to form the necessary reaction products that make cement. In buildings and roads cement is mixed with aggregate which can be sand and gravel or a mineral aggregate such as limestone or trap rock (basalt). If a lightweight concrete is needed, then a lightweight aggregate such as bloated shale is used. This bloated shale is formed by rapidly heating crushed shale particles to a temperature in which some volatiles in the shale are released at the same time the surface of the shale particle is melting. The volatiles then form internal voids because they expand when trapped by the viscous glassy outer melted surface. When cooled the shale particles are strong and light in weight and make a good lightweight aggregate for concrete. Other building materials also use industrial minerals. These include wallboard which is made from gypsum, bricks which are made from clays and shales, dimension or building stones which are limestone, marble, granite, and many other rocks, plaster which uses cement and perlite, and tile which uses kaolin, talc, wollastonite, and other minerals.

In making iron and steel, a flux is needed to lower the melting point of the iron ore. This flux is usually high-calcium limestone. In making metallic aluminum, the flux normally used is fluorite, another industrial mineral. In smelting copper an industrial mineral, silica sand, is used as a flux. Therefore, as can be seen, industrial minerals are necessary in the smelting of metallic ores.

In drilling oil wells, industrial minerals play a very important role. A drilling mud is necessary in order to successfully drill an oil well because the drilling mud is used to carry the tons of rock cuttings up the hole to the surface, it seals the hole by forming an impermeable layer of mud cake on the side of the hole thereby preventing the drilling fluid from escaping into the permeable and porous rock formations, and the mud cools and lubricates the drill bit. The drilling mud is comprised of industrial minerals, the major component of which is usually sodium bentonite. The material when mixed with water forms a material which is fluid when agitated or stirred and is gel-like when allowed to set. This property is called thixotropy. When the mud is pumped down the hole through the drill pipe to the bit it is fluid but is viscous enough to carry the rock cuttings up the hole. If the drilling stops for some reason, the drilling mud becomes a gel which prevents the rock cuttings from settling to the bottom and binding the drill bit so that it would twist off when the drilling was started again. Sodium bentonite is naturally thixotropic, so it is used for drilling oil wells all over the world and is an absolutely essential ingredient. In deep drilling to several thousands of feet the pressures of fluids in the rock formation are very high and if gas or oil is encountered that is under high pressure, the drilling mud could be blown out of the hole and dangerous fires would occur. In order to prevent these blow-outs, a weighting material is used. This finely pulverized heavy material is mixed with the drilling mud to make it heavy enough to contain the bottom hole pressures from blowing out the hole. The most common weighting agent is the industrial mineral barite which is very heavy.

Ceramic materials, which are used extensively in our everyday lives, are made from industrial minerals. Fine china is made using kaolin, silica, and feldspar in proper proportions. Pottery is made from clays and other industrial mineral ingredients. Refractories, for lining high temperature furnaces, use various refractory industrial minerals including kaolin, kyanite, dolomite, silica, and chromite. Glass, which is classed as a ceramic material, is made from high silica sand or quartzite along with a flux such as soda ash. Think of how much glass is used in your homes, cars, and office building. This means that it requires the availability of large quantities of high quality silica sand in various parts of the country to sustain our glass manufacturing industry.

Think of how much paper is used everyday in your homes and offices. Paper is made from wood pulp but also requires other ingredients which are mainly industrial minerals. Paper requires fillers to give it certain needed properties including whiteness, printability, smoothness, and opacity (hiding power). Industrial mineral fillers are used to give these properties and include kaolin, calcium carbonate, silica flour, talc, and titanium dioxide. For color printing, such as in National Geographic Magazine, the paper must be coated and this coating is normally kaolin and/or calcium carbonate along with a minor quantity of titanium dioxide. National Geographic Magazine is about 35 percent by weight of the industrial mineral kaolin.

In the foundry industry industrial minerals are essential in making the molds into which the molten metal is poured. The engine block in your automobile is made by pouring the hot gray iron into a mold which is made primarily of silica sand and bentonite clay. The mold must be strong enough to hold the molten metal, porous enough to release the gases from the molten metal as it cools, and refractory enough so that it doesn't melt and contaminate the iron. These two industrial minerals, high silica sand (which is the major component) and bentonite, have the needed properties to make these molds.

Industrial mineral fillers are used in many products including paint, plastics, rubber, and ink. These fillers are not inert fillers but they perform a needed function. Fillers or extenders in paint make the paint film glossy or dull, washable, and smooth, and give the paint film covering power and body so that it doesn't drip or run. In plastics, mineral fillers make the plastic less brittle so that it does not shatter on impact, make many plastic materials less vulnerable to infrared and ultra violet degradation, and improve the surface smoothness of the plastic. Some automobile bodies and many boat bodies are made of plastic and contain up to 40 percent mineral filler, usually calcium carbonate or kaolin. Without the mineral filler the plastic would shatter like glass upon impact. In rubber, industrial mineral fillers are used to improve abrasion resistance, to stiffen the rubber product, to reduce the elasticity, and to improve the hardness. In ink, fillers are used to control the viscosity and to improve the covering power. Thus, it can be seen that industrial mineral fillers are a necessary and functional part of many industrial products.

Another area where industrial minerals are used is when a sorptive material is used. Examples are in machine shops where oil spills commonly occur or in service stations where motor oil is spilled on the floor. An absorptive clay is used to soak up this oil. This type of clay is commonly referred to as fullers earth. These sorptive clays are also used in the agricultural area as carriers for insecticides and pesticides. The largest use by far of these sorptive clays is as a cat litter. More than a million tons of this clay are used annually in the United States by cat owners.

There are many, many other industrial uses for many other industrial minerals. The aforementioned examples are only to illustrate how important the industrial minerals have been and are to our industrial development.

LOCATION AND TRANSPORTATION

The location of the low price, high volume industrial minerals with respect to the markets is exceedingly important. In general, the transportation cost of these low priced industrial minerals such as sand and gravel controls their marketability because in many cases the transportation cost exceeds the cost of the sand or gravel at the mine. This is the reason that the mines must be located as close to the markets as possible. The urban growth around every major city in the United States has limited the expansion of operating mines and prohibited opening up new mines because of zoning restrictions. Thus, the delivered cost of sand and gravel has increased significantly in most major cities because of the increased cost of transportation. Policy development regarding this resource should begin to recognize (1) the necessity of having a viable sand and gravel industry, and (2) the need to locate and protect those resources needed for the future.

The medium to high price, large volume commodities and the high price, small volume industrial minerals are relatively insensitive to transportation costs although in many cases it is a marketing factor. The high cost of transportation gives considerable incentive to explore and locate resources nearer to markets. In many instances lower-quality materials are located and because of the lower transportation cost, additional money can be spent on processing to bring the product into market specifications.

SUMMARY

Industrial or non-metallic minerals are relatively unheralded materials but have a much higher value in the United States than the metallic minerals. They are essential ingredients in many process industries including paper, paint, plastics, rubber, ceramic, cement, glass, and many others. A high standard of living is largely due to the availability of an abundant supply of quality industrial minerals. Location and transportation costs are very important for the low cost, high volume industrial minerals and are relatively inconsequential for the higher cost, high volume and high cost, low volume industrial minerals. The economic development of a country is largely dependent on a supply of quality, reasonably priced industrial minerals.

In Central and South America, many countries should band together and form exploration and evaluation teams to locate quality industrial mineral resources. Many of these countries are small and would not have quality deposits. However, in the region many deposits could be found which would greatly help the region develop economically and industrially.

REFERENCES CITED

Industrial minerals many facets, 1988: London, Mining Journal, v. 310 (796B), p. 354-355.

Noestaller, R., 1987, Industrial minerals: Alton, Hampshire, Great Britain, World Bank Technical Paper 76, 117p.

Circum-Pacific Council for Energy and Mineral Resources
Earth Science Series, Vol. 16
R. L. Miller, G. Escalante, J. A. Reinemund, and M. J. Bergin (Eds.)
Energy and Mineral Potential of the Central American-Caribbean Regions
Springer-Verlag, Berlin Heidelberg 1995

THE OCCURRENCE, PRODUCTION, AND TRADE OF NON-METALLIC INDUSTRIAL MINERALS IN COSTA RICA

S. J. Mathers
Geological Survey, Keyworth, Nottingham, NG12 5GG, United Kingdom

N. Morales
Dirección de Geológia, Minas y Hidrocarburos, Apt. 10104, 1000 San José, Costa Rica

ABSTRACT

Costa Rica can be divided into a series of distinct geological units which contain characteristic assemblages of non-metallic industrial minerals. The most important are the Tertiary sediments, which include limestone and sandstone resources and the Quaternary volcanics which host deposits of sulfur, kaolin, diatomite, pumice, and aggregate.

Total production of non-metallic industrial minerals in 1988 was about 7.6 million MT with an unprocessed value of about US$ 40 million. On average Costa Rica imports about US$ 6 million of raw materials per year including large quantities of gypsum, glass sand, and fertilizer raw materials. Costa Rica has the resources to develop products such as high-purity limestone, diatomite, pumice, sulfur, aggregates, and kaolin-rich clays to both reduce its imports and develop valuable exports.

INTRODUCTION

This article briefly describes the geological occurrence, production, trade, and future potential of non-metallic industrial minerals in Costa Rica. It summarizes the first phase of a bilateral technical cooperation project between the British and Costa Rican governments, Proyecto Anglo-Costarricense de Minerales Industrializables (PACOMI), that is being implemented by the British Geological Survey on behalf of the Overseas Development Administration and the Dirección de Geológia Minas y Hidrocarburos (Ministerio de Recursos Naturales, Energía y Minas).

GEOLOGICAL FRAMEWORK

A simplified map of the geology of Costa Rica is presented in Figure 1 together with symbols showing the distribution of key non-metallic industrial minerals. The geology of Costa Rica can be divided conveniently into five principal units which provide a useful framework within which the occurrence of deposits can be considered.

The oldest unit comprises the Nicoya Complex which ranges in age from Late Cretaceous to Eocene. This unit crops out in several peninsulas along the Pacific coast of Costa Rica. These rocks are predominantly submarine tholeiitic basalts with subordinate pelagic cherts, limestones, and argillites. On the Peninsula de Santa Elena peridotites are developed. All these rocks represent segments of oceanic crust which have been obducted or thrust onto Costa Rica. They tend to be highly fractured, sheared, and altered.

In terms of industrial minerals the rocks of the Nicoya Complex and the Santa Elena peridotite are little exploited. Locally the basalts, where not heavily altered, have been dug, crushed, and graded as a source of roadstone.

The next unit comprises Tertiary igneous rocks that are differentiated into volcanics and intrusives (Figure 1). These represent the initial development of an island arc system in Costa Rica and today form the mountains which are the backbone of the central and south parts of the country. The extrusive rocks comprise both tuffs and lavas. They are predominantly andesitic in composition, but basaltic rocks are also widespread. These rocks are strongly faulted and folded and locally extensively altered by hydrothermal and

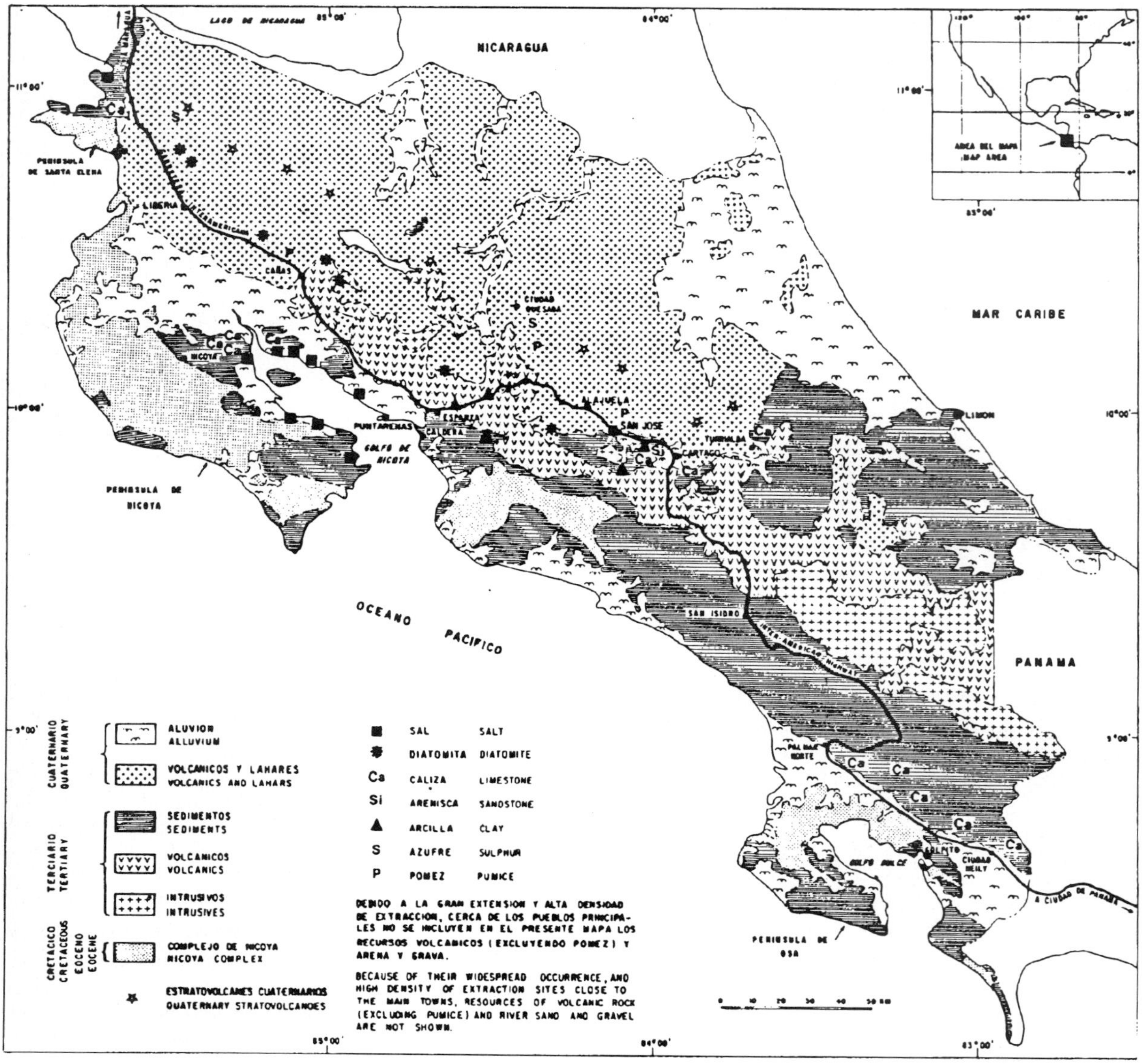

Figure 1. Map showing the simplified geology and location of key non-metallic industrial minerals in Costa Rica.

weathering processes. Where relatively fresh they are used as a source of aggregate, whereas in altered zones red kaolin-rich clays are exploited for brick making and artesanial pottery. Heavily silicified volcanics are used in the manufacture of refractory bricks. Numerous composite intrusive plutons ranging in composition from gabbro to adamellite (quartz monzonite) occur associated with the lavas.

During the emplacement of the oceanic rocks in the forearc and the development, emergence, and uplift of the inner volcanic arc through the Tertiary, numerous inter- and intra-arc basins were formed and infilled with marine Tertiary sediments. Deposition of these sediments continued into the Miocene, when marked uplift commenced causing emergence and the onset of erosion and terrestrial sedimentation. These rocks have been folded and faulted; locally they are also believed to be overthrust. The sedimentation in individual basins differs, but a general pattern of progression from deeper-water turbiditic sequences to shallower-water carbonates and sandstones can be discerned.

Argillaceous turbiditic sequences are being used for blending with limestone in the manufacture of cement. From an industrial minerals point of view, however, the most important rocks of the Tertiary sediments are the carbonates. Four deposits of limestone are of particular importance.

The Barra Honda limestone of Oligocene-Miocene age (Rivier, 1983) crops out around the head of the Golfo de

Nicoya (Figure 1). The limestone has a stratigraphic thickness of about 300 m and comprises stratified biomicrites overlain by massive algal biolithites (Mora, 1978). The deposits form a series of isolated knolls which may well reflect localized build-ups. Most of the limestone is of very high-purity, with analyses commonly indicating about 99.5% $CaCO_3$ (Garro, 1988). This limestone is currently extracted from about a dozen quarries and is used domestically to make cement; as a filler in paints, rubber, and plastics; in glass making; as dimension stone; for burning to make lime; and as an aggregate. The very high purity of the limestone, the lack of overburden and the coastal location near port facilities all suggest that the Barra Honda limestone could be a valuable export. A particular market that could be assessed is the potential to export the limestone as functional filler to the countries of the Pacific rim.

South of San José (Figure 1) a gray impure bioclastic limestone, the San Miguel limestone, crops out within a sequence of open-folded and faulted Tertiary sediments. The deposit comprises an eastward-thickening wedge, 15-190 m thick, which is believed to have been deposited as an offshore carbonate sand bar in the Miocene (Carballo and Fischer, 1978). The deposits are quarried together with interbedded shale for the manufacture of cement. Locally the limestone is also burned to make lime.

Farther east another limestone formation, the Las Animas limestone, crops out near Turrialba (Figure 1). This limestone comprises well-stratified biosparites about 100 m thick distributed in beds up to 1 m thick. The sequence is steeply dipping and locally vertically bedded. The deposit is probably Eocene in age (Rivier, 1973). Analyses of the limestones show that it varies between 92-98 percent $CaCO_3$. It is quarried principally for cement manufacture. Other uses are as a petro-fertilizer and in the manufacture of floor tiles.

In southeastern Costa Rica the Cajón limestone crops out between Palmar Norte and the Panamanian border (Figure 1). Several sub-parallel outcrops have been identified, the repetition being due to thrusting and tight folding. The formation is about 400 m thick, and consists of a lower algal biolithite and an upper detrital facies (Mora, 1979). The Cajón is thought to be Eocene in age (Malavassi, 1961). This limestone is of variable purity, ranging from 92-98.4 percent $CaCO_3$. It is presently little utilized, the only extraction being for use as a petro-fertilizer and for burning to make lime.

In addition to the resources offered by the carbonate rocks, the Tertiary sediments south of San José include the fine-grained Coris sandstone. It represents a regressive cycle of marginal sedimentation indicating the final phase of marine sedimentation and emergence in the Valle Central area prior to strong regional uplift. The Coris is of Miocene age and is up to 400 m thick. It consists of massive and thinly bedded sandstone with subordinate amounts of clay and conglomerate.

The Coris sandstone is used in several ways; massive beds within the sequence are dug and fashioned into grindstones, locally known as "mollejones," and into refractory bricks for kilns, whereas the more flaggy beds provide material for walling and facing stone. Silica flour is also produced from the sandstone and used in ceramic mixes. Investigations are in progress to assess the suitability of the Coris sandstone as a source of raw material for glass making. According to VICESA, the national glass manufacturer, the sandstone is not sufficiently pure to be used to make clear glass but may prove suitable for blending with imported materials to make colored container glass.

Much of Costa Rica's wealth of industrial minerals occurs within the deposits of the chain of active stratovolcanos which stretch from the Valle Central northwestward almost to the Nicaraguan border (Figure 1). These Quaternary volcanics and interbedded sediments contain deposits of pumice, sulfur, diatomite, kaolin, aggregate, and dimension stone.

Pumice-rich deposits occur as airfall and ash-flow tuffs. The pumice is utilized as a carrier for agrochemicals and as lightweight aggregate. Deposits of volcanogenic sulfur are being developed in Costa Rica at two principal sites, Chocoseula near Ciudad Quesada (Goméz and Samuels, 1988) and at Río Gongora north of Liberia (Figure 1). Production at both sites should begin within the next two years. Costa Rica also possesses good quality deposits of diatomite (Mathers, in press), in particular near Liberia in northwest Costa Rica. These deposits are presently utilized as an anti-caking agent for fertilizer pellets, but recent studies (Mathers, Inglethorpe, and Bloodworth, this volume) have shown that many of the deposits are likely to be suitable for use as filter-aids, the principal use of diatomite worldwide.

Widespread deposits of low-grade hydrothermal kaolin are associated with the volcanic chain. These deposits contain undesirable amounts of iron and sulfur and would need substantial benefication to manufacture commercial grades.

Substantial volumes of lavas and tuffs associated with the volcanic chain are dug for use in the construction industry, and locally lavas exhibiting good fissility are used as a dimension stone in walling.

Alluvial, fluvial, and littoral deposits are grouped in Figure 1 under the heading Quaternary alluvium. From these deposits Costa Rica obtains a substantial amount of good quality aggregate, derived mainly from clasts of fresh andesitic lavas. Large cobbles are used ornamentally in walling and as rip-rap, while small sizes are crushed and graded to provide aggregates for construction. Much of the material utilized is dug directly from river beds. This type of extraction minimizes environmental impact as the rivers with their high discharges and gradients quickly replenish the material removed. Elsewhere, river terrace, alluvial fan, and beach deposits provide local sources of aggregate.

Salt is produced by solar evaporation of seawater around the Golfo de Nicoya, which is favored because of its good tidal range and a predictable "dry" season.

PRODUCTION AND TRADE

The production of non-metallic industrial minerals in 1988 is shown in Table 1. The total value of the extracted raw material is about US$ 40 million. As one would expect, extraction is dominated by materials needed for the construction industry. These include tuffs, lavas, sand and gravel, limestone, and clay used collectively to provide aggregate and in the manufacture of cement, concrete, bricks, and tiles.

The agricultural industry consumes substantial amounts of limestone as petro-fertilizer and the high-purity Barra Honda limestone is additionally used in glass making and as a functional filler in paints, rubber, and plastics. Other products include salt, diatomite, pumice, and sandstone (Table 1).

Table 1. Costa Rican non-metallic industrial mineral extraction and end-use for 1988. All statistics are approximate and provisional.

PRODUCT	QUANTITY EXTRACTED (MT/yr)	USES
Tuff/Ignimbrite	3,300,000	Construction 51%, Roadstone 42%, Drainage 5%, Asphalt 2%
Lavas	1,500,000	Construction 57%, Roadstone 33%, Dimension Stone 6%, Asphalt 2%., Drainage 2%
Sand and Gravel (river)	1,175,000	Roadstone 49%, Construction 32%, Railway ballast 12%, Mosaic tiles 7%
Limestone	1,015,000	Cement 81%, Petro-fertilizer 14%, Construction 2%, Aglime 1%, Glass 1%, Paint filler 1%., Dimension Stone <<1%
Sand and Gravel	300,000	Roadstone 70%, Construction 30%
Clay	220,000	Cement 92%, Bricks and tiles 8%
Sand and Gravel (beach)	50,000	Roadstone 65%, Tiles 28%, Construction 7%
Salt	30,000	Human consumption 71%, Animal salt licks 29%
Diatomite	8,000	Anti-caking agent
Pumice	8,000	Carrier 75%, Lightweight aggregate 25%
Sandstone	1,000	Sanitaryware 80%, Dimension Stone 10%, Grindstones 10%
Total	7,607,000	

Approximate total value of raw materials US$ 40 million.

In recent years Costa Rica has imported substantially more raw non-metallic industrial minerals than it has exported. Exports are negligible, probably less than 5 percent of import value of US$ 5-8 million a year (Table 2). The only export of any significance has been the Barra Honda limestone which has been sold in Panama as a filler and ornamental stone.

Volumetrically, the principal imports since 1980 have been gypsum (for cement and glass), glass sand, phosphate rock and potash (for fertilizers), salt, clays (for ceramics and as fillers), and feldspar (for ceramics and glass). Relatively expensive commodities imported in smaller but significant amounts include talc, mainly used as a filler in paints, bentonite for drilling muds, asbestos for roofing and insulation, and diatomite for filtration.

The rapid growth of imports of fertilizer raw materials reflects the expansion of FERTICA SA which now supplies most of Costa Rica's agricultural industry with domestically manufactured products in place of importing ready-made foreign fertilizers. Imports of salt have declined due to an increased domestic production, and the banning of asbestos for use in roofing materials has led to the development of substitute technology using domestic materials.

Costa Rica has the resources to substitute domestic products for imports of diatomite, pumice, and sulfur and, to a lesser extent, kaolin-rich clays and glass sand. A principal difficulty, however, is the small size of the local market compared with the investment costs required. Therefore, is seems more prudent to continue importing while developing exports at the same time. Products that Costa Rica could be producing for export include diatomite, sulfur, pumice, aggregate, high-purity limestone, and cement. Markets exist for most of these commodities. These should be competitive as they have the advantage of relatively short distances to ports on both the Pacific and Caribbean coasts and a central geographical position within the North and South American markets. Given appropriate support from government, Costa Rica could readily reverse its present position and become a net exporter of industrial minerals.

ACKNOWLEDGMENTS

The authors wish to thank Dr. J. P. Berrangé for his comments on an earlier version of this paper. This paper is published with the permission of the Director of the British Geological Survey (Natural Environment Research Council).

REFERENCES

Carballo, M. A., and R. Fischer, 1978, La Formacíon San Miguel: Instituto Geográfico Nacional, Informe Semestral Enero-Junio 1978, p. 47-143.

Table 2. Costa Rican Imports of Non-Metallic Industrial Minerals 1980-1987

PRODUCT	1980	1981	1982	1983	1984	1985	1986	1987	1980-87
Gypsum	28,080	19,690	7,628	10,035	18,690	24,686	20,574	15,420	144,803
Sand (mainly for glass)	12,471	13,816	6,330	12,011	17,565	10,160	13,790	22,394	108,537
Phosphates and Potash	0	2,139	1,957	10,796	19,329	17,013	4,497	2,444	58,175
Salt	15,892	6,725	2,278	220	96	248	259	359	26,077
Ceramic Clays	3,113	2,488	2,493	2,481	4,393	2,729	2,782	3,178	23,657
Feldspar	3,441	2,055	1,813	2,235	2,622	2,229	2,954	3,907	21,256
Talc	991	713	667	969	912	808	1,095	990	7,145
Dolomite	0	0	0	0	0	0	144	4,938	5,082
Bentonite	367	507	850	372	98	807	784	563	4,348
Asbestos	1,317	761	1,191	381	326	3	0	10	3,989
Diatomite	255	431	361	342	211	271	264	290	2,425
Pumice	0	1	1	2	1	120	929	1,165	2,219
Dimension Stone	213	103	58	108	49	100	101	225	957
Sulfur	13	0	1	15	10	116	179	301	635
Mica	6	5	7	2	4	3	12	8	47
Graphite	10	7	4	3	4	7	6	3	44
Others [1]	505	254	788	3,705	2,085	1,920	6,220	1 869	17,346
TOTALS	66,674	49,695	26,427	43,677	66,395	61,220	54,590	58.064	426,742
VALUE (US$ millions)	7.7	5.5	4.3	5.2	7.8	5.8	5.4	5.0	46.7

Garro, M., 1988, Explotacíín de materias primas en cementos del Pacifico: ler Seminario, la actividad minera en Costa Rica, p. 100-111.

Goméz, M., and R. Samuels, 1988, Proyecto de azufre de Chocoseula, 1er Seminario, la actividad minera en Costa Rica, p. 87-99.

Malavassi, E., 1961, Some Costa Rican larger foraminiferal localities: Journal of Palaeontology, v. 35, p. 498-501.

Mathers, S. J., in press, Costa Rican diatomite—A review of existing knowledge and future potential: Revista Geológica de América Central.

Mathers, S. J., S. D. J. Inglethorpe, and A. J. Bloodworth, this volume, A provisional study of Costa Rican diatomites as raw materials for filter-aids.

Mora, S., 1978, Estudios geológicos de los cerros Barra Honda y alrededores, Nicoya, Guanacaste: Informe Final 6-5216, Campaña Geológica 1978, Escuela Centroamericana de Geológia.

Mora, S., 1979, Proyecto hidroeléctrico Boruca: San José, Instituto Costarricense de Electricidad, Informe de progreso 1, v. 5, 193p.

Rivier, F., 1973, Contribución estratigráfica sobre la geológia de la cuenca de Limón, zona de Turrialba, Costa Rica: Guatemala, ICAITI, Publicatión Geológica, v. 4, p 149-159.

Rivier, F., 1983, Sintesis geológica y mapa geológico del área del Bajo Tempisque, Guanacaste, Costa Rica: Instituto Geográfico Nacional, Informe Semestral Enero-Junio 1983, p. 7-30.

Circum-Pacific Council for Energy and Mineral Resources
Earth Science Series, Vol. 16
R. L. Miller, G. Escalante, J. A. Reinemund, and M. J. Bergin (Eds.)
Energy and Mineral Potential of the Central American-Caribbean Regions
Springer-Verlag, Berlin Heidelberg 1995

A PROVISIONAL STUDY OF COSTA RICAN DIATOMITES AS RAW MATERIALS FOR FILTER-AIDS

S. J. Mathers, S. D. J. Inglethorpe, and A. J. Bloodworth
British Geological Survey, Keyworth, Nottingham, NG12 5GG United Kingdom

ABSTRACT

Costa Rica possesses good quality deposits of diatomite which at present are used only as an anti-caking agent in the manufacture of fertilizer pellets. Diatomite-rich deposits are known from 10 locations, all in northwest Costa Rica. A provisional study of the suitability of some of these deposits for use as filter-aids has been carried out, including determination of bulk density, mineralogy, and filtration trials. The initial results are encouraging and indicate that calcined and flux-calcined diatomite comparable to commercially marketed grades can be manufactured from the Costa Rican deposits tested. This would enable Costa Rica to supply its own needs for filter-aids as well as creating a valuable export. Further studies are in progress.

INTRODUCTION

This paper describes a provisional laboratory study into the suitability of the principal Costa Rican diatomite deposits as raw materials in the manufacture of filter-aids. Costa Rica possesses significant resources of good quality diatomite which are presently used only as an anti-caking agent in coating fertilizer pellets. At the same time Costa Rica imports processed diatomite powders worth more than US$ 100,000 mainly for use in the filtration of sugar solutions, beers, and soft drinks. This study seeks to establish that some of the Costa Rican diatomite deposits are suitable as raw materials for producing filter-aids and that if Costa Rica installed a processing plant to manufacture filter-aids, it could become self-sufficient in diatomite and develop a valuable export business.

This work has been carried out in the laboratories of the British Geological Survey in conjunction with a bilateral technical cooperation project between the British and Costa Rican governments. This project, Proyecto Anglo-Costarricense de Minerales Industrializables (PACOMI) is an overall study of non-metallic industrial minerals in Costa Rica and is being carried out by the British Geological Survey on behalf of the Overseas Development Administration, in association with the Direcciín de Geología, Minas e Hidrocarburos (Ministerio de Recursos Naturales, Energía y Minas).

DIATOMITE DEPOSITS

There are 10 reported occurrences of diatomite in Costa Rica (Mathers, in press). They all occur in the northern half of Costa Rica (Figure 1) closely associated with the late Tertiary (Pliocene)-Quaternary volcanism and lie on the southwestern flank of the arc of active stratovolcanos which stretch northwestward from the Valle Central almost to the Nicaraguan border.

The diatomites are believed to be exclusively lacustrine in origin for the following reasons:

1. The limited identifications of diatoms all reveal genera and species typical of freshwater assemblages (for a compilation see Mathers, in press).

2. Cichlid fish, well preserved leaf impressions, and remains of terrestrial mammals (Alvarado, 1986) have all been recorded from a diatomite-rich sequence at Turrúcares (site 8, Figure 1). Similar mammal remains are also reported (Alvarado, 1986) from the Palmares area (site 7, Figure 1). Cichlid fish are today widespread in Costa Rica's rivers and lakes and are almost exclusively freshwater.

3. The deposits occur at heights up to 2,300 m above sea-level, and given their young age it is unlikely that marine sequences could have been uplifted to this extent.

4. The deposits lie within sequences of terrestrial volcanics and form parts of sequences which at many sites have been shown to be infilling small isolated basins and are

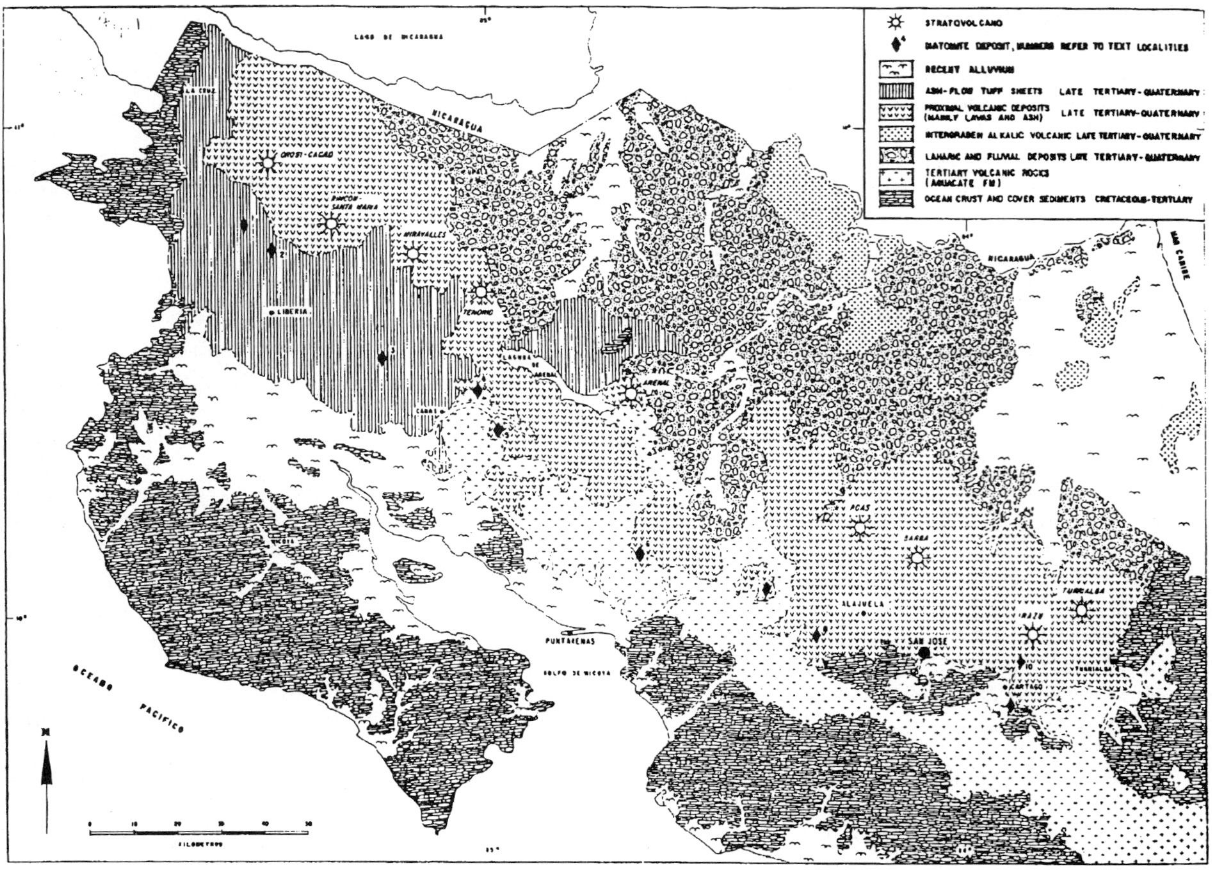

Figure 1. Geology of northwest Costa Rica showing the location of diatomite deposits.

interpreted as lacustrine in origin (sites 1, 2, 3, 7, 8, and 9, Figure 1).

INDIVIDUAL DEPOSITS

In this study grab samples were examined from the principal occurrences of diatomite (sites 1-8, Figure 1). This section briefly gives geological details and key references together with an idea of the resource potential of each site. The site numbers in the text correspond to those in Figure 1.

Loma Camastro (Site 1)

This site has been assessed geologically by Salazar (1978) and evaluated economically by Lacayo and Soto (1978). The diatomite deposits form part of a +100 m thick lacustrine sequence which infills a basin of at least 5 km^2. Surface evidence suggests that the diatomite may locally be 80-100 m thick (Salazar, 1978). The diatomite is overlain in part of the area by an andesitic lava flow, nevertheless areas exist where the diatomite can be dug directly without overburden. This deposit is thought to be the largest in Costa Rica and has reserves estimated at 5.8 million MT. Diatomite is currently being dug at Loma Camastro and used as an anti-caking agent for fertilizers. This deposit has considerable economic potential.

Cañas Dulces (Site 2)

The Cañas Dulces deposit was also investigated by Salazar (1978) and Lacayo and Soto (1978). The diatomite deposits infill a depression about 0.25 km^2 along the eastern margin of the relict Cañas Dulces volcanic center. There is no overburden and surface evidence indicates that up to 70 m of diatomite may be present, giving reserves of about 2.5 million MT (Salazar, 1978). The main physical properties of the diatomite are summarized by Salazar (1978, p. 319) based on the earlier work of Araya (1963). This site also appears to have economic potential, although drilling and pitting programs are needed to establish more precisely the available reserves.

Montano de Bagaces (Site 3)

This deposit was investigated in detail by Sandoval and Brenes (1967) and included a drilling program. Several small hills around Montano were found to contain erosional remnants of a lacustrine sequence which infilled a basin over an area of 1-1.5 km^2. Diatomite is well developed within the sequence in one of these hills. It is up to 7 m thick, and reserves were estimated at about 25,000 MT. Much of this material has, however, been subsequently worked-out.

Los Angeles (Site 4)

This deposit was identified during the reconnaissance study for the Corobicí hydroelectric project (Umaña, Bruce, and Mojica, 1975). It comprises a small lenticular outcrop. The deposit is 5-8 m thick and stretches for about 0.75 km along the side of the valley of the Río Santa Rosa. The diatomite is overlain by at least 100 m of younger volcanic rocks. A small quarry was opened to extract some of the readily accessible diatomite and further small quantities could probably still be dug. However, extensive extraction is unlikely to be feasible as this would involve working into the hillside which on the steep slopes would involve the stripping of very large amounts of overburden.

Líbano (Site 5)

Two deposits of diatomite are known from the area around Líbano, although their exact distribution and resources are not known. Limited extraction of material has taken place at one of the two sites.

Peñas Blancas (Site 6)

This deposit was reported by Madrigal (1960) and comprises a lacustrine sequence of 25-30 m within which at least 3 m of finely laminated diatomite is developed. The extent of the deposit is not known; however, the known outcrops lie within a nature reserve area, so any extraction at this site, if permitted, would be very tightly controlled.

Palmares (Site 7)

A large former lake-basin (ca 50 km^2) was developed during the Quaternary in the Palmares-San Ramón area and persists today as a marked depression. Within the basin a sequence of up to 90 m of lacustrine and fluvial sediments was deposited. These contain beds of diatomite up to a meter thick, which is too thin to be economic. Further investigation would be needed before finally writing-off this basin as a diatomite prospect, although the chances of encountering economic quantities of material must be viewed as slight.

Turrúcares (Site 8)

Lacustrine deposits containing diatomite crop out over an area of about 1 km^2 around Turrúcares (Sandoval, 1966; Montero, 1975). Within the sequence a unit of about 5 m of diatomite with interstratified thin volcaniclastic layers has been recognized. A detailed investigation, including the drilling of 16 boreholes, enabled the reserves to be estimated at about 300,000 MT (Sandoval, 1966). A small quarry was opened in the deposit during the 1960s and worked for several years for material for filter-aid and as an anti-caking agent.

Table 1. Mineralogy and Bulk Density of Costa Rican Diatomites.

SITE	Sample	Qtz	Crs	Fds	Smc	Kao	Mic	Bulk Density (kg/m^2)
Loma Camastro	JB15	**	*	ND	20	ND	ND	299
	JB 16	*	*	ND	24	ND	ND	242
Los Angeles	JB 24	*	tr	ND	16	*	ND	393
Cañas Dulces	JB 27	*	*	ND	20	*	ND	406
Montano de Bagaces	SM 20	*	*	*	7	ND	ND	430
Peñas Blancas	SM 89a	tr	tr	ND	5	ND	ND	368
	SM 89b	tr	tr	ND	5	ND	ND	346
Líbano	SM 133a	**	*	ND	15	*	tr	223
	SM 133b	*	*	ND	14	*	tr	200
Palmares	SM 135	**	*	*	24	**	tr	361
Turrúcares	SM 152	**	***	**	25	*	*	444
	'X'	*	*	**	9	ND	tr	292

Relative X-ray intensities
*** major
** minor
* present
tr trace
ND not detected

Minerals
Qtz Quartz
Crs Cristobalite
Fds Feldspar
Smc Smectite
Kao Kaolin
Mic Mica

Smectite content determined by the 2-ethoxyethanol method.

However, a substantial overburden over much of the area and the difficulties of extracting and processing a thinly interbedded sequence indicate that further extraction at the site of the quarry is unlikely.

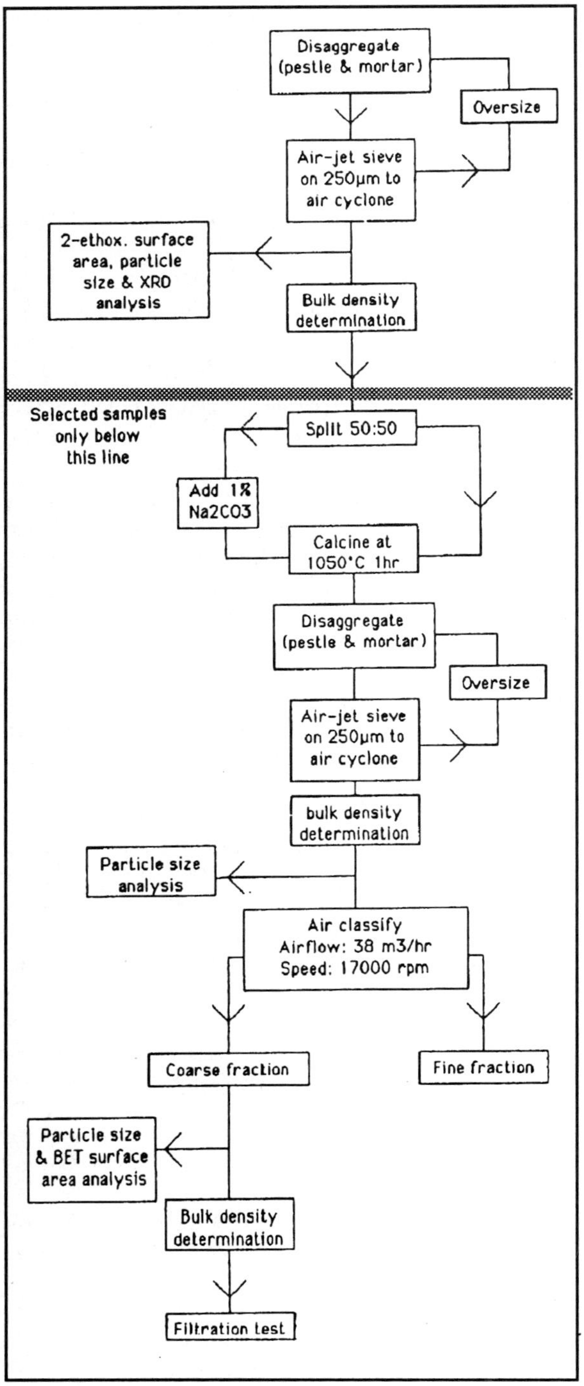

Figure 2. Processing flowsheet.

In summary, Costa Rica's best diatomite prospects in terms of ease of mining and reserves are at Loma Camastro and Cañas Dulces. Small quantities of diatomite could still be won from other sites and new occurrences will no doubt be found.

ASSESSMENT PROCEDURES

Eleven samples from the eight sites listed in Table 1 were submitted for examination. Initial characterization was carried out by determination of mineralogy and basic physical properties such as particle-size distribution and bulk density. Scanning electron microscopy was used to determine the form, size, and condition of individual diatom skeletons. The procedure for analyzing the samples is shown in Figure 2 and the results are presented in Table 1. The properties of the samples can be compared with values for a lacustrine diatomite sample from the western United States which is being dug commercially to manufacture filter-aids (Sample 'X').

X-ray diffraction was used to determine the mineralogical composition of the diatomites. This was carried out by analysis of randomly-oriented powder mounts prepared from micronized <250 μm material using a Philips Scientific PW 1760 x-ray diffractometer. Monochromated Co-Kα radiation was used over the scanning range 2-50°2θ at a scanning speed of 1-5°2θ/min. The silica which makes up the diatom skeletons has very little crystalline structure and this only gives a broad "hump" (Figure 3a) between 20 and 30°2θ on diffraction traces. Crystalline mineral "impurities" in the diatomites which generally show relatively sharp peaks (Figure 3b) are given in Table 1 along with a semi-quantitative estimate of proportions based on X-ray intensities. Quartz, cristobalite, feldspar, and smectite are the most abundant mineral components of the diatomites reflecting the considerable influence of volcanic processes on these sediments. The content of smectite in the samples was determined by the 2-ethoxyethanol surface area method.

Cumulative curves of the particle-size distribution of the raw diatomites show considerable variation. Of particular note sample SM 89 (Peñas Blancas, Figure 1) is very well sorted, whereas sample JB 15 (Loma Camastro) is very poorly sorted. Sample SM 133a (Líbano) shows a bimodal distribution and the very wide range of particle size shown by samples SM 135 (Palmares) and SM 152 (Turrúcares) probably reflects a high level of mineral impurities in these deposits.

Determination of bulk density also afforded a simple means of rapidly assessing the relative purity of the diatomites. In this study bulk density (Table 1) was calculated by measurement of the volume occupied by a known weight of diatomite following a controlled compaction procedure.

Theoretically "pure" diatomites will have a much lower bulk density than those which contain a higher proportion of mineral impurities. Complex diatom skeletons are composed

Table 2. Processing trials results.

SITE	Sample	% Na_2CO_3	Surface Area (m^2/g)	Bulk Density (Kg/m^3)	Filtration Flowrate (ml)
Loma Camastro	JB 16	0	8.1	383	228
		1	6.0	427	330
Peñas Blancas	SM 89b	0	NA	NA	NA
		1	3.7	450	320
Líbano	SM 133b	0	7.7	268	200
		1	5.2	308	195
	'X'	0	3.7	394	250
		1	2.4	463	325

NA = Not available, insufficient material for analysis.

of low density amorphous silica and therefore have a high void volume when packed. Crystalline minerals are relatively dense and form solid particles with simple shapes which have a high packing density. The relationship between bulk density and diatomite is, however, complicated by the influence of particle-size distribution. Diatomites with a coarser size distribution will tend to show higher bulk densities than finer material. Nevertheless, the samples examined show a broad correlation between bulk density and the quantity of mineral impurities. However, this method is only useful in rejecting samples with a high content of mineral impurities and cannot be used to assess the relative quality of fairly "pure" diatomites due to the effects of grain size.

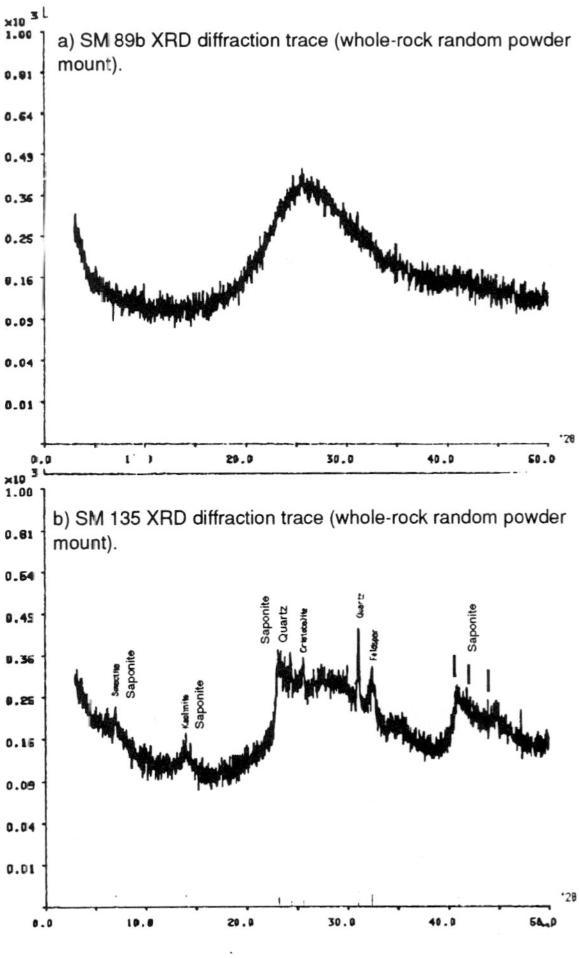

Figure 3.

Selected samples were examined under the scanning electron microscope (SEM). This technique provides useful visual information on the skeletal form of the diatoms. All samples examined from Costa Rica were dominated by centric diatom forms, mainly cylindrical types with subsidiary discoidal forms. Pennate forms occur in some of the samples but are less abundant. SEM photography also reveals a considerable contrast in the preservation of the diatom skeletons. In particular, forms from sample JB 27 (Cañas Dulces) are well preserved, whereas sample JB 15 (Loma Camastro) reveals a high percentage of broken-up material possibly indicative of re-working. Fragments of mineral impurities such as clay, quartz, and feldspar were observed in many samples. Clay tends to be present as discrete fine particles and aggregates, whereas the quartz and feldspar occur as coarser grains. Some samples reveal the presence of thin silica overgrowths on the diatom skeletons.

Based on the tests described above, three samples of relatively "pure" diatomite were selected for a simulated filtration test and compared to sample 'X', the commercial material. In assessing these samples it was necessary to process the ground diatomite in a manner analogous to a commercial operation. This was achieved by calcining the samples, with and without the addition of a flux, and then re-milling and classifying the material before performing the filtration test (Figure 2). The filtration test was performed using the methods developed by Bowman, Gajewski, and Sparrow (1983).

The results are shown in Table 2, and filtration flow rate curves are presented for two Costa Rican diatomite samples in Figure 4. The flowrate curves for the commercial material (sample 'X') are given on each graph for comparative purposes. The effect on basic physical properties of the addition of a sodium carbonate flux prior to calcination is clear. Flux calcination results in an increase in modal size and bulk

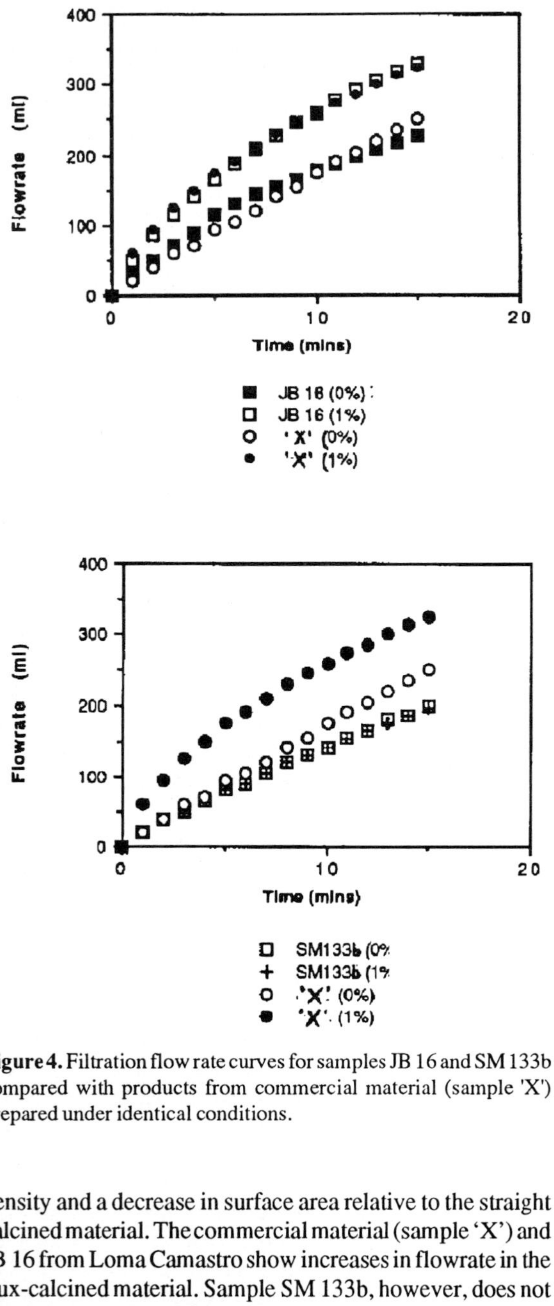

Figure 4. Filtration flow rate curves for samples JB 16 and SM 133b compared with products from commercial material (sample 'X') prepared under identical conditions.

density and a decrease in surface area relative to the straight calcined material. The commercial material (sample 'X') and JB 16 from Loma Camastro show increases in flowrate in the flux-calcined material. Sample SM 133b, however, does not follow this pattern (Figure 3b), possibly due to an overall finer particle size distribution.

DISCUSSION AND CONCLUSIONS

The aim of this study was to produce a preliminary assessment of the Costa Rican diatomites as raw materials for filter-aids. A small group of three samples was selected from the initial suite of 11 samples to carry out a simulated filtration test. The selection procedure was difficult because of the generally high quality of the diatomites submitted. The bulk density proved an inadequate means of grading a suite of relative high-purity diatomites, and the mineralogy was, therefore, used as the principal method of sample selection.

The performance of Costa Rican diatomite samples from Loma Camastro (JB 16) and Peñas Blancas (SM 89) is comparable to that of the commercial material (sample 'X'), and it is probable that many of the other samples examined with similar bulk density, particle-size, and mineralogy will also be suitable. Unlike the other samples, flux calcination of the material from Líbano (SM 133b) did not enhance its poor filtration characteristics.

In summary, Costa Rica has substantial resources of good-quality diatomite. On the basis of this preliminary study, it is considered likely that low flowrate filter-aids (and possibly other grades) could be produced from material similar to that of samples JB 16 (Loma Camastro) and SM 89 (Peñas Blancas). Costa Rica appears to have the resources to produce filter-aids to satisfy the domestic market and build an export business.

ACKNOWLEDGMENTS

This paper is published with the permission of the Director of the British Geological Survey (Natural Environment Research Council).

REFERENCES

Alvarado, G. E., 1986, Hallazgos de megamamíferos fósiles en Costa Rica: Revista Geológica América Central, v. 4, p. 1-46.

Araya, J. A., 1963, Estudio sobre el aprovechamiento industrial de la diatomita de Liberia: San José, Departamento de Química, Ciudad Universitaria "Rodrigo Facio," Lic. Tésis, 88p.

Bowman, R. G., P. J. Gajewski, and G. J. Sparrow, 1983, Composition and filtration properties of a drill core from the Conjuboy diatomite deposit, Queensland: Journal Australian Ceramic Society, v. 19, p. 19-27.

Lacayo, B., and G. Soto, 1978, Aspectos generales, técnicos y económicos del proyecto diatomita: Boletín Geológico y de Reservas Minerales, no. 1, CODESA.

Madrigal, R., 1960, Algunas localidades con diatomita de Costa Rica, informes técnicos y notas geológicas: Ministerio de Industrias (DGMP), no. 5, 13p.

Mathers, S. J., in press, Costa Rican diatomite: A review of existing knowledge and future potential: Revista Geológica América Central.

Montero, W., 1975, Estratigrafía del Cenozóico del área Turrúcares provincia de Alajuela, Costa Rica: San José, Escuela Centroamericana de Geológia, Universidad de Costa Rica, 40p.

Salazar, A. J., 1978, Geología de los depósitos de diatomita de las Brisas-Cañas Dulces y la Loma Camastro, Liberia, Guanacaste y su evolución preliminar: Boletín Geológico y de Reservas Minerales, no. 1, p. 296-319, CODESA.

Sandoval, F., 1966, Estudio geológico y evaluación de un yacimiento de diatomita localizado en la región Turrúcares de Alajuela, informes técnicos y notas geológicas: Dirección de Geología,

Minas y Petróleo, no. 23, 42p.
Sandoval, F., and M. Brenes, 1967, Estudio geológico y evaluación de un yacimiento de diatomita localizado en la región de Montano de Bagaces de Guanacaste, informes técnicos y notas geológicas, Dirección de Geológia, Minas y Petróleo, no. 25, 23p.
Umaña, J., E. Bruce, and E. Mojica, 1975, Informe de reconocimiento avanzado del proyecto hidroeléctrico de Corobicí: Instituto Costarricense de Electricidad, Dirección de Ingeniería de Energía, 32p.

Circum-Pacific Council for Energy and Mineral Resources
Earth Science Series, Vol. 16
R. L. Miller, G. Escalante, J. A. Reinemund, and M. J. Bergin (Eds.)
Energy and Mineral Potential of the Central American-Caribbean Regions
Springer-Verlag, Berlin Heidelberg 1995

SUMMARY AND RECOMMENDATIONS OF THE MINERALS WORKING GROUP

Orlando Martino
U.S Bureau of Mines

Stephen Ludington,
U.S. Geological Survey

The mineral endowment of Central America and the Caribbean area is significant. World-class deposits of gold, bentonite, silver, clay, copper, salt, manganese, diatomite, nickel, limestone, aluminum, sulfur, lead, and zinc make important contributions to local economies and to world supply. Potential for additional and new reserves appears promising.

The symposium brought to light both the excellence of the work being done by the local scientific and technological community, and the important role of international cooperation. Over half of the presentations during the minerals sessions were the direct result of local and international cooperative efforts. Development successes so far justify expanded local financial support for existing programs. Increased production of mineral commodities, particularly industrial minerals, will contribute significantly to easing the foreign debt crisis.

The working group identified the following specific needs to be filled within individual countries:

1. The acceleration of the acquisition of basic earth science data, including geologic mapping at both detailed and regional scales, and geophysical, geochemical, and remote sensing data;

2. Creation and maintenance of geologic, geophysical, and geochemical databases;

3. Wider distribution of collected and archived data by publication and other appropriate means;

4. Analytical facilities, both for chemical and physical properties;

5. Better production and trade statistics;

6. Better reserve estimates;

7. Long-term and multidisciplinary technical cooperation; and

8. Better earth science education.

To achieve sustained development that is environmentally sound, the Caribbean and Central American region will benefit from:

- Increased communication
- Increased local participation in the mining industry
- Rationalization of the efforts of governmental agencies
- Better logistic support facilities
- Long-term investment incentives
- Continuing cooperative efforts
- Programs designed to continue through to payback
- A clear legal environment

These activities should be carried forward within the context of a development strategy that takes into account both short and long-term goals.

Signed:
Fernando Alvarado (RECOPE)
Jevan Berrangé (British Geological Survey)
Siegfried Kussmaul (University of Costa Rica)
Stephen Ludington (U.S. Geological Survey)
Orlando Martino (U.S. Bureau of Mines)
Steve Mathers (British Geological Survey)
G. E. McKelvey (U.S. Geological Survey)
Carl Nelson (Consultant)
Donald Reichmuth (Geomax - Consultant)
Ricardo Vega (MINASA)

WATER RESOURCES

Circum-Pacific Council for Energy and Mineral Resources
Earth Science Series, Vol. 16
R. L. Miller, G. Escalante, J. A. Reinemund, and M. J. Bergin (Eds.)
Energy and Mineral Potential of the Central American-Caribbean Regions
Springer-Verlag, Berlin Heidelberg 1995

THE DISTINCTIVE HYDROLOGY OF TROPICAL ISLANDS

Allen L. Zack
Caribbean District Chief, Water Resources Division, U.S. Geological Survey,
GPO Box 4424, San Juan, Puerto Rico 00936 USA

ABSTRACT

The hydrology of tropical islands differs from that of temperate, continental land areas because of climate and geomorphology. The distinctive hydrologic cycle observed on tropical islands represents a history of orographic effects and thermal radiation on island geology and geometry, combined with the omnipresent influence of the sea.

Climate and geomorphology tend to alter the dimensions of time and space in the study and conceptualization of tropical islands hydrology. Islands and their hydrologic units such as stream basins and aquifers are smaller and have well-defined boundaries; floods and droughts are more frequent but have shorter duration; the time lag between rainfall and flooding is much less; aquifer recharge and discharge respond more immediately to rainfall and drought; rainfall and evapotranspiration are more variable in time and space; and average annual rainfall is more variable areally.

The distinctive hydrologic processes operating in tropical islands may be no more complex than in the temperate continents, but they are certainly less studied and documented. For example, (1) traditional empirical derivation of evapotranspiration from potential evaporation may be invalid in tropical islands; (2) indirect measurement of streamflow has been limited by an absence of hydraulic roughness values for stream channels in steep, tropical areas with banks vegetated with bananas, sugar cane, bamboo, pineapples, or other tropical plants; (3) the universal soil-loss equation may not be applicable in many tropical islands; (4) biological and chemical methodologies traditionally applied to determine sanitary conditions, study contamination, and measure the assimilative capacity of streams may not be applicable in the tropics; and (5) extreme chemical weathering found in the humid tropics is responsible for strong leaching of bedrock, high concentrations of clay in sediment, and high rates of surface denudation, but its effect on the occurrence, availability, and ionic composition of ground water has not been measured.

INTRODUCTION

Hydrologists conditioned to working in temperate, continental areas and relocated to a tropical-island environment must be reconditioned to hydrologic conceptualization given unfamiliar scales of time and space. The laws of physics and chemistry that define tropical-island hydrology remain the same, but they are cast under a distinctive set of constraints that alter the cause-and-effect relationships that hydrologists must interpret.

The hydrology of tropical islands differs from that of temperate, continental land areas because of dimension, climate, and geomorphology. The distinctive hydrologic cycle observed on tropical islands represents a long history of orographic effects and thermal radiation on island geology and geometry, combined with the omnipresent influence of the sea.

THE SCENE

The influence of climate on tropical island hydrology must be viewed in terms of water availability, in particular the extreme events caused by rainfall, and also the seasonal rhythms of wet and dry periods (Douglas and Spencer, 1984, p. 40). Wide ranges in runoff volumes, common to tropical islands, are attributable to the frequency and duration of rainfall. Flash flooding, extreme sediment transport, and long periods of drought are the hydrologic consequences. Discrete storm cells of intense activity are responsible for much of the local spatial and temporal variability of rainfall in tropical islands. In addition, rainfall magnitude and frequency in the mountainous interiors of the islands are dependent on relief and aspect where orographic uplift and instability result in increased rainfall. These local storms are in great contrast to frontal systems and cyclones which produce heavy rain over large areas. Cyclones develop only where sea surface temperatures are in excess of 26.5°C, and their frequency appears to be closely correlated with sea surface temperature (Spencer and Douglas, 1984, p. 29). In addition to cyclones, the tropics also incorporate such diverse climatic regimes as the doldrums, monsoons, trade winds, and incursions of polar air.

Along the coastlands of the mountainous islands of the West Indies, rainfall occurs chiefly in summer and autumn when the Intertropical Convergence Zone lies close to the islands and easterly waves and other tropical disturbances affect the region (Walsh, 1984, p. 95). The winter months are dry because of the proximity of the Azores subtropical anticyclone with its low-level inversion and descending stable air. These islands have climates that range from highly seasonal on the sheltered leeward coast, to weakly seasonal on the windward coast, to perennially wet in the central mountains. In general the elevated islands receive sufficient rainfall to endow them with dependable water supplies.

In contrast to the generally wet mountainous islands, many of the smaller, less elevated islands throughout the world's tropics have extremely limited water resources. Where orographic influences are minimal, rainfall in low-lying islands reflects that of the open ocean which, in most of the tropics, ranges from 380 to 635 mm (15 to 25 in.) of accumulation per year. Rooftop catchments, desalination, and reverse-osmosis facilities have historically been used to provide potable water. Drinking water in the relatively dry U.S. Virgin Islands is supplied principally by desalination, presently (1989) at the high cost of $14 per thousand gallons.

The tropical marine climate undoubtedly has been the greatest influence on the distinctive geomorphology and in some instances the dimensions of tropical islands. Climatic changes and associated changes in sea level have strongly influenced island size and tropical landform evolution (Douglas and Spencer, 1984, p. 11). In fact, the unique landform assemblages occurring in extratropical zones indicate the former existence of tropical climates. Coral reefs, mangrove forests, and beachrock—all tropical, shallow ocean, land-building phenomena—affect the geomorphology and dimensions of tropical islands. Whereas coral reefs and beachrock form limestone and calcite-cemented sandstone, respectively, mangroves trap and accumulate sediment. Through geologic time, this material provides the matrix for aquifers. The transmission and release of water to wells intended for water supply depends on the distinctive hydraulic conductivity and compressibility characteristics of the aquifer which, in turn, are attributable largely to tropical-island geomorphology.

Islands are natural hydrologic units inasmuch as their physical, economic, and hydrologic boundaries are effectively coincident (Matalas, 1987, p. 569). Owing to differences in dimension and geomorphology, islands and their hydrologic units such as stream basins and aquifers are smaller than those of the temperate continents and have well-defined boundaries. Accordingly, floods are more frequent and more responsive to rainfall events, but have shorter duration.

Surface-water development for water supply is more tenuous because of few sites to store water. Also, aquifers are less extensive and thus less dependable as a continuous water supply. Several different hydrologic disciplines can be examined in terms of the tropical-island environment and some suggestions for future studies and research initiatives can be made.

GROUND-WATER HYDROLOGY

Ground-water resources in the smaller, low-lying islands are generally limited because of low ground-water recharge from rainfall, restricted areal extent of aquifers, and the proximity to seawater. However, it has been demonstrated that even for small, dry islands, there are some methodologies and strategies available to optimize production from whatever limited ground-water resources are available. These include artificial ground-water recharge, scavenger wells, barrier wells, and statistical approaches to the development of low-yielding, fractured aquifers.

Ground-water availability from small, tropical-island aquifers is not as easy to quantify as that from extensive, continental aquifer systems. Features such as nearby impermeable boundaries, the saline-freshwater interface, ground-water recharge patterns, intimate ground-water/surface-water relationships, variable evapotranspiration, exaggerated anisotropy, and ambiguous storage characteristics make the conceptualization of these aquifers difficult when ground-water flow and solute-transport models are contemplated.

The distinct geomorphological history of near-surface, artesian aquifers in tropical islands provides unique opportunities to study ranges of specific storage rarely encountered in continental aquifer systems. Artesian aquifers occurring below impermeable deposits often contain a highly compressible aquifer matrix providing large values of compressible aquifer storage.

Saltwater intrusion into heavily developed coastal aquifers threatens water supplies in all tropical islands. The key in determining whether the saline-freshwater interface will migrate toward ground-water pumping centers is to identify the origin of water withdrawn by pumping wells. Water derived exclusively from aquifer storage may not initiate movement of the interface, whereas water derived from external sources adjacent to surface-water bodies such as the ocean may initiate migration of the saline-freshwater interface.

GEOCHEMISTRY

The extreme chemical weathering present in forested, volcanic tropical islands is responsible for strong leaching of bedrock, high concentrations of clay in sediment, and high rates of surface denudation. The deep weathered regolith on many Caribbean and Pacific islands results from high precipitation and the opportunity for percolating rainwater to incorporate carbon dioxide from decaying organic matter on the forest floor where temperatures are high (Douglas, 1984, p. 8). With heavy rainfall, erosion can be high and where slopes are steep, mass wasting of the unstable soils can occur. Rainfall of lesser magnitude and intensity infiltrates readily, contributes to ground-water resources, and limits the surface drainage density. The chemical processes and rates that produce deep saprolite aquifers have not been fully determined.

Calcite dissolution in the humid tropics has produced dramatic cone and tower karst. These landforms are different from temperate, continental karst because of climatic factors (rainfall intensity, year-round high temperature, and continuous production of carbon dioxide) and possibly lithologic factors (age and composition of the carbonates) (Troester, Back, and Mora-Castro, 1987; Peck, Troester, and Moore, 1988). In Puerto Rico, infiltration occurs through sinkholes, around the periphery of mogotes (haystack hills), providing important ground-water recharge. Dissolution continues within the carbonate aquifers contributing to secondary porosity. In general, excellent aquifers occur throughout tropical karst areas with a history of calcite dissolution. These areas usually have a poor surface drainage density with intermittent streams.

Dolomitization is known to be occurring in some tropical island carbonate aquifers. This process, which greatly affects aquifer porosity and permeability in all climates, is not well understood. Isla Mona, an uninhabited limestone platform west of Puerto Rico, serves as a natural laboratory for studying this process in an undisturbed environment in the midst of seawater and salt spray, both high in magnesium concentration.

Beachrock formation is a uniquely tropical phenomenon which, if understood and artificially induced, could be used to stabilize beach erosion. Monitoring of the near-shore, ground-water flow system and beach geometry would be required to observe the conditions under which it forms and erodes.

SURFACE-WATER HYDROLOGY

The tropical weather systems combine with orographic effects to promote frequent rainfall events of short duration and high intensity resulting in flash floods. Accurate measurement of streamflow using established methodologies in the small steep basins of tropical islands is often impossible during periods of high runoff for the following reasons:

1. Isolated, heavy rainfall events common to the tropics are difficult to predict.

2. Time lag from rainfall to runoff is short in small stream basins with steep slopes.

3. Most of the preferred data-collection sites are in remote areas without bridges or cable ways from which discharge measurements can be made.

4. Intense rainfall events of short duration promote rapid fluctuations in flow volumes and river stages. Normal streamflow-measuring techniques require that the river remain relatively stable for at least one hour. A rate-of-rise of 3 m/hr is not uncommon in the West Indies and a 6 m/hr rate has been recorded.

5. Stream velocities often exceed 3 m/sec, which is sufficient to dislodge large boulders. Use of conventional current meters under these conditions is impossible.

Therefore, it would seem that because of the problems associated with making direct current-meter measurements of streamflow in tropical islands, determination primarily by indirect methods is necessary. However, the accuracy of whatever method used is limited by the accuracy of estimation of channel roughness. Little work has been published relative to verification of roughness values for tropical areas. Accordingly, a need exists for research to evaluate the hydraulic roughness of stream channels in steep tropical areas with banks vegetated with bananas, sugar cane, bamboo, pineapples, marijuana, and other tropical plants.

Water-supply development on the smaller tropical islands is aggravated by the lack of perennial streams and few locations to store water. Frequent droughts cause water-supply emergencies. In many islands, cisterns fed by rooftop catchments are the most effective means to store water.

CLIMATOLOGY

Traditional empirical derivation of evapotranspiration from potential evaporation may be invalid in tropical islands according to recent studies conducted in Puerto Rico (Román-Más, 1988, p. 20). These studies reveal a poor correlation between evapotranspiration and potential evaporation, temperature, wind velocity, and solar radiation. These studies suggest that evapotranspiration is consistently higher than potential evaporation where rainfall is higher than measured pan-evaporation.

Accordingly, transpiration appears to be more important in the evapotranspiration process for tropical islands. Elements such as rainfall patterns, soil-water storage capacity, and vegetation seem to be critical to the understanding of the evapotranspiration process in tropical islands.

Transpiration from the rain forest canopy can have a marked impact on the rainfall-runoff relationship. Rain forest vegetation intercepts considerable portions of rainfall, up to 3 mm of light rain (F. Scatena, 1989, oral communication) but the proportion of rain that never reaches the ground decreases with the intensity and duration of the storm (Douglas and

Spencer, 1984, p. 51). Antecedent moisture conditions play a role in the rainfall-runoff relationship for tropical rain forests only after certain rainfall intensity and duration thresholds have been achieved.

There is mounting evidence of major long-term and short-term climatic fluctuations in several Caribbean islands. Mean annual rainfall in the islands has fluctuated considerably over the past hundred years at most stations with long records. The late 19th century appears to have been 15 to 25 percent wetter than at present (Walsh, 1984, p. 117). In addition, historical records of St. Croix, Isla Mona, Vieques, and several other relatively dry Caribbean islands indicate that rainfall has decreased over the past several hundred years.

GEOMORPHOLOGY AND SEDIMENT

Landslides are common in tropical islands having steep slopes. Rainfall intensities and accumulations can trigger landslide activity, causing loss of life and property, but quantitative assessments of landslide susceptibility in the tropics have not been made. In addition, the effect of landslides on sediment transport of tropical island streams has not been determined.

Erosion and sediment transport can be dramatically greater during storms in tropical islands than in the temperate continents because of the greater steepness of slopes, immature soils, and rainfall intensity. Only limited data-collection efforts have been conducted in tropical islands concerning sediment transport, land use, rainfall statistics, and soil types. It is suspected that mass wasting is responsible for the majority of the 125 MT/km^2/yr of suspended sediment load derived from a small, pristine watershed in the Caribbean National Forest, Puerto Rico (F. Scatena, 1989, oral communication). The universal soil-loss equation has not been tested and may not be applicable in many tropical islands.

BIOLOGY

Biological methodologies traditionally applied to determine sanitary conditions, to study contamination, and to measure the assimilative capacity of streams may not be applicable in the tropics. It has been demonstrated that *Escherichia coli* is widespread in pristine waters of El Yunque National Forest, Puerto Rico, suggesting the *E. coli* might be a naturally occurring bacteria in the tropics because of the naturally higher temperatures (Carillo-Estrada and Hazen, 1985, p. 470).

A survey of aquatic flora and fauna for rivers, lakes, lagoons, and estuaries does not exist for the majority of tropical islands. Also, tropical limnology in terms of populations, nutrients, function, productivity, eutrophication, sedimentation, and the hydrology budget has not been studied quantitatively. For example, the productivity of periphyton in controlling reaeration and deaeration for waste-load conveyance has not been determined for tropical streams.

Tropical-island wetlands are considerably different than wetlands occurring in the temperate continents not only because of the flora and fauna, but also because of differences in the hydrologic regime and water quality. The capacity of coastal wetlands in assimilating waste loads from sewage-treatment plants is unknown in tropical areas. Tropical islands with mountains of 900 m or more in elevation often have an unusual wetland environment occurring in the cloud forests. In these areas, it is believed that base flow conditions of streams originate from condensation rather than groundwater inputs.

SUMMARY

In summary, water-resource development in tropical oceanic islands must consider small, localized development and distribution rather than regional plants and systems. Mini-hydrology, in terms of time-scale and dimension, must be considered. The numerous hydrogeologic differences between tropical islands and continental land masses necessitate a more focused approach to problem solving in this setting.

REFERENCES

Carillo-Estrada, M. E., and T. C. Hazen, 1985, Survival and enumeration of the fecal indicators by bifidobacterium adolescentis *Escherichia coli* in a tropical rain forest watershed: Applied Environmental Microbiology, v. 50, p. 468-476.

Douglas, I., and T. Spencer, 1984, Present-day process as a key to the effects of environmental change, *in* I. Douglas and T. Spencer, eds., Environmental change and tropical geomorphology: London, George Allen and Unwin, p. 39-51.

Douglas, R. P. D., 1984, The influence of climate, lithology and time on drainage density and relief development in the tropical volcanic terrane of the Windward Islands, *in* I. Douglas and T. Spencer, eds., Environmental change and tropical geomorphology: London, George Allen and Unwin, p. 3-11.

Matalas, N. C., 1987, Hydrology in an island-continent context, *in* D. J. McLaren and B. J. Skinner, eds., Resources and world development: New York, John Wiley and Sons, Limited, p. 569-592.

Peck, D. L., J. W. Troester, and J. E. Moore, 1988, Karst hydrogeology in the United States of America: U.S. Geological Survey Open-File Report 88-0476, 19p.

Román-Más, A. J., 1988, Water required, water used, and potential water sources for rice irrigation, north coast of Puerto Rico: U.S. Geological Survey Water-Resources Investigations Report 86-4334, 35p.

Spencer, T., and I. Douglas, 1984, The significance of environmental change: Diversity, disturbance and tropical ecosystem, *in* I. Douglas and T. Spencer, eds., Environmental change and tropical geomorphology: London, George Allen and Unwin, p. 13-33.

Troester, J. W., W. Back, and S. C. Mora-Castro, 1987, Karst of the

Caribbean, *in* W. L. Graf, ed., Geomorphic systems of North America: Geological Society of America, Decade of North American Geology Centennial Special Volume No. 2, p. 347-357.

Walsh, R. P. D., 1984, The influence of climate, lithology and time on drainage density and relief development in the tropical volcanic terrane of the Windward Islands, *in* I. Douglas and T. Spencer, eds., Environmental change and tropical geomorphology: London, George Allen and Unwin, p. 93-164.

Circum-Pacific Council for Energy and Mineral Resources
Earth Science Series, Vol. 16
R. L. Miller, G. Escalante, J. A. Reinemund, and M. J. Bergin (Eds.)
Energy and Mineral Potential of the Central American-Caribbean Regions
Springer-Verlag, Berlin Heidelberg 1995

A QUATERNARY VOLCANIC AQUIFER SYSTEM IN CENTRAL AMERICA: KEY FACTORS IN GROUNDWATER POTENTIAL AND PROTECTION

S. S. D. Foster
British Geological Survey, Casilla Postal 4337, Lima 100 Peru

J. M. Parker
British Geological Survey, Hydrogeology Research Group, Maclean Building, Wallingsford 0X10 8BB, Great Britain

A. Gomez Cruz
Servicio Nacional de Aguas, Riego y Avenimiento (SENARA), Apartado Postal 5262, San José 1000 Costa Rica

ABSTRACT

In Central America, Quaternary volcanic aquifer systems play an important, and sometimes vital, role in public water-supply. For example, in Guatemala City, Managua, San Salvador, and San José de Costa Rica, high-yielding wellfields have been developed and/or major spring discharges directly or indirectly captured.

Over the past 15 years the Quaternary andesitic volcanic aquifer complex on the northern side of the Valle Central in Costa Rica has been rapidly exploited to provide groundwater supplies in excess of 250 Ml/d and scientifically investigated to identify its hydrogeological characteristics. It is now believed to be one of the best understood volcanic aquifer systems of the circum-Pacific. Interbedded lavas and pyroclasts combine to produce a multiaquifer system of very high transmissivity and major groundwater storage. Pyroclastic tuffs have extremely high porosity coupled with moderate permeability, permitting steady recharge to deeper lava aquifers by downward leakage. Perched water-tables are widely developed above aquitards formed by massive ignimbrites. The heavy rainfall of the area generates large groundwater resources but aquifer recharge and flow mechanisms are complex, and resultant problems in the evaluation of groundwater potential and protection requirements are highlighted.

HYDROGEOLOGICAL SETTING OF VALLE CENTRAL

The Valle Central of Costa Rica is situated between the southernmost active volcanos of the Middle America arc (Figure 1) and mountains of Tertiary sedimentary rocks. The northern flank of the valley is a Quaternary volcanic complex formed by emissions from the craters and vents of the Barba volcano (Figure 2A), with subordinate influences from the Irazu and Poas volcanos. This volcanic complex appears to be faulted against somewhat older and more consolidated volcanic and lacustrine deposits of the Irazu-Reventado Formation beneath the floor of the present valley.

The eroded volcanic cones comprise interbedded lava flows, exhibiting abrupt variations in thickness across pre-existing topographic depressions, and more extensive edge-flow tuffs and ignimbrites, deposited from gaseous clouds.

Geological correlation presents formidable problems because both lavas and pyroclasts vary little in chemical composition and show wide lateral variation in depositional texture and thickness. One pyroclastic deposit (the Puente de Mulas ignimbrite) has distinctive lithology and widespread occurrence, which has considerably aided geological interpretation (Figure 2B).

Groundwater levels throughout this multiaquifer sequence are complex, and correlation between boreholes can

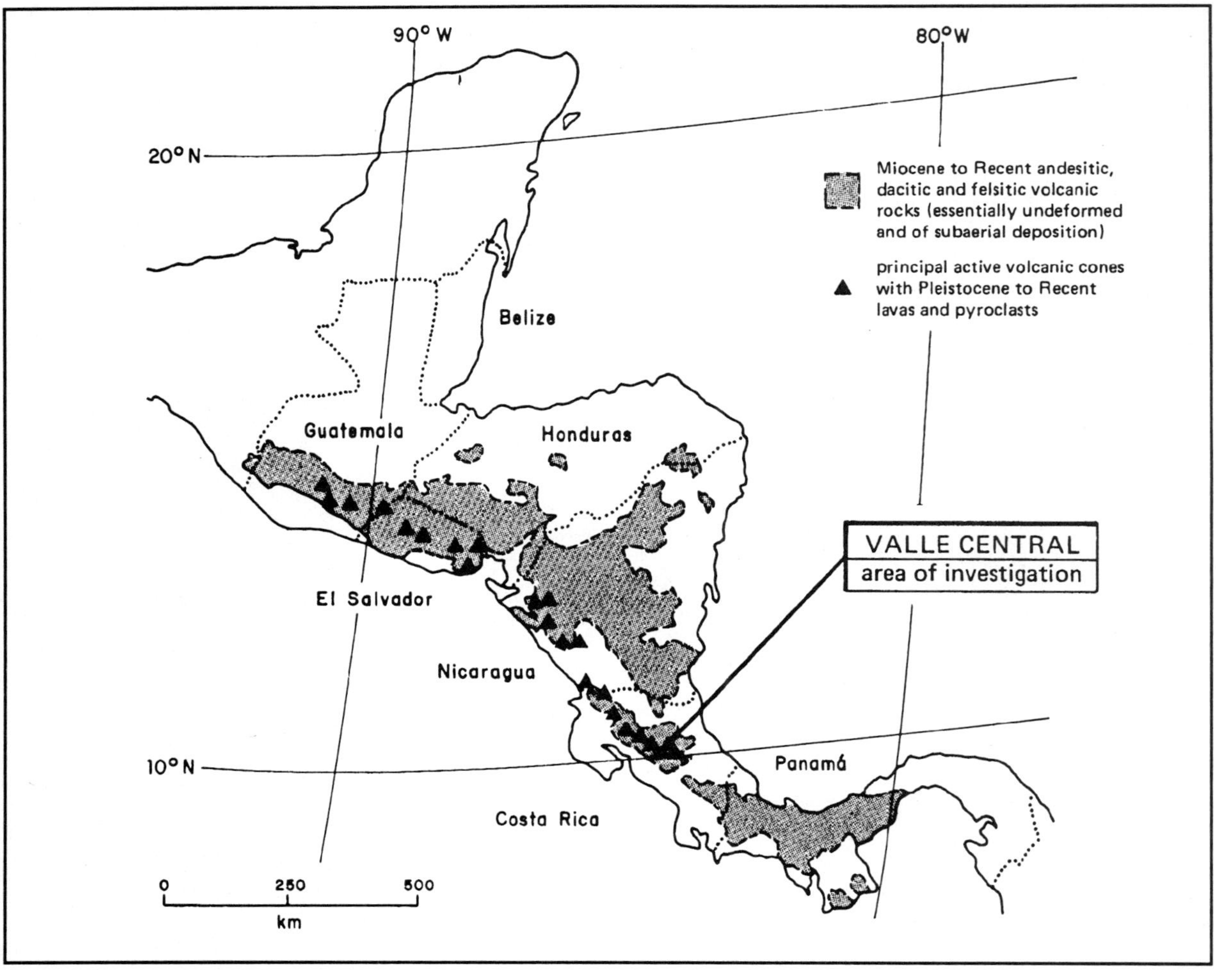

Figure 1. Volcanic province of Middle America (after Weyl, 1980; Circum-Pacific Council for Energy and Mineral Resources, 1984).

be difficult. The most prominent and extensive of the shallow lavas, known as the Barba (Inferior), forms a persistent perched aquifer (Figure 2B). The lavas of the deeper, prolific Colima aquifers have very limited outcrop and must be recharged by large-scale natural leakage, especially from the overlying perched aquifers on the higher ground.

The area is tropical in climate, with high, altitude-dependent rainfall (1,800-3,500 mm/a) falling mainly during May to November, and is drained by both infiltration to groundwater and by surface runoff. There are many small rivers with long, narrow catchments that exhibit complex influent-effluent relationships with underlying aquifers.

KEY FEATURES OF GEOHYDROLOGICAL REGIME

The aquifer system has a number of distinctive features:

1. Aquifers tend to have steeply-sloping bases, as a result of the conical form of volcanic deposition.

2. Although the shallowest perched water tables are subject to marked seasonal fluctuation (more than 5 or 8 m overall), monitoring boreholes in deeper aquifers exhibit little or no temporal variation in water level.

3. Hydraulic gradients in the deeper aquifers are both very steep (often exceeding 0.02) and essentially constant. The latter two features imply steady recharge rates, by implication from downward leakage. The explanation of this behavior was found in the hydraulic properties of the pyroclastic tuffs, which play a key role in the regional groundwater flow regime.

Laboratory testing of drill-core samples revealed extremely and consistently high effective porosity, coupled with moderate intergranular hydraulic conductivity (Figure 3), and pore-size distribution suggesting specific yields in the range 0.10-0.20. These results confirm that:

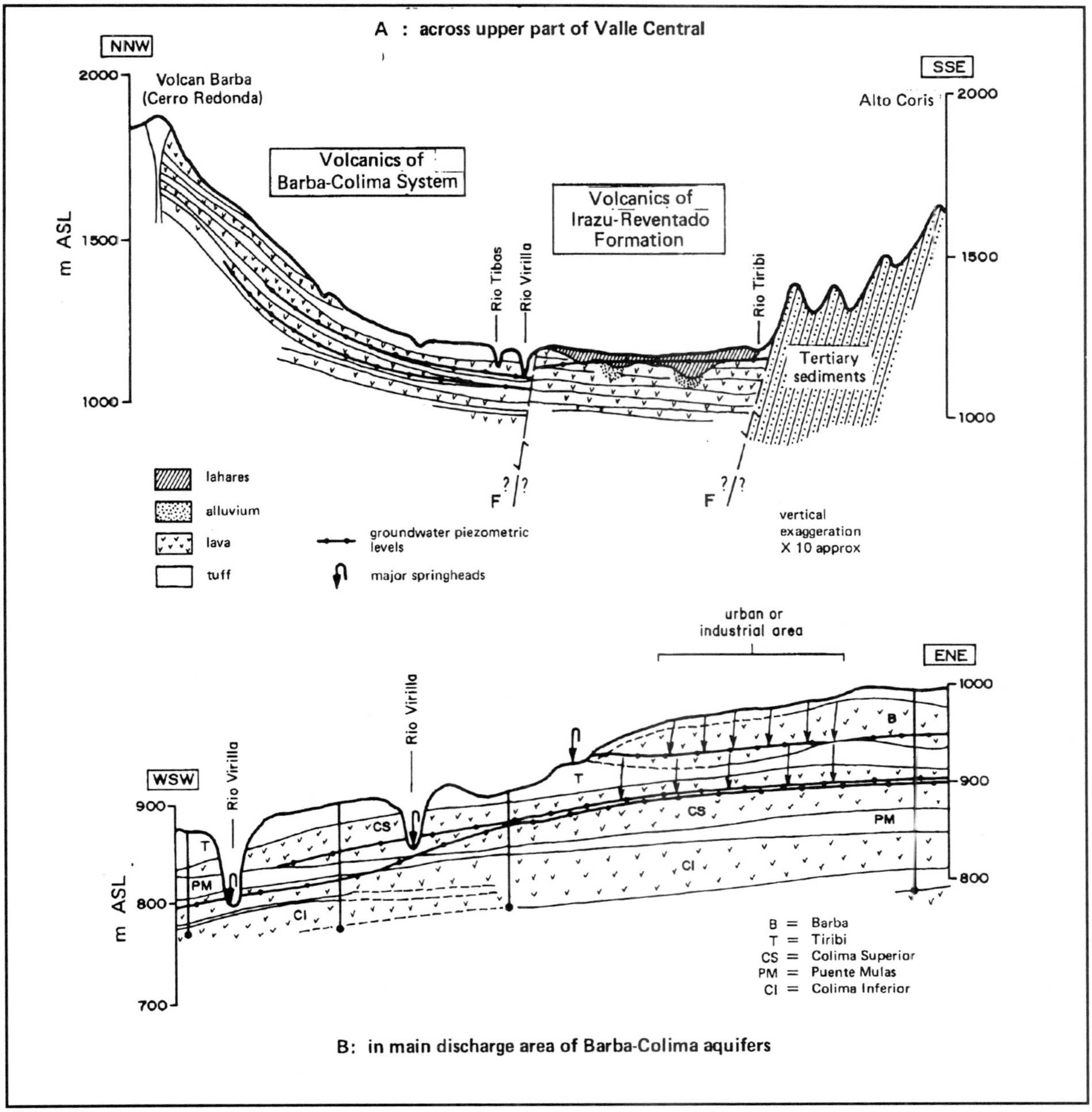

Figure 2. Simplified hydrogeological cross-sections in the Valle Central of Costa Rica (derived from BGS-SENARA, 1985).

1. The tuffs were not the limiting aquitards of the aquifer system responsible for perching; this role is more likely to be played by certain of the less-fractured ignimbrites.

2. The large storage in the various tuffaceous deposits overlying the Colima aquifer, which normally exceed 50 m aggregate thickness, provide steady recharge by natural downward leakage, damping-out seasonal and annual fluctuations in infiltration.

3. Where such tuffs occur immediately above lava aquifers they will increase their unconfined storage coefficient manyfold.

Although tuffaceous deposits unquestionably provide the bulk of the storage capacity of this volcanic aquifer system, it is the lava flows which possess high transmissivity. This is especially true where they are of brecciated or highly fractured texture. More massive sections may, on the other hand, act as aquitards.

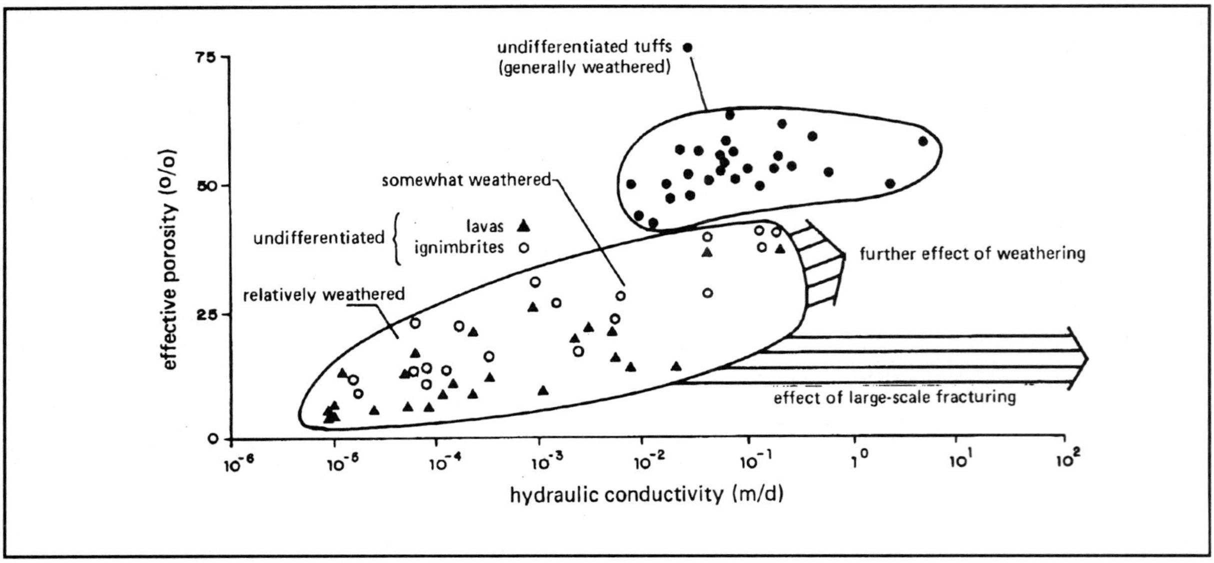

Figure 3. Laboratory hydraulic properties of Quaternary volcanic rocks from the Valle Central of Costa Rica (after Foster et al., 1985).

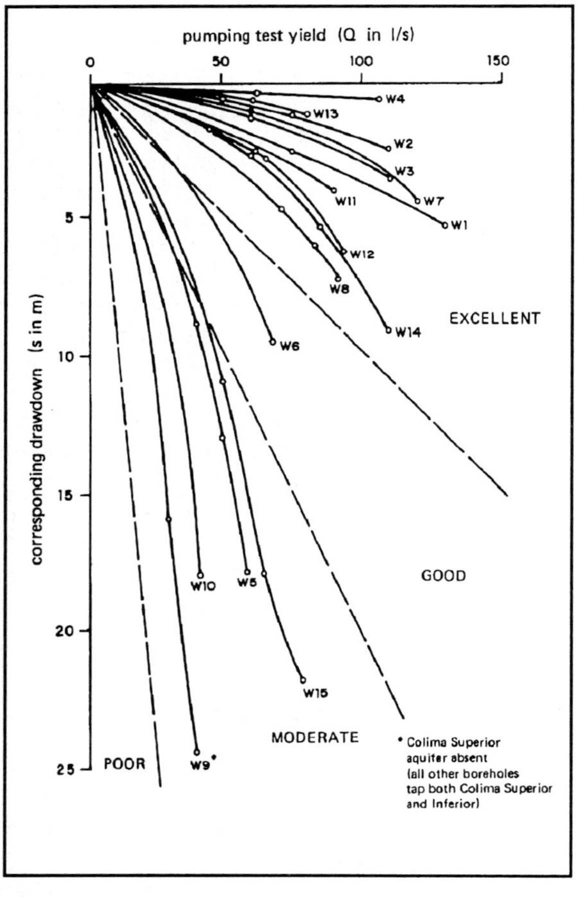

Figure 4. Yield-drawdown characteristics of boreholes in the Valencia wellfield of the Valle Central.

EVALUATION OF GROUNDWATER POTENTIAL

Aquifer pumping tests are difficult to conduct because of the very steep hydraulic gradients and rapid equilibration of individual production boreholes. Transmissivity values generally appear to be greater than 1,000 m^2/d, and in some parts of the Colima aquifers may exceed 5,000 m^2/d. This is witnessed by the excellent performance of production boreholes, which has been quite consistently achieved where both Colima Superior and Inferior aquifers are well developed, despite the gross heterogeneity exhibited by volcanic formations (Figure 4). It should be noted that much more variable performance is exhibited by production boreholes in the shallower Barba aquifer, and that boreholes in the older Irazu-Reventado Formation (Figure 2A) have performance in the poor, or moderate, category.

The variability and complexity of aquifer recharge make groundwater resources very difficult to quantify directly, and even the use of environmental isotopes (2H, 3H, and 18O), while confirming some hypotheses, has raised other questions about recharge mechanisms (Darling et al., 1987).

The evaluation can, however, be approached from the quantification of aquifer discharge because, in the cases of three principal aquifers (the Barba, Colima Superior, and Colima Inferior), the predominant discharge mechanism appears to be a series of major springheads in the lower parts of the valley and along the canyon of the Virilla River incised some 100 m below general land surface in this part of the valley (Figure 2B).

Analysis of baseflow on the Virilla River, together with springflow and differential riverflow gauging in the dry

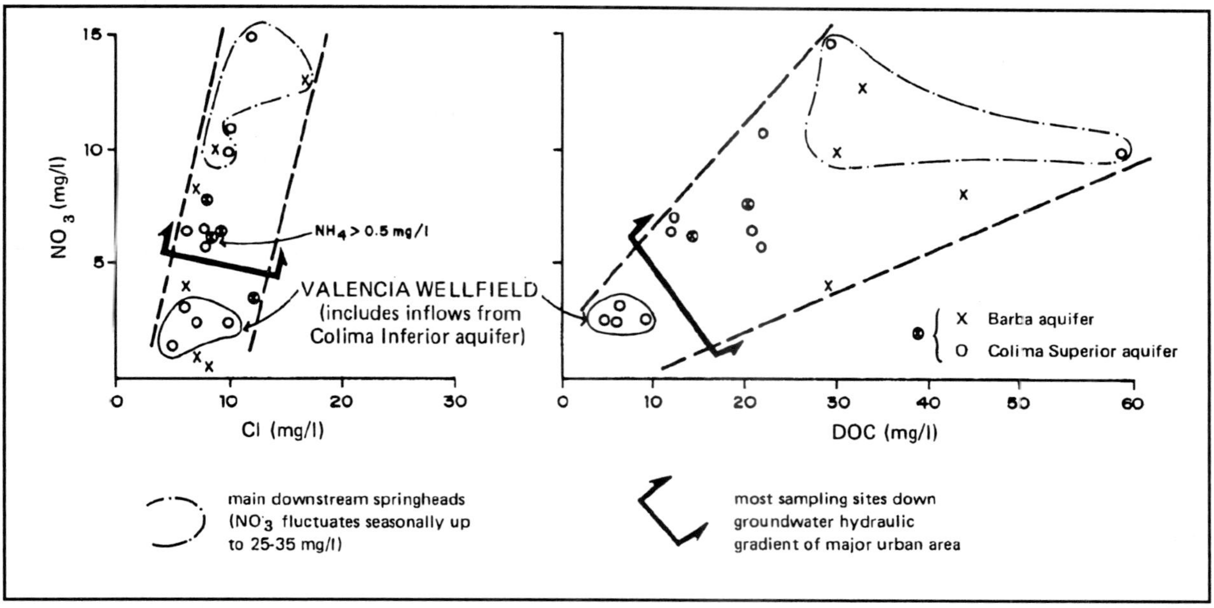

Figure 5. Correlation between NO_3, Cl, and DOC concentrations in groundwater of the Valle Central.

season, suggests that the resources of the Colima aquifers alone are at least 475 Ml/d and probably considerably more than 750 Ml/d.

A numerical two-layered aquifer simulation model, developed from the MODFLOW (McDonald and Harbough, 1983) program, proved useful in testing hypotheses about the behavior of the Colima aquifers and in estimating regional interference effects from possible new wellfields, although it presented much difficulty in formulation due to the complexity of the groundwater flow regime and the limitations of available hydrogeological data. Both regional steady-state and local non-steady-state calibration were attempted, using extrapolated average aquifer piezometric surfaces and drawndown data from a wellfield pumping up to 1,000 l/s over the period 1976-86, respectively.

AQUIFER POLLUTION VULNERABILITY

Field tests on weathered, brecciated Barba lavas recorded terminal infiltration capacities exceeding 200 mm/d. Locally where these lava formations are exposed at the surface, this aquifer is highly vulnerable to pollution even by impersistent contaminants, such as fecal pathogens, if they are discharged to the ground. However, the very high rates of annual infiltration, generally exceeding 1,000 mm/a, result in considerable dilution of nutrients and pesticides leached from coffee plantations, and serious diffuse groundwater pollution from this source appears unlikely to develop.

Tuffaceous deposits occupy much larger areas of the land surface than do lavas. Although less permeable, they exhibit field infiltration capacities in the range of 50-500 mm/d, which suggest they will be capable of accepting most rainfall. Surface infiltration of rainfall, therefore, appears to be the principal source of groundwater recharge to shallow aquifers, with an additional component from riverbed seepage on the higher ground.

The physical and mineralogical character of the tuffs means that they are likely to retard and/or to eliminate all but the most mobile and persistent types of contaminant. They thus offer a high degree of natural protection of underlying aquifers, especially the Colima formation. Indeed, the deeper groundwaters exhibit excellent natural quality, remarkably low in all dissolved constituents, both major ions and trace elements, with the exception of high silica (SiO_2 more than 60 mg/l) and moderate bicarbonate (80-120 mg HCO_3/l).

The subsurface contaminant load in parts of the Valle Central is considerable as a result of (1) soil discharge from on-site sanitation in almost all urban areas north of the Virilla, except for limited sectors of Heredia and Alajuela, and (2) industrial effluent discharge to the ground and to minor influent rivers, many of which already have a heavy sewage effluent load.

Monitoring of the Valle Central aquifers reveals a correlation between nitrate (NO_3) and chloride (Cl) concentrations (Figure 5), with levels increasing down hydraulic gradient of the main urban-industrial areas, suggesting that groundwater quality is being affected, directly or indirectly, by discharge of urban effluents. Moreover, NO_3 concentrations in the major downstream springheads of the shallower aquifers exhibit marked seasonal variation with maxima soon after the commencement of the wet season. This suggests leaching of nitrogenous material accumulating in the soil, which could be derived from agricultural practices, but in view of the spatial distribution of the higher values it is more likely to derive from urban, unsewered sanitation.

The implication of groundwater contamination from effluent discharge is compounded by the broad correlation between NO_3 and dissolved organic carbon (DOC) concentrations also. These values give grounds for concern about groundwater pollution, but limited follow-up sampling and analysis has not detected any of the toxic synthetic organic compounds in widespread industrial use.

CONCLUSIONS RELEVANT TO OTHER VOLCANIC AREAS

1. Under certain conditions of Quaternary subaerial volcanic deposition, which favor brecciation of lava flows, aquifers of high transmissivity can be formed.

2. Although these lava aquifers themselves have limited groundwater storage capacity, their storage characteristics are greatly enhanced if they are overlain directly or interbedded with non-welded tuffs, the net result being a high-yielding, drought-secure aquifer system of near-optimum hydraulic characteristics.

3. A surface cover of tuffaceous deposits offers a considerable degree of natural protection against groundwater pollution, but there are indications that heavy surface loading of persistent pollutants may still penetrate to underlying aquifers.

4. The hydrogeological characteristics of this andesitic volcanic complex differ widely from those of the Pacific ocean-ridge basaltic volcanos of Hawaii (Takasaki, Hirashima, and Lubke, 1969).

ACKNOWLEDGMENT

This paper is based on selected results of a long-term hydrogeological investigation program in the Valle Central of Costa Rica, carried out jointly by the British Geological Survey (BGS) and the Costa Rican Servicio Nacional de Aguas, Riego y Avenamiento (SENARA) for the Instituto Costarricense de Acueductos y Alcantarillados (ICAyA), and funded initially by the British Overseas Development Administration and later by the World Bank.

REFERENCES

BGS-SENARA, 1985, Hydrogeological map of the Valle Central of Costa Rica.

Circum-Pacific Council for Energy and Mineral Resources, 1984, Mineral resource map of circum-Pacific region: Tulsa, AAPG.

Darling, W. G., J. M. Parker, H. V. Rodriguez-Estrada, and A. J. Laudres, 1987, Investigations of a volcanic aquifer system in Costa Rica using environmental isotopes, in Proceedings, IAEA seminar "Application of isotope techniques in hydrology in Latin America," Mexico City, September, in press.

Foster, S. S. D., A. T. Ellis, M. Losilla-Penon, and H. V. Rodriguez-Estrada, 1985, Role of volcanic tuffs in groundwater regime of Valle Central, Costa Rica: Ground Water, v. 23, p. 795-801.

McDonald, M. G., and A. W. Harbaugh, 1983, A modular three-dimensional finite-difference ground water-water flow model (MODFLOW): U.S. Geological Survey Open-File Report 83-875.

Takasaki, K. J., G. T. Hirashima, and E. R. Lubke, 1969, Water resources of windward Oahu, Hawaii: U.S. Geological Survey Water-Supply Paper 1894, 119p.

Weyl, R., 1980, Geology of Central America: Stuttgart, Berlin, 371 p.

Circum-Pacific Council for Energy and Mineral Resources
Earth Science Series, Vol. 16
R. L. Miller, G. Escalante, J. A. Reinemund, and M. J. Bergin (Eds.)
Energy and Mineral Potential of the Central American-Caribbean Regions
Springer-Verlag, Berlin Heidelberg 1995

STRUCTURE AND DEPOSITIONAL PATTERNS AND THEIR INFLUENCE ON THE HYDRAULIC CONDUCTIVITY OF FAN-DELTAS IN SOUTHERN PUERTO RICO

Robert A. Renken
U.S. Geological Survey, Water Resources Division, 6481-B Peachtree Industrial Boulevard, Doraville, Georgia 30360 USA

Fernando Gómez-Gómez, Vicente Quiñones-Aponte, and Rafael Dacosta
U.S. Geological Survey, Water Resources Division, G.P.O. Box 4424, San Juan, Puerto Rico 00936 USA

ABSTRACT

Subsurface data and results of surface geologic mapping indicate that areal variations in hydraulic conductivity of water-bearing strata in the semiarid alluvial plain of southern Puerto Rico may be controlled by geologic structure and depositional patterns. Six fan-deltas coalesce to form a narrow, 3- to 8-km-wide alluvial plain that extends eastward 70 km from Ponce to Patillas. These unconsolidated coastal alluvial fan deposits consist of a moderately thick (50 to 1,000 m) clastic sequence of Holocene to Miocene(?) age. The fan-deltas are bordered to the north by foothills and mountains of the Cordillera Central and to the south by the Caribbean Sea. Bedrock hills protrude through thinner parts of the alluvial plain, and northwest-trending strike-slip faults and northeast-trending cross faults of the great southern Puerto Rico fault zone extend beneath the alluvial cover.

The reported hydraulic conductivity of the discrete water-bearing strata ranges from less than 5 to greater than 100 m per day. The more permeable strata of these fan-deltas consist of sheet-like to channelized deposits of massive to planar stratified and planar cross-stratified cobbles, gravel, and coarse to fine sand. These permeable sediments are interstratified with less permeable, planar stratified to massive beds of silt and some clay. Sand percentage maps correlate with hydraulic conductivity values estimated from specific-capacity data. The highest hydraulic conductivity values usually are associated with proximal and midfan areas and with buried stream channels. The lowest conductivity values usually are associated with distal or interfan areas.

A pull-apart basin tectonic mechanism helps explain both the horst and graben structures that underlie these fan-deltas and also the wide variation in aquifer thickness. Subsurface geologic facies maps provide an effective method to extrapolate aquifer characteristics. These maps can be used as an indirect method to verify or extrapolate hydraulic conductivity estimates needed to calibrate a digital ground-water flow model.

INTRODUCTION

The mineral resource potential of a fan-delta depositional system is not limited to hydrocarbon or placer deposits. Fan-deltas also are important as aquifers. Unconsolidated to poorly consolidated deposits of sand and gravel, which underlie Puerto Rico's southern fan-delta plain, function as a principal aquifer for agricultural, industrial, and municipal users. Approximately one-half of all ground water withdrawn on the island comes from this alluvial plain (Gómez-Gómez, Dacosta, and Orona, 1983).

Purpose and Scope

The water-resource potential of southern Puerto Rico's fan-delta plain is currently (1989) a part of the U.S. Geological Survey's Caribbean Islands Regional Aquifer-System Analysis (CI-RASA) study (Gómez-Gómez, 1987). The primary objectives are to (1) assess hydrologic conditions within the major aquifers, (2) identify and delineate the distribution of hydraulic conductivity, (3) describe the hydrochemistry, and (4) simulate the pattern of ground-water flow by use of a digital model.

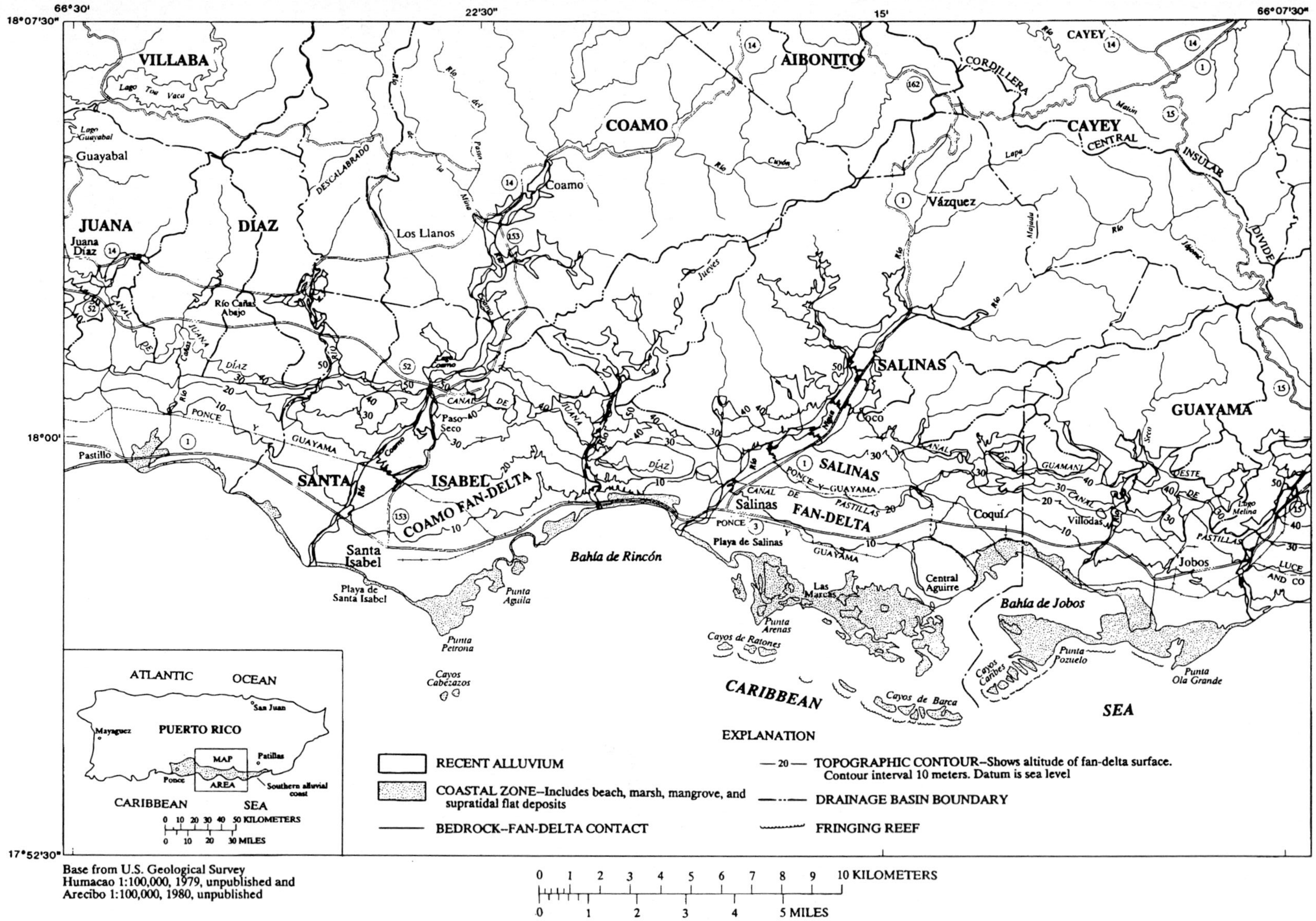

Figure 1. Surficial morphology of the Coamo and Salinas fan-deltas.

This article provides a framework that shows region-wide variations in thickness and hydraulic conductivity within the alluvial plain aquifer system. Subsurface geologic and hydrologic maps are presented for the Coamo and Salinas fan-deltas that are located in the central part of the southern fan-delta plain. A digital-flow model requires several types of hydrogeologic maps. Bedrock-surface and water-table maps are used to estimate the saturated thickness of the aquifer. Hydraulic conductivity maps are used to show variations in aquifer permeability.

General Setting

Six fan-deltas coalesce to form a narrow, but continuous, alluvial plain that extends eastward 70 km from Ponce to Patillas (Figure 1, inset map) on the southern Caribbean coast of Puerto Rico. The alluvial plain is bordered to the north by intensely faulted Cretaceous and Tertiary highlands of the Cordillera Central, which also form the principal insular drainage divide. Tributary streams in the mountains and foothills of the Cordillera Central transport the poorly-sorted clastic material that comprises the alluvial deposits. Fan-delta deposits (Figure 2) form a coastward-thickening sequence of bedded cobble, gravel, sand, and silt deposits of Quaternary age. Where found in the deeper subsurface, these deposits may be as old as Miocene age (Glover, 1971). Bedrock hills protrude through thinner parts of the alluvial plain, especially along the northern margin where the alluvial cover thins against the foothills.

The Coamo and Salinas fan-deltas are moderately large fan-deltas (10 km radii) (Figure 1). Both fan-deltas exhibit conical or radial surficial morphology which is a distinctive characteristic common to fan-deltas and alluvial fans. The Salinas and Coamo fan-deltas are drained by the Ríos Nigua and Coamo, respectively. The Ríos Nigua and Coamo are single, fairly straight, trunk streams that extend southwest from the fan apex. Prior to channelization, the confining channels of many streams crossing the southern fan-delta plain disappeared in the distal areas. This allowed surface water to spread out as sheetflow onto the coastal lowlands.

The subareal parts of the Coamo and Salinas fan-deltas are separated from the Caribbean Sea by a narrow coastal zone consisting of mangrove swamp, marsh, supratidal flat, and beach deposits. Terrigenous beaches of cobble-, gravel-, and sand-size material line the coast south of the Coamo fan-delta. However, beaches along the coastal margin of the Salinas fan differ lithologically and texturally, and they consist mostly of carbonate sand that has been derived from nearby fringing reefs.

The Cordillera Central mountains of Puerto Rico effectively cut off much of the precipitation from the northeast trade winds and create a rainshadow on the leeward southern side of the island. Rainfall averages less than 1,000 mm/yr on the southern side of the island. Near the coast and the westernmost part of the alluvial plain, precipitation averages about 500 mm/yr.

Ground water within the southern alluvial plain occurs under confined and unconfined conditions. Confined conditions are more prevalent in deeper parts of the alluvial section and near the Caribbean coast where water-bearing strata are overlain by less permeable, fine-grained materials. Water-table conditions dominate inland. Terrigenous deposits of cobbles, gravel, and sand, which comprise the principal water-bearing units, are interbedded with thick beds of silt and some clay. The silt and clay deposits impede the vertical movement of ground water and contribute to downfan confined conditions.

Under pre-development conditions, the aquifer was recharged principally by stream seepage mostly near the fan apex, and by infrequent rainfall. Discharge from the aquifer primarily occurred as seepage to the sea and along the lower reaches of stream channels, or as direct evapotranspiration in areas where shallow water-table conditions occurred. However, development of ground water and surface water has considerably modified the hydrologic system. Currently, recharge to the aquifer includes 30 percent of the water used for irrigation (Gómez-Gómez, 1987). The principal discharge from the aquifer is to wells, which are reported to yield 320,000 m^3/d.

GEOLOGY

Driller's and other well log data for more than 400 water wells were used to estimate the thickness of the alluvium, map the configuration of the bedrock surface, and quantify the textural and hydraulic character of water-bearing strata of the fan-delta plain. Many wells located in the study area do not reach the bedrock surface (reported as "hard, blue or gray" rock). These data are useful, however, as negative control points. In these wells, the altitude or the depth to the bedrock surface exceeds the total depth of the well. Some wells penetrate a sequence of weathered clay, silty clay, or silty limestone (reported as "Tosca"). McClymonds and Ward (1966) interpreted these deposits as a weathered zone overlying the bedrock and used the well data to construct their "buried ridges and valleys" contour maps. The structure map shown in this paper differs from McClymonds and Ward (1966). It is based, in part, on geologic maps for the adjacent highlands (Glover, 1971), and has resulted in an improved understanding of local and regional tectonics.

Structure

The great southern Puerto Rico fault zone (Glover, 1971) is a principal geologic structure of the island. It consists of a parallel series of northwest-striking, sinistral (left lateral), and northeast-striking cross faults that form a belt of horsts and grabens. More than 14 strike-slip faults have been mapped in the Cordillera Central highlands adjacent to the southern alluvial plain. Some extend to the southeast beneath

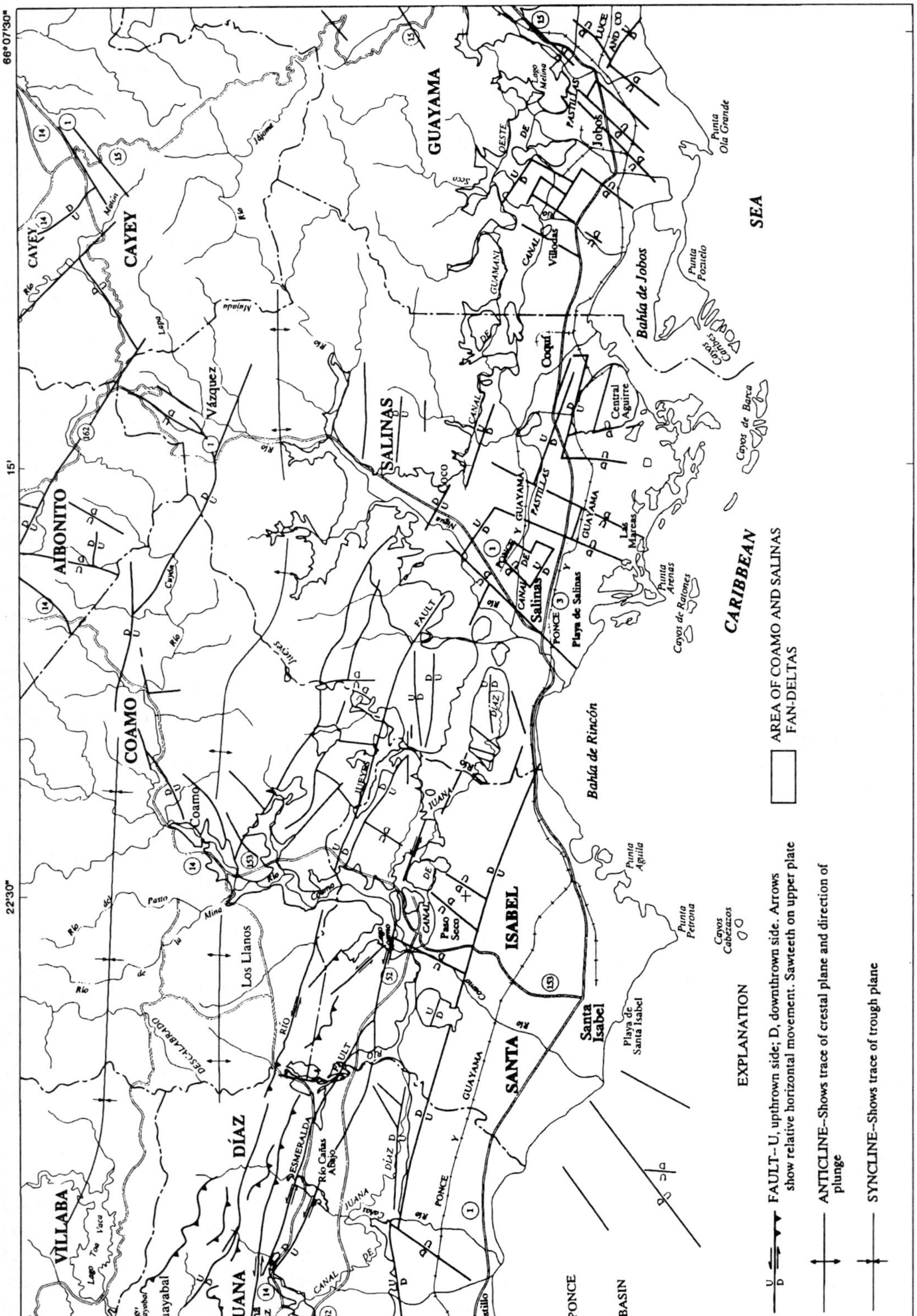

Base from U.S. Geological Survey
Humacao 1:100,000, 1979, unpublished and
Arecibo 1:100,000, 1980, unpublished

Figure 2. Location of major structural features in Coamo and Salinas areas.

the alluvial plain (Figure 2). A structure contour map of the base of the fan-delta deposits indicates that the fans are underlain by two different fault sets (Figure 3). One set of faults parallels, and is partly an expression of, the northwest-striking, sinistral faults. Southeastern extensions of the Rio Jueyes and the Esmeralda faults, the two northernmost left-lateral, strike-slip faults, do not underlie the alluvial plain as a continuous trace or resemble the mountainous branched pattern. Discontinuous faults extend southeast beneath the Salinas fan-delta possibly in a left-stepping, en echelon pattern. This fault set is complemented by a second set that strikes normal to them. These latter faults strike northeast and parallel a similar fault set mapped in the mountainous areas to the north. The pattern of faults and bedrock highs and lows suggests that a complex of horst and graben structures also lies beneath the Salinas fan-delta. Bedrock hills adjacent to the Salinas fan-delta are possible horst structures. A conceptual representation of the postulated structural pattern beneath the Salinas fan-delta is presented in Figure 4. Major horst and graben structures are shown as uplifted or downdropped blocks. Tensional stress between major en echelon strike-slip faults has resulted in bedrock low or basinal areas. Compressional stress at the ends of these same faults has resulted in bedrock highs or uplifted blocks. The pattern and the orientation of these en echelon faults and complementary cross faults is considered indicative of a pull-apart basin tectonic mechanism.

To the southwest, an unnamed fault parallels and lies south of the Rio Jueyes and Esmeralda faults, and it extends west-northwest across the Coamo fan-delta and beyond (Figure 3). Dip-slip movement is to the south, and vertical displacement exceeds 300 m in places. This fault may represent the landward limit of the offshore Ponce basin.

Depositional Patterns

On the basis of outcrop descriptions and core and lithologic well descriptions, subareal fan-delta deposits were separated into two types of lithofacies. The coarse-grained facies is herein defined as bedded deposits of boulder, cobble, gravel, and sand. The dominant bedforms found where the deposits crop out along the Río Coamo include massive, planar, and planar cross-stratified. Trough cross beds are less common, but are found in an outcrop area within the proximal (near the source) channel near Coamo. Core data indicate that upward-fining cycles occur within individual coarse-grained beds. Silt dominates the fine-grained facies, but some clay also is included. Beds along the Río Coamo range from massive to planar stratified, and locally may contain interbeds of fine to coarse sand and gravel. Core data collected west of the Coamo fan indicate an upward thickening of coarse-grained beds and an upward thinning of fine-grained beds. In combination, these data imply the fans are part of a coarsening-upward, prograding sequence.

Percentage facies maps were constructed to examine the spatial variability of the coarse-grained lithofacies. The coarse-grained lithofacies is expressed as a percent of the total section of alluvium; an assumption made was that the estimated percentage reflected the "coarseness" even for wells that did not penetrate bedrock. The resulting map helps illustrate downfan variations in grain size (Figure 5). Coarse-grained deposits dominate and extend downfan within a narrow proximal channel. Floodwaters from this channel cover the floodplain in the midfan areas. Where floodwaters diminish in depth and velocity, coarse detritus is deposited as a lobe. The orientation of the proximal channel and midfan lobe is largely perpendicular to the strike-slip faults and parallel to the cross faults. Percentage maps suggest that the coarse-grained facies constitute 40 percent or more of the entire alluvial section in the proximal and midfan areas. Conversely, the fine-grained facies is far more prevalent in the distal fan and interfan areas where coarse-grained beds constitute less than 20 percent of the total alluvial sequence. The lobate pattern associated with the Salinas fan-delta may suggest that the main depositional lobe extends beyond the modern coast, and distal deposits may lie buried in the offshore subsurface. Buried channel-fill deposits extend from the main Salinas depositional lobe as finger-like projections. These buried channels may once have fed fan segments that are now abandoned. There is a close correlation between the Esmeralda fault and an abandoned stream channel that lies north of the bedrock hill near Central Aguirre. It is possible that the channel shifted here in response to movement along the Esmeralda fault.

HYDRAULIC CONDUCTIVITY

The areal distribution of hydraulic conductivity or transmissivity is important to understanding the hydrogeologic system in constructing ground-water flow models and in simulating dissolved-mineral transport. Because of the high cost, few aquifer tests have been conducted in the Coamo and Salinas fan-deltas. Transmissivity estimates from available aquifer test results are representative of discrete water-bearing horizons within fairly localized areas. However, transmissivity estimates based on specific-capacity are available for most wells in the study area. Specific-capacity is defined as the yield per unit of water-level drawdown in a pumped well for a specific pumping rate. Estimates of transmissivity for the two fans were made from specific-capacity data using empirical equations developed by Theis (1963) and Brown (1963). The distribution of hydraulic conductivity shown in Figure 5 was computed by dividing the estimated transmissivity value by the thickness of the aquifer penetrated by the well. It should be noted, however, that the map showing the areal distribution of hydraulic conductivity is subject to qualification. It is based on specific-capacity methods which assume 100 percent well efficiency, full penetration of the aquifer, a 15 to 45 cm well diameter, a storage coefficient of

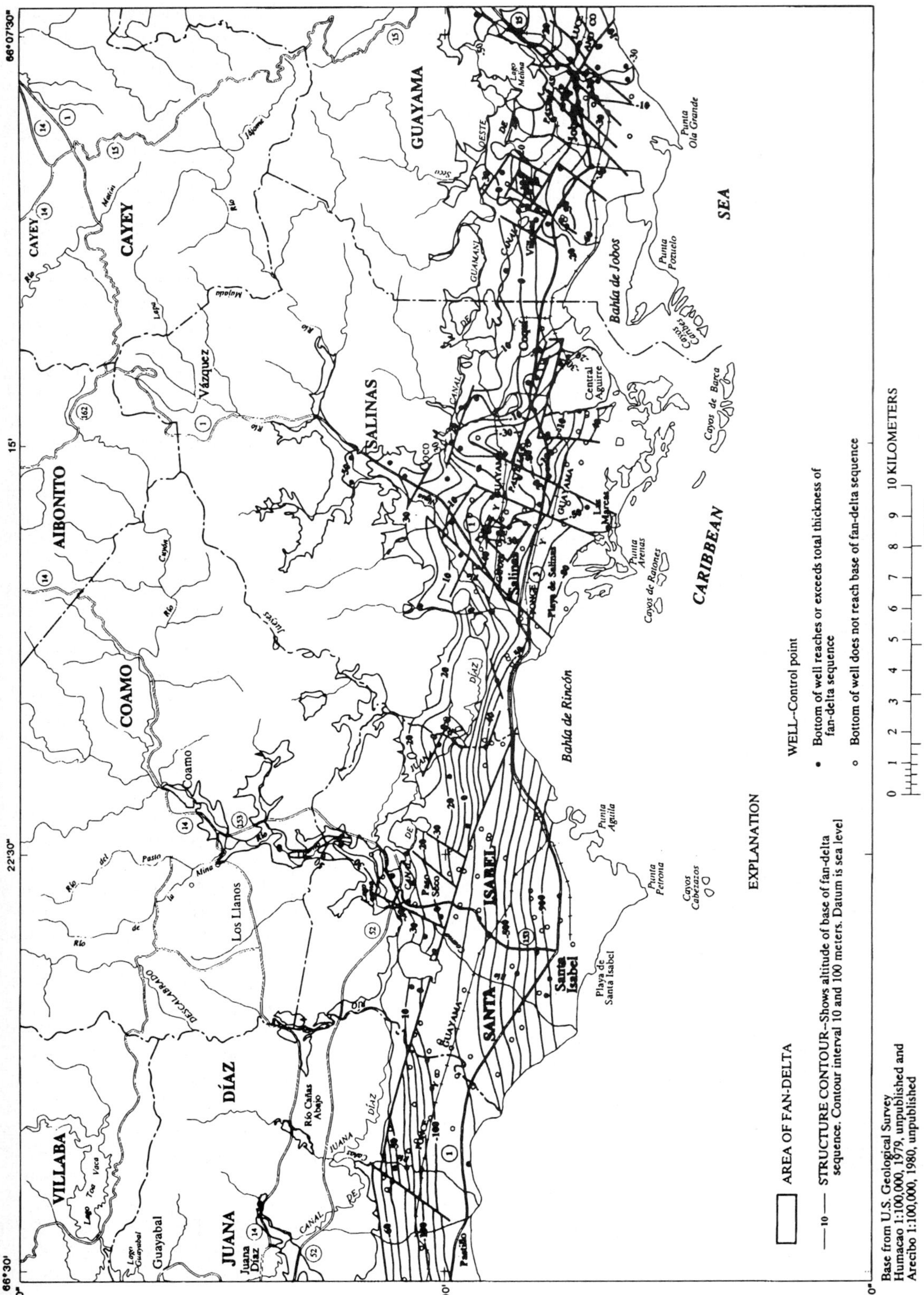

Figure 3. Configuration of the bedrock surface underlying the Coamo and Salinas fan-deltas.

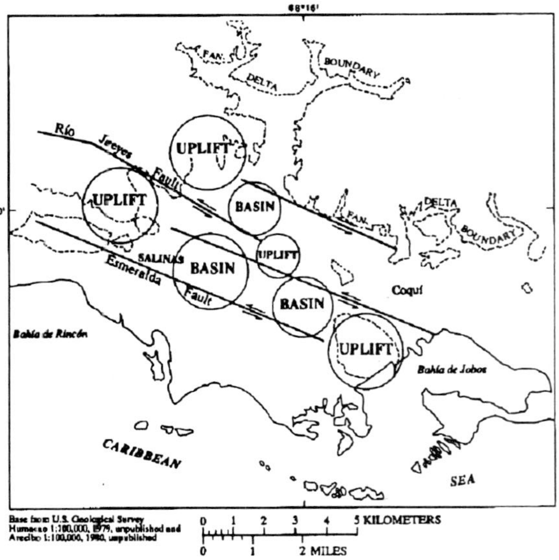

Figure 4. Idealized left-lateral pull-apart basins underlying the Salinas fan-delta produced by en echelon strike-slip faults.

1×10^{-4} for wells subject to artesian conditions or 0.15 for wells subject to water-table conditions, and a 24 hour pumping time.

DISCUSSION

Influence of Structure on Aquifer Thickness

The thickest part of the alluvial plain sequence is found where fan-delta deposits have infilled cross-cutting graben structures. The total thickness of alluvial deposits in the southern part of the Coamo fan-delta may, in places, exceed 900 m. The thickness of the alluvial sequence underlying the Salinas fan-delta exceeds 120 m just west of a bedrock hill at Central Aguirre. The section thins to 20 to 60 m where these deposits overlie the narrow ridge dividing the Salinas graben into two smaller blocks. Fan-delta deposits are thinnest in the proximal area and the narrow upland channel that extends into the foothills. In the Salinas and Coamo areas, the total thickness of deposits in the upland channel usually does not exceed 20 m. Fan-delta deposits also thin to a feather edge where they lie adjacent to upthrown bedrock hills.

Influence of Deposition on Hydraulic Conductivity

Lithofacies maps were used to indirectly corroborate the distribution of hydraulic conductivity shown for the central part of the southern fan-delta plain (Figure 5). The estimated conductivity for the Coamo and Salinas fan-deltas ranges from less than 1 to greater than 30 m/d. The sand and gravel percentage map was compared with the hydraulic conductivity map. The highest conductivities associated with the Coamo and Salinas fan-deltas are found in the proximal and midfan areas, which lie near the present position of the Río Coamo and Río Nigua and are underlain by sediments containing high percentages of sand and gravel. An area of moderate hydraulic conductivity (7.5 to 30 m/d) occurs parallel to the abandoned stream channel and the buried extensions of the Esmeralda fault north of the bedrock hill at Central Aguirre. The lowest hydraulic conductivity values (1 to 7.5 m/d) are found in distal fan and interfan areas. Less-permeable, distal fan deposits are important hydrologically because they minimize and/or retard saltwater intrusion from the deeper subsurface and from landward encroachment from the Caribbean Sea.

CONCLUSION

Understanding the physical framework of a fan-delta can prove useful in evaluating aquifer characteristics. In southern Puerto Rico, tectonic factors control aquifer thickness and the pattern of lithofacies influences hydraulic conductivity. Fan-deltas of southern Puerto Rico comprise an elongate depositional basin that is nearly parallel to the west-northwest structural grain of the great southern Puerto Rico fault zone. Strike-slip faults in the highlands extend southeast beneath the alluvial cover in an en echelon pattern, and they are complemented by cross faults that strike northeast. A left-lateral pull-apart basin tectonic framework is postulated to explain the horst and graben structures that underlie the southern coastal plain. Fan-deltas infill and occupy cross-cutting graben structures. Principal trunk streams that occur normal to the strike-slip faults concentrate coarse-grained material either as a narrow proximal channel near the fan apex or as a midfan lobe. Fine-grained deposits of silt and some clay are prevalent in the distal or interfan areas.

Coarse-grained deposits are the principal water-bearing units in the alluvial coastal plain. The hydraulic conductivity of these deposits ranges from 7.5 to greater than 30 m/d. Fine-grained, distal or interfan deposits impede the vertical movement of water, and they contribute to downfan artesian conditions. These deposits also restrict the landward and upward movement of saline waters from the ocean or from deeper parts of the alluvial section. The hydraulic conductivity of these deposits ranges from 1 to 7.5 m/d. Limited aquifer test data make it difficult to impossible to map regional variations in hydraulic conductivity accurately. Geologic facies maps can be used to help evaluate the spatial variation of coarse- and fine-grained strata, and, indirectly, can be used to help quantify and verify the distribution of hydraulic conductivity within a coastal alluvial fan or fan-delta sequence.

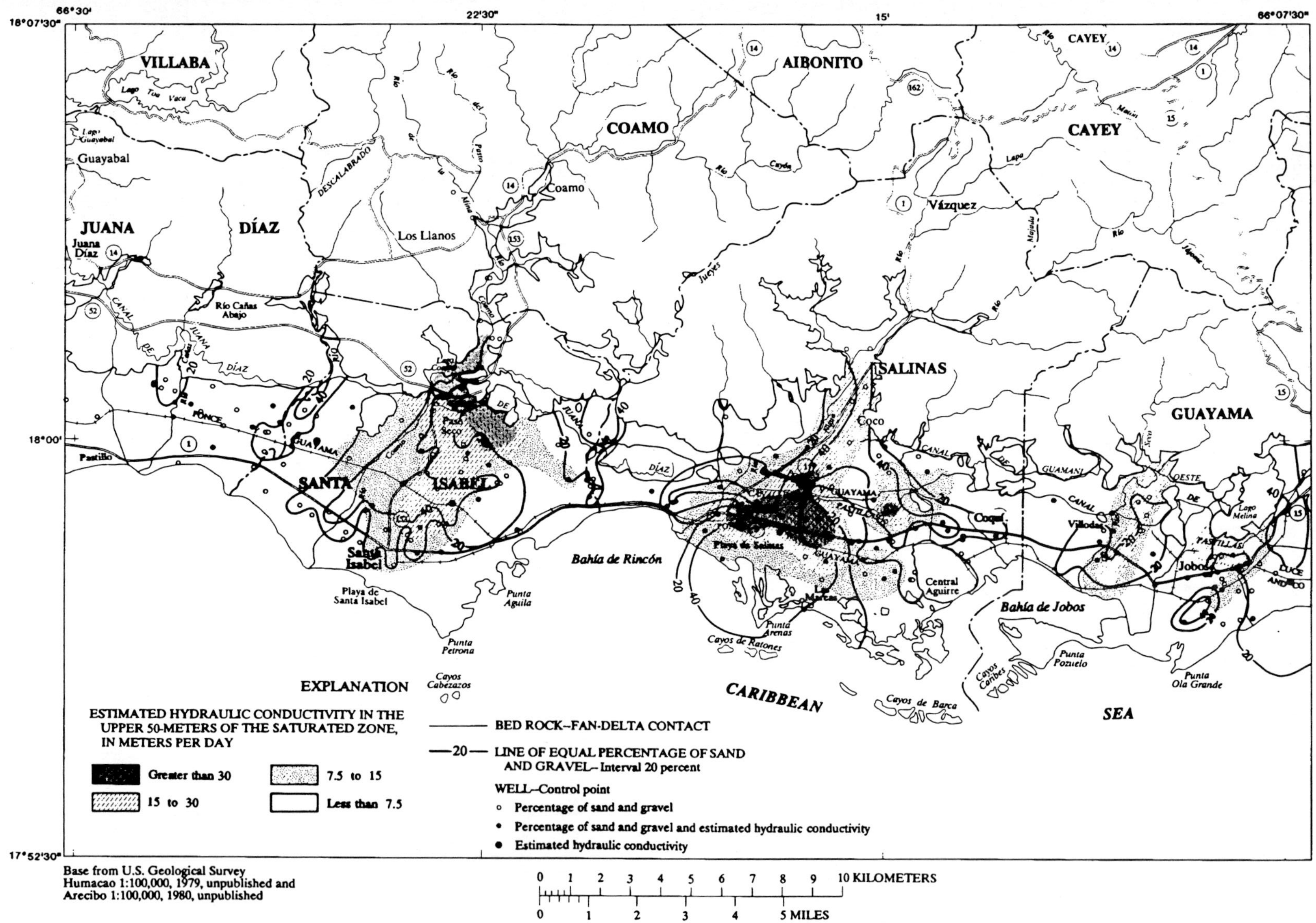

Figure 5. Comparison of estimated hydraulic conductivity and the percentage of sand and gravel in the Coamo and Salinas fan-deltas.

REFERENCES

Brown, R. H., 1963, Estimating the transmissibility of an artesian aquifer from specific capacity of a well, *in* R. Bentall, Methods of determining permeability, transmissibility and drawdown: U.S. Geological Survey Water-Supply Paper 1536-I, p. 336-338.

Glover, L., III, 1971, Geology of the Coamo area, Puerto Rico, and its relation to the volcanic arc-trench association: U.S. Geological Survey Professional Paper 636, 102p.

Gómez-Gómez, F., 1987, Planning report for the Caribbean Islands Regional Aquifer-System Analysis Project: U.S. Geological Survey Water-Resources Investigations Report 86-4074, 50p.

Gómez-Gómez, F., R. Dacosta, and M. Orona, 1983, Estimated water use in Puerto Rico, 1983: Puerto Rico Department of Natural Resources, Water Resources Division, Miscellaneous Map Series, 1 sheet.

McClymonds, N. E., and P. E. Ward, 1966, Hydrologic characteristics of the alluvial fan near Salinas, Puerto Rico, *in* Geological Survey Research: U.S. Geological Survey Professional Paper 550-C, p. 231-234.

Theis, C. V., 1963, Estimating the transmissibility of a water-table aquifer from the specific capacity of a well, *in* R. Bentall, Methods of determining permeability, transmissibility and drawdown: U.S. Geological Survey Water-Supply Paper 1536-I, p. 332-336.

Circum-Pacific Council for Energy and Mineral Resources
Earth Science Series, Vol. 16
R. L. Miller, G. Escalante, J. A. Reinemund, and M. J. Bergin (Eds.)
Energy and Mineral Potential of the Central American-Caribbean Regions
Springer-Verlag, Berlin Heidelberg 1995

KARST LOCALITIES IN COSTA RICA AND THEIR GEOLOGIC BACKGROUND

Sergio Mora C.
Eng. Geologist, Department of Geology, Instituto Costarricense de Electricidad
Escuela C. A. Geología-CIGEFI; INGEOSA.

ABSTRACT

Sixteen karst areas, developed in limestone that ranges in age from Cretaceous to Pleistocene, are known to exist in Costa Rica. Whereas their geologic setting is known as a result of regional stratigraphic reconnaissance, detailed exploration of most of the karst areas has not been accomplished, with one exception. Information on basic geologic conditions, geomorphology, speleology, ecology, and hydrogeology is available for the Barra Honda karst.

General development of karst systems in Costa Rica has reached the young to moderately mature stage. They exhibit many of the classic karst features. The available data suggest that many more discoveries are waiting to be made.

INTRODUCTION

Geologically, Costa Rica is a relatively young country. The oldest rocks yet found belong to the Late Mesozoic Nicoya Complex, an ancient sea floor/mantle-derived suite of rocks. Since the Late Mesozoic, strong tectonic events have resulted in accretion, faulting, and uplift of these oldest rocks, with subsequent deposition of oceanic and neritic sediments. Climatic and bathymetric conditions at times allowed the development of broad carbonate platforms in which the karst terrains later formed. An attempt will be made throughout this paper to describe the stratigraphy, structure, and geomorphology of the karst terrains.

GEOLOGIC AND TECTONIC SETTING

Several geotectonic events have shaped the landscape of Costa Rica, each of which is represented by individual rock units and by unique fault systems.

Pre-Campanian

In pre-Campanian time, sediments and ophiolitic basalts originated at an oceanic rift, and a primitive island arc was formed. This oceanic crust, as well as rocks derived from the oceanic crust and mantle (e.g., Nicoya Complex and Santa Elena ultramafics) were strongly deformed during the Santonian by north-south compression, under which large nappes were generated (Baumgardner, 1984). These units represent the underlying basement for karst-generating limestones in some areas.

Post-Santonian

By late Santonian, the Middle America Trench was established as an extensional-convergent margin (Aubouin et al., 1979), probably with no more accretion. During the Paleocene-Eocene, andesitic volcanism began in the arc area

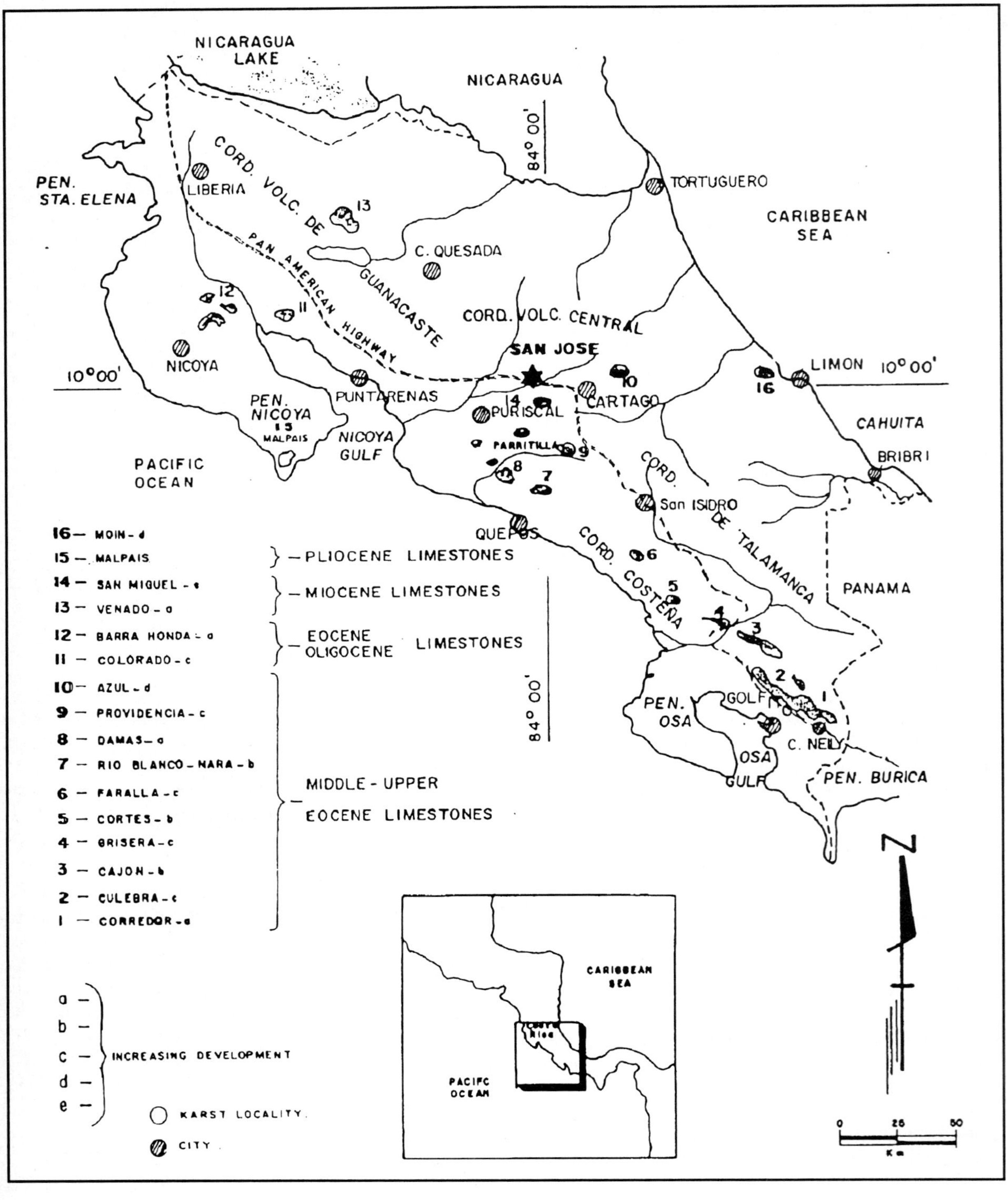

Figure 1. Karst localities in Costa Rica. Ranking reflects degree of karst development.

as a result of subduction. Some areas, later uplifted during strong tectonic events, were the sites upon which formed the carbonate platforms and reefs that became the source of limestones where karst would develop. Post-Oligocene up-

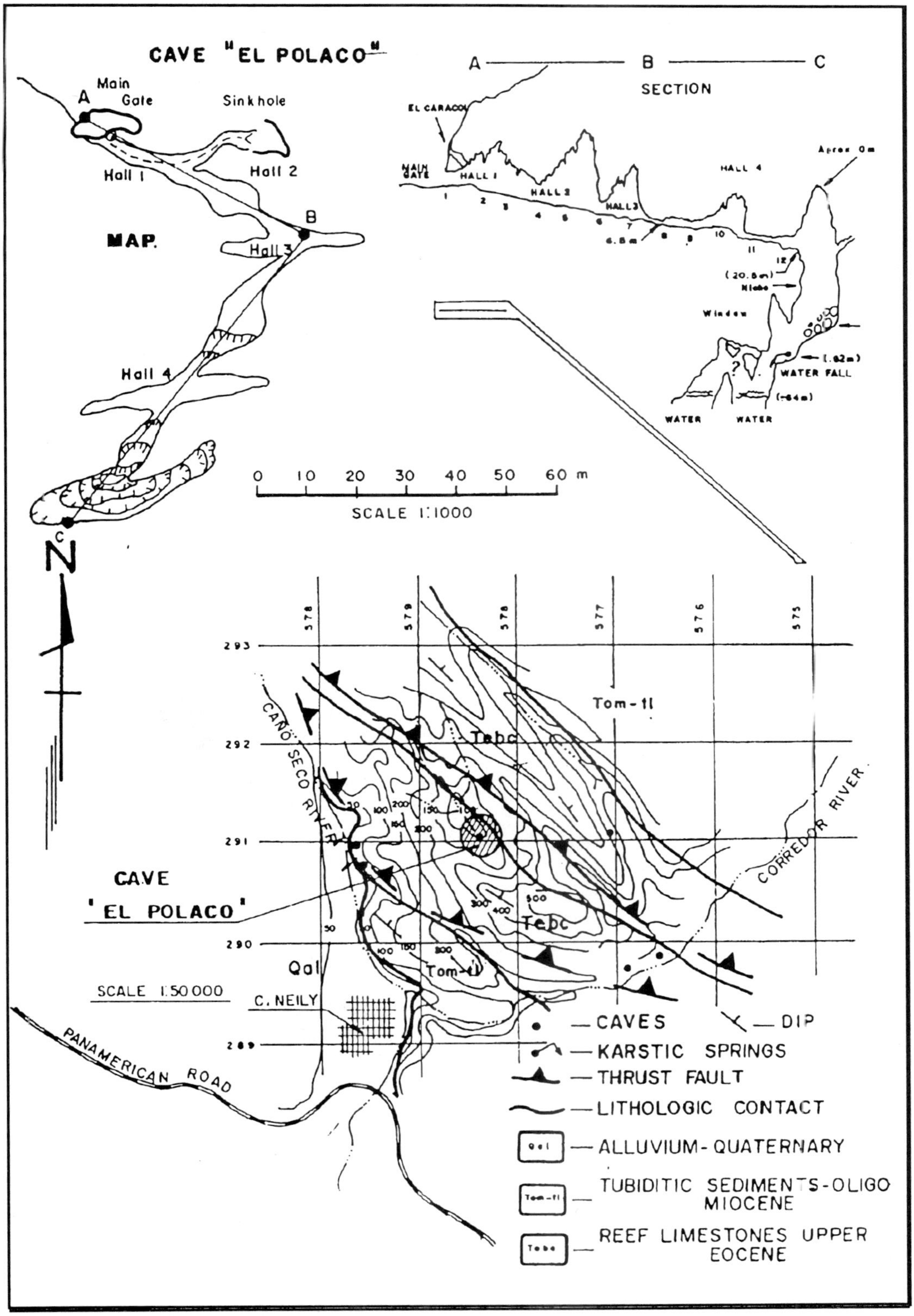

Figure 2. Geologic map of the Corredor karst site. Includes map and section of the El Polaco cave. Section after Asociacion Espeleologica Costarricenses, 1986.

lifting continued through the Pliocene and through the Quaternary, constructing an inner arc.

GEOLOGICAL ENVIRONMENT OF KARSTIC LIMESTONES

All limestones showing present-day evidence of karst features were deposited in shallow marine environments. Episodic and intensive tectonic-orogenic activity, accompanied by rapid rate of uplift, occurred mainly during the Eocene through Miocene. Some areas with favorable environmental conditions were hosts to reefs and other types of carbonate platforms, both of which have been subsequently uplifted.

Sedimentology, stratigraphy, and age data about most of the limestones of Costa Rica have been documented in the literature. However, the development and absolute age of karst landscapes are still poorly known. Those limestones that are hosts to karst development will be briefly described in the following paragraphs according to their age. Also, all known karst localities will be cited and described (Figure 1). Karst regions in Costa Rica have been ranked "a", representing the ones with the most well-developed features, to "e", the least developed, depending upon the occurrence and degree of development of caves, aven, aquifers, sinkholes, and karren (lapies).

Eocene Rocks

Reef limestones of late Eocene age are widespread throughout Costa Rica. Most facies of this kind are composed of orbitoid foraminifers, algal nodules (oncoids), and their synsedimentary erosional products (back or front reef facies). Boundstones, wackestones, grainstones, oosparite, and intrasparite are very common lithologies (Mora, 1979a; Sprechmann, 1984). All karst localities are shown in Figure 1. The best known karst localities in the figures are Corredor (1), Cajón (3), and Damas (8). Preliminary exploration has been conducted in those three (Mora, 1979a; Mora and Valdés, 1983; Asociación Espeleólogos Costarricenses, 1986).

The Corredor site is located near Ciudad Neilly, close to the Panama border. At Corredor, a blind valley drains into a cave. The cave water resurfacing as springs near Ciudad Neilly is intensively exploited for human uses, such as irrigation and potable water for domestic use. This site was briefly described by Asociación Espeleólogos Costarricenses (1986) and Mora and Alvarado (1985), and it has also been the subject of the work of a team from the National Speleological Society (Figure 2).

At the Cajón site, several sinkholes, karren areas, cave entrances, and springs are present, but no speleological exploration has yet been attempted, even though this site has been proposed for a major dam for hydroelectric power and detailed geologic and geotechnical surface and subsurface investigations have been carried out to determine technical feasibility of the project (Mora, 1979b; Grant, 1973; Consortium SNC-ACRES-TIL, 1979).

The Damas karst is well known to local people who associate the cave with traditions and with legends of lost people, appearances of ghosts, gateway to hell, etc. Scientific exploration of the cave has not been done so no technical information is available. It has an intricate system of galleries and halls, and several associated perennial springs yield considerable volumes of water (Mora and Valdés, 1983).

The Cortés karst area (Figure 1, 5) near Ciudad Cortés, has small outcrops of limestone and several important springs, some of which are exploited for human use. Other karst localities in Eocene limestones are Culebra (2), Grisera (4), Faralla (6), Nara (7), Providencia (9), and Azul (10). They all have similar geologic environments but little or no specific information is available.

Eocene-Oligocene Rocks

Two known outcrops of Eocene-Oligocene limestone occur in Costa Rica, of which the Barra Honda karst is certainly the best known and best developed of the country and was designated a National Park in 1974 (Figure 1, 12 and Figures 3 and 4). The other area of karst developed on Eocene-Oligocene limestone is Colorado (Figure 1, 11). This karst is not yet well explored. Both karst localities are developed in the Barra Honda Formation, composed of algal (corallinaceae, dasycladaceae) limestone, with biostroms and locally well developed bioherms (biolithite, biopelmicrite, boundstone), as well as back-reef deposits (biomicrite, biosparite, biopelmicrite, packstone). The age of this formation is still the subject of controversy because few guide fossils have been found. These, together with relations with underlying clastic units suggest an Eocene-Oligocene age (Mora 1978; Mora, 1981; Rivier, 1983). New tectonic and sedimentologic interpretations, however, indicate that an older age (Campanian-Maastrichtian) is possible. Evidence is not yet convincing, so I adhere to the earlier interpretation.

The maximum thickness of the Barra Honda Formation is 250 m. It forms a resistant layer that typically caps hills. The cap rocks represent remnants of the once continuous carbonate platform. The limestone cap rocks are no thicker than 90 m around the flanks of the hill, but caves as deep as 200 m are known from the interior of the Barra Honda hill. Apparently, the limestone is thicker within the interior of the hill than around the flanks (Figure 4). This suggests the development of a synsedimentary basin where continuous deposition caused subsequent weight-subsidence toward the center of the hill. Geomorphologically, this is a reversed landscape, in which the synsedimentary depression is now occupying the tops of the hills because of the greater resistance of the limestone to erosion than the surrounding and underlying units (Figure 3). The Barra Honda Formation is the youngest formation in the area, suggesting rather continuous uplift and emergence of the unit since deposition of the synsedimentary

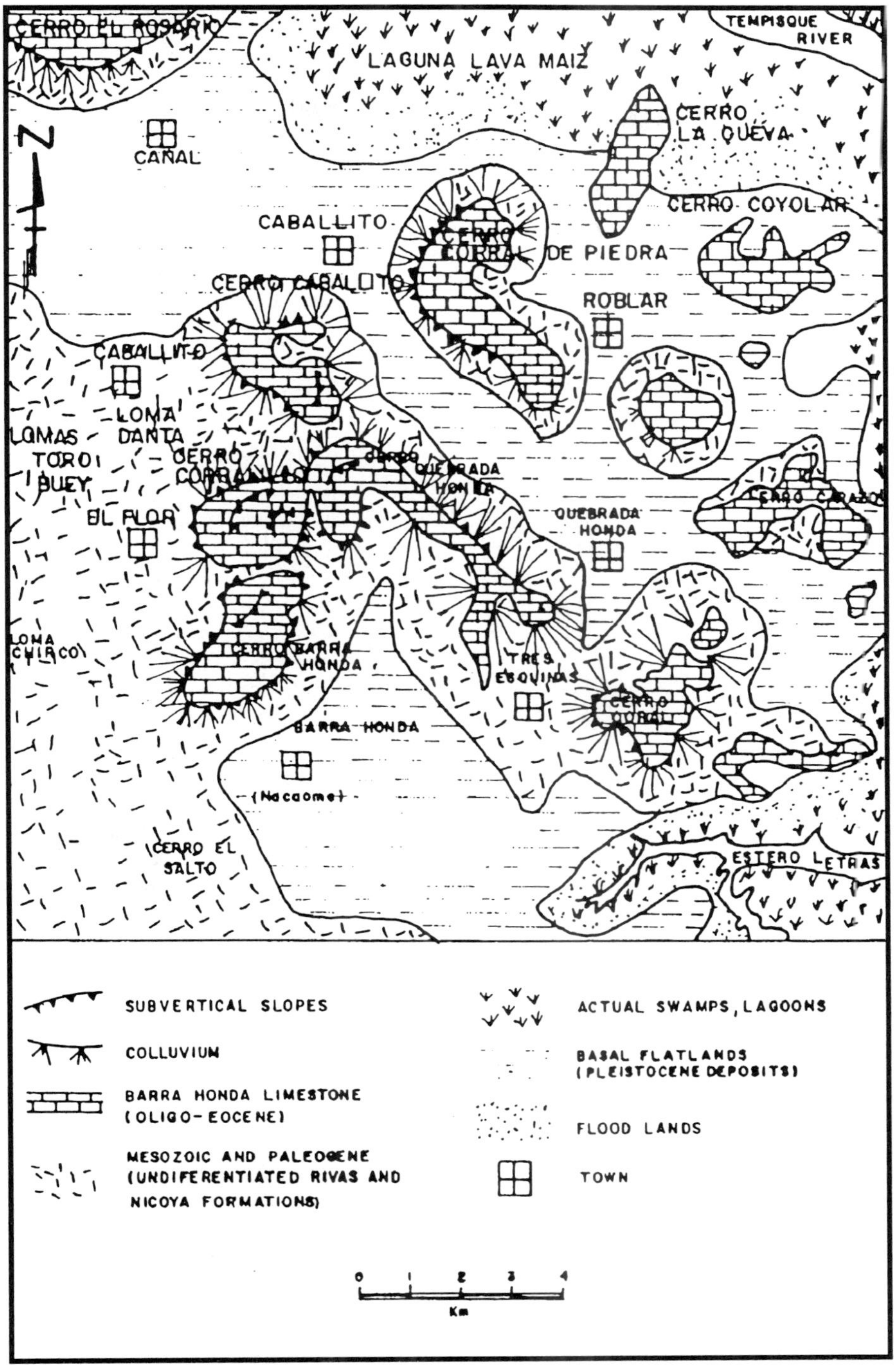

Figure 3. Geomorphological map of Cerros Barra Hondo, Corralillo, and surroundings. Cerro - hill, loma = long low hill, estero = swamp.

basin. It is interesting to note that the limestone seems to have been exposed for a long time to weathering and dissolution, probably more time than that needed to reach the early maturity degree of karst development (Figure 1).

The springs around the periphery yield important but still unknown volumes of water, even during long dry seasons. Many of the springs are extensively exploited as a water supply for surrounding communities. Although the cave

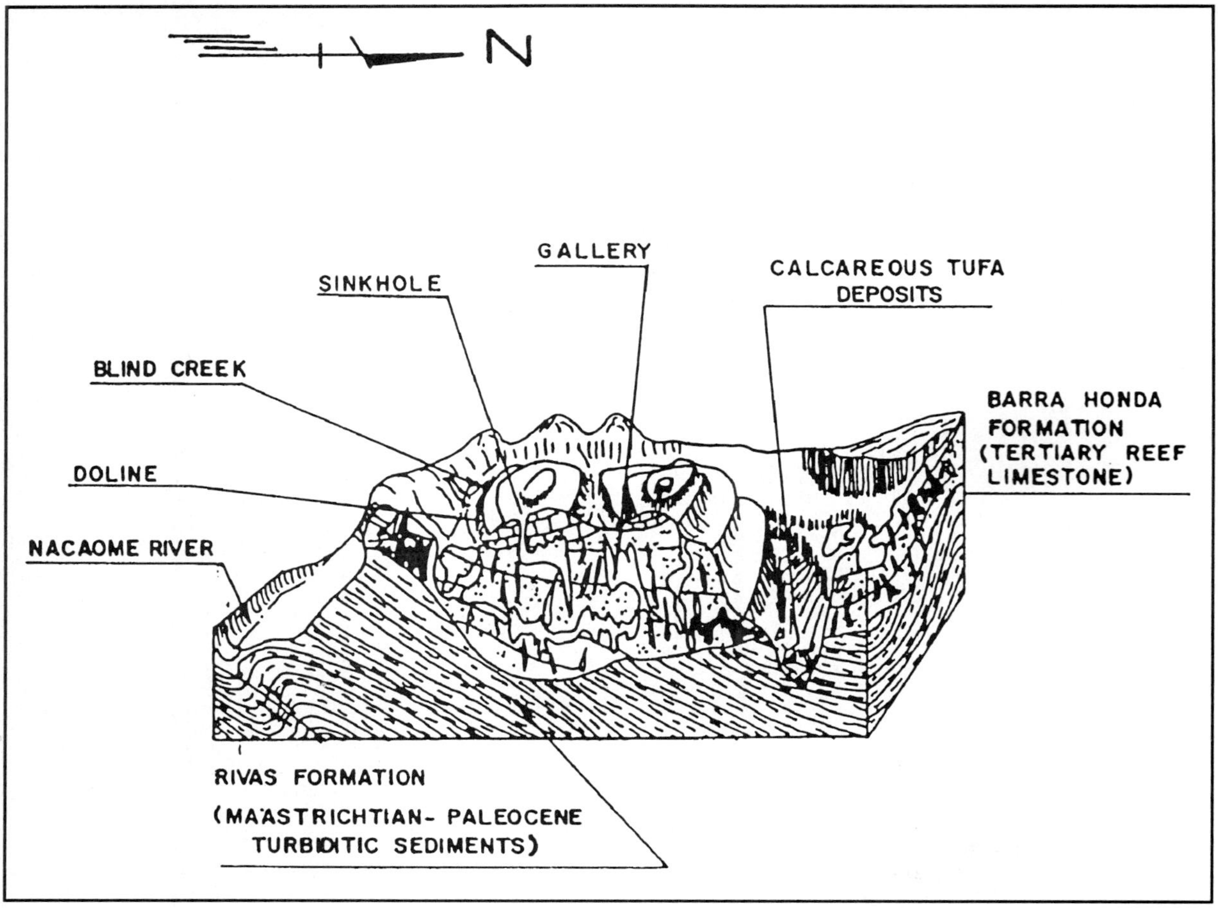

Figure 4. Barra Honda karst system (modified from Wells, 1973).

networks have been quite widely explored, and several springs occur at the base of the hills, no extensive groundwater reservoir nor conduit flow system has yet been identified in the Barra Honda area (Figure 4).

This karst area is also important from archaeological and anthropological viewpoints because human remains and artifacts have been found inside and outside of the Barra Honda caves (Mora, 1979a; Pereira, 1980; Mora, 1981). Occupational sites are thought to belong to cultures developed around 200 years B.C. to A.D. 250 years (Figure 5).

Very little is known about the Colorado karst (Figure 1, 11) even though its geologic setting is already well understood (Rivier, 1983). Behind its surficial appearance as a moderately developed karst, thickness of the limestone layers could allow for the development of larger solution features in addition to the conspicuous karren, small caves, and sinkholes. No big springs have been reported from this site, which is mainly exploited as a source of calcium carbonate for a cement factory and as ornamental stone for architectural purposes.

Miocene Rocks

Limestones and calcareous clastic units of Miocene age, as well as other carbonate facies are present on both Pacific and Caribbean sides of Costa Rica. Deltaic, transitional, and beach deposits are very widespread (Mora, 1979a; Fischer, 1980, 1981). Miocene limestones are composed mostly of calcareous sandstone, conglomerate, and shale in beds ranging from a few centimeters to a few meters thick. Pecten shells are the most abundant fossils, and ostracodes, gastropods, echinoids, bryozoans, and crustaceans are also common.

Three main karst localities are developed on Miocene limestone: Venado (Figure 1, 13), San Miguel (Figure 1, 14), and Malpaís (Figure 1, 15). Of these the most important site is Venado, which has large caverns, sinkholes, and blind creeks. No deep underground scientific exploration has yet been undertaken at this site.

The San Miguel area also has some incipient karst features. There the limestones are exploited for cement, fertilizers, and decorative stone.

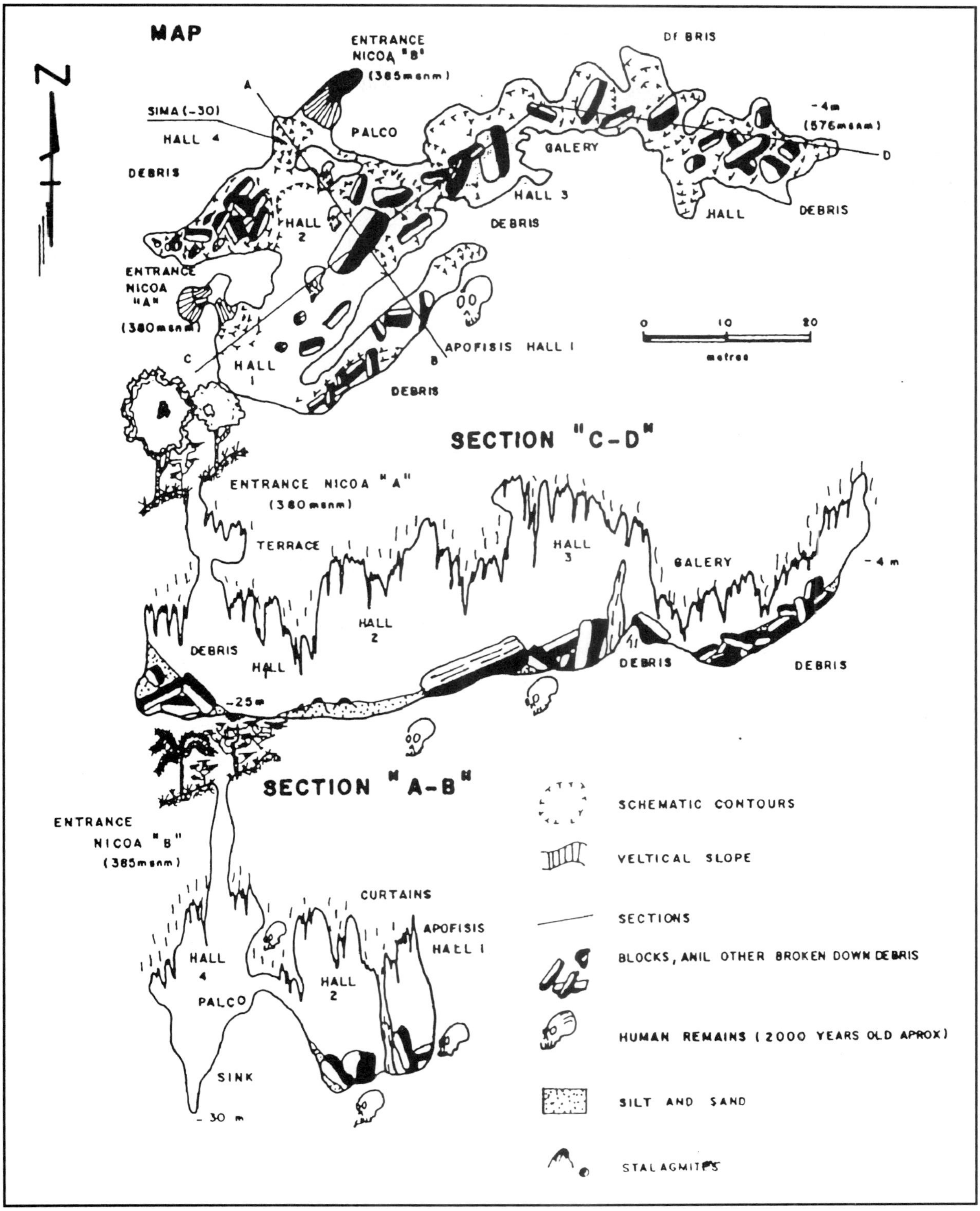

Figure 5. Nicoa cave (Barra Honda) map and sections.

Plio-Pleistocene Rocks

Due to the climatic setting, shore configuration, and local bathymetry, coral reefs began to develop during the late Pliocene and exist locally along the Caribbean coast. The biggest and best known karst locality in Plio-Pleistocene coral reef limestone is Moín (Figure 1, 16). The limestone contains abundant gastropod and other mollusk fossils. The thickness of the reef rock is about 40 m. Microkarst features have developed on it.

Some hydrogeological studies have been carried out in the Moín karst area due to its exploitation as an aquifer (Losilla, personal communication, 1979; Mora, Lucke, and Ramírez, 1980). It meets almost all the needs of potable water for the port of Limón. Fluorescein tests were conducted on certain of the main sinkholes. These tests showed rapid responses in peripheral springs (100 to 300 m in 90-160 min). A group of the springs yields an average 100 liters/sec during low rainfall periods, while more than 320 liters/sec have been measured after rainfalls of considerable intensity. Recently, some pollution has been detected in the spring water due to salt intrusion, quarry activities, and old nearby trash deposits.

CONCLUSIONS

There are several important karst sites in Costa Rica, most of them developed on Tertiary reef, platform, and shore limestones. They are present mainly in the southern and western parts of the country near the Pacific coasts. Some of the karst areas are very well developed, displaying classic features such as karren, sinkholes, aven, dolines, and caves. Present exploitation of and future plans for the associated springs for community water supplies make these karstic limestones very important. Detailed studies have not been made; studies to date have consisted of very preliminary reconnaissances. Data are sparse.

Barra Honda is the only exception to this situation, due to its designation as a National Park. Increasing economic interest, coupled with archeological and recreational importance urgently require that attention be given to the other karst localities in order to better understand their geology, hydrogeology, and mode of development.

REFERENCES

Asociación Espeleólogica Costarricenses, 1986, Descubren primer lago subterráneo del país: La República, January 27, p. 3.

Aubouin, J., et al., 1979, Prémiers résultats des forages profonds dans le Pacifique au niveau de la Fosse du Guatemala, Leg 67, Deep Sea Drilling Project: C. R. Académie des Sciences de Paris, v. 289, p. 1215-1220.

Baumgartner, P. O., 1984, Sedimentación y paleogeografía del Cretácico y Cenozoico del litoral pacífico de Costa Rica: Revista Geológica de América Central, Escuela Centroamericana de Geología, Universidad de Costa Rica, v. 1, p. 57-136.

Consortium SNC-ACRES-TIL, 1979, Feasibility study for the Boruca Hydroelectrical Project. Instituto Costarricense de Electricidad, 8 v., 5 sets of maps.

Fischer, R., 1980, Recent tectonic movements in the Costa Rica Pacific coast: Tectonophysics, v. 70, p. T25-T33.

Fischer, R., 1981, El desarrollo geográfico del Mioceno en Costa Rica: Porto Alegre, Brazil, Ann. 2nd Congreso Latinoamericano de Paleontología, p. 565-579.

Grant, L., 1973, Dye tests for Boruca limestone: Instituto Costarricense de Electricidad, Alcoa-Hensley-Schmidt, internal unpublished report, 9p.

Mora, D., O. Lucke, and O. Ramírez, 1980, Delimitación de la unidad hidrogeol<gica y zona de recarga de las fuentes de Moín: Instituto Costarricense de Acueductos y Alcantarillados, Informe interno, 12p.

Mora, S., 1978, Estudio geológico de los Cerros Barra Honda y alrededores: Universidad de Costa Rica, Escuela Centroamericana de Geología, Tesis de Bachillerato en Geología, 188p.

Mora, S., 1979a, Estudio geológico de una parte de la región sureste del Valle del General, Costa Rica: Universidad de Costa Rica, Escuela Centroamericana de Geología, Tesis de Licenciatura en Geología, 3 v., 275p.

Mora, S., 1979b, Estudios geológicos y geotécnicos para el Proyecto Hidroeléctrico Boruca: Instituto Costarricense de Electricidad, Departamento de Geología, Informe interno inédito, 2 v., 275p.

Mora, S., 1981, Barra Honda: San José, Costa Rica, Editorial EUNED, 92p.

Mora, S., and G. Alvarado, 1985, Análisis geológico preliminar de la Fila de la Cal, en los alrededores de Ciudad Neilly, Costa Rica: Universidad de Costa Rica, Escuela de Geografía, Informe interno inédito, 32p.

Mora, S., and R. Valdés, 1983, Estudio geológico-geotécnico para el Proyecto Hidroeléctrico Savegre. Instituto Costarricense de Electricidad, Departamento de Geología, 3 v., 375p.

Pereira, M., 1980, Le crâne No. 2 de Barra Honda: Paris, Museé de l'Homme, 18p.

Rivier, F., 1983, Síntesis geológica y mapa del área del Bajo Tempisque, Costa Rica. Instituto Geográfico Nacional, Informe Semestral Ene-Jun., p. 7-75.

Sprechmann, P., ed., 1984, Estratigrafía de Costa Rica, manual de geología de Costa Rica: San José, Editorial Costa Rica, 88p.

Wells, S., 1973, Report on the geologic reconnaissance of the Cerros Barra Honda karst, Costa Rica: Cincinnati, Case Research Foundation, University of Cincinnati, 36 p.

Circum-Pacific Council for Energy and Mineral Resources
Earth Science Series, Vol. 16
R. L. Miller, G. Escalante, J. A. Reinemund, and M. J. Bergin (Eds.)
Energy and Mineral Potential of the Central American-Caribbean Regions
Springer-Verlag, Berlin Heidelberg 1995

SUMMARY STATEMENT FOR HYDROLOGY SESSION WORKING GROUP

Allen L. Zack
Caribbean District Chief, Water Resources Division, U.S. Geological Survey,
GPO Box 4424, San Juan, Puerto Rico 00936 USA

The hydrology group identified three areas that would be of interest to the Circum-Pacific Council and participating nations.

First is a reminder that as mineral resources are developed, the pollution of surface waters by extraction processes and acid mine drainage is visibly impressive and emotionally distressing. Ground-water contamination, however, is less appreciated because it is hidden from view. The contamination of aquifers from the effects of mining and minerals development is generally more permanent and much more difficult and costly to correct. Therefore, aquifer protection and conservation must be practiced diligently.

Aquifers of the Circum-Pacific region are generally small and vulnerable. Once contaminated, their use is restricted. Every effort must be made to coordinate national and private interests to monitor ground-water quality continuously in order to preclude aquifer contamination. Because water is our most valuable mineral resource, we can no longer afford to take it for granted.

Second is a suggestion. Members of the Circum-Pacific Council, representing as they do diverse scientific disciplines, would benefit from a water-resources clearinghouse within the Council. It would be helpful knowing which institutions we can turn to for water-resources information. At present, it is difficult to determine what hydrologic data is available for specific areas. For example, it is a little-known fact that UNESCO's International Hydrological Program is presently assembling a series of hydrogeologic maps of the Caribbean Islands. Knowledge of the availability of hydrologic information such as geophysical logs, potentiometric maps, aquifer hydraulic data, streamflow and flood-frequency data, water-quality maps, landslide-vulnerability maps, and sediment-yield statistics would be extremely valuable to members of the Circum-Pacific Council.

The third is in the form of an open inquiry. How many of us would benefit from a real-time, data-collection network that would give early warning of floods, of landslides, and even possibly of volcanic activity? Satellite-transmission technology is presently available, providing real-time hydrologic information for making timely decisions.

Recent advances in computer technology will permit us to develop rainfall-runoff models that operate in real time. The U.S. Geological Survey Water-Resources Division (USGS-WRD) in Puerto Rico and the U.S. Virgin Islands maintains a flood-warning system for these islands. We will also be developing a landslide-alert network in cooperation with the Survey's Geologic Division. Both of these systems are invaluable for both public safety concerns and hydrologic research. The USGS-WRD suggests interested Circum-Pacific Council members consider developing programs for implementation of these types of systems.

GEOLOGIC HAZARDS

Circum-Pacific Council for Energy and Mineral Resources
Earth Science Series, Vol. 16
R. L. Miller, G. Escalante, J. A. Reinemund, and M. J. Bergin (Eds.)
Energy and Mineral Potential of the Central American-Caribbean Regions
Springer-Verlag, Berlin Heidelberg 1995

LANDSLIDES ON THE PANAMA CANAL

George Berman
Panama Canal Commission, Republic of Panama, APO Miami 34011-5000

ABSTRACT

Landslides were a dominant aspect of the construction of the Panama Canal. They resulted in 40 million m^3 of additional excavation, as much volume as contained in 15 Cheops pyramids. The landslides were a major factor in the bankruptcy of the French enterprise. The Americans also began to question the success of the undertaking because of the large number and size of slope failures during construction. Their resolve prevailed and they completed the 12 km navigation channel, called Gaillard Cut, through the isthmus' central mountain range. On August 15, 1914, the Canal opened to shipping. Shortly afterward, the two largest landslides, called the East and West Culebra landslides (13 and 10 million m^3 in motion, respectively), slipped simultaneously. These landslides closed the Canal for much of its second year of operation while dredges worked to reopen the channel.

Although landslides have not blocked the Canal since 1920, they still threaten navigation. They have partially obstructed it in three instances. Debris from the East Culebra Extension landslide (2.4 million m^3 in motion) restricted navigation in 1931 and 1974. The East Cucaracha landslide (4.6 million m^3 in motion) nearly closed the Canal in 1986. The impact of the landslides goes beyond the repair costs and the delayed transits. At a time of increasing competition, the Canal's reliability serves to sustain and increase its role in world commerce.

Through its Landslide Control Program, the Panama Canal Commission strives to prevent landslides from blocking the Canal. Rain, 215 cm/yr, and time, now 75 years, weaken the excavated slopes, but surface control monitoring of the slopes detects most potential landslides with enough lead time to prevent Canal encroachment. The remedial measures taken to increase stability after identifying problem slopes consist mainly of improving surface drainage and excavating the crest of the landslide. In operating the Canal, Gaillard Cut functions well simply by allowing continuous passage.

INTRODUCTION

The Panama Canal is a monumental example of the relation between solving slope stability problems and exploiting energy and mineral resources. Dr. D. J. Varnes (1981) said this at the second Circum-Pacific Council conference in Honolulu when he discussed the impact of landslides on major transportation corridors. The Panama Canal is an example of this relation not only because of past problems with landslides but also because of current ones.

The Panama Canal Commission, PCC, spends significant resources to mitigate the landslide hazard. These expenditures help the Canal sustain its impeccable record for reliability. Reliability of the waterway is critical for the Canal to maintain and increase its role in world commerce at a time of growing competition.

Slope stability will continue to be important to the Canal in the future. Beyond habitual stability problems, the Canal plans to widen Gaillard Cut from 151 m to 198 m, if traffic grows. This project will require excavating 25 million m^3 of earth. The widening will place new demands of considerable challenge on slope-stability engineering.

The Panama Canal crosses the Central American isthmus at the narrowest point, where it is only 80 km wide (Figure 1). At first, the French attempted to dig a sea level canal. Landslides, disease, and mismanagement caused the French enterprise to fail. Years later the Americans followed and succeeded. They decided on a lock canal instead of a sea

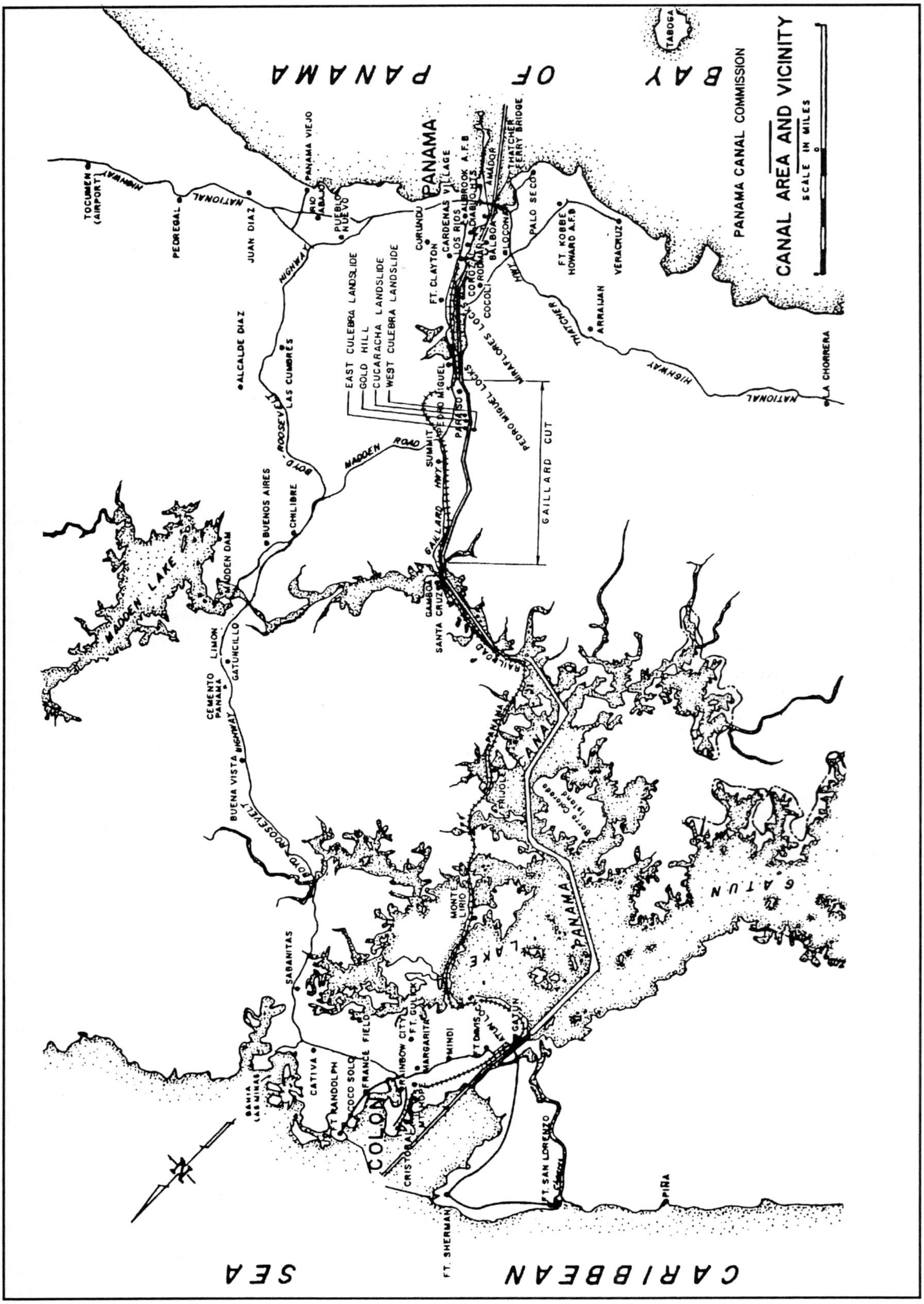

Figure 1. Map of the Panama Canal, showing salient features and location of landslides mentioned in text.

level canal. In doing so, they reduced the excavation and the depth of Gaillard Cut. Shallower excavations during the construction reduced the driving force for landslides, but not enough. Landslides in the cut were a constant hazard during the construction.

DESCRIPTIONS OF GAILLARD CUT

Gaillard Cut, 12 km long, is the part of the Canal which crosses the isthmus' central mountain range in a north-south direction. The water level in the cut is at elevation of 26 m above sea level, the level of Gatum Lake of which it is an arm. The highest peak in the cut, Gold Hill, is at elevation of 180 m. Other hills rise to a average elevation of 70 m.

For Canal operations, Gaillard Cut works well by allowing uninterrupted safe navigation. It does not, however, permit continuous two-way traffic of wide-beam ships. After the widening project, the Canal could lift this restriction.

The rocks in Gaillard Cut vary widely in strength. They range from extrusive igneous masses to clay shales. Several of the slopes have heavy igneous rocks overlying the weak clay shales, a serious detriment to their stability. Canal geologists refer to this particular landform as reverse topography. The Cucaracha Formation, consisting of slickensided clay shale strata, is infamous for its role in the mechanics of the major landslides.

The nature of the rocks makes the cut subject to landslides, but it is the rainfall which drives them into the navigation channel. The average rainfall in Gaillard Cut is 215 cm/yr, most of which falls during the rainy season from May through December. There is a direct relationship between rainfall and landslide motion. Rainfall induces instability by increasing the pore water pressure at depth and by increasing the overlying load. The increase in pore water pressure results from the cumulative seasonal effect. In contrast, the overlying load is the result of specific storms recharging standing water in cracks and saturating the surface layer. From our observations, landslides in Gaillard Cut occur when both factors are contributing during large storms at the height of the rainy season. Therefore, October is the month most prone to landslides.

LANDSLIDES DURING CONSTRUCTION

Because of the landslides, canal builders had to excavate 25 percent more (40 million m^3) than they planned. Gaillard Cut was the focus of most attention during the construction of the Canal. Other Canal construction problems were of secondary interest as long as the battle with landslides raged in the Cut. For example, Gatun Dam, at that time the highest earthfill dam ever attempted and which impounded the largest artificial lake, presented lesser problems.

Reports of the period describe more than 60 distinct landslides in Gaillard Cut during construction. Their size and number increased proportionately to the progress achieved in deepening the cut. The major ones were near the continental divide in segments of the Canal called Culebra and Cucaracha Reaches. Foremost of the landslides during the construction period were the East Culebra, West Culebra, and East Cucaracha. Respectively, their movement resulted in 13, 10, and 9 million m^3 of unplanned dredging. These three landslides alone yielded most of the slide debris removed during the original excavation of the Canal.

The Panama Canal opened to world commerce on August 15, 1914. In October, the East Culebra landslide moved into the Canal, but the dredges were able to keep the channel open until the next rainy season. In September, 1915, simultaneous movements of both the East Culebra and West Culebra slides closed the Canal. The bottom of the Canal rose from a depth of 10 m to above the surface in the center of the channel. Despite the combined work of all the dredges, the Canal remained closed for the next seven months. The dredges worked unceasingly until finally the rate of dredging exceeded the rate at which landslide debris was entering the navigation channel. The dredges, however, continued working for the next 30 years. In the end, they had widened the navigation channel from 91 m to 152 m in a 1.8 km-long segment of Gaillard Cut. This additional width served as an emergency repository for debris in the event of reactivations of the slides.

MODERN LANDSLIDES

Traffic through the Canal increased significantly after 1950. To safely accommodate the added demand, PCC widened the remainder of Gaillard Cut to the uniform width of 152 m. This major excavation project was completed in 1970 with only one landslide incident. In contrast to the original excavation of the cut, this incident added less than 2 percent to the planned project total of 22 million m^3.

In October, 1968, Canal geologists discovered cracks, 1.5 m wide, on a slope called Hodges Hill. It is near the West Culebra landslide. In response to this encroachment threat, PCC assembled a board of renowned consultants. Arthur Casagrande, a pillar of geotechnical engineering, headed the group. These consultants, together with PCC engineers, stabilized the threatened slope by improving the drainage. At this time they also launched PCC's Landslide Control Program.

The goal of the Landslide Control Program is clear: preventing landslides from blocking navigation in the Canal. Translating this statement into a successful course of action has not, however, been easy. In 1936, Karl Terzaghi, the father of geotechnical engineering, questioned the quantitative analysis of landslides. He said "the catastrophic descent of the slope of the deepest cut on the Panama Canal issued a warning that we are overstepping the limits of our ability." The problem still remains formidable despite advances in methods of analysis and in field instrumentation.

EAST CULEBRA SLIDE (STATION 1925)

On October 10, 1974, the slope called 1925 East Culebra plunged into Canal waters. This slope, located in Culebra Reach, is an extension of the slope on which the original East Culebra slide occurred. The slide moved on the gently dipping basal contact of the Cucaracha Formation, which surfaced at the toe of the slope. The mass broke up into three distinct blocks as it moved toward the Canal. The block closest to the Canal moved 20 m, and the material was projected across the channel's centerline. The second higher block moved only 2 m. Ground cracks corresponding to the third block had little movement and were 500 m from the Canal. The total mass in motion had a volume of 2.4 million m^3. Dredges had to remove 200,000 m^3 to restore fully the navigation channel.

In addition to clearing the landslide debris from the navigation channel, PCC took actions to stabilize this slope. Arthur Casagrande again assisted PCC engineers in the planning of remedial measures. The graben behind the forward block was full of loose rock fragments from the landslide, which naturally drained the water before it entered the block. Capping the loose material with clay delayed weathering and served to keep the water flowing laterally. Other remedial measures were the digging of an interceptor drain farther uphill and also unloading the top of the slope.

Because the 1925 East Culebra slide was an area of known instability, PCC engineers had selected it as one of the first slopes for the installation of surveillance instruments. Measurements of surface displacement before the slide showed 1.8 m of movement. PCC engineers also observed that the rate of movement had increased each of the two years before the slope's collapse. Remedial work before the encroachment had consisted only of drainage improvements. PCC engineers had considered major excavations to flatten the slope but these plans had been deferred. In retrospect, the implications of the observations of movement were not fully understood. New criteria have now emerged in dictating that the rate of movement is more important than its magnitude in triggering warning signals. Also, PCC can now accomplish stabilization work more readily in response to warning signals.

EAST CUCARACHA SLIDE

On October 13, 1986, the East Cucaracha slide nearly closed the Canal. Landslide debris filled the navigation channel leaving only a narrow passageway, 35 m wide, which is only slightly wider than the width of the lock chambers. Fortunately, experienced PCC pilots maneuvered ships through the constriction using modern tugs, and dredges were able to keep the narrow channel open.

This 1986 landslide is the latest reactivation of the major landslide cited before. Historical records show the slope began sliding as far back as 1886 when the French first excavated in this area. It was a major and continual menace to the American effort to excavate the cut from 1904 to 1914. Reactivations occurred in 1920, 1927, and as recently as 1972.

The mechanism of the East Cucaracha slide is complex. The entire landslide covers 10 ha. and the main body of the slide is 240 m wide. It reaches back 600 m from the Canal. The main body of the slide is next to Gold Hill on the Continental Divide. It is a gentle slope with an average rise of 20 percent. Gold Hill, an igneous block, confines the slide to the north and a basalt dike confines it to the south.

By itself, the main body did not pose a major threat to the Canal. The failure surface was relatively shallow, only about 20 m. Surface monitoring showed that the mass had moved more than 6 m in the following four years, but PCC engineers did not expect this mass to accelerate abruptly because of the weakened state of the materials. Because of its slow and persistent movements, Canal pioneers compared it to a glacier.

On the day of the slide, the rear part of Gold Hill broke off and moved 120 m onto the head of the main body of the slide. Partially weathered basalt composed the moving mass of bedrock but it slipped on the contact with the underlying clay shales. This mass, 153,000 m^3 in size, surcharged the active but slow moving Cucaracha slide. In addition to overloading the slide, a part of the bedrock mass broke up and tumbled down the slope. About 200,000 m^3 of this material entered Canal waters. Half of this material accumulated in the widened portion of the channel intended as a repository for slide debris, but the remainder almost blocked navigation.

One year earlier, PCC engineers had developed a plan to stabilize the block from Gold Hill that triggered the landslide. The project entailed the excavation of 120,000 m^3. PCC had awarded the contract, but the contractor had not yet started work when the landslide occurred. The urgency of the project was not well understood because of its intricate domino mechanism.

As a result of the remedial work accomplished after this 1986 slide, the East Cucaracha slope is now stable. The work has consisted of unloading 1.2 million m^3 from the head of the slide, including the block from Gold Hill, and improving the surface drainage. As an additional measure, hydrated lime has also been spread on the surface.

LANDSLIDE CONTROL PROGRAM

Following the landslide of 1986, PCC convened a new board of geotechnical engineers. This board meets on an annual basis for three days. PCC selected the members for the new board because of their recognized stature within the profession. The members of this board are Dr. Mike Duncan, Dr. Norbert Morgenstern, Professor George Sowers, and Dr. Robert Schuster. The role of the board is to be sure that the methodology followed in landslide control is consistent with sound engineering practice.

The major elements in landslide control are surveillance and stabilization of the slopes. The purpose of surveillance is to identify incipient landslides with enough anticipation to avert channel encroachment. Stabilization refers to the remedial work performed for this purpose.

The cornerstone of landslide surveillance in the cut is the surface control grid. There are about 250 active monuments throughout the cut. PCC surveyors measure the distance between the monitored point on one bank and two stable points on the opposite bank. From successive readings, PCC engineers calculate the magnitude, direction, and rate of any movement. A limited number of piezometers, to measure ground water pressure, and deflection boreholes, to observe movement at depth, complement the surface observations.

Stabilization methods used on the Canal include excavations to unload and flatten slopes, drainage trenches, horizontally drilled drainage holes, and, in an emergency, the application of lime. Controlling incipient landslides near collapse entails substantial expenditures.

Although the monitoring of landslide sites has been extensive, the knowledge of landslides in Gaillard Cut still remains imprecise in many aspects. PCC engineers, with the guidance of the advisory board, strive to better understand the intricate workings of this complex physical phenomena. PCC continuously enhances the sophistication of the surveillance coverage. However, the cost of enough measurements to resolve all the uncertainties throughout the cut by direct observation is prohibitive. Nevertheless, the available data evaluated with sound engineering criteria should yield a successful strategy for landslide control. The experience gained by studying the landslides in Gaillard Cut provides the most reliable means of assessing future risks.

Presently, our surface motion control grid indicates that all the slopes in Gaillard Cut are nearly quiescent. This is the result of multiple stabilization projects accomplished after the 1986 landslide. To avert future crises, PCC engineers recommend preventive maintenance. This work includes upgrading the surface drainage on the slopes and reforesting barren and grassy areas. The drainage efficiency and vegetation cover of a slope have an important bearing on its stability by reducing the negative impact of rain.

CONCLUSION

Much as been learned from observing the landslides in Gaillard Cut. Using this experience, PCC engineers conclude the following about the slides in the Cut.

1. New landslides are the result of the reactivation and extension, both backward and laterally, of old landslides. The major landslides occur in the deep cuts near the Continental Divide where the Cucaracha Formation controls the sliding mechanism.

2. Major landslides show substantial distress before accelerating catastrophically. This distress can be detected by measuring surface movements.

3. The strong correlation between landslide motion and month (time of year) emphasizes the multiple and progressive impact of rainfall on slope stability.

REFERENCES

Terzaghi, K., 1936, Stability of slopes of natural clay, in Proceedings, International Conference on Soil Mechanics and Foundation Engineering: ASCE, p. 161-165.

Varnes, D. J., 1981, Slope-stability problems of circum-Pacific region as related to mineral and energy resources, *in* M. T. Halbouty, ed., Energy resources of the Pacific Region: Tulsa, AAPG Studies in Geology No. 12, p. 489-505.

Circum-Pacific Council for Energy and Mineral Resources
Earth Science Series, Vol. 16
R. L. Miller, G. Escalante, J. A. Reinemund, and M. J. Bergin (Eds.)
Energy and Mineral Potential of the Central American-Caribbean Regions
Springer-Verlag, Berlin Heidelberg 1995

GEOLOGIC HAZARDS IN HONDURAS

Michael J. Kozuch
Peace Corps, A. P. 3158, Tegucigalpa, Honduras

ABSTRACT

Unlike most Central American countries Honduras is not susceptible to the same volcanic and seismic hazards that exist within its neighboring countries. That is not to say that the risk of geologic catastrophes does not exist. Several unlocated earthquakes dating as far back as 1539 caused significant damage. Similar earthquakes might be much worse in today's populated areas. Satellite photographs reveal volcanic centers of very recent activity that may still represent a future hazard.

Apart from the seismic and volcanic risks, the more immediate hazards in Honduras are landslides and flooding which may be in part due to severe deforestation. The major landslides present along the western highways and in Tegucigalpa tend to occur where there is water saturation in the contact zone between the Cretaceous Valle de Angeles redbeds and the Tertiary volcanic tuffs. The intersection of joint sets and bedding orientations occasionally produces ramps that carry overlying material downslope.

Much of the highway damage can be avoided if the geologic structures are studied beforehand and if adequate drainages are provided where highway routes through unstable areas are unavoidable. In urban centers, housing and other construction should be avoided in areas showing previous landslide activity and in areas where gradients exceeding 20-30 percent near the redbed-volcanic contact zones exist.

INTRODUCTION

Geologic hazards in developing countries are often neglected due to severe economic limitations which divert the attention of political leaders to the social and economic needs of the country. Unfortunately, not until the occurrence of a catastrophe, which may involve the loss of human life, is there sufficient attention devoted to geologic hazards. Honduras is not an exception to this situation. This paper attempts to summarize four prominent geologic hazards in Honduras, which include seismic, volcanic, flooding, and landslide hazards. Special attention is devoted to landslide problems which frequently afflict this country.

SEISMIC HAZARDS

Seismic risk to the general population of Honduras is relatively minor in comparison to its neighboring countries. However, it is also relatively unknown. A recent National Earthquake Information Service (NEIS) 20 year data set shows active areas along the Honduras depression and along the Swan Fracture Zone off the north coast (Figure 1). Activity in the south appears to be associated with the subducting Cocos slab. Reports of tremors felt by residents in the Comayagua valley substantiate the active faulting associated with this graben, the Comayagua graben being one of several associated with the Honduras depression. Historical earthquakes listed by Osiecki (1981) indicate that Honduras has recorded tremors as far back as 1539, with magnitudes up to 5 and 6. Activity of this magnitude could represent a significant hazard to today's populated areas. A study by Kiremidjian and Shah (1979) placed Puerto Cortez, a north coast port that lies close to the Guatemalan border, as a major city with the highest seismic risk.

VOLCANIC HAZARDS

Air photographs, LANDSAT, and Space Shuttle shots reveal numerous volcanic centers within Honduras. Two areas stand out as intraplate volcanic centers that are Quater-

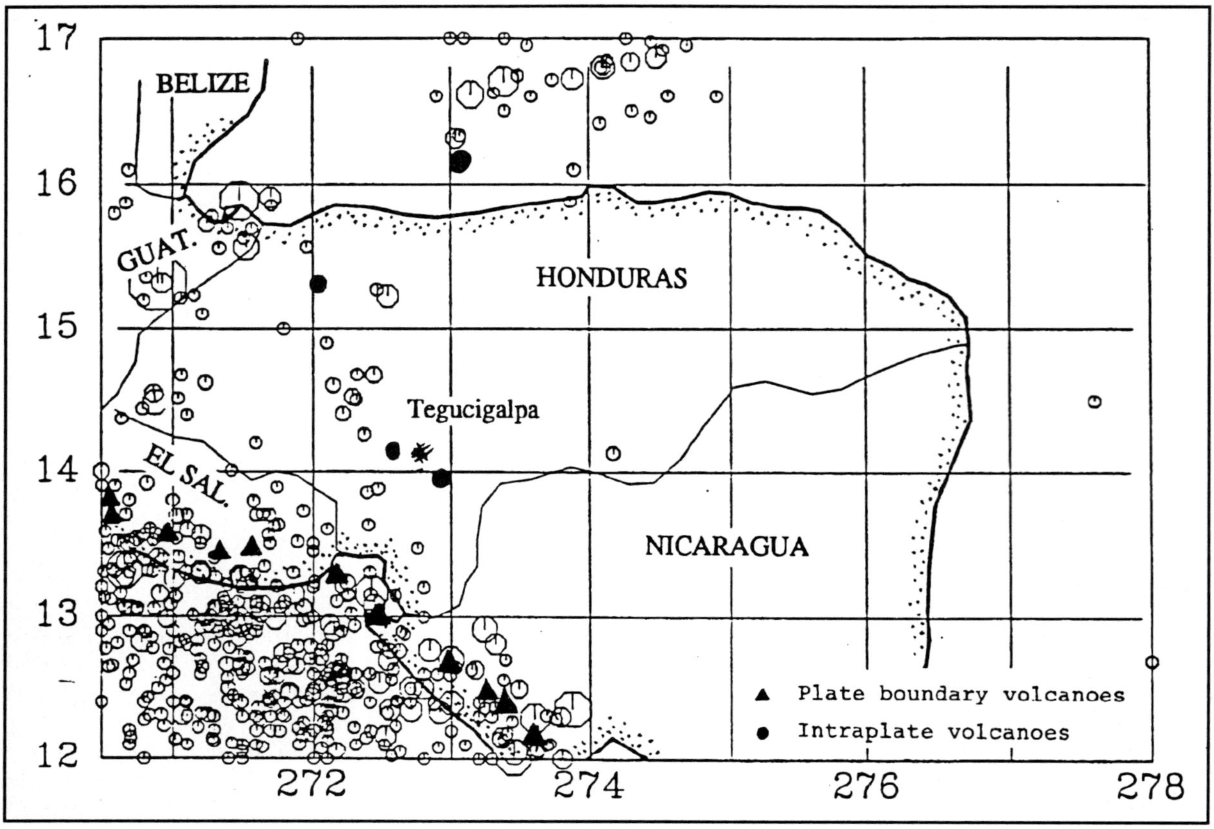

Figure 1. Plot of NEIS catalog of earthquake epicenters for Honduras from 1963 to 1983. Magnitudes within Honduras range from m_s and m_b magnitude 3.0-4.0. Also plotted are the intraplate and plate boundary volcanic centers.

nary in age (Figure 1). These volcanos may be the result of the extensional environment associated with the Honduras depression. These centers, which appear fairly recent, are inactive today. It appears that our greatest hazard is from volcanic eruptions in El Salvador, Nicaragua, and Guatemala. For example, recent eruptions of Pacaya volcano in Guatemala led to measurable ashfalls in Honduras.

Other eruptive centers that may exist, but which we do not recognize, may lie underneath the thick sequence of volcanic tuffs which cover a large portion of the country. Williams and McBirney (1969) claim that there may be as much as 5,000 km^3 of tuffaceous material in Honduras. The large number of thermal springs scattered across the country may be a clue to unrecognized volcanic centers or may be associated with intrusive bodies.

FLOODING HAZARDS

Flooding is a major problem in Honduras. It affects the country every year in the rainy season, especially from July to September. The north coast is affected by hurricanes and by runoff from the northern ranges, and low lying areas to the south may flood due to swelling rivers. Only two main rivers drain to the south into the Gulf of Fonseca but along these a large volume of water may bottleneck through small areas as in the Tegucigalpa basin. Tegucigalpa witnessed this phenomenon in September, 1988, when the Choluteca River swelled to over 30 feet above normal level and swept bridges, homes, and debris with it. The level of water experienced in the capital city will severely affect city planning of proposed landfills and development in the downtown area. Unless an adequate drainage system can be designed to handle flood waters, any such development projects are doomed to failure.

LANDSLIDE HAZARDS

Landslide activity has recently become more and more of a hazard in Honduras, especially in the marginal areas of Tegucigalpa and along some of the major highways. A comparison of these slide areas reveals a remarkable similarity among all of them. Nearly all are associated with the contact zone between the Cretaceous Valle de Angeles redbeds and the Tertiary Padre Miguel tuffs.

Geology

The Cretaceous redbeds are interbedded sandstones, siltstones, and mudstones. The predominant bonding cement is hematite, which gives the formation its distinctive color. Joints and gypsum veins are also common and act as conduits for water to percolate down to significant depths. Structurally, these beds are almost everywhere faulted and folded and lie with angular unconformity beneath the overlying volcanics. The Padre Miguel volcanics consist of massive ignimbrites with volcanic mudflows overlain by lacustrine beds and poorly indurated tuffs. A series of rhyolitic and andesitic flows caps this formation.

Causes

The surface of failure in these slides represents a plane of weakness which may have originally existed as a joint surface, intersecting joint sets, bedding surfaces, faults, karst cavities, or buried structures such as channels, basins, or anticlinal features. In all of these structural situations something happens to disrupt the state of equilibrium to where it becomes unstable. Landslide areas are subjected to additional factors which act to increase the stress in a rockmass while decreasing its strength (Sowers, 1989, unpublished report for the Dirección General de Caminos, Tegucigalpa, Honduras). These factors may be summarized as follows:

Increase in stress:
- external load*
- increase in moisture*
- undermining*
- water pressure increase*
- excavation
- earthquakes and blasting

Decrease in strength:
- increase in pore pressure*
- wetting, expansion*
- drying (tension)*
- progressive strain or creep
- hydrochemical weathering*

Much of the slide activity in Honduras is caused by a combination of these points and one cannot help but notice that water is involved in eight of these points (as indicated with an asterisk). A common feature to slides in Honduras is the saturation of the contact zone between the redbeds and the tuffs as a result of several of the previously mentioned factors. Water travels down the joints to the contact area where it saturates the zone. Some of it percolates farther down into the lower layers via the dipping bedding planes in the redbeds. This creates a situation where a fairly resistant layer is being supported by a weak layer. The characteristics and behavior of both of these rock types change completely when saturated and they cannot support the normal load.

Types of Failure

Of the various types of surfaces of weakness previously mentioned (under Causes), failure usually occurs as a plane or wedge failure. No surface is entirely flat, and there are features which determine whether a surface such as a joint will fail or move. One such feature is roughness, and this may consist of 1st, 2nd, and higher order surface irregularities which control how well two surfaces will grip each other on a hillslope (Figure 2a). It is the surface area that is in contact that determines how much shear stress a rock can handle before it fails. For example, the normal stress acting on an entire surface may be on the order of 100 bars; however, the asperities (or surfaces in contact) may be experiencing stresses up to one order of magnitude higher.

Below a certain normal stress level the existence of irregularities (especially 2nd order) on the joint surface will cause a dilation of the rockmass as irregularities try to lift and ride over each other (Nilsen, 1985) (Figure 2b). This will have a major effect on the shear strength of the rock joint (Figure 2c). From Figure 2c we see that the shear strength of a rough surface will be much greater than that of a relatively smooth surface. As the normal stress increases, the amount of shearing and breaking of irregularities also increases. In short, the rougher the surface the more resistant it is to sliding.

Other factors, which closely tie with roughness to determine when a rock joint will fail, are the thickness and type of filling in a joint (Goodman, 1976). The characteristics of the filling material (such as hardness) in combination with filling thickness and roughness determine the coefficient of friction between the filling and rock. Thicker fillings yield a drop in the coefficient of friction and the amount of drop is proportional to the roughness.

Finally, when two joint surfaces or a bedding plane and joint surface intersect they produce a wedge-like ramp which facilitates the transport of overburden material downhill (Figure 3). This feature was observed only in the redbed clay material in the Copán area of western Honduras. These wedges frequently evolved into new drainages.

Landslide Control

What can be done to control these slides or minimize the risk? There are at least four areas which could use improvement: reforestation, improvement in the design of retaining walls, improvement in drainages and maintenance, and the avoidance of construction in steep or known slide areas.

Deforestation has been a national concern not only for preserving our natural resources, but the severe erosion of topsoil directly affects landslide potential in the cities as well as in the field. The planting of deciduous trees in areas where shallow slides occur reduces the water content of clay soils down to a depth of 2-5 m (Stone and Lemmon, 1957). These trees have the largest consumption of water and the highest

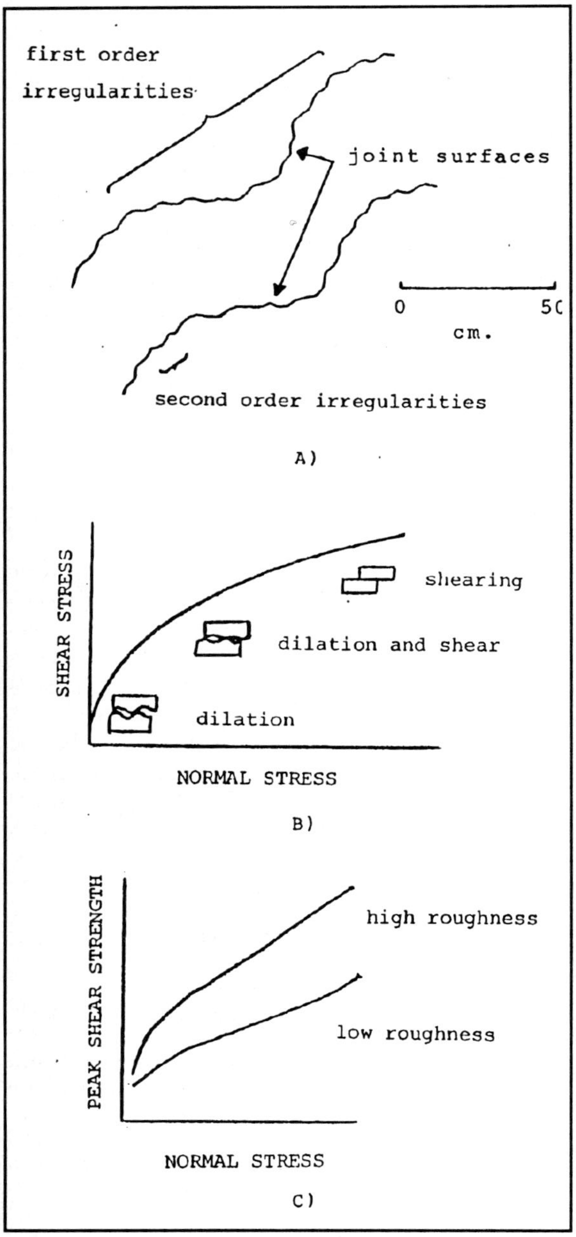

Figure 2. First and second order irregularities of a rough surface are shown in A (after Patton, 1966). Figure 2B illustrates the non-linear relationship for shear tests on rough joints (after Hoek, 1971), and finally, Figure 2C illustrates the extent to which roughness affects the shear test (Nilsen, 1985).

transpiration rates. It is also important to plant trees with deep roots to spread out the load of the trees themselves. A combination of broad leaf oaks and other trees native to the tropics, along with low lying bushes and grasses, is perhaps the best assortment to consider.

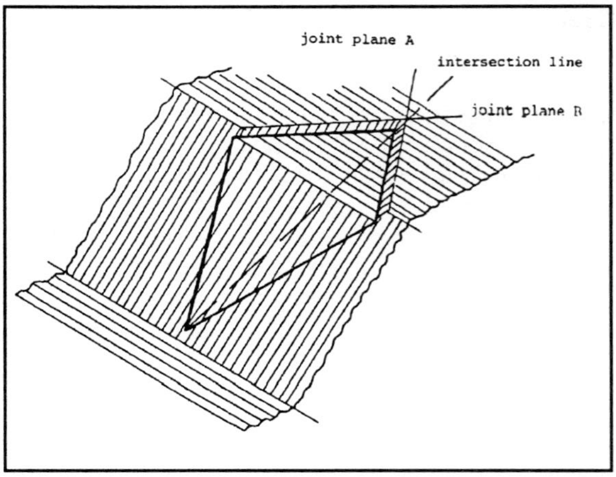

Figure 3. Illustration of wedge shaped ramps produced by intersecting joint surfaces.

The design of retaining walls desperately needs improvement in Honduras. Too many walls are built without adequate gravel filters and respirator drainages. The ones with drainages often lack a gravel or sand filter, the result being an impermeable wall which fails under stress (Figure 4).

Culvert designs must be improved to follow grades instead of remaining flat and susceptible to clogging (Figure 5). Inadequate drainage pipe designs have caused unnecessary erosion leading to landslides which could have been avoided.

Steep, hummocky topography is an excellent clue to previous slides, and construction should be avoided in these areas at all cost. In Tegucigalpa construction should be limited to areas with less than a 20° slope (or roughly a 30 percent grade).

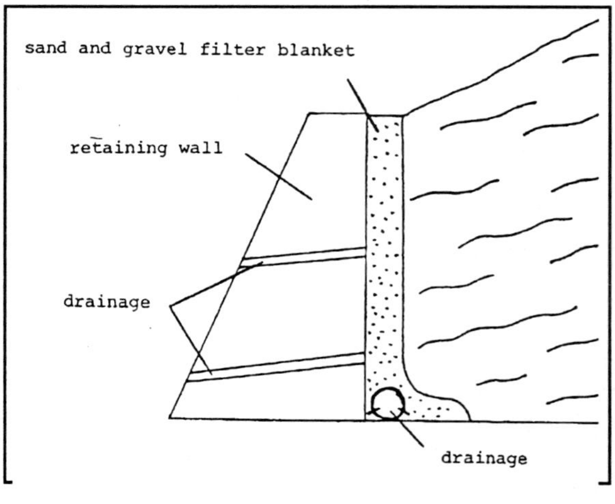

Figure 4. Ideal retaining wall setup with adequate drainages and a permeable blanket between the hillslope material and wall (after Sowers, 1989).

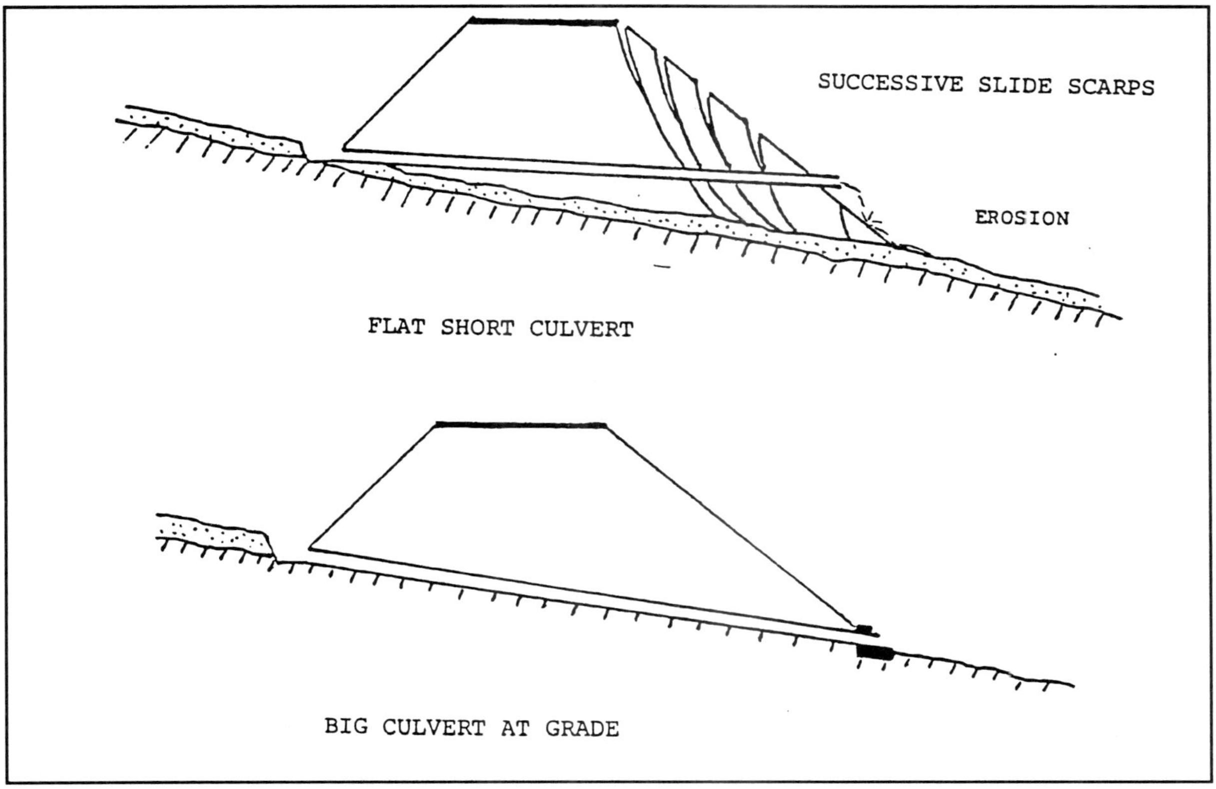

Figure 5. Construction of culverts at grade with the hillslope reduces the risk of erosion (after Sowers, unpublished report, 1989).

CONCLUSION

With respect to seismicity and volcanic hazards, the continuation of mapping efforts and the installation of an earthquake network are needed. Flood-prone areas need documentation and additional mapping to outline the extent of risk. Finally, when considering landslide potential, it should be noted that the safety of rock slopes is a function of time and depends largely on environmental factors. The safety of slopes also depends more on the nature of the failure plane contacts than on the overall strength of the rockmass. If we can actively pursue the four points of landslide control outlined in the previous section, then we may have found a cost-effective solution to landslide hazards in Honduras.

ACKNOWLEDGMENTS

Support for various aspects of this study was provided by Peace Corps, the Instituto Geográfico Nacional, the Dirección General de Minas e Hidrocarburos, and the Laboratorio de Suelos of SECOPT. I am also grateful to George F. Sowers for his educational insights into the nature of landslides.

REFERENCES CITED

Goodman, R. E., 1976, Methods of geological engineering in discontinuous rocks: St. Paul, Minnesota, West Publishing Company, p. 213-244.

Hoek, E., 1971, Estimating the stability of excavated slopes in opencast mines—Discussion and contributions: Transactions of the Institution of Mining and Metallurgy, v. 80, p. A72-A83.

Kiremidjian, A. S., and H. C. Shah, 1979, Seismic hazard analysis of Honduras: The John A. Blume Earthquake Engineering Center report no. 38: Stanford, California, Department of Civil Engineering, Stanford University, 323p.

Nilsen, B., 1985, Shear strength of rock joints at low normal stresses—A key parameter for evaluating rock slope stability: Bjorkliden, Proceedings of the International Symposium on Fundamentals of Rock Joints, p. 487-494.

Osiecki, P. S., 1981, Estimated intensities and probable tectonic sources of historic (pre-1898) Honduran earthquakes: Bulletin of the Seismological Society of America, v. 7, no. 3, p. 865-881.

Patton, F. D., 1966, Multiple modes of shear failure in rock and related materials: Thesis, University of Illinois.

Stone, E. L., Jr., and P. E. Lemmon, 1957, Soil and the growth of forests, in Soil: The yearbook of agriculture: Washington, U.S. Department of Agriculture, p. 721-732.

Williams, H., and A. R. McBirney, 1969, Volcanic history of Honduras: University of California Publications for Geological Sciences, v. 85, 101p.

Circum-Pacific Council for Energy and Mineral Resources
Earth Science Series, Vol. 16
R. L. Miller, G. Escalante, J. A. Reinemund, and M. J. Bergin (Eds.)
Energy and Mineral Potential of the Central American-Caribbean Regions
Springer-Verlag, Berlin Heidelberg 1995

EXTENT AND SOCIO-ECONOMIC SIGNIFICANCE OF SLOPE-INSTABILITY ON THE ISLAND OF HISPANIOLA (HAITI AND DOMINICAN REPUBLIC)

Sergio Mora C.
Departamento Geología, ICE, Escuela C.A. Geología-CIGEFI, Universidad de Costa Rica, INGEOSA, San José, Costa Rica

ABSTRACT

On the island of Hispaniola, unstable slopes are common phenomena, especially those associated with intensive soil erosion and landslides. In Haiti, soil erosion is extreme and has contributed to the problems of food supply and environmental degradation that currently afflict this country. Landslides are also common, such as those around the Peligre reservoir and the Massif de la Selle. In the Dominican Republic, intensive soil erosion is severe, as is the case of the neighborhood of Banica, where land utilization is completely gone. Large landslides are widespread throughout Septentrional and Central Mountain Ranges, examples of which are the well known case at the town of Carlos Díaz in the Septentrional Range and the many slides generated in the Central Range during the passage of the David and Frederic tropical storms in 1979. Climatic and geomorphic conditions have contributed to these phenomena, which in many cases are triggered by the frequent transit of hurricanes and by earthquakes. Irrational human activity, however, is steadily aggravating the situation.

INTRODUCTION

Unstable slopes are a common feature on the island of Hispaniola (Haiti and the Dominican Republic). This instability is the product of very high rates of erosion and the development of massive soil and/or rock slides, that can be subject to reactivation.

Added to the naturally unfavorable climatic and geomorphic conditions is the misuse of natural resources by human activities in both countries. The consequences are extreme environmental degradation and an increase in vulnerability to droughts, to tropical storms, and to earthquakes, all of which are very common in this part of the Caribbean.

Among the most detrimental consequences of slope instability and erosion-related phenomena, the following can be listed: degradation and loss of productivity of topsoils, increase of the relative albedo, progression of some areas toward desertification, and loss of efficiency of hydraulic (irrigation, hydropower) civil works. Unfortunately, very little or nothing is being done to prevent or reverse this rapid deterioration, which is one of the most important factors in the degradation of the environment and the quality of life in these two countries.

LOCATION

The island of Hispaniola is located in the Caribbean island arc, between the parallels of 17° and 20°N and the meridians 68° and 74°W. Bordering it are the Bahamas and the Atlantic Ocean to the north, Puerto Rico to the east, the Caribbean Sea to the south, and Cuba and Jamaica to the west (Figure 1).

RAINFALL

Rainfall averages vary locally from 500 to 3,000 mm/yr. The two periods of maximum precipitation are April-June and August-October, alternating with two drier periods in June-July and November-March. In the November-March period, rainfall is normally lacking. On the other hand, from September to November, the period of heaviest rains may

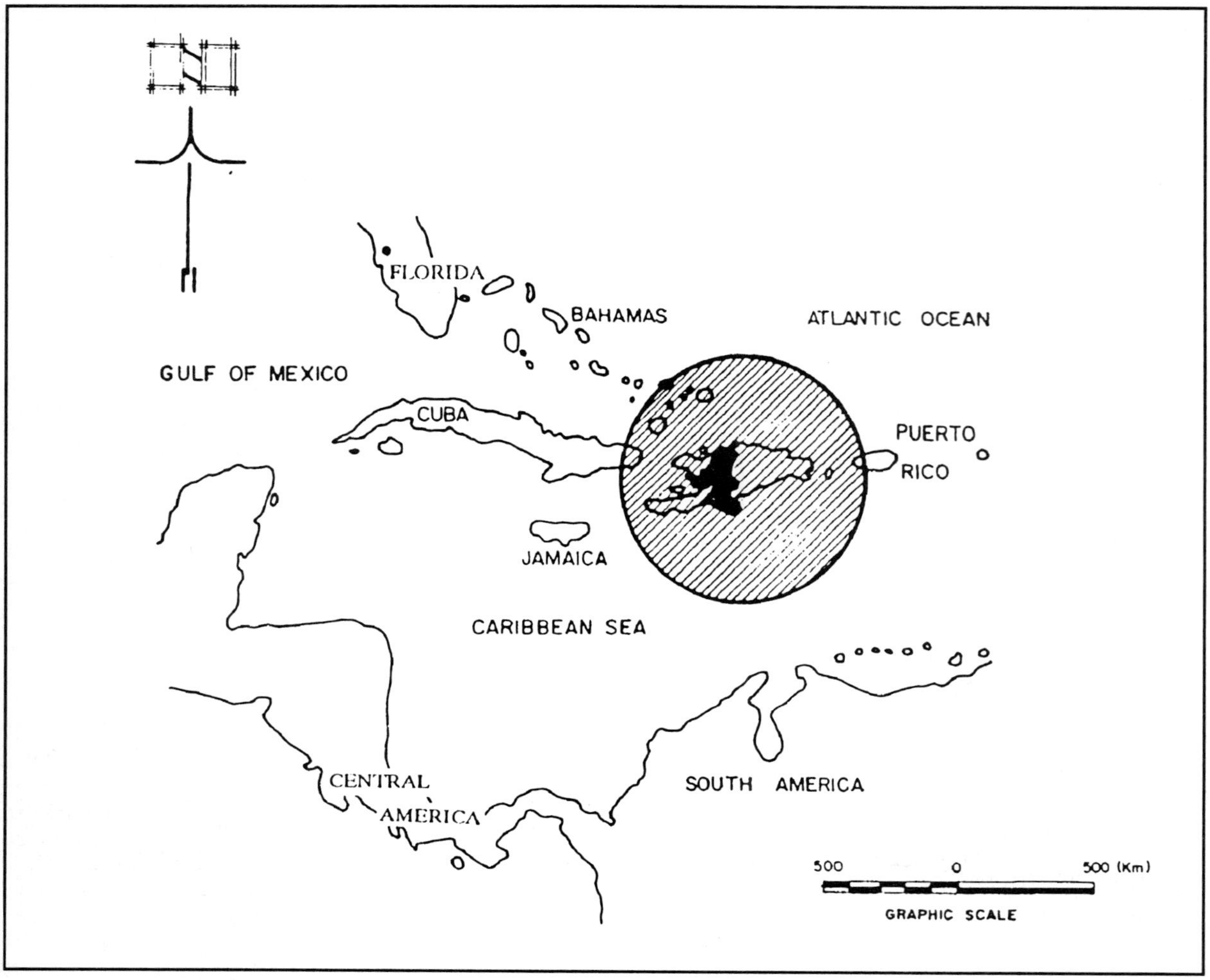

Figure 1. General location map of the island of Hispaniola.

total as much as 400 mm/month. A rain as much as 150 mm/ 24 hours is likely to occur at least once every five years (Alford, 1986; Mora, 1986b).

The "normal" climate is frequently disturbed by the traverse of tropical storms such as the David and Frederic hurricanes, which brought more than 700 mm of rain in five days to the Dominican Republic between August 30 and September 3, 1979. In Figure 2, time and space occurrence of most known historical storms passing through Hispaniola are shown.

PHYSIOGRAPHY

There are three major systems of mountain ranges (Figure 3) with which the steepest slopes of the island are associated. To the north is the Cordillera Septentrional (Dominican Republic)-Massif du Nord (Haiti), separated from the Cordillera Central (Dominican Republic)-Massif Central (Haiti) by the Cibao-Cap Haitien depression (in white on Figure 3). The central range, where the highest point on the island (Pico Duarte in the Dominican Republic, 3,247 m) is located, is separated from the Cordillera de Baoruco (Dominican Republic)-Massif de la Selle (Haiti) by the Enriquillo-Cul-de-Sac depression.

All these mountains, which represent 75 percent of the area of the island, have steep slopes that are frequently affected by landslides. Forty percent of the population of Hispaniola lives in the mountains, and because they have unwittingly contributed to the degradation of the land by erosion, they have been the victims of landslides.

PREVIOUS WORK

Very little research related to slope instability has been done, and is available in Hispaniola, although erosion problems were recognized in Haiti (and the Dominican Republic)

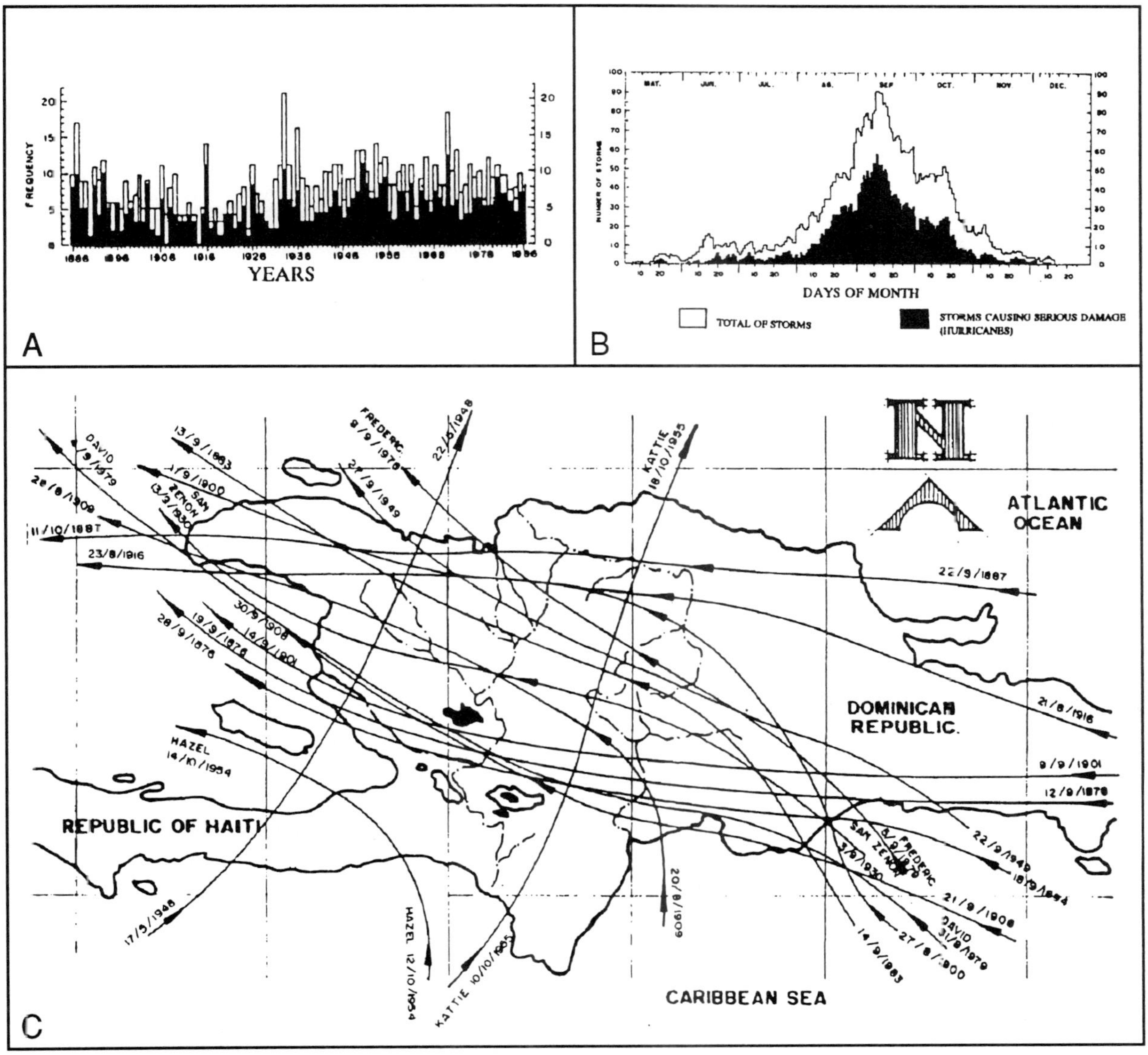

Figure 2. A. Histogram of the time distribution of 876 tropical storms traversing the Caribbean Sea since 1886. B. Profile of time-month occurrences of the same storms as A. C. Paths of main storms that have traversed the island of Hispaniola in the last 100 years.

for many years (Woodring, Brown, and Burbank, 1924). The phenomenon has been attributed both to land misuse, especially deforestation for timber and charcoal, and overgrazing. These are deteriorating and degradating the standard of living, particularly food availability. They have also provoked desertification. Publications on this subject are quite abundant: OEA/DDR (1972), Garvey (1982), Cohen (1984), Alford (1986), Mora (1986a), and others. On the other hand, information on landslides is sparse. Few publications are available. Information on other facets of the problems such as economic and social impact, when it exists, is included in different general reports, such as Nanita (1982), Mora (1986a,b), Prepetit and Joseph (1986), and others. The geology and some aspects of the unstable slopes are treated in the following publications: Woodring, Brown, and Burbank (1924), De la Fuente (1976), Case and Holcombe (1980), and Antonini (1986).

The number of people killed or injured in slope instability/erosion-related phenomena and the amount of socio-economic losses is almost impossible to assess at present, because no tabulations on these matters are available. Because most of the damaging events have occurred during earthquakes or hurricanes, the figures, if any, are usually included in general reports (De la Fuente, 1976; Mora,

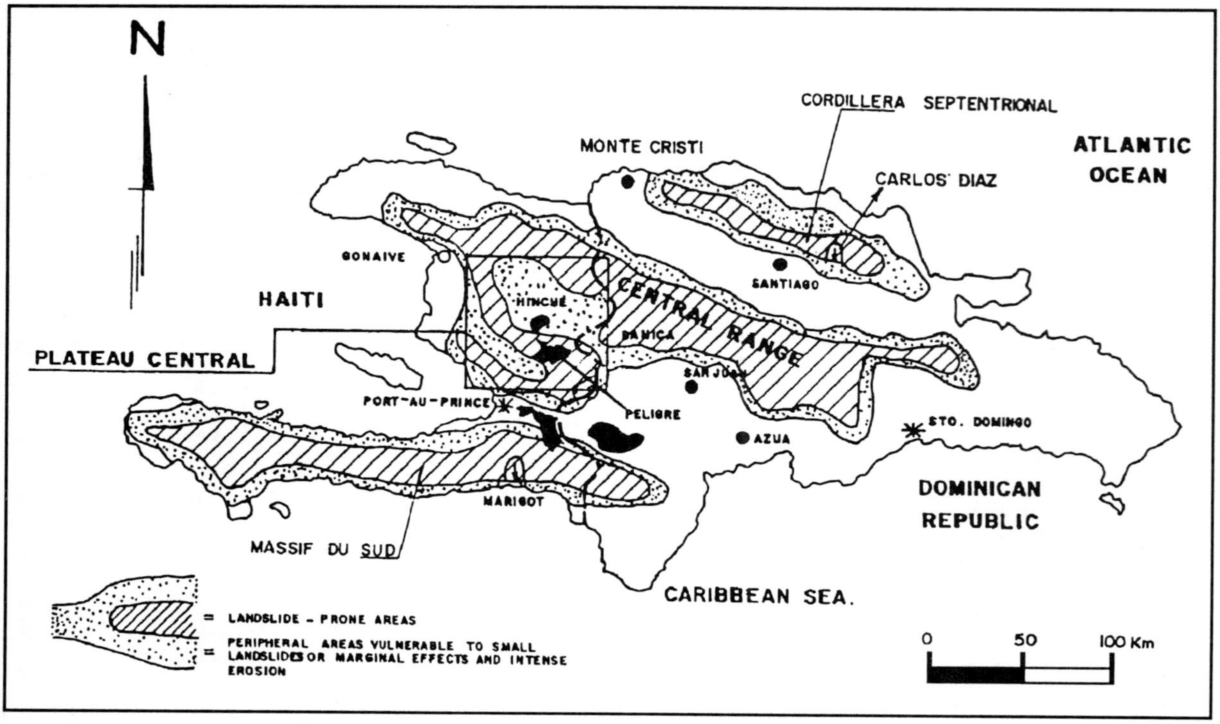

Figure 3. Schematic distribution of areas where intense and moderate slope instability is present in the island of Hispaniola and at the same time remain as the most prone and vulnerable. Areas of most damage from landslides are in black.

1986a,b; Mora, 1987). A quick overview, however, will show losses of at least several thousand lives and several tens of millions of U.S. dollars for this century only. Information on the best documented events that include data on landslides is in publications of De la Fuente (1976) and Mora (1986a,b). These are the Santiago earthquake of December 29, 1897; the San Severo storm of September 27, 1909; the Cibao earthquake of August 4, 1946; Hurricane Hazel on October 12, 1954; Hurricane Flora on October 3, 1963; and Hurricanes David and Frederic on August 31 and September 1, 1979. As the records show, the frequency of events that cause major damage is increasing at about the same rate as the environmental degradation due to human activities.

GEOLOGICAL ENVIRONMENT

The island of Hispaniola is located at the north edge of the Caribbean Plate. There the plate has interacted with the North American Plate, resulting in a complex system of subduction and transform displacements. A northwest-southeast trending group of thrust and transcurrent active faults has resulted.

Most landslides occur on residual and deeply weathered rock units, regardless of their age. Non-carbonate rocks are especially susceptible, as are ones structurally deformed. In the central range, many unstable slopes develop on metamorphic (schists and phyllites) or plutonic weathered rocks, especially on partially developed regolith. In other areas, intensive erosion and landslides usually occur on Tertiary-Neogene soft rocks such as sandstones and siltstones, or on unconsolidated Plio-Pleistocene tectonically uplifted alluvial deposits.

SLOPE INSTABILITY PROCESSES

High erosion rates are usually associated with alluvial sands and gravels that are unprotected by vegetation due to deforestation and overgrazing. Such is the case at Banica (Figures 3 and 4) in the Dominican Republic, where underground flowage also occurs because of fluctuating levels of the Artibonite River (Mora, 1987). The large quantity of sediment carried by the rivers is accelerating the loss of efficiency at several hydraulic works. The Peligre hydroelectric power reservoir is the best example of this adverse condition (Table 1).

Development of high pore pressures during heavy rainfalls and earthquakes (liquefaction) has proved to be a common trigger mechanism for landslides on the island of

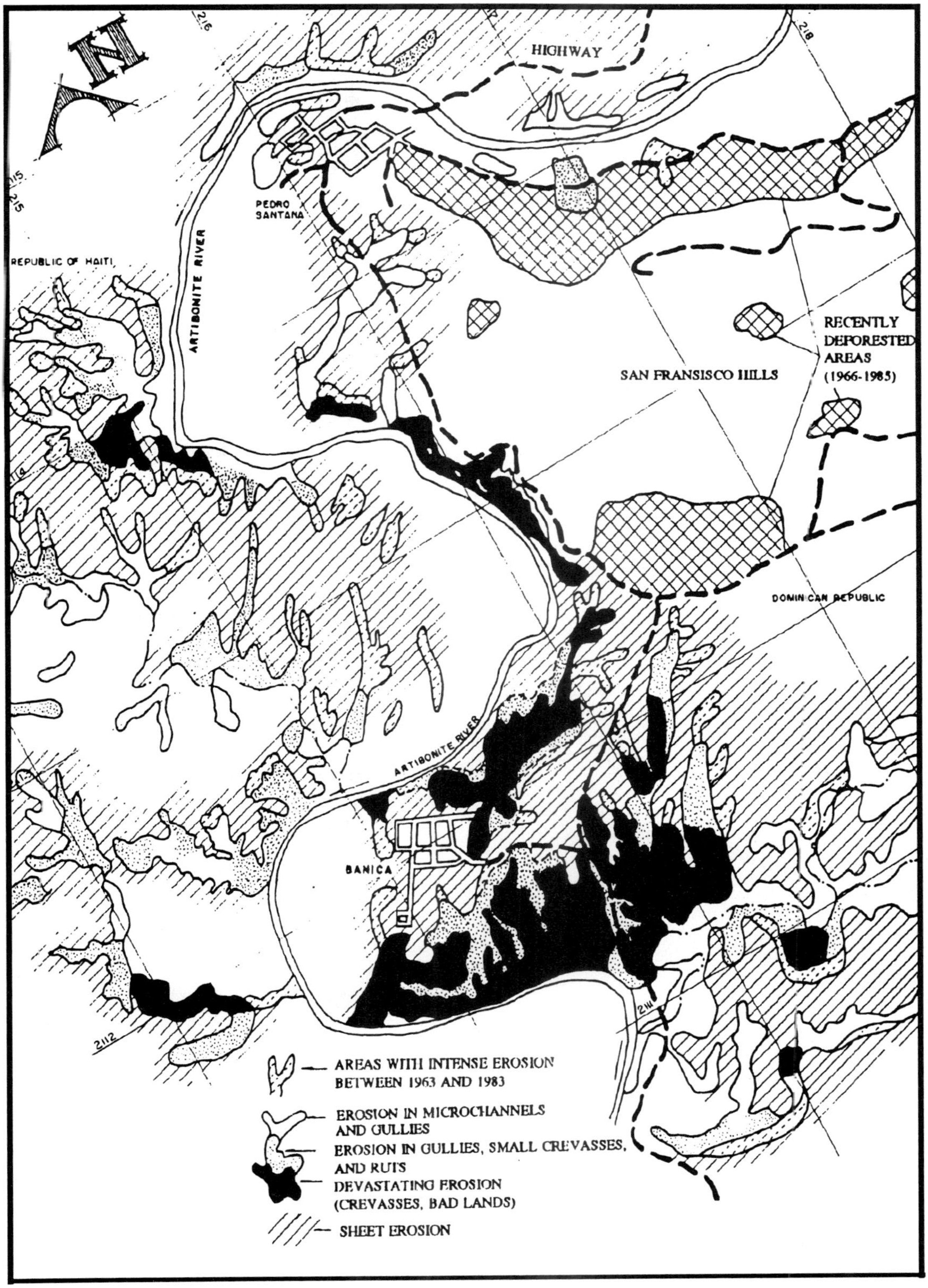

Figure 4. Distribution and extent of accelerated erosion phenomena in Banica and vicinity. Artibonite River is the international boundary. The heavy dashed lines are roads. Scale of grid is 100 m, after Mora (1986b).

Table 1. State of Silting of Peligre Hydropower Reservoir, Haiti.

Increase of Sediments (%) 1956-1980	Increase of Sediments (%) 1956-1984		Initial Reserve		1980 Reserve		1984 Reserve		Volume of Sediments 1956-1980		Volume of Sediments 1956-1984	
			10^6 m^3	&	10^6	m^3	10^6	m^3	10^6	m^3	10^6	m^3
45.8	52.7	Upstream	236	38.9	128	27.1	111.7	26.3	108	80.0	124.3	68.0
6.5	17.2	Center	123	20.2	115	24.4	101.9	24.0	8	5.9	21.1	11.5
7.7	15.2	Downstream	248	40.9	229	48.5	210.2	49.7	19	14.1	37.8	20.5
22.2	30.2	TOTAL	607	100.0	472	100.0	423.8	100.0	135	100.0	183.2	100.0

Yearly average increase in 24 years = 5.6 x 10 ^{6} m^3.
Yearly average increase in 28 years = 6.5 x 10 ^{6} m^3

Hispaniola. Fluvial slope-toe erosion and uncontrolled human activity are also responsible for precariously unstable slopes. In Haiti, Prepetit and Joseph (1985) reported a major landslide in Chemin Dieu Moune (Marigot) in the southern part of the country (Figure 5). The mobilized material was composed of a mixture of soil and blocks derived from a residual (regolith) profile that became saturated during a heavy rainstorm on June 3 and 4, 1985.

In the Plateau Central of Haiti (Figure 3) there are important areas suffering from intensive erosion and landslides. In the vicinity of the hydroelectric reservoir of Peligre (Morne la Yaille), several landslides of calcareous and clastic rocks have occurred. Some are the result of drawdown of reservoir level (30-50 m seasonally), adding to the already serious silting problem. Another landslide developed 4 km downstream from the dam, threatening the power house with potential flooding were it ever to dam the river.

Finally, a large number of landslides have occurred along the alluvial banks of almost all important rivers of Haiti, such as the Artibonite, Guayamouc, Bouyaha, Fer à Cheval, Canot, and others. In the Dominican Republic, the two most important landslide-prone areas are the Septentronial (Northern) and Central Cordilleras.

The Cordillera Septentronial was badly hit by the December 29, 1897, earthquake (Ms=7.1; MM max=IX). Several large landslides were reported on both (Cibao and Caribbean) sides of the range. A probable consequence of that earthquake is the Carlos Díaz (Tamboril) landslide (Figure 3) which damaged the village of that name and caused a forced relocation of about 1,500 people in 1982. The slope of the range is covered by colluvium overlying limestones that are interbedded with poorly consolidated marly sandstones and shales. High water tables and the presence of clay between layers have been the most significant factors causing instability, and frequent earthquakes such as the 1981 swarm (R. Acosta, 1986, personal communication) have been the trigger mechanism.

Tropical storms are also important phenomena that induce landslides, especially on the deforested and deteriorated slopes of the Central Range. During the David and Frederic storms (August-September, 1979; Figure 5), landslides were widespread in almost all mountains of the country. They also created ephemeral dams and ponds and subsequent mudflows. Combined losses provoked by the storms, floods, landslides, and other phenomena, kept the country's population in a state of shock for many weeks. About 2,000 died and 1,500,000 people were affected. Damage was estimated at about one billion U.S. dollars and the economy stagnated for at least eight years (Mora, 1986b).

INSTITUTIONAL ACTIONS TOWARD PREVENTION AND CONTROL

Erosion and landslide problems, consequences of land misuse as accelerators of socio-economic degradation, can only find solutions with integrated interdisciplinary management. The community should be actively involved in the development of the projects and guided by geoscientists, which should produce accessible technical solutions. At the present time there is almost no coherent nor coordinated policy of governmental, institutional, nor any individual action for prevention and control. In both countries there have been only a few local projects directed toward solving specific problems.

In Haiti, the ongoing famine problem motivated government and foreign groups to attempt to define a policy for soil protection (OEA/DDR, 1972; Pierre-Louis, 1985; and others). Other studies have been made in order to mitigate and control erosion and resultant silting related to the Pligre hydroelectric power reservoir, but so far no action. More recently, several other critical areas have been identified that should have priority for remedial action (OEA/DDR, 1985, 1986). As yet, no significant results have been obtained and, unfortunately, time is running out.

In the Dominican Republic, more regional studies have been made than in Haiti. These have taken into consideration the problems of natural hazards and their mitigation (Mora, 1986b, 1987; OEA/DDR, 1972, 1985, 1986; and others) but here again, no real action has yet been undertaken.

ENVIRONMENTAL AND SOCIO-ECONOMIC IMPACT

The natural geomorphologic cycle includes slope instability resulting from tectonic uplift with resultant regional

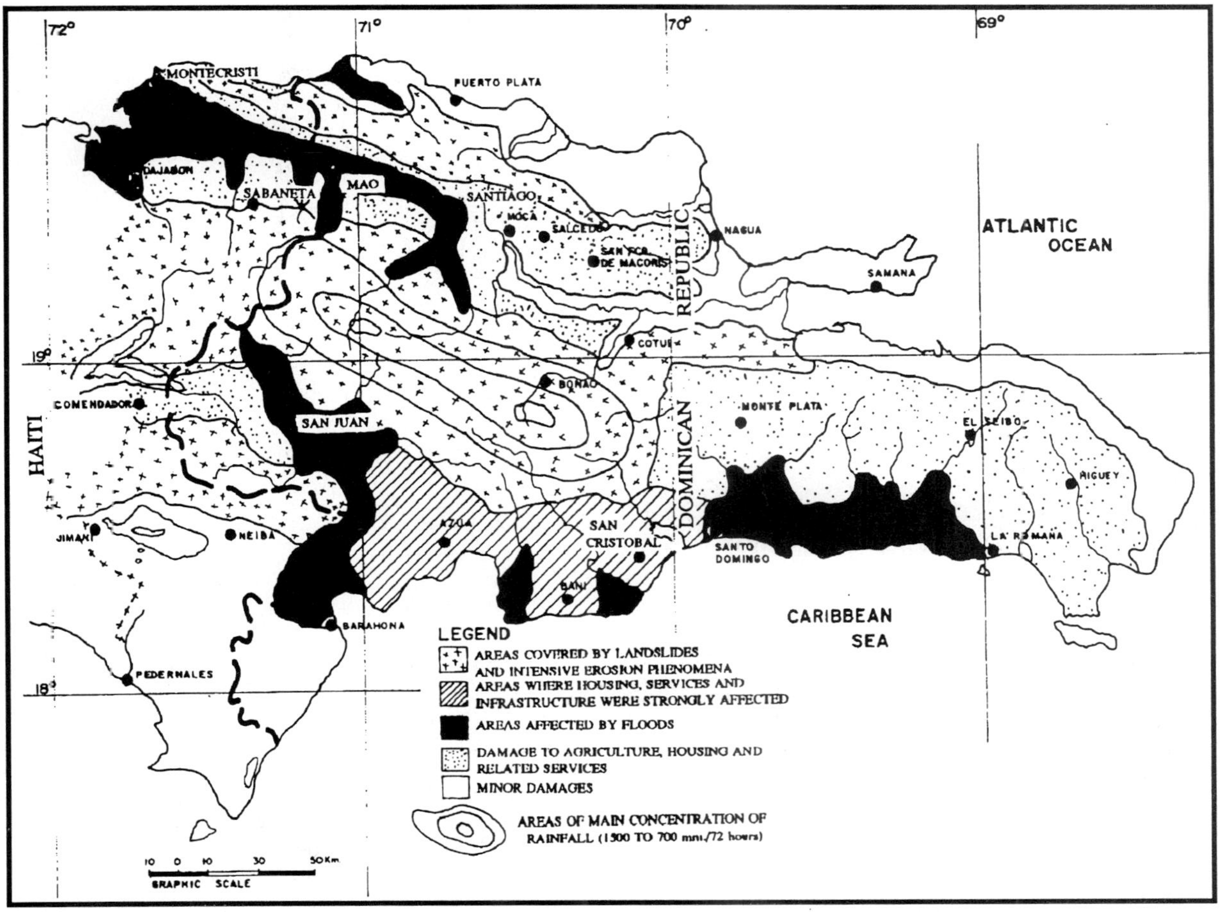

Figure 5. Schematic map of the main damage caused by tropical storms David and Frederic, August-September 1979, after Mora (1986b).

erosion. However, after equilibrium has been reestablished, the irrational use of natural resources by humans can be devastating. The situation appears to be a paradox in which men act both as the main cause of the erosion and instability and as the victims. Their works can be destroyed, including agriculture, vital lines of communication, infrastructure, and even lifestyles in general. Table 2 lists the degree of erosion in subbasins of the Artibonite River. Locations are not shown in figures because of the small scale of the maps. In both Haiti and in the Dominican Republic, the losses caused by erosion and landslides are huge. They constitute one of the major expenses required to maintain or replace infrastructure. In both countries the negative influence of these hazards on the economic status and quality of life, the foreign debt, and all other social indexes demonstrate how serious the situation is. It also points to the future difficulties in recovering a stable welfare for many communities. The link to be broken from this vicious circle is quite evident.

CONCLUSIONS

Slope instability, manifested as intensive and extensive erosion, and landslides are a widespread and major problem in the island of Hispaniola. Almost all mountain ranges are prone to slope instability and in large areas of the island intensive erosion patterns are present. Especially dramatic examples occur when diverse phenomena interact, such as unfavorable structures, deeply weathered rock units, heavy rainfalls, tropical storms, and earthquakes. At the same time, human activity is also a contributing factor because of the significant steady increase of land misuse, deforestation, overgrazing, and an overall irrational exploitation of natural resources. Because only a few studies have focused on the problem, and because almost none of them have resulted in real action, the immediate future does not permit an optimistic forecast.

Table 2. Classes of Erosion of the Artibonite River Valley by Subbasin in Hectares (after Snyder, 1985).

No.	Name of Subbasin	Light	Moderate	Strong	TOTAL
1	Bouyana	14,780	18,630	47,940	81,350
2	Libon	500	3,510	20,640	24,650
3	Locaine	1,230	2,110	27,060	30,400
4	Samana	640	2,680	10,480	13,800
5	Canot	13,810	21,620	41,620	77,050
6	Fond Bleu	0	4,980	12,500	17,480
7	Rio Frio	0	2,950	7,280	10,230
8	Guayamouc	4,820	26,910	34,790	66,520
9	Nord Artibonite	330	310	11,120	11,760
10	Sud Artibonite	170	310	4,750	5,230
11	Thomonde	850	4,530	14,890	20,270
12	Cange	0	1,890	3,250	5,140
13	Felician	0	220	4,930	5,150
14	Peligre*	1,500	1,950	15,300	18,750
15	Lascahobas	1,950	370	6,690	9,010
16	Roche Blanche	430	860	2,010	3,300
17	Onde Verde	2,240	6,640	11,870	20,760
	TOTAL	43,250	100,470	277,130	420,850
	PERCENTAGE	10	24	66	100

*3,000 hectares of reservoir of Peligre were not included.

The author strongly recommends the establishment of a policy and a subsequent plan involving local and foreign geoscientists and politicians. This plan should yield a better understanding of the factors involved in the generation, development, and occurrence of slope instability phenomena in the island. It will surely provide a foundation upon which a well-oriented proposal can be implemented to mitigate and finally to reverse the consequences of these problems of land degradation.

ACKNOWLEDGMENTS

María Laporte reviewed the first manuscript and significantly contributed to its improvement. Flory Abarca typed the first version. The Organization of the American States (OAS) funded the field work that produced most of the data contained in this paper.

REFERENCES

Alford, D., 1986, The water and sediment balance of the upper Artibonite basin, Haiti: Organización de los Estado Americanos/ Departamento de Desarrollo Regional, 33p.

Antonini, G., 1986, Processes and patterns of landscape change in the Línea Noreste, Dominican Republic: Columbia University, Ph.D. Thesis.

Case, J. E., and T. L. Holcombe, 1980, Geologic-tectonic map of the Caribbean Sea: U.S. Geological Survey, Miscellaneous Investigations Map Series Map I-1100, scale 1:2,500,000.

Cohen, W., 1984, Environmental degratation in Haiti, analysis of air photography: USAID, 4p.

De la Fuente, S., 1976, Geografía Dominicana: Santo Domingo, 232p.

Garvey, W., 1982, Disaster relief and preparedness evaluation, Haiti, Agriculture: USAID, 12p.

Mora, S., 1986a, Etude de reconnaissance des menaces naturelles dans le basin versant du haut Artibonite, Haiti: Organizaci<n de los Estados Americanos/Departamento de Desarrollo Regional/ Government of Haiti-INGEOSA, 70p.

Mora, S., 1986b, Estudio de reconocimiento de las amenazas naturales de la región fronteriza (occidental) de la República Dominicana: Organización de los Estados Americanos/Departamento de Desarrollo Regional/Goberno de la República Dominicana, Phase III, INGEOSA, 90p.

Mora, S., 1987, Análisis preliminar del fenómeno de erosión acelerada en Banica, Elías Piña, República Dominicana: Organización de los Estados Americanos/Departamento de Desarrollo Regional-INGEOSA, 160p.

Nanita, M., 1982, Disaster prevention in the Dominican Republic: Tokyo, Japan, National Research Center Disaster Prevention, SEOPC-JICA, 10p.

Organización de los Estado Americanos/Departamento de Desarrollo Regional, 1972, Mission d'assistance technique integrée: Haiti, Conseil National de Dévelopment et de Planification; Organisation des Etats Américains, 656p.

Organización de los Estado Americanos/Departamento de Desarrollo Regional, 1985, Programme de developpment de la zone frontaliOre dans le basin du haut Artiabonite: Plan de travail du projet d'aménagement et conservation des suclos ed des eaux, 73p.

Organización de los Estado Americanos/Departamento de Desarrollo Regional, 1986, Plan de sesarrollo de la zona fronteriza, fase I, diagnostico de los recursos didricos: Oficina Nacional de Planification, 36p.

Pierre-Louis, R., 1985, Soil degratation and declining argicultural productivity in Haiti: Min. Agriculture, 9p.

Prepetit, C., and Y. Joseph, 1986, Rapport sur le gissement de terrain produit a dieu moune (Marigot): Min. Mines et Press, Nat., 7p.

Snyder, R., 1985, Etudes de resources de la terre dans le bassin du Haut Artibonite, Haiti. Project OEA/DDR/60H, unpublished report, 31 p.

Woodring, W. P., S. S. Brown, and W. S. Burbank, 1924, Geology of Haiti: Port-au-Prince, Government of Haiti.

Circum-Pacific Council for Energy and Mineral Resources
Earth Science Series, Vol. 16
R. L. Miller, G. Escalante, J. A. Reinemund, and M. J. Bergin (Eds.)
Energy and Mineral Potential of the Central American-Caribbean Regions
Springer-Verlag, Berlin Heidelberg 1995

USE OF SEISMIC REFRACTION TECHNIQUES FOR INVESTIGATING RECENT LANDSLIDES IN A TROPICAL RAIN FOREST IN PUERTO RICO

Matthew C. Larsen
U.S. Geological Survey, P. O. Box 4424, San Juan, Puerto Rico 00936 USA

ABSTRACT

Seismic refraction surveys were conducted on two recent landslides (one translational slide and one rotational slump) that damaged a highway in the Caribbean National Forest, Puerto Rico. These surveys on recent landslides provided insight into the dimensions and depths of their failure planes. Field observations and drill hole data were used to verify the accuracy of refraction data. The seismic velocity measured in the overlying colluvium and saprolitic soils at a translational slide was 350 m/sec. The seismic velocity of the saprolite below the failure plane ranged from 1,000 to 1,500 m/sec. The velocity in the deeper weathered-bedrock measured about 7,300 m/sec. The thickness of colluvium ranged from 3 to 5 m at the toe of the slump, and the thickness of saprolitic soil above the head scarp ranged from 4 to 8 m. Estimated depth to the top of the quartz-diorite bedrock ranged from 12 to 24 m. The partially exposed failure plane was at the top of the saprolitic layer with velocities in the 1,000 to 1,500 m/sec range. A rotational slump at a second site had a measured seismic velocity of 200 m/sec in a 6-m thick layer of colluvium and saprolite which, based on drill-hole data, graded to saprolite at a depth of 1 to 2 m. A seismic velocity contact in the saprolite was found at this 6 m depth; the seismic velocity in the saprolite below this contact was 870 m/sec. The seismic velocity of weathered quartz-diorite bedrock underlying the saprolite was 2,040 m/sec. Depth to bedrock from seismic data was calculated to be 20 m, which compared reasonably well with the depth of exposed bedrock (15 m) in an adjacent slump. Drill-hole data obtained from other surveys confirmed the depth to bedrock at 20 m. The possible failure plane of this rotational slump was revealed in a velocity contact within the saprolite. The surface, as noted above, was 6 m deep and was confirmed with drill-hole data, which indicated a two-to three-fold increase in blow counts; this represents an increase in saprolite density below the contact. The presence of a perched water table was observed overlying this surface.

INTRODUCTION

Numerous naturally-occurring and road-related landslides are found in the Caribbean National Forest (CNF), a tropical rain forest in northeastern Puerto Rico (Guariguata and Larsen, 1989) (Figure 1). The Caribbean National Forest covers an 11,330 ha area with elevation ranging from 100 to 1,075 m above mean sea level. Mean annual temperature and rainfall range from 26.5°C and 2,450 mm in the lower elevations to 17.5°C and 4,000 mm at higher elevations (Brown et al., 1983). Intense precipitation may occur throughout the year and generally is associated with tropical depressions, hurricanes, or less intense tropical waves (Miller, 1965). Landslides commonly occur in response to these high rainfall events. The majority of failures observed in the CNF are translational slides, debris slides, and earth slumps (classification after Varnes, 1978). Also noted, but less common, are rock falls, which mainly occur along roadcuts. Many natural and road-related landslides are shallow, translational slides and occur along planar or concave slopes on planes of weakness in the saprolite, although unweathered bedrock is commonly exposed in large landslides of 2,000 m^3 or more (Simon and Larsen, 1988; in press). Headscarps range from 1 to 4 m in height except for the larger landslides where they may be as high as 8 m. The largest failures, located along a forest highway, are debris slides that include long and narrow

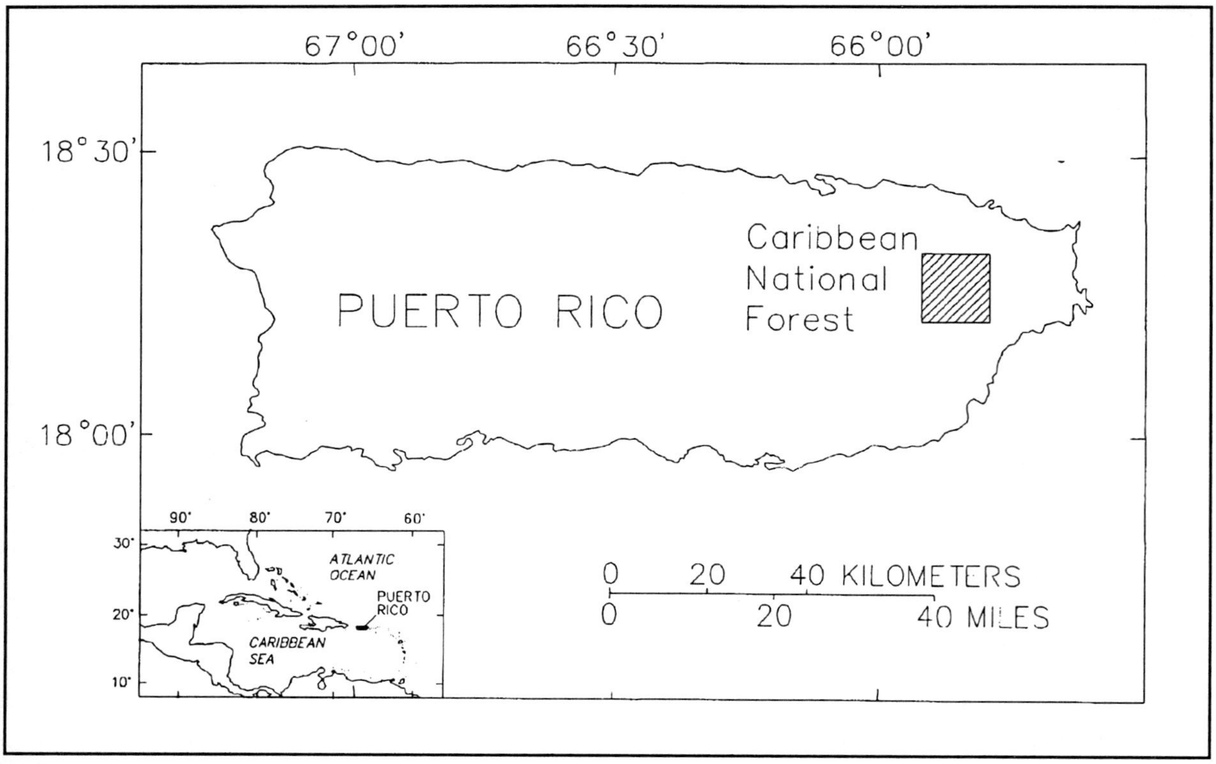

Figure 1. Location of the Caribbean National Forest, Puerto Rico.

run-out zones. Bedrock in the study area consists of marine-deposited andesitic to basaltic volcaniclastics of Cretaceous age (Seiders, 1971). These rocks have been intruded by the Rio Blanco stock, a light-gray, medium- to coarse-grained quartz diorite, that covers about 20 km^2 and is bounded by a zone of metamorphism extending 0.5 to 3.5 km into the surrounding volcaniclastics. Slopes in the CNF are mantled with silty-clay saprolitic soils, which are underlain by saprolite. This sequence of materials overlies weathered and unweathered bedrock and ranges in thickness from 3 to 24 m (Dames and Moore, 1980, unpublished report; T. Fennessey, U.S. Department of Transportation, 1988, personal communication). Landslides in the CNF cause periodic closures of highways, property damage, and they occasionally result in the loss of life. The U.S. Geological Survey, in cooperation with the Puerto Rico Planning Board, is evaluating landslide hazard potential in this area, and, as part of this study, is using seismic refraction techniques to estimate depth to bedrock, depth to water table, and types of material found beneath active landslides. One of the objectives of this work is to determine the feasibility of using seismic data to estimate the depths and dimensions of possible failure planes in active landslides. This seismic work will assist the Survey in defining geologic and hydrologic controls of landslide processes and will contribute to the evaluation of landslide hazard.

METHODS

Seismic refraction surveys generally were conducted along the elevation contour, using a signal-enhancing 12-channel seismograph in conjunction with an array of 12 geophones. Geophone spacing varied from 1.5 to 6 m; spread length ranged from 17 to 67 m; and shot offsets were 1.5, 3, 15, 30, and 46 m. The energy sources used were blows from a 3.6 kg sledge hammer on a steel plate, and a seismic shotgun using 8 and 12 gauge shells. Seismic data were analyzed using the delay-time method and a ray-tracing procedure described by Scott (1977). Drill-hole data was obtained from the U.S. Department of Transportation.

INVESTIGATIONS OF RECENT LANDSLIDES

Seismic refraction surveys were conducted on two recent landslides that damaged a highway in the Caribbean National Forest. During a two-week period of 1987, several storms produced a total of 790 mm of rainfall which caused numerous landslides in northeastern Puerto Rico (Guariguata and Larsen, 1989). One landslide, which blocked a highway, affected an area of approximately 740 m^3. The seismic velocities of overlying colluvium and saprolitic soil at this

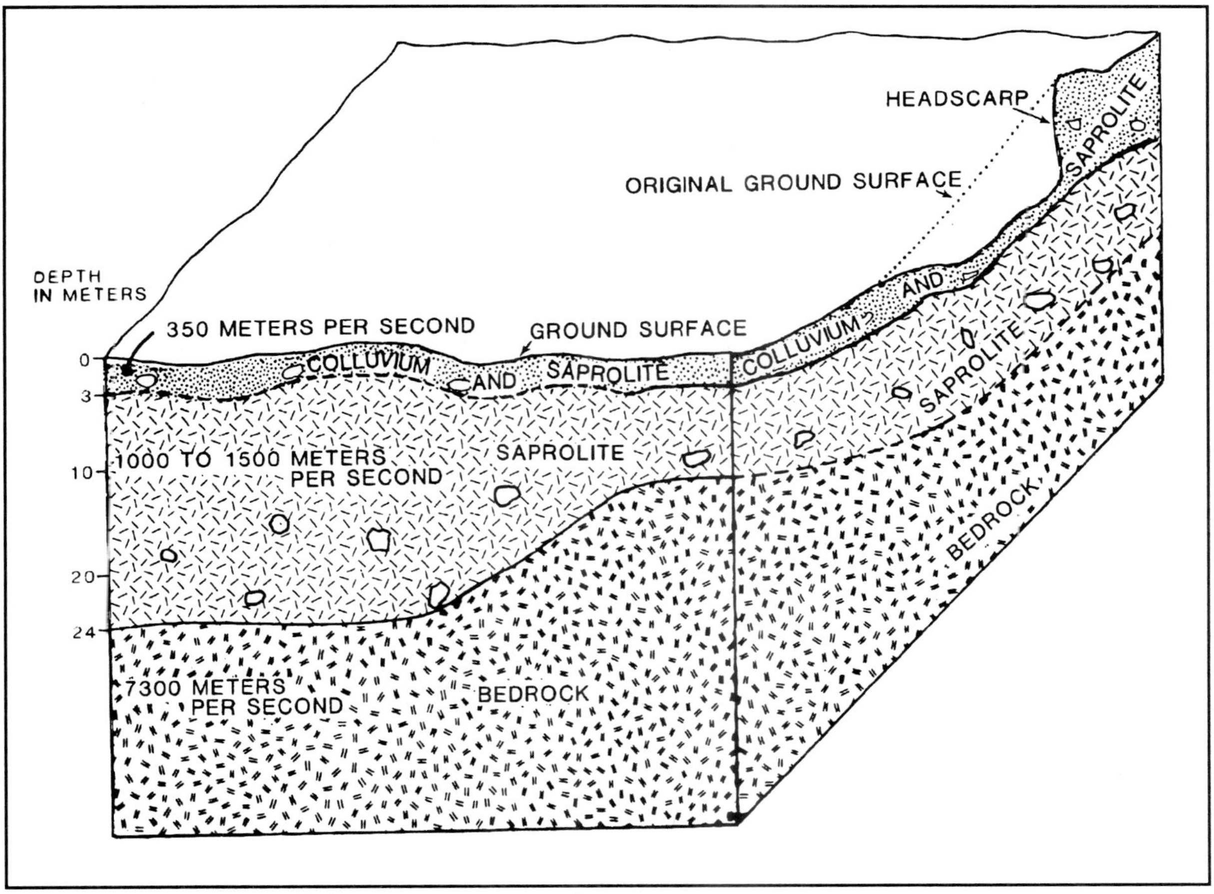

Figure 2. Block diagram showing seismic velocities and soil and rock types in the area of a translational slide.

translational slide were 350 m/sec (Figure 2). Saprolite below the colluvium soil had velocities ranging from 1,000 to 1,500 m/sec, and velocity in the underlying weathered bedrock was 7,300 m/sec. The thickness of colluvium ranged from 3 to 5 m at the toe of the slump at the time of the survey in 1988, and the thickness of saprolitic soil above the head scarp ranged from 4 to 8 m. The uneven bedrock refractor varied from 12 to 24 m below land surface. Drill hole data were not available for this site; however, depths of nearby bedrock exposures were used to verify these estimated depths. Along the same highway an active rotational slump, first observed in 1986 when 15 m long arcuate cracks appeared in the road surface, has caused a 60 cm drop in the road bed. Seismic velocities of 200 m/sec were measured in a 6 m thick sequence of colluvium, saprolitic soil and saprolite, lying immediately below the road bed (Figure 3). Seismic velocity in the saprolite below this contact was substantially higher (870 m/sec). The velocity contact within the saprolite is thought to be the possible failure plane of an active section of this slump. This 6 m deep contact was characterized by a two-to three-fold increase in blow counts and relatively sharp increase in saprolite density. A perched water table was also encountered at this horizon. This suggests that translational failures may occur primarily in saprolite zones of increased pore pressure immediately overlying saprolite layers with decreased permeability and increased density.

SUMMARY

A translational slide and a rotational slump were surveyed using seismic refraction techniques to estimate depth to bedrock and to delineate possible failure planes. Seismic velocity contacts, based on several surveys at each site, indicated that apparent failure planes in or at the top of the saprolite were 3 to 8 m below land surface. Bedrock depth ranged from 12 to 24 m. At the site in which drill-hole data were available, higher density and a perched water table were found at the seismic velocity contact within the saprolite, indicating that horizons of increased density and pore-pressure may exert a role in the movement of this active slump.

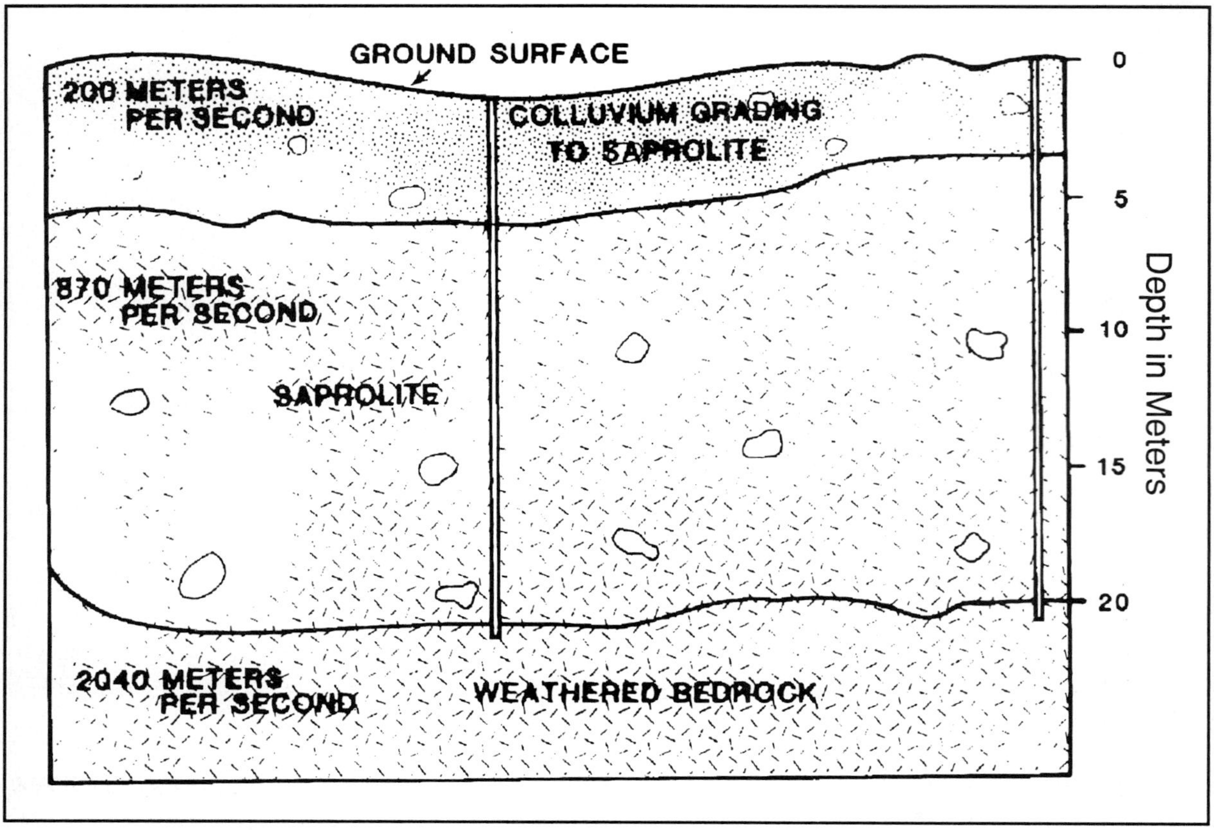

Figure 3. Generalized section showing seismic velocities and soil and rock types in the area of an earth slump.

REFERENCES CITED

Brown, S., A. E. Lugo, S. Silander, and L. Liegel, 1983, Research history and opportunities in Luquillo Experimental Forest: U.S. Department of Agriculture, Forest Service, General Technical Report 50-44, 128p.

Guariguata, M. R., and M. C. Larsen, 1989, Preliminary map showing landslides in El Yunque quadrangle, Puerto Rico: U.S. Geological Survey Open-File Report 89-257, scale 1:20,000, 1 sheet.

Miller, J. F., 1965, Two-to ten-day rainfall for return periods of 2 to 100 years in Puerto Rico and Virgin Islands: U.S. Department of Commerce, Weather Bureau, Technical Paper 53.

Scott, J. H., 1977, SIPB—A seismic inverse modeling program for batch computer systems: U.S. Geological Survey Open-File Report 77-366, 40p.

Seiders, V. M., 1971, Geologic map of the El Yunque quadrangle, Puerto Rico: U.S. Geological Survey, Miscellaneous Geologic Investigations Map I-658, scale 1:20,000, 1 sheet.

Simon, A., and M. C. Larsen, 1988, Shear strength and slope stability at sites in the Caribbean National Forest and Coamo area, Puerto Rico: Geological Society of America Abstracts with Programs, v. 20, no. 7, p. A-144.

Simon, A., and M. C. Larsen, in press, Slope-failure processes in saprolitic soils of a tropical rain forest, Puerto Rico (abs.): Proceedings of 12th Caribbean Geological Conference, St. Croix, U.S. Virgin Islands.

Varnes, D. J., 1978, Slope movement types and processes, *in* R. L. Schuster and R. J. Krizek, eds., Landslides, analysis and control: Washington, National Academy of Sciences, Transportation Research Board Special Report 176, p. 12-33.

Circum-Pacific Council for Energy and Mineral Resources
Earth Science Series, Vol. 16
R. L. Miller, G. Escalante, J. A. Reinemund, and M. J. Bergin (Eds.)
Energy and Mineral Potential of the Central American-Caribbean Regions
Springer-Verlag, Berlin Heidelberg 1995

REPORT OF THE WORKING GROUP ON NATURAL HAZARDS

Mary Ellen Williams
U.S. Geological Survey, Reston, Virginia, USA 22092

The Working Group on Natural Hazards considered their task in terms of three first-order questions that are shared by all natural hazards:

1. Are the impacts of natural hazards significant?
2. What are the major causes of losses due to natural hazards?
3. What can be done to mitigate the effects of natural hazards?

The Working Group then attempted to answer these major questions in general terms and with additional questions. Finally, the Working Group, within the constraints of the available time, assembled general recommendations that address mitigation of specific natural hazards presented during the conference.

ARE THE IMPACTS OF NATURAL HAZARDS SIGNIFICANT?

The Working Group concluded that impacts of natural hazards are significant in terms of lives, economy, and, possibly, regional politics. The figures at hand indicate that their negative effects are 100 times greater for developing countries than for developed countries. Moreover, the negative impact on developing countries is increasing at an alarming rate.

Since 1990, the number of persons affected by all kinds of natural disasters has increased at a rate of 6 percent per year, which is more than twice the rate of increase of the world population. The main causes of natural disasters are drought, floods, cyclones, earthquakes, volcanoes, and landslides. Some natural hazards occur due to degradation of the environment and uncontrolled population explosion in urban areas, especially in developing countries. During the last 20 years, property damage due to all natural disasters approached 100 billion U.S. dollars. The actual losses to the economy and to social organizations of the devastated areas are much lighter.

Examples: Loss Percentage of GNP due to natural disasters:

Nicaragua	Earthquake	1972	209.3%
Honduras	Hurricane	1974	46.5%
Guatemala	Earthquake	1976	31.8%
Bolivia	Flood/Drought (El Niño)	1983	28.2%

WHAT ARE THE MAJOR CAUSES OF LOSSES DUE TO NATURAL HAZARDS?

The hazards may not be detected for a variety of reasons. There may not have been a detectable warning (rare). Surveillance may not have been at an appropriate level. Technology might not have been available or funds weren't forthcoming. Decision-makers might have considered the problems too late, or a decision was made not to start at all. The warning signs were not correctly understood because of insufficient baseline data, or the observed pattern did not fit previous experience.

The reaction of the warning might have been too limited because of limited understanding of the hazards by the public, responsible agencies, and the government. There also might have been poor communication among the various groups responsible for disaster response. Appropriate mitigation measures might not be in place because of lack of warning, not enough funding, lack of coordination among involved groups, and because technical resources for mitigation were not available.

WHAT CAN BE DONE TO MITIGATE THE EFFECTS OF NATURAL HAZARDS?

Increased funding for studies of natural hazards, including surveillance and mitigation, is the first recommendation.

Better disaster mitigation, like any initiative, cannot be accomplished without appropriate funding.

It is important to increase communication and coordination before the next crises occur by forming, region by region, advisory panels composed of members from government and civil defense, and also social scientists, economic specialists, engineers, scientists with experience in each natural hazard, and, significantly, the press. In almost every natural disaster misunderstanding, distrust, turf protection, and consequent adversity may develop among the major participants. This is largely a matter of poor communication that occurs in the busy, often frantic, hours to weeks of the crisis. Panels should be organized to meet regularly to exchange information and educate the several functions before the next crisis, and thus develop the trust, understanding, and support required for proper delivery of hazards mitigation to the people during the next crisis.

Pre-crisis educational aids in the form of posters, mass mailings, public talks, and of paramount importance, video programs should be developed. Effective programs should be instituted that develop evaluation techniques, risk assessment, surveillance, and warning systems.

Central America is seismically very active. Recent work shows that volcanic chain earthquakes up to 6.5 on the Richter Scale have caused the major proportion of the damage and loss of life. The volcanic chain is particularly vulnerable to these earthquakes because some major population centers are located in seismically active areas where unconsolidated material can greatly amplify seismic waves. Vulnerability is increased because of the lack of engineering data needed for the aseismic design of buildings and emergency structures. As an example, the San Salvador earthquake of October 10, 1986, proved that the building codes that were based on California codes were not appropriate. Therefore, buildings thought to be of aseismic design were badly damaged.

WORKING GROUP RECOMMENDATIONS AND CONCLUSIONS

1. There is a need to continue the operation of networks of seismic instruments that will help to identify the location and extent of active faults and provide the needed engineering parameters for aseismic design of buildings. This is the basic data necessary for civil defense, city planners, and engineers who, through their positions, have the ability to implement disaster mitigation plans.

2. There is a need for a professional association for all Middle-American countries to pool knowledge and resources in order to focus on the problems and solutions while establishing continuity. The specific focus should be on the integration of regional data. There are not always scientists in each country with expertise in all hazards subjects, but on a regional basis natural hazards problems could be addressed adequately.

3. Generally each nation has an agency with a mandate to respond to catastrophes. Some work with high standards of efficiency. Most of these agencies, however, have not been created to assess and understand precursory events leading to disasters. Science and technology provide the basis for this understanding, and need to be supported before disasters occur.

4. A most important linkage must be made between the researcher and the decision-makers, who may be emergency managers, politicians, economists, or social scientists. This vital link is often lacking.

5. More attention should be given to surveying and monitoring areas that historically have shown disaster potential. As an example, in the case of landslides, surveying and monitoring may indicate that surficial stabilization is required.

6. To be effective, preventive planning must be incorporated into global development programs and into a long-term policy.

7. To be effective, risk management has to take into account not only scientific and technical data, but also political, sociological, economic, legal, and institutional factors. Risk management must be broadly based on information and on population participation.

8. To mitigate natural hazards, international cooperation (scientific, technical, or economic) has not been well organized to date, because it is too specific. The problem is not only a financial one, but is also due to the high specialization of the scientific and technical teams who generally work in an isolated manner.

9. It is very beneficial to participate in a symposium such as the Circum-Pacific, where there is an opportunity for regional interchange. More symposia should be scheduled where there can be a cross-fertilization of ideas, and an opportunity to hear what other countries in the Central American-Caribbean region are doing to effectively mitigate natural hazards.

INDEX

F

G

H

K

N

P

R